This is a text on the subject of superconductivity, an area of intense research activity worldwide. The book is in three parts: the first deals with phenomenological aspects of superconductivity, the second with the microscopic theory of uniform superconductors, and the third with the microscopic theory of nonuniform superconductors.

The first part of the book covers the London, Pippard, and Ginzburg–Landau theories, which are used to discuss a wide range of phenomena involving surface energies, vorticity, the intermediate and mixed states, boundaries and boundary conditions, the upper critical field in bulk, thin-film, and anisotropic superconductors, and surface superconductivity. The second part discusses the microscopic theory of Bardeen, Cooper, and Schrieffer. Finite temperature effects are treated using the Bogoliubov–Valatin transformation. The theory is used to discuss quasiparticle tunneling and the Josephson effects from a microscopic point of view. The final part of the book treats nonuniform superconductors using the Bogoliubov–deGennes approach with which it is possible to extract many important results without invoking Green's function methods.

This text will be of great interest to graduate students taking courses in superconductivity, superfluidity, many-body theory, and quantum liquids. It will also be of value to research workers in the field of superconductivity.

Superconductivity

Superconductivity

J. B. KETTERSON and S. N. SONG
Northwestern University

CAMBRIDGE
UNIVERSITY PRESS

PUBLISHED BY THE PRESS SYNDICATE OF THE UNIVERSITY OF CAMBRIDGE
The Pitt Building, Trumpington Street, Cambridge CB2 1RP, United Kingdom

CAMBRIDGE UNIVERSITY PRESS
The Edinburgh Building, Cambridge CB2 2RU, UK http://www.cup.cam.ac.uk
40 West 20th Street, New York, NY 10011-4211, USA http://www.cup.org
10 Stamford Road, Oakleigh, Melbourne 3166, Australia

First published 1999

Typeset in 9.5/13pt Times New Roman [VN]

A catalogue record for this book is available from the British Library

Library of Congress Cataloguing in Publication data

Ketterson, J. B. (John Boyd)
Superconductivity – J. B. Ketterson and S. N. Song.
p. cm.
ISBN 0–521–56295–3 (hardcover). – ISBN 0–521–56562–6 (pbk.)
1. Superconductivity. 2. Superconductors. I. Song, S. N.
(Shengnian N.) II. Title.
QC611.95.K48 1998
621.3′5—dc21 97–3060 CIP

~~ISBN 0 521 56295 3 hardback~~
ISBN 0 521 56562 6 paperback

Transferred to digital printing 2003

Contents

Preface

A preface is supposed to alert potential readers to the contents and style of what lies within so they can decide whether to proceed further. Superconductivity is now a vast subject extending from the esoteric to the very practical; the people who study or work on it have different preparations, goals, and talents. No treatment can or should address all these dimensions.

Part I is devoted to the phenomenological aspects of superconductivity, e.g., London's and Pippard's electrodynamics, the Ginzburg–Landau theory and the Landau Fermi liquid theory. These theories allow a discussion of the effects of magnetic fields, interfaces and boundaries, fluctuations, and collective response (which may all be thought of as different manifestations of inhomogeneities).

Since there is currently much interest in unconventional (non-s-wave) superconductivity, we have included a discussion of the associated Ginzburg–Landau theory (which then has a multidimensional, complex order parameter). ^{3}He is the only established example of an unconventional superfluid (triplet p-wave) and therefore our discussions of this substance are somewhat longer.

Part II is devoted to the microscopic theory of uniform superconductors: the theory of Bardeen, Cooper, and Schrieffer (BCS) and the Bogoliubov–Valatin canonical transformation, where the latter so greatly simplifies the discussion of excited states and finite temperature effects. Although not strictly a uniform-superconductor phenomenon, the theory of tunneling and the accompanying Josephson effects are also discussed in Part II.

Part III deals with the microscopic theory of nonuniform superconductors exclusively through the apparatus of the self-consistent Bogoliubov equations as developed extensively by deGennes and coworkers (and hence sometimes referred to as the Bogoliubov–deGennes theory). Inhomogeneities associated with a magnetic field, impurities, and boundaries are discussed; temporal 'inhomogeneities' (e.g., relaxation phenomena and collective modes) are also discussed via the time-dependent Bogoliubov equations. Bogoliubov theory is extended to include unconventional BCS superfluids, such as ^{3}He and (possibly) high temperature superconductors and heavy fermion superconductors (UPt_3). Part III ends with a simplified discussion of Green's functions starting from the Bogoliubov theory. This formalism serves as an appropriate departure point for those wishing to go deeper into the theory of superconductivity. Real time Green's functions, thermal or Matsubara Green's functions, as well as quasiclassical Green's functions are all discussed.

Our goal in this book is to focus primarily on the physics of superconductivity. The materials aspect is generally ignored; material properties enter the discussions only via idealized parameters. For this reason the book may not contain enough material on high temperature superconductors to suit some readers' taste, although idealized layered systems are treated in some detail. However, it is our view that much of the high T_c literature does, in fact, involve the systematics of the superconducting properties with regard to their chemical and physical make-up. On the other hand, the underlying physical pairing mechanism and the order parameter symmetry are still controversial at this time (1997).

The level of the treatment varies between sections. Much of Part I is accessible to fourth year undergraduate students or first year graduate students in physics who have had exposure to mechanics, electricity and magnetism, and elementary quantum mechanics. Parts II and III involve some use of second quantization at a level usually arrived at by the end of the first year of graduate school. With some omissions the book could form the basis of a one semester graduate course on superconductivity; if all topics are discussed it would likely extend to a year.

The microscopic origin of the attractive electron–electron interaction involving the exchange of phonons is discussed (in Part II) only in terms of the jellium model which is sufficient to bring out the basic physics of the electron–phonon interaction. A proper discussion, however, requires solving the coupled Green's function equations of motion for the electrons and phonons, which is beyond the scope of this book. A simplified discussion of the spin fluctuation pairing mechanism is included since pairing in superfluid ^{3}He likely arises from this effect. Other pairing mechanisms, as have been proposed for high T_c materials, are ignored entirely.

The character of many of our discussions was strongly influenced by deGennes' 1966 classic, *Superconductivity of Metals and Alloys*; any similarity of our discussions and his is likely intentional. Both of the present authors are primarily experimentalists; their involvement in writing a largely theoretical book about superconductivity was primarily a self-education exercise. We have tried to use the simplest mathematical methods we can while minimizing the (pedagogically useless) 'it-can-be-shown' approach. Our decision to use a uniform theoretical approach to the topics in Part III meant that in many cases we had to devise our own methods; in so doing we have strived to get things right but inevitably errors will creep in (here and elsewhere), for which we apologize in advance. Incidentally we would appreciate hearing about any errors detected.

This book is not meant to be a treatise on superconductivity (for this kind of treatment we recommend the well-known text edited by R. Parks); i.e., it is not our purpose to collect most of what is known and useful, but rather to present the nuts and bolts used to obtain some of the important results.

The literature on superconductivity is enormous; it is therefore not possible to be aware of anything but a small fraction of it (this situation was 'worsened' by the discovery of high temperature superconductivity). For this reason we decided to minimize the number of original citations to those which are particularly important historically or pedagogically (a difficult and dangerous decision) or to those where the reader may seek further details. We hope the many workers not cited will forgive us.

We have benefited from discussions with (or encouragement by) P. Auvil, K. Bennemann, S. Doniach, A. Patashinskii, P. Wölfle, and S. K. Yip. The text was typed (again and again) by Ms A. Jackson who deserves the most thanks.

Northwestern University

J. B. Ketterson
S. N. Song

PART I PHENOMENOLOGICAL THEORIES OF SUPERCONDUCTIVITY

1 Introduction

Many metals, alloys, and intermetallic compounds[1] undergo a phase transition at some (generally low) temperature, T_c, to a state having zero electrical resistance (see Fig. 1.1), a phenomenon which is called superconductivity. Superconductivity was discovered by Kamerlingh Onnes in 1911; it has been sensitively probed by observing the magnetic field produced by a circulating supercurrent using a sensitive technique such as nuclear magnetic resonance. Favorable materials exhibit no detectable decay of this current for periods limited by the patience of the observer (~ years). However, superconductivity is better defined by the nature of the associated phase transition and it then includes materials or measurement conditions where the resistance may not vanish; here we postpone the discussion of these exceptions and initially restrict ourselves to perfect superconductors.

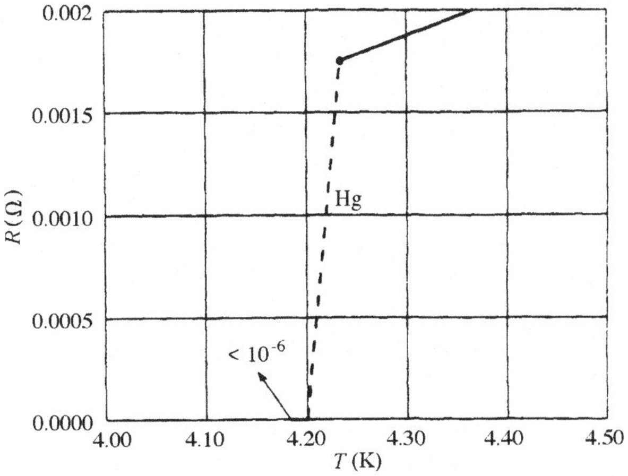

Figure 1.1 The original R vs T curve of Kamerlingh Onnes showing the superconducting transition in mercury. (After Kamerlingh-Onnes (1911).)

A second fundamental characteristic associated with a superconductor is the exclusion of magnetic flux, discovered by W. Meissner and R. Ochsenfeld in 1933 (which we will often refer to

1. An alloy is a solid solution of two or more elements at least one of which is a metal. An intermetallic compound involves a metal and one or more other elements which form a chemical compound having nearly precise ratios of its constituents.

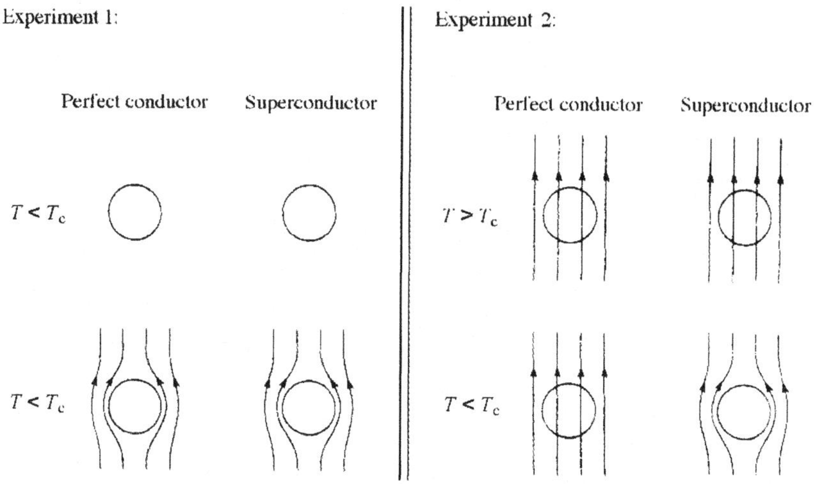

Figure 1.2 Hypothetical experiments showing the difference between a superconductor and a perfect conductor. Experiment 1: sample cooled in zero magnetic field after which a field is applied; experiment 2: sample cooled in applied magnetic field.

simply as the Meissner effect in what follows). The degree of flux exclusion can depend on the material or measurement conditions; we will also assume this property is nearly perfectly displayed for our initial discussion.

The combination of zero resistance and perfect diamagnetism results in a clear distinction between a superconductor (which as we will see is in a thermodynamic state) and a hypothetical 'perfect conductor' (which has the unique *transport property* of zero resistance); this difference is illustrated in Fig. 1.2 and involves the differing response each would have for different histories of cooling below the transition temperature and applying a magnetic field. If we start by cooling through T_c and then apply a magnetic field, both the superconductor and the perfect conductor would exclude the field. For the perfect conductor, induced currents arising from Faraday's law would screen the flux. Flux exclusion in the superconductor could be assigned to the same mechanism or (more fundamentally) to the Meissner effect itself. If we reverse the order by first applying the field and then cooling through T_c, the superconductor and the perfect conductor behave differently: the superconductor excludes the flux (the Meissner effect); the perfect conductor would remain fully permeated by the field.

These experimental observations argue that the transition associated with superconductivity is indeed a phase transition since an equilibrium thermodynamic state is defined by its independent thermodynamic variables (in this case T and H), and is *independent of its history* (which as we see is not true for the perfect conductor).

Superconductivity, and with it the Meissner effect, does not persist to arbitrarily high magnetic fields. For each temperature there is a well-defined critical field, $H_c(T)$, at which superconductivity disappears.[2] Fig. 1.3 shows a universal curve of the temperature dependence of H_c vs T.

2. This statement is restricted to so-called Type I materials having a shape for which there is no demagnetization effect, as will be discussed in later sections.

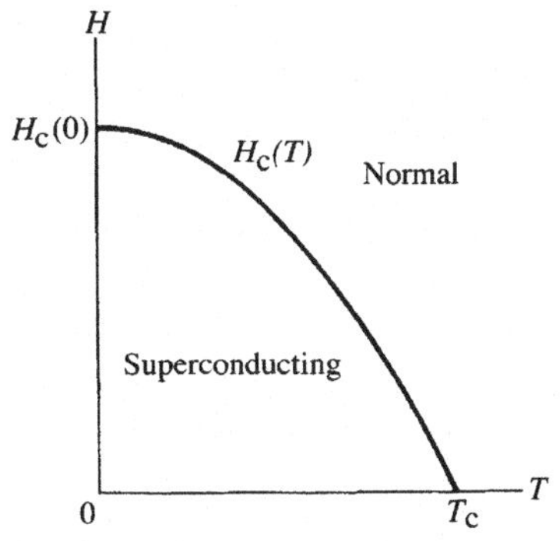

Figure 1.3 The temperature dependence of the critical field $H_c(T)$.

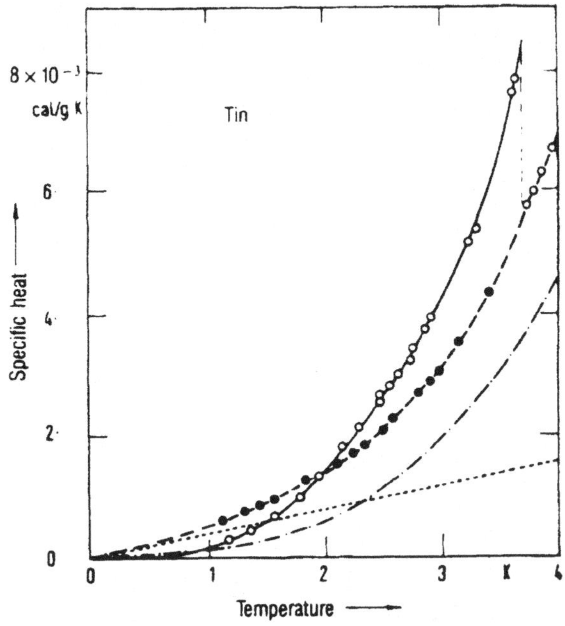

Figure 1.4 Specific heat of tin as a function of temperature: open circles without an external magnetic field; filled circles with an external field $H > H_c$. Shown also are the individual contributions from the electrons and the lattice for $H > H_c$: chain line – lattice contribution; dashed line – electron contribution. (After Keesom and van Laer (1938).)

A superconductor also exhibits a discontinuous increase in its specific heat at T_c (there is no latent heat at zero field) below which it drops rapidly (approaching an exponential dependence at low temperatures). A discontinuity in the specific heat is a signature of a second order phase transition providing added evidence that superconductivity is associated with a distinct thermodynamic phase. Fig. 1.4 shows the heat capacity of a typical superconductor.

There is a related low temperature phenomenon known as superfluidity which occurs in liquid helium. There are actually two such liquids involving the isotopes ^{3}He and ^{4}He having superfluid transitions near 2×10^{-3} K and 2 K, respectively, depending on the pressure. For some purposes a superconductor may be regarded as a superfluid having an electric charge. ^{4}He superfluidity involves a Bose condensation, a phenomenon which is related to superconductivity in a somewhat subtle or indirect way and will not be discussed in this book. The superfluidity of ^3He is intimately related to superconductivity and will be addressed in later sections. Another related superfluid is the neutron liquid in the interior of neutron stars.

2 The London–London equation

We now present two derivations of equations which are useful in describing many of the magnetic properties of superconductors. The older approach, used by F. London and H. London (1935) (see also London (1950)), starts with the Drude–Lorentz equation of motion for electrons in a metal, which is Newton's law for the velocity, $\mathbf{v}$, of an electron with mass, m, and charge, e, in an electric field, $\mathbf{E}$, with a phenomenological viscous drag proportional to $\mathbf{v}/\tau$:

$$m\left(\dot{\mathbf{v}} + \frac{1}{\tau}\mathbf{v}\right) = e\mathbf{E}. \tag{2.1}$$

For a perfect conductor $\tau \to \infty$. Introducing the current density $\mathbf{j} = ne\mathbf{v}$, where n is the conduction electron density, Eq. (2.1) can be written

$$\frac{\mathrm{d}\mathbf{j}}{\mathrm{d}t} = \frac{ne^2}{m}\mathbf{E}, \tag{2.2}$$

which is referred to as the first London equation. The time derivative of Maxwell's fourth equation (in cgs units) is

$$\nabla \times \frac{\partial \mathbf{H}}{\partial t} = \frac{4\pi}{c}\frac{\partial \mathbf{j}}{\partial t} + \frac{\varepsilon}{c}\frac{\partial^2 \mathbf{E}}{\partial t^2}, \tag{2.3}$$

where ε is the dielectric constant. Taking the curl of (2.3) and using (2.2) we have

$$\nabla \times \left(\nabla \times \frac{\partial \mathbf{H}}{\partial t}\right) = \left(\frac{4\pi ne^2}{mc} + \frac{\varepsilon}{c}\frac{\partial^2}{\partial t^2}\right)\nabla \times \mathbf{E}; \tag{2.4}$$

using $\nabla \times \mathbf{E} = -(1/c)\partial\mathbf{H}/\partial t$ we have

$$\nabla \times \left(\nabla \times \frac{\partial}{\partial t}\mathbf{H}\right) + \left(\frac{1}{\lambda_L^2} + \frac{\varepsilon}{c^2}\frac{\partial^2}{\partial t^2}\right)\frac{\partial}{\partial t}\mathbf{H} = 0, \tag{2.5}$$

where we introduced the London depth, λ_L, defined by

$$\frac{1}{\lambda_L^2} = \frac{4\pi ne^2}{mc^2}. \tag{2.6}$$

Eq. (2.5) has been obtained for a perfect conductor model. In order to conform with the experimentally observed Meissner effect, we must exclude time-independent field solutions arising from integrating (2.5) once with respect to time and we therefore write

$$\nabla \times (\nabla \times \mathbf{H}) + \left(\frac{1}{\lambda_L^2} + \frac{\varepsilon}{c^2}\frac{\partial^2}{\partial t^2}\right)\mathbf{H} = 0; \tag{2.7}$$

this is referred to as the second London equation. In what follows we will refer to Eq. (2.7) simply as the London equation.

An alternate derivation of (2.7) is motivated by the idea that some of the moving electrons behave collectively as a superfluid, a liquid possessing no viscosity. This concept is borrowed from the physics of liquid ^{4}He; below 2.19 K this system behaves as if it were composed of a mixture of two liquids: a superfluid, having no viscosity, and a normal liquid, having a finite viscosity. We assume that the free energy of a superfluid consists of three parts

$$F = F_N + E_{kin} + E_{mag}, \tag{2.8}$$

where F_N is the free energy associated with the normal liquid, E_{kin} is the kinetic energy of the moving superfluid, and E_{mag} is the magnetic field energy. We may write these latter two terms as

$$E_{mag} = (1/8\pi)\int H^2(\mathbf{r})\mathrm{d}^3r \tag{2.9}$$

and

$$E_{kin} = (1/2)\int \rho(\mathbf{r})v^2(\mathbf{r})\mathrm{d}^3r, \tag{2.10}$$

where $\rho(\mathbf{r})$ is the mass density associated with the superfluid. Writing $\rho = nm$ and $\mathbf{v} = (1/ne)\mathbf{j}$ and using the fourth Maxwell equation $\nabla \times \mathbf{H} = (4\pi/c)\mathbf{j}$, Eq. (2.10) becomes

$$E_{kin} = \frac{1}{8\pi}\int \frac{mc^2}{4\pi ne^2}(\nabla \times \mathbf{H})^2\mathrm{d}^3r; \tag{2.11}$$

n is now interpreted as the density of superconducting electrons. We will assume that the superconducting electrons adjust their motion so as to minimize the total free energy; this requires $\delta(E_{mag} + E_{kin}) = 0$ or

$$\int\left\{\mathbf{H}(\mathbf{r})\cdot\delta\mathbf{H}(\mathbf{r}) + \frac{mc^2}{4\pi ne^2}[\nabla \times \mathbf{H}(\mathbf{r})]\cdot[\nabla \times \delta\mathbf{H}(\mathbf{r})]\right\}\mathrm{d}^3r = 0, \tag{2.12}$$

where $\delta\mathbf{H}(\mathbf{r})$ is a variation of the (initially unknown) function $\mathbf{H}(\mathbf{r})$. Integrating the second term by parts (and placing the resulting surface outside the superconductor) we obtain

$$\int[\mathbf{H}(\mathbf{r}) + \lambda_L^2\nabla \times (\nabla \times \mathbf{H})]\cdot\delta\mathbf{H}(\mathbf{r})\mathrm{d}^3r = 0. \tag{2.13}$$

Since the variation $\delta\mathbf{H}(\mathbf{r})$ is arbitrary, the term in the square brackets must vanish; therefore

$$\nabla \times (\nabla \times \mathbf{H}) + \lambda_L^2\mathbf{H} = 0, \tag{2.14}$$

which is equivalent to (2.7) (including the displacement term in Maxwell's equation yields the last term in (2.7), which is negligible for most applications).

As a simple application of Eq. (2.14) we now discuss the behavior of a superconductor in a magnetic field near a plane boundary. Consider first the case of a field perpendicular to a superconductor surface lying in the x–y plane with no current flowing in the z direction. From the second Maxwell equation, $\nabla \cdot \mathbf{H} = 0$, we obtain $\partial H_z/\partial z = 0$ or $H = \text{const.}$ From the fourth Maxwell equation, $\nabla \times \mathbf{H} = (4\pi/c)\mathbf{j}$, the first term in (2.14) vanishes and hence $\mathbf{H} = 0$ is the only solution. Thus a superconductor exhibiting the Meissner effect cannot have a field component perpendicular to its surface.

As the second example consider a field lying parallel to the superconductor surface, e.g., $\mathbf{H} \parallel \hat{\mathbf{x}}$, which we may write as $\mathbf{H} = H(z)\hat{\mathbf{x}}$ (which satisfies $\nabla \cdot \mathbf{H} = 0$). Using the vector identity

$$\nabla \times (\nabla \times \mathbf{H}) = \nabla(\nabla \cdot \mathbf{H}) - \nabla^2 \mathbf{H}, \tag{2.15}$$

Eq. (2.14) becomes

$$\frac{\partial^2 H_x}{\partial z^2} - \frac{1}{\lambda_L^2} H_x = 0 \tag{2.16}$$

or (for a superconductor occupying the region $z > 0$)

$$\mathbf{H}(z) = \hat{\mathbf{x}} H_x(0) e^{-z/\lambda_L}. \tag{2.17}$$

A field parallel to the surface is therefore allowed; however, it decays exponentially, with a characteristic length, λ_L, in the interior; λ_L ($T = 0$) ranges from 500 to 10 000 Å, depending on the material. Accompanying this parallel field is a surface current density, which, from Maxwell's fourth equation, is

$$\mathbf{j}(z) = -\frac{c}{4\pi\lambda_L} H_x(0) e^{-z/\lambda_L} \hat{\mathbf{y}}. \tag{2.18}$$

This current density shields or screens the magnetic field from the interior of the superconductor.

3 Pippard's equation

At temperatures well below the superconducting transition temperature the heat capacity of a superconductor displays an exponential behavior $C \sim e^{-\Delta/k_B T}$ (see Fig. 1.4). This suggests that the conduction electron spectrum develops an energy gap, Δ (not to be confused with the gap in a semiconductor); recall that electrons in normal metals have a continuous (gapless) distribution of energy levels near the Fermi energy, μ. On dimensional grounds one can construct a quantity having the units of length from Δ and the Fermi velocity, v_F; we define the so-called coherence length by

$$\xi_0 = \frac{\hbar v_F}{\pi \Delta}. \tag{3.1}$$

This length bears no resemblance to the London depth, λ_L, and hence represents a different length scale affecting the behavior of a superconductor; it can be interpreted as a characteristic length which measures the spatial response of the superconductor to some perturbation (e.g. the distance over which the superconducting state develops at a normal metal–superconductor boundary). Length scales of this kind were introduced independently by Ginzburg and Landau (1950) and by Pippard (1953).[1] We first discuss Pippard's phenomenological theory (which semiquantitatively captures the main features of the microscopic theory to be discussed later). We begin by writing London's equation in an alternative form. Substituting $\nabla \times \mathbf{H} = (4\pi/c)\mathbf{j}$ (from the fourth Maxwell equation) in the London equation yields

$$\begin{aligned} \nabla \times \mathbf{j} &= -\frac{c}{4\pi\lambda_L^2}\mathbf{H} \\ &= -\frac{ne^2}{mc}\mathbf{H}. \end{aligned} \tag{3.2}$$

We next write $\mathbf{H} = \nabla \times \mathbf{A}$, where $\mathbf{A}$ is the magnetic vector potential, and restrict the gauge to satisfy

$$\nabla \cdot \mathbf{A} = 0 \qquad \text{(London gauge)} \tag{3.3}$$

and the boundary condition

1. These length scales are not identical, however; the Pippard length is temperature-independent while the Ginzburg–Landau length depends on temperature. The Pippard coherence length is related to the BCS coherence length.

$$\mathbf{A}_n = 0, \tag{3.4}$$

where $\mathbf{A}_n$ is the component of **A** perpendicular to the superconductor surface. London's equation may then be written

$$\mathbf{j} = -\frac{ne^2}{mc}\mathbf{A}. \tag{3.5}$$

Note that the condition (3.4) builds in the reasonable boundary condition that the normal component of the supercurrent, $\mathbf{j}_n$, vanishes at a boundary (this is a good boundary condition at a superconductor–insulator boundary but will require modification for metal–superconductor or superconductor–superconductor boundaries).

To generalize (3.5), Pippard reasoned that the relation between **j** and **A** should be nonlocal, meaning that the current **j**(**r**) at a point **r** involves contributions from **A**(**r**′) at neighboring points **r**′ located in a volume with a radius of order ξ_0 surrounding **r**. The mathematical form he selected was guided by the nonlocal relation between the electric field, **E**(**r**′), and the current, **j**(**r**), which had been developed earlier by Chambers (1952). The expression employed by Pippard was

$$\mathbf{j}(\mathbf{r}) = -C\int\frac{[\mathbf{A}(\mathbf{r}')\cdot\mathbf{R}]\mathbf{R}}{R^4}\mathrm{e}^{-R/\xi_0}\mathrm{d}^3r', \tag{3.6}$$

where $\mathbf{R} = \mathbf{r} - \mathbf{r}'$. The constant C is fixed by requiring (3.6) reduce to (3.5) in the quasiuniform limit where we may take **A** from under the integral sign; we then have

$$C\int\cos^2\theta\,\mathrm{d}\Omega\int\mathrm{e}^{-R/\xi_0}\mathrm{d}R = \frac{ne^2}{mc}, \tag{3.7}$$

from which we obtain $C = 3ne^2/4\pi\xi_0 mc$. Pippard's generalization of London's equation is then

$$\mathbf{j}(\mathbf{r}) = -\frac{3ne^2}{4\pi\xi_0 mc}\int\frac{[\mathbf{A}(\mathbf{r}')\cdot\mathbf{R}]\mathbf{R}}{R^4}\mathrm{e}^{-R/\xi_0}\mathrm{d}^3r'. \tag{3.8}$$

Since Eq. (3.8) involves two functions, **A**(**r**) and **j**(**r**), a complete description requires a second equation which is obtained by substituting $\mathbf{H} = \nabla\times\mathbf{A}$ in the fourth Maxwell equation to obtain

$$\nabla\times(\nabla\times\mathbf{A}) = \frac{4\pi}{c}\mathbf{j} \tag{3.9}$$

(resulting in an overall integrodifferential equation for **A**).

Eq. (3.8) applies only to a bulk superconductor. An important question we would like to examine is the behavior of a magnetic field near a surface, which will require a modification (or reinterpretation) of (3.8). To model the effect of the surface the integration over points **r**′ is restricted to the interior of the superconductor. If the surface is highly contorted, then it can happen that two points near the surface and separated by about a coherence length cannot be connected by a straight electron trajectory, without passing through the vacuum; one then has to account for this shadowing effect. We restrict ourselves here to plane boundaries which we take to be normal to the z direction.

In the limit $\lambda_L >> \xi_0$ Eq. (3.8) reduces to the London equation, as discussed above. (By expanding **A**(**r**′) in a power series in **R**, we may obtain corrections to the London equation due to nonlocality.) In the opposite limit, $\lambda_L << \xi_0$, **A**(**r**′) varies rapidly. Let us assume that **A**(**r**) falls off over a characteristic distance λ (which we will determine shortly through a self-consistency argument). When $\lambda < \xi_0$, the value of the integral (3.8) will be reduced roughly by a factor (λ/ξ_0); i.e.,

$$\mathbf{j}(\mathbf{r}) = -\frac{\lambda}{\xi_0}\frac{ne^2}{mc}\mathbf{A}(\mathbf{r}). \tag{3.10}$$

We may also write (3.10) in the 'London-like' form

$$\nabla \times (\nabla \times \mathbf{H}) + \frac{\lambda}{\xi_0}\cdot \lambda_L^{-2}\mathbf{H} = 0. \tag{3.11}$$

This equation has solutions which decay in a characteristic length $(\lambda_L^2\xi_0/\lambda)^{1/2}$; to achieve self-consistency we set this length equal to λ:

$$\lambda^3 = \lambda_L^2\xi_0. \tag{3.12}$$

(A more rigorous derivation from the microscopic theory carried out in Sec. 46 yields $\lambda^3 = 0.62\lambda_L^2\xi_0$.) We conclude that in the Pippard limit the effective penetration depth λ is larger than the London depth, λ_L: $\lambda/\lambda_L = (\xi_0/\lambda_L)^{1/3} > 1$. At the same time λ remains smaller than the coherence length: $\lambda/\xi_0 = (\lambda_L/\xi_0)^{2/3} < 1$.

If our metal has impurities it is natural to assume the relation between the current and vector potential will be altered. To account for the effects of electron scattering, Pippard modified the coherence length factor in the exponent of (3.8) as $1/\xi_0 \to (1/\xi_0) + (1/\ell)$ where ℓ is the electron mean free path;[2] the coefficient in front of the integral was not altered. Eq. (3.8) then becomes

$$\mathbf{j}(\mathbf{r}) = -\frac{3ne^2}{4\pi\xi_0 mc}\int \frac{[\mathbf{A}(\mathbf{r}')\cdot\mathbf{R}]\mathbf{R}}{R^4}\,\mathrm{e}^{-R\left(\frac{1}{\xi_0}+\frac{1}{\ell}\right)}\mathrm{d}^3r'. \tag{3.13}$$

In the limit $\lambda >> (\ell, \xi_0)$ we may again take $\mathbf{A}(\mathbf{r})$ from under the integral sign; carrying out the integration we obtain

$$\mathbf{j}(\mathbf{r}) = -\frac{ne^2}{mc\xi_0}\cdot\frac{1}{\frac{1}{\xi_0}+\frac{1}{\ell}}\mathbf{A}(\mathbf{r}). \tag{3.14}$$

For the case of a very dirty metal, where $\ell << \xi_0$,

$$\mathbf{j}(\mathbf{r}) = -\frac{ne^2}{mc}\frac{\ell}{\xi_0}\mathbf{A}(\mathbf{r}). \tag{3.15}$$

The effective penetration depth is then obtained from the expression

$$\lambda = \left(\frac{\lambda_L^2\xi_0}{\ell}\right)^{1/2}. \tag{3.16}$$

If $\lambda << (\ell, \xi_0)$ we continue to use Eq. (3.12).

The magnetic response of a superconductor depends on whether $\lambda \gtrsim \xi$ or $\lambda \lesssim \xi$, as will be developed later.[3] These regimes are designated in Table 3.1. A Type I superconductor displays a

2. Other expressions, such as $\xi = (\ell\xi_0)^{1/2}$, are sometimes used to estimate the effective coherence length. Such ambiguities are removed by the microscopic theory to be discussed in Part III.

Table 3.1 *Regimes defining Type I and Type II superconductors.*[3]

Type I (or Pippard)	$\lambda < \xi$
Type II (or London)	$\lambda > \xi$

complete Meissner effect (flux exclusion) up to some critical field H_c, above which it becomes normal. The magnetic response of a Type II superconductor is more complex and will be developed later.

3. The precise criterion separating the regimes is $\lambda = \xi/\sqrt{2}$.

4 Thermodynamics of a Type I superconductor

In this section we consider the thermodynamics of a Type I superconductor (Gorter and Casimir 1934a,b); our discussion follows that of London (1950). We restrict the geometry of the superconductor to a form for which the external field, **H**, is not distorted by the presence of the superconductor (examples being an infinitely long cylinder with **H** parallel to the axis or a plane slab of infinite extent with **H** parallel to its surface). Far inside the superconductor (i.e., several London depths from the surface) the magnetic field essentially vanishes in the superconducting state and is equal to **H** in the normal state. In the thermodynamic identities that follow we identify this interior field as **B**, the flux density. The **H** field will be taken as the applied external field. The relation between **B** and **H** is shown in Fig. 4.1.

We recall the thermodynamic identity for the response of a system in a magnetic field

$$\mathrm{d}\mathscr{E} = T\mathrm{d}\mathscr{S} + \frac{1}{4\pi}\mathbf{H}\cdot\mathrm{d}\mathbf{B}, \tag{4.1}$$

where $\mathscr{E}$ is the energy density and $\mathscr{S}$ is the entropy density. When T and **B** are the independent variables we use the Helmholtz free energy density, $\mathscr{F} = \mathscr{E} - T\mathscr{S}$, and when T and **H** are the independent variables we use the Gibbs free energy density, $\mathscr{G} = \mathscr{F} - (1/4\pi)\mathbf{B}\cdot\mathbf{H}$; taking the differential of these two quantities and using (4.1) yields

$$\mathrm{d}\mathscr{F} = -\mathscr{S}\mathrm{d}T + \frac{1}{4\pi}\mathbf{H}\cdot\mathrm{d}\mathbf{B} \tag{4.2}$$

and

$$\mathrm{d}\mathscr{G} = -\mathscr{S}\mathrm{d}T - \frac{1}{4\pi}\mathbf{B}\cdot\mathrm{d}\mathbf{H}. \tag{4.3}$$

A Type I superconductor displays the Meissner–Ochsenfeld properties:

$$B = 0, \qquad H < H_c; \tag{4.4}$$

$$B = H, \qquad H > H_c. \tag{4.5}$$

H_c is called the thermodynamic critical field. Since **H** and T will be our independent variables, we integrate (4.3) at constant T to obtain

$$\mathscr{G}(T,H) = \mathscr{G}(T,0) - \frac{1}{4\pi}\int \mathbf{B}\cdot\mathrm{d}\mathbf{H};$$

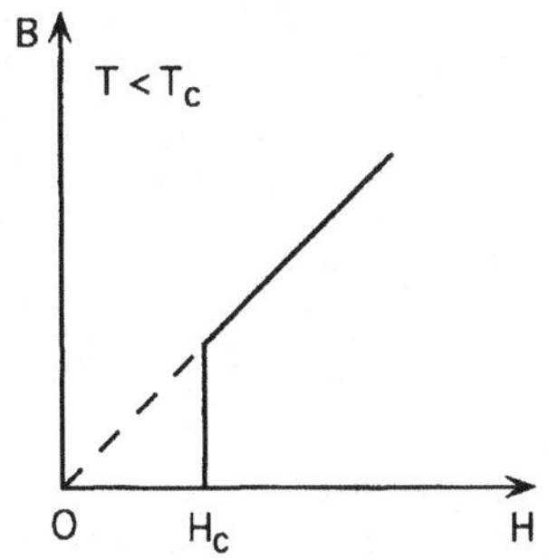

Figure 4.1 B vs H curve for a Type I superconductor.

therefore

$$\mathscr{G}(T,H) = \mathscr{G}(T,0) - \frac{1}{8\pi}(H^2 - H_c^2) \qquad (H > H_c) \tag{4.6}$$

or

$$\mathscr{G}(T,H) = \mathscr{G}(T,0) \qquad (H < H_c). \tag{4.7}$$

Note that $\mathscr{G}$ is continuous at the transition. We define a function

$$\mathscr{G}_0 \equiv \mathscr{G}(T,0) + \frac{1}{8\pi}H_c^2. \tag{4.8}$$

We may then write

$$\begin{aligned}\mathscr{G}(T,H) &= \mathscr{G}_0 - \frac{1}{8\pi}H^2 \qquad (H > H_c)\\ &\equiv \mathscr{G}_N\end{aligned} \tag{4.9}$$

and

$$\begin{aligned}\mathscr{G}(T,H) &= \mathscr{G}_0 - \frac{1}{8\pi}H_c^2 \qquad (H < H_c)\\ &\equiv \mathscr{G}_S,\end{aligned} \tag{4.10}$$

where $\mathscr{G}_N$ and $\mathscr{G}_S$ denote the normal and superconducting states, respectively. We may interpret $\mathscr{G}_0$ as the Gibbs free energy of the normal metal at zero field (were it stable); hence the Gibbs free energy density of the superconducting state is lower than that of the normal state by $(1/8\pi)H_c^2(T)$; this quantity is referred to as the *condensation energy*. Since $\mathscr{G} \equiv \mathscr{F} - (1/4\pi)\mathbf{H}\cdot\mathbf{B}$, $\mathscr{F}_S(T,0) = \mathscr{G}_0 - (1/8\pi)H_c^2$. The normal state Helmholtz free energy density is then

$$\mathscr{F}_N = \mathscr{F}_S + \frac{H_c^2}{8\pi}. \tag{4.11}$$

From Eq. (4.3),

$$\mathscr{S} = -\left(\frac{\partial \mathscr{G}}{\partial T}\right)_H; \tag{4.12}$$

thus

$$\mathscr{S}_{\mathrm{N}} = -\frac{\partial \mathscr{G}_0}{\partial T} \tag{4.13}$$

and

$$\mathscr{S}_{\mathrm{S}} = -\frac{\partial \mathscr{G}_0}{\partial T} + \frac{1}{4\pi} H_{\mathrm{c}} \left(\frac{\partial H_{\mathrm{c}}}{\partial T} \right)_H . \tag{4.14}$$

Note the entropy is discontinuous across the transition and hence we have a first order transition (when $H \neq 0$):

$$\mathscr{S}_{\mathrm{N}} - \mathscr{S}_{\mathrm{S}} = -\frac{1}{4\pi} H_{\mathrm{c}}(T) \left[\frac{\partial H_{\mathrm{c}}(T)}{\partial T} \right]_H . \tag{4.15}$$

The heat of the transition is

$$\mathscr{Q} = T(\mathscr{S}_{\mathrm{N}} - \mathscr{S}_{\mathrm{S}}) = -\frac{T}{4\pi} H_{\mathrm{c}}(T) \left[\frac{\partial H_{\mathrm{c}}(T)}{\partial T} \right]_H ; \tag{4.16}$$

this equation corresponds to the Clausius–Clapeyron equation of a (P,V,T) system. The specific heat (at constant H) is defined as

$$\mathscr{C}_H = T \left(\frac{\partial \mathscr{S}}{\partial T} \right)_H , \tag{4.17}$$

or

$$\begin{aligned} \Delta\mathscr{C}_H &\equiv \mathscr{C}_{\mathrm{N}} - \mathscr{C}_{\mathrm{S}} \\ &= -\frac{T}{4\pi} \left[H_{\mathrm{c}} \frac{\partial^2 H_{\mathrm{c}}}{\partial T^2} + \left(\frac{\partial H_{\mathrm{c}}}{\partial T} \right)^2 \right] . \end{aligned} \tag{4.18}$$

At $H = 0$, where the transition is second order,

$$\Delta\mathscr{C}_H |_{T=T_{\mathrm{c}}} = -\frac{T_{\mathrm{c}}}{4\pi} \left(\frac{\partial H_{\mathrm{c}}}{\partial T} \right)^2_{T=T_{\mathrm{c}}} ; \tag{4.19}$$

this is sometimes called Rutgers' formula (Rutgers 1933).

5 The intermediate state

If a superconducting body of arbitrary shape is placed in a magnetic field, the flux exclusion associated with the Meissner effect will in general distort the magnetic field. Exceptions are an infinite cylinder with the field parallel to the axis, or a sheet or half space with H_0 parallel to the plane of symmetry. For situations involving lower symmetry the local magnetic field can vary over the surface, being both higher and lower than the applied field, H_0. As a simple example consider the case of a spherical superconductor shown in Fig. 5.1. From magnetostatics the field will be highest at the equator (on the circle C in Fig. 5.1) where it is $H = \frac{3}{2}H_0$. Hence flux enters the sample, not at the thermodynamic critical field, H_c, but at a value $H_0 = \frac{2}{3}H_c$. For magnetic fields $H_c > H_0 > \frac{2}{3}H_c$ the sample consists of alternating domains of normal metal and superconductor. A superconductor in such a regime is said to be in the intermediate state (Landau 1937).

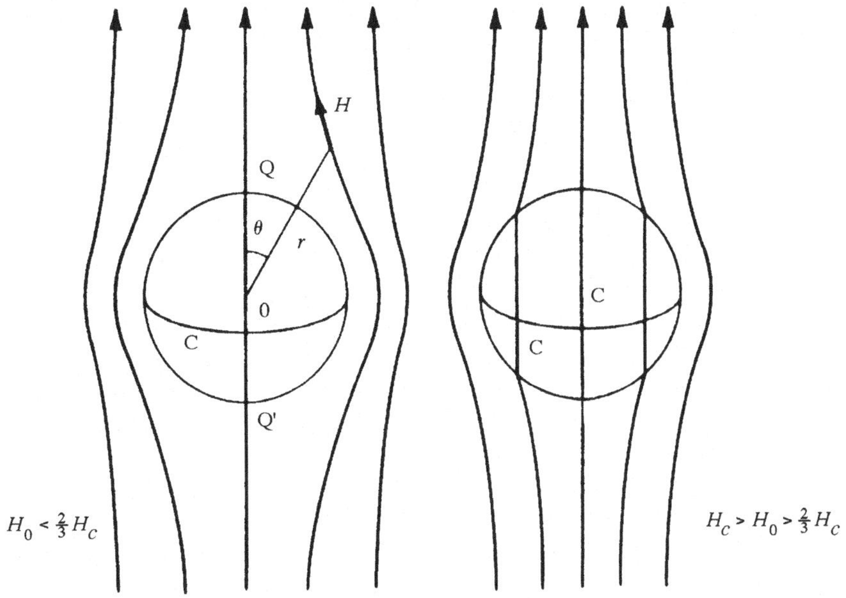

Figure 5.1 The magnetic field distribution about a superconducting sphere of radius a. For an applied field $H_0 < \frac{2}{3}H_c$, there is a complete Meissner effect and the field at the equator (at any point on circle C) is $\frac{3}{2}H_0$; the field at the poles (Q, Q′) is zero. For $\frac{2}{3}H_c < H_0 < H_c$, the sphere is in the intermediate state.

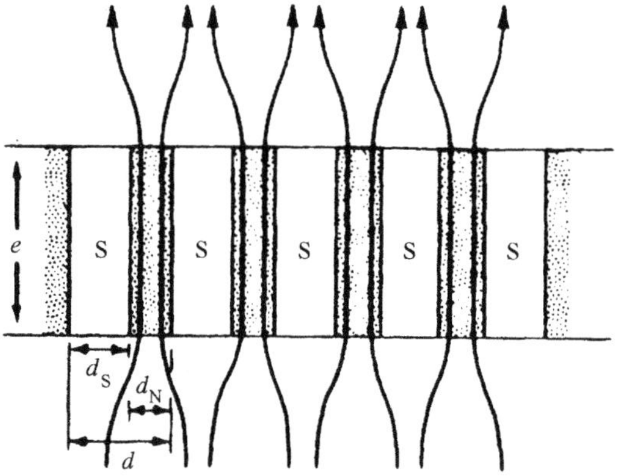

Figure 5.2 Schematic of the magnetic field distribution in a slab in a perpendicular field. For all fields $H_0 < H_c$, the sample is in an intermediate state having a laminar structure of normal and superconducting domains.

We will limit our discussion to the simple case of a plane superconducting sheet with H_0 parallel to the surface normal. From our earlier discussion we know that a superconductor cannot sustain a field component perpendicular to its surface. The field behavior is shown qualitatively in Fig. 5.2. It has the following features.

(i) For a magnetic field $0 < H_0 < H_c$ the sample consists of adjacent domains which are wholly superconducting (with no internal flux) or normal (with $H_{\text{local}} = H_c$).

(ii) In the interior of the superconductor and far from the surface, the domain walls are parallel to the applied field direction; the fraction of the cross section that is superconducting is fixed by H_0 and H_c such that the total flux through the sample is conserved

$$\frac{\text{normal cross section}}{\text{total cross section}} = \frac{H_0}{H_c}. \tag{5.1}$$

Hence the superconducting fraction is $1 - H_0/H_c$ (which vanishes, as it should, for $H_0 = H_c$).

(iii) Near the surfaces the flux sheets 'flare out' (which reduces the field curvature, which would otherwise raise the local fields at the interface). Were the local field to remain fixed at H_c, to sustain the average flux, needle-like superconducting domains would have to encroach on the normal domains in the vicinity of the surface. Such needles do not occur since the field profile is controlled by a minimization of the surface energy (between superconducting and free surface or the normal regions, see next section), the superfluid kinetic energy, and the total magnetic field energy.

A detailed treatment of the domain structure is mathematically complex and will not be dealt with here (see Landau, Lifshitz, and Pitaevskii (1984), deGennes (1966), London (1950)).

6 Surface energy between a normal and a superconducting metal

Consider a slab parallel to the x–y plane in a perpendicular magnetic field parallel to z. Assume we have a phase boundary perpendicular to the x axis with the superconductor occupying the region $x > 0$. The total free energy, F, in the London model is:

$$F = \mathscr{A}\int_0^\infty \mathrm{d}x\left[\mathscr{F}_\mathrm{S} + \frac{H^2(\mathbf{r})}{8\pi} + \frac{\lambda_\mathrm{L}^2}{8\pi}[\boldsymbol{\nabla}\times\mathbf{H}(\mathbf{r})]^2\right], \tag{6.1}$$

where $\mathscr{A}$ is the interface area, $\mathscr{F}_\mathrm{S}$ is the condensation energy density, $H_\mathrm{c}^2(T)/8\pi$, and the second and third terms are the magnetic field energy density and the superfluid electron-kinetic energy density, respectively. At our phase boundary in the intermediate state, where $H = H_\mathrm{c}$ for $x < 0$, we have (see Eqs. (4.6) and (4.7))

$$G_\mathrm{N} = G_\mathrm{S}. \tag{6.2}$$

By definition

$$G = F - \frac{\mathscr{A}}{4\pi}\int_0^\infty \mathbf{H}_\mathrm{c}\cdot\mathbf{B}(x)\mathrm{d}x. \tag{6.3}$$

According to the discussion at the beginning of Sec. 4 we set $\mathbf{B}(x) = \mathbf{H}(x)$; from Eq. (2.17) we have $\mathbf{H}(x) = H_\mathrm{c}\hat{\mathbf{z}}\mathrm{e}^{-x/\lambda_\mathrm{L}}$. Inserting this form in (6.3) and using (6.1) we obtain

$$G = \mathscr{A}\int_0^\infty\left[\mathscr{F}_\mathrm{S} + H_\mathrm{c}^2\left(\frac{1}{8\pi}\mathrm{e}^{-2x/\lambda_\mathrm{L}} + \frac{1}{8\pi}\mathrm{e}^{-2x/\lambda_\mathrm{L}} - \frac{1}{4\pi}\mathrm{e}^{-x/\lambda_\mathrm{L}}\right)\right]\mathrm{d}x$$

$$\equiv \int_0^\infty \mathscr{F}_\mathrm{S}\mathrm{d}^3r + \gamma\mathscr{A}, \tag{6.4}$$

where γ is the surface energy per unit area, i.e., the 'surface tension', given by

$$\gamma = \frac{H_\mathrm{c}^2}{4\pi}\left(-\frac{\lambda_\mathrm{L}}{2}\mathrm{e}^{-2x/\lambda_\mathrm{L}} + \lambda_\mathrm{L}\mathrm{e}^{-x/\lambda_\mathrm{L}}\right)\Bigg|_0^\infty$$

$$= -\lambda_\mathrm{L}\frac{H_\mathrm{c}^2}{8\pi}. \tag{6.5}$$

Note this surface energy is *negative*. This suggests the system can lower its energy by maximizing the interfacial area (i.e., the system is unstable to the formation of multiple domains with associated interfaces).

Type I superconductors (for geometries not possessing an intermediate state) display a single domain for $H \leq H_c$; hence they must have a positive domain wall energy. Physically this positive surface energy arises because superconductivity is destroyed over a region of order ξ perpendicular to the interface; i.e., we lose the condensation energy over a volume of order $\mathscr{A}\xi$, where ξ is a coherence length. This is equivalent to a positive contribution to the surface energy of order

$$\gamma = \xi \frac{H_c^2}{8\pi}. \tag{6.6}$$

In a Type I material $\xi > \lambda_L$ and hence the positive contribution (6.6) outweighs the negative contribution (6.5) and the interface is stable. In Sec. 9.3 we will continue this discussion and derive an expression for the surface tension.

For a Type II material the system does, in some sense, try to maximize the amount of internal interfacial area above some field (referred to as the lower critical field); however, it is subject to a constraint imposed by quantum mechanics, as we discuss in the next two sections.

7 Quantized vorticity

The previous discussion of the surface energy of a normal metal/superconductor interface suggests that Type II materials, where $\xi \lesssim \lambda$ are unstable to the formation of domain structures which in some way maximize the amount of interface area. Two possible domain geometries are: (i) an array of nested sheets (closed or open, depending on the geometry[1]) and (ii) a two-dimensional lattice of flux filaments. Calculations show the latter domain structure to be more stable.

Since the filaments (are presumed to) admit flux into the interior of the superconductor, we envision them as having a normal core with a diameter of order ξ, outside of which supercurrents flow in a diameter of order λ, which produce the internal field via Ampère's law.

As a primitive model of a single flux filament we consider the extreme limit $\xi \to 0$ for which the London approach should provide a good description. We recall Eq. (2.5) associated with our first derivation of the London equation

$$\frac{\partial}{\partial t}\left[\mathbf{\nabla} \times (\mathbf{\nabla} \times \mathbf{H}) + \frac{4\pi n e^2}{mc^2}\mathbf{H}\right] = 0; \tag{7.1}$$

we next integrate (7.1) over an area $\mathscr{A}$ intersecting the filament (for convenience we choose a plane perpendicular to its axis) and use Ampere's law (Maxwell's fourth equation) in the form

$$\frac{\partial}{\partial t}\int\left[\frac{mc}{ne^2}\mathbf{\nabla} \times \mathbf{j} + \mathbf{H}\right]\cdot \mathrm{d}\mathscr{A} = 0. \tag{7.2}$$

In integrating (7.2) with respect to time we now allow the possibility of a nonzero constant of integration (since the flux filament phenomen violates the Meissner behavior); thus

$$\int\left[\frac{mc}{ne^2}\mathbf{\nabla} \times \mathbf{j} + \mathbf{H}\right]\cdot \mathrm{d}\mathscr{A} = \phi. \tag{7.3}$$

Applying Stokes' law to the first term in (7.3) yields

$$\frac{mc}{ne^2}\oint \mathbf{j}\cdot \mathrm{d}\ell + \int \mathbf{H}\cdot \mathrm{d}\mathscr{A} = \phi. \tag{7.4}$$

If we choose the contour to enclose a large area, we may expect the first term to be exponentially

1. For a superconducting slab we envision an array of interfaces parallel to the surface and for a cylindrical sample an array of coaxial cylinders. Other shapes would have more complex structures.

small (since the currents fall off exponentially with a characteristic length λ); we can then identify the constant of integration, ϕ, as the total flux contained within the filament (most of which also falls inside a radius of order λ).

To gain further insight we substitute $\mathbf{H} = \nabla \times \mathbf{A}$ into (7.4) and again apply Stokes' law to obtain

$$\oint \left[\frac{mc}{ne^2} \mathbf{j} + \mathbf{A} \right] \cdot d\ell = \phi; \tag{7.5a}$$

writing $\mathbf{j} = ne\mathbf{v} = ne(\mathbf{p}/m)$, (7.5a) becomes

$$\oint \left[\mathbf{p} + \frac{e}{c} \mathbf{A} \right] \cdot d\ell = \frac{e}{c} \phi. \tag{7.5b}$$

We identify the integrand as the canonical momentum associated with the motion of a charged particle in the Hamiltonian formulation of mechanics. F. London correctly concluded that superconductivity was a macroscopic quantum phenomenon, and guided by this insight he suggested that Eqs. (7.5) must conform with the Bohr–Sommerfeld quantization rule for the (quasiclassical) motion of an electron, i.e. $(|e|/c)\phi = nh$ or $\phi = n(hc/|e|)$, where n is an integer. However, this assumes the orbiting entities are single electrons; Ginzburg and Landau allowed for a more general case where $e \to e^*$; we then have

$$\phi = n\phi_0, \tag{7.6a}$$

where

$$\phi_0 = \frac{hc}{|e^*|}. \tag{7.6b}$$

From the Bardeen–Cooper–Schrieffer (BCS) theory it is now known that $e^* = 2e$; i.e.,

$$\phi_0 = \frac{hc}{2|e|} = 2.07 \times 10^{-7}\,\mathrm{G\,cm^2} \tag{7.6c}$$

which is referred to as the flux quantum. Hence flux enters a Type II superconductor as an array of quantized flux filaments; the lowest energy situation corresponds to singly quantizied ($n = 1$) filaments each carrying a flux quantum ϕ_0.

In what immediately follows we adopt cylindrical coordinates (r, θ, z) and write the in-plane radius vector as $\mathbf{r}$. Let us next examine Eqs. (7.5) for a contour of radius $\lambda >> r >> \xi$; the amount of flux contained is then vanishingly small and the first term in the integrand of (7.5b) dominates yielding the condition $2\pi pr = h$ or $p = \hbar/r$; BCS theory also dictates that the mass of the orbiting entity is $m^* = 2m$; hence

$$\mathbf{v}(\mathbf{r}) = \frac{\hbar}{2mr} \hat{\boldsymbol{\theta}} \tag{7.7}$$

where $\hat{\boldsymbol{\theta}}$ is an azimuthal unit vector. This velocity profile corresponds to the large r behavior of a vortex in a fluid, although with the vorticity quantized. One then refers to the filaments as *quantized vortex lines* or *vortex lines* for short.

Vorticity involves a nonvanishing curl of the velocity; i.e., $\nabla \times \mathbf{v} \neq 0$; but the curl of $\mathbf{v}$ in (7.7) vanishes for $r \neq 0$. However, the circulation $\Gamma \equiv \oint \mathbf{v} \cdot d\ell \neq 0$ and thus all the vorticity is located in an infinitesimal region near the origin. Physically it would be spread out over a coherence length, ξ. In our $\xi \to 0$ model we can obtain an approximate description by adding a singular source term to the London equation in the form

$$\mathbf{H} + \lambda_L^2 \nabla \times (\nabla \times \mathbf{H}) = \phi_0 \hat{\mathbf{z}} \delta^{(2)}(\mathbf{r}), \tag{7.8}$$

where $\delta^{(2)}(\mathbf{r})$ is a two-dimensional δ function and $\hat{\mathbf{z}}$ is a unit vector (parallel to the vortex axis). Eq. (7.8) may be written in cylindrical coordinates as

$$H_z - \frac{\lambda_L^2}{r} \frac{d}{dr}\left(r \frac{dH_z}{dr}\right) = \phi_0 \delta^{(2)}(\mathbf{r}). \tag{7.9}$$

The left-hand side of (7.9) is a special case of Bessel's equation[2] and the solution having the required singular behavior near $r = 0$ is

$$H_z = \frac{\phi_0}{2\pi\lambda_L^2} K_0\left(\frac{r}{\lambda_L}\right), \tag{7.10}$$

where K_0 is the zeroth order modified Bessel function of imaginary argument. From Ampere's law, $\mathbf{j} = (c/4\pi)\nabla \times \mathbf{H} = -(c/4\pi)(dH_z/dr)\hat{\boldsymbol{\theta}}$ and, for small x, $K_0(x) \cong \ell n(1/x)$; hence $j_\theta = (c/4\pi) \cdot (\phi_0/2\pi\lambda_L^2 r) = neh/4\pi mr$ or $v = \hbar/2mr$, in agreement with (7.7). For large x we use the form $K_0(x) \cong (\pi/2x)^{1/2} e^{-x}$ or $H_z \cong (\phi_0/2\pi\lambda_L^2)(\pi\lambda_L/2r)^{1/2} e^{-r/\lambda_L}$ and the field drops off exponentially, as argued above.

The energy of a vortex line follows from Eqs. (2.8)–(2.11):

$$E = \frac{L}{8\pi} \int [\mathbf{H}^2(\mathbf{r}) + \lambda_L^2 (\nabla \times \mathbf{H})^2] d^2 r, \tag{7.11}$$

where L is the length of the vortex line. We integrate (7.11) by parts to obtain

$$\frac{E}{L} = \frac{\lambda_L^2}{8\pi} \oint \mathbf{H} \times (\nabla \times \mathbf{H}) \cdot d\ell + \frac{1}{8\pi} \int \mathbf{H} \cdot [\mathbf{H} + \lambda_L^2 \nabla \times (\nabla \times \mathbf{H})] d^2 r. \tag{7.12}$$

The quantity in square brackets in the second term of (7.12) is equal to the left-hand side of (7.8). Were we to simply replace it by $\delta^{(2)}(\mathbf{r})$ and integrate we would obtain $\mathbf{H}(\mathbf{r} \to 0)$, which is logarithmically divergent. This indicates that our simple model for the vortex core is not sufficiently accurate to evaluate the energy. To avoid the divergence problem we assume H is finite everywhere, thus eliminating $\delta^{(2)}(\mathbf{r})$ from the right-hand side of (7.8), which results in the second term in (7.12) vanishing. We must still account for the energy in the core of the vortex, however, which we do by separating the line integral in (7.11) into three parts: a circle at a very large radius (which makes a vanishingly small contribution due to the exponential fall-off of $\mathbf{H}(\mathbf{r})$ at large r), two counter-traversing radial paths from the outer circle to an inner circle of a very small radius (which cancel each other) and finally a path around the inner circle (which makes the only nonvanishing contribution). For small r the first term in (7.12) is

$$\frac{E}{L} = \frac{\lambda_L^2}{8\pi} \oint \frac{\phi_0^2}{(2\pi\lambda_L^2)^2} \frac{1}{r} \ell n \left(\frac{\lambda_L}{r}\right) d\ell \tag{7.13}$$

2. Here we adopt the definitions of Abramowitz and Stegun (1970), pp. 374ff.

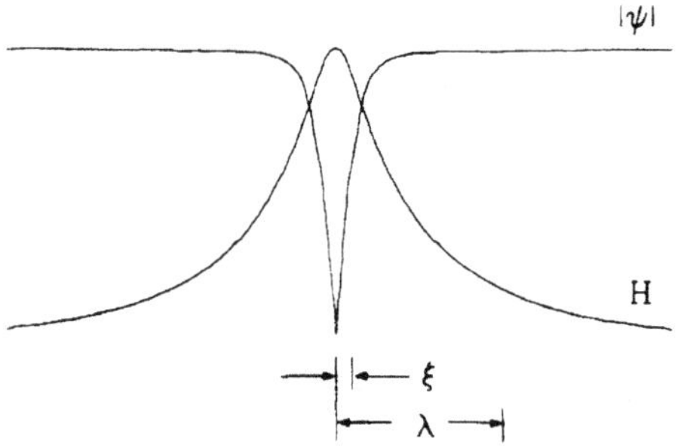

Figure 7.1 Structure of an isolated Abrikosov vortex line in a type II superconductor.

or

$$\frac{E}{L} = \frac{\phi_0^2}{(4\pi\lambda_L)^2}\ell n\frac{\lambda_L}{\xi}, \tag{7.14}$$

where we choose a radius $r = \xi$ for the (inner) line integral; this corresponds to the physically reasonable assumption that the field divergence is removed at the coherence length scale (as a more complete theory confirms). We expect that the $1/r$ divergence of the superfluid velocity ultimately destroys superconductivity in the vortex core; we may model this effect by assuming the density of superconducting electrons, $n(r)$, approaches zero (sufficiently rapidly) as $r \to 0$. Fig. 7.1 shows the qualitative behavior of $H(r)$ and $|\psi(r)|$ near the center (core) of a vortex filament where $|\psi(r)| \equiv [n(r)]^{\frac{1}{2}}$.

In the next section, where we discuss the mixed state of a Type II superconductor, we will require an expression for the interaction energy of two vortices. Returning to our London-like model, we generalize Eq. (7.8) to the case of two (parallel) vortices

$$\mathbf{H} + \lambda_L^2\nabla \times (\nabla \times \mathbf{H}) = \phi_0\hat{\mathbf{z}}[\delta^{(2)}(\mathbf{r} - \mathbf{r}_1) + \delta^{(2)}(\mathbf{r} - \mathbf{r}_2)]. \tag{7.15}$$

Since this equation is linear the resulting magnetic field will be the sum of two terms having the same form as (7.10):

$$\begin{aligned} H_z &= \frac{\phi_0}{2\pi\lambda_L^2}\left[K_0\left(\frac{|\mathbf{r} - \mathbf{r}_1|}{\lambda_L}\right) + K_0\left(\frac{|\mathbf{r} - \mathbf{r}_2|}{\lambda_L}\right)\right] \\ &= H_z^{(1)} + H_z^{(2)}. \end{aligned} \tag{7.16}$$

The total energy (which is quadratic in H) will involve three terms: two of these correspond to the 'self-energies' of the individual vortices (as given by (7.13)) and the third results from their interaction. The interaction energy in the extreme London limit can be evaluated from the second term in (7.12) along with (7.10) (we assuoe the path of the first integral involves a single circle of very large radius encircling both vortices); the interaction energy is then given by

$$\begin{aligned} \frac{E^{(12)}}{L} &= \frac{1}{8\pi}\int\{\mathbf{H}^{(1)}\cdot[\mathbf{H}^{(2)} + \lambda_L^2\nabla \times (\nabla \times \mathbf{H}^{(2)})] + \mathbf{H}^{(2)}\cdot[\mathbf{H}^{(1)} + \lambda_L^2\nabla \times (\nabla \times \mathbf{H}^{(1)})]\}\cdot \mathrm{d}^2\mathbf{r} \\ &= \frac{\phi_0}{8\pi}\int[\mathbf{H}^{(1)}(|\mathbf{r} - \mathbf{r}_1|)\delta^{(2)}(\mathbf{r} - \mathbf{r}_2) + \mathbf{H}^{(2)}(|\mathbf{r} - \mathbf{r}_2|)\delta^{(1)}(\mathbf{r} - \mathbf{r}_1)]\cdot \mathrm{d}^2\mathbf{r} \end{aligned}$$

$$= 2\left(\frac{\phi_0}{4\pi\lambda_L}\right)^2 K_0\left(\frac{|\mathbf{r}_1 - \mathbf{r}_2|}{\lambda_L}\right), \tag{7.17}$$

where, due to our assumptions, we require $|\mathbf{r}_1 - \mathbf{r}_2| \gtrsim \xi$. The sign of (7.17) is positive and hence the force (per unit length) between two vortices is *repulsive*. We may rewrite Eq. (7.17) in the form

$$\frac{E^{(12)}}{L} = \frac{\phi_0}{4\pi} H^{(12)}, \tag{7.18}$$

where $H^{(12)} \equiv H_z^{(1)}(\mathbf{r} = \mathbf{r}_2) = H_z^{(2)}(\mathbf{r} = \mathbf{r}_1)$ with the latter given by (7.10); i.e., $H^{(12)}$ is the contribution to the field at one vortex resulting from the presence of another. Regarding $(-\phi_0 L/4\pi)\hat{\mathbf{z}}$ as a magnetic moment, $\boldsymbol{\mu}$, we could write $E^{(12)} = -\boldsymbol{\mu}\cdot\mathbf{H}$, where $\mathbf{H} = H^{(12)}\hat{\mathbf{z}}$.

8 Type II superconductivity

8.1 Magnetic fields slightly greater than H_{c1}

Superconductors with $\xi < \lambda$, termed Type II materials, behave differently in a magnetic field from those with $\xi > \lambda$ (Type I materials). Fig. 8.1 shows the field dependence of the magnetization of an ideal Type II superconductor (the sample geometry is assumed to be a plane slab or a cylinder to avoid geometrically induced field inhomogeneities as noted earlier in our discussion of the intermediate state). For low fields the magnetization is $-H/4\pi$; i.e., the sample displays a Meissner-like behavior. However, at a field $H = H_{c1}$, called the *lower critical field*, flux abruptly enters the sample (the susceptibility $\chi = (\mathrm{d}M/\mathrm{d}H) \to +\infty$ for an ideal sample at this field). The magnetization increases (becomes less negative) continuously above H_{c1} and reaches zero at a field $H = H_{c2}$ called the *upper critical field*. The regime $H_{c1} < H < H_{c2}$ represents a new thermodynamic superconducting state called the mixed state or Shubnikov phase (Shubnikov *et al.* 1937).

We now show that the Meissner state of a Type II superconductor becomes unstable to the entry of vortex filaments at a field which we identify with H_{c1}. For magnetic fields which are only slightly above H_{c1} we may write the Gibbs free energy per unit volume as

$$\mathscr{G}(H) = \mathscr{G}(H = 0) + n_{\mathrm{L}} E/L + \frac{1}{L} \sum_{i<j} E^{(i,j)} - \frac{\mathbf{B} \cdot \mathbf{H}}{4\pi}. \tag{8.1}$$

The second term represents the self-energy of the individual lines where n_{L} is the number of lines per cm^2 with energy E/L per cm, the third term is the (repulsive) interaction energy between the lines (and is summed over a unit of area), and the last term is the usual field term relating the Gibbs and Helmholtz free energies. In the presence of a uniform array of flux lines the magnetic induction is

$$B = n_{\mathrm{L}} \phi_0. \tag{8.2}$$

Near H_{c1} (above which flux first enters), we will initially neglect the interaction term and, using (8.2), write

$$\mathscr{G}(H) = \mathscr{G}(H = 0) + n_{\mathrm{L}} \left(E/L - \frac{1}{4\pi} H \phi_0 \right). \tag{8.3}$$

For $H < 4\pi E/\phi_0 L$, (8.3) is minimized by setting $n_{\mathrm{L}} = 0$. However, for $H > 4\pi E/\phi_0 L$ Eq. (8.3)

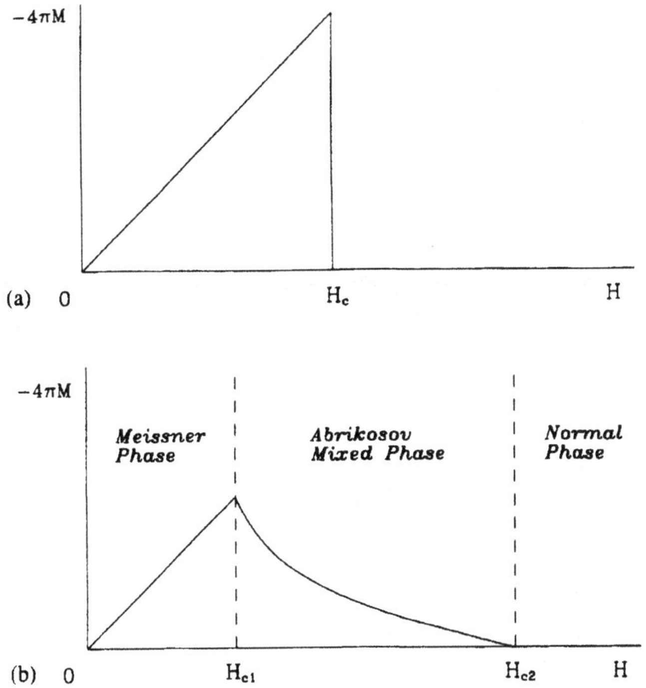

Figure 8.1 The reversible magnetization curves for a Type I (a) and a Type II (b) cylindrical superconductor. The magnetic field is applied parallel to the cylinder axis.

would suggest that $\mathcal{G} \propto -n_L$; i.e., the system can lower its free energy indefinitely simply by creating more flux lines. It is clear that we can identify H_{c1} as

$$H_{c1} = \frac{4\pi E}{\phi_0 L} \tag{8.4a}$$

or

$$H_{c1} = \frac{\phi_0}{4\pi\lambda_L^2} \ell n\left(\frac{\lambda_L}{\xi}\right), \tag{8.4b}$$

where we used (7.14) for the vortex self-energy in (8.4b).[1] The negative divergence of $\mathcal{G}$ for $H > H_{c1}$ is eliminated if we include vortex–vortex interaction (repulsion) effects; i.e., we must seek the minimum with respect to B of the quantity

$$\Delta\mathcal{G} = \frac{B}{4\pi}\left[H_{c1} - H + \frac{1}{2} z \frac{\phi_0}{2\pi\lambda_L^2} K_0\left(\frac{d}{\lambda_L}\right)\right]. \tag{8.5}$$

The first two terms are the same as in (8.3). The last term is the result of the vortex–vortex

1. We can now make contact with the comment made at the end of Sec. 6. Taking the classical limit as $h \to 0$, we see that the lower critical field would approach zero. At any finite field we would then have a divergent number of flux lines; i.e., the system would have a divergent internal interface area. Thus quantum mechanics imposes the constraint on the maximal amount of internal interfacial area.

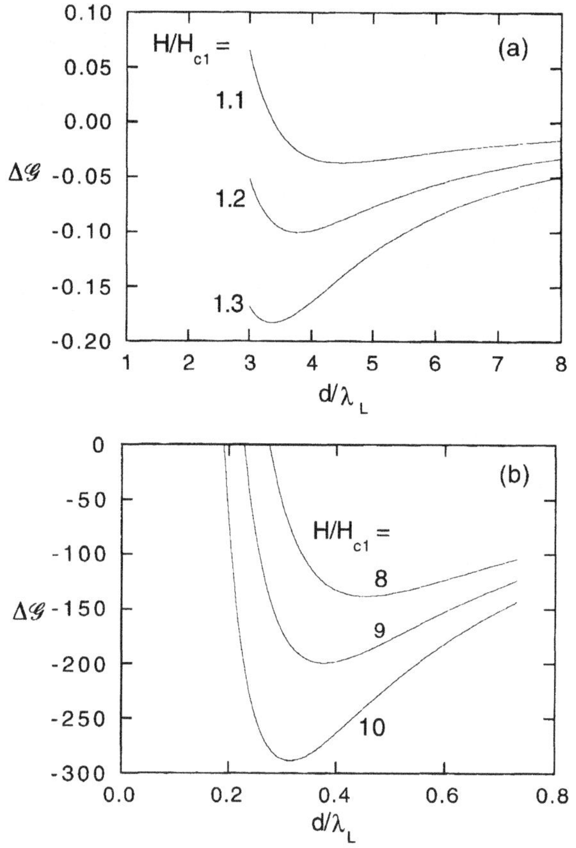

Figure 8.2 Normalized Gibbs free energy density calculated from Eq. (8.5) as a function of the reduced vortex lattice spacing in the low applied field regime (a) and high applied field regime (b).

repulsion; the factor 1/2 assigns half of the interaction energy to each vortex (in Eq. (7.18)), z is the number of nearest neighbors in the vortex lattice (the exponential fall-off of K_0 justifies including only nearest neighbor interactions at low fields where n_L is small), and d (appearing in the argument of K_0) is the vortex–vortex spacing, which depends on the symmetry of the vortex lattice. Calculations show that a triangular (centered hexagonal) arrangement of lines has the lowest energy. For this lattice $n_L = 2/\sqrt{3}d^2 = B/\phi_0$ or $d^2 = 2\phi_0/\sqrt{3}B$, and $z = 6$. On substituting these values of d and z in Eq. (8.5) we obtain $\Delta\mathcal{G} = \Delta\mathcal{G}(B)$; this function is shown qualitatively in Fig. 8.2 and the minimum value of $\Delta\mathcal{G}$ yields the magnetic induction, B, for a given external field, H. Carrying out the minimization process for each field H we can develop the function $B = B(H)$ (or $M = M(H)$ through $B = H + 4\pi M$). (Further analysis of this model shows that $\chi = (\partial M/\partial H)_T$ does diverge as $H \to H_{c1}$ from above.)

The above analysis breaks down as n_L increases (i.e., as B increases), and more powerful techniques are required. The simplest method involves the Ginzburg–Landau (G–L) theory, a discussion of which we begin in the next section.

In discussing the thermodynamics of a Type II superconductor the concept of the thermodynamic critical field, H_c, is retained in terms of the condensation energy, $H_c^2/8\pi$; however, $H_{c1} < H_c < H_{c2}$. From the discussion in Sec. 4 (where we showed $(1/4\pi)\int_0^{H_c} \mathbf{B}\cdot d\mathbf{H} = 0$) we identified $H_c^2/8\pi$ as the condensation energy. The fact that M has a different field dependence in a Type II superconductor does not alter this identity. Therefore

$$\int_0^{H_{c2}} M(H)\mathrm{d}H = -\frac{1}{8\pi}H_c^2. \tag{8.6}$$

H_{c2} greatly exceeds H_c in certain alloys and intermetallic compounds. This fact, coupled with the experimental observation that a zero resistance state often persists up to H_{c2}, makes Type II materials of great importance in the manufacture of magnets and related technologies.[2]

8.2 The region $H_{c1} << H << H_{c2}$

In the region $H_{c1} << H << H_{c2}$ there is a densely packed array of vortices; however, the spacing between vortices ($\cong n_L^{-1/2}$), d, satisfies the inequality $\xi << d << \lambda$ (here we assume that $\xi << \lambda$ so as to guarantee the existence of such a regime). When $d >> \xi$, $\mathbf{H}(\mathbf{r})$ is accurately given by the solution of the inhomogeneous London equation

$$\mathbf{H}(\mathbf{r}) + \lambda^2\nabla \times [\nabla \times \mathbf{H}(\mathbf{r})] = \phi_0\hat{\mathbf{z}}\sum_i \delta^{(2)}(\mathbf{r} - \mathbf{r}_i), \tag{8.7}$$

where i runs over all vortices in the lattice. Assuming the vortices lie on a periodic lattice[3] $\mathbf{H}(\mathbf{r})$ may be expanded in a two-dimensional Fourier series as

$$\mathbf{H}(\mathbf{r}) = \sum_{\mathbf{G}} \mathbf{H}_{\mathbf{G}} e^{-i\mathbf{G}\cdot\mathbf{r}} \tag{8.8a}$$

where

$$\mathbf{H}_{\mathbf{G}} = n_L \int_{\text{cell}} \mathbf{H}(\mathbf{r})e^{i\mathbf{G}\cdot\mathbf{r}}\mathrm{d}^2r; \tag{8.8b}$$

here $\mathbf{G}$ denotes all vectors of the two-dimensional reciprocal lattice associated with the real space lattice of the vortex array and n_L is the reciprocal of the area of the unit cell. Inserting (8.8a) into (8.7), using the fact that $\nabla \times [\nabla \times \mathbf{H}(\mathbf{r})] = -\nabla^2\mathbf{H}(\mathbf{r})$ (since $\nabla\cdot\mathbf{H}(\mathbf{r}) = 0$), multiplying by $e^{i\mathbf{G}'\cdot\mathbf{r}}$, and integrating over d^2r, and noting $\mathbf{G}'\cdot\mathbf{r}_i$ is a multiple of 2π we obtain

$$\mathscr{A}\sum_{\mathbf{G}}(1 + \lambda^2G^2)\mathbf{H}_{\mathbf{G}}\delta_{\mathbf{G}\mathbf{G}'} = \phi_0 N_L\hat{\mathbf{z}}$$

2. In the presence of a transport current (as one has in a magnet) the flux lines are subjected to a Lorentz force which would tend to make the flux lines move, which is a dissipative process. However, flux lines are usually pinned (immobilized) by various inhomogeneities present in the material (e.g., defects, grain boundaries, etc.).
3. In an inhomogeneous superconductor, vortices may be attached to so-called pinning sites, resulting in deviations from periodicity. On the other hand, thermal agitation may cause the lattice to melt in a high temperature superconductor.

or

$$H_G = \frac{\phi_0 n_L}{1 + \lambda^2 G^2} = \frac{B}{1 + \lambda^2 G^2}, \tag{8.9}$$

where $\mathscr{A}$ is the total cross sectional area of the vortex lattice, N_L is the total numbers of lines and $n_L \equiv N_L/\mathscr{A}$; henceforth we assume that $\mathbf{H}(\mathbf{r})$ and $\mathbf{H}_G$ are directed along the $\hat{z}$ axis and denote only their amplitudes, $H(\mathbf{r})$ and H_G.

The total energy of the vortex lattice follows from Eq. (7.11) as

$$\frac{E}{L} = \frac{1}{8\pi}\int \{H^2(\mathbf{r}) + \lambda^2[\nabla \times \mathbf{H}(\mathbf{r})]^2\} d^2 r$$

$$= \frac{\mathscr{A}}{8\pi}\sum_G (1 + \lambda^2 G^2) H_G H_{-G} = \frac{B^2 \mathscr{A}}{8\pi}\sum_G \frac{1}{1 + \lambda^2 G^2}.$$

Defining the energy density $\mathscr{E} = E/\mathscr{A}L$ we obtain

$$\mathscr{E} = \frac{B^2}{8\pi} + \frac{B^2}{8\pi}\sum_{G\neq 0}\frac{1}{1 + \lambda^2 G^2}. \tag{8.10}$$

Now the minimum nonzero value of $|\mathbf{G}| \cong 2\pi/d(\cong 2\pi n_L^{1/2})$ and with our assumption $d << \lambda$ we have $\lambda^2 G_{min}^2 >> 1$; therefore we may approximate $1/(1 + \lambda^2 G^2)$ by $1/\lambda^2 G^2$. The sum in (8.10) depends on the particular lattice adopted by the vortices. However, following deGennes (1966), we will limit ourselves to a semiquantitative estimate and replace the sum by an integral;

$$\sum_{G\neq 0}\frac{1}{G^2} \to \frac{1}{(2\pi)^2 n_L}\int_{G_{min}}^{G_{max}} \frac{2\pi G dG}{G^2} = \frac{1}{2\pi n_L}\ell n \frac{G_{max}}{G_{min}}. \tag{8.11}$$

From the above argument $G_{min} \cong 2\pi/d$; on the other hand we expect G_{max} to be cut off at the scale of an inverse coherence length and write $G_{max} = 2\pi/\xi$. We then obtain the total energy density of the lattice as

$$\mathscr{E} = \frac{B^2}{8\pi} + \frac{BH_{c1}}{4\pi}\frac{\ell n(\beta d/\xi)}{\ell n(\lambda/\xi)}; \tag{8.12}$$

here we used Eq. (8.4b) for H_{c1} and included a numerical factor β to offset partially the various approximations used to obtain (8.12). Matricon (see deGennes (1996)) finds $\beta = 0.381$ for a triangular lattice where $d^2 = 2\phi_0/\sqrt{3}B$. We note in passing that by considering the changes in the free energy arising from small distortions of the lattice from its equilibrium form, we may calculate the magnetic contribution to the elastic constants of the lattice. We will return to this topic in Sec. 20.7. Ignoring the entropy contribution we may replace $\mathscr{E}$ by $\mathscr{F}$; using the definition

$$\mathscr{G} = \mathscr{F} - \frac{1}{4\pi}BH \tag{8.13}$$

we may then calculate $B = B(H)$ from the condition $\partial G/\partial B = 0$ which is appropriate for a constant-H environment. Carrying out the calculation (including the contribution from the implicit dependence of d on H) we obtain

$$B = H - H_{c1}\frac{\ell n\left(\beta'\frac{d}{\xi}\right)}{\ell n\left(\frac{\lambda}{\xi}\right)}, \tag{8.14}$$

where $\beta' \equiv \beta e^{-1/2}$. From the definition $B = H + 4\pi M$ we have

$$M = -\frac{H_{c1}}{4\pi}\frac{\ell n\left(\beta'\frac{d}{\xi}\right)}{\ell n\left(\frac{\lambda}{\xi}\right)}. \tag{8.15}$$

Since $d \propto B^{-1/2}$ we see that M increases logarithmically (becomes less negative) with increasing magnetic field. This behavior agrees well with experimental data for materials with $\lambda >> \xi$.

8.3 Microscopic magnetic probes of the mixed state

We will now briefly discuss two experimental probes of the inhomogeneous magnetization in the mixed state: neutron diffraction and nuclear magnetic resonance. Other probes which have been applied or proposed include: decorating the surface with magnetic atoms (the Bitter technique) and magnetic force microscopy. (For more coarse grained magnetic structures magneto-optic and scanning Hall probe microscopy are useful.)

We begin with neutron diffraction. The neutron has a magnetic moment $\mu_n = (g_n/2)(e\hbar/2M_pc)$ where g_n is the anomalous g factor of the neutron ($g_n/2 = -1.91354$) and $e\hbar/2M_pc$ is the nuclear magneton with M_p the proton mass. The neutron interacts with a magnetic field via the usual spin Hamiltonian

$$\hat{H} = -\hat{\boldsymbol{\mu}}_n \cdot \mathbf{H}(\mathbf{r}), \tag{8.16}$$

where $\hat{\boldsymbol{\mu}}_n = \mu_n \underset{\sim}{\sigma}$ where $\underset{\sim}{\sigma}$ is the vector Pauli matrix. The coherent scattering cross section, σ_c, follows from the standard Born approximation expression

$$\sigma_c = |f(\mathbf{q})|^2,$$

where $f(\mathbf{q})$ is the scattering amplitude given by

$$f(\mathbf{q}) = \frac{M_n}{2\pi\hbar^2}\int \hat{H}(\mathbf{r})e^{i\mathbf{q}\cdot\mathbf{r}}d^3r; \tag{8.17}$$

here M_n is the neutron mass and $\mathbf{q} = \mathbf{k} - \mathbf{k}'$ with $\mathbf{k}$ and $\mathbf{k}'$ the wave vectors of the incident and scattered neutrons ($|\mathbf{k}| = |\mathbf{k}'|$ for a diffraction experiment). Inserting Eq. (8.8a), with H_G given by (8.9), into (8.17) we have

$$f(\mathbf{q}) = \frac{M_n}{2\pi\hbar^2}\mu_n B\sum_G \int \frac{e^{i(\mathbf{q}-\mathbf{G})\cdot\mathbf{r}}}{1+\lambda^2G^2}d^3r = \frac{M_n}{2\pi\hbar^2}\frac{\mu_n BV}{1+\lambda^2G^2}\delta_{\mathbf{Gq}}, \tag{8.18}$$

where V is the sample volume and the Kronecker delta function reflects the usual Ewald diffraction condition, $\mathbf{q} = \mathbf{G}$. Writing $B = n_L\phi_0$, $\phi_0 = 2\pi\hbar c/2e$ and $\mu_n \cong (g_n/2)(e\hbar/2M_n c)$ we have

$$f(\mathbf{q}) = \frac{1.91 n_L V}{4(1+\lambda^2 G^2)}\delta_{\mathbf{Gq}}. \tag{8.19}$$

Note that as discussed above $\lambda^2 G^2 >> 1$; hence $\sigma(\mathbf{q}) = |f(\mathbf{q})|^2 \propto (\lambda G)^{-4}$. For this reason it is very difficult to observe more than the lowest order peak. Experiments were first performed on Nb by Cribier *et al.* (1964). Since high T_c and heavy fermion materials have anomalously large London depths neutron diffraction has not been a useful probe of the vortex lattice in these materials.

We now discuss the expected behavior of the line width, $\Delta\omega$, measured in an nmr experiment. For many purposes the line width may be inferred from the root mean square deviation of the magnetic field from its average value[4]; i.e.,

$$\Delta H = [\overline{(H(\mathbf{r}) - \overline{H(\mathbf{r})})^2}]^{1/2} = [\overline{H^2(\mathbf{r})} - \overline{H(\mathbf{r})}^2]^{1/2} \tag{8.20}$$

and we take $\Delta\omega = \gamma\Delta H$, where γ is the gyromagnetic ratio of the nuclei which are in resonance. Here the bar implies an average over the sample volume. For simplicity we assume the external (applied) magnetic field is strictly uniform and that the sample has the shape of a long cylinder parallel to the applied field so that demagnetization effects may be ignored. Now $\overline{H(\mathbf{r})}$ in (8.20) is simply $B = n_L\phi_0$; $\overline{H^2(\mathbf{r})}$ is calculated as follows using (8.8a) and (8.9):

$$\overline{H^2(\mathbf{r})} = \frac{1}{\mathscr{A}}\int H^2(\mathbf{r})\mathrm{d}\mathscr{A} = \sum_{\mathbf{G}} H_{\mathbf{G}}H_{-\mathbf{G}} = (n_L\phi_0)^2\sum_{\mathbf{G}}\frac{1}{(1+\lambda^2G^2)^2}, \tag{8.21}$$

where $\mathscr{A}$ is the area of the sample perpendicular to the magnetic field. Separating off the $\mathbf{G} = 0$ term (which cancels the $\overline{[H(\mathbf{r})]}^2$ term in (8.20)), restricting to the limit $\lambda^2G^2 >> 1$, and again replacing the sum by an integral we obtain

$$[\Delta H]^2 = \frac{n_L\phi_0{}^2}{(2\pi)^2}\int_{G_{\min}}^{G_{\max}}\frac{2\pi G\mathrm{d}G}{\lambda^4G^4}. \tag{8.22}$$

The integral is convergent at the upper limit so we extend $G_{\max}$ to ∞ and we again take $G_{\min} \cong 2\pi/d$ (valid for a square lattice) yielding

$$\Delta H = \left(\frac{n_L\phi_0^2}{2\pi}\frac{1}{2G_{\min}^2\lambda^4}\right)^{1/2} = \frac{\phi_0}{2^{1/2}(2\pi)^{3/2}\lambda^2}. \tag{8.23}$$

Note that our result is independent of magnetic field and within a factor of order unity is equal to H_{c1}. Therefore nmr is readily observable in extreme Type II materials containing nuclei with sufficiently large moments. We emphasize that this line width is not to be interpreted as the reciprocal of a magnetic relaxation time; it is referred to as an inhomogeneous broadening. The relaxation of the magnetization will be discussed briefly in Subsec. 34.2.

4. For a discussion of moments in nmr see Abragam (1961), p. 106.

9 The Ginzburg–Landau theory

9.1 Basic equations

In 1937 Landau developed a model to describe second order phase transitions (those involving no latent heat). Central to this theory was the introduction of the concept of an order parameter. The order parameter is an appropriate quantity which vanishes in the high temperature phase ($T > T_c$) but which acquires a nonzero value below the transition ($T < T_c$). The identification of the order parameter is often obvious from the nature of the second order transition. Thus for the ferromagnetic transition it is natural to identify the spontaneous magnetization, **M**, as the order parameter.

The ferromagnet brings out a fundamental property of second order phase transitions: the development of a spontaneous magnetization is accompanied by a reduction in the symmetry of the system. Thus if the material is iron, where the high temperature (nonmagnetic) phase has a body centered cubic (bcc) structure, on passing through the transition (Curie) temperature ($T_c = 1043$ K) the material chooses one of several symmetry-equivalent crystal axes along which the magnetization develops; the choosing of one among several directions (for iron the directions of easy magnetization are the cube axes and there are six such directions, $\pm\hat{\mathbf{x}}$, $\pm\hat{\mathbf{y}}$, $\pm\hat{\mathbf{z}}$) lowers the symmetry of the crystal. The system is then said to have 'spontaneously broken symmetry'. It is a general property of second order phase transitions that the low temperature phase always has a lower symmetry.

Since the order parameter evolves continuously from zero below T_c, it is natural to expand the free energy as a power series in the order parameter. The free energy is a scalar but the order parameter may be a higher-dimensional object (e.g., a vector, tensor, complex number). For our example of the ferromagnet, the order parameter is the magnetization, **M**, a vector. Thus in making the expansion of $F(\mathbf{M})$ we can admit only symmetry-invariant combinations of the components, M_i; since the magnetization may develop along any of the (easy) crystal axes the free energy expansion must preserve the full symmetry of the high temperature phase. For a cubic ferromagnet the expansion of the free energy satisfying the above requirements has the form

$$F(\mathbf{M},T) = F(0,T) + \alpha(M_x^2 + M_y^2 + M_z^2) + \frac{1}{2}\beta_1(M_x^2 + M_y^2 + M_z^2)^2 + \frac{1}{2}\beta_2(M_xM_y + M_yM_z + M_zM_x)^2. \qquad (9.1)$$

Expression (9.1) is invariant under all the symmetry operations of a cube. However, to simplify

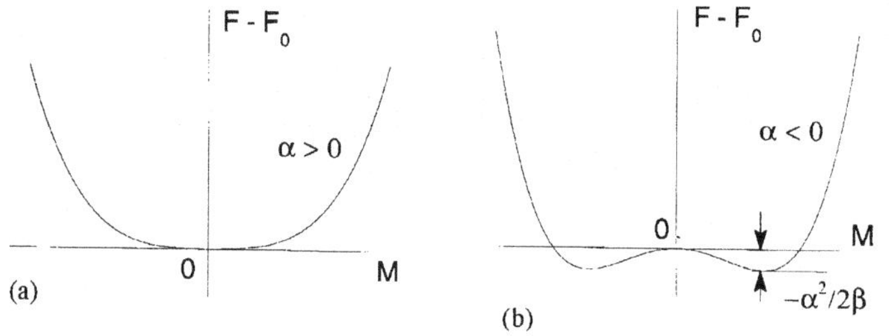

Figure 9.1 G–L free energy functions for $T > T_c$ ($\alpha > 0$) and for $T < T_c$ ($\alpha < 0$).

the discussion consider a (hypothetical) isotropic ferromagnetic liquid. The only rotationally invariant scalar that can be formed is then $\mathbf{M} \cdot \mathbf{M}$ and (9.1) simplifies to

$$F(M) = F(M = 0) + \alpha M^2 + \tfrac{1}{2}\beta M^4. \tag{9.2}$$

Limiting the expansion to fourth order will lead to an adequate description of ferromagnetism. Thermodynamic stability requires $\beta > 0$ (otherwise the system would seek a divergingly large magnetization). However, α may have either sign. If $\alpha > 0$ the minimum of (9.2) occurs at $M = 0$; if $\alpha < 0$ the minimum is for $M \neq 0$. These two situations are shown in Fig. 9.1. Hence we can model a second order phase transition simply by arranging for the sign of α to change at T_c which is easily done by writing

$$\alpha(T) = a(T - T_c); \tag{9.3}$$

it is sufficient to regard β as a constant. Setting $\partial F/\partial M = 0$ yields

$$[\alpha(T) + \beta M^2]M = 0. \tag{9.4}$$

This equation has the two solutions discussed above:

$$\alpha > 0, \qquad M = 0 \qquad (T > T_c); \tag{9.5a}$$

$$\alpha < 0, \qquad M^2 = -\alpha/\beta \qquad (T < T_c). \tag{9.5b}$$

In terms of Eq. (9.3) we have

$$M^2 = 0 \qquad (T > T_c); \tag{9.6a}$$

$$M^2(T) = -\frac{a(T - T_c)}{\beta} \qquad (T < T_c). \tag{9.6b}$$

It is important to recognize that the ferromagnetic state is degenerate. For the case of the crystal it is degenerate with respect to the number of independent crystal axes (or directions) the magnetization might orient along (six for our case of bcc iron with easy $\langle 100 \rangle$ axes). For our ferromagnetic liquid the ordered state is continuously degenerate: $\mathbf{M}$ can point in any direction.

There is an important aspect of the second order phase transition which is neglected in the above model: fluctuations. All thermodynamic systems undergo fluctuations. As an example the density of a liquid may fluctuate by an amount $\delta\rho$. However, this is at the expense of an increase in the local pressure, δp, governed by the bulk modules. This is a general feature: the 'restoring force' (pressure in our example) is linear in the 'displacement' (here the density). For a system which undergoes a second order phase transition the order parameter becomes a thermodynamic variable below the transition. For our magnetic system the generalized restoring force accompanying a change in the magnetization δM follows from expanding the free energy about its equilibrium value, M. However, as the transition is approached the linear restoring force goes to zero; just at the transition the restoring force is proportional to $(\delta M)^3$. We see that fluctuations can have a profound effect on the system near its transition temperature. The Landau model therefore assumes that the temperature is far enough from the transition temperature that fluctuation effects may be ignored. This is referred to as the *mean field* model.

We now take up the case of the superconductor. Our earlier discussion has brought out the idea that superconductivity is some kind of macroscopic quantum state. Ginzburg and Landau built this idea into the Landau second order phase transition theory by assuming the existence of a macroscopic 'wave function', ψ, which they took as the order parameter associated with superconductivity. Since wavefunctions can be complex, only the form $\psi\psi^*$ may enter the free energy expansion; we therefore write

$$F = F(T) + \alpha|\psi|^2 + \frac{1}{2}\beta|\psi|^4. \tag{9.7}$$

The minimization proceeds exactly as above, i.e.,

$$|\psi| = 0, \qquad T > T_c; \tag{9.8a}$$

$$|\psi| = \left[\frac{a(T_c - T)}{\beta}\right]^{1/2}, \quad T < T_c. \tag{9.8b}$$

To describe situations where the superconducting state is inhomogenous we must generalize (9.7). F is then interpreted as a free energy density, $\mathscr{F}(\mathbf{r})$, and we write

$$\begin{aligned} F &= F_0 + \int \mathscr{F}(\mathbf{r})\mathrm{d}^3r \\ &= F_0 + \int \mathrm{d}^3r\left[\alpha|\psi(\mathbf{r})|^2 + \frac{1}{2}\beta|\psi(\mathbf{r})|^4\right]; \end{aligned} \tag{9.9}$$

F is now the total free energy. Eq. (9.9) in its present form does not model the increase in energy associated with a spatial distortion of the order parameter, i.e., effects associated with a coherence length, ξ. To account for such effects Ginzburg and Landau added a 'gradient energy' term to (9.9) of the form

$$F_G = \int \frac{\hbar^2}{2m^*}|\nabla\psi(\mathbf{r})|^2\mathrm{d}^3r \tag{9.10}$$

with m^* as a parameter; the choice of the coefficient $\hbar^2/2m^*$ makes (9.9) mimic the quantum mechanical kinetic energy (introduced earlier in Eq. (2.10)). Ginzburg and Landau assumed that if (9.10) was to be regarded as the kinetic energy contribution to the Hamiltonian density of the superconducting electrons, then (as in Hamiltonian mechanics) the interaction of the electrons with an electromagnetic field would be accomplished by the Hamiltonian prescription

$$\mathbf{p} \to \mathbf{p} - \frac{e^*}{c}\mathbf{A}$$

or, since $\mathbf{p} \to \left(\frac{\hbar}{\mathrm{i}}\right)\nabla$ in quantum mechanics,

$$\nabla \to \nabla - \frac{\mathrm{i}e^*}{\hbar c}\mathbf{A}; \tag{9.11}$$

the use of e^* allows the superconducting entities to carry a different charge ($e^* = 2e$ in BCS theory). Combining the above we have

$$F_{\mathrm{G}} = \int \frac{\hbar^2}{2m^*}\left|\left[\nabla - \frac{\mathrm{i}e^*}{\hbar c}\mathbf{A}(\mathbf{r})\right]\psi(\mathbf{r})\right|^2 \mathrm{d}^3 r. \tag{9.12}$$

Finally we must add the contribution of the magnetic field to the energy density

$$\mathscr{F}_H = \frac{1}{8\pi}H^2(\mathbf{r}). \tag{2.9}$$

Combining the above we have

$$F = F_0 + \int \mathrm{d}^3 r\left\{\alpha|\psi(\mathbf{r})|^2 + \frac{1}{2}\beta|\psi(\mathbf{r})|^4 + \frac{\hbar^2}{2m^*}\left|\left[\nabla - \frac{\mathrm{i}e^*}{\hbar c}\mathbf{A}(\mathbf{r})\right]\psi(\mathbf{r})\right|^2 + \frac{1}{8\pi}H^2(\mathbf{r})\right\}. \tag{9.13}$$

The minimization of (9.13) must be carried out using the methods of the calculus of variations since F is a functional involving the free energy density $\mathscr{F}(\psi(\mathbf{r}), \psi^*(\mathbf{r}), \nabla\psi(\mathbf{r}), \nabla\psi^*(\mathbf{r}), \mathbf{H}(\mathbf{r}))$ which in turn involves the unknown functions $\psi(\mathbf{r})$, $\psi^*(\mathbf{r})$, and $\mathbf{H}(\mathbf{r})(= \nabla \times \mathbf{A}(\mathbf{r}))$.

Minimizing (9.13) with respect to $\psi^*(\mathbf{r})$ yields

$$\delta F = \int \mathrm{d}^3 r\left\{-\frac{\hbar^2}{2m^*}\left[\nabla - \frac{\mathrm{i}e^*}{\hbar c}\mathbf{A}(\mathbf{r})\right]^2\psi(\mathbf{r}) + \alpha\psi(\mathbf{r}) + \beta|\psi(\mathbf{r})|^2\psi(\mathbf{r})\right\}\delta\psi^*(\mathbf{r}) + \int \mathrm{d}^2\mathbf{r}\cdot\frac{\hbar^2}{2m^*}\left[\nabla - \frac{\mathrm{i}e^*}{\hbar c}\mathbf{A}(\mathbf{r})\right]\psi(\mathbf{r})\delta\psi^*(\mathbf{r}) \tag{9.14}$$

(variation with respect to ψ, which is an independent variable, yields the complex conjugate of (9.14)). To minimize $\mathscr{F}$ we set the integrand of the first part of (9.14) to zero; this yields the first G–L equation

$$-\frac{\hbar^2}{2m^*}\left[\nabla - \frac{\mathrm{i}e^*}{\hbar c}\mathbf{A}(\mathbf{r})\right]^2\psi(\mathbf{r}) + \alpha\psi(\mathbf{r}) + \beta|\psi(\mathbf{r})|^2\psi(\mathbf{r}) = 0. \tag{9.15}$$

The surface term (which was generated by an integration by parts) can be used (with caution) to establish certain boundary conditions and will be discussed later.

Variation of (9.13) with respect to $\mathbf{A}$ (with $\mathbf{H} = \nabla \times \mathbf{A}(\mathbf{r})$) yields Ampère's law

v006b;011

$$\nabla \times \mathbf{H}(\mathbf{r}) = \frac{4\pi}{c}\mathbf{j}(\mathbf{r}) \tag{9.16}$$

provided we identify $\mathbf{j}$ as

$$\mathbf{j}(\mathbf{r}) = \frac{e^*}{2m^*}\left\{\psi^*(\mathbf{r})\left[-i\hbar\nabla - \frac{e^*}{c}\mathbf{A}(\mathbf{r})\right]\psi(\mathbf{r}) + \psi(\mathbf{r})\left[+i\hbar\nabla - \frac{e^*}{c}\mathbf{A}(\mathbf{r})\right]\psi^*(\mathbf{r})\right\}, \tag{9.17a}$$

or equivalently,

$$\mathbf{j}(\mathbf{r}) = \frac{-ie^*\hbar}{2m^*}[\psi^*(\mathbf{r})\nabla\psi(\mathbf{r}) - \psi(\mathbf{r})\nabla\psi^*(\mathbf{r})] - \frac{e^{*2}}{m^*c}|\psi(\mathbf{r})|^2\mathbf{A}(\mathbf{r}). \tag{9.17b}$$

Eq. (9.17) is the second G–L equation; we note that (9.17) is the same as the expression for the current density in quantum mechanics. Note the current density satisfies the equation $j(\mathbf{r}) = c(\delta F/\delta\mathbf{A}(\mathbf{r}))$; i.e., it is the variable conjugate to $\mathbf{A}(\mathbf{r})$.

9.2 Gauge invariance

The simplest solution of (9.15) is for the case of a uniform superconductor, $\psi \neq \psi(\mathbf{r})$, with $\mathbf{A} = 0$, as given earlier in Eqs. (9.8). However, (9.15) possesses a continuum of other solutions having the same free energy, which we now show. As we can with any complex function, we write $\psi(\mathbf{r}) = a(\mathbf{r})e^{i\Phi(\mathbf{r})}$ where $a(\mathbf{r})$ and $\Phi(\mathbf{r})$ are the position-dependent amplitude and phase, respectively. Let us examine a class of solutions which satisfy the complex equation

$$\left(\nabla - \frac{ie^*}{\hbar c}\mathbf{A}(\mathbf{r})\right)\psi(\mathbf{r}) = 0, \tag{9.18}$$

which is equivalent to the two real equations

$$\nabla a(\mathbf{r}) = 0 \tag{9.19a}$$

and

$$\left[\nabla\Phi - \frac{e^*}{\hbar c}\mathbf{A}\right] = 0. \tag{9.19b}$$

From (9.19a) we see that the only allowed solutions of (9.18) involve a constant amplitude, a; Eq. (9.19b), on the other hand, has infinitely many solutions involving a vector potential and a position-dependent phase (which does not affect the free energy) related by

$$\mathbf{A} = \frac{\hbar c}{e^*}\nabla\Phi. \tag{9.20}$$

Any vector potential satisfying (9.20) results in a uniform free energy and (on substituting (9.18) into (9.17)) a vanishing current density (note that $\mathbf{H} = \nabla \times \mathbf{A} = 0$ for all $\mathbf{A}$ of the form (9.20)).

The above exercise shows that the symmetry broken in superconductivity is *gauge symmetry*, or equivalently, *phase symmetry*. Superconductors having different phase functions, $\Phi(\mathbf{r})$, are in a real sense physically distinct;[1] this arbitrariness of the phase is the analogue for a supercon-

1. Strictly speaking we cannot determine the absolute phase of a superconductor, but in our later discussion of the Josephson effects we will show that phase differences can be measured.

ductor of the property that the magnetization may point in any direction in an isotropic (liquid) ferromagnet.

9.3 Boundaries and boundary conditions

We first examine a simple case involving an inhomogeneous order parameter generated by the presence of a boundary, in the absence of a magnetic field. Assume we have a superconducting half space occupying the region $x > 0$. We further assume that the order parameter is driven to zero at this interface. Experimentally this can be accomplished by coating the surface of the superconductor with a film of ferromagnetic material.[2] We then seek a solution to the one-dimensional G–L equation

$$-\frac{\hbar^2}{2m^*}\frac{d^2\psi}{dx^2} + \alpha\psi + \beta\psi^3 = 0. \tag{9.21}$$

Noting α is negative in the superconducting state ($\alpha = -|\alpha|$), defining the G–L coherence length as

$$\xi^2 \equiv \frac{\hbar^2}{2m^*|\alpha|}, \tag{9.22}$$

and writing $(\beta/|\alpha|)\psi^2 = f^2$ we may rewrite (9.21) as

$$-\xi^2 f'' - f + f^3 = 0. \tag{9.23}$$

Multiplying by f' we may rewrite (9.23) as

$$\frac{d}{dx}\left[-\frac{\xi^2 f'^2}{2} - \frac{1}{2}f^2 + \frac{1}{4}f^4\right] = 0; \tag{9.24}$$

hence the quantity in square brackets must be a constant. Far from the boundary, $f' = 0$ and $f^2 = 1$ (equivalent to $\psi^2 = |\alpha|/\beta$); then (9.24) becomes

$$\xi^2(f')^2 = \frac{1}{2}(1 - f^2)^2 \tag{9.25}$$

which has the solution $f = \tanh(x/\sqrt{2}\xi)$ or

$$\psi = \left(\frac{|\alpha|}{\beta}\right)^{1/2} \tanh\left(\frac{x}{\sqrt{2}\xi}\right). \tag{9.26}$$

From (9.26) we see that ξ is a measure of distance over which the order parameter responds to a perturbation. Since $\alpha = a(T - T_c)$ we have

$$\xi(T) = \left(\frac{\hbar^2}{2m^* a T_c}\right)^{1/2}\left(1 - \frac{T}{T_c}\right)^{-1/2}. \tag{9.27}$$

2. Paramagnetic impurities (those bearing a spin in a host material) or interfaces with a ferromagnetic metal strongly depress superconductivity. A normal metal interface has a much smaller effect and an insulator or vacuum has a negligible effect for most purposes.

We see that the G–L coherence length diverges as $1/(1 - T/T_c)^{1/2}$; this divergence is a general property of the coherence length at all second order phase transitions (although the exponent differs in general from this 'mean field' value of 1/2 close to T_c).

We now examine Eq. (9.17b). In a limit where the first two terms are negligible, we recognize the remainder as the London equation, as written in the form (3.5), provided we identify $n = |\psi|^2$; this supports the identification of ψ as a (condensate) wave function associated with the superconducting electrons. We can immediately calculate the London penetration depth as

$$\lambda_L^2 = \frac{m^* c^2}{4\pi e^{*2} |\psi|^2} \tag{9.28a}$$

or

$$\lambda_L = \left(\frac{m^* c^2 \beta}{4\pi e^{*2} a T_c}\right)^{1/2} \left(1 - \frac{T}{T_c}\right)^{-1/2}. \tag{9.28b}$$

Comparing Eqs. (9.27) and (9.28) we see that in the G–L theory λ_L and ξ both diverge as $(1 - T/T_c)^{-1/2}$. Their ratio, called the G–L parameter, is therefore a constant which we write as

$$\kappa \equiv \frac{\lambda_L}{\xi} = \frac{m^* c}{e^* \hbar}\left(\frac{\beta}{2\pi}\right)^{1/2}. \tag{9.29}$$

Let us now return to the discussion of the surface tension of a normal–superconductor phase boundary begun in Sec. 6. From Eq. (6.3) the total Gibbs free energy, which is a constant, is given by

$$G = \mathscr{A} \int_{-\infty}^{+\infty} \mathrm{d}x \left[\mathscr{F}(x) - \frac{1}{4\pi} \mathbf{H}_c \cdot \mathbf{B}(x)\right]; \tag{9.30}$$

here $\mathscr{A}$ is the interface area, $\mathscr{F}$ is the G–L free energy density and the integral must now be extended into the region $x > 0$ since, due to the gradient term in the G–L model, the superconducting properties do not turn on abruptly at the interface (which we still locate, nominally, at $x = 0$). Inserting (9.13) into (9.30), employing a gauge where $\mathbf{A} = A(x)\hat{\mathbf{z}}$, and using the definition of γ given by (6.4), we obtain the one-dimensional equation

$$\gamma = \int_{-\infty}^{+\infty} \mathrm{d}x \left[\frac{\alpha^2}{2\beta} + \frac{\hbar^2}{2m^*}\left|\frac{\mathrm{d}\psi(x)}{\mathrm{d}x}\right|^2 + \frac{e^{*2}}{2m^* c^2} A^2(x) |\psi(x)|^2 \right.$$
$$\left. + \alpha |\psi(x)|^2 + \frac{1}{2}\beta |\psi(x)|^4 + \frac{B^2(x)}{8\pi} - \frac{H_c B(x)}{4\pi}\right]. \tag{9.31}$$

The vanishing of the cross term $\mathrm{i}\mathbf{A} \cdot \nabla$ in both (9.31) and the first G–L equation (9.15) allows us to choose ψ real; it then follows from (9.17) that

$$j_z(x) = -\frac{e^{*2}}{m^* c} \psi^2(x) A(x) \tag{9.32a}$$

and

$$j_x = j_y = 0. \tag{9.32b}$$

To compute γ we must simultaneously solve the first and second G–L equations, (9.15) and (9.16), for $\psi(x)$ and $B(x)(=-(\mathrm{d}A(x)/\mathrm{d}x))$, subject to boundary conditions which will be specified shortly, and insert the results in (9.31). To eliminate various parameters in the subsequent calculations we rewrite all equations in terms of the scaled variables:

$$\bar{x} = x/\lambda_L, \bar{\psi} = \left(\frac{\beta}{|\alpha|}\right)^{1/2}\psi, \bar{A} = \frac{A}{H_c\lambda_L}, \text{ and } \bar{B} = \frac{B}{H_c}. \tag{9.33}$$

In terms of these variables the first and second G–L equations become (where we now drop the bars over the variables)

$$\psi'' = \kappa^2\left[\left(\frac{1}{2}A^2 - 1\right)\psi + \psi^3\right] \tag{9.34}$$

and

$$A'' = A\psi^2. \tag{9.35}$$

The solutions of (9.34) and (9.35) for arbitrary κ must be obtained numerically. The appropriate boundary conditions are

$$\psi = 0, B = A' = 1 \text{ at } x = -\infty \tag{9.36a}$$

and

$$\psi = 1, A' = 0 \text{ at } x = +\infty \tag{9.36b}$$

(where $x > 0$ is nominally the superconducting side).[3] The behavior of $B(x)$ and $\psi(x)$ in the small κ regime where the field varies more rapidly than the order parameter is shown in Fig. 9.2. To obtain a first integral of these equations we multiply (9.34) by ψ', which leads immediately to

$$\frac{1}{\kappa^2}\frac{\mathrm{d}}{\mathrm{d}x}(\psi'^2) = \frac{\mathrm{d}}{\mathrm{d}x}\left[\left(\frac{1}{2}A^2 - 1\right)\psi^2 + \frac{1}{2}\psi^4\right] - \psi^2\frac{\mathrm{d}}{\mathrm{d}x}\left(\frac{A^2}{2}\right), \tag{9.37a}$$

and multiply (9.34) by A', which yields

$$\frac{\mathrm{d}}{\mathrm{d}x}\left(\frac{A'^2}{2}\right) = \psi^2\frac{\mathrm{d}}{\mathrm{d}x}\left(\frac{A^2}{2}\right). \tag{9.37b}$$

Combining (9.37a) and (9.37b) we have for our first integral

$$\frac{2}{\kappa^2}\psi'^2 + (2 - A^2)\psi^2 - \psi^4 + A'^2 = \text{const.} = 1; \tag{9.38}$$

the value of the constant was fixed by the boundary conditions at either $+\infty$ or $-\infty$. Using $\alpha^2/2\beta = H_c^2/8\pi$ and the scaled variables of Eq. (9.33) we may rewrite Eq. (9.31) as

$$\gamma = \frac{\lambda_L H_c^2}{8\pi}\int_{-\infty}^{+\infty}\left[\frac{2}{\kappa^2}\psi'^2 + (A^2 - 2)\psi^2 + \psi^4 + (A' - 1)^2\right]\mathrm{d}x.$$

3. From the structure of Eqs. (9.34) and (9.35) it follows from the boundary conditions (9.36a) and (9.36b) that $\psi' = 0$ at $x = \pm\infty$. Our restriction to real ψ requires that the constant A (which is equivalent to a phase Φ) vanishes at $x = +\infty$.

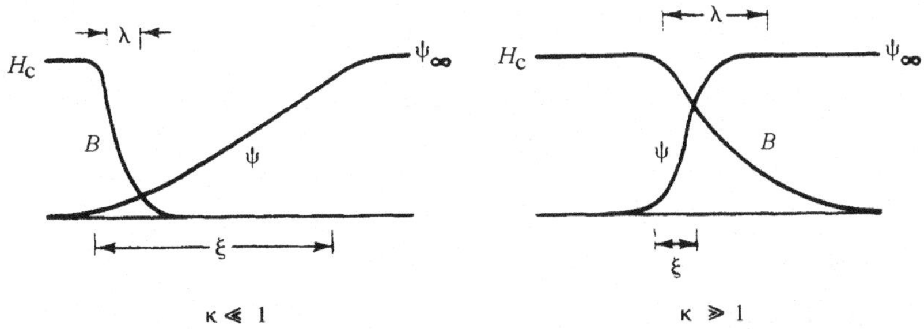

Figure 9.2 Schematic diagram of the variation of B and ψ in a domain wall. The case $\kappa << 1$ refers to a Type I superconductor with a positive surface energy; the case $\kappa >> 1$ refers to a Type II superconductor with negative surface energy.

Combining this expression with Eq. (9.38) we have

$$\gamma = \frac{\lambda_L H_c^2}{4\pi} \int_{-\infty}^{+\infty} \left[\frac{2}{\kappa^2} \psi'^2 + A'(A' - 1) \right] dx; \tag{9.39}$$

given $A'(x)(= - B(x))$ and $\psi'(x)$ we can compute γ.

Returning to Eq. (9.38) we verify that in the limit $A' = 0$ we recover the dimensionless form of Eq. (9.25);

$$\psi' = \frac{\kappa}{\sqrt{2}}(1 - \psi^2). \tag{9.40}$$

The solution of this equation is

$$\psi'(x) = \tanh \frac{\kappa(x - x_0)}{\sqrt{2}} \tag{9.41}$$

(which is the analogue of Eq. (9.26)); here x_0 is the nominal position of the boundary. The characteristic length scale of the order parameter variation associated with this normal/superconductor phase boundary is ξ (in unscaled units), as discussed earlier in connection with Eq. (9.26). This form does not satisfy the boundary conditions (9.36), as expected, since a stable phase boundary in an unbounded superconductor can exist only in the presence of a field, but rather satisfies the boundary conditions $\psi = +1$ or -1 as $x \to +\infty$. However, assuming the presence of an order parameter quenching mechanism at $x_0 = 0$ (e.g., a thin ferromagnetic plane embedded in an otherwise homogeneous superconductor) we may evaluate the associated surface tension by substituting (9.41) into (9.39); carrying out the integration (with $A = 0$) we obtain

$$\gamma = \frac{H_c^2 \lambda_L}{3\sqrt{2}\pi\kappa} = \frac{1.9\lambda_L}{\kappa} \frac{H_c^2}{8\pi}. \tag{9.42}$$

From the definition of κ given in (9.29) we see that $\gamma \sim \xi(H_c^2/8\pi)$ as anticipated earlier in Eq. (6.6). When A is nonzero the characteristic length scale of a field variation may be estimated from Eq. (9.35). In a Type I material at a vacuum/superconductor interface ψ may be regarded as constant over distances where the field varies. Since the length scale of Eq. (9.33) is unity this corresponds to the field varying over a distance λ_L (the London depth) in unscaled units. However, near a normal-metal–superconductor phase boundary the field variation occurs in a region where ψ is small. If we seek an approximate solution to (9.35) as an exponentially decaying form, $A \sim e^{-x/\delta}$ where δ is a characteristic length, then $1/\delta^2 \cong |\psi|^2$. From an expansion of Eq. (9.41) about $x = x_0$, $\psi \sim \kappa(x - x_0)$. The average value of ψ^2 in a region of width δ would be of order $(\kappa\delta)^2$. Combining these forms we see that the field decay would be governed by $(\kappa\delta)^2 \cong 1/\delta^2$ or $\delta \sim \kappa^{-1/2}$. This would result in a (negative) surface tension contribution of order $-\kappa^{1/2}\gamma$ which, although less than the expression (9.42) in a Type I material, decreases slowly with κ and hence limits the accuracy of this expression.

With increasing κ the surface tension continues to decrease and further analysis (see Lifshitz and Pitaevskii (1980), Sec. 46) shows that it passes through zero for $\kappa = 1/\sqrt{2}$.

We end this section with a discussion of the boundary conditions at an interface between two dissimilar materials at least one of which is superconducting. A boundary condition which is appropriate for the case when no current flows parallel to the surface normal, $\hat{\mathbf{n}}$, is (see Fig. 9.3(a))[4]

$$\left[-i\hbar\nabla - \frac{e^*}{c}\mathbf{A}\right]\cdot\hat{\mathbf{n}}\psi = -\frac{1}{b}\psi, \tag{9.43}$$

as is easily verified by substituting in Eq. (9.17a); here b is a parameter having the units of length. The superconductor insulator case corresponds to the limit $b \to \infty$.

Lastly we consider contact between two superconductors in the presence of a current normal to the boundary. In the most general case the boundary, which we will locate at $x = 0$, may have properties different from both superconductors, an example being a thin layer of a third (normal) metal, which is referred to as an SNS (superconductor–normal–superconductor) junction. The most general (Cauchy) boundary conditions connecting the two sides of the junction, denoted 1 ($x < 0$) and 2 ($x > 0$), have the form (deGennes 1964)[5]

$$\psi_1(0) = M_{11}\psi_2(0) + M_{12}\left(\frac{d}{dx} - \frac{ie^*}{\hbar c}A_x\right)\psi_2(0) \tag{9.44a}$$

and

$$\left(\frac{d}{dx} - \frac{ie^*}{\hbar c}A_x\right)\psi_1(0) = M_{21}\psi_2(0) + M_{22}\left(\frac{d}{dx} - \frac{ie^*}{\hbar c}A_x\right)\psi_2(0). \tag{9.44b}$$

If we choose a single gauge for regions 1 and 2 the vector potential will also be continuous. The coefficients M_{ij} are not independent but are constrained by the requirement of current conservation through the boundary:

$$j_1(0) = j_2(0); \tag{9.45}$$

as a result we may choose the M_{ij} to be real and they must satisfy the condition

$$M_{11}M_{22} - M_{12}M_{21} = 1. \tag{9.46}$$

The resulting expression for the current through the boundary, j_n, is

4. We emphasize that Eq. (9.43) is a *macroscopic* boundary condition. The behavior of the microscopic order parameter may differ substantially near the interface.
5. A boundary condition of this type will be discussed from the microscopic point of view in Part III, Sec. 42.

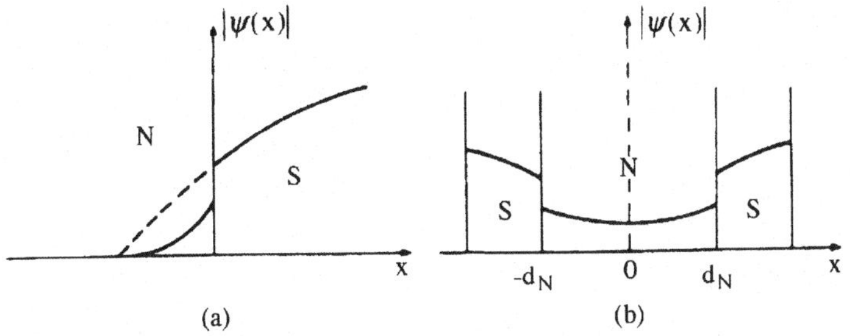

Figure 9.3 Schematic of the order parameter behavior: (a) near an NS boundary; (b) in an SNS junction.

$$j_n = -\frac{ie^*\hbar}{2m}\left\{\psi_2^*(0)\left[\frac{1}{M_{12}}\psi_1(0) - \frac{M_{11}}{M_{12}}\psi_2(0)\right] - cc\right\}$$

$$= -\frac{ie^*\hbar}{2m^*M_{12}}[\psi_1^*(0)\psi_2(0) - \psi_1(0)\psi_2^*(0)]$$

$$= \frac{e^*\hbar}{m^*M_{12}}|\psi_1(0)||\psi_2(0)|\sin[\Phi_2(0) - \Phi_1(0)]. \tag{9.47}$$

The maximum value of this current is

$$j_{\max} = \frac{e^*\hbar}{m^*M_{12}}|\psi_1(0)||\psi_2(0)|. \tag{9.48}$$

If $j_{\max}$ is much less than the bulk critical currents in the two superconductors then we may approximate the $\psi_i(x)$ by the $\mathbf{A} = 0$ forms analogous to (9.26)

$$\psi_1(x) = \psi_1(-\infty)\tanh\left(\frac{x + x_1}{\sqrt{2}\xi_1}\right), x \le 0 \tag{9.49a}$$

and

$$\psi_2(x) = \psi_2(+\infty)\tanh\left(\frac{-x + x_2}{\sqrt{2}\xi_2}\right), x \ge 0; \tag{9.49b}$$

here x_1 and x_2 are free parameters and $\xi_{1,2}$ are the G–L coherence lengths in the two media. From Eqs. (9.49) we have immediately

$$\psi_1(0) = \psi_1(-\infty)\tanh\left(\frac{x_1}{\sqrt{2}\xi_1}\right), \psi_2(0) = \psi_2(\infty)\tanh\left(\frac{x_2}{\sqrt{2}\xi_2}\right) \tag{9.50a}$$

and

$$\psi_1'(0) = \frac{\psi_1(-\infty)}{\sqrt{2}\xi_1}\left[1 - \tanh^2\left(\frac{x_1}{\sqrt{2}\xi_1}\right)\right],$$
$$\psi_2'(0) = -\frac{\psi_2(+\infty)}{\sqrt{2}\xi_2}\left[1 - \tanh^2\left(\frac{x_2}{\sqrt{2}\xi_2}\right)\right]. \tag{9.50b}$$

By substituting these quantities into the $\mathbf{A} = 0$ limit of Eqs. (9.44) we may numerically solve for the parameters x_1 and x_2, assuming the coupling coefficients M_{ij} are known. The behavior of $|\psi_i(x)|$ is shown schematically for an SNS junction in Fig. 9.3(b).

We will encounter Eq. (9.47) in Sec. 15 when we discuss the Josephson effects at an SIS junction (where I designates a thin insulating barrier). If the coupling is weak, as in the case of an SIS junction, we have $M_{11} = M_{22} \cong 0$ and $M_{21} = -1/M_{12}$. In this case the values of $|\psi_1(0)|$ and $|\psi_2(0)|$ are only slightly shifted from their equilibrium bulk values, $|\psi_1(-\infty)|$ and $|\psi_2(+\infty)|$.

10 The upper critical field of a Type II superconductor

As the magnetic field is raised above H_{c1}, and the density of flux lines n_L increases, a point is eventually reached where the distance between flux lines becomes of the order of the vortex core diameter; i.e. $n_L^{-1/2} \sim \xi$. One would then expect a transition to the normal state, and the field at which this occurs, called the upper critical field, is designated H_{c2}. In the region just below H_{c2} the superconducting order parameter must be small and this is the regime where the G–L approach should provide a good description.

We may calculate H_{c2} by linearizing the first G–L equation, since the order parameter is vanishingly small just at H_{c2}. Eq. (9.15) then becomes

$$-\frac{\hbar^2}{2m^*}\left(\nabla - \frac{ie^*}{\hbar c}\mathbf{A}\right)^2 \psi + \alpha\psi = 0; \tag{10.1}$$

this equation is the same as the Schrödinger equation for a particle with energy, $-\alpha$, mass, m^*, and charge, e^*, in a magnetic field, $\mathbf{H}$, with an associated vector potential, $\mathbf{A}$. We will assume a uniform field $\mathbf{H}_0 \parallel \hat{\mathbf{z}}$. The solution of (10.1) is easiest in the so-called Landau gauge where we write the vector potential as

$$\mathbf{A} = H_0\hat{\mathbf{y}}x; \tag{10.2}$$

the resulting G–L equation is then

$$-\frac{\hbar^2}{2m^*}\left[\frac{\partial^2}{\partial x^2} + \left(\frac{\partial}{\partial y} - \frac{ie^*H_0}{\hbar c}x\right)^2 + \frac{\partial^2}{\partial z^2}\right]\psi = -\alpha\psi. \tag{10.3}$$

We may separate the variables in (10.3) by writing $\psi_{n,k_z,k_y}(x, y, z) = e^{ik_z z + ik_y y}u_n(x)$ (we do not normalize ψ at this point as this property follows only from solving the nonlinear G–L equations, which we address in Sec. 14). Inserting this wavefunction in (10.3) we obtain:

$$\left[-\frac{\hbar^2}{2m^*}\frac{d^2}{dx^2} + \frac{1}{2}m^*\omega_c^2(x - x_0)^2\right]u_n(x) = \varepsilon_n u_n(x), \tag{10.4}$$

where $x_0 = (\hbar c/e^*H_0)k_y = (\phi_0/2\pi H_0)k_y$, and $\varepsilon_n = -\alpha - (\hbar^2 k_z^2/2m^*)$. Eq. (10.4) is the Schrödinger equation of a harmonic oscillator with frequency $\omega_c = |e^*|H_0/m^*c$ (the cyclotron frequency of the particle) with energies

$$\varepsilon_n = \left(n + \frac{1}{2}\right)\hbar\omega_c \tag{10.5}$$

having its origin at the point x_0. The full (unnormalized) wavefunctions are

$$\psi_{n,k_z,k_y}(x, y, z) = e^{i(k_y y + k_z z)} e^{-(x-x_0)^2/2a_H^2} H_n\left(\frac{x - x_0}{a_H}\right), \tag{10.6}$$

where $a_H^2 \equiv \hbar/m^*\omega_c = \hbar c/e^*H_0 = \phi_0/2\pi H_0$.

Only the smallest eigenvalue, $n = 0, k_z = 0$ solution (corresponding to the highest field at which superconductivity can nucleate in the interior of a large sample) is meaningful since our linearized theory is valid only as a description of the onset of superconductivity. Hence

$$-\alpha = \tfrac{1}{2}\hbar\omega_c \tag{10.7a}$$

or

$$\begin{aligned} H_{c2} &= -\frac{2m^*c\alpha}{\hbar e^*} \\ &= \frac{2m^*c}{\hbar e^*} a(T_c - T) \end{aligned} \tag{10.7b}$$

or, as it is more commonly written,

$$H_{c2} = \frac{\phi_0}{2\pi\xi^2}; \tag{10.7c}$$

in terms of the length a_H, Eq. (10.7c) may also be rewritten as $a_{H_{c2}} = \xi$. From (10.7b) we see that H_{c2} goes to zero at T_c and increases linearly below that temperature. Recalling $n_L = B/\phi_0$, we see that (10.7c) is in accord with the order-of-magnitude estimate of H_{c2} made at the beginning of this section. However, we are then struck with the fact that in our particle analogy, we have considered only a single quantum state ($n = 0$, the 'ground state'), and yet the material contains the largest possible number of flux lines. The resolution to this apparent contradiction is that our $n = 0$ quantum state is highly degenerate (as evidenced by the fact that the quantum number k_y does not affect the energy).

The relation between this high degeneracy and the high density of flux lines will be dealt with in Sec. 14, where we discuss the region of fields just below H_{c2}.

11 The anisotropic superconductor

Anisotropy may be incorporated in the phenomenological G–L theory simply by introducing an 'effective-mass tensor' into the 'kinetic energy' term in Eq. (9.15) in the form

$$\left[-i\hbar\left(\nabla - \frac{ie^*}{c}\mathbf{A}\right)\right]\cdot\frac{1}{2}\left[\frac{\mathbf{1}}{\mathbf{m}^*}\right]\cdot\left[-i\hbar\left(\nabla - \frac{ie^*}{\hbar c}\mathbf{A}\right)\right]\psi + \alpha\psi + \beta|\psi|^2\psi = 0, \tag{11.1}$$

where $\mathbf{1/m^*}$ is an effective-mass tensor.

For a system which may be regarded as uniaxial, the reciprocal effective-mass tensor may be written as

$$\left[\frac{1}{m^*}\right]_{ij} = \begin{bmatrix} \frac{1}{m_x} & 0 & 0 \\ 0 & \frac{1}{m_x} & 0 \\ 0 & 0 & \frac{1}{m_z} \end{bmatrix}, \tag{11.2}$$

where the z axis is the symmetry direction. The inverse of Eq. (11.2) yields the mass tensor $\mathbf{m}^*$.

We now obtain an expression for the upper critical field $H_{c2}(T)$. Near the field-dependent transition temperature the order parameter is vanishingly small and we may again neglect the nonlinear term. Our equation is then formally identical with the Schrödinger equation of a particle with charge e^* and an anisotropic mass tensor $\mathbf{m}^*$ in a uniform magnetic field $\mathbf{H}_0$ and the 'energy levels' have the harmonic oscillator form

$$-\alpha = (n + 1/2)\hbar\omega_c(\theta), \tag{11.3}$$

where $\omega_c(\theta)$ is the angle-dependent cyclotron frequency which is encountered in the effective-mass theory of cyclotron resonance in semiconductors. The latter can be worked out classically simply by solving Newton's equations of motion for our 'particle' moving under the influence of the Lorentz force:

$$\mathbf{m}^*\cdot\dot{\mathbf{v}} = \frac{e^*}{c}\mathbf{v}\times\mathbf{H}_0; \tag{11.4}$$

here $\mathbf{v}$ is the velocity. The solution of this equation involves elliptical orbits which are traversed at a frequency

$$\omega_c(\theta) = \frac{|e^*|H_0}{c}\left(\frac{\sin^2\theta}{m_x m_z} + \frac{\cos^2\theta}{m_x^2}\right)^{1/2}; \tag{11.5}$$

here θ measures the angle of the magnetic field from the z axis. The solution with the lowest free energy corresponds to $n = 0$ in Eq. (11.3) and the upper critical field is thus given by

$$H_{c2}(\theta) = \frac{2ca(T_c - T)}{\hbar e^*[\sin^2\theta/m_x m_z + \cos^2\theta/m_x^2]^{1/2}}. \tag{11.6}$$

If we define two coherence lengths,

$$\xi_x \equiv \left[\frac{\hbar^2}{2m_x a(T_c - T)}\right]^{1/2}$$

and

$$\xi_z \equiv \left[\frac{\hbar^2}{2m_z a(T_c - T)}\right]^{1/2},$$

and recall the 'flux quantum' $\phi_0 = hc/|e^*|$, then Eq. (11.6) may be written as

$$H_{c2}(\theta) = \frac{\phi_0}{2\pi(\sin^2\theta\,\xi_x^2\xi_z^2 + \cos^2\theta\,\xi_x^4)^{1/2}}; \tag{11.7a}$$

in particular, for fields parallel and perpendicular to the symmetry plane of the material we have, respectively,

$$H_{c2\parallel} = \frac{\phi_0}{2\pi\xi_x\xi_z} \tag{11.8a}$$

and

$$H_{c2\perp} = \frac{\phi_0}{2\pi\xi_x^2}. \tag{11.8b}$$

The angular dependence of the upper critical field is thus characterized by the two parameters ξ_x and ξ_z (Lawrence and Doniach 1971). Note that (11.8a) and (11.8b) both lead to a linear behavior of the upper critical field in $T_c - T$. In terms of the definitions (11.8a) and (11.8b) we may rewrite Eq. (11.7a) as

$$\frac{H_{c2}^2(\theta)}{H_{c2\perp}^2}\cos^2\theta + \frac{H_{c2}^2(\theta)}{H_{c2\parallel}^2}\sin^2\theta = 1, \tag{11.7b}$$

which is the equation of an ellipse.

Eqs. (9.17) can be generalized to the case of an anisotropic superconductor by replacing m^{*-1} by $\mathbf{1/m^*}$ from Eq. (11.2); an anisotropic London penetration depth then results.

12 Thin superconducting slabs

Consider an isolated superconducting slab which has a thickness d. As d is reduced, the effects of boundary scattering and reduced dimensionality usually cause T_c to fall (and ultimately to change the character of the transition); however, we shall here regard T_c and d as independent parameters. We consider the case of a parallel magnetic field and work in the gauge $\mathbf{A} = -H_0\hat{\mathbf{y}}z$. We seek a solution of Eq. (10.1) of the form

$$\psi = u(z)e^{ik_x x + ik_y y},$$

which yields the differential equation

$$-\frac{\hbar^2}{2m^*}u'' + \frac{1}{2}m^*\omega_c^2(z - z_0)^2 u + \left(\alpha + \frac{\hbar^2 k_x^2}{2m^*}\right)u = 0, \tag{12.1}$$

where $\omega_c = |e^*|H_0/m^*c$ and $z_0 \equiv c\hbar k_y/e^*H_0$. The lowest (most negative) eigenvalue corresponds to $k_x = 0$. In a bulk sample, z_0 may have any value since it simply refers to a change in the position of the origin. For a slab, we must solve Eq. (12.1) subject to boundary conditions at the surface of the slab. If we require the surface term to vanish in Eq. (9.14) we are led to the boundary condition

$$\left(\nabla - \frac{ie^*}{\hbar c}\mathbf{A}\right)\psi \cdot \hat{\mathbf{n}} = 0, \tag{12.2}$$

where $\hat{\mathbf{n}}$ is a unit vector normal to the surface. This is not a general boundary condition; it turns out to be correct only for a smooth nonmagnetic superconductor–insulator or the superconductor–vacuum interface. If we place the origin in the center of the slab, Eq. (12.2) requires that

$$\left.\frac{\partial u(z)}{\partial u}\right|_{z=\pm d/2} = 0.$$

We shall not seek an exact solution to Eq. (12.1) subject to this boundary condition. However, for $d \lesssim \xi$, we may regard the second term as a perturbation. The zero order solution satisfying the boundary condition is then $u =$ const. With this wavefunction and our perturbation, we have

$$-\alpha = \frac{1}{2}m^*\omega_c^2 \int_{-d/2}^{+d/2} u^2 z^2 \mathrm{d}z \bigg/ \int_{-d/2}^{+d/2} u^2 \mathrm{d}z = \frac{m^*\omega_c^2 d^2}{24}, \tag{12.3}$$

where we have set $z_0 = 0$ (which minimizes α). This leads immediately to

$$H_{c2\parallel} = \frac{(12)^{1/2}\phi_0}{2\pi\xi d}. \tag{12.4}$$

The behavior of Eq. (12.4) is radically different from that of Eq. (11.8a). The reduced dimensionality has the effect of replacing ξ_z by $d/(12)^{1/2}$. Since only one factor of ξ enters Eq. (12.4), the temperature dependence of $H_{c2\parallel}$ is altered from $H_{c2\parallel} \propto T_c - T$ to $H_{c2\parallel} \propto (T_c - T)^{1/2}$. We can refer to these two behaviors as 3D and 2D, respectively.

Let us now consider the case where the field is at an arbitrary angle with respect to the thin slab. We take the vector potential in the form

$$\mathbf{A} = \hat{\mathbf{y}}(x\cos\theta - z\sin\theta)H_0. \tag{12.5}$$

Writing $\psi(\mathbf{r}) = u(x,z)e^{ik_y y}$, the linearized G–L equation becomes

$$\mathscr{H}u + \alpha u = 0, \tag{12.6}$$

where

$$\mathscr{H} = \mathscr{H}_0 + \mathscr{H}' \tag{12.7}$$

with

$$\mathscr{H}_0 = \frac{-\hbar^2}{2m^*}\frac{\partial^2}{\partial x^2} + \frac{1}{2}m^*\omega_c^2 x^2\cos^2\theta \tag{12.8a}$$

and

$$\mathscr{H}' = \frac{-\hbar^2}{2m^*}\frac{\partial^2}{\partial z^2} + \frac{1}{2}m^*\omega_c^2(z - z_0)^2\sin^2\theta + m^*\omega_c^2 x(z - z_0)\sin\theta\cos\theta; \tag{12.8b}$$

here we have defined

$$z_0 = -\hbar c k_y/e^* H_0 \sin\theta.$$

Continuing our analogy to charged particle motion in a magnetic field, we note that there is no constraint on the in-plane motion and thus the term in x^2 would take on arbitrarily large values, were we to treat it as a perturbation on an x-independent trial wavefunction; we have therefore regarded this term as part of a zero order Hamiltonian, $\mathscr{H}_0$. The eigenvalues of $\mathscr{H}_0$ are simply harmonic oscillator levels with a frequency $\omega_c\cos\theta$. The requirement that $\partial u/\partial z$ vanishes at $z = \pm d/2$ again requires that (in lowest order) the wavefunction has no z dependence; i.e., $u(x,z) = u(x)$. Since particle motion in the z direction is restricted, we can treat $\mathscr{H}'$ as a perturbation. The eigenvalue is again minimized by setting $k_y = 0$ (the perturbation produced by the last term in Eq. (12.8b) then vanishes by symmetry). The resulting perturbation produced by $\mathscr{H}'$ is of the same form as Eq. (12.3). The total eigenvalue is then

$$-\alpha = \frac{1}{2}\hbar\omega_c\cos\theta + \frac{m^*\omega_c^2 d^2}{24}\sin^2\theta. \tag{12.9}$$

Defining $H_{c2\parallel} = (12)^{1/2}\phi_0/2\pi\xi d$ and $H_{c2\perp} = \phi_0/2\pi\xi^2$ we may rewrite Eq. (12.9) as

$$1 = \frac{H_{c2}(\theta)}{H_{c2\perp}}|\cos\theta| + \frac{H_{c2}^2(\theta)}{H_{c2\parallel}^2}\sin^2\theta; \tag{12.10}$$

note that the angular dependence of this expression, first suggested by Tinkham (1963), is radically different from that of Eq. (11.7b); in particular, it has a cusp at $\theta = \pi/2$.

We next treat the problem where the boundary condition is altered such that $u(z)|_{z=\pm d/2}$

$= 0$. In practice, this condition is approximated by a superconducting film which is sandwiched between two films containing a large concentration of paramagnetic impurities (such as ferromagnetic films) where 'pair breaking' effectively drives the order parameter to zero at the boundary. We return to Eq. (12.1). For $H_0 = 0$ ($\omega_c = 0$) the eigenfunction satisfying our boundary condition is

$$u = u_0 \cos(kz) \tag{12.11}$$

with $kd/2 = \pm \pi/2$ for the ground state. This leads to a suppression of the transition temperature governed by

$$-\alpha = \frac{\hbar^2\pi^2}{2m^*d^2}$$

or

$$T_c = T_{c0} - \frac{\hbar^2\pi^2}{2m^*d^2}; \tag{12.12}$$

here T_{c0} is the bulk transition temperature and T_c is the film transition temperature in the presence of the new boundary condition. Application of a parallel magnetic field induces an additional reduction in T_c which we calculate from perturbation theory (Wong and Ketterson 1986):

$$\begin{aligned}
-\delta\alpha &= \frac{\langle u(z)\,|\,\frac{1}{2}m^*\omega_c^2z^2\,|\,u(z)\rangle}{\langle u(z)\,|\,u(z)\rangle} \\
&= \frac{1}{2}m^*\omega_c^2 d^2\left(\frac{1}{12} - \frac{1}{2\pi^2}\right) \\
&= 0.0163m^*\omega_c^2d^2.
\end{aligned} \tag{12.13}$$

Combining Eqs. (12.12) and (12.13) gives

$$-\alpha = \frac{\hbar^2}{2m^*}\frac{\pi^2}{d^2} + 0.0163m^*\omega_c^2d^2. \tag{12.14}$$

Eq. (12.14) may be written in terms of T_c as

$$H_{c2\parallel}^2 = \left(\frac{m^*c}{|e^*|}\right)^2 \frac{2a(T_c - T)}{0.0327m^*d^2}. \tag{12.15}$$

If we define a coherence length involving T_c, rather than T_{c0}, by

$$\xi = \left(\frac{\hbar^2}{2ma(T_c - T)}\right)^{1/2}, \tag{12.16}$$

then we may write

$$H_{c2\parallel} = \frac{5.53\phi_0}{2\pi\xi d}, \tag{12.17}$$

which is to be compared with Eq. (12.4). The angular dependence follows from inserting (12.17) into (12.10).

13 Surface superconductivity

A superconducting half space in a parallel field may exhibit the phenomena of surface superconductivity (Saint-James and deGennes 1963). We take the interface to be the plane $z = 0$. The critical field is still determined by the eigenvalue of Eq. (12.1) ($k_x = 0$ with the requirement that $\partial u/\partial z = 0$ at $z = 0$). For values of z_0 (i.e., k_y) such that the 'orbit center' is deep in the bulk of the superconductor ($z_0 \to -\infty$) the corresponding eigenvalue yields the bulk critical field; the boundary condition is adequately satisfied since the ground-state wavefunction of our harmonic oscillator is a Gaussian (centered on z_0) having negligible amplitude at the boundary. For the opposite extreme, where we choose $z_0 = 0$, the same Gaussian wavefunction also satisfies the boundary condition, since the derivative at the peak of the Gaussian (now centered on $z_0 = 0$) vanishes. The question arises whether there is a smaller eigenvalue (than the bulk eigenvalue) for some value $0 > z_0 > -\infty$. To examine this situation, we reflect the harmonic oscillator potential $V = \frac{1}{2}m^*\omega_c^2(z - z_0)^2$ through the plane $z = 0$ such that

$$V = \begin{cases} \frac{1}{2}m^*\omega_c^2(z + |z_0|)^2, & z < 0, \\ \frac{1}{2}m^*\omega_c^2(z - |z_0|)^2, & z > 0. \end{cases} \tag{13.1}$$

Note that this is *not* a harmonic oscillator potential but rather has two minima (located at $z = \pm |z_0|$) and the slope $\partial V/\partial z$ is discontinuous at $z = 0$. If we seek a nodeless solution for the motion in this potential (which is symmetric under a sign change of z), the boundary condition at $z = 0$ will automatically be satisfied, and it should correspond to a possible ground state.

We can see immediately from Eq. (13.1) that an eigenvalue smaller than that of the bulk must exist, since the potential described by Eq. (13.1) is always less than or equal to that obtained by extrapolating the potential from either half space into the other. The above-defined eigenvalue problem admits an exact solution in terms of Weber functions (Saint-James, Thomas, and Sarma 1969). However, an adequate solution is obtained using a trial wavefunction $u \sim e^{-rz^2}$ (deGennes 1966). Minimization of α with respect to r and z_0 yields

$$-\alpha = \left(1 - \frac{2}{\pi}\right)^{1/2} \frac{\hbar\omega_c}{2} = 0.60\frac{|e^*|\hbar H_{c3}}{2m^*c}, \tag{13.2}$$

where we have defined the upper critical field associated with this surface superconducting state as H_{c3}. The exact solution yields 0.59 rather than 0.60 for the numerical factor. The upper critical fields for bulk and surface superconductivity are then related by

$$H_{c3} = 1.7H_{c2}. \tag{13.3}$$

Surface superconductivity persists to higher magnetic fields (in a Type II material) and thus may (in a conductivity measurement) mask the bulk transition. The state may be suppressed by the following:

(1) rendering the surface rough (by sand blasting);
(2) using sufficiently high measuring currents to destroy surface superconductivity;
(3) depositing a normal-metal overlayer.

We now briefly discuss the case in which we have a thick film, of thickness d (with the origin in the film center), rather than a superconducting half space. For films with $d \gtrsim \xi$, the amplitude of the G–L wavefunction at nucleation would be highest in the immediate vicinity of the surfaces at $z = \pm d/2$ for $H_{c2\parallel} < H < H_{c3}$. As d is reduced, the eigenvalues (satisfying the boundary conditions at $z = \pm d/2$ associated with bulk and surface superconductivity) approach each other and, below some distance $d = d_c$, only a state which has its wavefunction centered on $z = 0$ is stable (Saint-James *et al.* 1969).

As we saw above, when the G–L equation for the magnetic field at an arbitrary angle is written in terms of coordinates for which the boundary conditions are conveniently applied, it is nonseparable. Thus, for thick films where $u \sim u(x, z)$ (rather than thin films where $u \sim u(x)$), we must resort to a numerical, variational or perturbational technique. This complicated, and to some extent not fully explored, problem has been discussed by Minenko (1983), Saint-James (1965), Yamafuji, Kawashima, and Irie (1966) and Saint-James *et al.* (1969). However, we note that, at some angle $\theta = \theta_c$, we expect the eigenvalues associated with the surface and bulk superconducting states to merge.

14 The Type II superconductor for H just below H_{c2}

The discussion of the upper critical field given in Sec. 10 involved the linearized G–L theory and is strictly valid only for the transition line in the H–T plane where $H = H_{c2}(T)$. For fields below this value we must include the nonlinear terms in Eq. (9.15). In this section we obtain an approximate, but analytic, solution for the G–L equations which is useful for a small range of fields just below H_{c2}. This problem was first examined by Abrikosov (1957).

We recall the solution of the linearized G–L equation obtained in Sec. 10 for a 'nucleation wavefunction' at H_{c2} (for $\mathbf{H} \parallel \hat{\mathbf{z}}$) which follows from Eq. (10.6) with $n = 0$ and $k_z = 0$:

$$\psi(x, y) = C e^{iky} e^{-(x-x_0)^2/2\xi^2}, \tag{14.1}$$

where we dropped the subscript from k_y and $x_0 \equiv (\hbar c/e^* H_{c2})k = \phi_0 k/2\pi H_{c2} = \xi^2 k$; here C is a 'normalization constant', which vanishes for $H = H_{c2}$. As noted in Sec. 10, this wavefunction is highly degenerate in that k can have any value (so long as x_0 lies within the superconductor).

The wavefunction which is the solution to the full G–L equations for H just below H_{c2} is expected to resemble some particular solution of the linearized Eq. (10.3). At the same time, based on our earlier discussion, we also expect some sort of vortex lattice to be present. We therefore seek an approximate solution for ψ as a superposition of terms of the form (14.1) chosen so as to be appropriately periodic in both the x and y directions (thus forming the lattice). Eq. (14.1) is periodic in the y direction as it stands with a period $b = 2\pi/k$. We may retain this periodicity in the y direction while simultaneously forming a function which is quasiperiodic in the x direction by substituting nk for k in (14.1) and summing over all n to yield

$$\psi(x, y) = C \sum_{n=-\infty}^{n=+\infty} e^{inky} e^{-(x-x_n)^2/2\xi^2}, \tag{14.2}$$

where $x_n = n\xi^2 k = 2\pi n\xi^2/b$ and C is a (common) normalization constant. We define the period in the x direction as $a \equiv x_{n+1} - x_n = 2\pi\xi^2/b$, and the area of the unit cell is constrained by $ab = 2\pi\xi^2$. Eq. (14.2) is not, however, strictly periodic in x: if we substitute $x + a$ for x this is equivalent to replacing n by $n + 1$ *in the summation, which changes the phase of the wavefunction by* e^{iky}; i.e.,

$$\psi(x + a, y) = e^{-iky}\psi(x, y). \tag{14.3}$$

This behavior is inherent in our choice of the gauge $A_y = Hx$; under a coordinate displacement $x \to x + a$, $A_y \to A_y + Ha$ leading to a change in the canonical momentum $n\hbar k \to n\hbar k - (e^*/c)Ha$

which for $H \cong H_{c2}$ leads to $nk \to (n+1)k$. Eq. (14.3) shows that

$$|\psi(x+a, y)|^2 = |\psi(x, y|^2; \tag{14.4}$$

i.e., the probability density is strictly periodic.

Eq. (14.2) is not, however, the most general periodic function that can be formed by superimposing forms based on (14.1) since we may introduce a complex expansion coefficient associated with each term, provided we impose the periodicity condition $C_{n+\nu} = C_n$ where ν is an integer. Our wavefunction then has the form

$$\psi(x, y) = \sum_{n=-\infty}^{n=+\infty} C_n e^{inky} e^{-(x-x_n)^2/2\xi^2}, \tag{14.5}$$

where the quasiperiod in the x direction is given by $a = \nu(x_{n+1} - x_n) = 2\pi\nu\xi^2/b$ and the area of the unit cell is constrained by $ab = 2\pi\nu\xi^2$; ν corresponds to the number of vortices per unit cell.

In practice there are two lattices of interest: (i) the simple square lattice and (ii) the triangular (hexagonal) lattice. For the simple cubic case all the C_n are equal ($\nu = 1$) and $\psi(x, y)$ has the form (14.2) with $a = b = 2\pi/k$ and $x_n = n\xi^2 k = na$; hence

$$\psi_{\square}(x, y) = C \sum_{n=-\infty}^{n=+\infty} e^{2\pi i n y/a} e^{-(x-na)^2/2\xi^2}. \tag{14.6}$$

For the triangular case $C_{n+2} = C_n$ and $C_1 = iC_0$; Eq. (14.5) then has the form

$$\psi_{\triangle}(x, y) = C\left[\sum_{n=\text{even}} e^{inky} e^{-(x-x_n)^2/2\xi^2} + i \sum_{n=\text{odd}} e^{-inky} e^{-(x-x_n)^2/2\xi^2}\right]. \tag{14.7}$$

Writing $k = 2\pi/b$ and $a = 2\xi^2 k$, noting $b = \sqrt{3}a$ for a triangular lattice,[1] and combining the two sums, Eq. (14.7) becomes

$$\psi_{\triangle}(x, y) = C \sum_{n=\text{even}} [e^{2\pi i n y/b} e^{-(x-na/2)^2/2\xi^2} + i e^{2\pi i(n+1)y/b} e^{-[x-(n+1)a/2]^2/2\xi^2}]$$

or

$$\psi_{\triangle}(x, y) = C \sum_n e^{4\pi i n y/b} [e^{-(x-na)^2/2\xi^2} + i e^{2\pi i y/b} e^{-[x-(n+\frac{1}{2})a]^2/2\xi^2}]. \tag{14.8}$$

We fix the normalization constant, C (our only remaining internal free parameter), by requiring that the free energy be stationary with respect to a variation of ψ. If we temporarily ignore changes in the vector potential this is equivalent to setting $(\partial E/\partial C)_{\mathbf{A}} = 0$, which yields

$$\int \left[\frac{1}{2m^*}\left|\left(\frac{\hbar}{i}\nabla - \frac{e^*}{c}\mathbf{A}\right)\psi\right|^2 + \alpha|\psi|^2 + \beta|\psi|^4\right] d^3r = 0. \tag{14.9}$$

Denoting quantities of the form $(1/V)\int f(\mathbf{r})d^3r$ by a bar, where V is the volume, Eq. (14.9) may be written

$$\overline{\frac{1}{2m^*}\left|\left(\frac{\hbar}{i}\nabla - \frac{e^*}{c}\mathbf{A}\right)\psi\right|^2} + \overline{\alpha|\psi|^2} + \overline{\beta|\psi|^4} = 0. \tag{14.10}$$

1. A triangular lattice is a centered rectangular lattice with $b = \sqrt{3}a$ and two vortices per unit cell.

We must also include the effect of a change in the internal field, the magnetic induction $\mathbf{H}(\mathbf{r})$. We write this as a sum of two contributions arising from: (i) an explicit shift in the external field relative to H_{c2} (since we have an exact solution for the latter), and (ii) a (screening) shift $\mathbf{H}^{(s)}(\mathbf{r})$ arising from circulating supercurrents associated with the vortices (this latter effect is vanishingly small for $H = H_{c2}$ and was therefore neglected in Sec. 10). We denote the sum of these contributions by

$$\mathbf{H}(\mathbf{r}) = \mathbf{H}_{c2} + \mathbf{H}^{(1)}(\mathbf{r}) \tag{14.11}$$

with

$$\mathbf{H}^{(1)}(\mathbf{r}) = \mathbf{H}_0 - \mathbf{H}_{c2} + \mathbf{H}^{(s)}(\mathbf{r}). \tag{14.12}$$

Associated with the internal field $\mathbf{H}^{(1)}$ is a vector potential $\mathbf{A}^{(1)}$. Expanding (14.9) to first order in $\mathbf{A}^{(1)}$ yields (in a London gauge)

$$\beta\overline{|\psi|^4} + \overline{\frac{\hbar e^*}{2im^*c}\mathbf{A}^{(1)}(\mathbf{r})\cdot\left[\psi^*\left(\nabla - \frac{ie^*}{\hbar c}\mathbf{A}_0\right)\psi - \psi\left(\nabla + \frac{ie^*}{\hbar c}\mathbf{A}_0\right)\psi^*\right]} = 0, \tag{14.13}$$

where the remaining terms cancel (since they satisfy the linear Schrödinger-like equation at $H = H_{c2}$).

The last two terms may be written as $\overline{(1/c)\mathbf{j}^{(s)}(\mathbf{r})\cdot\mathbf{A}^{(1)}(\mathbf{r})}$ where we identify $\mathbf{j}^{(s)}(\mathbf{r})$ as the screening current density associated with the vortex lattice. Introducing $\nabla\times\mathbf{H}^{(s)} = (4\pi/c)\mathbf{j}^{(s)}(\mathbf{r})$, integrating by parts, and writing $\mathbf{H}^{(1)}(\mathbf{r}) = \nabla\times\mathbf{A}^{(1)}(\mathbf{r})$, Eq. (14.13) becomes

$$\beta\overline{|\psi|^4} - \frac{1}{4\pi}\overline{\mathbf{H}^{(1)}(\mathbf{r})\cdot\mathbf{H}^{(s)}(\mathbf{r})} = 0, \tag{14.14}$$

where $\mathbf{H}^{(s)}$ is parallel to $\mathbf{H}^{(1)}$ by symmetry. To the accuracy we are carrying the calculation, $\mathbf{H}^{(s)}$ is the diamagnetic response field ($\mathbf{H}^{(s)} = 4\pi\mathbf{M}^{(s)}$).

To obtain the field $\mathbf{H}^{(s)}$ we first calculate the associated magnetization current, $\mathbf{j}^{(s)}(\mathbf{r})$. The calculation of $\mathbf{j}^{(s)}$ is facilitated by noting that the contours of constant $|\psi(\mathbf{r})|$ and the (two-dimensional) streamlines of $\mathbf{j}^{(s)}$ coincide, which we now demonstrate. We introduce the canonical momentum operator $\hat{\mathbf{\Pi}}$ ($= m^*\hat{\mathbf{V}}$ where $\hat{\mathbf{V}}$ is the velocity operator)

$$\hat{\mathbf{\Pi}} = \frac{\hbar}{i}\nabla - \frac{e^*}{c}\mathbf{A}, \tag{14.15}$$

which has the commutation relations $[\Pi_x, \Pi_y] = i(e^*\hbar/c)H$. In analogy with the standard harmonic oscillator problem we introduce the raising (creation), $\Pi_+ = \Pi_x + i\Pi_y$, and lowering (destruction), $\Pi_- = \Pi_x - i\Pi_y$, operators.[2] The ground state has the property $\Pi_-\psi = 0$ from which we obtain

$$\left[\left(\frac{\hbar}{i}\frac{\partial}{\partial x} - \frac{e^*}{c}A_x\right) - i\left(\frac{\hbar}{i}\frac{\partial}{\partial y} - \frac{e^*}{c}A_y\right)\right]\psi = 0. \tag{14.16}$$

If we substitute the wavefunction in the form $\psi = |\psi|e^{i\Phi}$ into (14.16), and identify the current (in lowest order) as

$$\mathbf{j}^{(s)}(\mathbf{r}) = \frac{e^*}{m^*}|\psi|^2\left(\hbar\nabla\Phi - \frac{e^*}{c}\mathbf{A}_0\right), \tag{14.17}$$

2. The linearized Schrödinger-like G–L Eq. (10.4) may be written $\Pi_+\Pi_-\psi = 0$.

we obtain (on equating the real and imaginary parts separately to zero)

$$j_x^{(s)} = -\frac{e^*\hbar}{2m^*}\frac{\partial}{\partial y}|\psi|^2 \tag{14.18a}$$

and

$$j_y^{(s)} = \frac{e^*\hbar}{2m^*}\frac{\partial}{\partial x}|\psi|^2. \tag{14.18b}$$

From the structure of (14.18a) and (14.18b) we see that $\nabla|\psi|\perp\mathbf{j}^{(s)}$. The magnetization field, $\mathbf{H}^{(s)}$, follows from Maxwell's fourth equation, $\nabla\times\mathbf{H}^{(s)}(\mathbf{r}) = (4\pi/c)\mathbf{j}^{(s)}(\mathbf{r})$; comparing this equation with Eqs. (14.18) yields

$$\mathbf{H}^{(s)}(\mathbf{r}) = -4\pi\frac{e^*\hbar}{2m^*c}|\psi|^2\hat{\mathbf{z}}, \tag{14.19}$$

where the constant of integration was fixed such that $\mathbf{H}^{(s)}$ vanishes when $|\psi| = 0$. We may interpret $\mathbf{H}^{(s)}$ as a magnetization through the usual definition $\mathbf{B} = \mathbf{H} + 4\pi\overline{\mathbf{M}}^{(s)}$ where $\overline{\mathbf{M}^{(s)}(\mathbf{r})} = -(e^*\hbar/2m^*c)\overline{|\psi|^2}\hat{\mathbf{z}}$; note it is negative as expected (since superconductors are diamagnetic).

We now combine (14.12), (14.14), and (14.19) to obtain

$$\beta\overline{|\psi|^4} + \overline{\frac{e^*\hbar}{2m^*c}|\psi|^2\left[H_0 - H_{c2} - 4\pi\frac{e^*\hbar}{2m^*c}|\psi|^2\right]} = 0. \tag{14.20}$$

Writing $\psi(\mathbf{r}) = \psi_0 f(\mathbf{r})$, where $\psi_0^2 = |\alpha|/\beta$, and using the definitions (9.29) and (10.7) for κ and H_{c2} we may write (14.20) as

$$\overline{f^4}\left(1 - \frac{1}{2\kappa^2}\right) - \overline{f^2}\left(1 - \frac{H_0}{H_{c2}}\right) = 0. \tag{14.21}$$

The quantities $\overline{f^2}$ and $\overline{f^4}$ may be calculated provided we choose a symmetry ($\psi_\square$ or $\psi_\triangle$ from (14.6) or (14.8)) and a lattice spacing. It turns out that near H_{c2} the ratio

$$\beta_A \equiv \frac{\overline{f^4}}{\left(\overline{f^2}\right)^2} \tag{14.22}$$

(introduced by Abrikosov (1957)) is independent of H_0. Introducing this quantity into (14.21) we have

$$\overline{f^2} = \frac{1 - \dfrac{H_0}{H_{c2}}}{\beta_A\left(1 - \dfrac{1}{2\kappa^2}\right)}; \tag{14.23a}$$

$$\overline{f^4} = \beta_A\left(\overline{f^2}\right)^2 = \frac{\left(1 - \dfrac{H_0}{H_{c2}}\right)^2}{\beta_A\left(1 - \dfrac{1}{2\kappa^2}\right)^2}. \tag{14.23b}$$

From the above $\overline{|\psi|^2} = (|\alpha|/\beta)\overline{f^2}$ and $\overline{|\psi|^4} = \beta_A\left(\overline{|\psi|^2}\right)^2$.

We now proceed to a consideration of the thermodynamics of a Type II superconductor. We begin with the magnetic induction which is defined as the position average of the internal field, $\mathbf{H}(\mathbf{r})$: $B \equiv H_0 + \overline{H^{(s)}}$; using (14.19) and the definitions of κ and H_{c2} we have

$$B = H_0 - \frac{H_{c2}}{2\kappa^2}\overline{f^2} = H_0 - \frac{H_c}{\sqrt{2}\kappa}\overline{f^2}, \tag{14.24}$$

where we used the relation $H_{c2} = \sqrt{2}\kappa H_c$ in the second step. Substituting (14.23a) for $\overline{f^2}$ we obtain the thermodynamic magnetic induction $B = B(H_0)$ as

$$B = H_0 - \frac{H_{c2} - H_0}{\beta_A(2\kappa^2 - 1)}. \tag{14.25a}$$

We may invert this relation to obtain $H_0 = H_0(B)$ as

$$H_0 = B + \frac{H_{c2} - B}{1 + \beta_A(2\kappa^2 - 1)}. \tag{14.25b}$$

From (14.25a) we obtain the magnetization, $(B - H_0)/4\pi$, as

$$M = \frac{H_0 - H_{c2}}{4\pi\beta_A(2\kappa^2 - 1)}; \tag{14.25c}$$

note this is linear near H_{c2} as discussed earlier.

We now calculate the free energy. Comparing the variational expression (14.9) with Eq. (9.13) for the full free energy we have

$$F = F_0 + \int d^3r\left[-\frac{\beta}{2}|\psi|^4 + \frac{1}{8\pi}H^2(\mathbf{r})\right].$$

Rewriting the first integral and substituting the unaveraged form of (14.24) this equation becomes

$$F = F_0 + V\left[-\frac{H_{c2}^2}{8\pi}\frac{\overline{f^4}}{(2\kappa^2)} + \frac{1}{8\pi}\overline{\left(H_0 - \frac{H_{c2}}{2\kappa^2}f^2\right)^2}\right]$$

$$= F_0 + \frac{V}{8\pi}\left[-H_{c2}^2\left(1 - \frac{1}{2\kappa^2}\right)\frac{\overline{f^4}}{2\kappa^2} + H_0\left(H_0 - \frac{1}{\kappa^2}H_{c2}\overline{f^2}\right)\right]. \tag{14.26}$$

Substituting Eqs. (14.23) we obtain $F = F(H_0)$ as

$$F = F_0 + \frac{V}{8\pi}\left[H_0^2 - \frac{H_{c2}^2 - H_0^2}{\beta_A(2\kappa^2 - 1)}\right]. \tag{14.27}$$

Alternatively, using Eqs. (14.25), we may calculate $F = F(B)$ for which we find

$$F = F_0 + \frac{V}{8\pi}\left[B^2 - \frac{(B - H_{c2})^2}{1 + \beta_A(2\kappa^2 - 1)}\right]. \tag{14.28}$$

For the case of a long cylinder with an external field, H_{ext}, parallel to the axis or a thin film with

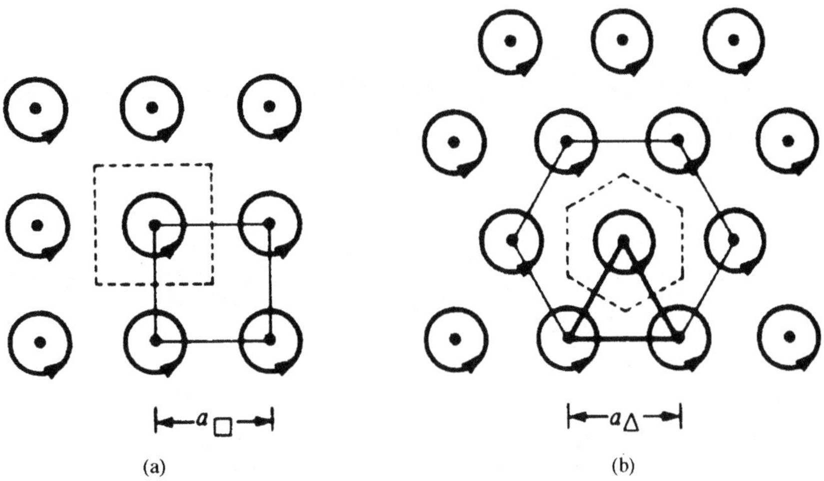

Figure 14.1 Schematic diagram of square and triangular vortex lattices. The dashed lines show the basic unit cells.

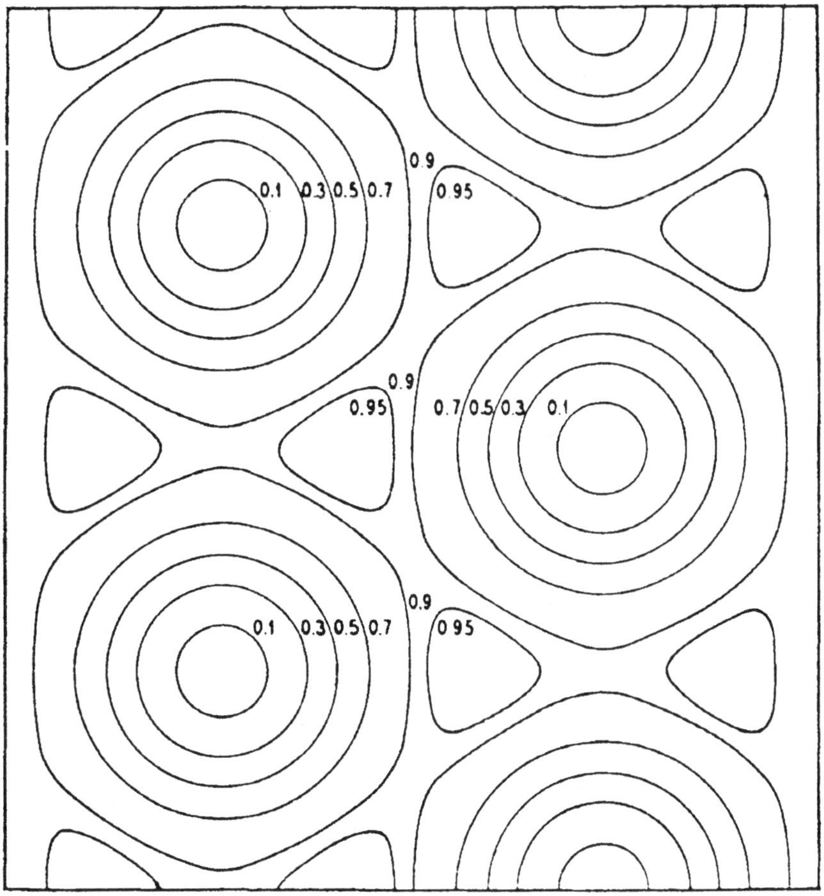

Figure 14.2 The spatial configuration of $|\psi|^2$ near H_{c2} for a triangular vortex lattice. The numbers labeling the contours specify the square of the reduced order parameter. (After Kleiner, Roth, and Autler (1964).)

H_{ext} lying in the plane (where $H_{ext} = H_0$) we would use (14.27); for the case where the external field is perpendicular to the film (where $H_{ext} = B$) we would use (14.28).

The only remaining problem is to calculate the Abrikosov parameter β_A. This calculation is somewhat tedious and we refer the reader to Saint-James *et al.* (1969) for the details. The results are

$$\beta_A^{\square} = 1.18 \qquad \text{(square lattice)} \tag{14.29a}$$

and

$$\beta_A^{\triangle} = 1.16 \qquad \text{(triangular lattice).} \tag{14.29b}$$

The triangular lattice is thus slightly more stable. A schematic diagram of the square and triangular vortex lattices is shown in Fig. 14.1. The spatial configuration of $|\psi|^2$ near H_{c2} for a triangular vortex lattice is shown in Fig. 14.2. The above treatment has neglected the effects of any in-plane anisotropy which, given that the structure is sensitive to β_A, can (and in some cases does) alter the observed symmetry.

15 The Josephson effects

15.1 The Josephson equations

Let us recall our expression for the G–L current density

$$\mathbf{j} = -\frac{ie^*\hbar}{2m^*}(\psi^*\nabla\psi - \psi\nabla\psi^*) - \frac{e^{*2}}{m^*c}|\psi|^2\mathbf{A}. \tag{9.17b}$$

We write ψ in the form $\psi = |\psi|e^{i\Phi}$ as before, obtaining

$$\mathbf{j} = \frac{e^*\hbar}{m^*}|\psi|^2\left[\nabla\Phi - \frac{e^*}{\hbar c}\mathbf{A}\right]; \tag{15.1}$$

i.e., the current in a superconductor involves the gradient of the (gauge-invariant) phase (recall that real G–L wavefunctions carry no current).

The G–L boundary condition at an insulator–superconductor interface was given earlier as

$$\hat{\mathbf{n}}\cdot\left(\frac{\hbar}{i}\nabla - \frac{e^*}{c}\mathbf{A}\right)\psi = 0, \tag{12.2}$$

which is equivalent to no current flow through the boundary. If we have a 'junction' which weakly couples two superconductors (formed, for example, by a thin insulating layer between the two superconductors through which electrons can tunnel), we must modify this condition. Taking $\hat{\mathbf{n}} \parallel \hat{\mathbf{x}}$, we replace (12.2) by more general (yet linear) boundary conditions[1]

$$\frac{\partial\psi_1}{\partial x} - \frac{ie^*}{\hbar c}A_x\psi_1 = \frac{\psi_2}{\lambda} \tag{15.2a}$$

and

$$\frac{\partial\psi_2}{\partial x} - \frac{ie^*}{\hbar c}A_x\psi_2 = -\frac{\psi_1}{\lambda}; \tag{15.2b}$$

the parameter λ is the same in both equations since it is a property of the boundary and not the superconductor designations, 1 and 2. We insert (15.2a) into Eq. (9.17b) yielding

$$j_x = -\frac{ie^*\hbar}{2m^*}\left[\psi_1^*\frac{\partial\psi_1}{\partial x} - \psi_1\frac{\partial\psi_1^*}{\partial x}\right] - \frac{e^{*2}}{m^*c}|\psi_1|^2A_x$$

1. This form is identical to Eq. (9.44) with $M_{11} = M_{22} = 0$, $-M_{12} = M_{21}^{-1} = \lambda$.

$$= -\frac{ie^*\hbar}{2m^*}\left[\psi_1^*\left(\frac{\psi_2}{\lambda}+\frac{ie^*}{\hbar c}A_x\psi_1\right)-\psi_1\left(\frac{\psi_2^*}{\lambda^*}-\frac{ie^*}{\hbar c}A_x\psi_1^*\right)\right]-\frac{e^{*2}}{m^*c}|\psi_1|^2A_x. \tag{15.3}$$

In the absence of magnetic atoms, the superconducting properties are invariant under time reversal, which results in $\psi \to \psi^*$, $\mathbf{j} \to -\mathbf{j}$, and $\mathbf{A} \to -\mathbf{A}$. Under these operations both sides of Eqs. (15.2) turn into their complex conjugates and hence λ must be real. We then obtain from (15.3)

$$j_x = \frac{-ie^*\hbar}{2m^*\lambda}(\psi_1^*\psi_2 - \psi_1\psi_2^*). \tag{15.4}$$

Writing $\psi_i = |\psi_i|e^{i\Phi_i}$, and assuming both sides are prepared from the same kind of superconducting material, $|\psi_1| = |\psi_2|$, we have

$$j = j_m \sin\Phi_{21} \tag{15.5}$$

where

$$j_m = \frac{e^*\hbar}{m^*\lambda}|\psi|^2, \tag{15.6a}$$

and

$$\Phi_{21} = \Phi_2 - \Phi_1. \tag{15.6b}$$

Note j_m is the maximum current density that may be carried by the junction.

In deriving Eq. (15.5) we have assumed that no electric field and magnetic flux density are present in the junction. When no electric field is present in the junction, the phase is time-independent. We generalize to the case in which a field is present by using a gauge invariance argument. Under a gauge transformation (recall $\mathbf{H} = \nabla \times \mathbf{A}$ and $\mathbf{E} = -\nabla V - (1/c)(\partial\mathbf{A}/\partial t)$)

$$\mathbf{A} \to \mathbf{A} + \nabla\chi \tag{15.7}$$

and

$$V \to V - \frac{1}{c}\frac{\partial\chi}{\partial t}, \tag{15.8}$$

where $\chi(\mathbf{r}, t)$ is an arbitrary single-valued function and V is the potential. Since the G–L equation contains the form $\nabla - (ie^*/\hbar c)\mathbf{A}$, we must change the phase of the G–L wavefunction by

$$\Phi \to \Phi + \frac{e^*}{\hbar c}\chi(t). \tag{15.9}$$

Comparing Eqs. (15.8) and (15.9) we see that the form

$$\frac{\partial\Phi}{\partial t} - \frac{e^*}{\hbar}V = 0 \tag{15.10}$$

is *gauge-invariant*. If V is (initially) assumed to be independent of time and denoted as V_{21}, then integration of (15.10) yields

$$\Phi_{21} = \Phi_{21}^{(0)} - \frac{e^*}{\hbar}V_{21}t \tag{15.11}$$

or

$$j = j_{\rm m}\sin\left[\Phi_{21}^{(0)} - \frac{e^*}{\hbar}V_{21}t\right]. \tag{15.12}$$

Introducing the frequency $\omega_{\rm J} = \partial\Phi_{21}/\partial t$ we see that (15.11) leads to

$$\begin{aligned}\omega_{\rm J} &= \frac{|e^*|}{\hbar}V_{21} \\ &= \frac{2|e|}{\hbar}V_{21}.\end{aligned} \tag{15.13}$$

The oscillating current $j(t)$ given by (15.12) will be associated with an oscillating voltage which will be superimposed on the static voltage, V_{21}.

We next examine the Josephson effects in the presence of a magnetic field. We will restrict ourselves to the case of a relatively weak field where a quasiclassical description is adequate; i.e., the dominant effect of a field, which is described by a vector potential **A**, is to make the phase position-dependent. From the discussion surrounding Eqs. (9.19) we know that for a 'pure' gauge field (one not involving a field **H**(**r**)) the *only* effect of the vector potential is to produce a position-dependent phase; this suggests that in the presence of a vector potential associated with a weak field, **H**(**r**), the effect may be approximately incorporated in the phase of the wavefunction. Comparing Eqs. (15.7) and (15.9) we have the gauge-invariant form analogous to (15.10) as

$$\nabla\Phi - \frac{e^*}{\hbar c}\mathbf{A} = 0. \tag{15.14}$$

Therefore, the gauge-invariant phase difference is given by

$$\Phi_{21} = \Phi_{21}^{(0)} + \frac{2\pi}{\phi_0}\int_1^2 \mathbf{A}\cdot d\ell. \tag{15.15}$$

where $\phi_0 \equiv hc/2e$ is the flux quantum.

The fundamental equations governing the behavior of Josephson junctions are the current–phase relation (15.5), the voltage–phase relation (15.11), and the gauge-invariant phase relation (15.15). They are believed to be exact. In the subsequent sections these equations are applied to some simple junction structures and circuits. For example, electromagnetic radiation is emitted from a Josephson junction in the presence of a potential (see Sec. 17). Eq. (15.13) now forms the basis for defining the standard volt in terms of a measured frequency and the fundamental constants, e and h (Taylor, Parker, and Langenberg 1969).

15.2 Magnetic field effects: the two-junction SQUID

Consider the superconducting circuit shown in Fig. 15.1 involving two Josephson junctions connected by superconducting leads. The loop formed by this circuit is assumed to contain a

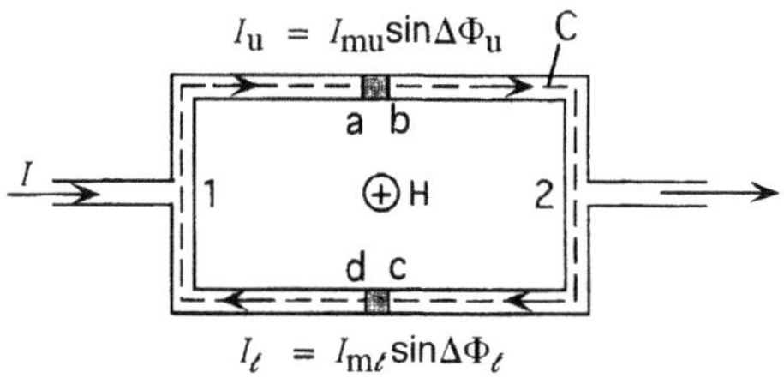

Figure 15.1 Schematic of a dc SQUID consisting of two Josephson junctions connected in parallel by superconducting links. The path of integration C is shown by the dashed line.

magnetic flux, Φ, arising from a position-dependent field, $\mathbf{H}(\mathbf{r})$, with some associated vector potential $\mathbf{A}(\mathbf{r})$.

On entering the loop on the left in Fig. 15.1 (point 1), the current I splits into two components I_u and I_ℓ, where the subscripts refer to the upper and lower paths, respectively. From Eq. (15.5) the total current arriving at point 2 is

$$\begin{aligned} I &= I_u + I_\ell \\ &= I_{mu} \sin \Delta\Phi_u + I_{m\ell} \sin \Delta\Phi_\ell, \end{aligned} \tag{15.16}$$

where I_{mu} and $I_{m\ell}$ correspond to the maximum currents associated with the upper and lower junctions, and $\Delta\Phi_u$ and $\Delta\Phi_\ell$ are the corresponding gauge-invariant phase shifts. Assuming a matched pair of junctions for simplicity ($I_{mu} = I_{m\ell} = I_m$), we may rewrite (15.16) as

$$I = 2I_m \sin\left(\frac{\Delta\Phi_u + \Delta\Phi_\ell}{2}\right) \cos\left(\frac{\Delta\Phi_u - \Delta\Phi_\ell}{2}\right). \tag{15.17}$$

The gauge invariant phase shifts may be obtained by integrating $\nabla\Phi$ around the closed path C shown in Fig. 15.1. Noting that Φ is a multivalued function that can change by $2\pi n$ upon completing the path we have

$$\oint_C \nabla\Phi \cdot d\ell = (\Phi_b - \Phi_a) + (\Phi_c - \Phi_b) + (\Phi_d - \Phi_c) + (\Phi_a - \Phi_d) = 2\pi n, \tag{15.18}$$

where n is an integer. The phase differences across the upper and lower Josephson junctions are given by Eq. (15.15) as

$$\Phi_b - \Phi_a = \Delta\Phi_u + \frac{2\pi}{\phi_0} \int_a^b \mathbf{A} \cdot d\ell$$

and

$$\Phi_d - \Phi_c = -\Delta\Phi_\ell + \frac{2\pi}{\phi_0} \int_c^d \mathbf{A} \cdot d\ell.$$

The second and fourth terms in Eq. (15.18) are phase differences in the superconducting leads themselves and are found by using the supercurrent equation (15.1) and the expression for the London penetration depth Eq. (2.6):

$$\Phi_c - \Phi_b = \int_b^c \nabla\Phi \cdot d\ell = \frac{2\pi}{\phi_0} \int_b^c \left(\mathbf{A} + \frac{4\pi\lambda_L^2}{c} \mathbf{j}\right) \cdot d\ell$$

and

$$\Phi_a - \Phi_d = \int_d^a \nabla\Phi \cdot d\ell = \frac{2\pi}{\phi_0}\int_d^a \left(\mathbf{A} + \frac{4\pi\lambda_L^2}{c}\mathbf{j}\right)\cdot d\ell.$$

Substituting the above four equations in Eq. (15.18) gives

$$\Delta\Phi_u - \Delta\Phi_\ell = 2\pi n + \frac{2\pi}{\phi_0}\oint_C \mathbf{A}\cdot d\ell + \frac{2\pi}{\phi_0}\frac{4\pi\lambda_L^2}{c}\int_{C'} \mathbf{j}\cdot d\ell. \tag{15.19a}$$

The integration of **A** is around a complete closed path C and is equal to the total flux ϕ inside the area enclosed by the contour. The integration of **j** follows a path C′ which excludes the integration over the insulators. If the superconducting leads are thicker than the London penetration depth, the integration path can be taken deep inside the superconductors where the integral involving the supercurrent density is negligible. The phase difference is then simply related to the total flux by

$$\Delta\Phi_u - \Delta\Phi_\ell = 2\pi n + \frac{2\pi\phi}{\phi_0}. \tag{15.19b}$$

Using this equation to eliminate $\Delta\Phi_u$ from Eq. (15.17), the total current is

$$I = 2I_m \sin\left(\Delta\Phi_\ell + \frac{\pi\phi}{\phi_0}\right)\cos\left(\frac{\pi\phi}{\phi_0}\right). \tag{15.20}$$

When the inductance L of the loop is taken into account, the total flux in Eq. (15.19b) consists of the externally applied flux ϕ_{ext} and the flux generated by the screening circulating current I_{cir}; i.e.,

$$\phi = \phi_{ext} + LI_{cir}. \tag{15.21}$$

For the identical junction case,

$$2I_{cir} = I_m(\sin\Delta\Phi_u - \sin\Delta\Phi_\ell). \tag{15.22}$$

In general, Eqs. (15.20), (15.21), and (15.22) must be solved self-consistently to describe the behavior of the two-Josephson-junction loop. For simplicity, we assume the loop inductance is negligible and consider only the effect of the externally applied flux on the characteristics of the loop. The maximum supercurrent density which can be carried by the loop is found by maximizing Eq. (15.20) with respect to $\Delta\Phi_\ell$; i.e.,

$$\Delta\Phi_\ell + \frac{\pi\phi_{ext}}{\phi_0} = (n + 1/2)\pi.$$

Hence the maximum supercurrent density, I_{max}, is given by

$$I_{max} = 2I_m\left|\cos\left(\frac{\pi\phi_{ext}}{\phi_0}\right)\right|, \tag{15.23}$$

which is periodic in the external flux. Since $\phi_0 \sim 2.07 \times 10^{-7}\,\mathrm{G\,cm^2}$, it is clear that the device pictured in Fig. 15.1 can be used to measure very small changes in magnetic field. It is sometimes

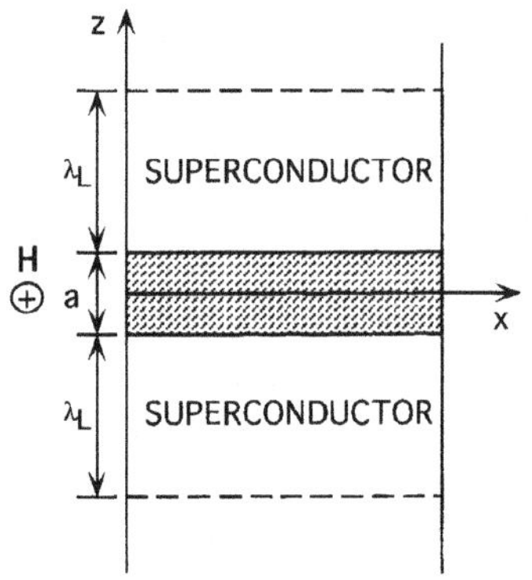

Figure 15.2 Schematic diagram of an SIS Josephson tunnel junction.

referred to as a two-junction SQUID where the latter is an acronym constructed from the words superconducting quantum interference device.

15.3 The extended Josephson junction

We next discuss the behavior of a single planar Josephson junction in a magnetic field. We refer to the junction cross section depicted in Fig. 15.2. The magnetic field is directed into the page along $\hat{\mathbf{y}}$. The middle of the junction is taken as the origin and the vector potential in the three regions is

$$A_z = \begin{cases} -Hx\mathrm{e}^{-(z-a/2)/\lambda_\mathrm{L}} & (z > a/2) \quad (15.24a) \\ -Hx & (a/2 > z > -a/2) \quad (15.24b) \\ -Hx\mathrm{e}^{(z+a/2)/\lambda_\mathrm{L}} & (z < -a/2) \quad (15.24c) \end{cases}$$

where the exponential dependencies arise from the Meissner effect in the two superconductors. The phase difference encountered in crossing the junction from $z = -\infty$ to $z = +\infty$ at a given horizontal coordinate x is

$$\Delta\Phi(x) = \Phi_{21}^{(0)} - \frac{e^*}{\hbar c}\int_{-\infty}^{\infty} A_z(x, z)\mathrm{d}z$$
$$= \Phi_{21}^{(0)} + \frac{e^*}{\hbar c} H(a + 2\lambda_\mathrm{L})x. \quad (15.25)$$

Assuming a rectangular junction of dimensions L_x and L_y, the total current, I, through the junction is

$$I = j_\mathrm{m} L_y \int_{-L_x/2}^{L_x/2} \sin \Delta\Phi(x)\mathrm{d}x$$
$$= j_\mathrm{m} L_y \int_{-L_x/2}^{L_x/2} \sin[2\pi H(a + 2\lambda_\mathrm{L})x/\phi_0 + \Phi_{21}^{(0)}]\mathrm{d}x$$

or

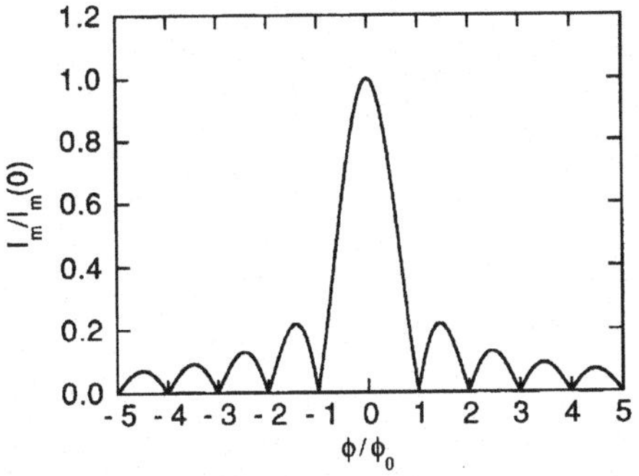

Figure 15.3 I_m vs ϕ characteristics for a short Josephson junction when the self-field effect is negligible.

$$I = I_m \frac{\sin(\pi\phi/\phi_0)}{\pi\phi/\phi_0} \sin\Phi_{21}^{(0)},$$

where $\phi \equiv H(a + 2\lambda_L)L_x$, and $I_m = j_m L_x L_y$. The maximum value of the junction current, I^{max}, occurs for $\sin\Phi_{21}^{(0)} = \pm 1$ (depending on the sign of $\sin(\pi\phi/\phi_0)$); thus

$$I^{max} = I_m \left|\frac{\sin(\pi\phi/\phi_0)}{\pi\phi/\phi_0}\right|, \tag{15.26}$$

We note we obtain a 'diffraction-like' ($\sin x/x$) pattern (see Fig. 15.3) involving the variable ϕ/ϕ_0 where ϕ is the flux contained within the junction.

The effective junction area entering the definition of ϕ involves an effective thickness (made up of the insulator thickness plus a contribution of one London depth in each superconductor) times the junction width, L_x.

15.4 Effect of an applied rf field

An interesting behavior results if an external rf voltage is applied to a dc voltage-biased Josephson junction, as first observed by Shapiro (1963). We take the applied voltage to have the form

$$V = V_0 + V_1 \cos\omega t. \tag{15.27}$$

From Eq. (15.12) the resulting current in the junction will be

$$I = I_m \sin\left[\frac{e^*}{\hbar}\int (V_0 + V_1 \cos\omega t)\mathrm{d}t\right]$$
$$= I_m \sin[\Phi(0) + \omega_J t + \delta \sin\omega t], \tag{15.28}$$

where $\Phi(0)$ is an arbitrary phase, ω_J is given by Eq. (15.13) and $\delta \equiv e^* V_1/\hbar\omega$ is called the modulation (or deviation) index. Using expressions which are well known in the theory of frequency modulation[2] we may rewrite (15.28) as

2. See Abramowitz and Stegun (1970), Eqs. (9.1.42) and (9.1.43), p. 361.

$$I(t) = I_\mathrm{m}\left\{\sin[\Phi(0) + \omega_\mathrm{J} t]\left[J_0(\delta) + 2\sum_{n=1}^{\infty} J_{2n}(\delta)\cos(2n\omega t)\right]\right.$$
$$\left. + \cos[\Phi(0) + \omega_\mathrm{J} t]2\sum_{n=0}^{\infty} J_{2n+1}(\delta)\sin[(2n+1)\omega t]\right\}, \qquad (15.29)$$

where the J_n are Bessel functions of first kind of integer order. Using $\sin a \cos b = \frac{1}{2}[\sin(a+b) + \sin(a-b)]$ we may rewrite (15.29) as

$$I(t) = I_m\left(J_0(\delta)\sin[\omega_\mathrm{J} t + \Phi(0)]\right.$$
$$\left. + \sum_{n=1}^{\infty} J_n(\delta)\{\sin[(\omega_\mathrm{J} + n\omega)t + \Phi(0)] + (-1)^n \sin[(\omega_\mathrm{J} - n\omega)t + \Phi(0)]\}\right). \qquad (15.30)$$

If we vary ω_J (by sweeping V_0) such that the condition $\omega_\mathrm{J} = \pm n\omega$ is met, then the time dependence associated with this term in (15.30), having the amplitude $J_n(\delta)$ and phase $\Phi(0)$, would be 'transformed to zero frequency' and would appear as a 'spike' in the current–voltage characteristics.

The impedance of a Josephson junction is usually much smaller than the resistance of the leads extending into the cryostat and hence a constant current source is a more accurate representation than the constant voltage source assumed above. Further analysis shows (and experiment confirms) that this results in a step-like (rather than a spike-like) current–voltage behavior.

15.5 The resistively shunted junction (RSJ) model

In analyzing the behavior of circuits involving small Josephson junctions, one may model the effects of various dissipative processes and the distributed junction capacity with so-called 'lumped' circuit parameters. Fig. 15.4 shows such a model; here C and R represent the capacity and effective resistance of the junction,[3] where the latter is represented by a cross. V_J is the voltage across the junction while I is the total current flowing through all three circuit elements which is given by[4]

$$I = I_\mathrm{J} + I_C + I_R \qquad (15.31a)$$

$$= I_\mathrm{m}\sin\Phi + \frac{\hbar C}{2e}\ddot{\Phi} + \frac{\hbar}{2eR}\dot{\Phi}, \qquad (15.31b)$$

where we have used the Josephson equations

3. The capacitance is largely determined by the geometry of the junction and may be regarded as constant. The resistance on the other hand may depend strongly on the junction voltage; for this reason an additional shunt resistance, which is much smaller than the junction resistance, is sometimes incorporated to bypass this effect.
4. We use MKS units in this subsection.

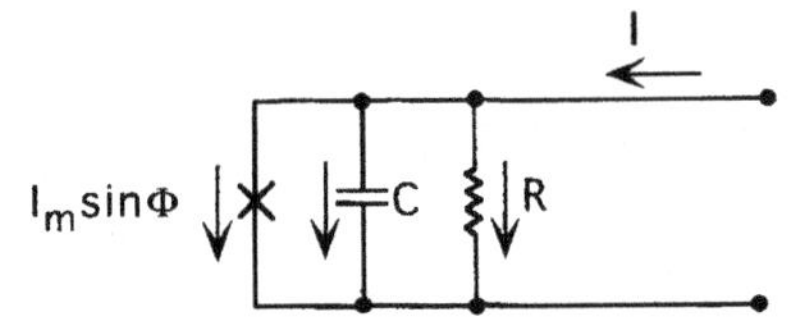

Figure 15.4 Equivalent circuit of a resistively shunted Josephson junction.

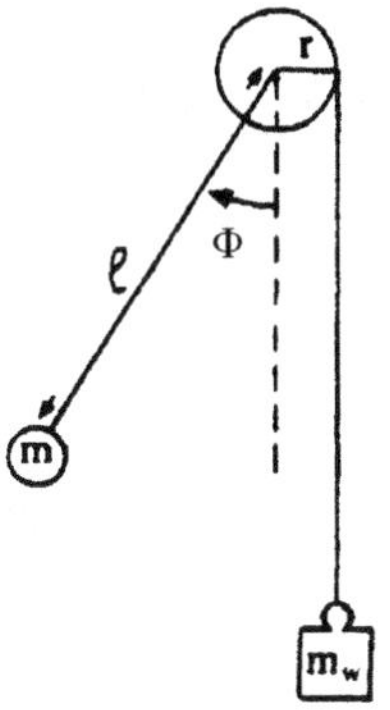

Figure 15.5 The simple pendulum with an applied torque.

$$2eV_J = \hbar\dot{\Phi} \tag{15.32}$$

and

$$I_J = I_m \sin \Phi \tag{15.33}$$

in obtaining (15.31b).

Let us examine the behavior of an RSJ to which a constant current is applied. If I is slowly increased the voltage across the junction will remain zero ($V_J = 0$), implying $\Phi = \sin^{-1}(I/I_m)$, until $I = I_m$. Above this point a voltage develops and the junction becomes resistive with a time-dependent phase. To examine the behavior in this regime it is useful to note that the phase, $\Phi(t)$, of a junction described by Eq. (15.31b) is in one-to-one correspondence with the angle of rotation (which we will also call Φ) of a *damped pendulum*, driven by a constant torque, in a constant gravitational field (see Fig. 15.5): the applied torque, $\mathcal{T} = m_w gr$, gravitationally induced torque, $mg\ell \sin \Phi$, moment of inertia, $\mathcal{I} = m\ell^2$, and damping coefficient, k, are identified with I, $I_m \sin \Phi$, $\hbar C/2e$, and $\hbar/2eR$, respectively. The regime $I < I_m$ corresponds to a situation in which the applied torque $\mathcal{T}$ is less than a critical torque $\mathcal{T}_m$ necessary to raise the pendulum to an angle $\Phi = \pi/2$ (where the opposing gravitational torque is maximal). For $\mathcal{T} > \mathcal{T}_m$ the pendulum rotates in a manner such that the average energy dissipated per rotation is equal to the average work per rotation (were there no damping the average angular velocity $\dot{\Phi}$ would increase without limit in this regime). We now go to a rotating reference frame by writing

$$\Phi(t) = \omega t + \Phi'(t), \tag{15.34}$$

where

$$\omega \equiv \bar{\dot{\Phi}} \tag{15.35}$$

is the average angular velocity and Φ' is periodic (with an average value of zero); Eq. (15.31b) then becomes

$$\frac{\hbar C}{2e}\ddot{\Phi}' + \frac{\hbar}{2eR}\dot{\Phi}' + I_m \sin[\omega t + \Phi'(t)] = 0. \tag{15.36}$$

In the limit where the departure of Φ from a steady state rotation, ωt, is small, i.e. $\Phi'(t) << 1$, we may neglect the Φ' contribution to the last term in (15.36). We then have the equation for a driven harmonic oscillator.

Writing $\Phi'(t) = \Phi'_0 \sin(\omega t + \theta)$ leads to the pair of equations

$$I_m - \frac{\hbar C}{2e}\omega^2 \Phi'_0 \cos\theta - \frac{\hbar}{2eR}\omega\Phi'_0 \sin\theta = 0 \tag{15.37}$$

and

$$-\frac{\hbar C}{2e}\omega^2 \Phi'_0 \sin\theta + \frac{\hbar}{2eR}\omega\Phi'_0 \cos\theta = 0. \tag{15.38}$$

From the second condition we obtain

$$\tan\theta = \frac{1}{\omega RC} \tag{15.39a}$$

or

$$\sin\theta = \frac{1}{(1+\omega^2\tau^2)^{1/2}}, \tag{15.39b}$$

where $\tau \equiv RC$. Substituting (15.39) into (15.37) yields

$$\Phi'_0 = \frac{2e}{\hbar}\frac{I_m R}{\omega}\frac{1}{(1+\omega^2\tau^2)^{1/2}}. \tag{15.40}$$

Now

$$\dot{\Phi} = \frac{2e}{\hbar}\bar{V} = \frac{2e}{\hbar}IR. \tag{15.41}$$

In the limit, $\omega\tau << 1$, $\Phi'_0 = (I_m/I)$. In the opposite limit, $\omega\tau >> 1$, (15.40) becomes $\Phi'_0 = (1/\omega\tau)I_m/I$. The requirement $\Phi'_0 << 1$ is therefore best satisfied in the limit of large I and large $\omega\tau$.

The limit $\omega\tau \to 0$ may be obtained by eliminating the shunting capacitor in Fig. 15.4; the differential equation (15.31b) then reduces to

$$I = I_m \sin\Phi + \frac{\hbar}{2eR}\dot{\Phi} \tag{15.42}$$

or

$$t = \frac{\hbar}{2eRI_m}\int_{\Phi(0)}^{\Phi(t)} \frac{d\Phi}{I/I_m - \sin\Phi},$$

which upon integration yields

$$\Phi(t) = 2\tan^{-1}\left[\left(\frac{\alpha^2-1}{\alpha^2}\right)^{\frac{1}{2}} \tan\left(\frac{\pi t}{T}\right) - \alpha\right], \tag{15.43}$$

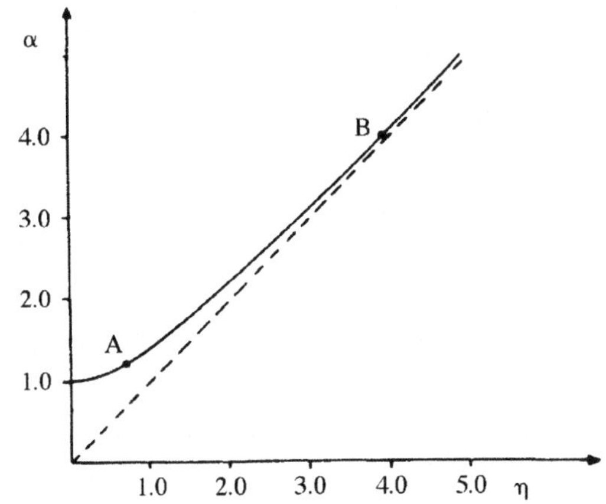

Figure 15.6 Time-averaged nonhysteretic I–V characteristic computed in the RSJ model in the zero capacitance limit; $\alpha = I/I_m$, $\eta = \bar{V}/RI_m$. Points A and B give the time-averaged voltages corresponding to $\alpha = 1.2$ and $\alpha = 4$, respectively.

where we have defined

$$\alpha \equiv \frac{I}{I_m}, \quad T = \frac{\hbar}{2eRI_m}\frac{2\pi}{(\alpha^2 - 1)^{1/2}}, \tag{15.44}$$

where α is a parameter and T is the period (which is the minimum time required for $\Phi(t)$ to return to its initial value $\Phi(0)$). The voltage across the junction from Eq. (15.5) is $(\hbar\dot{\Phi}/2e)$. If we assume the voltage is sensed with a device which records the time average of $V(t)$ (over many periods T) then

$$\bar{V} = \frac{\hbar}{2e}\bar{\dot{\Phi}} = \frac{\hbar}{2e}\frac{1}{T}\int_0^T \frac{\mathrm{d}\Phi}{\mathrm{d}t}\mathrm{d}t = \frac{\hbar}{2e}\frac{2\pi}{T} = RI_m(\alpha^2 - 1)^{1/2}. \tag{15.45}$$

Note that for $I >> I_m$ ($\alpha >> 1$) (15.45) approaches (15.41). Fig. 15.6 shows the I–V characteristics from Eq. (15.45); note that the I–V characteristic is nonhysteretic.

We now give a qualitative discussion of the general case when $C \neq 0$. We begin by rewriting Eq. (15.31b) in dimensionless form. We measure time in units of ω_0 and introduce a parameter β where

$$\omega_0 \equiv \left(\frac{2e}{\hbar}\frac{I_m}{C}\right)^{1/2}, \qquad \beta = \frac{1}{\omega_0 RC}; \tag{15.46}$$

ω_0 is the natural frequency for small oscillations (referred to as the Josephson plasma frequency). Then Eq. (15.31b) may be rewritten as

$$\ddot{\Phi} + \beta\dot{\Phi} + \sin\Phi = \alpha, \tag{15.47}$$

where the derivatives are now with respect to the dimensionless time variable. We define an angular velocity $\Omega(t) \equiv \dot{\Phi}$; the angular acceleration is then $\ddot{\Phi} = \dot{\Omega} = (\mathrm{d}\Omega/\mathrm{d}\Phi)\dot{\Phi} = (\mathrm{d}\Omega/\mathrm{d}\Phi)\Omega = (\mathrm{d}/\mathrm{d}\Phi)(\Omega^2/2)$ and Eq. (15.47) becomes

$$\frac{\mathrm{d}}{\mathrm{d}\Phi}\left(\frac{\Omega^2}{2}\right) + \beta\Omega + \sin\Phi = \alpha. \tag{15.48}$$

The 'orbits' associated with the solutions of (15.48) for various initial conditions and parameter

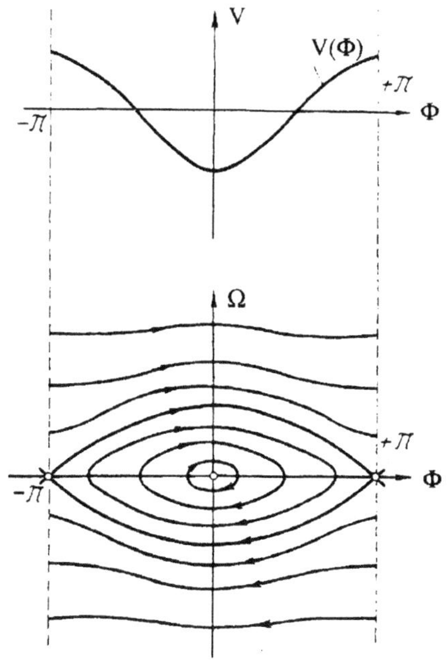

Figure 15.7 The Ω–Φ plane trajectories (lower panel) associated with the solution of Eq. (15.49). There are two kinds of orbits: an open orbit for $(1/2)\Omega^2(\Phi = 0) > 2$ and a closed orbit for $(1/2)\Omega^2(\Phi = 0) < 2$. The upper panel shows the potential as a function of Φ.

values (α and β) are trajectories in the Ω–Φ plane; we may think of this plane as the two-dimensional 'phase space' associated with the motion. We now discuss the nature of these orbits.

We begin by discussing the orbits when $\alpha = \beta = 0$. Our equation is then

$$\frac{d}{d\Phi}\left(\frac{\Omega^2}{2}\right) + \sin\Phi = 0. \tag{15.49}$$

This equation may be integrated analytically in terms of Euler elliptic functions, as discussed in mechanics textbooks. We will restrict ourselves to some qualitative statements. Since (15.49) is invariant under the transformation $\Phi \to -\Phi$, we expect extrema at $\Phi = 0$ and (when they exist) $\pm\pi$. For the rotating case, Φ advances continuously in time with Ω having a maximum at $\Phi = 0$ (where the kinetic energy of the pendulum is largest) and a minimum at $\Phi = \pm\pi$ (where the potential energy is maximal); the phase space trajectory is shown in Fig. 15.7. Such 'open' orbits require that the kinetic energy always be larger than the potential energy; i.e.,

$$\frac{1}{2}\Omega^2(\Phi = 0) > 2. \tag{15.50}$$

When this condition is not satisfied the motion is oscillatory and the orbit in phase space is closed as shown in Fig. 15.7.

We now turn on a small damping ($\beta > 0$), but still keeping the drive current (torque) zero ($\alpha = 0$); we will then have $\Omega(\Phi) > \Omega(\Phi + 2\pi) > \Omega(\Phi + 4\pi)$, etc., i.e., the rate of rotation slows down. Eventually a point is reached where Ω passes through zero before the system (pendulum) reaches the 'top' after which it reverses its motion. This point marks the transition between rotary

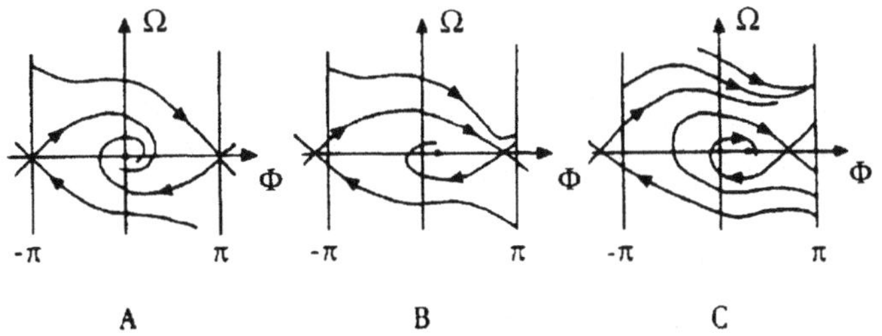

Figure 15.8 The qualitative phase-plane trajectories, corresponding to the three different situations (A, B, C) in the β–α plane in Fig. 15.10. (After Belykh, Pedersen, and Soerensen (1977).)

and oscillatory motion. After the transition the phase point proceeds along a spiral, asymptotically approaching the origin.

The above described sequence of events is shown as a phase space plot in Fig. 15.8. (Note that since dissipation is present the arrow of time is relevant.)

We now consider the general case: $\alpha > 0, \beta > 0$. When $\alpha >> 1$ ($I >> I_m$), Ω is practically constant and given by $\Omega = \alpha/\beta$. As α is lowered through unity from above the behavior of the system depends on the magnitude of β. Since α is starting at a value greater than 1, the pendulum motion must evolve from a rotary state. If we now decrease α to a value less than 1 ($I < I_m$) the pendulum can continue to rotate (i.e., we can maintain a steady state motion) provided the work performed per cycle by the external torque, $2\pi\alpha$, can compensate for the energy dissipated per cycle; the latter is given by

$$\beta \int_0^T \Omega^2(t)\mathrm{d}t = \beta \int_0^{2\pi} \Omega^2(\Phi)\frac{\mathrm{d}t}{\mathrm{d}\Phi}\mathrm{d}\Phi$$

$$= \beta \int_0^{2\pi} \Omega(\Phi)\mathrm{d}\Phi. \tag{15.51}$$

The behavior of $\Omega(\Phi)$ is not known without actually integrating the equation of the orbit (15.48). However, we can estimate the critical torque, α_c, by assuming $\Omega(\pi) = 0$ and $\Omega(0)$ is the angular velocity resulting from 'free-fall' rotation under the influence of gravity, which from Eq. (15.50) is $\Omega(\Phi = 0) = 2$; we linearly interpolate between 0 and π by writing $\Omega(\Phi) = \Omega(0)\Phi/\pi$. We then have

$$2\pi\alpha_c \cong 2\beta\Omega(0) \int_0^{\pi} \frac{\Phi}{\pi}\mathrm{d}\Phi$$

or

$$\alpha_c \cong \beta.$$

A more accurate expression can be obtained from the exact treatment of the free pendulum which yields

$$\alpha_c = \frac{4}{\pi}\beta, \tag{15.52}$$

which is valid for $\beta \lesssim 0.2$ (Stewart 1968, 1974). We see that as $\beta \to 0, \alpha_c \to 0$; i.e., the junction is hysteretic, switching from a constant-phase, zero-voltage state to a phase-precessing, finite-

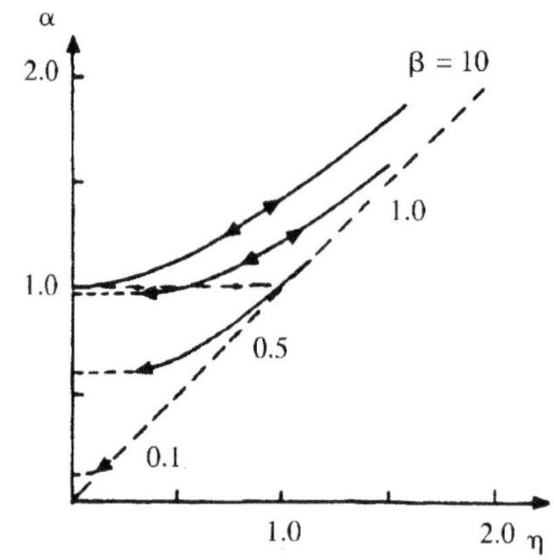

Figure 15.9 Analog computer results for the time average of $\alpha(= I/I_m)$ vs $\eta(= \bar{V}/RI_m)$ characteristics for different values of the parameter $\beta = 1/\omega_0 RC$. (After Johnson (1968).)

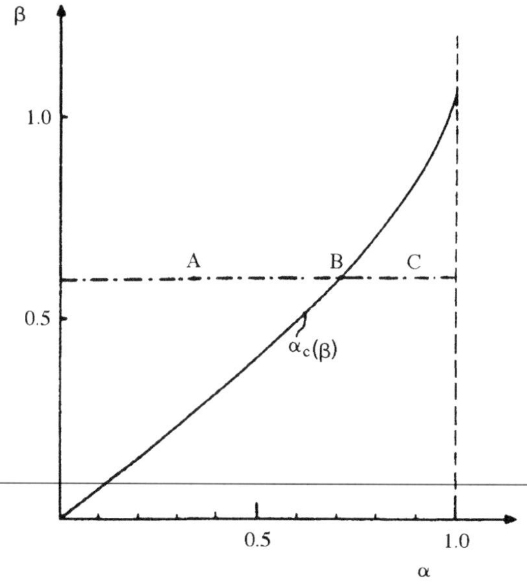

Figure 15.10 Analog computer results for the hysteresis parameter α_c as a function of β. In the region above the $\alpha_c(\beta)$ curve only one stable solution exists. In the region below the curve two stable solutions exist. (After Johnson (1968).)

voltage state as α is increased above 1, but returning to the original (constant-phase) state at a value $\alpha < 1$. The I–V behavior of a hysteretic junction is shown in Fig. 15.9. The junction instability line in the α–β plane obtained by computer is shown in Fig. 15.10 (Johnson 1968).

For a discussion of the case in which the junction resistance is voltage-dependent see Barone and Paterno (1982).

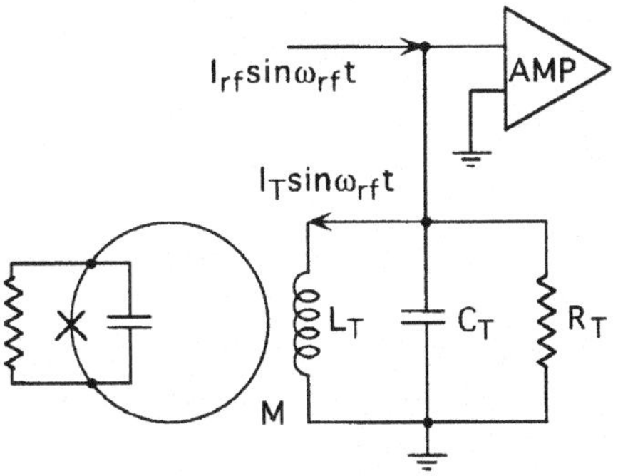

Figure 15.11 A single junction SQUID coupled to a tank circuit via a mutual inductance M.

15.6 The rf biased SQUID

Rather than measuring a magnetic flux with a two-junction interferometer (a dc SQUID), as discussed in Sec. 15.2, an alternative method involves the use of only a single junction (Silver and Zimmerman 1967).

The device involves three circuits (see Fig. 15.11): (i) the primary SQUID, which is a superconducting loop connected to a resistively shunted Josephson junction; (ii) a second loop (not shown in the figure), which is inductively coupled to the first loop and introduces an external flux $\phi_{ext} = M_{ext} I_{ext}$ as a result of a mutual inductance, M_{ext}, coupling the two loops and a current, I_{ext}, applied to the second loop; and (iii) a third loop which is driven by an external oscillator (typically operating at a frequency of order 10^7 Hz) which applies a periodically changing flux, ϕ_{rf}, through a mutual inductance, M.

To explain the operation of the device, which is somewhat subtle, we must examine the action of ϕ_{ext} and ϕ_{rf} on the response of the primary SQUID loop. We begin by discussing the equation of motion when $\phi_{ext} = \phi_{rf} = 0$. The equivalent circuit consists of an RSJ connected to an inductor, L (represented by the loop in Fig. 15.11). The equations of motion are[5]

$$I = I_L + I_J + I_C + I_R = 0 \tag{15.53a}$$

where

$$\dot{I}_L L = V_J = \frac{\hbar\dot{\Phi}}{2e}. \tag{15.53b}$$

Combining these equations with the forms in Eq. (15.31b) we have (where we multiply through by L)

$$\frac{\hbar\Phi}{2e} + LI_m \sin\Phi + \frac{\hbar LC}{2e}\ddot{\Phi} + \frac{\hbar L}{2eR}\dot{\Phi} = 0. \tag{15.54}$$

The phase variable, Φ, in Eq. (15.54) may be replaced by the internally generated flux, ϕ, according to the relation $\phi = (\Phi/2\pi)\phi_0$. Incorporating the contribution of the external flux, ϕ_{ext}, Eq. (15.54) then becomes

5. We use MKS units in this subsection.

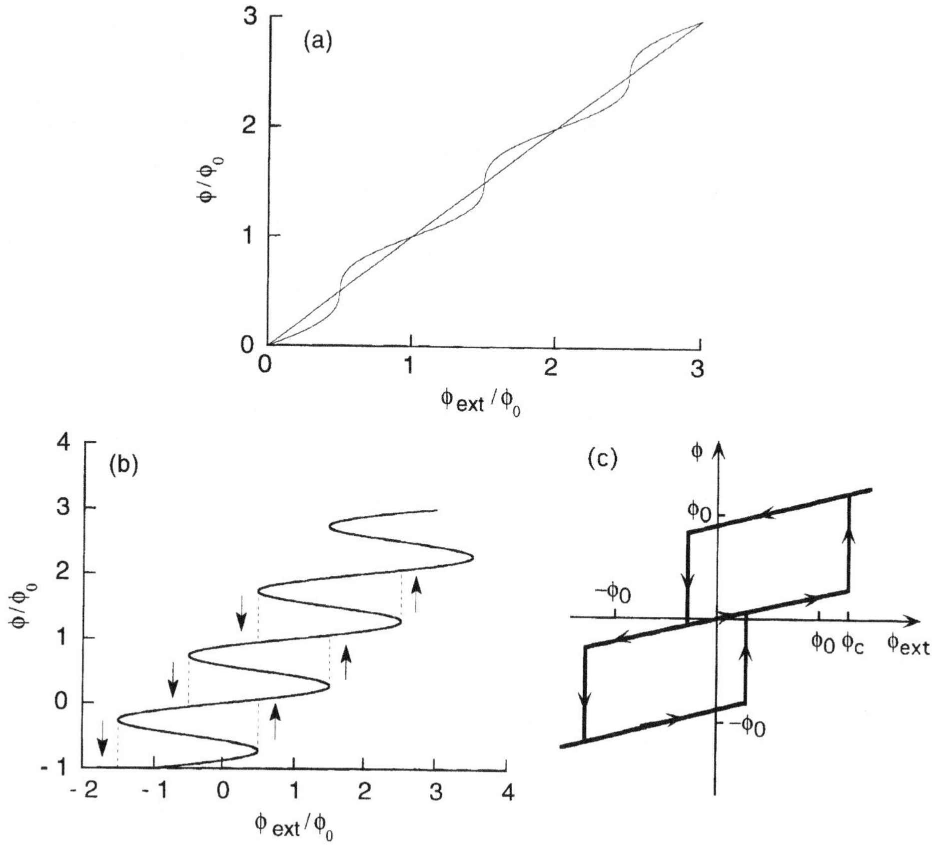

Figure 15.12 The static behavior of a single-junction superconducting loop in an applied external flux ϕ_{ext}: (a) $\phi_m/\phi_0 = 0.25$; (b) $\phi_m/\phi_0 = 1.25$; (c) a 'linearized' version of Fig. 15.12(b) showing the path traced out by ϕ and ϕ_{ext} for the case $\phi = n\phi_0$ involving three branches.

$$\phi_{ext} = \phi + \phi_m \sin\frac{2\pi\phi}{\phi_0} + \frac{\ddot{\phi}}{\omega_0^2} + \tau\dot{\phi}, \tag{15.55}$$

where $\phi_m \equiv LI_m$, $\omega_0^2 = 1/LC$, and $\tau = L/R$. We examine the static solutions of this equation for the responses $\phi = \phi(\phi_{ext})$, which are shown in Fig. 15.12(a) and (b). Fig. 15.12(a) shows the behavior when $\phi_m/\phi_0 = 0.25$, where a continuous, single-valued evolution of ϕ with ϕ_{ext} is obtained; a line with a slope of 1 is shown for reference. Fig. 15.12(b) on the other hand shows the behavior when $\phi_m/\phi_0 = 1.25$. We now have a multi-valued behavior. Between successive points ϕ_c at which $d\phi/d\phi_{ext}$ diverges (indicated by the dashed vertical lines in Fig. 15.12(b)) there are alternating regions with positive and negative slopes; the former are stable while the latter are unstable and the system spontaneously oscillates. However, the frequency involved is typically in the microwave regime and if τ is sufficiently short the oscillations damp quickly (relative to the period associated with ϕ_{rf}). The system then, effectively, switches from one stable (positive slope) region to an adjacent one. (By Taylor expanding in the region $\phi/\phi_0 \cong n$ (an integer) one sees that the slope of the plateau at this point is $(2\pi\phi_m/\phi_0 + 1)^{-1}$.) The upward and downward directed arrows in Fig. 15.12(c) indicate how the system switches for increasing and decreasing ϕ_{ext}, respectively.

To use these characteristics to make measurements we need a means of locating a stable branch and the position along it. To do this we introduce the third loop, which is driven sinusoidally by an rf generator, and induces an additional flux $\phi_{\text{rf}}(t) = \phi_{\text{rf}} \sin \omega_{\text{rf}} t$ into the primary loop. We will assume that a capacitor is associated with this additional loop and that the pair resonate at the rf frequency. The quality factor, Q, of this resonant circuit along with the resonant frequency fixes its bandwidth, $\Delta\omega = \omega_{\text{rf}}/Q$, which in turn fixes the response time $\tau_{\text{rf}} \cong (\Delta\omega)^{-1}$. Suppose that at $t = 0$ the rf generator is connected to the resonant circuit. The amplitude, ϕ_{rf}, of the oscillatory flux will then increase with a characteristic time τ_{rf}. When the total externally induced flux, $\phi_{\text{ext}} + \phi_{\text{rf}}(t)$, exceeds one of the thresholds, ϕ_c, (associated with the primary SQUID flux, ϕ, depicted in Fig. 15.12(b)), the SQUID loop becomes resistive, since it is then in a finite voltage state (oscillating at some characteristic frequency much higher than ω_{rf}, which we ignore). This results in a rapid dissipation leading to a partial quenching of the amplitude, ϕ_{rf}; however, the process then repeats itself. The peak amplitude of ϕ_{rf}, denoted $\bar{\phi}_{\text{rf}}$ (which can be measured with the appropriate electronic circuitry), is then a measure of ϕ_{ext} as we will now discuss.

Referring to Fig. 15.12(b), we identify two 'special' points: (i) the midpoint of a stable branch, where $\phi_{\text{ext}} = n\phi_0$ (with n an integer), and (ii) the midpoint of an unstable branch, where $\phi_{\text{ext}} = (n + \frac{1}{2})\phi_0$. We first examine case (i). Assuming sufficient rf drive, as the oscillations build up in time, the thresholds, ϕ_u, connecting the $n, n + 1$ and $n, n - 1$ branches are encountered and the SQUID switches symmetrically between the three branches. If the rf drive is further increased the thresholds connecting the $n + 1, n + 2$ and the $n - 1, n - 2$ branches also are encountered and the SQUID then oscillates symmetrically between five branches and so on. An example of a path traversed by the system for a circuit involving three stable regions is shown in Fig. 15.12(c).

For case (ii), where $\phi_{\text{ext}} = (n + \frac{1}{2})\phi_0$, the path traversed by the SQUID is nominally centered at the midpoint of an unstable region. As the rf drive is increased from zero, a level is reached where a complete circuit involves traversing the portion of the two stable branches lying directly above, $n + 1$, and below, n, the unstable $n + \frac{1}{2}$ section. The threshold amplitude corresponds to the extent of the unstable branch. Since this extent is smaller than that of the stable region, the threshold for first completing circuits in which the SQUID switches irreversibly is lower for case (ii) than for case (i).

Fig. 15.13 shows the peak amplitude of the rf tank circuit, V_T, as a function of the rf drive current, I_{rf}. The slope is initially high but drops abruptly (nominally to zero for a peak reading detector) when the thresholds for SQUID transitions are encountered, involving $3, 5, \ldots$ stable branches for case (i) and $2, 4, 6, \ldots$ for case (ii). From the symmetry it follows that cases (i) and (ii) represent the maximum and minimum values of V_T (or $\bar{\phi}_{\text{rf}}$) for a given drive level, I_{rf}.

It is clear from the above discussion that if ϕ_{ext} is continuously varied (at fixed I_{rf}), the peak amplitude $\bar{\phi}_{\text{rf}}$ will move back and forth (in a triangular fashion) between maxima, at $\phi_{\text{ext}} = n\phi_0$, and minima, at $\phi_{\text{ext}} = (n + \frac{1}{2})\phi_0$. By observing the change in the number of flux quanta, one can accurately measure a change in a current, I_{ext}, associated with the loop generating ϕ_{ext}.

Fig. 15.14(a) shows the characteristic 'staircase' dependence on I_{rf} for an rf SQUID operated at 27 MHz that was shown schematically in Fig. 15.13. The two curves correspond to the limiting cases for the dc external flux. Fig. 15.14(b) shows the triangular dependence of the detected rf voltage V_T vs I_{ext} characteristics. The different curves refer to different values of the rf drive current, I_{rf}.

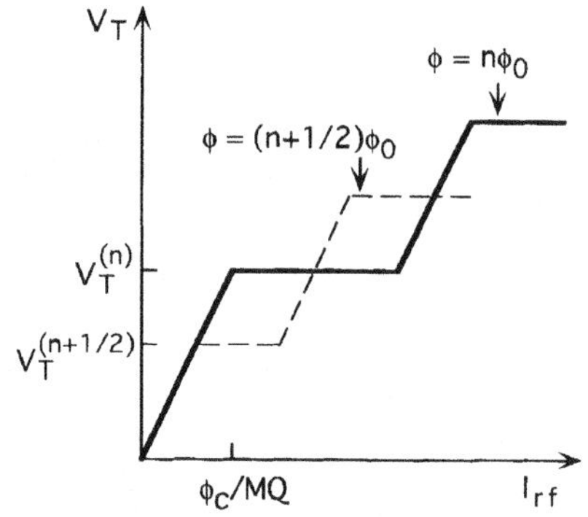

Figure 15.13 Relation between the tank circuit voltage V_T and the rf current amplitude I_{rf} for the cases of integral and half integral numbers of flux quanta. For intermediate values of flux, the voltage steps occur at intermediate values of V_T.

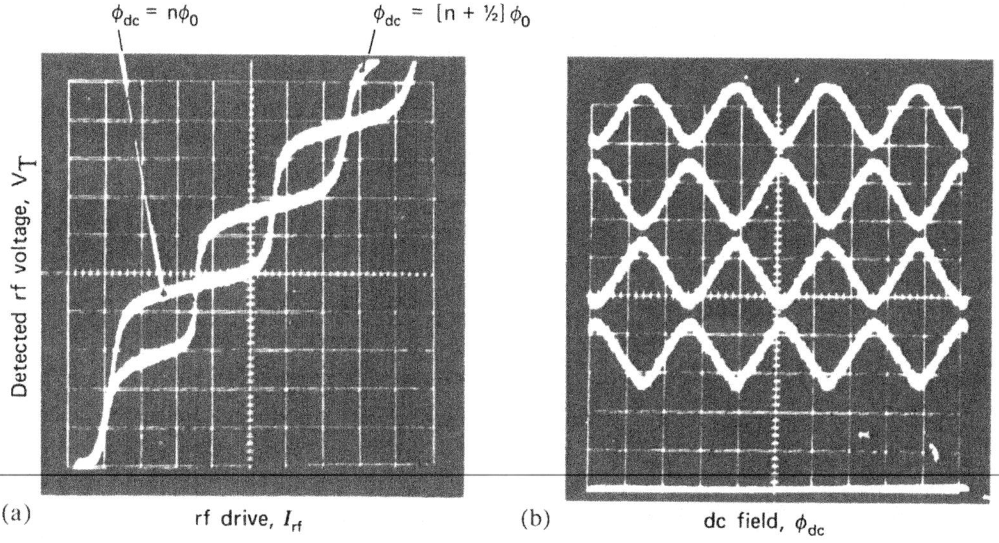

Figure 15.14 Experimental responses of an rf SQUID in the dissipative mode: (a) detected tank circuit voltage V_T vs rf current amplitude I_{rf} (the two curves are the two limiting cases for external dc flux $\phi_{dc} = n\phi_0$ and $\phi_{dc} = (n + \frac{1}{2})\phi_0$); (b) V_T vs applied magnetic flux curves at different values of I_{rf}. (After Zimmerman (1972).)

If the loop generating ϕ_{ext} is coupled to another loop which, in turn, is coupled to a magnetic sample in a fixed field (via an all superconducting circuit), changes in the magnetization with temperature or some other parameter may be observed. (Note that current changes induced through magnetization changes are persistent.) The self-inductance of the two loops, together with the mutual inductances coupling them to the SQUID and the sample, make up a 'flux transformer' the characteristics of which can be optimized for maximal sensitivity.

For further discussion of the rf SQUID see Zimmerman (1972) or Barone and Paterno (1982), where many additional references may be found.

16 The Josephson lattice in 1D

Certain intermetallic compounds have a layered structure in which the in-plane conductivity greatly exceeds the conductivity normal to the layers. This applies for most of the high temperature superconductors. Such structures may also be made artificially by depositing alternating layers of a superconductor and a low conductivity material.

The simplest model is to assume infinitesimally thin superconducting layers which are coupled via *order parameter tunneling* (Josephson coupling) through insulating layers of thickness s. Following Lawrence and Doniach (1971), we introduce a modified free-energy functional of the form

$$F = F_{\mathrm{n}} + F_{\mathrm{s}}, \tag{16.1}$$

where

$$F_{\mathrm{s}} = \sum_n s \int \mathrm{d}^2 r \left[\frac{\hbar^2}{2m^*} \left| \left(\nabla - \frac{\mathrm{i}e^*}{\hbar c} \mathbf{A}_\perp \right) \psi_n \right|^2 + \alpha |\psi_n|^2 + \frac{1}{2} \beta |\psi_n|^4 \right.$$
$$\left. + \frac{\hbar^2}{2m_z^* s^2} \left| \psi_{n+1} \exp\left(-\frac{\mathrm{i}e^*}{\hbar c} \bar{A}_z s \right) - \psi_n \right|^2 \right] + \frac{1}{8\pi} \int B^2 \mathrm{d}^3 r; \tag{16.2}$$

here we have defined $\bar{A}_z = (1/s) \int_{ns}^{(n+1)s} A_z \mathrm{d}z$. The structure of this equation is similar to Eq. (9.13) with respect to the in-plane components while the interplane coupling is seen (on expanding the exponent) to be a generalization of the operator

$$\frac{\partial}{\partial z} - \frac{\mathrm{i}e^*}{\hbar c} A_z$$

to the case of finite differences. Our order parameter $\psi_n(\mathbf{r})$ has a discrete dependence on the index n and a continuous dependence on $\mathbf{r} = x\hat{\mathbf{x}} + y\hat{\mathbf{y}}$; the total vector potential $\mathbf{A}_\perp + A_z \hat{\mathbf{z}}$ is, however, defined at all points. Variation with respect to ψ^* yields

$$-\frac{\hbar^2}{2m^*} \left(\nabla - \frac{\mathrm{i}e^*}{\hbar c} \mathbf{A}_\perp \right)^2 \psi_n$$
$$- \frac{\hbar^2}{2m_z^* s^2} \left[\psi_{n+1} \exp\left(-\frac{\mathrm{i}e^*}{\hbar c} \bar{A}_z s \right) - 2\psi_n + \psi_{n-1} \exp\left(+\frac{\mathrm{i}e^*}{\hbar c} \bar{A}_z s \right) \right]$$
$$+ \alpha \psi_n + \beta \psi_n |\psi_n|^2 = 0. \tag{16.3}$$

Variation with respect to $\mathbf{A}_\perp$ and A_z yields Eq. (9.16) with

$$\mathbf{j}_\perp = -\frac{\mathrm{i}e^*\hbar}{2m^*}(\psi_n^*\nabla\psi_n - \psi_n\nabla\psi_n^*) - \frac{e^{*2}}{m^*c}\mathbf{A}_\perp|\psi_n|^2 \tag{16.4}$$

and

$$j_{\|,n+1,n} = -\frac{\mathrm{i}e^*\hbar}{2m_z^*s}\left[\psi_{n+1}\exp\left(-\frac{\mathrm{i}e^*}{\hbar c}\bar{A}_z s\right)\psi_n^* - \psi_{n+1}^*\exp\left(+\frac{\mathrm{i}e^*}{\hbar c}\bar{A}_z s\right)\psi_n\right]. \tag{16.5}$$

Eq. (16.5) for the tunneling current flowing between planes n and $n+1$ is equivalent to Eq. (15.4) (written in a gauge-invariant form).

Solutions of Eq. (16.3) were first examined in detail by Klemm, Luther, and Beasley (1975). As before, we confine our interest to the evaluation of H_{c2}, in which case we may neglect the last (nonlinear) term. The external magnetic field will be assumed to lie in the x–z plane. In order to take advantage of the layer periodicity, we choose a gauge for which the vector potential has no z dependence:

$$\mathbf{A}_\perp + A_z\hat{\mathbf{z}} = Hx(\cos\theta\hat{\mathbf{y}} - \sin\theta\hat{\mathbf{z}}). \tag{16.6}$$

We seek a solution of the form

$$\psi_n = \mathrm{e}^{+\mathrm{i}k_z ns + \mathrm{i}k_y y}u_n(x);$$

noting that $\bar{A}_z = -Hx\sin\theta$, we find that

$$\left\{-\frac{\hbar^2}{2m^*}\frac{\mathrm{d}^2}{\mathrm{d}x^2} + \frac{1}{2}m^*\omega_c^2(x-x_0)^2\cos^2\theta - \frac{\hbar^2}{m_z^*s^2}\left[\cos\left(k_z s + \frac{e^*Hs}{\hbar c}x\sin\theta\right) - 1\right] + \alpha\right\}u_n(x) = 0, \tag{16.7}$$

where $x_0 = \hbar ck_y/e^*H\cos\theta$ is an orbit center.

For the field perpendicular to the layers, we again have a harmonic oscillator problem and we recover our earlier result $-\alpha = \frac{1}{2}\hbar\omega_c(0)$. For H parallel to the planes Eq. (16.7) becomes

$$\left\{-\frac{\hbar^2}{2m^*}\frac{\mathrm{d}^2}{\mathrm{d}x^2} - \frac{\hbar^2}{m_z^*s^2}\left[\cos\left(k_z s + \frac{e^*Hs}{\hbar c}x\right) - 1\right] + \alpha\right\}u_n = 0. \tag{16.8}$$

Changing k_z has the effect of shifting the origin on the x axis. We shall initially fix the origin such that the argument of the cosine vanishes when $x = 0$. Near the zero-field transition temperature, where α is small, we expect the critical magnetic field, $H_{c2\|}$, also to be small and further that the lowest eigenfunction u, which by Floquet's theorem is periodic (since the potential is periodic), is concentrated near the minimum of the '*potential*'

$$V(x) = \frac{\hbar^2}{m_z^*s^2}\left[1 - \cos\left(\frac{e^*Hs}{\hbar c}x\right)\right]. \tag{16.9}$$

Expanding Eq. (16.9) to second order, we obtain a harmonic oscillator potential which, on substitution into Eq. (16.8), yields

$$-\frac{\hbar^2}{2m^*}u'' + \left[\frac{1}{2}m^*\omega_c^2\left(\frac{\pi}{2}\right)x^2 + \alpha\right]u = 0, \tag{16.10}$$

where

$$\omega_c\left(\frac{\pi}{2}\right) \equiv \frac{|e|^*H}{(m^*m_z^*)^{1/2}c}.$$

The lowest eigenvalue is $-\alpha = \frac{1}{2}\hbar\omega_c(\pi/2)$, yielding an upper critical field $H_{c2\parallel} = \phi_0/2\pi\xi\xi_z$ where $\xi \equiv (\hbar^2/2m^*|\alpha|)^{1/2}$ and $\xi_z \equiv (\hbar^2/2m_z^*|\alpha|)^{1/2}$. For temperatures further from T_c, where the eigenvalue $-\alpha$ and the associated parallel critical field are both large, u becomes less concentrated at the potential minimum and our expansion of $V(x)$ breaks down. To examine this limit, we write Eq. (16.8) in its '*canonical*' form (Abramowitz and Stegun 1970) by introducing new variables

$$2v \equiv \frac{e^*Hs}{\hbar c}x + \pi,\ 2q \equiv \frac{2\phi_0^2}{\pi^2H^2s^4}\frac{m^*}{m_z^*}$$

and

$$a' = \frac{4\phi_0^2}{(2\pi)^2H^2s^4}\frac{m^*}{m_z^*}\left(\frac{s^2}{\xi_z^2} - 2\right),$$

which yields

$$\frac{d^2u}{dv^2} + (a' - 2q\cos 2v)u = 0. \tag{16.11a}$$

For large H, both a' and q are small and are related by the expansion (Abramowitz and Stegun 1970, p. 722)

$$a' = -\frac{q^2}{2} + \frac{7q^4}{128} + \ldots. \tag{16.11b}$$

Inserting our expressions for a' and q into Eq. (16.11b) we obtain

$$H_{c2\parallel} = \frac{\phi_0}{2\pi s^2(1 - s^2/2\xi_z^2)^{1/2}}\left(\frac{m^*}{m_z^*}\right)^{1/2}. \tag{16.12}$$

Since this is a high field expansion, Eq. (16.12) is valid only for temperatures such that $\xi_z \approx s/\sqrt{2}$; however, $H_{c2\parallel}$ is real only for $\xi_z > s/\sqrt{2}$. No (real) solution exists for $\xi_z < s/\sqrt{2}$ and hence the critical field becomes formally infinite (in this model) at a temperature T^* such that $\xi_z(T^*) = s/\sqrt{2}$. In a real system, this divergence would be removed by the effect of paramagnetic limiting (see Part III, Eq. (41.45)).

Deutcher and Entin–Wohlman (1978) generalized the above model to the case of a superlattice of thin slabs of thickness d separated by Josephson-coupled insulating layers of thickness s. We shall continue to take the z axis normal to the layers, and we place $z = 0$ at the interface with the lower surface of a metallic layer. The free energy is given by

$$F = \sum_n \int \mathrm{d}x\mathrm{d}y \left\{ \int_{nD}^{nD+d} \mathrm{d}z \left[\frac{\hbar^2}{2m^*} \left| \left(\nabla - \frac{\mathrm{i}e^*}{\hbar c} \mathbf{A} \right) \psi_n(x, y, z) \right|^2 \right. \right.$$

$$+ \alpha \,|\, \psi_n(x, y, z) \,|^2 + \frac{\beta}{2} \,|\, \psi_n(x, y, z) \,|^4$$

$$+ \frac{\hbar^2}{2m_z^* s^2} \left| \psi_{n+1}(x, y, (n+1)D) \exp\left(-\frac{\mathrm{i}e^*}{\hbar c} \bar{A}_{n+1} s \right) \right.$$

$$\left. \left. \left. - \psi_n(x, y, nD + d) \right|^2 \right] \right\} + \frac{1}{8\pi} \int \mathrm{d}x\mathrm{d}y\mathrm{d}z B^2(x, y, z); \qquad (16.13)$$

here $\bar{A}_{n+1} \equiv (1/s) \int_{nD+d}^{(n+1)D} A_z(x, y, z)\mathrm{d}z$ and $D = d + s$. We seek a solution to Eq. (16.13) in the limit $d << \lambda_L$ and $d << \xi$. The first of these assures us that we may neglect the diamagnetic screening currents and hence we shall take the magnetic field as uniform and given by the external field, H. We consider only the case where $\mathbf{H}$ is parallel to $\hat{\mathbf{y}}$ and work in the gauge $\mathbf{A} = -Hx\hat{\mathbf{z}}$. The limit $d << \xi$ implies that the magnitude of the order parameter is constant along z for a given x; however, we do allow the phase to vary. We thus choose a solution of the form

$$\psi = u(x)\mathrm{e}^{\mathrm{i}\Phi(x,z)}.$$

The z component of the gradient energy in Eq. (16.13) is

$$\left| \left(\frac{\partial \Phi}{\partial z} + \frac{e^*}{\hbar c} Hx \right) \psi \right|^2.$$

To avoid an arbitrarily large growth in this term for large x, we choose the phase in the form

$$\Phi = -\frac{e^*}{\hbar c} Hx(z - z_n);$$

z_n is chosen to minimize the overall free energy. For an isolated slab it has the value $z_n = nD + d/2$, i.e., the origin must be taken to be in the center of the nth slab. For the ground state, we assert that, by symmetry, z_n will have the same value in the coupled superlattice. We include no variation in the phase along y since this results in an increase in the free energy. Varying Eq. (16.13) with respect to ψ^* yields the linearized equation

$$\int_{-d/2}^{+d/2} \left\{ -\frac{\hbar^2}{2m^*} \left(\frac{\mathrm{d}^2 u}{\mathrm{d}x^2} - 2\frac{\mathrm{i}e^*}{\hbar c} Hz \frac{\mathrm{d}u}{\mathrm{d}x} - \frac{e^{*2}}{\hbar^2 c^2} H^2 z^2 u \right) \right.$$

$$\left. - \frac{\hbar^2}{m_z^* s^2} \left[\cos\left(\frac{e^* HDx}{\hbar c} \right) - 1 \right] u + \alpha u \right\} \mathrm{d}z = 0. \qquad (16.14)$$

Integration over z (noting that the second term in the first parentheses vanishes by symmetry) yields

$$-\frac{\hbar^2}{2m^*} \frac{\mathrm{d}^2 u}{\mathrm{d}x^2} - \frac{\hbar^2}{m_z^* s^2} \left[\cos\left(\frac{e^* HD}{\hbar c} x \right) - 1 \right] u + \left(\alpha + \frac{m^* \omega_c^2 d^2}{24} \right) u = 0, \qquad (16.15)$$

where we have introduced the notation $\omega_c = e^* H/m^* c$. Using Eqs. (12.3) and (12.4) we may rewrite Eq. (16.15) as

$$-\frac{\hbar^2}{2m^*}\frac{d^2u}{dx^2}-\frac{\hbar^2}{m_z^*s^2}\left[\cos\left(\frac{e^*HD}{\hbar c}x\right)-1\right]u+\alpha\left(1-\frac{H^2}{H_{\|}^2}\right)u=0, \tag{16.16}$$

where $H_{\|}$ is the parallel critical field of the isolated slab. The equation is identical with Eq. (16.8) except for the additional term multiplying α.

To obtain the low-field behavior, we again expand the cosine term; this time we shall include the H^4 term, the effect of which we calculate by perturbation theory. We require the matrix element of x^4 with respect to the unperturbed Gaussian ground-state wavefunction u_0 which is given by (Landau and Lifshitz 1977, p. 136)

$$\langle u_0|x^4|u_0\rangle=\frac{3}{4}\frac{\hbar^2}{m^{*2}\omega_c^2}\frac{s^2}{D^2}\frac{m_z^*}{m^*}. \tag{16.17}$$

Our eigenvalue equation is then

$$-\alpha=\frac{1}{2}\hbar\omega_c\frac{D}{s}\left(\frac{m^*}{m_z^*}\right)^{1/2}+\frac{m^*\omega_c^2d^2}{24}\left(1-\frac{3}{4}\frac{D^2}{d^2}\right)+\dots \tag{16.18}$$

or

$$1=\frac{2\pi}{\phi_0}\frac{D}{s}\left(\frac{m^*}{m_z^*}\right)^{1/2}\xi^2H+\frac{1}{12}\left(\frac{2\pi}{\phi_0}\right)^2\left(1-\frac{3}{4}\frac{D^2}{d^2}\right)\xi^2d^2H^2. \tag{16.19}$$

For $H\to 0$ ($T\to T_c$), we have the limiting behavior

$$H_{c2\|}=\frac{\phi_0}{2\pi\xi\xi_z(D/s)}; \tag{16.20}$$

this we refer to as 3D behavior with $H_{c2\|}\propto T_c-T$.

To obtain the high-field behavior, we must write Eq. (16.15) in the canonical form of Eq. (16.11). The constants now become

$$2v=\frac{e^*HD}{\hbar c}x+\pi,\ 2q=\frac{2\phi_0^2}{\pi^2H^2s^2D^2}\frac{m^*}{m_z^*}$$

and

$$a'=\frac{4\phi_0^2}{(2\pi)^2H^2D^2\xi^2}\left(1-\frac{H^2}{H_{\|}^2}-2\frac{\xi^2}{s^2}\frac{m^*}{m_z^*}\right). \tag{16.21}$$

Using Eq. (16.12), we have

$$1-\frac{H^2}{H_{\|}^2}-2\frac{\xi^2}{s^2}\frac{m^*}{m_z^*}=-\frac{\phi_0^2}{2\pi^2H^2s^4}\frac{\xi^2}{D^2}\left(\frac{m^*}{m_z^*}\right)^3. \tag{16.22}$$

From Eq. (16.22), as $H\to\infty$, we obtain $H_{c2\|}=H_{\|}$; i.e., we obtain the upper critical field of an isolated slab, where $H_{c2\|}\propto(T_c-T)^{1/2}$, which we refer to as 2D behavior. Thus as the temperature is lowered a 3D to 2D '*crossover*' occurs. Note that the divergence encountered with the Lawrence–Doniach superlattice model involving infinitesimally thin superconducting layers is avoided in the Deutcher–Entin-Wohlman model where $d\neq 0$.

17 Vortex structure in layered superconductors

Dichalcogenides[1] of transition metals such as $NbSe_2$, superconductor/insulator superlattices, and high T_c oxide superconductors all have a layered structure. Most of these systems are Type II superconductors with relatively large κ values. One of the fascinating characteristics of the layered superconductors is their strongly anisotropic magnetic properties. Usually, the coherence length perpendicular to the layer plane ($\xi_\perp$) is much smaller than that parallel to the layer ($\xi_\parallel$). The anisotropy can be characterized by an anisotropy ratio, γ, defined as $\gamma = \xi_\parallel/\xi_\perp$. Other length scales of interest are the penetration length, λ, and the scale of intrinsic inhomogeneities.

Depending on the relative size of $\xi_\perp$ and the layer spacing, s, we may identify three different regimes for the vortex structure in the layered, high-κ superconductors: (1) If $\xi_\perp(T) >> s$, then the layered structure is largely irrelevant and the superconductor may be regarded as three-dimensional: anisotropic, but uniform. Since the coherence length diverges as T_c is approached, this regime will always occur sufficiently close to T_c. The vortex structure can be described using the London or G–L theories by introducing an anisotropic mass tensor as discussed in Sec. 11. (2) With decreasing temperature, we may have a regime where $\xi_\perp(T) << s$ (especially in high T_c oxides). If both regimes can be entered by sweeping the temperature a 3D–2D crossover will occur at some temperature. Below this temperature we have a quasi-2D regime; the layered structure is then relevant and the Lawrence–Doniach-like model discussed in Sec. 16 is adequate to describe the vortex structure. (3) For the case of extreme anisotropy, as in Bi- and Tl-based oxide superconductors, with $\gamma \gtrsim 10$, the interlayer Josephson coupling is very weak. The flux 'lines' are then better viewed as stacks of 2D vortices residing in the superconducting layers (so-called pancake vortices).

In this section we will give a brief account of the vortex structure in each of these three regimes. We will focus on the weak field region and derive expressions for H_{c1}, which contain information on the anisotropy in addition to that contained in H_{c2}.

17.1 3D anisotropic London model

As discussed in Sec. 11, the upper critical field, H_{c2}, in layered Type II superconductors can be described by the G–L equations with a phenomenological anisotropic mass tensor. Due to the linearity of the equations near H_{c2}, a relatively simple solution can be found in this region (see

1. The chalcogenides are the group VI elements S, Te, and Se.

Sec. 11). However, close to the lower critical field, H_{c1}, it is difficult to solve the nonlinear G–L equations to obtain H_{c1} in the anisotropic case. Furthermore, the G–L expansion itself is problematic in this regime since we are far from T_c. For this reason, we will use the less accurate London model, which provides a reasonable description, at least for a large G–L parameter κ (see, for example, Kogan (1981)). For $\kappa(\equiv \lambda/\xi) >> 1$, the amplitude of the order parameter, $|\psi|$, varies considerably only for distances smaller than ξ; one may then assume $|\psi|$ to be constant everywhere except in a narrow core of radius ξ.

As in Sec. 11, we introduce a phenomenological effective mass tensor. In the reference frame aligned with the principal axes of a crystal with orthorhombic symmetry or higher, this mass tensor is diagonal with elements which we now write as M_i. For convenience we define a geometric mean mass $\bar{M} = (M_1 M_2 M_3)^{1/3}$ and normalize the mass tensor such that $m_i = M_i/\bar{M}$. In terms of $\bar{M}$ we define a mean penetration depth, λ, and mean coherence length, ξ, by using Eqs. (2.6) and (9.22), respectively. The penetration depths, $\lambda_i = \lambda m_i^{1/2}$, describe the decay of the components of the screening supercurrent along the principal directions, i. The coherence lengths, $\xi_i = \xi/m_i^{1/2}$, characterize the spatial variation of the order parameter along these directions. For a uniaxial crystal, $m_3 = m_\perp, m_1 = m_2 = m_\parallel$ (in high T_c oxides, the anisotropy within the Cu–O layers is very weak). The anisotropy ratio is defined as $\gamma \equiv \xi_\parallel/\xi_\perp = \lambda_\perp/\lambda_\parallel = (M_\perp/M_\parallel)^{1/2}$.

The free energy, given in Eq. (2.8) for the isotropic case, may be generalized to the anisotropic case as

$$F = \frac{1}{8\pi}\int [\mathbf{B}^2 + \lambda^2(\nabla \times \mathbf{B})\cdot \mathbf{m}\cdot(\nabla \times \mathbf{B})]\mathrm{d}^3 r, \tag{17.1}$$

where $\mathbf{B}$ is the local field. Straightforward minimization of the energy (17.1) yields the anisotropic London equation

$$\mathbf{B} + \lambda^2 \nabla \times [\mathbf{m}\cdot(\nabla \times \mathbf{B})] = 0. \tag{17.2}$$

Taking into account the boundary condition, the local magnetic field around a single flux line directed along $\hat{\mathbf{z}}$ and carrying a flux quantum ϕ_0 is determined by

$$\mathbf{B} + \lambda^2 \nabla \times (\mathbf{m}\cdot \nabla \times \mathbf{B}) = \phi_0 \hat{\mathbf{z}}\delta(x)\delta(y). \tag{17.3}$$

In the isotropic case where $m_{ij} = \delta_{ij}$, Eq. (17.3) coincides with the usual London equation (2.14). To deal with an arbitrarily directed flux line, we define two reference frames as shown in Fig. 17.1. The vortex frame $(\hat{\mathbf{x}}, \hat{\mathbf{y}}, \hat{\mathbf{z}})$ is obtained from the crystal frame $(\hat{\mathbf{x}}_0, \hat{\mathbf{y}}_0, \hat{\mathbf{z}}_0)$ by rotating by an angle θ from $\hat{\mathbf{z}}_0$ about $\hat{\mathbf{y}}_0$ axis. Restricting ourselves to the case of uniaxial symmetry, the components m_{ij}, transformed to the vortex frame, are

$$\mathbf{m}(\theta) = \begin{pmatrix} m_1\cos^2\theta + m_3\sin^2\theta & 0 & (m_1 - m_3)\sin\theta\cos\theta \\ 0 & m_1 & 0 \\ (m_1 - m_3)\sin\theta\cos\theta & 0 & m_1\sin^2\theta + m_3\cos^2\theta \end{pmatrix} \tag{17.4}$$

The algebra is greatly simplified by working in the vortex frame, since both $\mathbf{B}$ and $\mathbf{j}$ are independent of z. Substituting (17.4) into Eq. (17.3) results in a set of equations for B_x, B_y, and B_z. These equations lead to a more complicated vortex structure in the anisotropic superconductor than the isotropic Abrikosov vortex. For example, transverse fields, $B_{x,y}$, are present even when

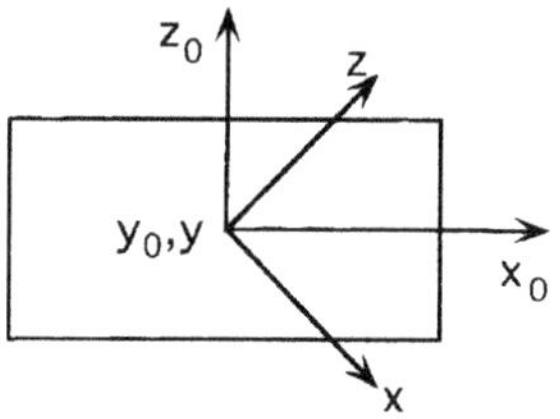

Figure 17.1 In the crystal frame $(\hat{\mathbf{x}}_0, \hat{\mathbf{y}}_0, \hat{\mathbf{z}}_0)$ the plane x_0y_0 coincides with the basal crystal plane. The vortex frame $(\hat{\mathbf{x}}, \hat{\mathbf{y}}, \hat{\mathbf{z}})$ is obtained from the crystal frame by rotating by an angle θ from $\hat{\mathbf{z}}_0$ about the $\hat{\mathbf{y}}_0$ axis. The axis $\hat{\mathbf{z}}$ is parallel to the vortex axis.

the applied field is along the z direction. For simplicity, we consider two special cases in which the magnetic field is applied perpendicular to or parallel to the layer plane.

For the perpendicular field case, $\theta = 0$. The mass tensor is diagonal; $m_{xx} = m_{yy} = m_1, m_{zz} = m_3$. Eq. (17.3) then becomes

$$B_z - \lambda^2\left(m_1\frac{\partial^2 B_z}{\partial x^2} + m_1\frac{\partial^2 B_z}{\partial y^2}\right) = \phi_0\delta(x)\delta(y). \tag{17.5}$$

This equation is equivalent to Eq. (7.8) and the lower critical field is given by (8.4b) as

$$H_{c1\perp} \approx \frac{\phi_0}{4\pi\lambda_\parallel^2}\ell n\left(\frac{\lambda_\parallel}{\xi_\parallel}\right). \tag{17.6}$$

When the applied magnetic field is parallel to the layer plane, $\theta = \pi/2$. The mass tensor is also diagonal: $m_{xx} = m_3, m_{yy} = m_{zz} = m_1$. Eq. (17.3) now reads

$$B_z - \lambda^2\left(m_1\frac{\partial^2 B_z}{\partial x^2} + m_3\frac{\partial^2 B_z}{\partial y^2}\right) = \phi_0\delta(x)\delta(y); \tag{17.7a}$$

i.e.,

$$B_z - \left(\lambda_\parallel^2\frac{\partial^2 B_z}{\partial x^2} + \lambda_\perp^2\frac{\partial^2 B_z}{\partial y^2}\right) = \phi_0\delta(x)\delta(y). \tag{17.7b}$$

By interchanging $z_0 \leftrightarrow x$, this equation can be transformed back to the crystal frame with B along the $\hat{\mathbf{x}}_0$ axis:

$$B - \left(\lambda_\parallel^2\frac{\partial^2 B}{\partial z_0^2} + \lambda_\perp^2\frac{\partial^2 B}{\partial y_0^2}\right) = \phi_0\delta(z_0)\delta(y_0). \tag{17.8}$$

Eq. (17.8) can be solved by a Fourier transform method. Alternatively, we may map Eq. (17.8) to the isotropic case by defining

$$\tilde{z} \equiv z_0/\lambda_\parallel,\ \tilde{y} \equiv y_0/\lambda_\perp,\ \tilde{\rho}^2 \equiv \tilde{z}^2 + \tilde{y}^2. \tag{17.9}$$

From Sec. 7 we know that the solution of Eq. (17.8) at large distances is given by

$$B(z_0, y_0) = \frac{\phi_0}{2\pi\lambda_\parallel\lambda_\perp}K_0(\tilde{\rho}), \tag{17.10}$$

where $K_n(\tilde{\rho})$ is a modified Bessel function of the second kind of order n. The free energy per unit length of the flux line, E_1, can be calculated from Eq. (17.1). To logarithmic accuracy it is given by

$$E_1 = \frac{\phi_0^2}{16\pi^2\lambda_{\parallel}\lambda_{\perp}} \ln\left(\frac{\lambda_{\parallel}}{\xi_{\perp}}\right). \tag{17.11}$$

Note for this case the normal core radius $\xi_{\perp}$ serves as the cut-off scale. From the usual thermodynamic relation $E_1 = \phi_0 H_{c1}/4\pi$, the parallel lower critical field is

$$H_{c1\parallel} = \frac{\phi_0}{4\pi\lambda_{\parallel}\lambda_{\perp}} \ln\left(\frac{\lambda_{\parallel}}{\xi_{\perp}}\right). \tag{17.12}$$

Comparing Eqs. (17.6) and (17.12), we see that important information about the anisotropy can be obtained by measuring H_{c1} in both parallel and perpendicular fields.

17.2 Lawrence–Doniach model

In Sec. 16 we discussed the upper critical fields for layered superconductors which can be modeled as Josephson lattices in 1D. We saw that when the temperature is lowered to a value T^* such that $\xi_z(T^*) = s/\sqrt{2}$, a 3D–2D crossover occurs. Above T^*, the layered superconductor behaves as an anisotropic 3D superconductor. At temperatures lower than T^*, the barriers of the layered structure dominate, resulting in a 2D behavior of the temperature dependence of $H_{c2\parallel}(T)$. In this subsection we will show that below T^* the vortex structure in layered superconductors is also drastically modified.

We will start with Eqs. (16.4) and (16.5) which were derived from the Lawrence–Doniach free energy functional (16.2). For high κ superconductors in a weak magnetic field, the influence of the field on the value of $|\psi_n|$ can be neglected, and we can regard $|\psi_n|$ as a constant; i.e., we use the London approximation. We confine ourselves to the parallel field case. We take the field along the $\hat{\mathbf{x}}$ direction and choose a gauge such that $A_z = 0$. Writing $\psi_n = |\psi_n| e^{i\Phi_n}$, from Eqs. (16.4) and (16.5) we obtain

$$\mathbf{A}_{\perp} = -\frac{4\pi\lambda_L^2}{c}\mathbf{j}_{\perp} + \frac{\phi_0}{2\pi}\nabla\Phi_n \tag{17.13}$$

and

$$j_{z,n+1,n} = j_m \sin(\Phi_{n+1} - \Phi_n), \tag{17.14}$$

where λ_L is the bulk London penetration depth of the superconducting layers, Φ_n is the phase in the nth superconducting layer and $j_m \equiv e^*\hbar|\psi|^2/m_z^* s$ is the maximum Josephson supercurrent density. Note that in this model it is assumed, for simplicity, that each superconducting layer is isotropic with an intrinsic bulk penetration depth λ_L. We use the rectangular contour C shown in Fig. 17.2 to compute the phase difference across one unit cell: $\Phi \equiv \Phi_{n+1} - \Phi_n$. From Eq. (17.13) we have

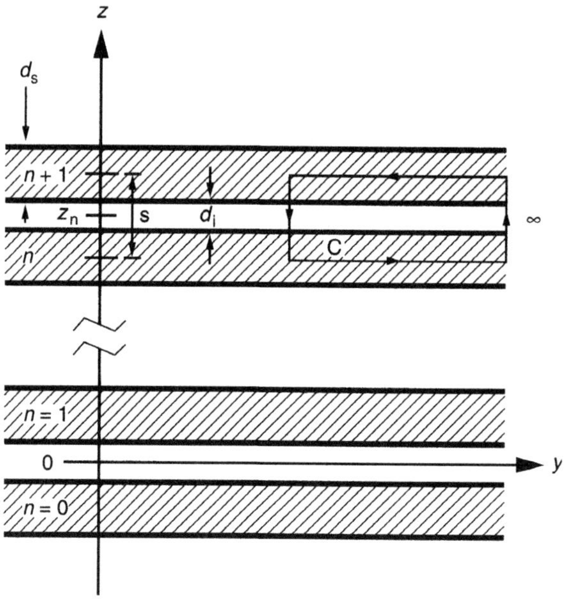

Figure 17.2 Schematic of a Josephson-coupled superconductor–insulator superlattice. The insulating layers of thickness d_i alternate with superconducting layers (crosshatched) of thickness d_s. The modulation wavelength $s = d_i + d_s$. The rectangular contour C is used to compute the magnetic flux in Eq. (17.15).

$$\int_y^{y+\Delta y} dy' \int_{ns}^{(n+1)s} dz' B(y', z') = \oint_C \mathbf{A}_\perp \cdot d\ell$$

$$= -\frac{4\pi\lambda_L^2}{c}\left(\int_y^{y+\Delta y} j_{y,n} dy' + \int_{y+\Delta y}^{y} j_{y,n+1} dy'\right)$$

$$+ \frac{\phi_0}{2\pi}[\Phi(y) - \Phi(y + \Delta y)].$$

The left-hand side in the above equation can be approximated as $s\Delta y B(y, z)$, and we obtain

$$s\Delta y B(y, z) \approx \frac{4\pi\lambda_L^2}{c}(j_{y,n+1} - j_{y,n})\Delta y + \frac{\phi_0}{2\pi}[\Phi(y) - \Phi(y + \Delta y)].$$

Hence

$$\frac{4\pi\lambda_L^2}{c}\frac{\partial j_y}{\partial z} - \frac{\phi_0}{2\pi s}\frac{\partial \Phi}{\partial y} \approx B(y, z), \tag{17.15}$$

where we have used the following approximations: $\partial j_y/\partial z \approx (j_{y,n+1} - j_{y,n})/s$ and $\partial\Phi/\partial y \approx [\Phi(y + \Delta y) - \Phi(y)]/\Delta y$. For small Φ the Josephson current relation (17.14) may be approximated as $j_z \approx j_m\Phi$ and substituting this form into Eq. (17.15) we obtain

$$\frac{4\pi\lambda_L^2}{c}\frac{\partial j_y}{\partial z} - \frac{4\pi\lambda_z^2}{c}\frac{\partial j_z}{\partial y} = B(y, z), \tag{17.16}$$

where we have defined

$$\lambda_z^2 = \lambda_J^2 \equiv \frac{c\phi_0}{8\pi^2 s j_m}; \tag{17.17}$$

λ_J is the corresponding Josephson penetration depth. Using the Maxwell equation $\nabla \times \mathbf{B} = 4\pi\mathbf{j}/c$ we obtain from (17.16) an anisotropic London equation

$$\lambda_J^2 \frac{\partial^2 B}{\partial y^2} + \lambda_L^2 \frac{\partial^2 B}{\partial z^2} = B(y, z). \tag{17.18}$$

In comparison with Eq. (17.8) we see that in the above Lawrence–Doniach model, the Josephson penetration depth λ_J plays the role of $\lambda_\perp$ while $\lambda_\parallel = \lambda_L$.

The single-vortex solution of Eq. (17.18) and the lower critical field can be computed by the procedures outlined in Subsec. 17.1. Note that in the above Lawrence–Doniach model, the vortex *resides in the insulating layer*. There is no normal core since no variation in the amplitude of the order parameter is associated with the vortex. To calculate the flux line energy and H_{c1}, we assume that the natural cut-off scale is the periodicity of the superlattice, s. The parallel lower critical field is then

$$H_{c1\parallel} = \frac{\phi_0}{4\pi\lambda_\parallel\lambda_J} \ell n\left(\frac{\lambda_\parallel}{s}\right). \tag{17.19}$$

The above model may be extended to take into account the effect of a finite superconducting layer thickness (Clem and Coffey 1990).

17.3 Vortex structure in a 2D film

In this subsection we examine the magnetic field generated by the so-called 2D pancake vortex in an isolated thin superconducting film, which may serve as a basis to study the vortex structure in extremely anisotropic layered superconductors. The problem was first discussed by Pearl (1964).

For simplicity, we employ the London approximation and regard $|\psi(\mathbf{r})|$ as a constant except in the small normal-core region. The second G–L equation (9.17b) can then be written as

$$\mathbf{j} = -\frac{c}{4\pi\lambda^2}\left(\mathbf{A} - \frac{\phi_0}{2\pi}\nabla\Phi\right), \tag{17.20}$$

where Φ is the phase of the order parameter, and

$$\lambda^2 = \frac{m^* c^2}{4\pi e^{*2} |\psi(\mathbf{r})|^2}$$

is the penetration depth and $n_{3D}^* = |\psi(\mathbf{r})|^2$ is the density of Cooper pairs. Consider a superconducting film of thickness d and infinite extent, located in the x–y plane (see Fig. 17.3). When $d \to 0$, we can define a sheet current density, $\mathbf{K}$, as $\mathbf{K} = \mathbf{j}d$ and replace Eq. (17.20) by

$$\mathbf{K} = -\frac{c}{2\pi\Lambda}\left(\mathbf{A} - \frac{\phi_0}{2\pi}\nabla\Phi\right); \tag{17.21}$$

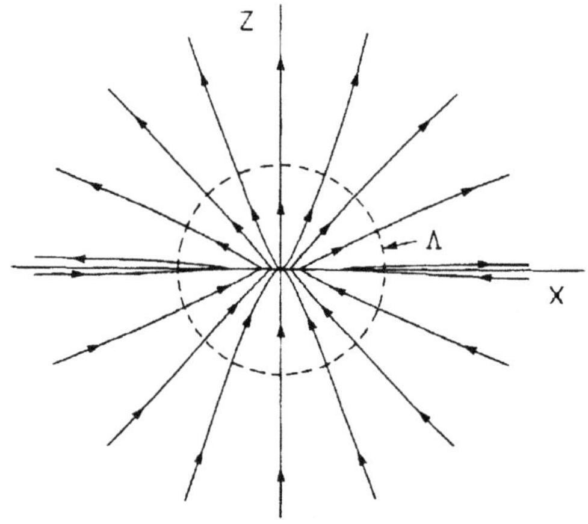

Figure 17.3 Sketch of the magnetic field generated by a 2D pancake vortex in an isolated superconducting layer with thin-film screening length Λ. Note the differing behavior for $\rho < \Lambda$ and $\rho > \Lambda$. The field for $z > 0$ and $\rho > \Lambda$ resembles that of a magnetic monopole such that $B = \phi_0/2\pi\rho^2$. (After Clem (1991).)

here we introduced a 2D screening length defined as

$$\Lambda = \frac{2\lambda^2}{d} = \frac{m^* c^2}{2\pi e^{*2} n^*_{2D}} \tag{17.22}$$

and the sheet density of Cooper pairs as $n^*_{2D} = dn^*_{3D} = d|\psi(\mathbf{r})|^2$. We consider the case of a superconducting film containing a single pancake vortex with one fluxoid ($\hat{\mathbf{z}}\phi_0$) at the origin. The Maxwell equation is

$$\nabla \times \mathbf{B} = 0, \qquad z \neq 0. \tag{17.23}$$

Using $\mathbf{B} = \nabla \times \mathbf{A}$, we obtain

$$\nabla \times (\nabla \times \mathbf{A}) = 0, z \neq 0. \tag{17.24}$$

In cylindrical coordinates (ρ, θ, z) with the unit vectors $\hat{\boldsymbol{\rho}} = \hat{\mathbf{x}}\cos\theta + \hat{\mathbf{y}}\sin\theta, \hat{\boldsymbol{\theta}} = \hat{\mathbf{y}}\cos\theta - \hat{\mathbf{x}}\sin\theta$, and $\hat{\mathbf{z}}$, the vector potential becomes $\mathbf{A} = \hat{\boldsymbol{\theta}} A_\theta(\rho, z)$, and Eq. (17.24) takes the form

$$\left(\frac{\partial^2}{\partial\rho^2} + \frac{1}{\rho}\frac{\partial}{\partial\rho} + \frac{\partial^2}{\partial z^2} - \frac{1}{\rho^2}\right) A_\theta(\rho, z) = 0, \qquad z \neq 0. \tag{17.25}$$

Eq. (17.25) can be solved by separating variables. The z-dependent part is proportional to e^{-qz} for $z > 0$ and e^{qz} for $z < 0$; the ρ-dependent part is proportional to the Bessel function of order one, $J_1(q\rho)$. With mirror symmetry about the $z = 0$ plane, the general solution of Eq. (17.25) is

$$A_\theta(\rho, z) = \int_0^\infty a(q) J_1(q\rho) e^{-q|z|} dq, \tag{17.26}$$

where $a(q)$ is the Fourier expansion coefficient. From Eq. (17.21) the discontinuity in the magnetic field's radial component across the film is

$$K_\theta(\rho, z=0) = (c/4\pi)[B_\rho(\rho, 0_+) - B_\rho(\rho, 0_-)] \tag{17.27a}$$

or

$$K_\theta(\rho, z=0) = -(c/2\pi\Lambda)\left[A_\theta(\rho, z=0) - \frac{\phi_0}{2\pi\rho}\right]. \tag{17.27b}$$

From Eqs. (17.26) and (17.27) we have

$$\int_0^\infty dq a(q)(q + 1/\Lambda)J_1(q\rho) = \phi_0/2\pi\Lambda\rho.$$

Using the expression

$$\frac{1}{\rho} = \int_0^\infty J_1(q\rho)dq,$$

we find that

$$a(q) = \frac{\phi_0}{2\pi(q\Lambda + 1)}. \tag{17.28}$$

Combining Eqs. (17.28) and (17.27a) and using the Hankel transformation, the resulting integral yields

$$K_\theta(\rho) = (c\phi_0/8\pi\Lambda^2)[H_1(\rho/\Lambda) - Y_1(\rho/\Lambda) - 2/\pi], \tag{17.29}$$

where H_1 is the first order Struve function and Y_1 the first order Bessel function of the second kind (Gradshteyn and Ryzhik 1994); Abramowitz and Stegun 1965). Limiting expressions for the sheet current density $K_\theta(\rho)$ are

$$K_\theta(\rho) \approx c\phi_0/4\pi^2\Lambda\rho, \qquad \rho << \Lambda, \tag{17.30a}$$

and

$$K_\theta(\rho) \approx c\phi_0/4\pi^2\rho^2, \qquad \rho >> \Lambda. \tag{17.30b}$$

From Eq. (17.27a), the magnetic flux $\phi_z(\rho, 0)$ through a circle of radius ρ is found to be

$$\phi_z(\rho) \approx \phi_0(\rho/\Lambda), \qquad \rho << \Lambda \tag{17.31a}$$

and

$$\phi_z(\rho) \approx \phi_0(1 - \Lambda/\rho), \qquad \rho >> \Lambda. \tag{17.31b}$$

The magnetic field

$$\mathbf{B} = \hat{\rho}B_\rho(\rho, z) + \hat{\mathbf{z}}B_z(\rho, z)$$

is sketched in Fig. 17.3.

From Eq. (17.30) we see that while for small values of ρ the current is identical with the Abrikosov solution for bulk superconductors, the behavior at large distances is drastically different; instead of an exponential cut-off, we now find a slow $1/\rho^2$ decay of currents. The reason for this is that in films distant regions are electromagnetically coupled through free space, while in bulk superconductors the fields are diamagnetically screened by circulating superconducting

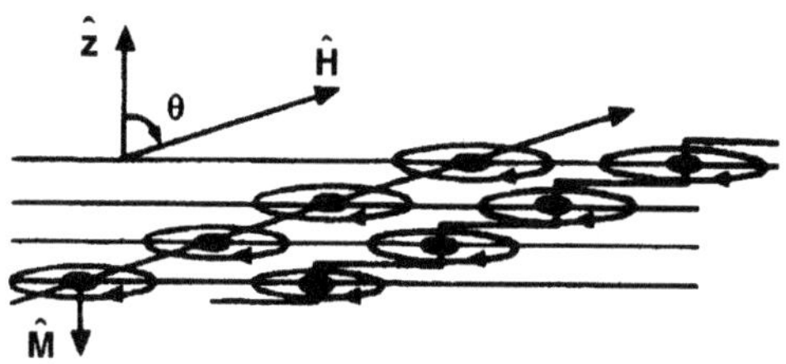

Figure 17.4 Schematic view of two representations of flux lines in a superconductor–insulator superlattice when the field **H** makes an angle θ with the z axis. The ellipses (lying in the layer plane) depict superconducting screening currents. To the left is the continuous flux-line picture for very anisotropic superconductors; to the right, the segmented flux-line picture with pancake vortices valid for Josephson-coupled superconducting layers. Note that the magnetization is directed along the z axis.

electrons. Knowing the current distribution around a single vortex, we can calculate the free energy associated with a single vortex $E_1 d$. It turns out that $E_1 d$ has almost the same form as Eq. (7.14) and is given by $E_1 d \approx (\phi_0/4\pi\lambda_{\parallel})^2 \ell n(\lambda_{\parallel}/\xi_{\parallel})$. However, the magnetization associated with a single vortex in thin films is larger than that in a bulk superconductor by a factor R/d; in particular, $M \approx \phi_0 R/4\pi d$ where R is the film radius and d is the film thickness. Hence, the lower critical field $H_{c1\perp}$ for a thin-film superconductor is

$$H_{c1\perp} \approx \frac{\phi_0}{4\pi\lambda_{\parallel}^2} \frac{d}{R} \ell n \frac{\lambda_{\parallel}}{\xi_{\parallel}}, \tag{17.32}$$

which is a factor of d/R smaller than the bulk value.

The above treatment has been extended by Clem (1991) to study the vortex structure within an infinite stack of thin superconducting layers in the limit of zero interlayer Josephson coupling. For this case 3D vortex lines can be built up by superposing the contributions of stacks of 2D pancake vortices as schematically shown in Fig. 17.4.

When the magnetic field is perpendicular to the layer plane, the lower critical field, $H_{c1\perp}$, is again given by the usual London expression (17.6). However, if the applied field is exactly parallel to the layers, with the approximation $\lambda_{\perp} \sim \lambda_J = \infty$, the energy cost for magnetic flux to penetrate parallel to the layers vanishes; thus $H_{c1\parallel} = 0$. A more sophisticated model, taking into account both electromagnetic and Josephson interlayer coupling, is required in this case.

18 Granular superconductors: the Josephson lattice in 2D and 3D

Granular superconductors are formed, for example, by embedding superconducting grains in an insulating matrix. For low metal-volume fractions, there is no direct metallic contact between the grains, and the grains are coupled only by the Josephson tunneling. In such a system the superconducting transition (if it occurs at all) appears in two steps. As the temperature is lowered the superconducting order parameter first develops in each grain at a temperature T_{c0}, but because the thermal energy is higher than the intergrain Josephson coupling, E_J, the phases of the order parameter are uncorrelated due to thermal fluctuations. Only at a lower temperature $k_B T_c \sim E_J$ will phase locking take place, leading to long-range phase coherence and zero resistivity. Granular superconductors have been an active field of research for many years. The discovery of high T_c oxide superconductors has renewed interest in granular superconductivity since these oxides often appear to have an intrinsic 'granularity', which is commonly found in single crystals and films.

At first glance, a granular superconductor constitutes a very complex system having random grain size and random intergrain coupling. In principle, one can write down a G–L-like free energy functional by taking into account the intergrain coupling energy. After an ensemble average, the system is then characterized by an effective grain size and an effective coupling energy, E_J; both the amplitude and the phase of the order parameter are macroscopic quantum variables. However, if the two stages of the ordering process are separated by a large temperature interval, we can neglect the influence of Josephson tunneling on the amplitude of the order parameter, and we consider only the phase degrees of freedom. In this sense, a granular superconductor may be modeled by a two-dimensional or three-dimensional Josephson junction array as depicted in Fig. 18.1.

To derive an effective Hamiltonian for a granular system, following Anderson (1964), we first examine the Hamiltonian (energy) for a single Josephson junction. Including the Coulomb charging energy, and using the Josephson relation $\dot{\Phi} = 2e\mathscr{V}/\hbar$, the Hamiltonian is

$$H = \frac{1}{2}C\mathscr{V}^2 + V(\Phi) = \frac{\hbar^2}{8e^2}C\dot{\Phi}^2 + E_J(1 - \cos\Phi), \tag{18.1}$$

where C is the capacitance of the junction, $\mathscr{V}$ is the electric potential, and $V(\Phi)$ is the Josephson barrier energy. This Hamiltonian can be transformed to a suitable Lagrangian L:

$$L = -\frac{\hbar^2}{8e^2}C\dot{\Phi}^2 - E_J(1 - \cos\Phi), \tag{18.2}$$

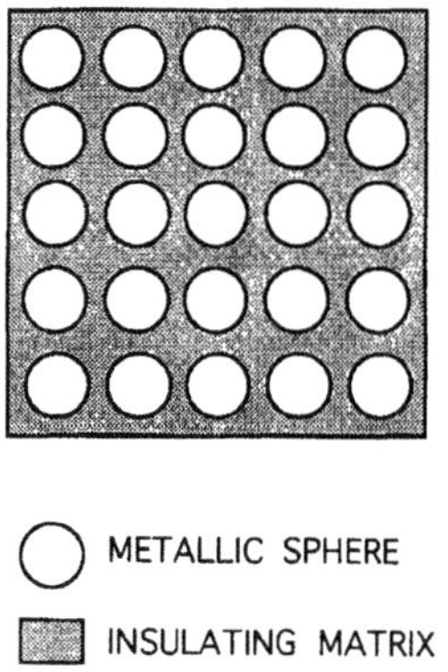

Figure 18.1 Schematic of a model system for a granular superconductor consisting of a lattice of metallic spheres embedded in an insulating matrix.

provided that we define a canonical momentum, p, such that

$$p = \frac{\hbar^2}{4e^2} C\dot{\Phi}. \tag{18.3}$$

Eq. (18.1) may then be quantized by setting

$$[\Phi, \hat{p}] = i\hbar, \tag{18.4}$$

or equivalently

$$\hat{p} = -i\hbar \frac{\partial}{\partial \Phi}. \tag{18.5}$$

If we define $\hat{n} \equiv \hat{p}/\hbar = -i(\partial/\partial\Phi)$, then Eq. (18.1) may be written in operator form as

$$\hat{H} = -2U\hat{n}^2 + E_J(1 - \cos\Phi), \tag{18.6}$$

where $U \equiv e^2/C$ is just the single-electron charging energy. We see that $\hat{n}$, as the variable conjugate to the phase, measures the number of Cooper pairs transferred across the junction. The Hamiltonian (18.6) describes a quantum Josephson junction. Neglecting dissipation the eigenstates of Eq. (18.6) are Mathieu functions for which there is an energy gap between the ground state and first excited state. The charging energy leads to zero-point phase fluctuations which tend to destroy the long-range superconducting order. Therefore, in contrast to the classical junction, the quantum junction behaves as a kind of insulator. Experimentally, it is found that large clusters of grains are more stable against zero-point phase fluctuations than isolated pairs of grains. The intergrain Josephson coupling, as a competitive mechanism to the thermal fluctuations and Coulomb charging, may restore the long-range superconducting order. In what follows we will extend the single-junction model to treat an array of grains in a collective manner.

If we neglect the Coulomb interaction between charged neighboring grains, we may extend Eq. (18.6) to a three-dimensional ordered array as

$$\hat{H} = -2U\sum_i \hat{n}_i^2 + \sum_{ij} E_{ij}[1 - \cos(\Phi_i - \Phi_j)], \tag{18.7}$$

where $\hat{n}_i = -\mathrm{i}(\partial/\partial\Phi_i)$ is the operator describing the deviation from the average number of Cooper pairs on the ith grain and $E_{ij} = E_J\delta_{i,i\pm 1}$; we only consider the nearest neighbor coupling, since the long-range coupling is effectively cut off by the strong screening effect. For a more rigorous treatment, see, for example, Efetof (1980). From the temperature dependence of the critical current (see Sec. 30) it follows that

$$E_J(T) = \frac{\pi\hbar\Delta(T)}{4e^2R_n}\tanh\frac{\Delta(T)}{2k_BT}, \tag{18.8}$$

where R_n is the normal state resistance of the junction. Here we assume that the individual grains are large enough to sustain an energy gap $\Delta(T)$.

To calculate the transition temperature, T_c, we use a mean field approximation which replaces the second term of Eq. (18.7) by $-2zE_J\langle\cos\Phi\rangle\Sigma_i\cos\Phi_i$ (Simáneck 1979) to obtain

$$\hat{H}_{MF} = 2U\sum_i \hat{n}_i^2 - 2zE_J\langle\cos\Phi\rangle\sum_i\cos\Phi_i, \tag{18.9}$$

where z is the number of nearest neighbors in the array and $\langle\cos\Phi\rangle$ is the average order parameter of the array given by

$$\langle\cos\Phi\rangle = \frac{\sum_n \mathrm{e}^{-E_n/k_BT_c}\langle\psi_n|\cos\Phi|\psi_n\rangle}{\sum_n \mathrm{e}^{-E_n/k_BT_c}}; \tag{18.10}$$

here ψ_n and E_n are the eigenfunctions and eigenvalues of the following Schrödinger equation:

$$\left[2U\frac{\mathrm{d}^2}{\mathrm{d}\Phi^2} + 2zE_J\langle\cos\Phi\rangle\cos\Phi\right]\psi_n(\Phi) = -E_n\psi_n(\Phi). \tag{18.11}$$

Eq. (18.11) can be cast into the standard form of Mathieu's equation:

$$\mathrm{d}^2\psi_n/\mathrm{d}y^2 + [a_n - 2q\cos(2y)]\psi_n = 0, \tag{18.12}$$

where $y = \Phi/2$, the eigenvalues $a_n = 2E_n/U$, and $q = -2zE_J\langle\cos\Phi\rangle/U$. Near the global phase transition temperature, T_c, q is a small parameter. Both the eigenfunctions and eigenvalues, a_n, may be expanded in a power series in q. To obtain the eigenvalues we keep only the zeroth order in q while for eigenfunctions we restrict ourselves to expanding to order q. We keep only wavefunctions that are 2π periodic in the phases of the superconducting order parameter of the grains; i.e., we retain only those eigenstates with $n = 0, \pm 2, \pm 4, \ldots$. In terms of the parameters $x \equiv U/2k_BT_c$ and $\alpha \equiv zE_J/U$, from Eq. (18.10), the global mean field transition temperature, T_c, is determined by the following equation:

$$\alpha\frac{1 - \frac{5}{3}\mathrm{e}^{-4x} - 2\sum_{m=2}^{\infty}\mathrm{e}^{-(2m)^2x}/[(2m)^2-1]}{1 + 2\sum_{m=1}^{\infty}\mathrm{e}^{-(2m)^2x}} = 1, \tag{18.13}$$

where m is an integer. The phase diagram calculated from Eq. (18.13) is shown in Fig. 18.2. Note that the classical limit, $k_BT_c = zE_J$, is correctly obtained when $\alpha \to \infty$; T_c is depressed for finite α

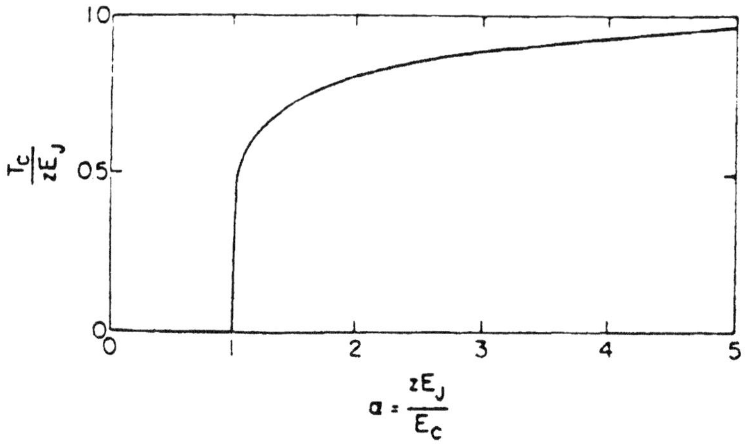

Figure 18.2 T_c/zE_J plotted as a function of $\alpha = zE_J/E_c$ from Eq. (18.13). Note that $T_c/zE_J \to 1$ as $\alpha \to \infty$ and $T_c/zE_J \to 0$ as $\alpha \to 1$. (After Imry and Strongin (1981).)

and vanishes when $\alpha \to 1$. Since α is directly proportional to the intergrain Josephson coupling, E_J, and hence inversely proportional to the normal state resistivity, ρ, the extremely fast drop of T_c to zero around $\alpha = 1$ is consistent with the sharp dependence of T_c on ρ observed experimentally.

19 Wave propagation in Josephson junctions, superlattices, and arrays

19.1 Wave propagation in a junction

In this section we derive the equations governing the propagation and generation of electromagnetic waves in a Josephson junction (Swihart 1961). The electric and magnetic fields in the bulk superconductor and insulator obey the Maxwell equations:[1]

$$\varepsilon \nabla \cdot \mathbf{E} = 4\pi\rho, \tag{19.1}$$

$$\nabla \cdot \mathbf{H} = 0, \tag{19.2}$$

$$\nabla \times \mathbf{E} = -\frac{\mu}{c}\frac{\partial \mathbf{H}}{\partial t}, \tag{19.3}$$

and

$$\nabla \times \mathbf{H} = \frac{4\pi}{c}\mathbf{J} + \frac{\varepsilon}{c}\frac{\partial \mathbf{E}}{\partial t}. \tag{19.4}$$

The metal charge fluctuations are screened out in a distance of the order of the Thomas–Fermi screening length, and we can safely set $\rho = 0$. Assuming no free charges in the insulators, we will set $\rho = 0$ everywhere in the bulk material.

The G–L expression for the current density applies in the superconductor

$$\mathbf{J} = \frac{e^*\hbar}{m^*}|\psi_0|^2\nabla\Phi - \frac{e^{*2}}{m^*c}|\psi_0|^2\mathbf{A}; \tag{19.5}$$

here $|\psi_0|$ and Φ are the magnitude and phase of the G–L order parameter.

The junction is depicted in Fig. 15.2. The z axis is normal to the junction and we will assume that the geometry and the external magnetic field are such that waves propagate only in the x direction. In this case the fields will be independent of y. At the interfaces between the superconducting and insulating layers we apply the usual Maxwell boundary conditions involving the continuity of the tangential **E** and **H** fields. In addition, the normal component of the current density through the insulator obeys the first Josephson relation,

$$\mathbf{J}_z = \mathbf{J}_m \sin \Delta\Phi, \tag{19.6}$$

1. We use uppercase **J** in this section to distinguish it from its Fourier transform which will be introduced shortly.

where $\mathbf{J}_m$ is a constant characterizing the junction. The time dependence of $\Delta\Phi$ is governed by the second Josephson relation:

$$\frac{\partial \Delta\Phi}{\partial t} = \frac{|e^*|}{\hbar} V. \tag{19.7}$$

Here V is the voltage across the junction,

$$V = -\int_{-a/2}^{a/2} E_z(x, z, t) \mathrm{d}z, \tag{19.8}$$

and $\Delta\Phi$ is the gauge-invariant phase difference between opposite sides of the junction:

$$\Delta\Phi(x, t) \equiv \Phi[x, (a/2), t] - \Phi[x, -(a/2), t] - \frac{e^*}{\hbar c}\int_{-a/2}^{+a/2} A_z(x, z, t)\mathrm{d}z. \tag{19.9}$$

We assume for the present that waves propagate only in the x direction and seek a solution where all magnetic fields, static and dynamic, have the form which satisfies Eq. (19.2):

$$\mathbf{H} = H(x, z, t)\hat{\mathbf{y}}. \tag{19.10}$$

This allows us to choose a gauge where $\mathbf{A} = \mathbf{A}(x, z, t)\hat{\mathbf{x}}$. Note that this eliminates the gauge term integral from Eq. (19.9).

It is convenient to perform a Fourier transform on the variables x and t entering the various fields:

$$G(x, z, t) = \int \mathrm{d}k\mathrm{d}\omega \mathrm{e}^{\mathrm{i}kx - \mathrm{i}\omega t} g(k, z, \omega). \tag{19.11}$$

We first consider the *fields in the insulator*. In the insulator, the electric field has the form

$$\mathbf{E} = E_x(x, z, t)\hat{\mathbf{x}} + E_z(x, z, t)\hat{\mathbf{z}}. \tag{19.12}$$

In terms of the transformed variables, the x component of Eq. (19.4) and Eq. (19.3) yields

$$e_x = -\frac{\mathrm{i}c}{\varepsilon_\mathrm{I}\omega}\frac{\partial h}{\partial z} \tag{19.13}$$

and

$$e_z = -\frac{\omega}{ck}\left(h + \frac{c^2}{\varepsilon_\mathrm{I}\omega^2}\frac{\partial^2 h}{\partial z^2}\right), \tag{19.14}$$

where ε_I is the dielectric constant in the insulator. Eqs. (19.13) and (19.14) together with Eq. (19.1) then yield

$$\frac{\partial^3 h}{\partial z^3} = \left(k^2 - \frac{\varepsilon_\mathrm{I}\omega^2}{c^2}\right)\frac{\partial h}{\partial z},$$

which has the solution

$$h = h_0(k, \omega) + h_+(k, \omega)\mathrm{e}^{\gamma(z - z_c)} + h_-(k, \omega)\mathrm{e}^{-\gamma(z - z_c)}, \tag{19.15}$$

where $\gamma^2 \equiv k^2 - \varepsilon_\mathrm{I}\omega^2/c^2$ and z_c is an arbitrary constant that we shall choose to be the center of the insulator layer. Combining Eqs. (19.14) and (19.15) with (19.6) and the z component of (19.4), we have

$$\int \mathrm{d}k\mathrm{d}\omega \mathrm{e}^{ikx - i\omega t}\frac{\gamma^2}{ik}h_0(k,\omega) = -(4\pi/c)J_{\mathrm{m}} \sin \Delta\Phi(x,t). \tag{19.16}$$

Combining Eqs. (19.14) and (19.15) with (19.7) and (19.8) gives

$$\frac{\phi_0}{2\pi c}\frac{\partial \Delta\Phi}{\partial t} = \int \mathrm{d}k\mathrm{d}\omega \mathrm{e}^{ikx - i\omega t}\left(-\frac{\omega}{ck}\right)\left\{ah_0 + \frac{2c^2k^2}{\varepsilon_{\mathrm{I}}\omega^2}\frac{\sinh[\gamma(a/2)]}{\gamma}(h_+ + h_-)\right\}, \tag{19.17}$$

where $\phi_0 \equiv hc/e^*$ is the flux quantum.

We now examine the *fields in the superconductor*. Consistent with Eqs. (19.5), (19.6), and (19.7), we write the electric field as

$$\mathbf{E} = \frac{4\pi\lambda_{\mathrm{L}}^2}{c^2}\frac{\partial}{\partial t}\mathbf{J}, \tag{19.18}$$

which is London's equation; here $1/\lambda_{\mathrm{L}}^2 \equiv 4\pi e^{*2}|\psi_0|^2/m^*c^2$. Inserting Eq. (19.18) into the fourth Maxwell equation yields

$$\nabla \times \mathbf{H} = \frac{4\pi}{c}\mathbf{J} + \frac{\varepsilon_{\mathrm{s}}}{c}\frac{4\pi\lambda_{\mathrm{L}}^2}{c^2}\frac{\partial^2 \mathbf{J}}{\partial t^2},$$

and on Fourier transforming

$$j_x = -\frac{c}{4\pi}\frac{1}{1 - \varepsilon_{\mathrm{s}}\omega^2\lambda_{\mathrm{L}}^2/c^2}\frac{\partial h}{\partial z} \tag{19.19a}$$

and

$$j_z = \frac{ikc}{4\pi}\frac{1}{1 - \varepsilon_{\mathrm{s}}\omega^2\lambda_{\mathrm{L}}^2/c^2}h. \tag{19.19b}$$

In what follows we will require the phase variation which is obtained by integrating Eq. (19.5). As noted in Eq. (19.10), we choose **H** parallel to $\hat{\mathbf{y}}$ and the gauge $\mathbf{A} = A\hat{\mathbf{x}}$. This yields

$$A(x,z,t) = \int H(x,z,t)\mathrm{d}z. \tag{19.20}$$

Combining Eqs. (19.5), (19.19), and (19.20) yields

$$\frac{\phi_0}{2\pi}\frac{\partial \Phi(x,z,t)}{\partial z} = \int \mathrm{d}\omega\mathrm{d}k \mathrm{e}^{ikx - i\omega t} ik\overline{\lambda}_{\mathrm{L}}^2 h(k,z,\omega) \tag{19.21a}$$

and

$$\frac{\phi_0}{2\pi}\frac{\partial \Phi(x,z,t)}{\partial x} = -\int \mathrm{d}\omega\mathrm{d}k \mathrm{e}^{ikx - i\omega t}\left(\overline{\lambda}_{\mathrm{L}}^2\frac{\partial h(k,z,\omega)}{\partial z} - \int_0^z h(k,z,\omega)\mathrm{d}z\right), \tag{19.21b}$$

where $\bar{\lambda}_L^2 \equiv \lambda_L^2/(1 - \varepsilon_s \lambda_L^2 \omega^2/c^2)$. Requiring that the cross partial derivatives of Φ with respect to x and z be equal yields

$$\frac{\partial^2 h}{\partial z^2} - k^2 h = \frac{1}{\bar{\lambda}_L^2} h, \tag{19.22}$$

which has the general solution

$$h(k, z, \omega) = h_>(k, \omega) e^{-\Gamma(z - z_c)} + h_<(k, \omega) e^{\Gamma(z - z_c)}, \tag{19.23}$$

where

$$\Gamma^2 \equiv k^2 + 1/\bar{\lambda}_L^2 \tag{19.24}$$

and an origin shift z_c has been included for later convenience. The electric fields follow from the third Maxwell equation and are given by

$$e_x = -(i\omega \bar{\lambda}_L^2 \Gamma/c)(h_> e^{-\Gamma(z - z_c)} - h_< e^{\Gamma(z - z_c)}) \tag{19.25a}$$

and

$$e_z = -(i\omega \bar{\lambda}_L^2/c) ik (h_> e^{-\Gamma(z - z_c)} + h_< e^{\Gamma(z - z_c)}). \tag{19.25b}$$

Having derived the general form of the field amplitudes in the insulators and superconductors, we must now satisfy the boundary conditions at the two interfaces for the sandwich geometry of Fig. 15.2. We fix the origin by setting $z_c = 0$ in the expressions for the insulator. The requirement that the field amplitudes be finite at $z = \pm\infty$ restricts the allowed solutions in the superconductor to

$$h(k, z, \omega) = \begin{cases} h_> e^{-\Gamma z}; & z > a/2 \\ h_< e^{\Gamma z}; & z < -a/2. \end{cases} \tag{19.26}$$

Matching h (Eqs. (19.15) and (19.26)) and e_x (Eqs. (19.13) and (19.25a)) for the interfaces at $z = \pm a/2$ yields the conditions

$$h_> e^{-\Gamma(a/2)} = h_0 + h_+ e^{\gamma(a/2)} + h_- e^{-\gamma(a/2)}, \tag{19.27a}$$

$$h_< e^{-\Gamma(a/2)} = h_0 + h_+ e^{-\gamma(a/2)} + h_- e^{\gamma(a/2)}, \tag{19.27b}$$

$$\frac{i\omega \bar{\lambda}_L^2 \Gamma}{c} h_> e^{-\Gamma(a/2)} = \frac{ic\gamma}{\varepsilon_I \omega}(h_+ e^{\gamma(a/2)} - h_- e^{-\gamma(a/2)}) \tag{19.28a}$$

and

$$\frac{i\omega \bar{\lambda}_L^2 \Gamma}{c} h_< e^{-\Gamma(a/2)} = -\frac{ic\gamma}{\varepsilon_I \omega}(h_+ e^{-\gamma(a/2)} - h_- e^{\gamma(a/2)}). \tag{19.28b}$$

Solving these four equations in terms of h_0 yields

$$h_+ = h_- = \frac{-1}{(1 - p)e^{\gamma(a/2)} + (1 + p)e^{-\gamma(a/2)}} h_0, \tag{19.29}$$

and

$$h_> = h_< = \frac{-2p e^{\Gamma(a/2)} \sinh[\gamma(a/2)]}{(1 - p)e^{\gamma(a/2)} + (1 + p)e^{-\gamma(a/2)}} h_0, \tag{19.30}$$

where $p \equiv c^2\gamma/\varepsilon_I\omega^2\bar{\lambda}_L^2\Gamma$.

The value of h_0 is fixed by Eq. (19.16); in order to determine this value we must evaluate the phase difference $\Delta\Phi$ as defined by Eq. (19.9). The phase itself is obtained by integrating Eq. (19.21b). Taking the phase as zero at the point $x = 0, z = 0$, and being sure to include the contribution arising from the magnetic field in the insulator (arising from the integral in the second term), we obtain in the region $z \geq a/2$:

$$\frac{\phi_0}{2\pi}\Phi(x,z,t) = \int dk d\omega e^{ikx - i\omega t}\frac{1}{ik}\left[\bar{\lambda}_L^2\Gamma h_> e^{-\Gamma z} - \frac{1}{\Gamma}h_>(e^{-\Gamma z} - e^{-\Gamma a/2}) + \frac{a}{2}h_0 + h_+\frac{(e^{\gamma(a/2)} - 1)}{\gamma} - h_-\frac{(e^{-\gamma(a/2)} - 1)}{\gamma}\right], \tag{19.31a}$$

and similarly for $z \leq -a/2$,

$$\frac{\phi_0}{2\pi}\Phi(x,z,t) = \int dk d\omega e^{ikx - i\omega t}\frac{1}{ik}\left[-\bar{\lambda}_L^2\Gamma h_< e^{\Gamma z} + \frac{1}{\Gamma}h_<(e^{\Gamma z} - e^{-\Gamma a/2}) - \frac{a}{2}h_0 + h_+\frac{(e^{-\gamma(a/2)} - 1)}{\gamma} - h_-\frac{(e^{\gamma(a/2)} - 1)}{\gamma}\right]. \tag{19.31b}$$

The phase difference $\Delta\Phi$ defined in Eq. (19.9) is thus given by

$$\frac{\phi_0}{2\pi}\Delta\Phi = \int dk d\omega e^{ikx - i\omega t}\frac{1}{ik}\left\{\bar{\lambda}_L^2\Gamma e^{-\Gamma(a/2)}(h_> + h_<) + \frac{2\sinh[\gamma(a/2)]}{\gamma}(h_+ + h_-) + a h_0\right\}. \tag{19.32}$$

Substituting $h_>$, $h_<$, h_+, and h_- from Eqs. (19.29) and (19.30) into Eq. (19.32) we have

$$\frac{\phi_0}{2\pi}\Delta\Phi = \int dk d\omega e^{ikx - i\omega t}\frac{h_0}{ik} \times\left\{a - \frac{k^2}{\varepsilon_I\omega^2\gamma}\frac{\sinh[\gamma(a/2)]}{(1 + p)e^{\gamma(a/2)} + (1 - p)e^{-\gamma(a/2)}}\right\}. \tag{19.33}$$

Eqs. (19.33) and (19.16) form a set of simultaneous equations for $\Delta\Phi$ and h_0. If we construct E_z from Eq. (19.14) and V from Eq. (19.8), Eq. (19.7) for the time variations of $\Delta\Phi$ is satisfied exactly.

For cases of physical interest, these equations may be considerably simplified by noting $\gamma a << 1$ and $\Gamma^2 \cong 1/\bar{\lambda}_L^2 \cong 1/\lambda_L^2$, in which case Eq. (19.33) becomes

$$\frac{\phi_0}{2\pi}\Delta\Phi = \int dk d\omega e^{ikx - i\omega t}\frac{h_0}{ik} \times\left\{(a + 2\lambda_L)\gamma^2\Big/\left[k^2 - \left(1 + \frac{2\lambda_L}{a}\right)\frac{\varepsilon_I\omega^2}{c^2}\right]\right\}. \tag{19.34}$$

Combining this with Eq. (19.16) yields the Kulik equation (Kulik 1965a,b, 1966)

$$\left(\nabla_\perp^2 - \frac{1}{\bar{c}^2}\frac{\partial^2}{\partial t^2}\right)\Delta\Phi = \frac{1}{\lambda_J^2}\sin\Delta\Phi, \tag{19.35a}$$

where

$$\frac{1}{\bar{c}^2} \equiv \left(1 + \frac{2\lambda_L}{a}\right)\frac{\varepsilon_I}{c^2} \tag{19.35b}$$

and

$$\frac{1}{\lambda_J^2} \equiv \frac{8\pi^2 J_m}{c\phi_0}(a + 2\lambda_L), \tag{19.35c}$$

and we have allowed for wave propagation for any direction in the x–y plane by writing $\partial^2/\partial x^2 \to \nabla_\perp^2$ where $\nabla_\perp^2 = \partial^2/\partial x^2 + \partial^2/\partial y^2$. Note that since $\lambda_L >> a, \bar{c} << c$, Eq. (19.35a) will be recognized as the sine-Gordon equation which possesses soliton solutions. In the limit where the phase variations $\Delta\Phi << 1$, Eq. (19.35a) may be linearized as

$$\left(\nabla_\perp^2 - \frac{1}{\bar{c}^2}\frac{\partial^2}{\partial t^2}\right)\Delta\Phi = \frac{1}{\lambda_J^2}\Delta\Phi, \tag{19.36}$$

which is the Klein–Gordon equation. Plane-wave solutions of the form $\Delta\Phi = \mathrm{Re}\Delta\Phi_0 e^{ikx - i\omega t}$ have the dispersion relation

$$\omega^2 = \omega_0^2 + \bar{c}^2 k^2, \tag{19.37}$$

where $\omega_0^2 = \bar{c}^2/\lambda_J^2$ is referred to as the 'Josephson plasma frequency' and defines a frequency below which small amplitude electromagnetic waves will not propagate in the junction.

Electromagnetic waves may be excited in a junction via oscillating Josephson supercurrents associated with a dc voltage applied to the junction. The amplitude of such waves will be largest if two conditions are satisfied: (i) The first condition is that the length of the junction, L, be such that an integral number of half wavelengths fit into the junction, which then becomes a cavity; i.e., $L = n(\lambda/2)$ and the resulting resonant frequencies are given by $\omega_n = (n\pi\bar{c}/L)$. Combining this with the Josephson relation $2eV = \hbar\omega$ we obtain $V_n = (h/2e)(\bar{c}/2L)n$. (ii) The second condition is that the phase of the oscillating Josephson supercurrent vary across the junction. The largest coupling occurs when the supercurrent phase matches that of the electromagnetic wave. The second condition can be achieved by applying a magnetic field in the plane of the junction and perpendicular to the direction of wave propagation. Above H_{c1} for the junction there will then be a linear array of Josephson vortices (long junction limit). The number of such vortices is given by ϕ/ϕ_0 where ϕ is the flux in the junction, $(2\lambda_L + a)LH$. Each fluxon is associated with a phase shift of 2π and the length over which a 2π phase shift occurs is then $\phi_0/[(2\lambda_L + a)H]$. The amplitude of the excited electromagnetic wave will be largest if this length is equal to the electromagnetic wavelength; i.e., if $\lambda = \phi_0/(2\lambda_L + a)H$.

Cavity self-resonances of the above kind were first observed by Fiske (1964) as steps in the I–V characteristics of a junction as the current in the junction was swept; they are referred to as Fiske steps. For a more quantitative discussion, including the effect of the cavity Q and other parameters on the step size, see Barone and Paterno (1982) and the theoretical and experimental references therein.

We now give a brief discussion of some solutions to the sine-Gordon equation. If we measure length in units of λ_J and time in units of $\lambda_J/\bar{c}$, then we may write the one-dimensional form of Eq. (19.35a) in the dimensionless form

$$\frac{\partial^2 \Phi}{\partial x^2} - \frac{\partial^2 \Phi}{\partial t^2} - \sin \Phi = 0. \tag{19.38}$$

For simplicity, we have denoted the gauge-invariant phase difference as Φ. We seek solutions of the form $\Phi(x, t) = \Phi(x - ut) = \Phi(\xi)$ where $\xi \equiv x - ut$. The form chosen is that of a shape and amplitude preserving pulse propagating with a constant velocity, u; such solutions are called solitary waves (or solitons if multiple traveling wave solutions exist). The constant u is a dimensionless velocity and may be written $u = v/\bar{c}$ where v is the pulse velocity in the original variables. Inserting this form into Eq. (19.38) we obtain the total differential equation

$$(1 - u^2)\frac{\mathrm{d}^2 \Phi}{\mathrm{d}\xi^2} = \sin \Phi. \tag{19.39}$$

A first integral of (19.39) is obtained by multiplying by $\mathrm{d}\Phi/\mathrm{d}\xi$:

$$\frac{1}{2}(1 - u^2)\frac{\mathrm{d}}{\mathrm{d}\xi}\left(\frac{\mathrm{d}\Phi}{\mathrm{d}\xi}\right)^2 = -\frac{\mathrm{d}}{\mathrm{d}\xi}(\cos \Phi)$$

or

$$\frac{\mathrm{d}\Phi}{\mathrm{d}\xi} = \left[\frac{2(E - \cos \Phi)}{1 - u^2}\right]^{1/2}, \tag{19.40}$$

where E is a constant of integration. This equation has the following second integral for the special case $E = 1$

$$\Phi = 4 \tan^{-1}\left\{\exp\left[\pm \frac{(x - ut)}{(1 - u^2)^{\frac{1}{2}}}\right]\right\}; \qquad |u| < 1. \tag{19.41}$$

(This solution may be verified by substitution into (19.39).) This is the soliton solution (another solution with $E = -1$, corresponding to $\Phi(x \to \pm \infty) = \pi$, is unstable to a global rotation by π). Eq. (19.41) describes a behavior where the phase increases $(+)$ or decreases $(-)$ by 2π as x passes from $-\infty$ to $+\infty$; these two cases may be referred to as a kink (soliton) or antikink (antisoliton), respectively.

The solutions for $E \neq 1$ are periodic; for the case $E > 1$ and $|u| < 1$

$$\Phi = 2 \sin^{-1}\left[\pm \mathrm{cn}\left(\frac{x - ut}{k(1 - u^2)^{1/2}}\middle| k\right)\right] \tag{19.42}$$

where $\mathrm{cn}(z)$ is a Jacobi elliptic function of modulus k, where $0 < k < 1$. The limit $k = 1$ is the single-soliton solution. For $-1 < E < 1$ and $|u| > 1$ the solution is

$$\Phi = 2 \sin^{-1}\left[k \mathrm{sn}\left(\frac{x - ut}{(u^2 - 1)^{\frac{1}{2}}}\middle| k\right)\right] \tag{19.43}$$

where the modulus $k > 1$ and $\mathrm{sn}(z)$ is another Jacobi elliptic function.

In certain limits we can obtain solutions which have simple forms. We first consider the case where the phase increases nearly linearly with $x - ut$. We seek a solution of the form $\Phi = \Phi^{(0)}(x - ut) + \Phi^{(1)}(x - ut) + \ldots$, where $\Phi^{(0)} = k(x - ut)$ and $\Phi^{(1)}$ is a perturbation. Inserting this form in Eq. (19.39) yields

$$(1 - u^2)\frac{d^2}{d\xi^2}\Phi^{(1)}(\xi) = \sin[k\xi + \Phi^{(1)}(\xi)].$$

If we assume $\Phi^{(1)} << 1$ and neglect the $\Phi^{(1)}$ term in the argument of the sine function (to lowest order) we obtain

$$\Phi^{(1)} = -\frac{1}{(1 - u^2)k^2}\sin[k(x - ut)]$$

or

$$\Phi(x - ut) = k(x - ut) - \frac{1}{k^2(1 - u^2)}\sin[k(x - ut)]. \quad (19.44)$$

By differentiating this form and comparing the result with (19.40) we obtain the constant E (to the same accuracy) as $E = (1 - u^2)k^2/2$. This solution corresponds to an array of flux quanta moving with velocity u, or $u\bar{c}$ in unscaled units. The limit $u \to 0$ corresponds to an array of uniformly distributed static flux quanta with a spatial period $\lambda = 2\pi/k$. These forms are limiting cases of the exact solution (19.42).

Another simple form arises in the limit when the variation of the phase with $x - ut$ is much less than 2π. We may then expand the argument of the sine in Eq. (19.39) yielding the equation

$$(1 - u^2)\frac{d^2\Phi}{d\xi^2} = \Phi \quad (19.45)$$

having solutions of the form

$$\Phi = \Phi_0 e^{-\kappa(x - ut)}, \qquad x > 0; \qquad u^2 < 1 \quad (19.46a)$$

or

$$\Phi = \Phi_0 \sin[k(x - ut) + \theta], \qquad u^2 > 1, \quad (19.46b)$$

where $\kappa = 1/(1 - u^2)^{1/2}$ and $k = 1/(u^2 - 1)^{1/2}$. By differentiating these forms and again comparing them with (19.40) we obtain $E = 1 - \Phi_0^2/2$ (k and E being independent constants in this limit). These solutions correspond to low amplitude waves (referred to as plasmons). They are limiting cases of the exact solution (19.43). The form (19.46a) corresponds to flux exclusion from the junction. The static limit $u \to 0$, leading to $\kappa = 1$, corresponds to an exponential decay of the form e^{-x/λ_J}, in unscaled units. For this reason λ_J is called the Josephson penetration depth.

Up to this point, our discussion of the long Josephson junction has ignored the effects of dissipation as well as an externally applied current bias. Referring to our earlier discussion of the RSJ model, it is clear that these two effects may be incorporated by altering Eq. (19.38) to have the form

$$\Phi'' - \ddot{\Phi} - \beta\dot{\Phi} - \alpha = \sin\Phi, \quad (19.47)$$

where we have measured length in units of λ_J and β and α are as in Eq. (15.47). (An additional dissipative term of the form $\gamma\dot{\Phi}''$ resulting from current flow parallel to the junction may also be included but will be ignored in our discussion.)

An exact solution to (19.47) is not known. A scheme for including dissipation and a bias current perturbatively has been given by McLaughlin and Scott (1977, 1978); we will limit ourselves to expressions which incorporate these effects into the limiting forms (19.44) and (19.46). We also restrict to the case where the damping is small. We seek a perturbative solution of the form $\Phi(x,t) = \Phi^{(0)}(x,t) + \Phi^{(1)}(x,t)\ldots$, where $\Phi^{(0)} = k(x - ut)$. Inserting this in (19.47) and assuming a traveling wave form $\Phi^{(1)} = \Phi^{(1)}(x - ut) = \Phi^{(1)}(\xi)$ yields, in first order,

$$(1 - u^2)\frac{\partial^2\Phi^{(1)}}{\partial\xi^2} - \sin k\xi = -\beta k u + \alpha. \tag{19.48}$$

In steady state the right-hand side vanishes (in our pendulum analogy, the applied torque, α, is compensated by the frictional torque, $\beta k u$); i.e.,

$$u = \frac{\alpha}{\beta k} \tag{19.49}$$

and therefore u and k are no longer independent parameters. Note we recover the uniform junction precession frequency $\Omega = \alpha/\beta$ ($\alpha >> 1$ in Eq. (15.47)) by taking the limit $k \to 0, u \to \infty$, with $ku = \Omega$. The remainder of (19.48) can be integrated to yield

$$\Phi(x,t) = k(x - ut) - \frac{1}{k^2(1 - u^2)}\sin[k(x - ut)], \tag{19.50}$$

which is the same as (19.44). In the opposite limit of small phase variations we have the linear equation (with $\alpha << 1$)

$$\Phi'' - \ddot{\Phi} - \beta\dot{\Phi} - \alpha - \Phi = 0. \tag{19.51}$$

The solutions of (19.51) are damped plasmons

$$\Phi(x,t) = \Phi_0 e^{-t/\tau}\sin[k(x - ut) + \theta] - \alpha, \tag{19.52}$$

where $\tau \cong \beta^{-1}$ and $k \cong (u^2 - 1)^{-1/2}$.

We now turn to the behavior of a soliton in the presence of dissipation and a current bias. In the steady state approximation, the damping is governed by $\beta\dot{\Phi} + \alpha = 0$. Writing $\dot{\Phi} = -u\Phi'$ we have from (19.40) (with $E = 1$)

$$\frac{2(1 - \cos\Phi)}{1 - u^2} = \left(\frac{\alpha}{\beta u}\right)^2$$

or

$$1 - \cos\Phi = \frac{\alpha^2(1 - u^2)}{2\beta^2 u^2}. \tag{19.53}$$

The left-hand side of (19.53) is nonzero only in the vicinity of the kink where Φ increases or decreases by 2π. The width of the kink is, from (19.41), of order $(1 - u^2)^{1/2}$. We average (19.53) over a junction of length x_0 to obtain

$$C(1 - u^2)^{1/2} = \frac{\alpha^2(1 - u^2)x_0}{2\beta^2 u^2},$$

where the constant C is given by

$$C \equiv \int_{-\infty}^{+\infty} \{1 - \cos[4\tan^{-1}(e^{\pm\xi})]\} d\xi.$$

The soliton velocity (in the limit of small u) is given by

$$u = \frac{\alpha}{\beta}\left(\frac{x_0}{2nC}\right)^{1/2},$$

where we included a factor n to account for the possibility of multiple soliton propagation.

19.2 Wave propagation in a superlattice

In this section we consider an infinite set of superconductor/insulator layers of widths b and a, respectively (Auvil and Ketterson 1987). We look for a solution to our basic equations which is periodic in the z direction with period $a + b$. To produce $E_z(x, z, t)$ in the insulators with this periodicity, the potential biasing of the superconducting layers must be a monotonic voltage sequence $0, V, 2V, 3V, \ldots$. Fig. 19.1 shows this periodic structure and the expected wave propagation direction in the insulating layers.[2]

We utilize the same equations and solutions as in Subsec. 19.1; however, we must keep both pieces of h (Eq. (19.23)) in each superconducting layer. Also, where z_c occurs in an equation involving a superconductor, we interpret it as the center of that respective layer, and similarly for z_c in an insulator equation. For example, $z - z_c$ at the top of a superconducting layer is $b/2$, and $z - z_c$ at the bottom of an insulating layer is $-(a/2)$.

Using Eqs. (19.13), (19.15), (19.23), and (19.25a) and matching at the top of any insulating layer and the bottom of the next superconducting layer yields the conditions

$$h_> e^{\Gamma(b/2)} + h_< e^{-\Gamma(b/2)} = h_0 + h_+ e^{\gamma(a/2)} + h_- e^{-\gamma(a/2)} \tag{19.54a}$$

and

$$\frac{-i\omega\bar{\lambda}_L^2\Gamma}{c}(h_> e^{\Gamma(b/2)} - h_< e^{-\Gamma(b/2)}) = \frac{-ic\gamma}{\varepsilon_I\omega}(h_+ e^{\gamma(a/2)} - h_- e^{-\gamma(a/2)}). \tag{19.54b}$$

Using our periodic boundary conditions, $f(z + a + b) = f(z)$, we now match the bottom of an insulating layer to the top of the next superconductor layer. For example, at $z = -(a/2)$, $f(a/2 + b) = f[-(a/2)]$. Using the same equations, the results are

$$h_> e^{-\Gamma(b/2)} + h_< e^{\Gamma(b/2)} = h_0 + h_+ e^{-\gamma(a/2)} + h_- e^{\gamma(a/2)} \tag{19.55a}$$

and

$$\frac{-i\omega\bar{\lambda}_L^2\Gamma}{c}(h_> e^{-\Gamma(b/2)} - h_< e^{\Gamma(b/2)}) = \frac{-ic\gamma}{\varepsilon_I\omega}(h_+ e^{-\gamma(a/2)} - h_- e^{\gamma(a/2)}). \tag{19.55b}$$

Solving these four equations in terms of h_0 yields

2. Note that the most general solution to this problem would involve a Bloch wave vector with a component k_z along the z direction.

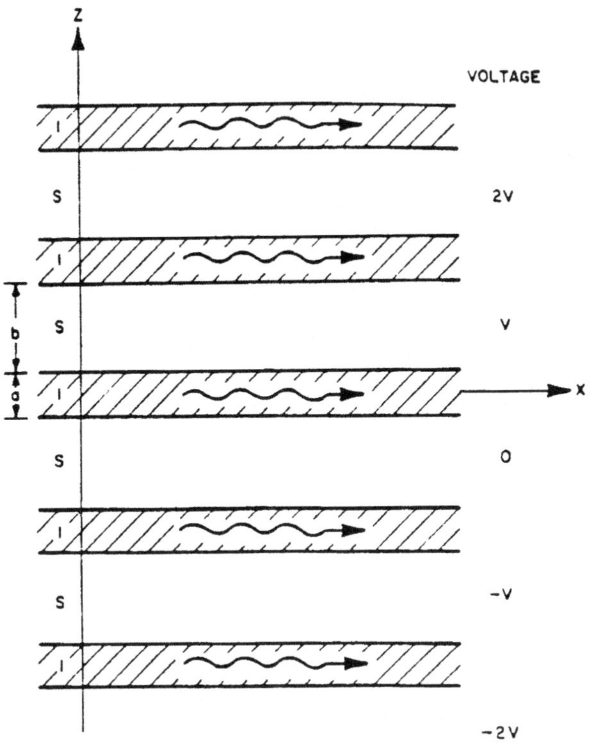

Figure 19.1 Schematic diagram of a series-driven superconductor–insulator superlattice. The wavy lines show the direction of the waves generated via the Josephson effect.

$$h_+ = h_- = \frac{\sinh[\Gamma(b/2)]}{2\{p\sinh[\gamma(a/2)]\cosh[\Gamma(b/2)] - \cosh[\gamma(a/2)]\sinh[\Gamma(b/2)]\}} h_0 \tag{19.56}$$

and

$$h_> = h_< = \frac{p\sinh[\gamma(a/2)]}{2\{p\sinh[\gamma(a/2)]\cosh[\Gamma(b/2)] - \cosh[\gamma(a/2)]\sinh[\Gamma(b/2)]\}} h_0, \tag{19.57}$$

where, as in Subsec. 19.1, $p = c^2\gamma/\varepsilon_I\omega^2\bar{\lambda}_L^2\Gamma$. If we take into account the shift in z_c for the superconducting layers as used in this section, Eqs. (19.56) and (19.57) reduce exactly to Eqs. (19.29) and (19.30) in the limit that b tends to infinity.

To complete our solution, we must now construct $\Phi(x, z, t)$ and $\Delta\Phi(x, t)$. We again choose the phase to be zero at $x = 0$ and $z = 0$. The actual expressions for $\Phi(x, z, t)$ obtained by integrating Eqs. (19.21a) and (19.21b) are quite lengthy, but because all quantities are periodic in z, $\Delta\Phi(x, t)$ is independent of the insulating layer. A straightforward calculation yields

$$\frac{\phi_0}{2\pi}\Delta\Phi(x, t) = \int dk d\omega e^{ikx - i\omega t}\frac{1}{ik}\left\{2\bar{\lambda}_L^2\Gamma\sinh[\Gamma(b/2)](h_> + h_<) + \frac{2\sinh[\gamma(a/2)]}{\gamma}(h_+ + h_-) + ah_0\right\}. \tag{19.58}$$

Substituting $h_>$, $h_<$, h_+, and h_- from Eqs. (19.56) and (19.57) into Eq. (19.58) we have

$$\frac{\phi_0}{2\pi}\Delta\Phi(x,t) = \int dk d\omega e^{ikx - i\omega t}\frac{h_0}{ik}$$

$$\times\left(a + \frac{c^2k^2}{\varepsilon_I\omega^2\gamma}\frac{\sinh[\gamma(a/2)]\sinh[\Gamma(b/2)]}{\varepsilon_I\omega^2\gamma\{p\sinh[\gamma(a/2)]\cosh[\Gamma(b/2)] - \cosh[\gamma(a/2)]\sinh[\Gamma(b/2)]\}}\right). \quad (19.59)$$

As in Subsec. 19.1 we construct E_z and V. Eq. (19.7) for the time variation of $\Delta\Phi$ is then satisfied exactly, so that now Eqs. (19.59) and (19.16) form a simultaneous set of equations for $\Delta\Phi$ and h_0.

For cases of physical interest, we simplify by assuming that $\gamma a << 1$ and $\Gamma b << 1$. It follows that $\Gamma \sim \bar{\lambda}_L^{-1}$ and $\bar{\lambda}_L \sim \lambda_L$, in which case $\Delta\Phi$ simplifies to

$$\frac{\phi_0}{2\pi}\Delta\Phi(x,t) = \int dk d\omega e^{ikx - i\omega t}\frac{h_0}{ik}\left\{(a+b)\gamma^2\Big/\left[k^2 - \left(1 + \frac{b}{a}\right)\frac{\varepsilon_I\omega^2}{c^2}\right]\right\}. \quad (19.60)$$

Combining this with Eq. (19.16) yields the Kulik equation (Eq. (19.35a). In place of Eqs. (19.35b) and (19.35c) we now have

$$\frac{1}{\bar{c}^2} = \left(1 + \frac{b}{a}\right)\frac{\varepsilon_I}{c^2} \quad (19.61)$$

and

$$\frac{1}{\lambda_J^2} = (8\pi^2 J_m/c\phi_0)(a+b); \quad (19.62)$$

note that the phase velocity involves the ratio b/a instead of λ_L/a.

In this same approximation we can relate $\Delta\Phi$ to H in a simple way as follows (note that H is independent of z in this approximation):

$$\frac{\partial\Delta\Phi}{\partial x} = \frac{2\pi(a+b)}{\phi_0}H \quad (19.63)$$

and $\Delta\Phi$ is exactly related to E_z in Eq. (19.7)

$$\frac{\partial\Delta\Phi}{\partial t} = \frac{2\pi c}{\phi_0}\int_{\text{bottom}}^{\text{top}} E_z dz. \quad (19.64)$$

For constant external H_0 and increasing voltages $0, V, 2V, \ldots$, a zeroth order solution for $\Delta\Phi$ is

$$\Delta\Phi^{(0)}(x,t) = \left(\frac{2\pi(a+b)}{\phi_0}H_0\right)x - \left(\frac{2\pi c}{\phi_0}V\right)t, \quad (19.65)$$

which, when substituted into Eq. (19.34), yields traveling waves together with harmonics which resonate at

$$\frac{2\pi(a+b)}{\phi_0}H_0 = \frac{1}{\bar{c}}\frac{2\pi c}{\phi_0}V; \quad (19.66)$$

that is

$$H_0 = [\varepsilon_I/a(a+b)]^{1/2}.$$

Of course, we expect that relaxing our assumptions of a perfect superconductor and a perfect insulator will give this response a width and will damp the traveling wave.

In the same spirit we may also consider the parallel biasing case,[3] i.e., to bias the superconducting layers alternately as $0, V, 0, V, 0, \ldots$. This produces an E_z which is alternate in the plus and minus z direction in the insulating layers. Such a structure has periodicity $2(a+b)$, rather than $a+b$ as in the series case. The details of the analysis have been described by Auvil and Ketterson (1987).

As mentioned above, the wave propagation and generation in a Josephson coupled superconductor/insulator superlattice are governed by the same form of the sine-Gordon equation but with a different phase velocity (Eq. (19.61)) and Josephson penetration depth (Eq. (19.62)). The efficient production of submillimeter microwave radiation by a conventional Josephson junction is severely limited by the fact that there is an enormous mismatch between the impedances of the junction and the free space. The electrical impedance of a medium is directly proportional to the phase velocity, which is given by (19.35b) for a single junction and by Eq. (19.61) for a superconductor/insulator superlattice, respectively. As long as $b << 2\lambda_L$, the phase velocity in the superconductor/insulator superlattice will be greatly increased, leading to an enhanced output of Josephson radiation. Furthermore, the area of the radiating surface scales with the number of series junctions in the superlattice, which further increases the radiated power. For the case of phase-locked operation, the field amplitude scales with N and the radiation power as N^2, where N is the number of coherently radiating junctions. Hence superconductor/insulator superlattice Josephson radiators should have greatly increased power output.

The fluxon dynamics in Josephson coupled superconductor/insulator superlattices is governed by the nonlinear sine-Gordon equation. Proceeding in the same way as in Subsec. 19.1, we may seek a single fluxon solution in the static case; this yields a lower critical field, H_{c1}, of the superlattice given by

$$H_{c1} = 8\left(\frac{\hbar J_m}{2\pi e d}\right)^{1/2}, \tag{19.67}$$

where the magnetic length $d = a + b$. For a single junction, $d = a + 2\lambda_L$. Therefore, the critical field of a superconductor/insulator superlattice is higher than that of a single junction, provided that $b < 2\lambda_L$.

For a small single junction, the maximum supercurrent as a function of a weak applied magnetic field is given by

$$I_c(\phi) = I_c(0)\left|\frac{\sin(\pi\phi/\phi_0)}{\pi\phi/\phi_0}\right|,$$

where ϕ_0 is the flux quantum, $\phi = dwH$ is the flux linking the junction, and w the width of the junction. For a superconductor/insulator superlattice, if $b < 2\lambda_L$, then $d = a + b$ is smaller than $a + 2\lambda_L$, and the oscillation period in the magnetic field will be larger by a factor of $(a + 2\lambda_L)/(a + b)$. All these observations have been confirmed experimentally.

3. Experimentally, the series-biased structure is easier to fabricate. However, it has the disadvantage that the resistances between successive layers must closely match, otherwise the waves may not phase lock. To our knowledge a parallel-biased structure has not yet been fabricated.

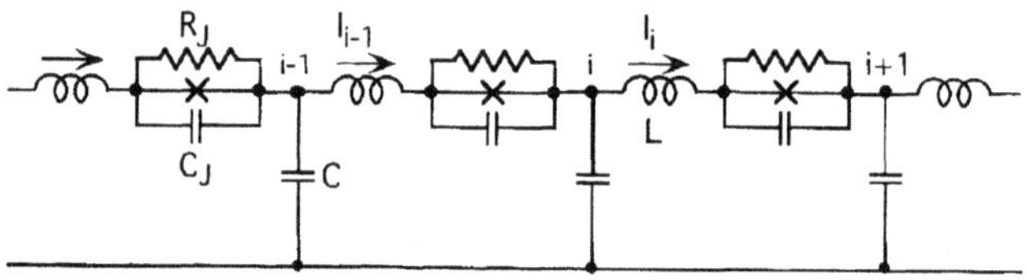

Figure 19.2 Equivalent circuit of a series-connected one-dimensional Josephson array.

19.3 The Josephson transmission line

Using thin-film patterning techniques (originally developed for technological applications involving semiconductors), one- and two-dimensional (periodic or aperiodic) arrays (or lattices) can be fabricated where each unit cell involves at least one Josephson junction along with various (lumped or distributed) inductors, capacitors, and resistors. We begin our discussion with 1D arrays.

Figs. 19.2 and 19.3 show two examples of 1D Josephson transmission lines incorporating, respectively, a series and parallel arrangement of the junctions, indicated by a cross; C_J and R_J represent the junction capacitance and resistance, C a distributed capacitance, and L a series inductance (all elements other than the junction are assumed to be lossless).

Today, series arrays of Josephson junctions (as shown in Fig. 19.2) are widely used as voltage standards. For this purpose, a large number of junctions are integrated into a suitable microwave distribution circuit, i.e., a transmission line consisting of a top layer of series-connected junctions, an insulating dielectric layer, and a superconducting ground plane (to ensure the uniformity of the microwave field sensed by the junctions). Arrays of over 1000 Nb/Al–Al0_x/Nb junctions have been successfully used to operate a 1 V reference (Yoshida *et al.* 1993).

A one-dimensional array of Josephson junctions connected in parallel by superconducting wires (as shown in Fig. 19.3) may be viewed as a discrete version of a single continuous junction. This kind of structure has potential applications as: a coherent Josephson radiator, elementary logic gates, and memory cells (which can be integrated into a computing network). It is also an ideal nonlinear system to study soliton (fluxon) dynamics. The dynamics of the (discrete) array can differ considerably from that of a single junction, where the latter is governed by the sine-Gordon equation.

Consider the discrete Josephson transmission line shown in Fig. 19.3. For simplicity, we assume that the junctions are sufficiently small that they never contain an appreciable fraction of a flux quantum. The critical currents, the capacitances, the normal-state junction resistances, and the self-inductances are taken to be identical for all unit cells. We denote the phase difference across the ith junction as Φ_i and the current through the self-inductance in the ith unit cell as I_i. From the equivalent circuit shown in Fig. 19.3, we see that in the absence of external magnetic flux the phase differences Φ_{i+1}, Φ_i, and Φ_{i-1} are restricted by the gauge-invariant phase relation Eq. (15.19b); i.e.,[4]

$$\Phi_i - \Phi_{i+1} = 2\pi\phi_{i,i+1}/\phi_0 = 2\pi L I_i/\phi_0, \tag{19.68a}$$

and

4. We use MKS units in this subsection.

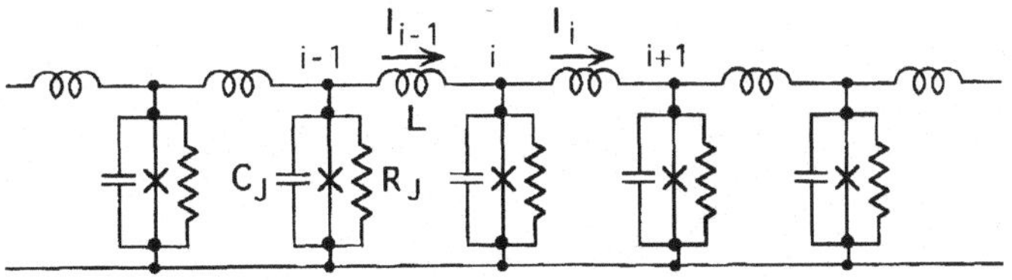

Figure 19.3 Equivalent circuit of a parallel-connected one-dimensional Josephson array.

$$\Phi_{i-1} - \Phi_i = 2\pi\phi_{i-1,i}/\phi_0 = 2\pi L I_{i-1}/\phi_0, \tag{19.68b}$$

where $\phi_{i,i+1}$ is the net magnetic flux whtihin the unit cell consisting of the ith and $(i+1)$th junctions. Using the current relation (15.31b) for a resistance-shunted junction, current conservation at the ith node gives

$$I_i = I_{i-1} - \frac{\hbar C}{2e}\ddot{\Phi}_i - \frac{\hbar}{2eR_J}\dot{\Phi}_i - I_m \sin\Phi_i. \tag{19.69}$$

Combining Eqs. (19.68a), (19.68b), and (19.69), the differential equation coupling Φ_{i+1}, Φ_i, and Φ_{i-1} takes the form

$$\frac{\phi_0}{2\pi L}(\Phi_{i+1} - 2\Phi_i + \Phi_{i-1}) - \frac{\hbar C}{2e}\ddot{\Phi}_i - \frac{\hbar}{2eR_J}\dot{\Phi}_i = I_m \sin\Phi_i. \tag{19.70}$$

When time is measured in units of ω_0^{-1} given by Eq. (15.46), and a bias current I_b is applied to every junction (from external sources not shown in Fig. 19.3), Eq. (19.70) can be written

$$\frac{\phi_0}{2\pi L I_m}(\Phi_{i+1} - 2\Phi_i + \Phi_{i-1}) - \frac{d^2\Phi_i}{d\tau^2} - \beta\frac{d\Phi_i}{d\tau} = \sin\Phi_i - \alpha_i, \tag{19.71}$$

where

$$\tau \equiv \omega_0 t, \qquad \beta \equiv 1/\omega_0 R_J C, \qquad \alpha = I_b/I_m.$$

Single fluxon solutions can be obtained from Eq. (19.71) using numerical methods. In Eq. (19.71), $(2\pi L I_m/\phi_0)^{1/2}$ is a discreteness parameter which determines the spatial spread of the fluxon. We note that a Josephson junction can be viewed as a nonlinear inductance element with a characteristic inductance given by $L_J = \phi_0/2\pi I_m$; this is easily seen from the expression $V = d(L_J(I)I)/dt$. Therefore, the discreteness parameter is defined as

$$1/\Lambda_J = (L/L_J)^{1/2} = (2\pi L I_m/\phi_0)^{1/2}. \tag{19.72}$$

When $1/\Lambda_J << 1$, the single fluxon spreads spatially over several unit cells. In the opposite limit, $1/\Lambda_J >> 1$, the single fluxon is essentially confined to one unit cell. Numerical calculations also show that single fluxons cannot propagate on discrete Josephson junction transmission lines if the bias currents are smaller than some threshold value.

20 Flux pinning and flux motion

20.1 Nonideal Type II superconductors

In previous sections we have discussed the static properties of magnetic flux structures in ideal (homogeneous) superconductors. In a real superconductor, various kinds of structural imperfections or defects are present resulting in a nonideal (or inhomogeneous) superconductor. These defects interact with the Abrikosov vortices or other flux structures which, in turn, influence the magnetization curves.

Fig. 20.1 shows typical magnetization curves of an ideal superconductor and a nonideal superconductor. The superconductors were cooled in zero field to a temperature $T < T_c$. As the applied magnetic field increases from zero to $H_{c1}(T)$, the nonideal superconductor shows perfect diamagnetism; i.e., $-4\pi M = H$ (line OC_1). When $H > H_{c1}(T)$, the magnetization deviates from line OC_1. With increasing magnetic field it reaches a maximum at a characteristic field H_p and then decreases to zero at $H = H_{c2}(T)$. Above $H_{c2}(T)$ the superconductor is in the normal state. If H decreases from above $H_{c2}(T)$, the magnetization of the nonideal superconductor does not follow curve C_2PC_1O; instead, it follows curve C_2O'; i.e., it is hysteretic and hence irreversible. At $H = 0$ there is a remnant magnetization. The magnetization of the ideal superconductor, on the other hand, follows curve C_2C_1O; it is nonhysteretic and hence reversible.

It is easy to demonstrate that the hysteresis in the magnetization of nonideal superconductors is caused by defects. As an example we consider a lead 8.23 wt% indium alloy in which the diameter was reduced from 1.91 cm to 0.317 cm by cold working at LN_2 temperature; the resulting magnetization curve, curve A in Fig. 20.2, shows a very large hysteresis and remnant magnetization. The specimen was then annealed at room temperature. Curves B, C, D, and E in Fig. 20.2 represent the magnetization curves obtained after annealing times of 30 min, 1 day, 18 days, and 46 days, respectively. Clearly, the longer the annealing time, the smaller the hysteresis and the remnant magnetization. Since cold work creates a large dislocation density in the specimen, which is reduced by the annealing processes, the above experiment demonstrates convincingly that the irreversibility arises from defects.

Fig. 20.1 shows that as soon as $H > H_{c1}(T)$ a magnetic field penetrates into a nonideal superconductor. The superconductor then enters the mixed state, which does not display perfect diamagnetism. We know that in the mixed state of an ideal Type II superconductor, the flux lines form a periodic triangular lattice. Fig. 20.3 is a photograph taken by the Bitter method[1] on a

1. In the Bitter method a magnetic material is deposited on a sample while it is exposed to a field. The magnetic material tends to concentrate in regions where the magnetic field is greater.

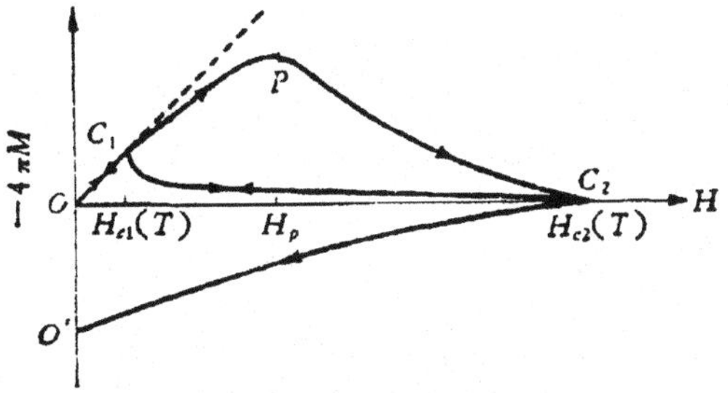

Figure 20.1 Schematic of the magnetization curves for an ideal (curve OC_1C_2) and nonideal (curve OC_1PC_2O') Type II superconductor.

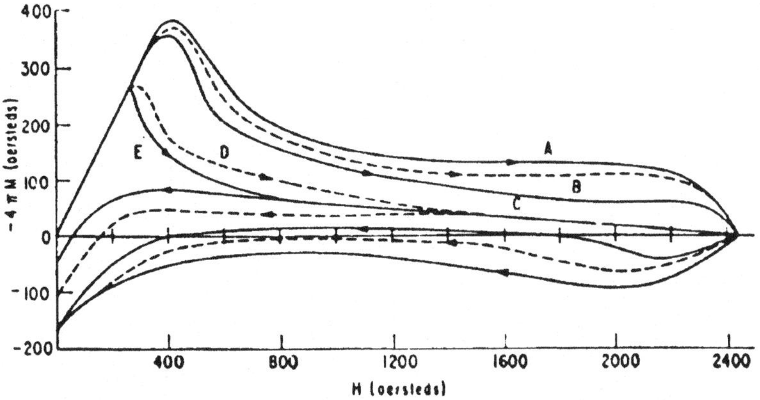

Figure 20.2 Results for the successive room-temperature annealing of a lead–8.23 wt% indium alloy: as cold swaged (A); and as annealed for 30 min (B), 1 day (C), 18 days (D) and 46 days (E). (After Livingston (1963).)

deformed cylindrical lead 6.33 at% indium specimen which shows how flux lines are distributed in a nonideal Type II superconductor. The Bitter pattern shows the flux-line distribution over a relatively large area. We see that the density of the flux increases from right to left. For any small region, the flux lines form an approximately triangular lattice, but the lattice constant, a, increases from left to right. The dislocations in the flux-line lattice are marked by the symbol $\perp$ in Fig. 20.3.

From the above discussion, we see that even though the magnetic properties of a nonideal Type II superconductor are more complicated than those of an ideal superconductor, microscopically the difference is not large. The main difference is that in the former the flux lines are not uniformly distributed but in the latter they distribute uniformly in the form of a triangular lattice. Those concepts which do not connect directly to the flux-line distribution (e.g., the vortex structure, the energy of a single flux line, and the interaction between vortices) are still applicable to nonideal superconductors.

20.2 Microscopic description

To describe a periodic flux-line lattice, it is sufficient to know the symmetry of the lattice and its lattice constant. However, a complete description of a nonuniformly distributed flux-line system

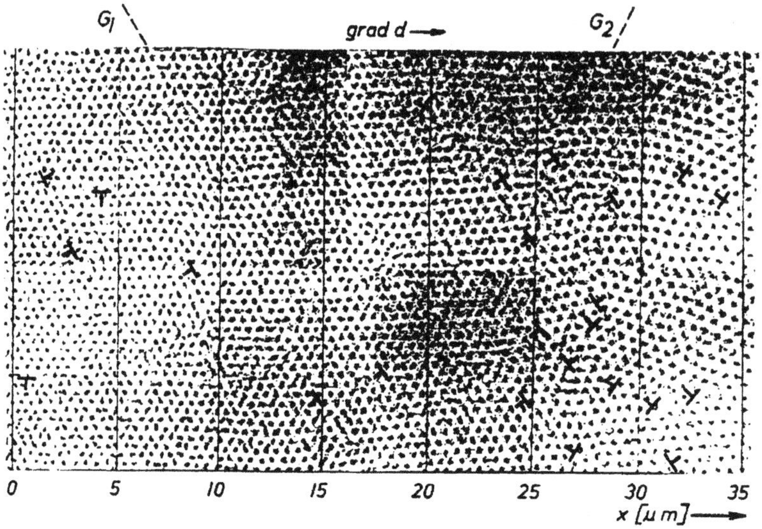

Figure 20.3 A Bitter photograph showing a section of the flux-line lattice with an approximately constant flux-line density gradient from right to left. The dislocation cores are indicated by the symbol ⊥. (After Träuble and Essmann 1968).)

requires, in principle, a knowledge of the spatial position of each vortex which, of course, is not practical. From Fig. 20.3 we see that on a macroscopically small but microscopically large scale (with respect to the lattice constant a), the flux lines in the nonideal Type II superconductor are arranged, approximately, in a triangular lattice; however, the lattice constant varies spatially on a length scale much larger than a. A spatially slowly varying flux-line system can be described using continuum methods which provide an adequate description of the macroscopic properties of the nonideal Type II superconductor.

We begin our discussion by introducing the flux-line density, $n_{\mathrm{L}}(\mathbf{r})$, at a point $\mathbf{r}$ as

$$n(\mathbf{r}) = \lim_{\Delta S \to 0} \frac{\Delta N(\mathbf{r})}{\Delta S}, \tag{20.1}$$

where ΔS is a macroscopically small but microscopically large area element and $\Delta N(\mathbf{r})$ is the number of flux lines crossing ΔS. A given distribution of flux lines is always associated with a corresponding magnetic induction, $\mathbf{H}(\mathbf{r})$, and supercurrent, $\mathbf{j}_{\mathrm{s}}(\mathbf{r})$. Note that both $\mathbf{H}(\mathbf{r})$ and $\mathbf{j}_{\mathrm{s}}(\mathbf{r})$ are microscopic quantities. Through superposition we have the relations

$$\mathbf{H}(\mathbf{r}) = \sum_i \mathbf{H}_i(\mathbf{r}), \tag{20.2}$$

and

$$\mathbf{j}_{\mathrm{s}}(\mathbf{r}) = \sum_i \mathbf{j}_{\mathrm{s}i}(\mathbf{r}), \tag{20.3}$$

where $\mathbf{H}_i(\mathbf{r})$ and $\mathbf{j}_{\mathrm{s}i}(\mathbf{r})$ are the magnetic induction and supercurrent created at a point $\mathbf{r}$ produced by the ith flux line. The macroscopic magnetic induction is then defined as

$$\mathbf{B}(\mathbf{r}) = \lim_{\Delta V \to 0} \frac{1}{\Delta V} \int_{\Delta V} \mathbf{H}(\mathbf{r}) \mathrm{d}^3 r, \tag{20.4}$$

where ΔV is a macroscopically small but microscopically large volume element. Within ΔV, the flux lines are parallel and form a lattice; therefore,

$$\mathbf{B}(\mathbf{r}) = n(\mathbf{r})\phi_0. \tag{20.5}$$

In the same way, we can write

$$\mathbf{J}_s(\mathbf{r}) = \lim_{\Delta V \to 0} \frac{1}{\Delta V}\int_{\Delta V} \mathbf{j}_s(\mathbf{r})\mathrm{d}^3 r; \tag{20.6}$$

$\mathbf{J}_s(\mathbf{r})$ is the macroscopic volume current density induced by the static magnetic field; $\mathbf{B}(\mathbf{r})$ and $\mathbf{J}_s(\mathbf{r})$ are related by the macroscopic Maxwell equation

$$\nabla \times \mathbf{B}(\mathbf{r}) = \frac{4\pi}{c}\mathbf{J}_s(\mathbf{r}), \tag{20.7}$$

which may be obtained by averaging the microscopic Maxwell equation $\nabla \times \mathbf{H}(\mathbf{r}) = (4\pi/c)\mathbf{j}_s(\mathbf{r})$ over ΔV.

The induced current density, $\mathbf{J}_s(\mathbf{r})$, is an important quantity. It arises from the inhomogeneous distribution of the flux lines. To see this we consider a one-dimensional superconductor. Assuming B and J_s vary with x only ($B(x)$ is along the $\mathbf{z}$ direction and $J_s(x)$ is along the $\mathbf{y}$ direction) we have

$$-\frac{\mathrm{d}B(x)}{\mathrm{d}x} = \frac{4\pi}{c}J_s(x). \tag{20.8}$$

Eq. (20.8) clearly shows that $J_s(x)$ is related to the gradient of the flux-line density. Once $\mathbf{J}_s(x)$ is known, the flux-line distribution in the superconductor can be determined from Eq. (20.8) with the proper boundary conditions. Therefore, in nonideal Type II superconductors,[2] $\mathbf{B}(\mathbf{r}) \neq 0$, and $\mathbf{J}_s(\mathbf{r}) \neq 0$. In the mixed state of an ideal Type II superconductor, the magnetic field also penetrates and hence $\mathbf{B}(\mathbf{r}) \neq 0$; however, $\mathbf{J}_s(\mathbf{r}) \equiv 0$, since $\mathbf{B}(\mathbf{r})$ is independent of $\mathbf{r}$. The presence of the bulk induced current $\mathbf{J}_s(\mathbf{r})$ is one of the remarkable properties of the nonideal Type II superconductors. It forms the basis for many important technological applications.

20.3 The Lorentz force

In Sec. 7 we showed that the interaction energy per unit length between two flux lines is given by

$$\frac{E_{12}}{L} = \frac{\phi_0^2}{8\pi^2\lambda_L^2}K_0\left(\frac{|\mathbf{r}_1 - \mathbf{r}_2|}{\lambda_L}\right) = \frac{\phi_0}{4\pi}H_1(\mathbf{r}_2). \tag{20.9}$$

This interaction is repulsive for the usual case where the vortices are parallel. The force, $\mathbf{F}_2$, acting on the second flux line as a result of the magnetic field produced by the first is determined by the gradient of the interaction energy:

2. If the sample has been previously magnetized we may have $\mathbf{B}(\mathbf{r}) \neq 0$ and $\mathbf{J}_s(\mathbf{r}) \neq 0$ in the absence of an external field.

$$\frac{\mathbf{F}_2}{L} = -\frac{\partial}{\partial \mathbf{r}_2}(E_{12}|\mathbf{r}_1 - \mathbf{r}_2|)/L. \tag{20.10}$$

Using Maxwell's equation $\nabla \times \mathbf{H}_1(\mathbf{r}_2) = (4\pi/c)\mathbf{j}_{1s}(\mathbf{r}_2)$, the force per unit length on vortex 2 is

$$\frac{\mathbf{F}_2}{L} = \frac{1}{c}\mathbf{j}(\mathbf{r}_2) \times \boldsymbol{\phi}_0, \tag{20.11}$$

where $\boldsymbol{\phi}_0$ is parallel to the flux density.

Using the procedure outlined in Subsec. 20.2, the above expression can be volume averaged for an arbitrary flux array and written as

$$\frac{\mathbf{F}}{L} = \frac{1}{c}\mathbf{J}_s \times \boldsymbol{\phi}_0, \tag{20.12}$$

where $\mathbf{J}_s$ is now the total volume-averaged supercurrent density (due to all the remaining vortices) at the position of the flux line under consideration. Introducing the flux-line density $n_L = B/\phi_0$, and using Eq. (20.5), the Lorentz force per unit volume of the flux-line lattice (which we refer to as the force density or volume force) is

$$\mathbf{f}_L = \frac{1}{c}\mathbf{J}_s \times \mathbf{B}. \tag{20.13}$$

In the above derivation we have assumed $\kappa >> 1$, since Eq. (20.9) is valid in this limit. For the general case, Eq. (20.13) may be obtained from a thermodynamic argument (Friedel, deGennes, and Matricon 1963).

To see the origin of the Lorentz force, we use the vector identity

$$\frac{1}{2}\nabla(\mathbf{B}\cdot\mathbf{B}) = (\mathbf{B}\cdot\nabla)\mathbf{B} + \mathbf{B} \times (\nabla \times \mathbf{B}) \tag{20.14}$$

and the Maxwell equation, (20.7), to obtain

$$\mathbf{f}_L = \frac{1}{c}\mathbf{J}_s \times \mathbf{B} = -\frac{1}{8\pi}\nabla(\mathbf{B}\cdot\mathbf{B}) + \frac{1}{4\pi}(\mathbf{B}\cdot\nabla)\mathbf{B}. \tag{20.15}$$

Therefore, the Lorentz force on a unit volume of the flux lattice arises from the magnetic pressure and an additional term describing a tension along the lines of force.

For a long cylindrical superconductor with the magnetic field applied parallel to its axis, the demagnetization effect is absent and the flux lines are essentially parallel to the $\hat{\mathbf{z}}$ direction. Hence

$$\mathbf{f}_L = \frac{-B}{4\pi}\nabla B; \tag{20.16}$$

i.e., the Lorentz force density is solely determined by the flux density gradient and is directed along the density gradient. We will come back to this point later.

Eq. (20.13) implies that a flux line can be in a static equilibrium at any given position only if the total supercurrent arising from all other sources is zero at that point. This condition can be fulfilled if a vortex is surrounded by a symmetrical array, as with a *uniform* (undistorted) lattice.

Even for this case, however, an applied transport current will produce a transverse driving force on the flux lattice, and if this force is sufficient to move the flux lines a phenomenon called flux flow results. The situation is quite different for the *nonuniform* vortex lattice. In this latter case the net force exerted on a vortex by the remaining vortices is nonzero, even in the absence of a transport current.

20.4 Pinning centers and pinning forces

In a nonuniformly distributed flux-line system the lattice is subjected to a Lorentz volume force, $\mathbf{f}_L$, as discussed above. This volume force tends to drive the vortices down the flux density gradient, leading to flux flow. However, measurements of the time dependence of the magnetization, M, in nonideal Type II superconductors show that M does not change significantly with time (the phenomenon of flux creep will be discussed later). We thus conclude that there exists another force which balances the Lorentz force. Since the only difference between a nonideal and an ideal superconductor is the presence or absence of defects, respectively, this balancing force must be associated with defects. We call this force the *pinning force* and call the defects the *pinning centers*. We denote the pinning force per unit volume as $\mathbf{f}_p$. The pinning force density and pinning centers are two important concepts in understanding the complex properties of the nonideal Type II superconductors. In an inhomogeneous superconductor many kinds of imperfections may exist, such as second-phase precipitates, interstitial impurity atoms, voids, grain boundaries, and dislocations. The changes of material parameters at these defects (e.g., electronic density, of states, pairing interaction, and electron mean free path) give rise to a local variation of T_c, ξ, and λ. As a consequence the free energy varies at defect locations. The defects perturb the coefficients of α, β, and $1/2m^* \equiv \gamma$ as well as the order parameter, ψ, and the vector potential, $\mathbf{A}$, in the G–L free energy, Eq. (9.13). Because F is supposed to be minimized with respect to ψ and $\mathbf{A}$, the pinning energy can, in first order, be obtained by taking into account the variations of the coefficients. This approximation is valid if the relevant size of the defect is small compared to the distance over which ψ changes. The pinning potential of a defect is thus given by

$$\delta F_p = \int d^3r \left[\delta\alpha(\mathbf{r} - \mathbf{r}_d)|\psi|^2 + \frac{1}{2}\delta\beta(\mathbf{r} - \mathbf{r}_d)|\psi|^4 + \delta\gamma(\mathbf{r} - \mathbf{r}_d)\left|\left(-i\hbar\nabla - \frac{e^*}{c}\mathbf{A}\right)\psi\right|^2\right]. \tag{20.17}$$

To account for an array of defects Eq. (20.17) must be summed over defect positions, $\mathbf{r}_d$. The variations $\delta\alpha$, $\delta\beta$, and $\delta\gamma$ differ from zero in the vicinity of the position of the defect, $\mathbf{r}_d$. From the microscopic derivation of the G–L free energy (see Part III) we know that the coefficient γ in Eq. (20.17) is related to ℓ_{tr}, the transport mean free path, in the dirty limit. It is evident from Eq. (20.17) that the T_c perturbations couple to the terms containing α and β, while the ℓ_{tr} fluctuations lead to a perturbation of the third term. They are referred to as δT_c-pinning and $\delta\kappa$-pinning effects, respectively. The latter pinning mechanism arises from a change in the mean free path of the electrons at the defects resulting in a variation in both κ and γ. It follows from Eq. (20.17) that the pinning potential depends on several length scales: the size of the pin, and the length scales over which the order parameter and the vector potential change. For isolated vortices ($H < 0.2H_{c2}$) the relevant length scales are ξ and λ_L, respectively; for overlapping vortices ($H > 0.2H_{c2}$), it is the flux-line lattice parameter, a. If the pin size is less than ξ, λ_L, or a, the pinning potential can be determined from Eq. (20.17) using the solutions for ψ and $\mathbf{A}$ obtained for the homogeneous superconductor. For large pins, the boundary conditions at the interface become important. Therefore to evaluate the vortex–defect interaction, one has to decide which of the above considerations is most relevant for each pinning mechanism.

For high flux densities or for materials with a relatively low κ-value, we may use Eq. (20.17) to evaluate the pinning potential. Recalling that $H_c^2/8\pi = \alpha^2/2\beta$ and $|\psi_0|^2 = |\alpha|/\beta$, Eq. (20.17) may be written

$$\delta F_p = \int \delta \mathscr{F}_p \mathrm{d}^3 r, \tag{20.18a}$$

where

$$\delta \mathscr{F}_p = \frac{H_c^2}{4\pi}\left[-\frac{\delta H_{c2}}{H_{c2}}\left|\frac{\psi}{\psi_0}\right|^2 + \frac{1}{2}\frac{\delta\kappa^2}{\kappa^2}\left|\frac{\psi}{\psi_0}\right|^4 + \frac{\delta\gamma}{\gamma}\xi^2\left|\nabla\left(\frac{\psi}{\psi_0}\right)\right|^2\right]. \tag{20.18b}$$

The elementary pinning force density, $\mathbf{f}_p$, is obtained by determining the maximum value of $\nabla\delta\mathscr{F}(\mathbf{r})$. The final result depends on how $\psi(\mathbf{r})$ varies with the magnetic field. From the discussion in Sec. 14, when $H \to H_{c2}$ we may write[3]

$$\left|\frac{\psi}{\psi_0}\right|^2 \approx 1 - h, \tag{20.19}$$

where $h = H/H_{c2}$. For a triangular lattice, the lattice constant is $a_0 = 1.075(\phi_0/H)^{1/2}$ and the related wave number is $k_0 = 2\pi/a_0$. Therefore by ignoring the (higher order) second term in Eq. (20.18) we have

$$\delta\mathscr{F} \approx \left[\frac{2\pi}{a_0}(1-h)\frac{\delta H_{c2}}{H_{c2}} + \left(\frac{2\pi}{a_0}\right)^3 \xi^2(1-h)\frac{\delta\gamma}{\gamma}\right]\Omega\frac{H_c^2}{4\pi}, \tag{20.20}$$

where Ω is some mean volume fraction of the defects. Because $\delta H_{c2}/H_{c2}$ and $\delta\gamma/\gamma$ behave differently, the temperature dependence of the δT_c- and $\delta\kappa$-pinning effects will differ accordingly.

20.4.1 *The core interaction*

We first consider a simple pinning mechanism involving nearly isolated flux lines. The contribution to the condensation energy per unit length of a flux line resulting from the formation of a normal core, E_{core}/L, is

$$E_{core}/L = \frac{H_c^2}{8\pi}\pi\xi^2, \tag{20.21}$$

where H_c is the thermodynamic critical field. If a second phase, having a thermodynamic critical

3. This result follows from Eq. (14.23a) in the limit of large κ and approximating β_A by unity (see Eqs. (14.29)).

field H'_c, precipitates within the host superconductor, then if the core resides completely within the second phase the resulting difference in the energy, i.e., the pinning potential, is

$$\delta E = \left(\frac{H_c^2}{8\pi} - \frac{H_c'^2}{8\pi}\right)\pi\xi L. \tag{20.22}$$

For the maximum pinning case where $H'_c = 0$, the pinning potential per unit length of vortex line is given by Eq. (20.21). The magnitude of the maximum pinning force, $|\mathbf{F}_{\text{core}}^{\max}|$, per unit length L is then

$$\frac{|\mathbf{F}_{\text{core}}^{\max}|}{L} = \left|\nabla\left(\frac{E_{\text{core}}}{L}\right)\right| \approx \frac{E_{\text{core}}}{L\xi} \approx \frac{H_c^2\xi}{8} = \frac{1}{16}\left(\frac{\phi_0}{2\pi}\right)^{1/2}\frac{H_{c2}^{3/2}}{\kappa^2}, \tag{20.23}$$

where we have used $H_{c2} = \sqrt{2}\kappa H_c = \phi_0/2\pi\xi^2$. If the normal inclusions extended throughout the whole of the superconductor (in the same direction as the applied magnetic field) the maximum pinning force density, $|\mathbf{f}_{\text{core}}^{\max}|$, would be $n_L|\mathbf{F}_{\text{core}}^{\max}|/L$ or

$$|\mathbf{f}_{\text{core}}^{\max}| = BH_c^2\xi/8\phi_0, \tag{20.24}$$

where $n_L = B/\phi_0$. (Such a matching arrangement of pinning sites may be fulfilled by a lattice of holes etched into the superconductor.) By using the Lorentz force density, Eq. (20.13), we obtain

$$J_c^{\max} = cH_c^2\xi/8\phi_0 = cH_c/16\pi\sqrt{2}\lambda_L. \tag{20.25}$$

20.4.2 *Surface magnetic interaction*

In a parallel field, the sample surface or a grain boundary can act as a pinning center, since it forms an energy barrier for flux entry and exit. For $H \approx 0$ this can be understood in terms of the boundary conditions for the screening currents surrounding the vortex core when the core is within a distance of order λ_L from the surface. The boundary conditions may be simulated by assuming an *antivortex* is present outside the superconductor at the image position. The vortex and its image attract each other while the London surface currents arising from the applied field repel the vortex. The net effect for the equilibrium case is a potential barrier, called the Bean–Livingston barrier, which leads to initial flux entry at a field H_s which is larger than H_{c1}.

Assuming $H \approx H_{c1}$, the repulsive force exerted by the surface current on the vortex at x is given by $(\phi_0 H_{c1}/4\pi\lambda_L)e^{-x/\lambda_L}$. The force from the image vortex takes the form Ce^{-2x/λ_L}, where C is a constant which we determine by requiring that in thermal equilibrium, the total work done by quasistatically displacing the vortex from the surface ($x = 0$) to $x = \infty$ should be zero,

$$\int_0^\infty dx[(\phi_0 H_{c1}/4\pi\lambda_L)e^{-x/\lambda_L} - Ce^{-2x/\lambda_L}] = 0. \tag{20.26}$$

Hence

$$C = \phi_0 H_{c1}/2\pi\lambda_L. \tag{20.27}$$

The force per unit length arising from the magnetic surface interaction reaches its maximum at $x = 0$ given by

$$\frac{|\mathbf{F}_{\text{surf}}^{\text{max}}|}{L} = \phi_0 H_{c1}/4\pi\lambda_L. \tag{20.28}$$

Using Eqs. (8.4b) and (10.7c), this result can be written as

$$\frac{|\mathbf{F}_{\text{surf}}^{\text{max}}|}{L} = \frac{1}{4}\left(\frac{\phi_0}{2\pi}\right)^{1/2} \frac{H_{c2}^{3/2}}{\kappa^3} \ell n\kappa. \tag{20.29}$$

Comparing Eq. (20.29) with Eq. (20.23), we find

$$\frac{|\mathbf{F}_{\text{core}}^{\text{max}}|}{|\mathbf{F}_{\text{surf}}^{\text{max}}|} = \frac{\kappa}{4\ell n\kappa}. \tag{20.30}$$

Eq. (20.30) indicates that the core interaction dominates in high κ superconductors. The main reason for this is the different ranges involved in the two interactions; the relevant length scales are approximately ξ for the core interaction and λ_L for the magnetic interaction. Since the pinning force is related to the spatial derivative of the interaction energy, $\mathbf{F}_{\text{core}}$ will be larger than $\mathbf{F}_{\text{surf}}$ in high κ materials. From Eqs. (20.23) and (20.29) we also see that the temperature dependences of the core and magnetic interactions are quite similar. Hence, it is difficult to distinguish the two interactions experimentally.

20.4.3 *Summation of the pinning forces*

For inhomogeneous superconductors in thermal equilibrium, the flux-line lattice will adapt to the distribution of defects by assuming a configuration of minimum free energy determined by the sum of the deformation energy and the interaction energy with the defects. The net effect is that the flux-line lattice is pinned by the defects. Lattice deformation is crucial here in obtaining a finite net pinning force on summing over all random pinning centers. Consider the case of a 'rigid' flux-line lattice (i.e., the limit of infinite line tension) shown in Fig. 20.4(a). Because the pinning centers are randomly distributed the average numbers of the pinning centers located on the two sides of a flux line are equal and hence the resultant (in plane) pinning force vanishes. In contrast, in a 'soft' flux-line lattice, the flux lines can deform (bend) such that the total pinning force in a particular direction is maximized (see Fig. 20.4(b)). Therefore, the elastic properties of the flux-line lattice are very important in determining the overall effect of the pinning; this will be discussed in Sec. 20.7.

20.5 The equation of motion

For various technological applications, the dynamics of the vortices is important, since it governs their response to an electromagnetic field and hence determines the transport properties. To treat such problems we must introduce dissipation (or damping). Therefore, in addition to the various pinning forces per unit length, discussed above, which we denote collectively as $\mathbf{F}_p/L$, we must introduce a viscous damping force per unit length, $\mathbf{F}_\eta/L$. We will assume that for small velocities

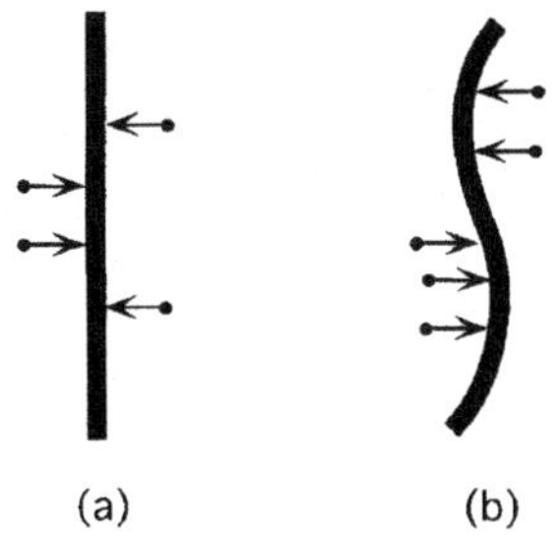

Figure 20.4 The interaction between a flux line (shown as a thick black line) and pinning centers (shown as small solid circles): (a) a rigid flux line; (b) a 'soft' flux line.

this damping is described by $\mathbf{F}_\eta/L = -\eta\mathbf{v}$, where $\mathbf{v}$ is the velocity (perpendicular to the vortex axis) and η is a friction coefficient. In the continuum limit, where many flux lines pass through the averaging volume V_c, we may introduce a viscous volume force through

$$\mathbf{f}_\eta = -n_L\eta\mathbf{v}, \tag{20.31}$$

where $\mathbf{v}$ is now the volume-average flux-line velocity, $n_L = B/\phi_0$ is again the number of flux lines per unit area, and η is the friction coefficient.

In addition to the internal volume force associated with the pinning, it is convenient to introduce a volume force associated with the distortion of the flux lattice, by the pinning and other volume forces, from its equilibrium form. In the continuum limit and for small displacements this force may be modeled using linear elasticity theory in terms of a two-dimensional lattice displacement vector, $\mathbf{u}(\mathbf{r})$. If we introduce an elastic free energy density, $\mathscr{F}_{el}$, then the elastic restoring force density is given by

$$\mathbf{f}_{el} = -\frac{\partial \mathscr{F}_{el}}{\partial \mathbf{u}}; \tag{20.32}$$

we will return to this point in Subsec. 20.7 where the associated elastic constants will be calculated. We note in passing that the velocity entering (20.31) is clearly given by $\mathbf{v} = \partial\mathbf{u}(\mathbf{r}, t)/\partial t$.

We are now in a position to write the continuum limit equation of motion for a flux lattice. Requiring that the sum of all the contributing force densities vanishes we have, on assembling the above,

$$\mathbf{f}_L + \mathbf{f}_\eta + \mathbf{f}_{el} + \mathbf{f}_p = 0. \tag{20.33}$$

In the next two subsections we discuss models used to describe the pinning and the elastic force densities. The quantity which is most difficult to determine theoretically and experimentally is the pinning force, $\mathbf{f}_p$. Usually, the microscopic parameters associated with pinning centers are not well known. In some cases strong pinning even drives the lattice deformation into a plastic regime and makes the quantitative calculations even more difficult. The most difficult problem in determining the pinning effects is that the statistical summation of $\mathbf{f}_p$ depends on the relative positions of the pinning centers within the lattice. When the interaction force with one pinning center reaches a maximum, the interaction force with other centers is generally smaller and may even be in the opposite direction. As a result, the mean pinning force is less than the sum of the maximum values of the individual forces discussed earlier.

20.6 The critical state

Consider a nonideal Type II superconductor in zero field. On applying a magnetic field the flux first enters near the surface, in the London penetration layer. When $H > H_{c1}$ (we ignore the surface barrier for the moment), flux lines penetrate into the bulk of the superconductor and adopt some density gradient. In the static state, the flux-line distribution does not vary with time; i.e., the condition

$$\mathbf{f}_L = -\mathbf{f}_p$$

is fulfilled everywhere inside the superconductor. Such a state is called the critical state and was first discussed by Bean (1962, 1964).

The concept of the critical state plays a central role in understanding the irreversible magnetization in hard Type II superconductors. In the critical state, a superconductor responds to a change in the applied field via shielding (or trapping) currents that flow at the level of the critical current density, J_c. As the applied field increases from zero and reaches a value corresponding to the Bean–Livingston barrier, the flux lines start to penetrate into the superconductor. The lines adjust themselves so that the driving force arising from the flux density gradient is equal to the maximum pinning force. The field penetration profile in the superconductor is determined, in conjunction with proper boundary conditions, by the Maxwell equation

$$\nabla \times \mathbf{B} = \frac{4\pi}{c}\mathbf{J}_c, \tag{20.34}$$

and a constitutive or critical state equation

$$J_c = J_c(T, B). \tag{20.35}$$

The critical state equation (20.35) is, in general, very complicated. In practice various simplified versions have been proposed. It was originally assumed by Bean that $J_c(T, B)$ was a constant in the critical state. Later, based on experimental results, Kim, Hempstead, and Strand (1963) proposed the following critical state equation

$$J_c(T, B) = \frac{J_0(T)}{1 + (B/B_0)}. \tag{20.36}$$

Here $J_0(T)$ is the critical current density at zero field and at temperature T, B is the local flux density and B_0 is a model-dependent parameter. Other models, e.g., where J_c depends linearly or exponentially on B, have also been proposed.

To determine the field penetration profile and to demonstrate how the current density, J_c, may be determined based on the measured magnetization, we consider a superconducting slab with a width $2d$ in the x direction and with the field applied parallel to the slab (and taken as the z direction). For this case the demagnetization effect can be neglected and the vortices are not curved. Eq. (20.34) then reduces to

$$\frac{\mathrm{d}B}{\mathrm{d}x} = \pm\frac{4\pi}{c}J_c(T, B), \tag{20.37}$$

where the signs depend on the field sweep history. In order to simplify the boundary conditions,

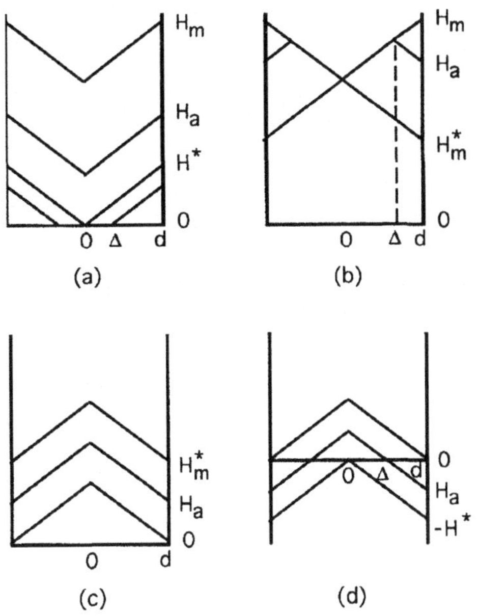

Figure 20.5 Schematic of the magnetic field penetration profiles at various stages of the applied field for the zero-field-cooled case: (a) $0 \le H_a \le H_m$; (b) $H_m \ge H_a \ge H^*_m$; (c) $H^*_m \ge H_a \ge 0$; (d) $0 \ge H_a \ge -H^*$.

we ignore the effect of the lower critical field and any possible role of a surface current or an equilibrium magnetization. We use a generalized critical state equation of the form

$$J_c(T, B) = \frac{J_0(T)}{(1 + B/B_0)^n}. \tag{20.38}$$

When the exponent $n = 0$ and 1, Eq. (20.38) reduces to Bean's model and Kim's critical state equation (20.36), respectively.

We consider the magnetization for the zero-field-cooled case. As the magnetic field changes magnitude (or direction), a shielding (or trapping) current will be induced. The resulting field penetration profiles and current distributions at various stages of the field evolution are shown schematically in Fig. 20.5. In this figure H_m is the maximum field applied and Δ is the penetration depth. We define two characteristic fields: the full penetration field, H^*, which depends only on J_c and the sample dimension, and H^*_m, which is the field at which all shielding currents reverse direction; note H^*_m is related to H_m. To simplify the expression, we define

$$B_J = \frac{4\pi J_0(T) d}{c}. \tag{20.39}$$

Solving Eqs. (20.37) and (20.38) determines the field penetration profiles in the slab. We distinguish the following regions of the applied field, H_a:

(a) $\quad 0 \le H_a \le H_m$

In this region the local field is given by

$$B(x) = B_0\left[(1 + H_a/B_0)^{n+1} - \frac{(n+1)B_J}{B_0}\left(1 - \frac{x}{d}\right)\right]^{1/(n+1)} - B_0. \tag{20.40}$$

The full penetration field H^* is determined by

$$H^* = B_0\left[1 + \frac{(n+1)B_T}{B_0}\right]^{1/(n+1)} - B_0. \tag{20.41}$$

The penetration depth Δ is

$$\frac{\Delta}{d} = \begin{cases} \dfrac{B_0}{(n+1)B_J}[1 - (1 + H_a/B_0)^{n+1}] + 1, & 0 \le H_a \le H^*; \\ 0, \ H^* \le H_a \le H_m. \end{cases} \tag{20.42}$$

(b) $\quad H_m \ge H_a \ge H_m^*$

When the applied field is decreased from H_m, a portion of the shielding current reverses direction as shown in Fig. 20.5(b). The penetration depth Δ is

$$\frac{\Delta}{d} = 1 - \frac{B_0}{2(n+1)B_J}\left[\left(1 + \frac{H_m}{B_0}\right)^{n+1} - \left(1 + \frac{H_a}{B_0}\right)^{n+1}\right]. \tag{20.43}$$

At $H_a = H_m^*$, $\Delta = 0$, and we have

$$H_m^* = B_0\left[\left(1 + \frac{H_m}{B_0}\right)^{n+1} - \frac{2(n+1)B_J}{B_0}\right]^{1/(n+1)} - B_0, \tag{20.44}$$

$$B(x) = \begin{cases} B_1(x) = B_0\left[\left(1 + \dfrac{H_m}{B_0}\right)^{n+1} - \dfrac{(n+1)B_J}{B_0}\left(1 - \dfrac{x}{d}\right)\right]^{1/(n+1)} - B_0, & 0 \le x \le \Delta; \\ B_2(x) = B_0\left[\left(1 + \dfrac{H_a}{B_0}\right)^{n+1} + \dfrac{(n+1)B_J}{B_0}\left(1 - \dfrac{x}{d}\right)\right]^{1/(n+1)} - B_0, & \Delta \le x \le d. \end{cases} \tag{20.45}$$

(c) $\quad H_m^* \ge H_a \ge 0$

As shown by Fig. 20.5(c), $\Delta = 0$ and therefore

$$B(x) = B_2(x), \qquad 0 \le x \le d. \tag{20.46}$$

(d) $\quad 0 \ge H_a \ge -H_m^*$

When the applied field changes direction, the resulting field profile is shown by Fig. 20.5(d) with H^* still given by Eq. (20.41). We have

$$\frac{\Delta}{d} = -\frac{B_0}{(n+1)B_J}\left[\left(1 - \frac{H_a}{B_0}\right)^{n+1} - 1\right] \tag{20.47}$$

and

$$B(x) = \begin{cases} B_1(x) = B_0\left[2 - \left(1 - \dfrac{H_a}{B_0}\right)^{n+1} + \dfrac{(n+1)B_J}{B_0}\left(1 - \dfrac{x}{d}\right)\right]^{1/(n+1)} - B_0, & 0 \le x \le \Delta; \\ B_2(x) = B_0 - B_0\left[\left(1 - \dfrac{H_a}{B_0}\right)^{n+1} - \dfrac{(n+1)B_J}{B_0}\left(1 - \dfrac{x}{d}\right)\right]^{1/(n+1)}, & \Delta \le x \le d. \end{cases} \tag{20.48}$$

The rest of the hysteresis loop can be completed by a symmetry consideration. The magnetization is related to $B(x)$ and H_a by the expression

$$\langle 4\pi\mathbf{M}\rangle = \langle\mathbf{B}\rangle - \mathbf{H}_a, \tag{20.49}$$

where $\langle\,\rangle$ denotes an average over the whole sample volume. Using the local flux density $B(x)$ given above, the hysteresis loop can then be calculated. The calculated magnetization loop is quite different from the reversible magnetization curve predicted by the Abrikosov theory. Using Bean's model ($n = 0$ in Eq. (20.38)) to estimate J_c from the magnetization, we can obtain useful expressions. For $H^* < H_a < H_m^*$, the magnetization corresponding to the increasing and decreasing field branches is

$$\langle 4\pi M^-\rangle = -\frac{1}{2}B_J \tag{20.50}$$

and

$$\langle 4\pi M^+\rangle = +\frac{1}{2}B_J, \tag{20.51}$$

respectively. Therefore

$$J_c = \frac{c}{d}\langle\Delta M\rangle, \tag{20.52}$$

where $\Delta M = M^+ - M^-$. In addition to deducing J_c and its field dependence, magnetization curves are directly related to the ac losses in Type II superconductors.

For the general case, where the demagnetization effect is significant, one has to use Eq. (20.34) to account for the curved nature of the vortices.

20.7 The elastic constants of a flux-line lattice

In Sec. 14 we presented the Abrikosov solution for the mixed state. In the mixed state of an ideal Type II superconductor the flux lines form a regular triangular array. The interaction between the vortices can be expressed in terms of an elasticity tensor. The elastic properties play a central role in understanding the pinning mechanisms in a Type II superconductor. The elastic constants of a flux-line lattice can be calculated in terms of the reversible magnetization curve by taking the appropriate derivatives of the free energy of the vortex assembly.

Consider a flux-line lattice consisting of vortices parallel to the z axis. In the x–y plane the vortices form a regular array with a triangular unit cell. In the usual engineering (Voigt) notation, the stresses, σ_i, are related to the strains, ε_i, by

$$\sigma_i = c_{ij}\varepsilon_i, \tag{20.53}$$

where c_{ij} are the corresponding elastic constants and i takes the values 1–6 which denote the xx, yy, zz, yz, zx, and xy components of the stress/strain tensors, respectively, and

$$\varepsilon_1 = \varepsilon_{xx} = \frac{\partial u_x}{\partial x} \text{ and } \varepsilon_6 = \varepsilon_{xy} = \left(\frac{\partial u_x}{\partial y} + \frac{\partial u_y}{\partial x}\right), \tag{20.54}$$

where **u** is the displacement.

Since the length of the vortex line is infinite in the z direction, the line tension is independent of length; i.e., the forces cannot depend on $\varepsilon_3\,(\varepsilon_{zz})$. Consequently, the third column and by symmetry the third row of the c_{ij} tensor are zero. The generalized Hooke's law thus takes the form (Campbell and Evetts 1972)

$$\begin{pmatrix} \sigma_1 \\ \sigma_2 \\ \sigma_4 \\ \sigma_5 \\ \sigma_6 \end{pmatrix} = \begin{pmatrix} c_{11} & c_{12} & 0 & 0 & 0 \\ c_{12} & c_{11} & 0 & 0 & 0 \\ 0 & 0 & c_{44} & 0 & 0 \\ 0 & 0 & 0 & c_{44} & 0 \\ 0 & 0 & 0 & 0 & c_{66} \end{pmatrix} \begin{pmatrix} \varepsilon_1 \\ \varepsilon_2 \\ \varepsilon_4 \\ \varepsilon_5 \\ \varepsilon_6 \end{pmatrix} \tag{20.55}$$

with the usual uniaxial anisotropy condition $2c_{66} = c_{11} - c_{12}$. For a complete description of the elastic properties, we only need to calculate the moduli c_{11}, c_{44}, and c_{66} or any three independent linear combinations of these. Note c_{44} and c_{66} are the tilt and shear moduli, respectively, and $c_{\mathrm{L}} \equiv c_{11} - c_{66}$ is the compressional modulus.

For a flux bundle with a volume V, and keeping the number of the vortices in the bundle constant, the modulus c_{L} is given by

$$c_{\mathrm{L}} = V\left[\frac{\partial^2(V\mathscr{F})}{\partial V^2}\right]_{N=\mathrm{const}}, \tag{20.56}$$

where $\mathscr{F}$ is the free energy per unit volume of the flux lattice. Because $BV = \text{const.}$, we have $(\partial/\partial V) = -(B/V)(\partial/\partial B)$ and Eq. (20.56) becomes

$$c_{\mathrm{L}} = B^2\frac{\partial^2\mathscr{F}(B)}{\partial B^2} = \frac{B^2}{4\pi}\frac{\partial H(B)}{\partial B}, \tag{20.57}$$

since $\partial\mathscr{F}/\partial B = H/4\pi$.

The constant c_{44} is determined by tilting the lattice in the x–z plane through an angle θ. Since flux is conserved, $B\cos\theta$ is constant. For small tilt angles $\delta B \approx \theta B\delta\theta$, so the work done is

$$\delta W = \sigma_5\delta\theta = \frac{V}{4\pi}H\delta B = \frac{VHB}{4\pi}\theta\delta\theta = Vc_{44}\theta\delta\theta;$$

therefore

$$c_{44} = \frac{BH}{4\pi}. \tag{20.58}$$

To find c_{66} we shear the lattice in the x–y plane. In Sec. 14 we derived an expression (Eq. (14.28)) for the free energy density near H_{c2}:

$$\mathscr{F} = \mathscr{F}_{\mathrm{n}} - \frac{H_{\mathrm{c}}^2}{4\pi}\frac{\kappa^2}{1 + (2\kappa^2 - 1)\beta_{\mathrm{A}}}\left(1 - \frac{B}{H_{\mathrm{c2}}}\right)^2 + \frac{B^2}{8\pi}. \tag{20.59}$$

The general solution of the linearized G–L equation introduced in Sec. 14 has the form

$$\psi_{\mathrm{L}} = \sum_n C_n e^{inqy} \exp\left[-\frac{1}{2}\left(\frac{x - x_n}{\xi(T)}\right)^2\right] \tag{20.60}$$

with

$$x_n = \frac{n\hbar qc}{e^* H_{c2}}. \tag{20.61}$$

The function ψ_{L} of Eq. (20.60) is periodic in the y direction with a period

$$\Delta y = \frac{2\pi}{q}. \tag{20.62}$$

As discussed in Sec. 14, the periodicity in the x direction can also be established, if the coefficients C_n are periodic functions of n, such that $C_{n+v} = C_n$, where v is some integer. For the triangular lattice $v = 2$. From Eqs. (20.61) and (20.62), the periodicity in the x direction is

$$\Delta x = \frac{\hbar c}{e^* H_{c2}} \frac{2\pi}{\Delta y} = \frac{\hbar cq}{e^* H_{c2}}. \tag{20.63}$$

Therefore $\Delta x \Delta y H_{c2} = \phi_0$. Combining Eqs. (20.62) and (20.63) we see that a variation of q by δq introduces a shear strain $\varepsilon_6 = 2(\delta q/q)$ and an elastic energy density

$$\delta \mathscr{F} = \frac{1}{2} c_{66} \varepsilon_6^2 = 2c_{66}\left(\frac{\delta q}{q}\right)^2. \tag{20.64}$$

Shearing in the x–y plane causes a variation of β_{A} in Eq. (20.59) which can be expanded as a Taylor series in δq. For a triangular lattice, $\partial\beta_{\mathrm{A}}/\partial q = 0$, hence we must go to second order:

$$\delta\beta_{\mathrm{A}} \approx \frac{1}{2}\frac{\partial^2 \beta_{\mathrm{A}}}{\partial q^2}(\delta q)^2. \tag{20.65}$$

From Eq. (20.59) we have

$$\delta \mathscr{F} = \left[\frac{H_c^2}{4\pi} \frac{\kappa^2(2\kappa^2 - 1)}{[1 + (2\kappa^2 - 1)\beta_{\mathrm{A}}]^2}\left(1 - \frac{B}{H_{c2}}\right)^2 \frac{1}{2} q^2 \frac{\partial^2 \beta_{\mathrm{A}}}{\partial q^2}\right]\left(\frac{\delta q}{q}\right)^2. \tag{20.66}$$

For the triangular lattice $\beta_{\mathrm{A}} = 1.16$ and a numerical calculation yields

$$q^2 \frac{\partial^2 \beta_{\mathrm{A}}}{\partial q^2} = 1.921.$$

Comparing Eqs. (20.64) and (20.66) we obtain

$$c_{66} = \frac{0.48 H_c^2}{4\pi} \frac{\kappa^2(2\kappa^2 - 1)}{[1 + (2\kappa^2 - 1)\beta_{\mathrm{A}}]^2}\left(1 - \frac{B}{H_{c2}}\right)^2, \tag{20.67}$$

or in terms of the applied field, H_a (rather than the internal flux density B),

$$c_{66} = \frac{0.48 H_c^2}{4\pi} \frac{\kappa^2}{(2\kappa^2 - 1)\beta_{\mathrm{A}}^2}\left(1 - \frac{H_a}{H_{c2}}\right)^2. \tag{20.68}$$

Since $H_{c2} = 2^{1/2}\kappa H_c$, for high κ materials

$$c_{66} \approx 0.1 \frac{H_{c2}^2}{4\pi\kappa^2}\left(1 - \frac{B}{H_{c2}}\right)^2. \tag{20.69}$$

Near H_{c1} the expression for c_{66} is (Labusch 1967, 1969b)

$$c_{66} = \frac{1}{4\pi}\int_0^B \xi^2 \frac{d^2H(\xi)}{d\xi^2} d\xi. \tag{20.70}$$

For intermediate values of the applied field, a useful interpolation formula has been given by Brandt and Essmann (1987), which for large κ is

$$c_{66} \approx \frac{H_c^2}{16\pi} b(1 - 0.29b)(1 - b)^2, \qquad \kappa >> 1, \tag{20.71}$$

where $b = B/H_{c2}$.

A comparison of c_{66} with other moduli gives $c_{66}/c_{11} \approx c_{66}/c_{44} \sim 1/10\kappa^2$. For large κ materials it is clear that shear deformations are predominant, since they involve less energy. Note that the compressional modules, c_L, and the tilt modules, c_{44}, are finite and increase up to H_{c2}. In contrast, the shear modulus, c_{66}, goes quadratically to zero. Therefore near H_{c2} the flux-line lattice is soft for shear deformations.

In the above derivations a local elasticity theory has been assumed. It is valid in the long wavelength (slowly varying) limit of the lattice deformations (i.e. in the limit of $k \to 0$). In the case of short wavelength deformations, elastic nonlocality has to be taken into account. This can be understood as follows. The deformations are defined by the displacements of the vortex cores. The origin of the elastic energy of the flux-line lattice is the magnetic interactions; both the magnetic field and supercurrent associated with the flux lines vary on the scale of λ_L. For high κ superconductors (such as high T_c oxides), λ_L may be very large. Therefore, for deformations with $k > 1/\lambda_L$ (i.e., small wavelength), the displacement of the vortex cores is decoupled from the field and current profiles of the vortices. Consequently, the flux-line lattice is softer than for a homogeneous strain. The k-dependent tilt and compressional moduli, c_{44} and $c_{11} - c_{66}$, have been derived by Brandt (1977) within the framework of the G–L theory. The moduli have been rederived from microscopic theory (Larkin and Ovchinnikov 1978, 1979) and shown to apply even at low temperature. The derivation of Brandt is very lengthy and will not be repeated here. We only quote his final results:

$$c_{11}(k) = \frac{B^2}{4\pi}\frac{\partial H}{\partial B}\frac{1}{1 + k^2/k_h^2}\frac{1}{1 + k^2/k_\psi^2} + c_{66}, \tag{20.72}$$

$$c_{44}(k) = \frac{B^2}{4\pi}\frac{1}{1 + k^2/k_h^2}\frac{B(H - B)}{4\pi}, \tag{20.73}$$

where

$$k_h^2 \approx (1 - B/H_{c2})/\lambda_L^2$$

and

$$k_\psi^2 = 2\kappa^2(1 - B/H_{c2})/\lambda_L^2.$$

The spatial dispersion of c_{66} is small, since the induction B does not change for any wavelength of shear deformation.

From elementary elasticity theory, the elastic deformation can be uniquely defined by the elastic moduli $c_L \equiv c_{11} - c_{66}, c_{44}, c_{66}$, and a two-dimensional displacement, $\mathbf{u}$. The correction to the free energy resulting from elastic deformations of a vortex lattice is equal to (Labusch 1969a)

$$\delta F_{\mathrm{el}} = \frac{1}{2}\int \mathrm{d}^3 r \left[(c_{11} - c_{66}) \left(\frac{\partial \mathbf{u}}{\partial r}\right)^2 + c_{66} \sum_{\alpha,\beta} \left(\frac{\partial u_\beta}{\partial r_\alpha}\right)^2 + c_{44} \left(\frac{\partial \mathbf{u}}{\partial z}\right)^2 \right] \tag{20.74a}$$

and the associated elastic force density is

$$\mathbf{f}_{\mathrm{el}} = -\frac{\delta \mathscr{F}_{\mathrm{el}}}{\partial \mathbf{u}} = (c_{11} - c_{66}) \frac{\partial}{\partial r}\left(\frac{\partial \mathbf{u}}{\partial r}\right) + \left(c_{66} \frac{\partial^2}{\partial \mathbf{r}^2} + c_{44} \frac{\partial^2}{\partial z^2} \right) \mathbf{u}, \tag{20.74b}$$

where $\mathbf{r}$ is a two-dimensional vector in the x–y plane and the z axis is chosen in the direction of the magnetic field. In cartesian coordinates Eq. (20.74b) can be cast in the form

$$\mathbf{f}_{\mathrm{el}} = \underset{\sim}{\mathbf{D}}_2 \cdot \mathbf{u},$$

where the elastic matrix is given by

$$\underset{\sim}{\mathbf{D}}_2 \equiv \begin{bmatrix} c_{11}\dfrac{\partial^2}{\partial x^2} + c_{66}\dfrac{\partial^2}{\partial y^2} + c_{44}\dfrac{\partial^2}{\partial z^2} & (c_{11} - c_{66})\dfrac{\partial^2}{\partial x \partial y} \\ (c_{11} - c_{66})\dfrac{\partial^2}{\partial x \partial y} & c_{66}\dfrac{\partial^2}{\partial x^2} + c_{11}\dfrac{\partial^2}{\partial y^2} + c_{44}\dfrac{\partial^2}{\partial z^2} \end{bmatrix}. \tag{20.75}$$

20.8 Collective flux pinning

As argued earlier, a rigid flux-line lattice cannot be pinned by random pinning sites (pins); it is the elastic response of the flux-line lattice to these pinning forces which causes a finite volume pinning force, $\mathbf{f}_\mathrm{p}$, by correlating the positions of the pinned flux lines with the pin positions. The calculation of the average pinning force exerted by some distribution of pins is a complicated problem, and for most real systems the pins are either too strong, too extended or too correlated for a simple statistical theory to apply. In the limit of weak pinning and the assumption of random pins a model based on the collective pinning concept can be employed (Larkin and Ovchinnikov 1978, 1979).

The interaction between flux lines is electromagnetic in origin and hence has a range of order λ_L which can be very large, especially in high κ materials. On the other hand in a real system of pins the concentration of pinning centers, n_p, is often very large, resulting in $n_\mathrm{p}^{-1/3} << \lambda_\mathrm{L}$. This suggests that a collective treatment of the pinning force is appropriate. The collective nature of pinning is apparent from the fact that pinning occurs for both attractive and repulsive pins.

One of the striking features of collective pinning arising from weak pins is the absence of long-range order (Larkin and Ovchinnikov 1979). In an ordinary solid body, point defects do not break the long-range order, since both the defects (unoccupied lattice sites) and the atoms (occupied lattice sites) shift when the body is displaced. On the other hand, when the vortex lattice in a superconductor is displaced it moves independently of the pinning sites. The shift of the positions of the vortices caused by a defect decreases slowly with the distance from the defect. At large distances the shifts caused by the ensemble of defects accumulate in a random-like fashion, resulting in a gradual breakdown of the long-range order in the flux-line lattice. However, the local displacements due to weak pins are very small. Therefore, short-range order still exists over regions of volume V_c. If a current is flowing, the flux-line lattice within this correlation volume V_c (containing many pinning centers) displaces as a whole. Each such region may be considered as one large pinning center. Our approach will be to determine the total pinning force exerted on V_c and to estimate the relevant correlation lengths.

The average volume-pinning force, $\mathbf{f}_p$, may be estimated from the *fluctuation* of the total force exerted on V_c: within a volume V_c there are $n_p V_c$ pins and the statistical summation of these forces is therefore

$$V_c f_p = \langle n_p V_c \mathbf{F}_i^2 \rangle_{\text{pin}}^{1/2}, \tag{20.76}$$

where $\mathbf{F}_i$ is the actual force exerted by the ith pin and $\langle\,\rangle_{\text{pin}}$ represents the average over the pins. The square of the pinning forces averaged over a unit volume is defined as

$$w = \langle \mathbf{F}_i^2 \rangle_{\text{volume}} \approx n_p \langle \mathbf{F}_i^2 \rangle_{\text{pin}}, \tag{20.77}$$

and eliminating $\langle \mathbf{F}_i^2 \rangle^{1/2}$ using (20.76) we have

$$f_p \approx (w/V_c)^{1/2}. \tag{20.78}$$

The volume V_c may be estimated by studying the growth with $\mathbf{r}$ of the displacement correlation function,

$$g(\mathbf{r}) = \langle |\mathbf{u}(\mathbf{r}) - \mathbf{u}(0)|^2 \rangle_{\text{ensemble}}. \tag{20.79}$$

The average is taken over an ensemble of pin configurations. Qualitatively, when $g(\mathbf{r})$ is of order a^2 (where a is the flux-line lattice constant), the spatial long-range order is destroyed. More precisely we may define V_c from the condition

$$g(\mathbf{r}) \leq r_c^2, \tag{20.80a}$$

where r_c is the distance over which the pinning force *changes sign*, which is given approximately by $r_c \approx \xi$ for $b \equiv B/H_{c2} < 0.2$ for isolated vortices and by $r_c \approx a/2$ for $b > 0.2$ for overlapping vortices. Writing $\mathbf{r}$ as $(\mathbf{r}_\perp, z)$ we may define correlation lengths parallel, L_c, and perpendicular, R_c, to the local axis of the lattice as

$$g(R_c, 0) = r_c^2 \tag{20.80b}$$

and

$$g(0, L_c) = r_c^2. \tag{20.80c}$$

A crude estimate of R_c and L_c will be given below; more precise values will be developed later in this subsection (see Eqs. (20.93) and (20.94)). In 3D the correlation volume is an ellipsoid having a volume $V_c = (4\pi/3)R_c^2 L_c$; in 2D and 1D the relevant 'volumes' are the area πR_c^2 and the length $2L_c$.

The order of magnitude of V_c can be found from simple energy considerations. Suppose the flux-line lattice undergoes an elastic deformation due to the inhomogeneities. The energy density arising from the interaction between the flux line and the random pins in V_c is of order $a(w/V_c)^{1/2}$, where a is the lattice constant. Including the elastic energy due to the shear and tilt deformations the total energy change per unit volume, $\delta\mathcal{F}$, is of order

$$\delta\mathcal{F} \approx c_{66}(a/R_c)^2 + c_{44}(a/L_c)^2 - a(w/V_c)^{1/2}. \tag{20.81}$$

Setting $V_c \approx R_c^2 L_c$, and minimizing Eq. (20.81) with respect to R_c and L_c, we find

$$R_c \approx \frac{32^{1/2} c_{66}^{3/2} c_{44}^{1/2} a^2}{w}, \qquad L_c \approx \frac{8 c_{66} c_{44} a^2}{w}, \qquad V_c \approx \frac{256 c_{66}^4 c_{44}^2 a^6}{w^3}. \tag{20.82}$$

The critical current density follows from the critical state equation

$$\frac{1}{c} j_c B = F_p = (w/V_c)^{1/2} = \frac{w^2}{16 c_{66}^2 c_{44} a^3}. \tag{20.83}$$

From the above discussion we see that a collective pinning model characterizes the pins in terms of their 'strength' w and 'range' r_c and the flux-line lattice in terms of its linear elastic response. We now examine the problem in more detail within the framework of G–L theory as described by Brandt (1977, 1986a,b).

For a 3D isotropic medium the balance between the elastic and pinning force densities is expressed by

$$\underset{\sim}{\mathbf{D}}_2 \cdot \mathbf{u} + \mathbf{f}_p = 0, \tag{20.84}$$

where the elastic matrix $\underset{\sim}{\mathbf{D}}_2$ is given by Eq. (20.75); here we neglect for the moment viscous and external driving forces. We Fourier transform the pinning force density as

$$\mathbf{f}_p(\mathbf{r}) = \int \frac{d^3k}{(2\pi)^3} \mathbf{f}_p(\mathbf{k}) e^{i\mathbf{k}\cdot\mathbf{r}}; \tag{20.85}$$

the integration is over Brillouin zone of the 2D flux-line lattice and over $-\infty < k_z < \infty$ (the applied field is along the z direction). It follows from Eq. (20.75) that the Fourier transform of $\underset{\sim}{\mathbf{D}}_2$ is

$$\underset{\sim}{\mathbf{D}}_2(\mathbf{k}) = \begin{bmatrix} c_{11}k_x^2 + c_{66}k_y^2 + c_{44}k_z^2 & (c_{11} - c_{66})k_x k_y \\ (c_{11} - c_{66})k_x k_y & c_{11}k_y^2 + c_{66}k_x^2 + c_{44}k_z^2 \end{bmatrix}. \tag{20.86}$$

Using Eq. (20.84), the displacement $\mathbf{u}(\mathbf{r})$ is given by

$$\mathbf{u}(\mathbf{r}) = -\int \frac{d^3k}{(2\pi)^3} [\underset{\sim}{\mathbf{D}}_2(\mathbf{k})]^{-1} \cdot \mathbf{f}_p(\mathbf{k}) e^{i\mathbf{k}\cdot\mathbf{r}}. \tag{20.87}$$

For simplicity we calculate the displacement correlation function, Eq. (20.79), for point pins exerting forces $\mathbf{F}_j$ at random positions $\mathbf{r}_j$ on the flux-line lattice. Thus $\mathbf{f}_p(\mathbf{r})$ in Eq. (20.85) has the form

$$\mathbf{f}_p(\mathbf{r}) = V_c^{-1} \sum_j \mathbf{F}_j(\mathbf{r} - \mathbf{r}_j). \tag{20.88}$$

From Eq. (20.79) we obtain

$$g(\mathbf{r}) = \int \frac{d^3k\, d^3k'}{(2\pi)^6} (e^{i\mathbf{k}\cdot\mathbf{r}} - 1)(e^{-i\mathbf{k}'\cdot\mathbf{r}} - 1)[\underset{\sim}{\mathbf{D}}_2(\mathbf{k})]_{\alpha\gamma}^{-1} [\underset{\sim}{\mathbf{D}}_2(\mathbf{k}')]_{\beta\gamma}^{-1}$$

$$\times \left\langle V_c^{-2} \sum_{jj'} F_{j\alpha} F_{j'\beta} e^{-i\mathbf{k}\cdot\mathbf{r}_j} e^{i\mathbf{k}'\cdot\mathbf{r}_{j'}} \right\rangle. \tag{20.89}$$

Here α, β, γ denote the in-plane (x, y) components and the summation convention is used. After ensemble averaging, only the terms with $j = j'$ in the double sum are nonzero since the pin positions are assumed to be uncorrelated. The resulting sum over the pins may be replaced by an integral over the pin position with the F_j^2 replaced by their average, i.e.,

$$V_c^{-1}\left\langle \sum_j F_{j\alpha}F_{j\beta}e^{-i(\mathbf{k}-\mathbf{k}')\cdot\mathbf{r}_j}\right\rangle \approx \delta_{\alpha\beta}n_pF_j^2\int d^3r_j e^{-i(\mathbf{k}-\mathbf{k}')\cdot\mathbf{r}_j}$$

$$\approx (2\pi)^3 w\delta(\mathbf{k}-\mathbf{k}')\delta_{\alpha\beta}. \tag{20.90}$$

Substituting (20.90) into Eq. (20.89) and carrying out the integration over d^3k' we obtain

$$g(\mathbf{r}) = \int\frac{d^3k}{(2\pi)^3}[1-\cos(\mathbf{k}\cdot\mathbf{r})]w\mathrm{Tr}\{[\underset{\sim}{\mathbf{D}}_2(\mathbf{k})]^{-1}\cdot[\underset{\sim}{\mathbf{D}}_2(\mathbf{k})]^{-1}\}$$

$$= w\int\frac{d^3k}{(2\pi)^3}[1-\cos(\mathbf{k}\cdot\mathbf{r})]\left[\frac{1}{(c_{66}k_\perp^2+c_{44}k_z^2)^2}+\frac{1}{(c_{11}k_\perp^2+c_{44}k_z^2)^2}\right], \tag{20.91}$$

where $k_\perp^2 \equiv k_x^2 + k_y^2$.

As mentioned in the previous subsection, in general the elastic moduli c_{11} and c_{44} depend on the wave number of the displacement field $\mathbf{u}$. In the long wavelength limit, the stress at position $\mathbf{r}$ (i.e., the elastic response) is 'local'. We evaluate Eq. (20.91) in the local limit with the k-independent tilt modulus c_{44} given by Eq. (20.58). We also neglect the term $1/(c_{11}k_\perp^2 + c_{44}k_z^2)^2$ in Eq. (20.91), since $c_{11} >> c_{66}$. For the 3D infinite flux-line lattice we may make the transforms

$$\mathbf{k}' = \hat{\mathbf{x}}k_x + \hat{\mathbf{y}}k_y + \hat{\mathbf{z}}(c_{44}/c_{66})^{1/2}k_z$$

and

$$\mathbf{r}' = \hat{\mathbf{x}}x + \hat{\mathbf{y}}y + \hat{\mathbf{z}}(c_{66}/c_{44})^{1/2}z$$

such that $\mathbf{k}'\cdot\mathbf{r}' = \mathbf{k}\cdot\mathbf{r}$. Introducing spherical coordinates and extending the integral to infinity Eq. (20.91) becomes

$$g(\mathbf{r}) = \frac{wr'}{8\pi c_{66}^{3/2}c_{44}^{1/2}} = \frac{w}{8\pi c_{66}^{3/2}c_{44}^{1/2}}\left[x^2+y^2+\left(\frac{c_{66}}{c_{44}}z^2\right)\right]^{1/2}. \tag{20.92}$$

Combining Eqs. (20.80) and (20.92) the corresponding correlation lengths R_c and L_c can be estimated using the definitions (20.80b) and (20.80c) which yields

$$R_c \approx r_c^2 8\pi c_{66}^{3/2}c_{44}^{1/2}/w \tag{20.93}$$

and

$$L_c = (c_{44}/c_{66})^{1/2}R_c. \tag{20.94}$$

We see that the correlation lengths R_c and L_c, and hence the correlation volume V_c, are determined by the pinning strength w, the 'range' r_c of the pinning force, and the elastic moduli of the flux-line lattice. The critical current density can be calculated by combining Eqs. (20.78), (20.93), and (20.94) and using $V_c = (4\pi/3)R_c^2L_c$:

$$\frac{1}{c}J_cB = \frac{(3/2)^{1/2}w^2}{32\pi^2r_c^3c_{66}^2c_{44}}. \tag{20.95}$$

For a more complete treatment using the nonlocal elastic theory, we refer the reader to Larkin and Ovchinnikov (1979) and Brandt (1977, 1986a,b). These authors also studied the correlation function $g(\mathbf{r})$ in superconducting films. Since w is small and $L_c >> R_c$, L_c may be larger than the thickness, d, of the film when the field is perpendicular to the film surface. In this case the flux-line deformations in the field direction may be ignored, so that only the 2D order in the x–y plane is relevant. A more rigorous analysis shows that the 3D–2D crossover in the collective flux pinning occurs at $L_c = d/2$.

20.9 Mechanisms of flux motion

Flux-line lattices are not absolutely static. At finite temperatures thermally activated flux lines can hop between neighboring pinning positions in response to the driving force of a flux-density gradient. The resulting flux creep, as first pointed out by Anderson (1962), manifests itself in two ways: (1) it leads to a slow relaxation of the magnetization (or trapped magnetic fields); and (2) it leads to a measurable resistive voltage.

Flux creep is observed both in the conventional low T_c superconductors and in the high T_c oxides. Flux creep is especially pronounced in high T_c superconductors since usually high temperatures and low pinning potentials are involved. This is demonstrated, for instance, by the 'giant' flux creep and by the broadening of the resistive transition in magnetic fields which is a direct consequence of thermally activated flux creep. Significant relaxation of the magnetization (and hence the critical current density) is a very important factor limiting the current-carrying capacity of these materials.

We will discuss (and parametrize) the thermally activated flux hopping in terms of the flux creep model of Anderson and Kim (1964). In a Type II superconductor in the mixed state, the flux lines will usually be pinned due to interactions with various imperfections. These flux lines may correlate to form a flux bundle (especially in the weak pinning limit where collective flux pinning is justified) having a finite volume. The flux bundle can be regarded as an entity located at the bottom of a pinning-potential well of depth U_0. The activation energy, U_0, is the difference in the Gibbs free energy of the system between a situation in which the flux-line bundle is trapped in the well and one in which it can move. The model assumes that in the absence of a driving force, a flux-line bundle can be thermally activated out of a well at a rate (both forward and backward)

$$\nu = \nu_0 e^{-U_0/k_B T}, \tag{20.96}$$

where ν_0 is some attempt frequency. As the hopping rates are the same in both directions, there is no net motion of the flux-line bundle. When a flux-density gradient is present, the flux-line bundle experiences a Lorentz force. We can construct an energy, U_L, from the product of the Lorentz force density $\mathbf{f}_L = (1/c)\mathbf{J} \times \mathbf{B}$, the volume of the flux bundle that moves independently of the other flux bundles, V_c, and the range of the pinning potential r_p as

$$U_L = \frac{1}{c} J B V_c r_p. \tag{20.97}$$

When the effect of U_L is added, the original pinning potential becomes a washboard potential as

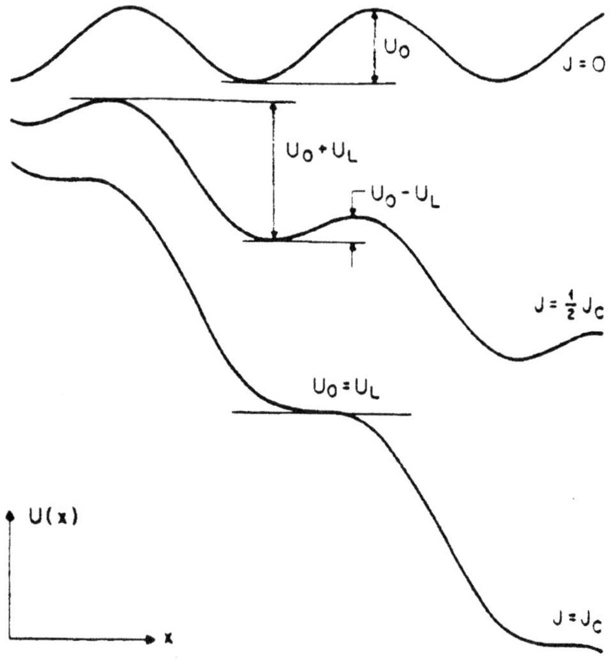

Figure 20.6 Schematic diagram of the condensation energy along a direction parallel to the Lorentz force. The highest curve shows the unperturbed potential, U_0, and the lowest curve shows depinning ($U_0 = U_L$).

shown in Fig. 20.6. The rate of forward hopping (in the direction of the Lorentz force) is then given by

$$\nu_f = \nu_0 e^{-(U_0 - U_L)/k_B T}, \tag{20.98}$$

while the rate of backward hopping (in the direction opposite to that of the Lorentz force) is given by

$$\nu_b = \nu_0 e^{-(U_0 + U_L)/k_B T}]. \tag{20.99}$$

There is now a net forward rate of hopping given by the difference between these two expressions

$$\nu_{net} = 2\nu_0 e^{-U_0/k_B T} \sinh(U_L/k_B T). \tag{20.100}$$

The average flux-drift velocity is $v = \nu_{net}\ell$, where ℓ is the average distance over which hopping takes place. As the flux bundles move they generate an electric field $E = vB/c = \rho J$. For a current density, J, the electric field, E, arising from thermally activated flux hopping is thus

$$E = \frac{2\nu_0 B\ell}{c} e^{-U_0/k_B T} \sinh(U_L/k_B T). \tag{20.101}$$

For simplicity, we employ the pinning potential, U_0, of the critical state model, i.e., $U_0 = J_{c0} B V_c r_p / c$; here J_{c0} is the critical current density without flux creep. Depending on the relative amplitude of the quantities U_0, U_L and $k_B T$, we can distinguish three interesting regimes in the $E(J)$ relation:

(1) Thermally assisted flux flow (TAFF)

For current densities and temperatures such that $J << J_{c0}$ and $U_L << k_B T$, we may expand the $\sinh(U_L/k_B T)$ term in Eq. (20.101) to first order yielding

$$E = \frac{2v_0 B\ell J U_0}{cJ_{c0}k_B T} e^{-U_0/k_B T}. \tag{20.102}$$

In this regime we have ohmic behavior with a thermally activated resistivity $\rho_{TAFF} \propto e^{-U_0/k_B T}$. This kind of ohmic behavior has been unambiguously observed in high T_c oxides in the small current limit. We will refer to this process as the TFF (or TF for short) regime.

(2) Flux creep (FC)

Flux creep occurs when $J \approx J_{c0}$, i.e., when the superconductor is in the critical state. In this regime we have $U_L >> k_B T$ and $U_0 >> k_B T$. For this case, as shown in Fig. 20.6, backward hopping is negligible. The induced electric field is given by

$$E = \frac{v_0 B\ell}{c} e^{-U_0(1 - J/J_{c0})/k_B T}. \tag{20.103}$$

It is worth noting that in Eqs. (20.102) and (20.103) the attempt frequency, hopping distance, flux-bundle volume, etc., are phenomenological parameters. All these quantities depend on the specific mechanisms governing the flux dynamics and pinning. The related microscopic models are still controversial. However, by defining a voltage criterion E_c for the critical current density (say, $E_c \cong 1\ \mu$V/cm), the flux creep can be described by directly measured macroscopic quantities. For a chosen E_c, Eq. (20.103) becomes

$$E = E_c \exp\left[-\frac{U_0}{k_B T}\left(1 - \frac{J}{J_{c0}}\right)\right] \tag{20.104}$$

and

$$J = J_{c0}\left[1 + \frac{k_B T}{U_0}\ell n\left(\frac{E}{E_c}\right)\right]. \tag{20.105}$$

(3) Flux flow (FF)

When a transport current **J** is present in a Type II superconductor, the flux lines are subject to a Lorentz driving force exerted by the flux-density gradient. In an ideal Type II superconductor, where the flux pinning is absent, the Lorentz force density $\mathbf{J} \times \mathbf{B}/c$ will cause the flux lines to move with a velocity v_L. This motion dissipates energy owing to a viscous-drag force density $-\eta v_L$ with η being a phenomenological viscosity coefficient of the medium. In practice lattice defects of various kinds are present in the material which effectively pin the flux lines and enable the specimen to sustain a finite transport current density **J**, which in the critical state is just the critical current density, $\mathbf{J}_c$. However, when the Lorentz driving force exceeds the pinning force the flux lines undergo a viscous flow and the specimen enters a 'flux flow' state. The associated energy dissipation can sometimes lead to a thermal instability (or flux jumps) when the flux lines are forced to move against a large and irregular pinning force. The flux flow state is characterized by the following force balance equation (see Subsec. 20.5)

$$\eta \mathbf{v}_L = \mathbf{f}_L + \mathbf{f}_P. \tag{20.106}$$

When a steady state (with a constant $\mathbf{J}$ and $\mathbf{v}_L$) is maintained the electric field generated by the flux motion is

$$\mathbf{E} = \mathbf{B} \times \mathbf{v}_L/c. \tag{20.107}$$

Combining Eqs. (20.13), (20.106), and (20.107), for the ideal case where there is no pinning we may define a differential flux flow resistivity as

$$\rho_f = \frac{dE}{dJ} = B^2/\eta c^2. \tag{20.108}$$

The measurement of ρ_f thus yields the viscosity coefficient η. The motion of flux lines clearly involves a time variation of the order parameter and a more complete analysis of the problem requires time-dependent G–L theory, as discussed by Schmid (1966), Caroli and Maki (1967a,b, 1968), Thompson and Hu (1971), and Hu and Thompson (1972). Some simplified models have been proposed by Bardeen and Stephen (1965) and by Tinkham (1964). The problem was reviewed in some detail by Tinkham (1975). Here we will derive an empirical expression for ρ_f using an interpolation scheme based on simple phenomenological arguments. Assuming the electrical resistance is contributed only by the normal cores, and that the current does not bypass them, we expect the viscosity coefficient, η, to be proportional to the flux density, B; i.e., $\eta = kB$ for $b = B/H_{c2} << 1$. For $b = 1$ we must obtain the normal resistivity, ρ_n; the constant is then given by $k = H_{c2}/\rho_n c^2$ and the flux flow resistivity is given by

$$\rho_f = \rho_n \frac{B}{H_{c2}}. \tag{20.109}$$

The flux flow in conventional low T_c Type II superconductors has been extensively studied experimentally by Kim *et al.* (1963, 1965) and Kim and Stephen (1969). They found that for $f_L > f_P$, E increases rapidly and becomes linear in J, indicating that $E \propto v_L$ as was assumed in Eq. (20.107). The flux-flow resistivity is determined by the bulk superconducting properties of the material and is independent of pinning. This can be clearly seen in Fig. 20.7. Thus we can write the E vs J relation in the flux-flow regime as

$$E = \rho_f J_c \left(\frac{J}{J_c} - 1 \right), \qquad J > J_c. \tag{20.110}$$

The general E vs J characteristics for a Type II superconductor are shown schematically in Fig. 20.7(a). For comparison, the approximate E vs J curve according to the critical state model is shown in Fig. 20.7(b).

20.10 Relaxation of the magnetization with time

Flux creep in superconductors can be formulated in terms of a nonlinear diffusion of magnetic flux through a sample (Beasley, Labusch, and Webb 1969). The process is described by the Maxwell equations,

$$\nabla \times \mathbf{E} = -\frac{1}{c}\frac{\partial \mathbf{B}}{\partial t} \tag{20.111}$$

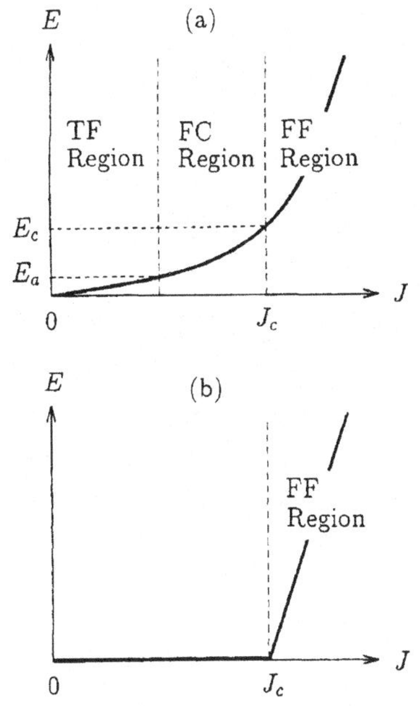

Figure 20.7 (a) Schematic illustration of the E vs J curve in different regimes. E_a is the onset field for the FC regime. (b) Approximate E vs J curve according to the critical state model.

and

$$\nabla \times \mathbf{H} = \frac{4\pi}{c}\mathbf{J}, \tag{20.112}$$

in conjunction with the constitutive equation $\mathbf{E} = \mathbf{E}(\mathbf{J})$, which may involve various resistivity mechanisms. To illustrate the basic idea, we consider a slab of thickness $2a$ along the x axis and infinite in the y–z plane with the external magnetic field $\mathbf{H}$ parallel to the z axis (see Fig. 20.8). We confine ourselves to the field region $H_{c1} << H << H_{c2}$ so that $B \approx H$. For the constitutive equation we use Eq. (20.104). For this planar case the electric field $\mathbf{E} = \hat{\mathbf{y}}E$ has only a y component. Differentiating Eq. (20.111) with respect to x and using (20.112) and (20.104) we have

$$\frac{\partial^2 E}{\partial x^2} = g(E)\frac{\partial E}{\partial t}, \tag{20.113}$$

where

$$g(E) = \frac{4\pi}{c^2}\left(\frac{\partial E}{\partial J}\right)^{-1} = \frac{4\pi J_{c0}}{c^2}\frac{k_B T}{U_0}\frac{1}{E}. \tag{20.114}$$

A complete solution of the nonlinear partial differential equation (20.113) would involve the temporal and spatial boundary conditions as determined by the field ramp rate and the field configuration. We will seek a special solution which is independent of the initial conditions and gives the asymptotic (long-time) behavior of $E(x, t)$. For the usual flux creep measurements, either the external field is ramped to a finite value, after which the relaxation of the magnetization with

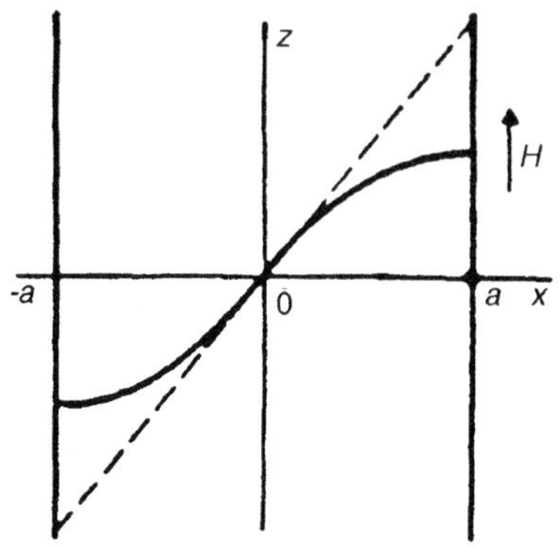

Figure 20.8 A slab of thickness $2a$ in a parallel magnetic field. The solid and dashed curves show the electric field profile $E(x, t)$ and the initial distribution $E(x, 0)$, respectively.

time is measured, or the applied external field is switched off, after which the time decay of the trapped field is measured. In both these cases the proper boundary condition is

$$E'(\pm a, t) = 0. \tag{20.115}$$

The corresponding electric field profile $E_y(x, t)$ is shown in Fig. 20.8. A solution which obeys both Eq. (20.113) and the boundary condition (20.115) is

$$E(x, t) = \left(\frac{4\pi J_{c0}}{c^2}\frac{k_B T}{U_0}\right)\frac{2ax - x^2 \operatorname{sgn} x}{2t}. \tag{20.116}$$

Substituting (20.116) into (20.113) we have

$$-\frac{1}{t} = \frac{1}{E}\frac{\partial E}{\partial t}. \tag{20.117}$$

Upon integrating (20.117) and using (20.104) we finally obtain

$$J = J_{c0}\left[1 - \frac{k_B T}{U_0}\ell n\left(\frac{t}{\tau_0}\right)\right], \tag{20.118}$$

where $\tau_0 \sim 1/E_c$ depending on the particular criterion chosen for E_c. Eq. (20.118) describes the long-time relaxation behavior of the current in the framework of the Anderson–Kim model (1964). In actual experiments, it is the time decay of the magnetization that is measured. The magnetization, $M(t)$, is a macroscopic quantity depending on the sample geometry. Using the Bean critical state model, for the slab geometry considered here, the magnetization for the case of full field penetration (see Subsec. 20.6) is simply proportional to the critical current and is given by

$$M_0 = \frac{4\pi J_{c0} a}{c}.$$

Hence the relaxation of the magnetization is governed by the following expression

$$M(t) = M_0\left[1 - \frac{k_B T}{U_0}\ell n\left(\frac{t}{\tau_0}\right)\right]. \tag{20.119}$$

Eq. (20.119) predicts the well-known logarithmic decay of the magnetization with time. The effective pinning potential and its temperature dependence can be deduced from flux creep measurements. However, M_0 in Eq. (20.119) is not a well-defined quantity. It is convenient to evaluate a normalized relaxation rate defined as the logarithmic derivative of the magnetization which in the case of Eq. (20.119) yields

$$- \mathrm{d}\ell\mathrm{n}[M(t)]/\mathrm{d}\ell\mathrm{n}(t) = k_B T/[U_0 - k_B T \ell\mathrm{n}(t/\tau_0)]. \qquad (20.120)$$

In the low temperature limit, this is just $k_B T/U_0$. Therefore a plot of the logarithmic derivative of the magnetization with time vs the temperature can be used to determine U_0.

20.11 Phase diagram of high T_c oxide superconductors

The behavior of superconductors in the presence of a magnetic field has been the subject of much scientific, as well as practical, interest. On the practical side, the design and performance of superconducting magnets are often limited by the suppression of superconductivity in large magnetic fields. The equilibrium theory for the behavior of Type II superconductors was developed by Abrikosov (1957). From discussions in the previous sections we know that when the applied magnetic field is smaller than the lower critical field H_{c1}, an ideal conventional (low T_c) Type II superconductor is in the Meissner state; the superconductor shows perfect diamagnetism (except within a London depth of the surface). When the magnetic field is greater than H_{c1} and less than the upper critical field, H_{c2}, it penetrates the superconductor in the form of quantized flux lines each carrying exactly one quantum of magnetic flux, ϕ_0. The superconductor is thus in the mixed or the Abrikosov vortex lattice phase. In the mixed phase, the flux lines, which run parallel to the magnetic field, are arranged in a regular triangular crystalline lattice. The mixed phase has two spontaneously broken symmetries. First, there is phase coherence in the pairing field. Second, there is crystalline long-range of the vortex lattice, which breaks the translational symmetry. In the absence of flux pinning, this phase is not truly superconducting; it will have a nonzero ohmic resistivity due to flux flow. When the magnetic field reaches H_{c2}, the superconductor undergoes a second order phase transition and enters the normal state. The mean field phase diagram in the H–T plane is shown in Fig. 20.9.

Beyond the mean-field picture for an ideal Type II superconductor, the dynamic properties of the vortex system are modified by the existence of various disorders. The static disorder (crystalline defects) affecting the superconducting order parameter will contribute to a finite pinning force density, f_p, which counteracts the driving Lorentz force. As a result, the technological usefulness of the Type II superconductors is reestablished. The critical depinning current density, defined by the critical state equation, $J_c = cf_p/B$, is always bounded by the depairing current density, $J_0 = cH_c/3\sqrt{6}\pi\lambda_L$, determined by the thermodynamic critical field H_c. The dimensionless ratio J_c/J_0 is a parameter characterizing the strength of the volume pinning force.

Thermal fluctuations are another source of disorder having significant consequences both for the phase diagram of the vortex system and for its dynamic properties (for a review see Blatter *et al.* (1994)). For example, the smoothing of the pinning potential due to thermal fluctuations may lead to thermal depinning of the flux lines, resulting in flux creep. As the fundamental

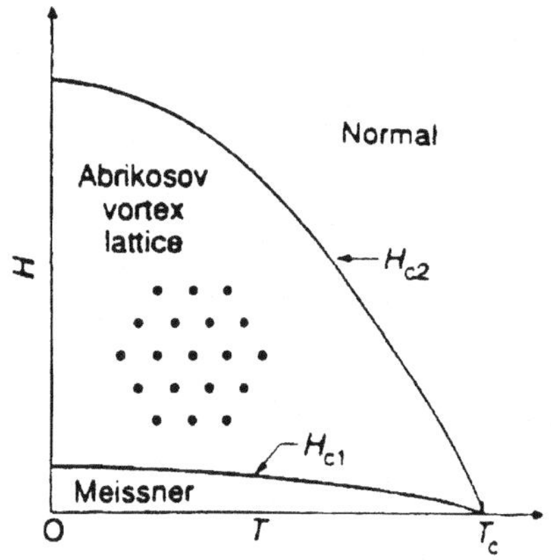

Figure 20.9 Schematic phase diagram of a conventional low T_c Type II superconductor.

parameter governing the strength of thermal fluctuations, we may introduce the Ginzburg number, $G = [k_B T_c/H_c^2(0)\xi_{\parallel}^2\xi_{\perp}]^2/2$, which measures the relative size of the transition temperature and the condensation energy within a coherence volume.

Fig. 20.10 shows a phenomenological phase diagram for the new high temperature oxide superconductors. There are three factors (not completely unrelated) that make the high temperature superconductors drastically different from the conventional low T_c superconductors: (1) The transition temperature, T_c, is very large and hence the coherence length $\xi \propto \hbar v_F/k_B T_c$ is small. (2) The penetration depth, λ, is very large. (3) The layered structure of the oxides introduces a large uniaxial anisotropy in the system, and the resulting anisotropy parameter $\gamma \equiv \xi_{\parallel}/\xi_{\perp}$ is much larger than unity. The difference may be quantified by comparing the corresponding parameters for the pinning strength and the thermal fluctuation. In conventional superconductors pinning is usually strong, $J_c/J_0 \sim 10^{-2}$–10^{-1}, while thermal fluctuations are weak, $G \sim 10^{-8}$. In high T_c superconductors such as $YBa_2Cu_3O_{7-\delta}$ (YBCO), pinning is usually weak, $J_c/J_0 \sim 10^{-3}$–10^{-2}, while thermal fluctuations are significant, $G \sim 10^{-2}$. As a result, a wealth of novel phenomena have been revealed experimentally in high temperature superconductors, including a broadening of the resistive transition in a finite magnetic field (Tinkham 1988), the existence of a distinct irreversibility line far below the mean field H_{c2}, and the presence of a giant flux creep (Yeshurun and Malozemoff 1988) even at temperatures far below T_c.

We now discuss the irreversibility line in some detail. Experimentally it is found that high T_c superconductors show a wide region in the H–T plane where diamagnetism is present but $J_c = 0$. In the H–T plane, there is a distinct line $H_{irr}(T)$, called the irreversibility line, which separates the regions of irreversible (low H, T) and reversible magnetic behavior. The irreversibility line can be determined from dc magnetization measurements, where the temperature is swept at constant field and the point where the zero-field-cooled and the field-cooled branches merge is identified, or from ac susceptibility measurements, in which the loss peak is usually the criterion for irreversibility.

A consistent theory for the irreversibility line is still lacking at present. A simple explanation is that it arises from thermally activated depinning. Above the irreversibility line thermally activated flux flow is present. At still higher magnetic fields and temperatures (up to H_{c2}) flux flow occurs resulting in completely reversible magnetization curves. For this reason, the irreversibility

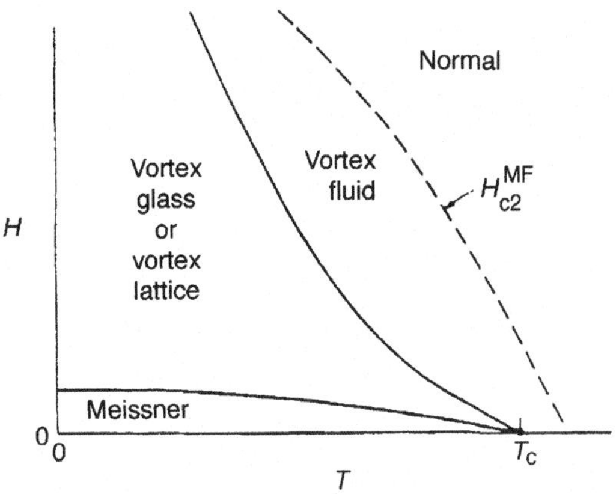

Figure 20.10 Schematic phase diagram of a Type II superconductor with strong thermal fluctuations, where H_{c2}^{MF} is the mean field behavior of the upper critical field.

line is also called the depinning line. In this picture, the resistivity is given by Eq. (20.102), which is nonzero at all nonzero temperatures. This is a fundamentally single-particle view of the flux dynamics.

A different (and more controversial) point of view is that the irreversibility line actually reflects a vortex lattice melting transition from a low temperature ordered phase into a high temperature vortex liquid. This transition is driven by the strong thermal fluctuations characterized by the large Ginzburg number, G. The vortex lattice melting line is usually determined by the Lindemann criterion, $\langle u^2(T)\rangle_{th}^{1/2} \simeq c_L a_0$; i.e., the spatial extent of the thermal displacement is of order the lattice constant, a_0, where the Lindemann number $c_L \sim 0.1$–0.4 (Houghton *et al.* 1989).

In the melting or phase transition picture, many-body effects are important. Taking into account the pinning disorder, Fisher *et al.* (1991) postulated a vortex glass transition. In the vortex glass phase, the vortices at low temperatures are frozen into a particular random configuration, as determined by the details of the pinning centers, and thus are not free to move, so that the ohmic linear resistivity is strictly zero. In the vortex glass model there is a true thermodynamic phase transition at T_g, between the vortex liquid and vortex glass phases. Associated with this transition there is a diverging correlation length given by $\xi_G \sim (T - T_g)^{-\nu}$. The theory predicts that both the linear response resistivity, R, and the current scale for nonlinear response vanish as powers of $(T - T_g)$ near T_g. Measurements by Gammel, Schneemeyer, and Bishop (1991) with picovolt sensitivity on YBCO single crystals provide compelling support for the vortex glass phase transition.

The layered structure of high T_c superconductors leads to additional novel features. Whereas the YBCO compound still can be described by the anisotropic London model, the Bi- and Tl-based compounds, where the interlayer coupling is weaker, are better described by the discrete Lawrence–Doniach model. Superconductivity in these materials can then become quasi-2D over a large portion of the H–T plane. As shown in Fig. 17.4, the simple rectilinear vortex is replaced by a more complicated object consisting of an array of 2D 'pancake vortices' interconnected by coreless Josephson vortices running within the 'insulating layers' between two successive superconducting planes.

21 Time-dependent G–L theory

A system which is disturbed from its equilibrium state will relax toward equilibrium when the associated perturbation is removed. This relaxation process has two characteristic forms in general: (i) damped oscillations, or (ii) monotonic decay. In this section we will be concerned primarily with the latter type of decay (oscillatory responses of bulk superconductors will be discussed in Sec. 54 and are termed collective modes).

Thermodynamic systems have characteristic microscopic (rapid) relaxation times[1] and the relaxation of the vast majority of their internal degrees of freedom proceeds on these time scales. There are two important exceptions: (i) modes involving degrees of freedom for which conservation laws exist; and (ii) additional modes involving a broken symmetry of the system. The conservation laws are those involving mass (or charge), energy (or entropy) and momentum (which, being a vector quantity, involves an equation for each component). For a liquid this leads to the existence of five low frequency (also called hydrodynamic) modes involving: a heat conduction mode (1), transverse viscous shear waves (2), and longitudinal sound[2] (2). The number in parentheses denotes the number of such modes, there being five in all, corresponding to the five conservation laws. For a derivation of this mode structure see Landau and Lifshitz (1987) or Miyano and Ketterson (1979). Additional equations of motion exist when the system spontaneously breaks some symmetry, and we now list some examples. When a liquid freezes a lattice is formed and the macroscopic specification of the lattice sites results in three additional equations of motion resulting in a total eight modes (longitudinal sound (2), fast and slow shear waves (2 + 2), heat diffusion (1), and vacancy diffusion (1)). Liquid crystals, which have various degrees of orientational and translational order, have an intermediate number of modes. In superfluid ^{4}He the order parameter has the same form, $\psi = |\psi| e^{i\Phi}$, as the superconductor. The superfluid velocity obeys the Josephson-like equation[3] $\mathbf{v}_s = (\hbar/M_4)\nabla\Phi$; the resulting six equations lead to six modes in superfluid ^{4}He: longitudinal sound (2), second sound (2), and viscous shear modes (2) (note the diffusive heat conduction mode has been replaced by a pair of propagating second sound or temperature waves). The second sound mode is a counter-oscillation of the normal and superfluid densities with a constant total density.

1. Examples are elastic and inelastic scattering times associated with electrons and phonons.
2. Each sound wave counts as two modes since the equations have the solutions $\omega = \pm V_s k$ where V_s is the velocity of sound; the two viscous shear waves involve the two orthogonal polarizations.
3. If we combine this with the first Josephson equation $\hbar\dot{\Phi} = \mu'$, where μ' is the chemical potential per particle, we obtain $\partial \mathbf{v}_s/\partial t = (1/M_4)\nabla\mu' = \nabla\mu$ where μ is the chemical potential per unit mass. This latter equation is Landau's acceleration equation for the superfluid.

In a charged superfluid (a superconductor) the longitudinal sound modes do not exist since variations in the charge density oscillate at the plasma frequency. The second sound mode also does not exist since it requires that the electron–lattice scattering time be longer than the electron–electron scattering time, a condition which is difficult to achieve in a metal.[4]

Note that while we had an equation of motion involving the phase of the order parameter, we did not introduce an equation of motion for the magnitude, $|\psi|$. The time scale for achieving phase variations is necessarily slow (hydrodynamic) whereas $|\psi|$ may change rapidly (by the conversion of superfluid into normal fluid and vice versa). An exception is near the second order phase transition where the free energy becomes 'soft' with respect to a variation of the magnitude of the order parameter and the relaxational dynamics slows down. This intuitive idea was quantified by Landau and Khalatnikov (1954) in their discussion of relaxation phenomena in superfluid ^{4}He near the lambda point; they made the ansatz

$$\frac{\partial|\psi|}{\partial t} = -\gamma\frac{\partial F}{\partial|\psi|}, \tag{21.1}$$

where γ is a rate constant ($\gamma > 0$). For an inhomogeneous superfluid we would replace $\partial F/\partial|\psi|$ with the functional derivative $\delta F/\delta|\psi|$. Note the static case then corresponds to the usual minimization condition of the G–L free energy functional, $\delta F = 0$. When applying (21.1) to a charged superfluid, a superconductor, the question immediately arises as to whether the time derivative should be written in its gauge-invariant form

$$\frac{\partial}{\partial t} \to \frac{\partial}{\partial t} + \frac{\mathrm{i}e^*}{\hbar c}\phi \tag{21.2}$$

where ϕ is the electrostatic potential (which may be nonzero in a dynamic state). We may answer this question by imposing the general thermodynamic requirement that the rate of generation of entropy must be positive definite. The rate of entropy generation, $\dot{S}$, is given by

$$\dot{S} = \frac{\mathrm{d}}{\mathrm{d}t}\int \rho s\,\mathrm{d}^3r = \int\frac{\partial}{\partial t}(\rho s)\mathrm{d}^3r, \tag{21.3}$$

where s is the entropy per unit charge and ρ is the charge density. We recall the differential form of the combined first and second laws of thermodynamics which in the present context takes the form[5]

$$\mathrm{d}\varepsilon_{\mathrm{in}} = T\mathrm{d}(\rho s) + \mu'\mathrm{d}\rho; \tag{21.4}$$

4. Hydrodynamic modes involve frequencies $\omega << 1/\tau$, where τ is a characteristic time for equilibrium within the locally moving excitation gas, and for the normal fluid this would involve the electron–electron scattering time, τ_{ee}. If the electron–lattice scattering time, $\tau_{e\ell}$, is much shorter than τ_{ee}, the energy of the second sound mode would be rapidly transferred to lattice degrees of freedom; i.e., it would be rapidly damped. In superfluid ^{4}He there is no lattice and this relaxation mechanism is absent. There is another mode in superfluid ^{4}He which occurs when the liquid is dispersed in a solid matrix which clamps the normal fluid (the matrix functions loosely as a lattice). We then have a density wave in the superfluid alone which is referred to as fourth sound. The analogue of this mode in a superconductor has been reported by Carlson and Goldman (1975).
5. When writing our equations in terms of quantities per unit volume, the pressure (which enters the usual form of the first law as $p\mathrm{d}V$) will not appear.

here μ' is the chemical potential per unit charge ($\equiv \mu/e$) and ε_{in} is the internal energy per unit volume. In the presence of a moving superfluid background, a term involving the vector potential must be added to the right-hand side of (21.4); if in addition we assume that the superfluid is not equilibrated with respect to the order parameter we must include terms to account for this effect. The thermodynamic identity then becomes

$$\delta\varepsilon_{\text{in}} = T\delta(\rho s) + \mu'\delta\rho - \frac{1}{c}\mathbf{j}_s\cdot\delta\mathbf{A} + \frac{\delta F}{\delta\psi}\delta\psi + \frac{\delta F}{\delta\psi^*}\delta\psi^*, \tag{21.5}$$

where $\mathbf{j}_s$ is the supercurrent (a thermodynamic quantity), $\mathbf{A}$ is the vector potential, and we used the relation $(\delta\varepsilon/\delta\psi)_{\mathbf{A},\rho s,\rho,\psi*} = (\delta F/\delta\psi)_{\mathbf{A},T,\rho,\psi*}$; the derivatives involved are assumed to be functional derivatives. Inserting (21.5) into (21.3) we obtain

$$\dot{S} = \int \mathrm{d}^3r\frac{1}{T}\left[\dot{\varepsilon}_{\text{in}} - \mu'\dot{\rho} + \mathbf{j}_s\cdot\frac{\dot{\mathbf{A}}}{c} + \frac{\partial F}{\partial\psi}\dot{\psi} + \frac{\partial F}{\partial\psi^*}\dot{\psi}^*\right]. \tag{21.6}$$

We now invoke the conservation laws of energy,

$$\frac{\partial\varepsilon}{\partial t} + \nabla\cdot\mathbf{q} = 0, \tag{21.7}$$

and charge,

$$\frac{\partial\rho}{\partial t} + \nabla\cdot\mathbf{j} = 0, \tag{21.8}$$

where $\mathbf{q}$ is the energy flux vector, $\mathbf{j}$ is the total current, and $\varepsilon = \varepsilon_{\text{in}} + \varepsilon_{\text{em}}$ is the total energy, where ε_{em} is the electromagnetic energy. From electromagnetic theory we have the equation

$$\mathbf{E} = -\frac{1}{c}\frac{\partial\mathbf{A}}{\partial t} - \nabla\phi, \tag{21.9}$$

where ϕ is the electric potential. We also require an expression for $\dot{\varepsilon}_{\text{em}}$, which follows from Poynting's theorem. Assuming the radiated energy flux is negligible we may write $\dot{\varepsilon}_{\text{em}} = \mathbf{j}\cdot\mathbf{E}$ (i.e., it is given by the ohmic heating). Inserting Eqs. (21.7), (21.8), and (21.9) into Eq. (21.6) we have

$$\dot{S} = \int \mathrm{d}^3r\frac{1}{T}\left[-\nabla\cdot\mathbf{q} + \mathbf{j}\cdot\mathbf{E} + \mu'\nabla\cdot\mathbf{j} - (\mathbf{E} + \nabla\phi)\cdot\mathbf{j}_s - \frac{\delta F}{\delta\psi}\dot{\psi} - \frac{\delta F}{\delta\psi^*}\dot{\psi}^*\right]. \tag{21.10}$$

Performing an integration by parts on terms involving the ∇ operator, converting divergences to surface integrals, and assuming all fluxes vanish at ∞, we obtain

$$\dot{S} = \int \mathrm{d}^3r\Bigg[(\mathbf{q} - \mu'\mathbf{j}_n + \phi\mathbf{j}_s)\nabla\cdot\frac{1}{T} + \frac{\mathbf{j}_n}{T}\cdot(\mathbf{E} - \nabla\mu') - \frac{1}{T}\left\{\frac{\partial\psi}{\partial t} + \frac{ie^*}{\hbar}(\phi + \mu')\psi\right\}\frac{\delta F}{\delta\psi} - \frac{1}{T}\left\{\frac{\partial\psi^*}{\partial t} - \frac{ie^*}{\hbar}(\phi + \mu')\psi^*\right\}\frac{\partial F}{\partial\psi^*}\Bigg], \tag{21.11}$$

where we used the equation

$$\frac{ie^*}{\hbar}\left(\psi^*\frac{\partial F}{\partial\psi^*} - \psi\frac{\delta F}{\delta\psi}\right) = \nabla\cdot\mathbf{j}_s. \tag{21.12}$$

and we introduced the normal current through the relation $\mathbf{j}_n = \mathbf{j} - \mathbf{j}_s$. From the structure of Eq. (21.11) it is apparent that, if we wish to ensure that the entropy production is positive definite, the time derivative in Eq. (21.1) should be introduced as

$$\frac{\partial}{\partial t} \to \frac{\partial}{\partial t} + \frac{ie^*}{\hbar}\left(\phi + \frac{\mu}{e}\right); \tag{21.13}$$

in this way the theory will be gauge-invariant and will allow for shifts in the chemical potential from its equilibrium value (taken to be zero). From (21.1) and (21.13) we have immediately the time-dependent G–L equation (TDGL) as

$$\gamma^{-1}\left[\frac{\partial}{\partial t} + ie^*\left(\phi + \frac{\mu}{e}\right)\right]\psi = \alpha\psi + \beta|\psi|^2\psi - \frac{\hbar^2}{2m^*}\left(\nabla - \frac{ie}{\hbar c}\mathbf{A}\right)^2\psi. \tag{21.14}$$

Substituting (21.14) into (21.11) we obtain the entropy production rate arising from the last term as

$$\dot{S} = \frac{2}{\gamma}\int\left|\left\{\frac{\partial}{\partial t} + \frac{ie^*}{\hbar}\left(\phi + \frac{\mu}{e}\right)\right\}\psi\right|^2 d^3r. \tag{21.15}$$

Eq. (21.14) is somewhat controversial; in particular the nonlinear form is valid only for a gapless superconductor (to be discussed in Part III). The linearized form (in which the term involving β is ignored) has wider applicability.

As an application of the linearized form of (21.14) we consider a Type II superconductor for $H = H_{c2}$ in the presence of a transport current, which exerts a force on the Abrikosov lattice. In the absence of pinning the lattice moves and the resulting dissipation creates an electric field resulting in a potential, $\phi = -Ex$. In the steady state the vortices drift with a velocity $u = (E/H_{c2})c$. It is easy to verify that the linearized form of Eq. (21.14) has the solution

$$\psi_k(x, y) = \exp[ik(y + ut)]\exp\left[-\frac{e^*H'}{2\hbar c}\left(x - \frac{\hbar kc}{e^*H'} - \frac{ium^*c}{e^*H'}\right)\right], \tag{21.16}$$

where

$$H' \equiv H_{c2}\left(1 - \frac{m^{*2}cu^2}{e^*\hbar H_{c2}}\right). \tag{21.17}$$

As in our previous discussion the wavefunction of the vortex lattice would be constructed as a superposition of the forms (21.16).

22 Fluctuation effects

22.1 The Ginzburg criterion

The G–L theory presented in Sec. 9 neglects the effects of thermodynamic fluctuations. However, near T_c the added energy associated with a change in the order parameter, $\delta\psi(\mathbf{r})$, is small (vanishing to the first order in $|\psi|^2$ at T_c for a uniform variation) and hence fluctuation effects are expected to be important for some interval of temperatures about the transition. The key idea in examining fluctuation effects is to reinterpret the G–L free energy as a Hamiltonian (see Patashinskii and Pokrovskii (1979)), which we write as

$$H_{GL} = H_0 + \Delta H, \tag{22.1a}$$

where $\Delta H = \int \mathscr{H} \mathrm{d}^3 r$ with

$$\mathscr{H} = \frac{\hbar^2}{2m^*}\left|\left(\nabla - \frac{\mathrm{i}e^*\mathbf{A}}{\hbar c}\right)\psi\right|^2 + \alpha|\psi|^2 + \frac{\beta}{2}|\psi|^4 + \frac{1}{8\pi}H^2(\mathbf{r}); \tag{22.1b}$$

here H_0 is the Hamiltonian associated with all the nonsuperconducting degrees of freedom. System properties are then calculated by evaluating the partition function, Z, as a sum involving all the states associated with this Hamiltonian. Since each function $\psi(\mathbf{r})$ corresponds to a possible state of the system, the partition function involves a functional integral (Feynman and Hibbs 1965); i.e.,

$$Z = \mathrm{e}^{-F/k_B T} = \mathrm{e}^{-F_0/k_B T}\mathrm{e}^{-\Delta F/k_B T}, \tag{22.2a}$$

where

$$\mathrm{e}^{-\Delta F/k_B T} = \int \mathscr{D}\psi(\mathbf{r})\mathrm{e}^{-(1/k_B T)\int \mathrm{d}^3 r \mathscr{H}(\psi(\mathbf{r}))} \tag{22.2b}$$

and F_0 is the normal state free energy arising from H_0. We may convert the functional integral into a discrete set of integrations by first assuming the system is in a large (but finite) cubical box of volume L^3 with $\psi(\mathbf{r})$ satisfying boundary conditions on the free surfaces. We next expand an arbitrary state $\psi(\mathbf{r})$ as a Fourier series

$$\psi(\mathbf{r}) = \sum_{\mathbf{k}} \psi_{\mathbf{k}} \mathrm{e}^{\mathrm{i}\mathbf{k}\cdot\mathbf{r}}. \tag{22.3}$$

Finally, to account for the contribution from all such states, we integrate over the expansion coefficients, $\psi_{\mathbf{k}}$. However the presence of the fourth order term precludes an exact evaluation of Eq. (22.2). For our present purposes it is sufficient to expand $\mathscr{H}$ to second order in $\delta\psi(\mathbf{r}) \equiv \psi(\mathbf{r}) - \psi_e$, where ψ_e is the equilibrium order parameter ($|\psi_e|^2 = 0$, $T > T_c$; $|\psi_e|^2 = -\alpha/\beta, T < T_c$) obtaining

$$\Delta H = \Delta H_e + \Delta H_f, \tag{22.4}$$

where

$$\Delta H_e = \frac{\alpha}{2}|\psi_e|^2 L^3 = -\frac{\alpha^2}{2\beta}L^3 \tag{22.5a}$$

and

$$\Delta H_f = L^3 \sum_{\mathbf{k}\neq 0}\left[\frac{\hbar^2 k^2}{2m^*} + \alpha + 3\beta|\psi_e|^2\right]|\psi_{\mathbf{k}}|^2; \tag{22.5b}$$

Eqs. (22.5a) and (22.5b) are the contributions, respectively, from the uniform ($\mathbf{k} = 0$) superconductor and all the fluctuations[1] having wave vectors $\mathbf{k}$. The mean square amplitude of a fluctuation of wave vector $\mathbf{k}$ can then be calculated as

$$\begin{aligned}\langle|\psi_{\mathbf{k}}|^2\rangle &= \left(\int_0^\infty \mathrm{d}|\psi_{\mathbf{k}}|^2 |\psi_{\mathbf{k}}|^2 \mathrm{e}^{-\Delta H_f/k_B T}\right)\Big/\int_0^\infty \mathrm{d}|\psi_{\mathbf{k}}|^2 \mathrm{e}^{-\Delta H_f/k_B T}\\ &= \frac{2m^*}{L^3\hbar^2}\frac{k_B T}{k^2 + \xi^{-2}(T)},\end{aligned} \tag{22.6}$$

where $\xi^2(T) = \hbar^2/2m^*\alpha(T)$ for $T > T_c$ and $\hbar^2/4m^*|\alpha(T)|$ for $T < T_c$; Eq. (21.6) is referred to as the Ornstein–Zernicke form, and we note that it diverges at $k = 0$ as $T \to T_c$ from either side.[2] From this calculation we see that long wavelength fluctuations are clearly going to be important near the transition.

A physical quantity which is especially sensitive to the fluctations is the specific heat. In the Landau theory it is constant but undergoes a discontinuity at the transition given by $\Delta C = T\mathrm{d}S/\mathrm{d}T = -T(\partial^2/\partial T^2)_V\Delta F = T(\partial^2/\partial T^2)(\alpha^2/2\beta)L^3 = (a^2/\beta)L^3$. As a measure of the importance of fluctuations we will evaluate the fluctuation contribution to the heat capacity and compare it to the discontinuity occurring in the Landau theory. The fluctuation contribution to the partition function may be written

$$\exp\left(\frac{-(1}{k_B T}\Delta F_f\right) = \prod_{\mathbf{k}}\int_0^\infty 2\pi\mathrm{d}|\psi_n|^2\,\mathrm{e}^{-\Delta H_f(\psi_{\mathbf{k}},\psi_{\mathbf{k}}^*)/k_B T},$$

where ΔF_f is the fluctuation contribution to the free energy. Thus

$$\begin{aligned}\Delta F_f &= -k_B T\sum_{\mathbf{k}}\ell n\left\{2\pi\int_0^\infty \mathrm{d}|\psi_{\mathbf{k}}|\exp\left[-\frac{L^3}{k_B T}\sum_{\mathbf{k}}\left(\frac{\hbar^2 k^2}{2m^*} + \alpha + 3\beta|\psi_e|^2\right)|\psi_{\mathbf{k}}|^2\right\}\right.\\ &= -k_B T\sum_{\mathbf{k}}\ell n\left[\frac{4\pi 2k_B T m^*}{L^3\hbar^2}\frac{1}{k^2 + \xi^{-2}(T)}\right].\end{aligned} \tag{22.7}$$

1. Here we used the property $\int \mathrm{d}^3 r\, \mathrm{e}^{\mathrm{i}(\mathbf{k},\mathbf{k}')\cdot\mathbf{r}} = L^3\delta_{\mathbf{kk}'}$.
2. If we insert Eq. (22.6) into Eq. (22.5b) the latter reduces to a sum over all modes, each having an energy $k_B T$.

The sum is divergent at the upper limit (referred to as an ultraviolet divergence) and its evaluation requires the introduction of a cut-off momentum, k_m. However, our present interest is in the heat capacity in which this divergence does not enter. The sum is also divergent to the lower limit as $\xi \to 0$ (referred to as an infrared divergence) which we now examine. Ignoring all temperature dependencies except that arising from $\xi(T)$, which results in a divergence as $T \to T_c$, we obtain (for $T < T_c$)

$$S_f = -\frac{\partial F}{\partial T} = \frac{2m^* k_B T_c a}{\hbar^2} \sum_{\mathbf{k}} \frac{1}{k^2 + \xi^{-2}} \tag{22.8a}$$

(the sum still being divergent at large k), and

$$\Delta C_f = T\frac{\partial S_f}{\partial T} = k_B \left(\frac{2m^* T_c a}{\hbar^2}\right)^2 \sum_{\mathbf{k}} \frac{1}{(k^2 + \xi^{-2})^2}$$

$$= k_B \left(\frac{2m^* T_c a}{\hbar^2}\right)^2 \frac{L^3}{(2\pi)^3} 4\pi\xi \int_0^\infty \frac{x^2 \mathrm{d}x}{(x^2+1)^2}, \tag{22.8b}$$

(where the sum is now convergent at large k). The integral in (22.8b) is given by $I = \frac{1}{2}\Gamma(\frac{3}{2})\Gamma(\frac{1}{2})/\Gamma(2) = \pi/4$. We see that the heat capacity is proportional to ξ and hence diverges as $1/|T - T_c|^{1/2}$ as we approach the transition temperature (from either side). The ratio of the fluctuation heat capacity to the Landau heat capacity discontinuity is given by

$$\frac{\Delta C_f}{T_c(a^2/\beta)L^3} = \frac{k_B I}{2\pi^2 \xi_0^3 (a^2/\beta)} \frac{1}{(T_c|T - T_c|)^{1/2}}$$

$$\sim \frac{T_F}{T_c(\xi_0 k_F)^3}\left(\frac{T_c}{|T - T_c|}\right)^{1/2};$$

here $\xi_0 = \xi(T = 0)$ and we substituted $(a^2/\beta) \sim (nk_B/T_F) \sim (k_B k_F^3/T_F)$, where n and T_F are the electron density and Fermi temperature, respectively. For typical superconductors this ratio is of order unity for

$$\frac{T - T_c}{T_c} \sim \begin{cases} 10^{-16} & \text{(clean Al)} \\ 10^{-6} & \text{(dirty Nb)} \end{cases}$$

and hence fluctuations make a negligible contribution for this type of second order phase transition. For a 2D superconductor Eq. (22.8b) becomes

$$\Delta C_f = k_B \left(\frac{2m^* T_c a}{\hbar^2}\right)^2 \frac{L^2}{(2\pi)^2} \pi\xi^2.$$

The heat capacity now diverges as $1/|T - T_c|$. This is a general property; i.e., fluctuation effects are larger for lower dimensionality.

22.2 The diamagnetic susceptibility for $T > T_c$

The magnetic susceptibility is given by the usual thermodynamic expression

$$\chi = -\frac{1}{L^3}\frac{\partial^2 F}{\partial H^2} \tag{22.9}$$

and hence we must evaluate the fluctuation contribution to the field-dependent free energy, $F(H)$. Above T_c the order parameter fluctuations are accurately described by the linearized G–L functional

$$\Delta F_f = \int d^3r\left[\frac{\hbar^2}{2m^*}\left|\left(\nabla - \frac{ie^*}{\hbar c}\mathbf{A}\right)\psi(\mathbf{r})\right|^2 + \alpha|\psi(\mathbf{r})|^2\right]; \tag{22.10}$$

here we have neglected the fluctuation contribution to the field energy and we will use the Landau gauge $\mathbf{A} = Hx\hat{\mathbf{y}}$ where $\mathbf{H}$ is the external field. Rather than representing $\psi(\mathbf{r})$ as a Fourier series, as in the previous subsection, it is now appropriate to use the eigenfunctions of the Schrödinger-like equation

$$-\frac{\hbar^2}{2m^*}\left(\nabla - \frac{ie^*}{\hbar c}\mathbf{A}\right)^2\psi_{n,k_y,k_z} = \varepsilon_{n,k_z}\psi_{n,k_y,k_z}, \tag{22.11}$$

where

$$\varepsilon_{n,k_z} = (n+\tfrac{1}{2})\hbar\omega_c + \frac{\hbar^2k_z^2}{2m^*} \tag{22.12}$$

and ψ_{n,k_yk_z} is given in Eq. (10.6). Thus

$$\psi(\mathbf{r}) = \sum_q c_q\psi_q(\mathbf{r}), \tag{22.13}$$

where q stands for the three quantum numbers n, k_y, k_z and the c_q are complex expansion coefficients. We may then write the fluctuation free energy as

$$\Delta F_f = \sum_q(\varepsilon_q + \alpha)|c_q|^2 \tag{22.14}$$

and we assume the ψ_q are normalized by $\int|\psi_q|^2d^3r = 1$. The sum over q involves two continuous quantum numbers (k_y and k_z) and one discrete one (n) and we require a prescription for counting the states; this same problem is encountered in the theory of the de Haas van Alphen effect and the required rule is[3]

$$\sum_q \to \frac{1}{(2\pi)^2}\frac{e^*HL^3}{\hbar c}\sum_n\int dk_z \tag{22.15}$$

(note k_y does not appear).

3. Using the well-known quasiclassical Onsager–Lifshitz quantization the area of an orbit in k space is $A_k = (2\pi e^*H/\hbar c)n$ and the area of an annulus between neighboring orbits (n and $n+1$) is then $\Delta A_k = 2\pi e^*H/\hbar c$; the associated volume element in k space is $\Delta A_k dk_z$. Hence

$$\frac{L^3}{(2\pi)^3}\int d^3k \to \frac{L^3}{(2\pi)^3}\frac{2\pi e^*H}{\hbar c}\sum_n\int dk_z = \frac{1}{(2\pi)^2}\frac{e^*HL^3}{\hbar c}\sum_n\int dk_z.$$

Analogous to Eq. (22.6) we obtain for the mean square value of the expansion coefficients in (22.13)

$$\langle |c_q|^2 \rangle = \frac{k_B T}{\varepsilon_q + \alpha} \tag{22.16}$$

and the equilibrium superconducting free energy, ΔF, follows from Eq. (22.2) as

$$\Delta F = -k_B T \sum_q \ell n \left\{ \int dc_q' dc_q'' \exp\left[-\frac{1}{k_B T} |c_q|^2 (\varepsilon_q + \alpha) \right] \right\}, \tag{22.17}$$

where c_q' and c_q'' are the real and imaginary parts of c_q (which count as separate modes). Carrying out the integrations we obtain

$$\begin{aligned} \Delta F &= -k_B T \sum_q \ell n \left(\frac{\pi k_B T}{\varepsilon_q + \alpha} \right) \\ &= -L^3 \frac{e^* k_B T H}{(2\pi)^2 \hbar c} \sum_n \int \ell n \left[\frac{\pi k_B T}{(n + \frac{1}{2}) \hbar \omega_c + \hbar^2 k_z^2 / 2m^* + \alpha} \right] dk_z, \end{aligned} \tag{22.18}$$

where we used (22.12) and (22.15) in the last step. The above expression is divergent, again due to the fact that the G–L theory is limited to low energy fluctuations and slow variations involving wavelengths longer than ξ_0. Hence we can assume that the sum is to be cut off at some maximum energy, $\hbar\omega_m = N\hbar\omega_c$ (where N is some large integer) and the integration over dk_z at some maximum wave vector, $k_m \sim \xi_0^{-1}$. For low fields the summation approaches an integration; the first correction to the integral is obtained using the Poisson summation formula which states that

$$\sum_{n=0}^{N} f(n + \tfrac{1}{2}) \cong \int_0^N f(x) dx - \frac{1}{24} f'(x) \Big|_0^N, \tag{22.19}$$

where f is some function. On writing the integral on the right in dimensionless form it is easy to see that the magnetic field cancels out; the free energy arising from the second term involves only the lower limit for large N and we have

$$\begin{aligned} \Delta F &= L^3 H^2 \frac{e^{*2} k_B T_c}{96\pi^2 m^* c^2} \int_{-\infty}^{+\infty} \frac{dk_z}{\hbar^2 k_z^2 / 2m^* + \alpha} \\ &= L^3 H^2 \frac{e^{*2} k_B T_c}{48\pi \hbar c^2 (2m^* \alpha)^{1/2}}. \end{aligned} \tag{22.20}$$

Using Eq. (22.9) we obtain the susceptibility as

$$\chi = -\frac{e^{*2} k_B T_c}{24\pi c^2 (2m^* a)^{1/2} (T - T_c)^{1/2}} \tag{22.21}$$

(Schmidt 1968, Schmid 1969); note it is negative (diamagnetic) as expected and diverges as $(T - T_c)^{-1/2} \propto \xi(T)$ as $T \to T_c$.

Let us also consider the magnetic response of a thin film of thickness d. For the case where the magnetic field is perpendicular to the plane of the film the only change in the above treatment is to remove the integration over dk_z (the film is assumed thin enough that the excited states, having wave vectors of order d^{-1} and larger, make a negligible contribution). Replacing the sum over states $(L/2\pi)\int dk_z$, by unity and setting $k_z = 0$, we have from Eq. (22.20)

$$\Delta F = L^2 H^2 \frac{e^{*2} k_B T_c}{48\pi m^* c^2 \alpha}. \tag{22.22}$$

The magnetic moment per unit volume of the film is given by $M = -(1/L^2 d)/(\partial F/\partial H)$ or

$$M = \frac{H}{d} \frac{e^{*2} k_B T_c}{24\pi m^* c^2 a(T - T_c)}. \tag{22.23}$$

We see that the divergence is strong, $\propto (T - T_c)^{-1}$ in two dimensions; i.e., the fluctuation effect is enhanced by the lower dimensionality as discussed earlier.

22.3 Paraconductivity for $T > T_c$

The fluctuations of the order parameter for $T > T_c$ can also contribute to the electrical conductivity, and the pretransitional rise in the conductivity is referred to as paraconductivity. The Fourier transform of the nonlocal electrical conductivity can be calculated with the aid of the Kubo formula

$$\sigma(\mathbf{q}) = \frac{1}{2k_B T} \int_{-\infty}^{+\infty} dt \langle \hat{\mathbf{j}}_{\mathbf{q}}(t) \hat{\mathbf{j}}_{\mathbf{q}}(0) \rangle, \tag{22.24}$$

where the bracket implies a combined quantum mechanical and statistical average. Using the plane wave decomposition of $\psi(\mathbf{r})$ given in Eq. (22.3) and the definition of the G–L current operator given in Eq. (9.17) we have

$$\mathbf{j}(\mathbf{r}) = \frac{e^*\hbar}{2m^*} \sum_{\mathbf{k},\mathbf{k}'} (\mathbf{k} + \mathbf{k}') \psi_{\mathbf{k}} \psi_{\mathbf{k}'}^* e^{i(\mathbf{k}-\mathbf{k}')\cdot\mathbf{r}} \tag{22.25}$$

Fourier transforming (22.25) in $e^{-i\mathbf{q}\cdot\mathbf{r}}$ yields

$$\mathbf{j}(\mathbf{q}) = \frac{e^*\hbar L^3}{2m^*} \sum_{\mathbf{k}} (2\mathbf{k} + \mathbf{q}) \psi_{\mathbf{k}}^* \psi_{\mathbf{k}+\mathbf{q}}. \tag{22.26}$$

Substituting (22.26) into (22.24) and limiting ourselves to the $q = 0$, we obtain

$$\sigma(0) = \frac{\hbar^2 e^{*2} L^6}{2k_B T m^{*2}} \int_{-\infty}^{+\infty} dt \sum_{\mathbf{k},\mathbf{k}'} \mathbf{k}\mathbf{k}' \langle |\psi_{\mathbf{k}}(t)|^2 |\psi_{\mathbf{k}'}(0)|^2 \rangle.$$

We assume terms with $\mathbf{k} \neq \mathbf{k}'$ are statistically independent and average to zero; thus

$$\sigma_{ij}(0) = \frac{\hbar^2 e^{*2} L^6}{2k_B T m^{*2}} \int_{-\infty}^{+\infty} dt \sum_{\mathbf{k}} \mathbf{k}_i \mathbf{k}_j |\langle \psi_{\mathbf{k}}^*(t) \psi_{\mathbf{k}}(0) \rangle|^2 \tag{22.27}$$

If we assume an exponential decay, the 'correlation function' entering (22.27) may be written

$$\langle \psi_{\mathbf{k}}^*(t) \psi_{\mathbf{k}}(0) \rangle = \frac{2m^*}{L^3 \hbar^2} \frac{k_B T}{k^2 + \xi^{-2}} e^{-t/\tau_k}, \tag{22.28}$$

which reduces to the Ornstein–Zernicke form (22.6) for $t = 0$. The relaxation time, τ_k, will be calculated from the Landau–Khalatnikov model discussed in Sec. 21:

$$\frac{\partial \psi_{\mathbf{k}}}{\partial t} = -\gamma \frac{\partial \Delta F_{\mathrm{f}}/L^3}{\partial \psi_{\mathbf{k}}^*}, \tag{22.29}$$

where, for the present situation, ΔF_{f} is given by Eq. (22.5b); we then obtain the relaxation equation

$$\frac{\partial \psi_{\mathbf{k}}}{\partial t} = -\gamma \frac{\hbar^2}{2m^*}(k^2 + \xi^{-2})\psi_{\mathbf{k}},$$

which has the solution

$$\psi_{\mathbf{k}}(t) = \psi_{\mathbf{k}}(0)\mathrm{e}^{-t/\tau_k} \tag{22.30a}$$

where

$$\tau_k = \frac{2m^*}{\gamma\hbar^2}\frac{1}{k^2 + \xi^{-2}}. \tag{22.30b}$$

Inserting (22.28) with (22.30b) into the $\mathbf{q} = 0$ limit of (22.27), carrying out the integration over time,[4] and writing $k_i^2 = \frac{1}{3}k^2$, we obtain

$$\begin{aligned}\sigma(0) &= \frac{4}{3}k_{\mathrm{B}}T\frac{e^{*2}m^*}{\gamma\hbar^4(2\pi)^3}\int \mathrm{d}^3k\frac{k^2}{(k^2 + \xi^{-2})^3} \\ &= \frac{2}{3}k_{\mathrm{B}}T\frac{e^{*2}m^*\xi}{\gamma\hbar^4\pi^2}\int_0^\infty \frac{x^4\mathrm{d}x}{(x^2+1)^3},\end{aligned} \tag{22.31}$$

where the integral is given by $\frac{1}{2}\Gamma(\frac{5}{2})\Gamma(\frac{1}{2})/\Gamma(3) = 3\pi/16$. The divergence in 3D is again rather weak: $\sigma(T) \sim (T - T_{\mathrm{c}})^{1/2}$. Inserting the expression for γ obtained in Sec. 49,

$$a\gamma = \frac{8k_{\mathrm{B}}}{\hbar\pi} \tag{2.32}$$

we may write Eq. (22.31) as

$$\sigma(0) = \frac{e^2}{32\hbar\xi_0}\left(\frac{T_{\mathrm{c}}}{T - T_{\mathrm{c}}}\right),$$

where we substituted $e^* = 2e$.

For a thin film with $d << \xi$ we may neglect the fluctations along the film normal and the system becomes effectively 2D. The resulting conductivity is

$$\sigma(0) = k_{\mathrm{B}}T_{\mathrm{c}}\frac{e^{*2}m^*\xi^2}{\gamma\hbar^2\pi d}\int_0^\infty \frac{x^3\mathrm{d}x}{(x^2+1)^3}, \tag{22.34}$$

where the integral is given by 1/4 and our final G–L results for the temperature dependence of the conductivity for a thin film of thickness d is

$$\sigma(0) = \frac{e^{*2}k_{\mathrm{B}}\mathrm{T}_{\mathrm{c}}}{2\hbar\pi d\gamma a(T - T_{\mathrm{c}})}.$$

4. The integration over negative time is accomplished by replacing e^{-t/τ_k} with $\mathrm{e}^{-|t|/\tau_k}$.

Again inserting our expression for γ we have

$$\sigma = \frac{e^2}{16\hbar d}\frac{T_c}{T - T_c}. \tag{22.35}$$

These results were obtained by Aslamasov and Larkin (1968) using field theoretic methods; the derivation given above is due to Schmidt (1968).

23 G–L theory of an unconventional superfluid

Although our discussion in this section will be largely phenomenological we must anticipate the major result of the microscopic theory which will be developed in Parts II and III; in particular we must make use of the fact that superfluidity in Fermi systems arises from the formation of a special kind of bound state in which electrons (or ^{3}He quasiparticles) on (or near) the Fermi surface having *oppositely* directed momenta, $+\mathbf{p}$ and $-\mathbf{p}$, form (collectively) bound pairs.

For the vast majority of superfluid electron systems (metals), the pairs form in a state with no net orbital angular momentum, an $\ell = 0$ or s-wave state which we will refer to as conventional fermion superfluidity; by the Pauli principle the electrons must then be in a singlet spin state ($s = 0$).

In the decade or so following the development of the BCS theory, and especially after the experimental discovery of the superfluid phases of liquid ^{3}He, theorists explored various types of pairing in systems with a more complicated orbital and spin structure in which the gap can vary, in both magnitude and phase, with position on the Fermi surface. In the case of odd ℓ pairing, the pairs form (again due to the Pauli principle) in a triplet $s = 1$ rather than a singlet spin state. The energy gap associated with such states may have zeros (i.e., vanish) at points or lines on the Fermi surface, and as a result the number of excitations at low temperatures varies as a power of the temperature, rather than exponentially. After a rather hectic two-year period of exploration following the discovery of superfluidity in liquid ^{3}He below ~ 0.002 K by Osheroff, Richardson, and Lee (1972), it was established that the pairing occurs in two distinct p-wave states. The A phase (also called the Anderson–Brinkman–Morel (ABM) state, Anderson and Brinkman (1973)) has point nodes on the Fermi surface, and the B phase (or Balian–Werthamer (BW) state, Balian and Werthamer (1963)) has a gap with a constant magnitude over the Fermi surface. These p-wave triplet states possess magnetic and other properties which are very different from singlet paired states (see Leggett (1975)).

Evidence has been accumulating that certain (so-called) heavy fermion intermetallic compounds may have unconventional superconducting order parameters (see Sarma, Levy, Adenwalla, and Ketterson (1992) and Hess, Riseborough, and Smith (1993) for reviews). These systems involve a rare earth or actinide metal (typically Ce or U) and, due to the presence of narrow f-derived conduction bands, have enormous effective masses ($\sim 100m_e$–$1000m_e$). In particular the compound UPt_3, with $m^* \cong 100m_e$, appears to display three distinct superconducting phases in the H–T plane as shown in Fig. 23.1 (here we do not include the H_{c1} phase boundary). The data shown are consistent with experiments involving: ultrasound propagation, heat capacity, thermal expansion, etc. One interpretation of these phase transitions is that the nature of the order

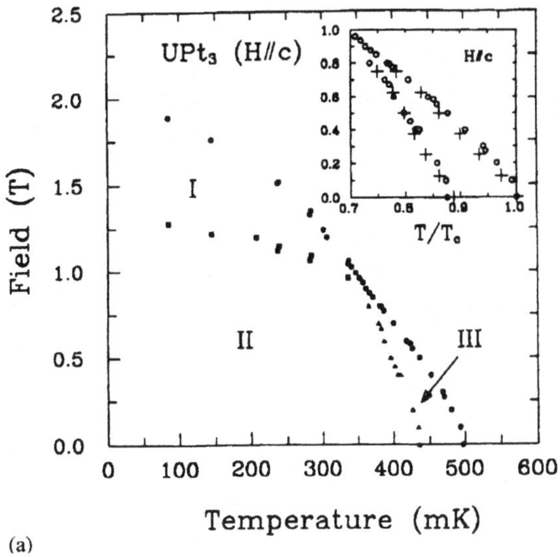

(a)

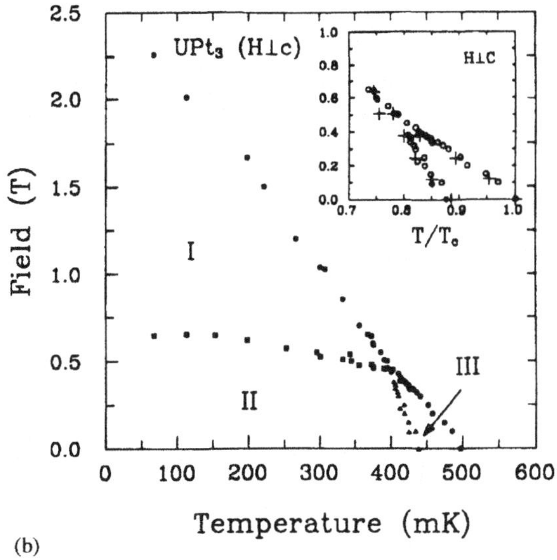

(b)

Figure 23.1 Phase diagram of superconducting UPt_3 obtained from ultrasonic velocity measurements for $H \| c$ (a) and $H \perp c$, (b). The insets show the comparison of the velocity data (open circles) with the heat capacity data (crosses). (After Adenwalla *et al.* (1990).)

parameter changes from one symmetry class to another (some of the various symmetry classes will be discussed shortly).

Finally, the possibility that high temperature superconductivity involves a (crystal-field split) d-wave-like state appears to be supported by microwave absorption and certain tunneling experiments (Wollman *et al.* 1993); there is also support from some microscopic calculations for this type of pairing (Monthouk, Balatsky, and Pines 1992).

23.1 The order parameter of an unconventional superfluid

In our earlier discussion of the G–L theory of (conventional) superconductivity in Sec. 9, we assumed the existence of a wavefunction associated with the superconducting electron pairs. Such a simple description is possible only when the condensate formed from the pairs has no internal degrees of freedom (other than the phase). As discussed above, in the BCS pairing theory of an isotropic electron liquid, the electrons are bound in a state with no orbital angular momentum (an $\ell = 0$ or s-wave state) and no net spin (an $s = 0$ or singlet state) and hence the pair wavefunction may be characterized by a single complex amplitude. In this subsection we consider an unconventional superconductor (superfluid) where the pairs form with a net orbital angular momentum. It is then necessary to generalize the form of the order parameter. For an isotropic liquid an appropriate form is $\psi_{\alpha\beta}(\mathbf{k})$, where the subscripts α and β ($= \uparrow$ or $\downarrow$) arise from the spin variables of the two electrons and $\mathbf{k}$ accounts for a possible angular dependence over the Fermi surface.[1] Order parameters of the above form divide naturally into two classes: singlet and triplet states which are respectively antisymmetric and symmetric under exchange of the spin indices. By the Pauli principle the *total* wavefunction (orbital $\times$ spin) must be antisymmetric under interchange of the particles making up the pair. The parity of the spherical harmonics associated with pairing in a given angular momentum 'channel' is $(-1)^{\ell}$; hence singlet pairing will always be associated with even angular momentum and triplet pairing with odd angular momentum.

23.1.1 Superfluid ^{3}He: isotropic p-wave pairing

A firmly established example of an unconventional fermion superfluid is liquid ^{3}He where the ^{3}He quasiparticles form in an $\ell = 1$ (p-wave), $s = 1$ (triplet) state (see Anderson and Brinkman (1978) for a review). The microscopic theory of ^{3}He superfluidity will be discussed in Sec. 53. The spin dependence of the ^{3}He–^{3}He quasiparticle interaction, leading to the triplet pairing, has its origin in a combination of the hard-core repulsion and exchange effects (spin–orbit effects are negligible).[2] The angular dependence of the order parameter arising from the $\ell = 1$ character of the pairing may be represented by the Cartesian components of the unit vector $\mathbf{k}$ (which involve the polar angles on the Fermi surface)

$$k_1 = \sin\theta\cos\phi, \tag{23.1a}$$

1. Microscopically the order parameter is defined as $\psi_{\alpha\beta}(\mathbf{k}) \equiv \langle \hat{c}_{\mathbf{k}\alpha}\hat{c}_{-\mathbf{k}\beta}\rangle$, where $\hat{c}_{\mathbf{k}\alpha}$ is a destruction operator for a quasiparticle with momentum $\mathbf{k}$ and spin α and the brackets denote an ensemble average as well as an integration over the magnitude of $\mathbf{k}$.
2. The usual qualitative argument given is that the hard core repulsion between ^{3}He atoms, which tends to raise the energy, can be reduced by forming a paired state with $\ell > 0$; the centrifugal barrier then tends to keep the atoms from seeing the repulsive force. In addition, if the spins are aligned parallel, the Pauli principle keeps them from occupying the same region which also suppresses the effect of the hard core interaction and, simultaneously, favors a triplet state. Finally if one calculates the phase shifts for scattering by a phenomenological '6–12' ^{3}He interaction potential, higher ℓ states turn out to be attractive, as opposed to the s state (note the bare interaction is not the same as the quasiparticle–quasiparticle interaction although both are repulsive in the singlet channel).

$$k_2 = \sin\theta \sin\phi, \tag{23.1b}$$

and

$$k_3 = \cos\theta; \tag{23.1c}$$

these forms replace the usual spherical harmonics, $Y_{1,1}$, $Y_{1,-1}$, and $Y_{1,0}$, and clearly transform as a vector in real space. For the spin wavefunctions we choose

$$\chi_1 = \frac{1}{\sqrt{2}}[-(\uparrow)(\uparrow) + (\downarrow)(\downarrow)], \tag{23.2a}$$

$$\chi_2 = \frac{i}{\sqrt{2}}[(\uparrow)(\uparrow) + (\downarrow)(\downarrow)], \tag{23.2b}$$

and

$$\chi_3 = \frac{1}{\sqrt{2}}[(\uparrow)(\downarrow) + (\downarrow)(\uparrow)], \tag{23.2c}$$

which transform as a vector under rotations in spin space. These functions have the further property that the expectation values $\mathrm{Tr}\chi_\nu^\dagger S_\nu \chi_\nu$ vanishes, where S_ν are the components of the total spin operator $\mathbf{S} = \frac{1}{2}(\sigma_1 + \sigma_2)$ and σ_1 and σ_2 denote the vector Pauli matrices associated with the two quasiparticles. Any linear combination of (23.2a)–(23.2c) with real amplitudes (together with an overall complex amplitude) will correspond to a superfluid with no net spin polarization. Such states are called unitary.

The most general order parameter for ^{3}He (which includes both polarized and unpolarized states) may be written as the superposition

$$\psi_{\alpha\beta}(\mathbf{k}) = \sum_{\nu=1}^{3} \sum_{i=1}^{3} D_{\nu i} k_i \chi_\nu \tag{23.3}$$

and the nine complex amplitudes $D_{\nu i}$ (18 real amplitudes) define 18 different superfluid degrees of freedom.

In the absence of a magnetic field superfluid ^{3}He has no net spin polarization. From the above discussion it follows that there are two possible forms the $D_{\alpha i}$ matrix may take:[3]

$$D_{\alpha i} = n_{\alpha i}\Delta \qquad \text{(case I)} \tag{23.4a}$$

and

$$D_{\alpha i} = n_\alpha \Delta_i \qquad \text{(case II)}; \tag{23.4b}$$

in the first case $n_{\alpha i}$ is a real matrix and Δ is a complex constant while in the second case n_α is a (real) unit vector and Δ_i is a complex vector. The vector n_α allows a linear superposition of the three spin wavefunctions (with real amplitudes) while the vector Δ_i allows a superposition of the three orbital wavefunctions (with complex coefficients). The two observed ground states, A and B, correspond to cases I and II respectively. As we will show later in Subsec. 23.3.1, the equilibrium B

3. In Part III we will use $\alpha, \beta, \ldots$ to denote spin components $(\uparrow,\downarrow)$ of a spin wavefunction and $\nu, \mu, \ldots$ to denote spin eigenfunctions. For the remainder of the section, however, we will follow the literature convention and use α, β (in place of ν, μ) to denote the spin eigenfunctions.

phase can be characterized by the amplitude matrix

$$\mathbf{D} = \frac{\Delta}{\sqrt{3}}\begin{pmatrix} 1 & 0 & 0 \\ 0 & 1 & 0 \\ 0 & 0 & 1 \end{pmatrix}, \tag{23.5}$$

where Δ is the arbitrary complex amplitude (in (23.4a)) and the normalization chosen is such that $\mathrm{Tr}\mathbf{D}\cdot\mathbf{D} = |\Delta|^2$; more generally the unit matrix is replaced by a rotation matrix accounting for the fact that the axes defining orbital and spin space are in general different. The order parameter of the B phase corresponding to Eq. (23.5) is

$$\psi(\mathbf{k}) = \Delta(k_1\chi_1 + k_2\chi_2 + k_3\chi_3), \tag{23.6}$$

where we have adopted the normalization $\int \Psi^*\Psi \mathrm{d}\Omega = |\Delta|^2$.

The A phase order parameter may be written as

$$\mathbf{D} = \frac{\Delta}{\sqrt{2}}\begin{pmatrix} 1 & \mathrm{i} & 0 \\ 0 & 0 & 0 \\ 0 & 0 & 0 \end{pmatrix}, \tag{23.7}$$

corresponding to the wavefunction

$$\psi = \sqrt{\frac{3}{2}}\Delta(\mathbf{k}_1 + \mathrm{i}\mathbf{k}_2)\chi_1; \tag{23.8}$$

$\hat{n}_\alpha$ was arbitrarily chosen parallel to the $\hat{\mathbf{x}}$ axis in spin space in Eq. (23.4b). In so-called weak coupling theory, where the pairing interaction remains unaltered on entering the superfluid state, only the B phase would be stable. For the case of ^{3}He the relative stability of the possible phases is determined by the *fourth* order terms in the G–L expansion; at the level of the second order terms all possible $\ell = 1$ superfluids would have the same transition temperature.

By applying the total angular momentum operator $J^2 = L^2 + S^2 + 2\mathbf{L}\cdot\mathbf{S}$, it is easy to verify that Eq. (23.5) corresponds to a $J = 0$ state. Although Eq. (23.5) has a structure that corresponds to case I above, there are eight other independent $n_{\alpha i}$ matrices one can construct. A real 3×3 matrix has nine independent components. It is convenient to classify these according to their total angular momentum. One can easily verify that the following set of matrices corresponds to $J = 1$ states

$$\underset{\text{(a)}}{\frac{1}{\sqrt{2}}\begin{pmatrix} 0 & 0 & 0 \\ 0 & 0 & -1 \\ 0 & 1 & 0 \end{pmatrix}}; \quad \underset{\text{(b)}}{\frac{1}{\sqrt{2}}\begin{pmatrix} 0 & 0 & 1 \\ 0 & 0 & 0 \\ -1 & 0 & 0 \end{pmatrix}}; \quad \underset{\text{(c)}}{\frac{1}{\sqrt{2}}\begin{pmatrix} 0 & 1 & 0 \\ -1 & 0 & 0 \\ 0 & 0 & 0 \end{pmatrix}}, \tag{23.9}$$

while a set of $J = 2$ states is

$$\underset{\text{(a)}}{\frac{1}{\sqrt{2}}\begin{pmatrix} 0 & 1 & 0 \\ 1 & 0 & 0 \\ 0 & 0 & 0 \end{pmatrix}}; \quad \underset{\text{(b)}}{\frac{1}{\sqrt{2}}\begin{pmatrix} 1 & 0 & 0 \\ 0 & -1 & 0 \\ 0 & 0 & 0 \end{pmatrix}}; \quad \underset{\text{(c)}}{\frac{1}{\sqrt{2}}\begin{pmatrix} 0 & 0 & 0 \\ 0 & 0 & 1 \\ 0 & 1 & 0 \end{pmatrix}};$$

$$\frac{1}{\sqrt{2}}\begin{pmatrix}0 & 0 & 1\\ 0 & 0 & 0\\ 1 & 0 & 0\end{pmatrix}; \quad \frac{1}{\sqrt{6}}\begin{pmatrix}1 & 0 & 0\\ 0 & 1 & 0\\ 0 & 0 & -2\end{pmatrix}. \tag{23.10}$$

(d) (e)

To verify that the $J = 0$ state has the lowest energy we must minimize the appropriate free energy expression, which will be carried out later in this section. Excited $J = 0$ states are also possible. The amplitudes associated with (23.9) or (23.10) may be either real or imaginary; with the two $J = 0$ states this results in a total of 18 excited states which are the so-called collective modes (of the B phase) which will be discussed phenomenologically later in this section and microscopically in Sec. 54.

23.1.2 *Isotropic d-wave pairing*

For a singlet superfluid the spin wavefunction would be

$$\chi_0 = \frac{1}{\sqrt{2}}[(\uparrow)(\downarrow) - (\downarrow)(\uparrow)]. \tag{23.11}$$

Thus an s-wave superfluid has the order parameter

$$\psi = \chi_0 \Delta, \tag{23.12}$$

where Δ is again a complex number. Singlet pairing in a higher angular momentum state involves a more complicated orbital wavefunction. For the case of singlet d-wave pairing the order parameter can be written

$$\psi = \chi_0 \sum_{i=1}^{3} \sum_{j=1}^{3} k_i D_{ij} k_j = \chi_0 \mathbf{k} \cdot \mathbf{D} \cdot \mathbf{k}, \tag{23.13}$$

where $\mathbf{D}$ is a traceless symmetric matrix and $\mathbf{k}$ is again given by Eq. (23.1). (Tr$\mathbf{D} \neq 0$ would imply an s-wave admixture). A representation of the allowed states (which have $J = L = 2$) has the same form as Eq. (23.10)

$$\frac{1}{\sqrt{2}}\begin{pmatrix}0 & 1 & 0\\ 1 & 0 & 0\\ 0 & 0 & 0\end{pmatrix}; \quad \frac{1}{\sqrt{2}}\begin{pmatrix}1 & 0 & 0\\ 0 & -1 & 0\\ 0 & 0 & 0\end{pmatrix}; \quad \frac{1}{\sqrt{2}}\begin{pmatrix}0 & 0 & 0\\ 0 & 0 & 1\\ 0 & 1 & 0\end{pmatrix};$$

(a) (b) (c)

$$\frac{1}{\sqrt{2}}\begin{pmatrix}0 & 0 & 1\\ 0 & 0 & 0\\ 1 & 0 & 0\end{pmatrix}; \quad \frac{1}{\sqrt{6}}\begin{pmatrix}1 & 0 & 0\\ 0 & 1 & 0\\ 0 & 0 & -2\end{pmatrix}; \tag{23.14}$$

(d) (e)

here each state can have an independent complex amplitude. These forms correspond to the usual (unnormalized) Cartesian d-wave functions:

$$2k_1k_2; \quad k_1^2 - k_2^2; \quad 2k_2k_3; \quad 2k_1k_3; \quad k_1^2 + k_2^2 - 2k_3^2, \\ \text{(a)} \qquad \text{(b)} \qquad \text{(c)} \qquad \text{(d)} \qquad \text{(e)} \tag{23.15}$$

where the k_i are defined in Eq. (23.1). By taking linear combinations of (a) and (b) and of (c) and (d) we may form the d-wave spherical harmonics $Y_{2\pm2}$ and $Y_{2,\pm1}$; form (e) corresponds to $Y_{2,0}$. We will use the s-, p-, and d-wave states discussed above as a starting point for discussing some allowed electron-superconducting states.

23.2 Crystal-field and spin–orbit effects

A metal is not an isotropic electron liquid and the classification of various allowed electron superfluids must take account of the lowered symmetry of the host crystal lattice; to be specific we will focus on hexagonal and tetragonal crystals (examples being the intermetallic compounds UPt_3 and URu_2Si_2). A rigorous discussion involves group theory (see Volovik and Gorkov (1985), but most of the relevant physics can be extracted by considering the effect of various perturbations on the states of an isotropic system. All known heavy fermion compounds involve rare earths or actinides and, due to their high atomic numbers, Z, we must also include spin–orbit effects. We will deal with this latter effect first.

In the L–S coupling scheme of atomic physics, the perturbing Hamiltonian arising from spin–orbit coupling may be written as

$$V_{so} = A\mathbf{L}\cdot\mathbf{S}, \tag{23.16}$$

where A is a constant. The eigenvalues of the operator $\mathbf{L}\cdot\mathbf{S}$ acting on a state of a given j, ψ_j, are given by

$$\mathbf{L}\cdot\mathbf{S}\psi_j = \tfrac{1}{2}[j(j+1) - \ell(\ell+1) - s(s+1)]\psi_j. \tag{23.17}$$

Hence states with the same ℓ and s, but differing j will be shifted; however, the degeneracy within a multiplet of a given j is not lifted. For our p-wave superfluid the degeneracy of the states with $j = 0$, $j = 1$, and $j = 2$ (Eqs. (23.5), (23.9), and (23.10)) would be lifted in the *leading order* of the G–L expansion; i.e., they would each have different transition temperatures. The spin–orbit energy shifts for this case are: $-2A(j=0)$; $-A(j=1)$; and $+1A(j=2)$.

We next consider the effect of the crystal structure. We begin by recalling the independent symmetry operations (classes) associated with the hexagonal D_{6h} (UPt_3) and tetragonal D_{4h} (URu_2Si_2) point groups. For D_6 we have: the identity (E); two (C_2), three ($2C_3$), and six ($2C_6$) fold rotations about the hexagonal axis; and two sets of two-fold ($3U_2$ and $3U_2'$) rotations about axes in the basal plane; the notation within the parenthesis here is the standard group theory designation. (The number preceding the symmetry designation denotes the number of such operations.) For D_4 we have: the identity (E); two (C_2) and four (C_4) fold rotations about the tetragonal axis; and two sets of two-fold rotations about axes in the basal plane ($2U_2$ and $2U_2'$). The inclusion of inversion symmetry (which is equivalent in the present case to reflection through the basal plane) doubles the number of classes and the corresponding point group designations are D_{6h} and D_{4h}.

The perturbing effect of the crystal potential for our cases of interest may be represented by a function $V(\mathbf{k})$

$$V(\mathbf{k}) = \sum_{\ell,m} V_{\ell,m} \cos \ell\theta \cos m\phi, \tag{23.18}$$

where $m = 4p$ for D_4 and $6p$ for D_6, and $\ell = 2q$; here p and q denote all positive integers. We will generally limit ourselves to the leading-order, symmetry-breaking effect: $p = q = 1$.

An important aspect of the theory of point groups is concerned with the transformation properties of (wave)functions. Wavefunctions (in our case order parameters) may have two kinds of behaviors: either (i) a wavefunction transforms into itself under all symmetry operations (other than a phase factor), or (ii) a minimal set of wavefunctions transforms into linear combinations of each other; (i) is said to be a 1D (nondegenerate) representation and the dimensionality (degeneracy) of (ii) is equal to the minimum number of required wavefunctions. As an example, consider atomic states corresponding to the total angular momentum quantum number j. Such states are $(2j + 1)$-fold degenerate; for a singlet ($\ell = j$), they would correspond to spherical harmonics, $Y_{\ell,\ell_z}(\theta, \phi)$, and on rotating a system in a given state ℓ about an arbitrary axis these functions transform into linear combinations involving the quantum number ℓ_z. Since there is no limit on j, spherical symmetry admits arbitrarily high degeneracies. Nonspherical perturbations remove most of the degeneracy.

There is an upper limit on the degeneracies of discrete point groups. For D_{6h} or D_{4h} it is two and hence we can have only 1D and 2D representations. The 1D representations are denoted by the letters A and B; they are even and odd respectively under the minimum rotation about the principal symmetry axis (i.e., a 60° or 90° rotation for D_{6h} and D_{4h}). The 2D representations are denoted by the letter E (not to be confused with the identity). If there is more than one symmetry distinct representation, then subscripts 1, 2, etc., are included. Some of the representations have the same symmetry as the various crystal axes (x, y, or z) and the relevant axis is then included as a subscript. Finally in the presence of inversion symmetry a representation may be either even or odd and the subscripts g and u, respectively, are used to denote these symmetries. Note there is no limit on the number of times a given symmetry type may occur; however, barring an accidental degeneracy, they will have different energies. The ground state corresponds to the globally lowest energy state (across all representations), which in the absence of other information could have any of the allowed symmetries (and hence degeneracies).

We now employ some simple heuristic arguments to arrive at the important representations for our chosen point groups (D_{6h} and D_{4h}) starting from s-, p-, and d-wave pairing states of the isotropic quasiparticle superfluid. We will begin by assuming the largest energy shifts are associated with a transition from spherical to uniaxial symmetry (with an axis involving the principal four- or six-fold crystal axis). This symmetry is broken by the $V_{2,0}$ term in Eq. (23.18); as far as the effect on symmetry breaking is concerned, it is equivalent to the quadratic Stark effect in atomic physics: $2j + 1$ degenerate states of a given j split into $j + 1$ levels with j doubly degenerate $\pm |j_z|$ levels and a singly degenerate $j_z = 0$ level. Note that at this elementary stage (uniaxial symmetry + spin–orbit coupling) we have already lifted most of the relevant degeneracy. Let us examine the effect of crystal-field splitting on the previously enumerated s-, p-, and d-wave states. We write the wavefunctions of the various states in the form $\psi = \psi^{\ell,s}_{j,j_z}(\mathbf{k})$. Clearly the singlet s-wave state is unaffected and has the symmetry A_{1g}. The p-wave states split into a 1D and a 2D representation having the structure (where the quantity in parenthesis is the group theoretical designation of the state)

$j = 0$:

$$\mathbf{D} = \frac{1}{\sqrt{3}}\mathbf{1}, \tag{23.19}$$

$$\psi^{1,1}_{0,0}(\hat{\mathbf{k}}) = k_1\chi_1 + k_2\chi_2 + k_3\chi_3 \quad (\mathrm{A}_{1\mathrm{u}}); \tag{23.20}$$

$j = 1; j_z = 0$:

$$\mathbf{D}_z = \frac{1}{\sqrt{2}}\begin{pmatrix} 0 & 1 & 0 \\ -1 & 0 & 0 \\ 0 & 0 & 0 \end{pmatrix}, \tag{23.21}$$

$$\psi^{1,1}_{0,0}(\hat{\mathbf{k}}) = \sqrt{\frac{3}{2}}(k_2\chi_1 - k_1\chi_2) \quad (\mathrm{A}_{2\mathrm{u};z}); \tag{23.22}$$

$j = 1; j_z = \pm 1$:

$$\mathbf{D}_x = \frac{1}{\sqrt{2}}\begin{pmatrix} 0 & 0 & 0 \\ 0 & 0 & -1 \\ 0 & 1 & 0 \end{pmatrix}, \tag{23.23}$$

$$\mathbf{D}_y = \frac{1}{\sqrt{2}}\begin{pmatrix} 0 & 0 & 1 \\ 0 & 0 & 0 \\ -1 & 0 & 0 \end{pmatrix}, \tag{23.24}$$

$$\psi^{1,1}_{1,\pm 1}(\hat{\mathbf{k}}) = \sqrt{\frac{3}{2}}\begin{pmatrix} -k_3\chi_2 + k_2\chi_3 \\ k_3\chi_1 - k_1\chi_3 \end{pmatrix} \quad (\mathrm{E}_{1\mathrm{u};x,y}); \tag{23.25}$$

$j = 2; j_z = 0$:

$$\mathbf{D}_{r^2-3z^2} = \frac{1}{\sqrt{6}}\begin{pmatrix} 1 & 0 & 0 \\ 0 & 1 & 0 \\ 0 & 0 & -2 \end{pmatrix}, \tag{23.26}$$

$$\psi^{1,1}_{2,0}(\hat{\mathbf{k}}) = \frac{1}{\sqrt{2}}(k_1\chi_1 + k_2\chi_2 - 2k_3\chi_3) \quad (\mathrm{A}_{1\mathrm{u}}); \tag{23.27}$$

$j = 2; j_z = \pm 1$:

$$\mathbf{D}_{yz} = \frac{1}{\sqrt{2}}\begin{pmatrix} 0 & 0 & 0 \\ 0 & 0 & 1 \\ 0 & 1 & 0 \end{pmatrix}, \tag{23.28}$$

$$\mathbf{D}_{xz} = \frac{1}{\sqrt{2}}\begin{pmatrix} 0 & 0 & 1 \\ 0 & 0 & 0 \\ 1 & 0 & 0 \end{pmatrix}, \tag{23.29}$$

$$\psi^{1,1}_{2,\pm 1}(\hat{\mathbf{k}}) = \sqrt{\frac{3}{2}}\begin{pmatrix} k_3\chi_2 + k_2\chi_3 \\ k_3\chi_1 + k_1\chi_3 \end{pmatrix} \quad (\mathrm{E}_{1u;x,y}); \tag{23.30}$$

$j = 2; j_z \pm 2$:

$$\mathbf{D}_{xy} = \frac{1}{\sqrt{2}}\begin{pmatrix} 0 & 1 & 0 \\ 1 & 0 & 0 \\ 0 & 0 & 0 \end{pmatrix}, \tag{23.31}$$

$$\mathbf{D}_{x^2-y^2} = \frac{1}{\sqrt{2}}\begin{pmatrix} 1 & 0 & 0 \\ 0 & -1 & 0 \\ 0 & 0 & 0 \end{pmatrix}, \tag{23.32}$$

$$\psi^{1,1}_{2,\pm 2}(\hat{\mathbf{k}}) = \sqrt{\frac{3}{2}}\begin{pmatrix} k_2\chi_1 + k_1\chi_2 \\ k_1\chi_1 - k_2\chi_2 \end{pmatrix} \quad (\mathrm{E}_{2u}). \tag{23.33}$$

(Note that under a 90° rotation about the z axis $k_1 \to k_2, k_2 \to -k_1$; $\chi_1 \to \chi_2, \chi_2 \to -\chi_1$ and similarly for rotations about the other two axes.) For the $\ell = j = 2$ (d-wave) states we have

$\ell = 2; \ell_z = 0$:

$$\psi^{2,0}_{2,0}(\hat{\mathbf{k}}) = \sqrt{\frac{5}{6}}(k_1^2 + k_2^2 - 2k_3^2)\chi_0 \quad (\mathrm{A}_{1g}); \tag{23.34}$$

$\ell = 2; \ell_z = \pm 1$:

$$\psi^{2,0}_{2,\pm 1}(\hat{\mathbf{k}}) = \sqrt{15}\begin{pmatrix} k_2k_3 \\ k_1k_3 \end{pmatrix}\chi_0 \quad (\mathrm{E}_{1g;x,y}); \tag{23.35}$$

$\ell = 2; \ell_z = \pm 2$:

$$\psi^{2,0}_{2,\pm 2}(\hat{\mathbf{k}}) = \sqrt{\frac{15}{2}}\begin{pmatrix} k_1k_2 \\ \frac{1}{2}(k_1^2 - k_2^2) \end{pmatrix}\chi_0 \quad (\mathrm{E}_{2g}). \tag{23.36}$$

One may, of course, choose other relative amplitudes for the components of the 2D representations in the above.

We next examine the effect of removing the continuous (uniaxial) rotational symmetry about the $\hat{z}$ axis and replacing it with four- or six-fold symmetry. In lowest order this results in a perturbing potential involving the V_{04} or V_{06} terms in Eq. (23.18) for the D_4 and D_6 symmetries, respectively. (The two components of an E_2 representation transform as $k_1^2 - k_2^2(\sim \cos 2\phi)$ and $2k_1k_2(\sim \sin 2\phi)$.) The matrix elements of the V_{04} terms of Eq. (23.18) (associated with D_4 symmetry) with the components of an E_2 representation are different and hence E_2 states split into two nondegenerate levels (the symmetry of which we will discuss shortly). For the case of D_6 symmetry (involving the V_{06} terms), the two levels making up an E_2 representation remain degenerate.

With the understanding that much of the degeneracy is lifted, it is still common to talk of pairing in $\ell = 0, \ell = 1, \ell = 2$ channels, or s-wave (singlet), p-wave (triplet), and d-wave (singlet) superconductors. We will assume that pairing effects arising from channels with $\ell > 2$ are negligible. Conventional s-wave superconductors correspond to a nodeless A_{1g} representation and the gap has the full point group symmetry of the crystal.

Note that so far we have not designated any B representations. For principal axes of fourth or sixth order (D_4 or D_6) this would require wavefunctions containing the terms $\cos m\phi$ or $\sin m\phi$ where $m = 2(D_4)$ or $3(D_6)$, which change sign under a rotation of $\pi/2(D_4)$ or $\pi/3(D_6)$. (The $\cos m\phi$ term peaks on the U axis and the $\sin m\phi$ on the U' axis.) For D_6, such terms cannot occur for $\ell \leq 2$; for D_4 the $j = 2$, E_2 representation of the uniaxial symmetry case splits into a $B_1(\cos 2\phi)$ and a $B_2(\sin 2\phi)$ representation.

In the absence of spin–orbit coupling, the (Stark-like) potential producing the uniaxial symmetry, which may be represented by the term V_{20} in Eq. (23.18), has a matrix element connecting the $j = 0, j_z = 0$ state (Eq. (23.20)) and the $j = 2, j_z = 0$ state (Eq. (23.27)) (there are no matrix elements connecting $j = 1, j_z = 0$ with either of these two states). We must therefore perform a *degenerate* perturbation calculation in this subspace. The resulting wavefunctions and energy shifts, ΔE, are:

$$\psi^{1,1}_{\text{polar}}(\hat{\mathbf{k}}) = \sqrt{3}k_3\chi_3; \quad \Delta E = \frac{3}{5}V_{20} \tag{23.37}$$

and

$$\psi^{1,1}_{\text{planar}}(\hat{\mathbf{k}}) = \sqrt{\frac{3}{2}}(k_1\chi_1 + k_2\chi_2); \quad \Delta E = \frac{1}{5}V_{20}. \tag{23.38}$$

The angular dependence of the nodes of the gap follows from $\psi^\dagger(\hat{\mathbf{k}})\psi(\hat{\mathbf{k}})$. Form (23.37) is analogous to the polar phase of ^{3}He and has a node in the x–y (basal) plane; form (23.38) is analogous to the planar phase and has a node along the $\hat{z}$ axis. (Note neither of these phases is stable in ^{3}He.) The addition of spin–orbit coupling restores forms (23.20) and (23.25) as eigenstates. Form (23.20) has no nodes and may be regarded as the analogue of the ^{3}He B phase; form (23.27) has the dependence $\sin^2\theta$ and retains the planar-like behavior, while the polar-like state disappears.

The A or ABM phase of ^{3}He (see (23.8)) may be regarded as a superposition of $j_z = \pm 1$ states of the $j = 1$ and $j = 2$ multiplets, both of which are $E_{1u;xy}$ representations. However, in the presence of spin–orbit coupling, $j = 1$ and $j = 2$ states have different energies (i.e., are non-degenerate) and hence a coherent superposition is not possible; thus the A phase cannot exist for D_6 or D_4.

Neutron scattering experiments suggest that there is an antiferromagnetic ordering in this basal plane of UPt_3; formally, this lowers the symmetry from D_{6h} to D_{2h} (or D_{4h} to D_{2h} for URu_2Si_2); however, the weakness of the perturbation suggests that any degeneracies lifted involve only a very small splitting. We may impose this lowered symmetry on the representations discussed earlier by examining the effect of a term V_{02} in Eq. (23.18). We have already shown that E_2 is split by a V_{04} term (which is also present when the symmetry is reduced to D_2). A short calculation shows that V_{02} splits E_1. The surviving representations have the symmetries and designations: A, $B_{3;x}$, $B_{1;z}$ and $B_{2;y}$ in addition to either u or g symmetry for the triplet or singlet cases respectively. These splittings proceed as follows

$$E_1 \sim \begin{pmatrix} \sin\phi \\ \cos\phi \end{pmatrix} \rightarrow B_{2;y}; B_{3;x}. \tag{23.39}$$

For our split E_2 representations

$$E_2 \sim \begin{pmatrix} \sin 2\phi \\ \cos 2\phi \end{pmatrix} k_z \rightarrow A_{1;z}, A_{1;z}. \tag{23.40}$$

Table 23.1 *Wavefunction symmetries for* D_2, D_4, *and* D_6 *classes.*

D_2	A	$B_{1;z}$	$B_{2;y}$	$B_{3;x}$		
D_4	A_1	$A_{2;z}$	B_1	B_2	$E_{1;x,y}$	
D_6	A_1	$A_{2;z}$	B_1	B_2	E_2	$E_{1;x,y}$

The allowed wavefunction symmetries for D_2, D_4, and D_6 are summarized in Table 23.1. The various allowed symmetry types are called 'irreducible representations', because they characterize the allowed wavefunction symmetries using the minimum number of dimensions (i.e., wavefunctions) possible.

23.3 The G–L theory of an unconventional superfluid

As with the general Landau theory of a second order phase transition, it is assumed that the free energy may be expanded in terms of an order parameter, and that this order parameter can be written as a linear combination of the allowed wavefunction symmetries (or irreducible representations) discussed in Subsecs. 23.1 and 23.2. It will further be assumed that only one wavefunction symmetry (that form having the highest T_c) will be important near the transition temperature. We begin with the case of ^{3}He.

23.3.1 G–L theory for ^{3}He

We now construct a G–L theory which is applicable to the p-wave superfluid ^{3}He (Mermin and Stare 1973, Brinkman and Anderson 1973). As discussed in the previous subsection the A and B ground states of ^{3}He correspond to unitary phases which are described by Eqs. (23.4a) and (23.4b), respectively; unitary states satisfy the property

$$\sum_i D^*_{\alpha i} D_{\beta i} = S_{\alpha\beta}, \tag{23.41}$$

where $S_{\alpha\beta}$ is a symmetric matrix.

Even if we restrict our discussion to unitary states, there are still several possible independent structures for the gap matrix. The quantities $D_{\alpha i}$ form a representation of the superfluid order parameter. $D_{\alpha i}$ is a 'bivector' which under a simultaneous rotation of the spin and momentum space coordinates transforms as

$$D'_{\alpha i} = \sum_\beta \sum_j R_{\alpha\beta}(\theta) R_{ij}(\theta) D_{\beta j}; \tag{23.42}$$

i.e., it transforms as a tensor.

For the most general case the matrix $\mathbf{D}$ involves 9 complex (or 18 real) numbers. To

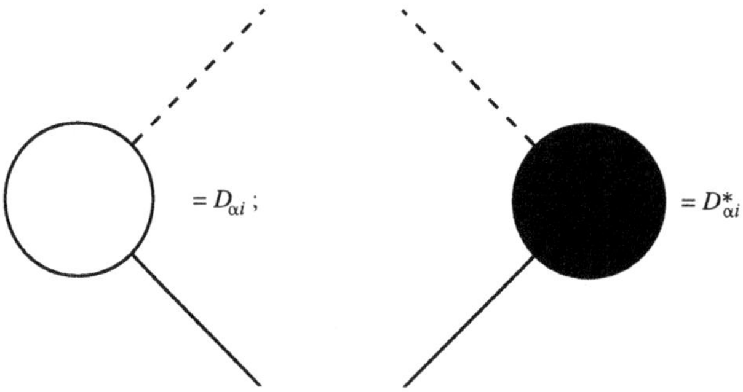

Figure 23.2 Diagrams representing $D_{\alpha i}$ and $D^*_{\alpha i}$.

construct a G–L theory we must determine all the second and fourth order invariants that can be formed from $D_{\alpha i}$. If the very weak nuclear dipolar (and still weaker spin–orbit) energy is neglected, the free energy must be invariant under *separate* rotations of the coordinate systems associated with spin space (α) and real space (i); i.e., it must be invariant under the operations

$$D'_{\alpha i} = \sum_{\beta} R_{\alpha\beta}(\theta) D_{\beta i} \tag{23.43}$$

or

$$D'_{\alpha i} = \sum_{j} R_{ij}(\theta) D_{\alpha j}. \tag{23.44}$$

The rotational invariance will be assured if we form pairs of matrices $\mathbf{D}$ and $\tilde{\mathbf{D}}$ and sum over the spin and momentum (orbit) space indices. In addition we must include an equal number of $\mathbf{D}$ and $\mathbf{D}^*$ matrices to insure that the result is real. There is only one second order invariant which may be written as

$$\mathrm{Tr}\,\mathbf{D}\cdot\mathbf{D}^\dagger = \sum_{\alpha,i} D^*_{\alpha i} D_{\alpha i}. \tag{23.45}$$

In fourth order there are five invariants. They are most easily enumerated by introducing the diagrams shown in Fig. 23.2. Here the dashed and solid lines refer to the spin and orbit indices, respectively. The five fourth order invariants correspond to the diagrams shown in Fig. 23.3. Connecting a line between two circles corresponds to summing over an index. Since we exclude terms mixing the spin and orbital degrees of freedom we connect a solid line only to a solid line and a dashed line only to a dashed line. If we think of the dashed and solid lines as strings connected to light and dark balls then we may rotate or twist the structure without altering the corresponding mathematical forms. For example we could also draw Fig. 23.3(d) in a 'twisted' form shown in Fig. 23.4. The five diagrams shown in Fig. 23.3 correspond to the following quadratic terms:

$$|\mathrm{Tr}(\mathbf{D}\cdot\tilde{\mathbf{D}})|^2 = \left(\sum_{\alpha,i} D_{\alpha i} D_{\alpha i}\right)\left(\sum_{\alpha,i} D^*_{\alpha i} D^*_{\alpha i}\right); \tag{23.46a}$$

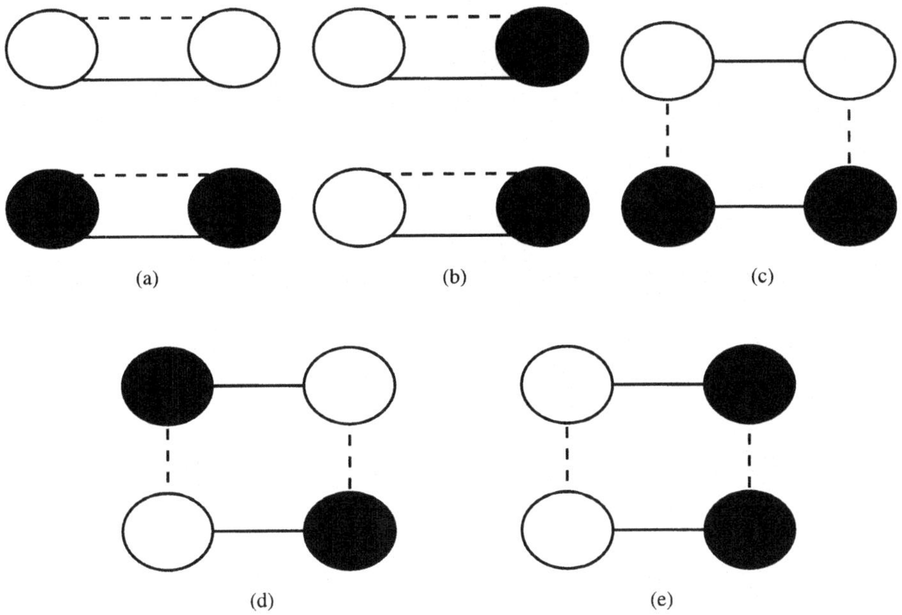

Figure 23.3 The five quadratic invariants involving $D_{\alpha i}$ and $D^*_{\alpha i}$.

$$[\mathrm{Tr}(\mathbf{D}\cdot\mathbf{D}^\dagger)]^2 = \left(\sum_{\alpha,i} D_{\alpha i}D^*_{\alpha i}\right)^2; \tag{23.46b}$$

$$\mathrm{Tr}[(\mathbf{D}\cdot\tilde{\mathbf{D}})(\mathbf{D}\cdot\tilde{\mathbf{D}})^*] = \sum_{\substack{\alpha,\beta\\ i,j}} D_{\alpha i}D_{\beta i}D^*_{\alpha j}D^*_{\beta j}; \tag{23.46c}$$

$$\mathrm{Tr}[(\mathbf{D}\cdot\mathbf{D}^\dagger)(\mathbf{D}\cdot\mathbf{D}^\dagger)] = \sum_{\substack{\alpha,\beta\\ i,j}} D_{\alpha i}D_{\beta j}D^*_{\alpha j}D^*_{\beta i}; \tag{23.46d}$$

$$\mathrm{Tr}[(\mathbf{D}\cdot\mathbf{D}^\dagger)(\mathbf{D}\cdot\mathbf{D}^\dagger)^*] = \sum_{\substack{\alpha,\beta\\ i,j}} D_{\alpha i}D_{\alpha j}D^*_{\beta i}D^*_{\beta j}; \tag{23.46e}$$

Combining the above we may write the G–L free energy as

$$\begin{aligned}\mathscr{F} = \mathscr{F}_0 &- \alpha\mathrm{Tr}\mathbf{D}\cdot\mathbf{D}^\dagger + \beta_1|\mathrm{Tr}\mathbf{D}\cdot\tilde{\mathbf{D}}|^2 + \beta_2[(\mathrm{Tr}\mathbf{D}\cdot\mathbf{D}^\dagger]^2\\ &+ \beta_3\mathrm{Tr}[(\mathbf{D}\cdot\tilde{\mathbf{D}})(\mathbf{D}\cdot\tilde{\mathbf{D}})^*] + \beta_4\mathrm{Tr}[(\mathbf{D}\cdot\mathbf{D}^\dagger)(\mathbf{D}\cdot\mathbf{D}^\dagger)]\\ &+ \beta_5\mathrm{Tr}[(\mathbf{D}\cdot\mathbf{D}^\dagger)(\mathbf{D}\cdot\mathbf{D}^\dagger)^*],\end{aligned} \tag{23.47}$$

where near the transition temperature $\alpha = a(T_c - T)$ and the β_i are constants.

For case I, Eq. (23.4a), $D_{\alpha i}$ is given by $n_{\alpha i}$, apart from the complex constant. All of the matrix products involving $\mathbf{D}$ and $\tilde{\mathbf{D}}$ or $\mathbf{D}^\dagger$ in Eq. (23.4a) have the form $S_{\alpha\beta} = \Sigma_i n_{\alpha i}n_{\beta i}$; S is therefore a real symmetric matrix, and such a matrix has three real, nonnegative eigenvalues, λ_1, λ_2, and λ_3. In terms of these eigenvalues we may write the free energy (which involves $\mathrm{Tr}S$, $(\mathrm{Tr}S)^2$, and $\mathrm{Tr}(SS)$) as

$$\mathscr{F} = \mathscr{F}_0 - \alpha\sum_i\lambda_i + (\beta_1+\beta_2)\left(\sum_i\lambda_i\right)^2 + (\beta_3+\beta_4+\beta_5)\sum_i\lambda_i^2. \tag{23.48}$$

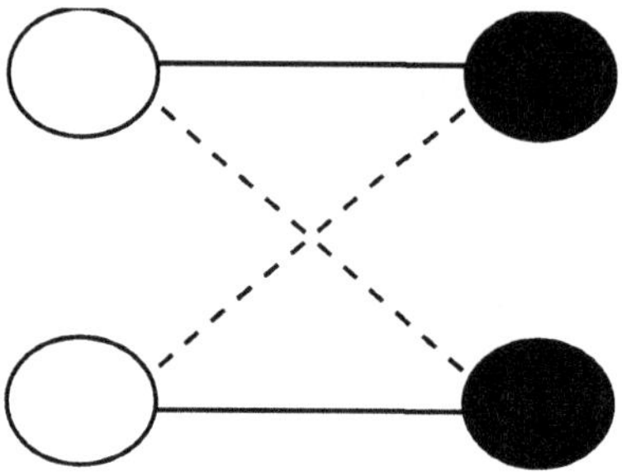

(d')

Figure 23.4 An alternative form for Fig. 23.3(d).

Thus, although many parameters (involving the degeneracy with respect to rotations in spin and real space and a phase factor) are required to specify $D_{\alpha i}$ completely, the free energy is governed by only three parameters. The system will spontaneously seek a minimum with respect to these parameters, i.e., we must minimize $\mathscr{F}$ with respect to the λ_i. There are three cases depending on whether one, two, or three of the λ_i differ from zero. These cases are referred to as the 1D (polar), 2D (planar), and 3D (Balian–Werthamer) cases respectively. Minimizing $\mathscr{F}$ with respect to λ_i (the minimum occurs when the nonzero λ_i are equal) yields an expression for λ_i which when substituted into $\mathscr{F}(\lambda_i)$ yields the following expressions

$$\mathscr{F} - \mathscr{F}_0 = \begin{cases} -\dfrac{3}{2}\dfrac{\alpha^2}{6(\beta_1+\beta_2)+2(\beta_3+\beta_4+\beta_5)}, & \text{Balian–Werthamer;} \quad (23.49a) \\ -\dfrac{\alpha^2}{2[2(\beta_1+\beta_2)+(\beta_3+\beta_4+\beta_5)]}, & \text{planar;} \quad (23.49b) \\ -\dfrac{\alpha^2}{4(\beta_1+\beta_2+\beta_3+\beta_4+\beta_5)}, & \text{polar.} \quad (23.49c) \end{cases}$$

The form of the order parameter for the three cases is:

(*a*) 3D (Balian–Werthamer):

$$S_{\alpha\beta} = \delta_{\alpha\beta}, \tag{23.50a}$$

$$D_{\alpha i} = \Delta R_{\alpha i} = \Delta \sum_{\lambda=1}^{3} \ell_\alpha^{(\lambda)} m_i^{(\lambda)}; \tag{23.50b}$$

(*b*) 2D (planar):

$$S = P_2, \tag{23.51a}$$

$$D_{\alpha i} = \Delta \sum_{\lambda=1}^{2} \ell_\alpha^{(\lambda)} m_i^{(\lambda)}; \tag{23.51b}$$

(*c*) 1D (polar):

$$S = P_1, \tag{23.52a}$$

$$D_{\alpha i} = \ell_\alpha m_i \Delta. \tag{23.52b}$$

Here Δ is the complex gap amplitude, $\mathbf{R}$ is a 3D rotation matrix, and P_1 and P_2 are 1D and 2D projection operators respectively. The $\ell^{(\lambda)}$ and $\mathbf{m}^{(\lambda)}$ are each sets of mutually perpendicular unit vectors; i.e., $\ell^{(1)} \cdot \ell^{(2)} = 0$, $\mathbf{m}^{(1)} \cdot \mathbf{m}^{(2)} = 0$, etc.

We see that within each class there is a large degeneracy in that the two coordinate system associated with the sets of vectors $\ell^{(\lambda)}$ and $\mathbf{m}^{(\lambda)}$ may each be arbitrarily oriented.

We now examine case II, Eq. (23.4b), where $D_{\alpha i} = n_\alpha \Delta_i$, where $\mathbf{\Delta}$ is a vector order parameter. The matrix $\mathbf{D}$ has the structure

$$\mathbf{D} = \begin{bmatrix} n_1\Delta_1 & n_1\Delta_2 & n_1\Delta_3 \\ n_2\Delta_1 & n_2\Delta_2 & n_2\Delta_3 \\ n_3\Delta_1 & n_3\Delta_2 & n_3\Delta_3 \end{bmatrix}.$$

Inserting this form into Eq. (23.47) we obtain

$$\mathscr{F} = \mathscr{F}_0 - \alpha \mathbf{\Delta} \cdot \mathbf{\Delta}^* + (\beta_1 + \beta_3) |\mathbf{\Delta} \cdot \mathbf{\Delta}|^2 + (\beta_2 + \beta_4 + \beta_5)(\mathbf{\Delta} \cdot \mathbf{\Delta}^*)^2. \tag{23.53}$$

If we write $\mathbf{\Delta} = \mathbf{\Delta}_1 + i\mathbf{\Delta}_2$, where $\mathbf{\Delta}_1$ and $\mathbf{\Delta}_2$ are real vectors, then

$$\mathbf{\Delta} \cdot \mathbf{\Delta}^* = \Delta_1^2 + \Delta_2^2$$
$$|\mathbf{\Delta} \cdot \mathbf{\Delta}| = [(\Delta_1^2 - \Delta_2^2)^2 + 4(\mathbf{\Delta}_1 \cdot \mathbf{\Delta}_2)^2]^{1/2}.$$

We clearly have the following inequality

$$0 \le |\mathbf{\Delta} \cdot \mathbf{\Delta}| \le \mathbf{\Delta} \cdot \mathbf{\Delta}^*;$$

$|\mathbf{\Delta} \cdot \mathbf{\Delta}|$ is largest, and equal to $\mathbf{\Delta} \cdot \mathbf{\Delta}^*$, for $\mathbf{\Delta}_1$ parallel or antiparallel to $\mathbf{\Delta}_2$; it has the smallest value, 0, for $\mathbf{\Delta}_1 \perp \mathbf{\Delta}_2$ and $|\mathbf{\Delta}_1| = |\mathbf{\Delta}_2|$. If $\beta_1 + \beta_3 > 0$ we want to minimize the contribution of this term and therefore choose this latter situation yielding

$$\mathscr{F} = \mathscr{F}_0 - 2\alpha\Delta_1^2 + 4(\beta_2 + \beta_4 + \beta_5)\Delta_1^4. \tag{23.54}$$

Minimizing (23.54) with respect to the magnitude of $\mathbf{\Delta}$ yields

$$\mathscr{F} = \mathscr{F}_0 - \frac{\alpha^2}{4(\beta_2 + \beta_4 + \beta_5)}. \quad \text{(axial)} \tag{23.55}$$

(For $\beta_1 + \beta_3 < 0$ we obtain the 1D case already discussed.) From the above-determined structure of our $\mathbf{\Delta}$ vector we may write the order parameter as

$$\mathbf{D} = \Delta \begin{bmatrix} n_1(\ell_1^{(1)} + i\ell_1^{(2)}) & n_1(\ell_2^{(1)} + i\ell_2^{(2)}) & n_1(\ell_3^{(1)} + i\ell_3^{(2)}) \\ n_2(\ell_1^{(1)} + i\ell_1^{(2)}) & n_2(\ell_2^{(1)} + i\ell_2^{(2)}) & n_2(\ell_3^{(1)} + i\ell_3^{(2)}) \\ n_3(\ell_1^{(1)} + i\ell_1^{(2)}) & n_3(\ell_2^{(1)} + i\ell_2^{(2)}) & n_3(\ell_3^{(1)} + i\ell_3^{(2)}) \end{bmatrix}, \tag{23.56}$$

where $\ell^{(1)}$ and $\ell^{(2)}$ are a pair of orthogonal unit vectors. The expressions for the free energy are compared in Table 23.2. Note that the 2D state is never stable. In ^{3}He only the axial and 3D states occur, the A and B phases respectively.

We emphasize again that the above analysis of phase stability was limited to unitary phases. A complete analysis of all possible phases for an isotropic p-paired triplet superfluid has not been carried out to date (1997), although some more general classes have been examined (Barton and Moore 1975).

Table 23.2 *Stability criteria.*

$\beta_1 + \beta_3$	$\beta_3 + \beta_4 + \beta_5$	Stable state
> 0	$< \frac{3}{2}(\beta_1 + \beta_3)$	axial
> 0	$> \frac{3}{2}(\beta_1 + \beta_3)$	3D
< 0	< 0	1D
< 0	> 0	3D

23.3.2 *G–L theory for an isotropic d-paired superfluid*

Mermin (1974) has studied the case of an isotropic d-paired superfluid. The free energy is constructed as follows. The resulting expression must be gauge and rotationally invariant. In second order there is only one invariant, $\mathrm{Tr}\mathbf{D}\cdot\mathbf{D}^*$ (note that $\mathbf{D}^* = \mathbf{D}^\dagger$ since $\mathbf{D}$ is a symmetric matrix). In fourth order there are four: $(\mathrm{Tr}\mathbf{D}\cdot\mathbf{D}^*)^2$, $|\mathrm{Tr}\mathbf{D}\cdot\mathbf{D}|^2$, $\mathrm{Tr}(\mathbf{D}\cdot\mathbf{D}^*)^2$, and $\mathrm{Tr}\mathbf{D}^2\cdot\mathbf{D}^{*2}$ (note that $\mathrm{Tr}\mathbf{D} = 0$ so $\mathbf{D}$ or $\mathbf{D}^*$ cannot be contracted with itself). One of these may be eliminated using the identity

$$\mathrm{Tr}(\mathbf{D}\cdot\mathbf{D}^*)^2 = \tfrac{1}{2}|\mathrm{Tr}\mathbf{D}^2|^2 + (\mathrm{Tr}\mathbf{D}\cdot\mathbf{D}^*)^2 - 2\mathrm{Tr}(\mathbf{D}^2\cdot\mathbf{D}^{*2}).$$

A suitable form for the free energy is then

$$\mathscr{F} = \mathscr{F}_0 + \alpha\mathrm{Tr}\mathbf{D}\cdot\mathbf{D}^* + \beta_1|\mathrm{Tr}\mathbf{D}^2|^2 + \beta_2(\mathrm{Tr}\mathbf{D}\cdot\mathbf{D}^*)^2 + \beta_3\mathrm{Tr}\mathbf{D}^2\cdot\mathbf{D}^{*2}. \quad (23.57)$$

We will not carry out the minimization here. Mermin finds that there are three stable phases, two of which are degenerate in the weak coupling limit; his phase diagram is shown in Fig. 23.5.

23.3.3 *Unconventional G–L theory in metals*

If the order parameter is dominated by one of the 1D representations discussed earlier then the G–L expansion takes the same form as for an ordinary (A_{1g}) superconductor. The observation of the various phenomena discussed earlier, which imply the existence of more than one 'transition temperature' in UPt_3, strongly suggests that the superconductivity is unconventional. One possibility is that it is associated with one of the 2D representations, E_1 or E_2, and that the otherwise-degenerate transition temperatures (associated with the two components making up a 2D representation) are split by a weak symmetry breaking, brought on by, e.g., the antiferromagnetic ordering mentioned earlier. It is, of course, also possible that two symmetry-distinct 1D representations accidentally have nearly the same transition temperatures (within a small percentage for UPt_3). In BCS theory, the transition temperature depends exponentially on the pairing potential associated with a given representation; therefore such a situation would represent an 'accident'. In what follows we use the E_1 and E_2 symmetries as discussion examples[4] (Tokuyasu, Hess, and Sauls 1990, Tokuyasu 1990).

4. At present there are no confirmed E state superconductors.

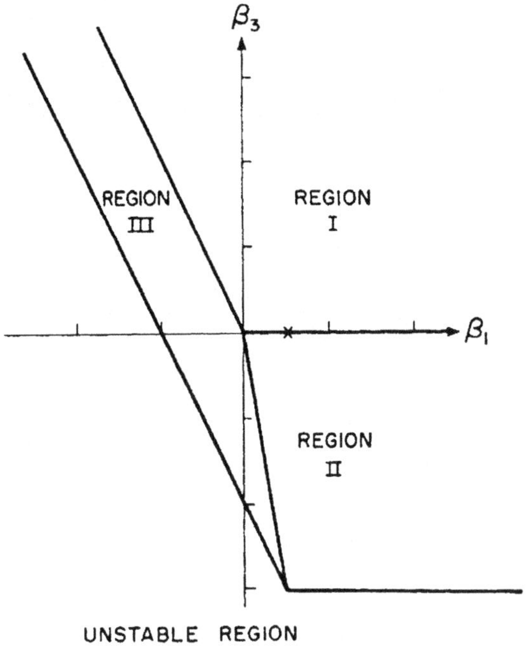

Figure 23.5 Parameter space for the general fourth order d-wave free energy. The scale on the β_1 and β_3 axes is in units of β_2, which stability requires to be positive. The point $\beta_1 = \beta_2/2$, $\beta_3 = 0$, corresponding to the elementary BCS case, is indicated by an '×'. In regions I and II the equilibrium order parameters are unique (to within a constant factor and a rotation) and are given by $\Delta \propto (\hat{k}_1 + i\hat{k}_2)^2$ (region I) and $\Delta \propto \hat{k}_1^2 + e^{2\pi i/3}\hat{k}_2^2 + e^{4\pi i/3}\hat{k}_3^2$ (region II). In region III the free energy is minimized by any suitably normalized order parameter that is real (to within a constant phase factor). (After Mermin (1974).)

E_1 case

The two basis wavefunctions making up an E_1 representation transform as k_x and k_y. The most general E_1 state must therefore be written as a linear superposition of the two basis functions with complex coefficients Δ_x and Δ_y:

$$\psi = \Delta_x k_x + \Delta_y k_y = \mathbf{\Delta} \cdot \mathbf{k}_\perp. \tag{23.58}$$

The free energy is expanded in terms of the parameters contained in $\mathbf{\Delta}$; these parameters must be combined such that the free energy is: (i) real, (ii) invariant under all crystal symmetry operations, and (iii) restricted to even powers of $|\Delta_x|$ and $|\Delta_y|$. At the level of the quadratic terms the only invariant function involving k_x and k_y is $(k_x^2 + k_y^2)$ with the associated invariant $(\mathbf{\Delta} \cdot \mathbf{\Delta}^*)$. The allowed quartic terms differ from D_4 and D_6. For D_6 the only invariant function is $(k_x^2 + k_y^2)^2$ leading to two fourth order invariants: $(\mathbf{\Delta} \cdot \mathbf{\Delta}^*)^2$ and $|\mathbf{\Delta} \cdot \mathbf{\Delta}|^2$; for D_4 we have the additional invariant $(k_x^4 + k_y^4)$ leading to the invariant $\Delta_x^2 \Delta_x^{*2} + \Delta_y^2 \Delta_y^{*2}$. The resulting G–L expansions are

$$\mathscr{F} = \mathscr{F}_0 + \alpha(\mathbf{\Delta} \cdot \mathbf{\Delta}^*) + \beta_1(\mathbf{\Delta} \cdot \mathbf{\Delta}^*)^2 + \beta_2 |\mathbf{\Delta} \cdot \mathbf{\Delta}|^2 \quad (D_6) \tag{23.59a}$$

and

$$\mathscr{F} = \mathscr{F}_0 + \alpha(\mathbf{\Delta}\cdot\mathbf{\Delta}^*) + \beta_1(\mathbf{\Delta}\cdot\mathbf{\Delta}^*)^2 + \beta_2|\mathbf{\Delta}\cdot\mathbf{\Delta}|^2 + \beta_3(\Delta_x^2\Delta_x^{*2} + \Delta_y^2\Delta_y^{*2}) \quad (\mathrm{D}_4). \tag{23.59b}$$

We will restrict ourselves to the case of D_6 (UPt_3). Minimizing Eq. (23.59a) with respect to $\mathbf{\Delta}^*$ we obtain

$$\alpha\mathbf{\Delta} + 2\beta_1(\mathbf{\Delta}\cdot\mathbf{\Delta}^*)\mathbf{\Delta} + 2\beta_2(\mathbf{\Delta}\cdot\mathbf{\Delta})\mathbf{\Delta}^* = 0. \tag{23.60}$$

Near T_c we write $\alpha = a(T - T_c)$. Since the quantity $\mathbf{\Delta}\cdot\mathbf{\Delta}^*$ is positive definite, thermodynamic stability requires $\beta_1 > 0$; β_2 may have either sign, but thermodynamic stability requires $\beta_2 > -\beta_1$. If $\beta_2 > 0$ we achieve the lowest energy when $\mathbf{\Delta}\cdot\mathbf{\Delta} = 0$, which requires the form

$$\mathbf{\Delta} = \frac{\Delta_0}{\sqrt{2}}(\hat{\mathbf{x}} \pm \mathrm{i}\hat{\mathbf{y}}) \qquad (\beta_2 > 0). \tag{23.61}$$

Inserting this in the remaining terms in (23.60) we obtain

$$|\Delta_0|^2 = -\frac{\alpha}{2\beta_1}, \qquad T < T_c \qquad (\beta_2 > 0) \tag{23.62a}$$

or

$$|\Delta_0|^2 = 0, \qquad T > T_c; \tag{23.62b}$$

Δ_0 contains an arbitrary phase.

For the case $\beta_2 < 0$, the free energy is minimized when $\mathbf{\Delta}\cdot\mathbf{\Delta}$ is maximal; i.e., when

$$\mathbf{\Delta} = \Delta_0(\cos\phi_0\hat{\mathbf{x}} + \sin\phi_0\hat{\mathbf{y}}) \tag{23.63}$$

where ϕ_0 is an arbitrary angle of rotation in the x–y plane and Δ_0 again contains an arbitrary phase. The value of $|\Delta_0|$ minimizing Eq. (23.59a) is

$$|\Delta_0|^2 = -\frac{\alpha}{2(\beta_1 + \beta_2)} \qquad (T < T_c) \tag{23.64a}$$

or

$$|\Delta_0|^2 = 0 \qquad (T > T_c). \tag{23.64b}$$

The two signs associated with Eq. (23.61) correspond to two different states which cannot be continuously deformed into each other and hence the superfluid state can consist of a mixture of domains of each type. The state described by Eq. (23.63) is continuously (infinitely) degenerate with respect to the arbitrary angle, ϕ_0.

E_2 case

The E_2 representations, associated with p- or d-wave pairing, involve the forms $k_x^2 - k_y^2$ and k_xk_y. Writing $k_x = \cos\phi$ and $k_y = \sin\phi$ and taking the appropriate linear combinations we may schematically write the wavefunction of our E_2 representation as

$$\psi \sim \begin{pmatrix} \mathrm{e}^{2\mathrm{i}\phi} \\ \mathrm{e}^{-2\mathrm{i}\phi} \end{pmatrix}. \tag{23.65}$$

The order parameter must be a linear combination of the two basis functions which we may write as $\Delta_+ e^{2i\phi} + \Delta_- e^{-2i\phi}$. If we define a vector $\mathbf{\Delta} = \begin{pmatrix}\Delta_+ \\ \Delta_-\end{pmatrix}$ and note that the second and fourth order invariants can only involve $(k_x + k_y)^2$ and $(k_x + k_y)^4$, then our free energy takes a form identical to Eq. (23.59a):

$$\mathscr{F} = \mathscr{F}_0 + \alpha(\mathbf{\Delta}\cdot\mathbf{\Delta}^*) + \beta_1(\mathbf{\Delta}\cdot\mathbf{\Delta}^*)^2 + \beta_2|\mathbf{\Delta}\cdot\mathbf{\Delta}|^2; \tag{23.66}$$

the minimization of Eq. (23.66) is completely analogous to the E_1 case. In the ordered state we have the following two possibilities

$$\mathbf{\Delta} = \Delta_0\begin{pmatrix}1 \\ \pm\mathrm{i}\end{pmatrix} \qquad (\beta_2 > 0) \tag{23.67a}$$

or

$$\mathbf{\Delta} = \Delta_0\begin{pmatrix}\cos 2\phi_0 \\ \sin 2\phi_0\end{pmatrix} \qquad (\beta_2 < 0) \tag{23.67b}$$

with $|\Delta_0|$ given by (23.62a) and (23.64a), respectively.

23.4 Inhomogeneities in the order parameter

To treat the effects of boundaries or finite fields we must incorporate the gradient terms in the G–L free energy. The case of an isotropic conventional superconductor was discussed in Sec. 9 and that for an anisotropic superconductor in Sec. 11. Here we will limit ourselves to the case of an isotropic p-wave superfluid (^{3}He) and the case of a uniaxial E representation in D_6.

For the isotropic p-wave superfluid there are three independent ways to combine the spatial derivatives of the order parameter, $\partial D_{\alpha i}/\partial x_\beta$, which are rotationally invariant with respect to separate rotations in spin space and real space (Ambegaokar, deGennes, and Rainer 1974, Anderson and Brinkman 1978):

$$\mathscr{F}_{G1} = K_1 \sum_{\alpha\beta}\sum_i \frac{\partial D_{\beta i}}{\partial x_\alpha}\frac{\partial D^*_{\beta i}}{\partial x_\alpha}, \tag{23.68a}$$

$$\mathscr{F}_{G2} = K_2 \sum_i \left|\sum_\alpha \frac{\partial D_{\alpha i}}{\partial x_\alpha}\right|^2, \tag{23.68b}$$

$$\mathscr{F}_{G3} = K_3 \sum_{\alpha\beta}\sum_i \frac{\partial D_{\alpha i}}{\partial x_\beta}\frac{\partial D^*_{\beta i}}{\partial x_\alpha}. \tag{23.68c}$$

If these terms are to result in a bulk contribution to the free energy only, then integrations by parts using Green's theorem must result in any surface contributions vanishing identically; application of this constraint leads to the condition $K_1 = K_2$. The full expression for the gradient energy is

$$\mathscr{F}_G = K_1 \sum_{\alpha\beta}\sum_i \frac{\partial D_{\beta i}}{\partial x_\alpha}\frac{\partial D^*_{\beta i}}{\partial x_\alpha} + K_2 \sum_i \left|\sum_\alpha \frac{\partial D_{\alpha i}}{\partial x_\alpha}\right|^2 + K_3 \sum_{\alpha\beta}\sum_i \frac{\partial D_{\alpha i}}{\partial x_\beta}\frac{\partial D^*_{\beta i}}{\partial x_\alpha}. \tag{23.69}$$

For the E_1 representation the gradient energy must be rotationally invariant in the $x, y(1,2)$plane. The vector $\mathbf{\Delta}$ and the perpendicular component of the gradient, $\nabla_\perp(\partial_x\hat{\mathbf{x}}' + \partial_y\hat{\mathbf{y}})$, each behave as 2D vectors and we can form the following 2D, rotationally invariant combinations which contribute to the free energy (Tokuyasu 1990, Tokuyasu *et al.* 1990)

$$\mathscr{F}_1 = \kappa_1 \sum_{i,j=1}^{2} \frac{\partial \Delta_i}{\partial x_j}\frac{\partial \Delta_i^*}{\partial x_j}, \tag{23.70a}$$

$$\mathscr{F}_2 = \kappa_2 \sum_{i,j=1}^{2} \frac{\partial \Delta_i}{\partial x_i}\frac{\partial \Delta_j^*}{\partial x_j}, \tag{23.70b}$$

and

$$\mathscr{F}_3 = \kappa_3 \sum_{i,j=1}^{2} \frac{\partial \Delta_i}{\partial x_j}\frac{\partial \Delta_j^*}{\partial x_i}. \tag{23.70c}$$

In addition we must add the (single) contribution associated with variations along the z axis

$$\mathscr{F}_4 = \kappa_4 \sum_{i=1}^{2} \frac{\partial \Delta_i}{\partial x_3}\frac{\partial \Delta_i^*}{\partial x_3}. \tag{23.70d}$$

In the presence of a magnetic field we use the standard gauge invariance prescription

$$\frac{\partial}{\partial x_i} \to \frac{\partial}{\partial x_i} - \frac{2ie}{\hbar c}A_i, \tag{23.71}$$

where A_i is the vector potential and $i = 1,2,3$. Finally we must add the energy density of the magnetic field

$$\mathscr{F}_H = \frac{1}{8\pi}H^2. \tag{23.72}$$

To obtain the expressions for the current density we use the relation $\mathbf{j}(\mathbf{r}) = c(\delta F/\delta \mathbf{A}(\mathbf{r}))$ as mentioned at the end of Subsec. 9.1. The resulting expressions are also valid for a neutral ($e = 0$) superfluid.

23.5 Collective modes in an unconventional superfluid

One of the most striking features of an unconventional superfluid is the appearance of collective modes associated with the order parameter. Qualitatively such modes represent oscillations of various components (or linear combinations of components) with respect to their equilibrium values. We may write this variation in the form

$$D_{\alpha i} = D_{\alpha i}^{(\mathrm{eq})} + D'_{\alpha i}, \tag{23.73}$$

where the $D_{\alpha i}^{(\mathrm{eq})}$ are the equilibrium values and the $D'_{\alpha i}$ represent a small departure from equilibrium.

A proper treatment of these modes requires the solution of the time-dependent Bogoliubov equations, the generalized Boltzman equation, or some other such approach, which we take up in Sec. 54. Our discussion here will involve a heuristically developed time-dependent G–L model from which many of the predictions of the more formal approaches appear relatively simply. We will limit ourselves to the cases of the superfluid ^{3}He B phase and an E state, uniaxial superconductor.

23.5.1 Collective modes of ^{3}He B

For the case of ^{3}He B it turns out that we may factor $D_{\alpha i}$ in Eq. (23.73) into real and imaginary parts as (where we drop the prime on the fluctuating part)

$$D_{\alpha i} = D_{\alpha i}^{(\mathrm{R})} + \mathrm{i} D_{\alpha i}^{(\mathrm{I})}, \tag{23.74a}$$

where

$$D_{\alpha i}^{(\mathrm{R})} = \frac{1}{2}(D_{\alpha i} + D_{\alpha i}^*) \tag{23.74b}$$

and

$$D_{\alpha i}^{(\mathrm{I})} = -\frac{\mathrm{i}}{2}(D_{\alpha i} - D_{\alpha i}^*). \tag{23.74c}$$

We may further decompose the real (but unsymmetric) matrices $\mathbf{D}^{(\mathrm{R})}$ and $\mathbf{D}^{(\mathrm{I})}$ into two traceless symmetric, $\mathbf{D}^{(\mathrm{RS})}$ and $\mathbf{D}^{(\mathrm{IS})}$, and two antisymmetric, $\mathbf{D}^{(\mathrm{RA})}$ and $\mathbf{D}^{(\mathrm{IA})}$, matrices together with unit matrices with amplitudes proportional to the traces: $D^{(\mathrm{RT})}\,\mathbf{1}$ and $D^{(\mathrm{IT})}\,\mathbf{1}$. The number of complex independent elements of these matrices is 5, 3, and 1 for each traceless symmetric matrix, antisymmetric matrix, and trace for a total of 18 (9 real and 9 imaginary) amplitudes; each of these amplitudes corresponds to an independent excited state (or collective mode) of ^{3}He B. The appropriate forms for the matrices were given earlier in Eqs. (23.10) ($j = 2$ or traceless symmetric), Eq. (23.9) ($j = 1$ or traceless antisymmetric), and Eq. (23.5) ($j = 0$ unit matrix).

The G–L theory of the collective modes assumes the existence of a Lagrangian density, $\mathscr{L}$, having the form[5]

$$\mathscr{L}\left(\mathbf{D}, \dot{\mathbf{D}}, \frac{\partial \mathbf{D}}{\partial x_\alpha}, \mathbf{D}^*, \dot{\mathbf{D}}^*, \frac{\partial \mathbf{D}^*}{\partial x_\alpha}\right) \equiv \mathscr{T} - \mathscr{V}, \tag{23.75}$$

where the 'kinetic energy', $\mathscr{T}$, is given by

$$\mathscr{T} = \Lambda \sum_{\alpha,i} \dot{D}_{\alpha i} \dot{D}_{\alpha i}^* \tag{23.76a}$$

and the potential $\mathscr{V}$ is formed from expanding the G–L free energy about its equilibrium value to

5. This model has been employed by Koch and Wölfle (1981), Volovik and Khazan (1984), Salomaa and Volovik (1989), Theodorakis (1988), and Zhao, Adenwalla, Sarma, and Ketterson (1992).

second order[6] in the variations of the order parameters, $D_{\alpha i}$, from their equilibrium values, $D_{\alpha i}^{(\mathrm{eq})}$; this yields

$$\begin{aligned}\mathscr{V}^{(2)} = &-\alpha\sum_{\alpha i} D_{\alpha i}D_{\alpha i}^{*} + \beta_1\Delta^2\left[4\left|\sum_{\alpha i} D_{\alpha i}\delta_{\alpha i}\right|^2 + 3\sum_{\alpha i}(D_{\alpha i}D_{\alpha i} + D_{\alpha i}^{*}D_{\alpha i}^{*})\right]\\ &+ \beta_2\Delta^2\left[\left(\sum_{\alpha i}(2D_{\alpha i} + D_{\alpha i}^{*})\delta_{\alpha i}\right)^2 + 6\sum_{\alpha i} D_{\alpha i}D_{\alpha i}^{*}\right]\\ &+ (\beta_3 + \beta_5)\Delta^2\sum_{\alpha i}(2D_{\alpha i}D_{\alpha i}^{*} + 2D_{\alpha i}D_{i\alpha}^{*} + D_{\alpha i}D_{\alpha i} + D_{\alpha i}^{*}D_{\alpha i}^{*})\\ &+ \beta_4\Delta^2\sum_{\alpha i}(4D_{\alpha i}D_{\alpha i}^{*} + D_{\alpha i}^{*}D_{i\alpha}^{*} + D_{\alpha i}D_{i\alpha}).\end{aligned} \tag{23.76b}$$

The equations of motion for the various components of the order parameter follow from the Euler–Lagrange equation, which for our case is given by

$$\frac{\partial}{\partial t}\frac{\partial \mathscr{L}}{\partial \dot{D}_{\alpha i}} + \frac{\partial}{\partial x_\beta}\frac{\partial \mathscr{L}}{\partial\left(\dfrac{\partial D_{\alpha i}}{\partial x_\beta}\right)} - \frac{\partial \mathscr{L}}{\partial D_{\alpha i}} = 0 \tag{23.77}$$

(the second term in this equation contributes only when the gradient energies are included). We may phenomenologically include the effects of dissipation by adding a term $\partial\mathscr{R}/\partial\dot{D}_{\alpha i}$ on the left-hand side of (23.77) where $\mathscr{R}$ is a Rayleigh dissipative function, which we write in the form

$$\mathscr{R} = \sum_{\alpha i} f_{\alpha i}\dot{D}_{\alpha i}\dot{D}_{\alpha i}^{*}. \tag{23.78}$$

Here the $f_{\alpha i}$ are friction coefficients which measure the damping frequency of the nonequilibrium amplitudes, $D_{\alpha i}$. We define the relaxation times through $\tau_{\alpha i} = f_{\alpha i}^{-1}$.

The equation of motion resulting from (23.77) and (23.76) is

$$\begin{aligned}\Lambda\ddot{D}_{\alpha i} - \alpha D_{\alpha i} + \Delta^2\Bigg\{&\beta_1\left[4\delta_{\alpha i}\sum_{\beta j} D_{\beta j}\delta_{\beta j} + 6D_{\alpha i}^{*}\right]\\ &+ \beta_2\left[2\delta_{\alpha i}\sum_{\beta j}(D_{\beta j} + D_{\beta j}^{*})\delta_{\beta j} + 6D_{\alpha i}\right]\\ &+ 2(\beta_3 + \beta_5)[D_{\alpha i} + D_{\alpha i}^{*} + D_{i\alpha}] + 2\beta_4(2D_{\alpha i} + D_{i\alpha}^{*})\Bigg\} = 0.\end{aligned} \tag{23.79}$$

In terms of the variables defined in Eqs. (23.74), the equations of motion are

$$\begin{aligned}\Lambda\ddot{D}_{\alpha i}^{(\mathrm{R})} - \alpha D_{\alpha i}^{(\mathrm{R})} + \Delta^2\Bigg[&\beta_1\left(4\delta_{\alpha i}\sum_{\beta j} D_{\beta j}^{(\mathrm{R})}\delta_{\beta j} + 6D_{\alpha i}^{(\mathrm{R})}\right)\\ &+ \beta_2\left(4\delta_{\alpha i}\sum_{\beta j} D_{\beta j}^{(\mathrm{R})}\delta_{\beta j} + 6D_{\alpha i}^{(\mathrm{R})}\right)\\ &+ 2(\beta_3 + \beta_5)(2D_{\alpha i}^{(\mathrm{R})} + D_{i\alpha}^{(\mathrm{R})}) + 2\beta_4(2D_{\alpha i}^{(\mathrm{R})} + D_{i\alpha}^{(\mathrm{R})})\Bigg] = 0.\end{aligned} \tag{23.80}$$

6. The first order variation vanishes since this is the equilibrium condition for minimizing the free energy.

and

$$\Lambda \dot{D}^{(I)}_{\alpha i} - \alpha D^{(I)}_{\alpha i} + \Delta^2 \left\{ \beta_1 \left[4\delta_{\alpha i} \sum_{\beta j} D^{(I)}_{\beta j} \delta_{\beta j} - 6D^{(I)}_{\alpha i} \right] + 6\beta_2 D^{(I)}_{\beta i} \right.$$
$$\left. + 2(\beta_3 + \beta_5) D^{(I)}_{i\alpha} + 2\beta_4 (2D^{(I)}_{\alpha i} - D^{(I)}_{i\alpha}) \right\} = 0. \tag{23.81}$$

We will assume the fluctuating quantities vary as $e^{iqz - i\omega t}$. Forming the symmetric (S), antisymmetric (A), and trace (T) contributions of the matrices $D^{(R)}_{\alpha i}$ and $D^{(I)}_{\alpha i}$ we obtain

$$\omega^2_{RS} = \frac{\alpha}{\Lambda} \frac{2(\beta_3 + \beta_4 + \beta_5)}{3(\beta_1 + \beta_2) + (\beta_3 + \beta_4 + \beta_5)}, \tag{23.82a}$$

$$\omega^2_{RA} = 0, \tag{23.82b}$$

$$\omega^2_{RT} = \frac{2\alpha}{\Lambda}, \tag{23.82c}$$

$$\omega^2_{IS} = -\frac{\alpha}{\Lambda} \frac{6\beta_1}{3(\beta_1 + \beta_2) + (\beta_3 + \beta_4 + \beta_5)}, \tag{23.82d}$$

$$\omega^2_{IA} = -\frac{\alpha}{\Lambda} \frac{6\beta_1 + 2\beta_3 - 2\beta_4 + 2\beta_5}{3(\beta_1 + \beta_2) + (\beta_3 + \beta_4 + \beta_5)}, \tag{23.82e}$$

and

$$\omega^2_{IT} = 0. \tag{23.82f}$$

At this point it is convenient to utilize a result from the microscopic theory. For the case in which the pairing is dominated by a single, temperature-independent, p-wave-pairing potential, the β_i of the G–L theory are related as[7]

$$\beta_2 = \beta_3 = \beta_4 = -\beta_5 = -2\beta_1 \equiv \beta; \tag{23.83}$$

the expression for Δ in this case is

$$|\Delta|^2 = \frac{1}{5}\frac{\alpha}{\beta}. \tag{23.84}$$

If we are to obtain numerical agreement with the microscopic theory (Sec. 54) in this limit we must set $\alpha/\Lambda = 2\omega_g^2$ where we have introduced a gap frequency, ω_g, defined by $\hbar\omega_g = |\Delta|$. We then have

$$\hbar^2 \omega^2_{RS} = \frac{8}{5}|\Delta|^2, \tag{23.85a}$$

$$\hbar^2 \omega^2_{RA} = 0, \tag{23.85b}$$

$$\hbar^2 \omega^2_{RT} = 4|\Delta|^2, \tag{23.85c}$$

$$\hbar^2 \omega^2_{IS} = \frac{12}{5}|\Delta|^2, \tag{23.85d}$$

7. This is the so-called weak coupling BCS approximation.

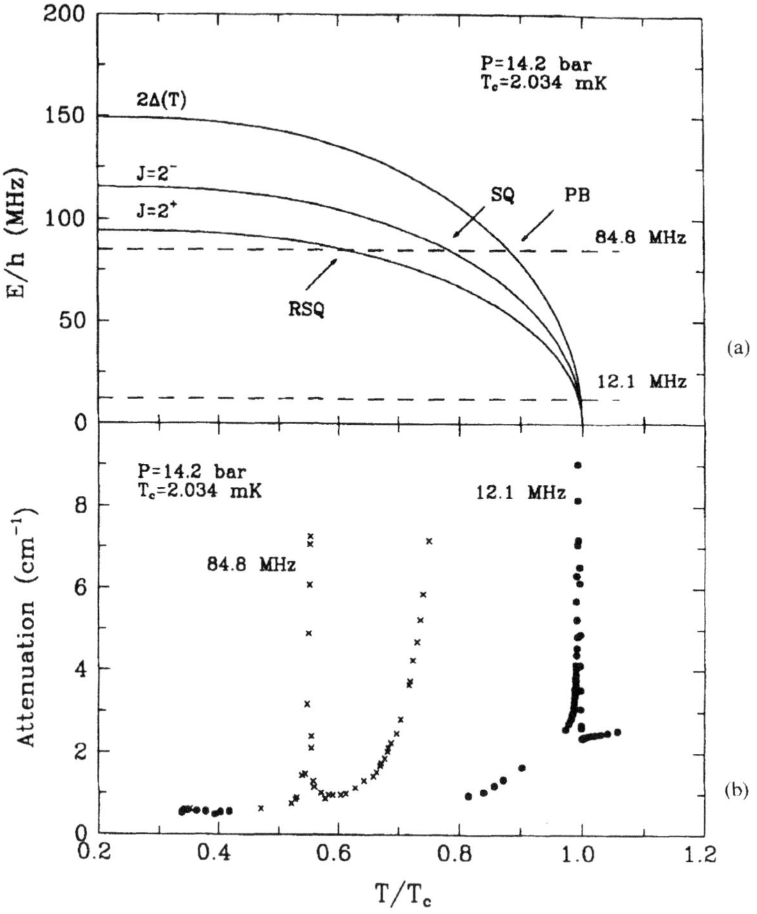

Figure 23.6 (a) A plot of the temperature dependence of the frequencies ($f = \omega/2\pi$) corresponding to $2|\Delta|$ and the RS ($J = 2^+$) and IS ($J = 2^-$) modes as a function of temperature. The dashed horizontal lines denote two frequencies (12.1 MHz and 84.8 MHz) used in the experiments. Mode–mode coupling occurs where the sound frequency is equal to a mode frequency. (b) The experimental data. The attentuation peaks at the right and left arise from the $J = 2^-$ (12.1 MHz) and $J = 2^+$ (84.8 MHz) modes respectively (the $J = 2^-$ coupling at 84.8 MHz is so strong that the peak is off scale). (After Mast *et al.* (1980).)

$$\hbar^2\omega_{\text{IA}}^2 = 4|\Delta|^2, \tag{23.85e}$$

and

$$\hbar^2\omega_{\text{IT}}^2 = 0. \tag{23.85f}$$

In Subsec. 54.3 we will rederive these results from the microscopic theory.

For those collective modes having a frequency proportional to $\Delta(T)$, we may use the temperature dependence of the gap to 'sweep' the collective mode frequency. It is relatively easy to propagate high frequency sound waves in superfluid ^{3}He and when the sound frequency matches a collective mode frequency there is a possibility of observing a mode–mode coupling.

Fig. 23.6 shows the effect of this coupling for the case described by Eqs. (23.85a) and (23.85d) in which $\hbar\omega_{\mathrm{RS}} = \sqrt{\frac{8}{5}}|\Delta(T)|(J = 2^+)$ and $\hbar\omega_{\mathrm{IS}} = \sqrt{\frac{12}{5}}|\Delta(T)|(J = 2^-)$; note the very sharp peaks in the attenuation at the temperature where the modes are degenerate with the sound mode $(J = 0^-)$.

By including the gradient terms of the G–L theory in our phenomenological Lagrangian it is possible to compute the shifts of the collective mode frequencies at finite wave vector.

23.5.2 *E state collective modes*

As an example of a possible collective mode spectrum of an unconventional metallic system we will apply the theory to the case of an E_1 state superfluid having D_6 symmetry. Expanding the free energy expression (23.59a) to second order we obtain

$$\begin{aligned}\mathscr{V}^{(2)} = \alpha\boldsymbol{\Delta}'\cdot\boldsymbol{\Delta}'^* + \beta_1[2(\boldsymbol{\Delta}\cdot\boldsymbol{\Delta}^*)(\boldsymbol{\Delta}'\cdot\boldsymbol{\Delta}'^*) + (\boldsymbol{\Delta}\cdot\boldsymbol{\Delta}'^* + \boldsymbol{\Delta}^*\cdot\boldsymbol{\Delta}')^2] \\ + \beta_2[(\boldsymbol{\Delta}\cdot\boldsymbol{\Delta})(\boldsymbol{\Delta}'\cdot\boldsymbol{\Delta}')^* + (\boldsymbol{\Delta}^*\cdot\boldsymbol{\Delta}^*)(\boldsymbol{\Delta}'\cdot\boldsymbol{\Delta}') + 4(\boldsymbol{\Delta}\cdot\boldsymbol{\Delta}')(\boldsymbol{\Delta}^*\cdot\boldsymbol{\Delta}'^*)].\end{aligned} \tag{23.86}$$

The equation of motion for the collective modes again follows from Lagrange's equation

$$\frac{\partial}{\partial t}\frac{\partial\mathscr{L}}{\partial\boldsymbol{\Delta}'^*} - \frac{\partial\mathscr{L}}{\partial\boldsymbol{\Delta}'^*} = 0, \tag{23.87}$$

where

$$\mathscr{L} = \mathscr{T} - \mathscr{V}. \tag{23.88}$$

The resulting equation is

$$\begin{aligned}\Lambda\ddot{\boldsymbol{\Delta}}' + \alpha\boldsymbol{\Delta}' + \beta_1\{2(\boldsymbol{\Delta}\cdot\boldsymbol{\Delta}^*)\boldsymbol{\Delta}' + 2\boldsymbol{\Delta}[(\boldsymbol{\Delta}\cdot\boldsymbol{\Delta}'^*) + (\boldsymbol{\Delta}^*\cdot\boldsymbol{\Delta}')]\} \\ + \beta_2\{2(\boldsymbol{\Delta}\cdot\boldsymbol{\Delta})\boldsymbol{\Delta}'^* + 4(\boldsymbol{\Delta}\cdot\boldsymbol{\Delta}')\boldsymbol{\Delta}^*\} = 0.\end{aligned} \tag{23.89}$$

To proceed further we must incorporate one of the equilibrium forms discussed earlier (in general the collective mode frequencies are a function of the form and amplitude of the equilibrium order parameter).

For the case of an E_1 superconductor with $\beta_2 < 0$ we may write the equilibrium order parameter as $\boldsymbol{\Delta} = \Delta_0\hat{\mathbf{x}}$ (case Δ_0 real and $\phi_0 = 0$). The distortions of the order parameter, $\boldsymbol{\Delta}'$, can be written as $\boldsymbol{\Delta}' = \Delta'_x\hat{\mathbf{x}} + \Delta'_y\hat{\mathbf{y}}$ (here Δ'_x and Δ'_y may be complex) and our equations of motion are (where we again drop the prime)

$$\Lambda\ddot{\Delta}_x + \alpha\Delta_x + 2\beta_1\Delta_0^2[2\Delta_x + \Delta_x^*] + 2\beta_2\Delta_0^2[\Delta_x^* + 2\Delta_x] = 0 \tag{23.90a}$$

and

$$\Lambda\ddot{\Delta}_y + \alpha\Delta_y + 2\beta_1\Delta_0^2\Delta_y + 2\beta_2\Delta_0^2\Delta_y^* = 0. \tag{23.90b}$$

Note we actually have four equations of motion involving Δ_x, Δ_x^*, Δ_y, and Δ_y^*; hence there will be four independent collective modes. We define the combinations

$$\boldsymbol{\Delta}^{(\mathrm{R})} = \frac{1}{2}(\boldsymbol{\Delta} + \boldsymbol{\Delta}^*) \tag{23.91a}$$

and

$$\Delta^{(\mathrm{I})} = -\frac{\mathrm{i}}{2}(\Delta - \Delta^*). \tag{23.91b}$$

Taking the complex conjugate of Eqs. (23.90a) and (23.90b) and adding and subtracting we obtain

$$\Lambda\ddot{\Delta}_x^{(\mathrm{R})} + \alpha\Delta_x^{(\mathrm{R})} + 6\beta_1\Delta_0^2\Delta_x^{(\mathrm{R})} + 6\beta_2\Delta_0^2\Delta_x^{(\mathrm{R})} = 0, \tag{23.92a}$$

$$\Lambda\ddot{\Delta}_x^{(\mathrm{I})} + \alpha\Delta_x^{(\mathrm{I})} + 2\beta_1\Delta_0^2\Delta_x^{(\mathrm{I})} + 2\beta_2\Delta_0^2\Delta_x^{(\mathrm{I})} = 0, \tag{23.92b}$$

$$\Lambda\ddot{\Delta}_y^{(\mathrm{R})} + \alpha\Delta_y^{(\mathrm{R})} + 2\beta_1\Delta_0^2\Delta_y^{(\mathrm{R})} + 2\beta_2\Delta_0^2\Delta_y^{(\mathrm{R})} = 0, \tag{23.92c}$$

and

$$\Lambda\ddot{\Delta}_y^{(\mathrm{I})} + \alpha\Delta_y^{(\mathrm{I})} + 2\beta_1\Delta_0^2\Delta_y^{(\mathrm{I})} - 2\beta_2\Delta_0^2\Delta_y^{(\mathrm{I})} = 0. \tag{23.92d}$$

Assuming a solution of the form $\Delta \sim \mathrm{e}^{-\mathrm{i}\omega t}$ we obtain the four frequencies

$$\omega_{x\mathrm{R}}^2 = -\frac{2\alpha}{\Lambda}, \tag{23.93a}$$

$$\omega_{x\mathrm{I}}^2 = 0, \tag{23.93b}$$

$$\omega_{y\mathrm{R}}^2 = 0, \tag{23.93c}$$

$$\omega_{y\mathrm{I}}^2 = \frac{2\alpha}{\Lambda}\frac{\beta_2}{\beta_1 + \beta_2}. \tag{23.93d}$$

Given our assumption $\beta_2 < 0$ and noting $\alpha < 0$ for $T < T_c$, we verify that (23.93a) and (23.93d) have real frequencies. On the basis of our experience with ^{3}He we presume that (23.93a) corresponds to a 'gap mode' where $\hbar\omega = 2\Delta(T)$ which means we identify Δ^2 as $-\alpha\hbar^2/2\Lambda$. For small $|\beta_2|$, (23.93d) lies within the gap. Forms (23.93b) and (23.93c), which are called Goldstone modes, would have finite frequencies in the presence of dispersion (caused by the gradient terms in the G–L expansion). The Goldstone mode (23.93b) associated with Eq. (23.92b), arises from fluctuations of the phase of the order parameter (since it involves the imaginary component and we chose our equilibrium order parameter to be real); for small $\phi(\mathbf{r}, t)$

$$\mathrm{e}^{\mathrm{i}\phi(\mathbf{r},t)} = 1 + \mathrm{i}\phi(\mathbf{r}, t).$$

It is known that for conventional superconductors the frequency of this mode will be lifted to the plasma frequency; this follows from the fact that time- and position-dependent phase fluctuations will have an associated superfluid velocity $\mathbf{v} = (\hbar/m)\nabla\phi$ (the Josephson equation) which, through the equation of continuity, will produce charge density fluctuations, which oscillate at the plasma frequency in a metal.

The Goldstone mode (23.93c) is associated with a (soft) oscillation of the direction of the order parameter about its equilibrium direction $\hat{\mathbf{x}}$, in the x–y plane. Eqs. (23.92d) and (23.93d) directly involve an infinitesimal rotation by an imaginary angle, which changes the magnitude of the gap and hence has a finite frequency.

We now consider the E_1 superconductor with $\beta_2 > 0$. Eq. (23.89) takes the form (assuming Δ_0 to be real)

$$\Lambda\ddot{\Delta}_x + \alpha\Delta_x + \beta_1\Delta_0^2[3\Delta_x + \Delta_x^* - \mathrm{i}(\Delta_y - \Delta_y^*)] + 2\beta_2\Delta_0^2(\Delta_x + \mathrm{i}\Delta_y) = 0 \tag{23.94a}$$

$$\Lambda\ddot{\Delta}_y + \alpha\Delta_y + \beta_1\Delta_0^2[3\Delta_y - \Delta_y^* + i(\Delta_x + \Delta_x^*)] + 2\beta_2\Delta_0^2(\Delta_y - i\Delta_x) = 0. \quad (23.94b)$$

We introduce the variables

$$\Delta_\pm = \Delta_x \pm i\Delta_y \quad (23.95)$$

and Eqs. (23.90a) and (23.90b) become

$$\Lambda\ddot{\Delta}_+ + \alpha\Delta_+ + 2\Delta_0^2(\beta_1 + 2\beta_2)\Delta_+ = 0 \quad (23.96a)$$

and

$$\Lambda\ddot{\Delta}_- + \alpha\Delta_- + 2\Delta_0^2\beta_1(2\Delta_- + \Delta_-^*) = 0. \quad (23.96b)$$

Eq. (23.96a) has the solution

$$\omega^2 = -\frac{2\alpha}{\Lambda}\frac{\beta_2}{\beta_1}; \quad (23.97a)$$

Since the complex conjugate equation has the same solution this mode is doubly degenerate. Solving the coupled set of equations involving (23.96b) and its complex conjugate we obtain

$$\omega^2 = \begin{cases} -2\alpha/\Lambda, & (23.97b) \\ 0. & (23.97c) \end{cases}$$

We identify (23.97b) as the gap mode. Eq. (23.97c) is the Goldstone mode corresponding to the phase mode of an ordinary superconductor and it will be lifted to the plasmon frequency. Since $\alpha < 0$ and $\beta_2 > 0$, Eq. (23.97a) has a real solution for the frequency ω.

Using perturbation theory it is relatively straightforward to evaluate the dispersion of the collective modes from the expressions for the gradient energy given earlier.

24 Landau Fermi liquid theory

24.1 Basic equations

The Pauli–Sommerfeld free electron model of a metal is remarkably successful in describing many of the properties of metals such as the linear behavior of the specific heat ($C = \gamma T$) and the temperature-independent paramagnetic susceptibility. However, the energy associated with the Coulomb repulsion between electrons is of the same order as the kinetic energy and it is therefore not at all obvious that such a model should work at all.

Liquid ^{3}He (at temperatures below 0.1 K) is another Fermi system which has some features in common with a Fermi gas. (^{3}He has no electronic spin and its statistics are governed by its $I = 1/2$ nuclear spin.) It may be thought of as having an even larger particle–particle interaction since the 'empty volume' for the atoms to move in is small; most of the liquid volume is occupied by the essentially impenetrable 'hard core' associated with the filled $1s^2$ shell of the He atoms.

The long range of the Coulomb potential adds an initially unnecessary feature to understanding the success of the independent (or free) electron model so we will use (neutral) liquid ^{3}He as our model Fermi 'liquid'.[1]

The success of the Pauli–Sommerfeld model rests on there being a sharp discontinuity separating the occupied from the unoccupied states at $T = 0$ which is associated with the Fermi occupation factor,[2] $n(\varepsilon)$. A sharp discontinuity, in turn, requires that the energy of the 'one-particle' states at ε_F be precisely defined; i.e., there can be no uncertainty in the energy of these states, $\delta\varepsilon$, caused by scattering; from the uncertainty principle $\delta\varepsilon \sim \hbar/\tau$, where τ is the ^{3}He–^{3}He scattering time. This says we must have $\tau \to \infty$ at $T = 0$.

Following Landau's (Landau 1957a,b) argument let us consider a gas of ^{3}He atoms at absolute zero with the interatomic potential 'switched off'. All states with an energy less than the Fermi energy (or chemical potential), $\mu = \hbar^2 k_F^2/2m$ where $k_F = (3\pi^2 N/V)^{1/3}$, will be occupied and all states of higher energy will be empty;[3] here N and V are the number of particles and the volume of the system, respectively, and N/V is the particle number density. At a temperature

1. One sometimes refers to the conduction electrons in a metal as the electron liquid, since they are bound to the metal yet free to move within it.
2. In this section we will use $n(\varepsilon)$, rather than $f(\varepsilon)$ which is used elsewhere in the book, for the Fermi distribution function, consistent with most of the literature.
3. The number of carriers, N, is given by the usual expression $N = 2(2\pi)^{-3}L^3 V_k$ where the factor 2 accounts for the spin degeneracy, L^3 is the real space volume, and $V_k = (4/3)\pi k_F^3$ is the occupied region of momentum space at $T = 0$; the expression $k_F = (3\pi^2 N/V)^{1/3}$ then follows immediately.

$T << T_F$ (where $k_B T_F \equiv \mu$) the excitations will consist of a few 'particles' with $\varepsilon > \mu$ and 'holes' with $\varepsilon < \mu$. These particle and hole states have energies in an interval of order $k_B T$ about μ. If we now slowly (adiabatically) turn on the interatomic potential two things happen: (i) The nature of the excited states 'deforms', and rather than involving a single particle an excitation really involves a particle and a 'coherent or adiabatic' motion of the surrounding particles called *back-flow* as they 'get out of the way' of the excited particle. The total entity (particle + backflow) is called a quasiparticle (where, from the Latin, quasi means 'almost'). (ii) The second thing that happens is that the quasiparticles may scatter from each other; i.e., transitions occur between quasiparticle states. From the uncertainty principle this results in a broadening of the quasiparticle energies, $\delta\varepsilon$, by an amount of the order of $\hbar/\tau$, where τ is the collision time. By considering the transition rate for the scattering of particles in the background of a Fermi sea one can show (see Pines and Nozières (1966)) that

$$\tau^{-1} \propto \begin{cases} (\varepsilon - \mu)^2; \ k_B T << |\varepsilon - \mu|, \\ (k_B T)^2; \quad k_B T >> |\varepsilon - \mu|. \end{cases} \tag{24.1}$$

This behavior has its origin in the Pauli exclusion principle. At low temperatures the majority of the particles in the Fermi sea cannot scatter since the final states (which energetically they could scatter into) are filled. Crudely speaking, one says that the number of excited particles is proportional to T and the probability of finding an empty state in the vicinity of ε_F for the particles to scatter into also scales as T giving an overall T^2 behavior of the scattering rate. However, a convincing argument can only be made by examining the structure of the collision integral.

Based on the above one can make the following very important observation: as $T \to 0$ the ratio of the energy level uncertainty to the energy itself approaches zero; i.e., $\delta\varepsilon/\varepsilon \to 0$. In this sense the few excited states of the system behave as *independent* 'particles'. However, as noted above, the motion of the excited particles really involves a coordinated motion in the background of the remaining (unexcited) particles.

The quasiparticle energies must be associated with various quantum numbers. A fundamental postulate of Landau's Fermi liquid theory is that energies are a function of the momentum, p (or wave vector k) and that the momentum p_F corresponding to the Fermi energy continues to be given by the *free particle* expression

$$p_F = \hbar(3\pi^2 N/V)^{1/3}. \tag{24.2}$$

This postulate was later proved formally (Landau 1959) using Green's function methods. Thus the quasiparticle states are in one-to-one correspondence with those of an ideal gas. In addition, each quasiparticle has a spin quantum number, $m_s = \pm\frac{1}{2}$.

Since the quasiparticle states are well defined only in the immediate vicinity of μ, it is not possible to write the total energy of the liquid as a sum over the quasiparticle energies (even in the Hartree–Fock sense); this is to be contrasted with the free particle case where $E = \frac{3}{5}N\mu$. However, it is possible to express the *change* in the energy density of the system $\delta(E/V)$ as a function of a *change* of occupation of a few states near μ, $\delta n(\mathbf{p})$, which we write in the form

$$\delta\left(\frac{E}{V}\right) = \int \varepsilon(\mathbf{p})\delta n(\mathbf{p})\mathrm{d}\tau, \tag{24.3}$$

where $\mathrm{d}\tau = [2/(2\pi\hbar)^3]\mathrm{d}^3 p$, V is the volume of the system and $\varepsilon(\mathbf{p})$ is defined as the quasiparticle energy. To understand this definition we consider a change in the distribution function $\delta n(\mathbf{p}) \sim \delta^{(3)}(\mathbf{p} - \mathbf{P})$ (with the δ function normalized such that $\int \delta^{(3)}(\mathbf{p} - \mathbf{P})\mathrm{d}\tau = 1$). If we excite one quasiparticle with momentum $\mathbf{P}$, we then have immediately that the change in system energy $\delta(E/V)$ is $\varepsilon(\mathbf{P})$, as it must be.

In a many-body system the excitation energies are themselves a functional of the distribution function. We write this dependence in the form

$$\delta\varepsilon(\mathbf{p}) = \int f(\mathbf{p}, \mathbf{p}')\delta n(\mathbf{p}')\mathrm{d}\tau', \tag{24.4}$$

where $f(\mathbf{p}, \mathbf{p}')$ is a function characteristic of the system.

The corrections to $\varepsilon(\mathbf{p})$ represented by (24.4) are essential in considering collective modes (e.g., sound waves) which result in a position- and time-dependent distribution function; in general both the shape and volume of the Fermi surface change. The change in the quasiparticle energies resulting from the change in the distribution function is often referred to as a molecular field effect. That E and ε are both functionals follows from the fact that they involve integrals over the unspecified function, $\delta n(\mathbf{p})$. Formally we write

$$\varepsilon(\mathbf{p}) = \frac{\delta(E/V)}{\delta n(\mathbf{p})} \tag{24.5}$$

and

$$f(\mathbf{p}, \mathbf{p}') = \frac{\delta\varepsilon(\mathbf{p})}{\delta n(\mathbf{p}')} = \frac{\delta^2(E/V)}{\delta n(\mathbf{p})\delta n(\mathbf{p}')}; \tag{24.6}$$

the latter property shows f is symmetrical in the variables $\mathbf{p}$ and $\mathbf{p}'$.

The quasiparticle velocity is given by the usual quantum mechanical expression for the group velocity

$$\mathbf{v} = \frac{\partial\varepsilon}{\partial\mathbf{p}}. \tag{24.7}$$

We define an effective mass as the ratio of the momentum to the velocity

$$m^* \equiv \frac{p}{v}. \tag{24.8}$$

This quantity is 3–6 times heavier than the mass of a bare ^{3}He atom (depending on the pressure) and this large enhancement is another manifestation of the strongly interacting character of ^{3}He.

We now consider the form of the equilibrium distribution function, $n(\mathbf{p})$, which follows from maximizing the entropy subject to the constraints that the total energy, E, and particle number, N, of the system are constant. From statistical mechanics the variation of the entropy is given by

$$\delta\left(\frac{S}{k_\mathrm{B}V}\right) = \delta\sum_{\mathbf{p}}\{-n(\mathbf{p})\ell n n(\mathbf{p}) - [1 - n(\mathbf{p})]\ell n[1 - n(\mathbf{p})]\}. \tag{24.9}$$

Imposing the constraints by introducing the usual Lagrange multipliers μ and β we have

$$\delta\left(\frac{S}{k_\mathrm{B}V}\right) - \beta\mu\delta\left(\frac{N}{V}\right) + \beta\delta\left(\frac{E}{V}\right) = 0, \tag{24.10}$$

which leads to the distribution function

$$n(\mathbf{p}) = \frac{1}{e^{\beta[\varepsilon(\mathbf{p}) - \mu]} + 1}, \tag{24.11}$$

where $\beta = 1/k_B T$. Note that since $\varepsilon(\mathbf{p})$ is a functional of $n(\mathbf{p})$, Eq. (24.11) is more complicated than it initially appears.

Information concerning the function $f(\mathbf{p}, \mathbf{p}')$ must, at present, be obtained experimentally. Landau derived two 'sum rules' which fixed two moments of this function which are referred to as the compressibility and effective-mass sum rules, which we now derive.

The first of these is derived from the expression for the sound velocity $c^2 = \partial p/\partial \rho$, where p and ρ are the pressure and density, respectively. From the differential relation for the Gibbs free energy, $d(N\mu) = V dp - S dT + \mu dN$, we obtain (at constant T) $N(\partial\mu/\partial N) = V(\partial p/\partial N)$; noting $\rho = Nm/V$ we have $c^2 = (N/m)(\partial\mu/\partial N)$. (Strictly speaking sound propagates adiabatically; however, the distinction between isothermal and adiabatic sound is negligible at low temperatures.) In evaluating the derivative of μ with respect to N we must include: (i) the explicit dependence resulting from the density dependence of p_F, Eq. (24.2), and (ii) an implicit dependence arising from Eq. (24.4). The change in p_F associated with the first effect causes the second effect, in that a shift in the occupation of states in the vicinity of p_F alters the average molecular field associated with a quasiparticle excitation. Combining the two effects we have

$$\delta\mu = \frac{\partial \varepsilon_F}{\partial p_F}\frac{\partial p_F}{\partial N}\delta N + \int f(\mathbf{p}_F, \mathbf{p}')\delta n' d\tau' = \frac{p_F}{3N}\left(v_F + \frac{p_F^2}{\hbar^3\pi^2}\bar{f}\right)\delta N, \tag{24.12}$$

where we have written $\delta n(\mathbf{p}') = (\partial/\partial p_F)\theta(p_F - p')(\partial p_F/\partial N)\delta N = \delta(p_F - p')(\partial p_F/\partial N)\delta N$ and have defined $\bar{f} = (1/4\pi)\int f(\hat{\mathbf{p}}_F, \hat{\mathbf{p}}_F')d\Omega'$. Thus

$$\bar{f} = \frac{3m\pi^2\hbar^3}{p_F^3}\left(c^2 - \frac{p_F^2}{3mm^*}\right), \tag{24.13}$$

which is our first sum rule.

The second sum rule follows from the fact that the laws of mechanics must be invariant under a transformation to a uniformly moving reference frame, which is called Galilean invariance in nonrelativistic mechanics. For reference we give the resulting transformation laws: each excitation of momentum $\mathbf{p}_0$ and energy $\varepsilon_0(\mathbf{p}_0)$ in a frame which is at rest relative to the fluid has, in a frame moving with velocity $\mathbf{V}$ relative to the fluid, a momentum $\mathbf{p}$ and energy ε which are given by

$$\mathbf{p} = \mathbf{p}_0 + M\mathbf{V} \tag{24.14}$$

and

$$\varepsilon = \varepsilon_0(\mathbf{p}_0) + \mathbf{p}_0 \cdot \mathbf{V} + \tfrac{1}{2}MV^2, \tag{24.15}$$

where M is the total mass of the fluid. The latter relation follows simply by replacing the momentum operator by (24.14) in the Hamiltonian of the liquid.

The momentum of the liquid per unit volume may be written as $\int \mathbf{p}n(\mathbf{p})d\tau$; alternatively the quasiparticle velocity is $\partial\varepsilon/\partial\mathbf{p}$ and the momentum per unit volume transported by the liquid is therefore[4] $m\int(\partial\varepsilon/\partial\mathbf{p})n(\mathbf{p})d\tau$. Equating these two forms we have

$$\int \mathbf{p} n(\mathbf{p}) \mathrm{d}\tau = m \int \frac{\partial \varepsilon}{\partial \mathbf{p}} n(\mathbf{p}) \mathrm{d}\tau. \qquad (24.16)$$

Taking a variation of both sides, and recalling that ε is a functional of n, yields

$$\int \mathbf{p} \delta n(\mathbf{p}) \mathrm{d}\tau = m \int \frac{\partial \delta\varepsilon}{\partial \mathbf{p}} n(\mathbf{p}) \mathrm{d}\tau + m \int \frac{\partial \varepsilon}{\partial \mathbf{p}} \delta n(\mathbf{p}) \mathrm{d}\tau. \qquad (24.17)$$

Using (24.4) we obtain

$$\frac{1}{m} \int \mathbf{p} \delta n \mathrm{d}\tau = \int \int \frac{\partial}{\partial \mathbf{p}} f(\mathbf{p}, \mathbf{p}') \delta n(\mathbf{p}') n(\mathbf{p}) \mathrm{d}\tau \mathrm{d}\tau' + \int \frac{\partial \varepsilon}{\partial \mathbf{p}} \delta n(\mathbf{p}) \mathrm{d}\tau. \qquad (24.18)$$

Since (24.18) holds for an arbitrary variation we have, on integrating the first term by parts,

$$\frac{\mathbf{p}}{m} = -\int \frac{\partial n(\mathbf{p}')}{\partial \mathbf{p}'} f(\mathbf{p}', \mathbf{p}) \mathrm{d}\tau' + \frac{p_{\mathrm{F}}}{m^*}, \qquad (24.19)$$

where we used Eqs. (24.7) and (24.8) to rewrite the second term on the right. We again write $n(p') = \theta(p_{\mathrm{F}} - p')$ from which we obtain $\partial n(\mathbf{p}')/\partial \mathbf{p}' = -\delta(p_{\mathrm{F}} - p')\mathbf{p}'/p$. Inserting this form into Eq. (24.19), carrying out the integration with respect to $\mathrm{d}p'$, and canceling a common factor of p_{F} we obtain

$$\frac{1}{m} = \frac{1}{m^*} + \frac{2p_{\mathrm{F}}}{(2\pi\hbar)^3} \int f(\mathbf{p}_{\mathrm{F}}, \mathbf{p}'_{\mathrm{F}}) \cos\theta' \mathrm{d}\Omega', \qquad (24.20)$$

which is our second sum rule.[5] We have used the fact that the δ function and limiting ourselves to infinitesimal variations δn about p_{F} restrict the values of $\mathbf{p}$ and $\mathbf{p}'$ to the Fermi surface. Furthermore, rotational invariance, and the fact that $f(\mathbf{p}, \mathbf{p}')$ is symmetric, result in the form $f(\cos\chi)$, where χ is the angle between the momenta of the two quasiparticles on the Fermi surface.

The physical idea implicit in the above derivation is that a moving reference frame results in a displacement of the entire Fermi sphere by a momentum $M\mathbf{V}$. However, the displaced sphere may be regarded as the original undisplaced sphere with the addition of a 'crescent' of particle-like excitations on the 'half sphere' in the direction of motion and a crescent of holes on the opposite 'half sphere', as shown in Fig. 24.1. These crescents clearly represent a macroscopic change in the distribution function, $\delta n(\mathbf{p})$, which, in turn, causes a corresponding shift in the quasiparticle energies, $\delta\varepsilon(\mathbf{p})$. This energy shift results in $m^* \neq m$.

In order to parameterize $f(\cos\chi)$ we first introduce a dimensionless function $F(\cos\chi)$ defined as

$$F(\cos\chi) = \mathcal{N}_{\mathrm{F}} f(\cos\chi), \qquad (24.21)$$

where $\mathcal{N}_{\mathrm{F}}$ is the density of states at the Fermi energy (including both spins): $\mathcal{N}_{\mathrm{F}} = m^* p_{\mathrm{F}}/\pi^2\hbar^3$. Second we expand this function in Legendre polynomials

4. Eq. (24.8) would suggest that we should have m^* rather than m in this expression; however, (24.8) applies to the case of an excitation relative to a stationary fluid.
5. The derivation presented here can be criticized in that (24.6) involves an integration over $n(\mathbf{p})$ which is not strictly valid; for a more rigorous derivation see Nozières (1964).

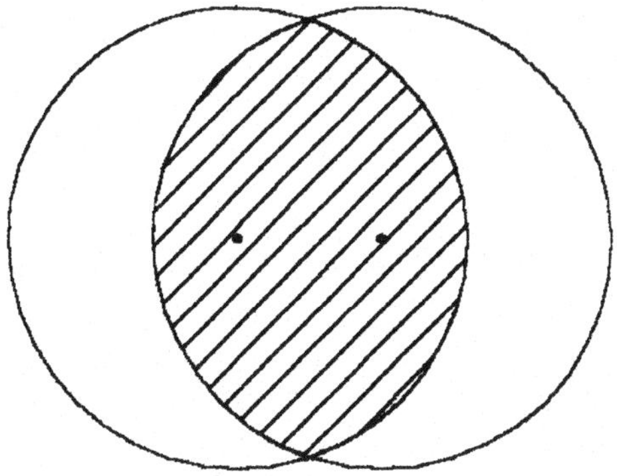

Figure 24.1 The displaced Fermi sphere in a moving reference frame.

$$F(\cos\chi) = \sum_n F_n P_n(\cos\chi) \tag{24.22}$$

$$= F_0 + F_1 \cos\chi + \ldots. \tag{24.23}$$

In terms of this expansion our two sum rules, Eqs. (24.13) and (24.20), become

$$F_0 = \frac{3mm^*c^2}{p_F^2} - 1 \tag{24.24}$$

and

$$F_1 = 3\left(\frac{m^*}{m} - 1\right). \tag{24.25}$$

These two sum rules exhaust the information that may be obtained from purely thermodynamic measurements (at zero field). Additional information concerning F may be obtained from zero sound measurements, as will be discussed in Subsec. 24.2.

Up to this point we have ignored the spin degree of freedom associated with the quasiparticles. In order to treat the effects of a magnetic field we must generalize our distribution function $n(\mathbf{p})$ to a two by two matrix, $n_{\alpha\beta}(\mathbf{p})$, having the form

$$\underset{\sim}{n}(\mathbf{p}) = \begin{pmatrix} n_{11}(\mathbf{p}) & n_{12}(\mathbf{p}) \\ n_{21}(\mathbf{p}) & n_{22}(\mathbf{p}) \end{pmatrix}, \tag{24.26}$$

where 1 and 2 refer to spin up (↑) and spin down (↓), respectively; since our distribution function must be Hermitian, we have $n_{12} = n_{21}^*$. In equilibrium $\underset{\sim}{n}$ will be diagonal;[6] however, when treating dynamic situations (e.g., spin waves) $\underset{\sim}{n}$ will have off diagonal elements.

When account is taken of the spin degree of freedom, Eq. (24.3) must be written

$$\delta\left(\frac{E}{V}\right) = \sum_{\alpha\beta} \int \varepsilon_{\alpha\beta}(\mathbf{p})\delta n_{\beta\alpha}(\mathbf{p})d\tau, \tag{24.27}$$

where the quasiparticle energies, $\varepsilon_{\alpha\beta}(\mathbf{p})$, must also be written as a two by two matrix (like $\underset{\sim}{n}$, this

6. Here we assume spin is a good quantum number (i.e., we neglect spin–orbit coupling) and we quantize along the z axis.

matrix is diagonal in equilibrium). Eq. (24.4), describing a change in the quasiparticle energies due to a change in the distribution function, generalizes to

$$\delta\varepsilon_{\alpha\beta}(\mathbf{p}) = \sum_{\delta\gamma}\int f_{\alpha\beta,\gamma\delta}(\mathbf{p},\mathbf{p}')\delta n_{\delta\gamma}(\mathbf{p}')\mathrm{d}\tau', \tag{24.28}$$

where, in the absence of spin–orbit coupling, the Landau interaction function has the form[7]

$$f_{\alpha\beta,\gamma\delta}(\mathbf{p},\mathbf{p}') = f^{(\mathrm{s})}(\mathbf{p},\mathbf{p}')\delta_{\alpha\beta}\delta_{\gamma\delta} + f^{(\mathrm{a})}(\mathbf{p},\mathbf{p}')\sigma_{\alpha\beta}\cdot\sigma_{\gamma\delta}; \tag{24.29}$$

$f^{(\mathrm{s})}(\mathbf{p},\mathbf{p}')$ is identical to the function $f(\mathbf{p},\mathbf{p}')$ of our earlier discussion and $f^{(\mathrm{a})}(\mathbf{p},\mathbf{p}')$ is a new function.

The one thermodynamic property of ^{3}He which we have not discussed is the magnetic susceptibility, and we now show that it is related to an average of $f^{(\mathrm{a})}(\mathbf{p},\mathbf{p}')$. In the presence of a magnetic field, which we will take along the **z** axis, each quasiparticle will have its energy shifted according to

$$\delta\varepsilon_{\alpha\beta} = -\tfrac{1}{2}\hbar\gamma_0 H(\sigma_z)_{\alpha\beta} + \sum_{\gamma\delta}\int \mathrm{d}\tau' f_{\alpha\beta,\gamma\delta}(\mathbf{p},\mathbf{p}')\delta n_{\delta\gamma}(\mathbf{p}'). \tag{24.30}$$

The first term on the right is a direct shift of the quasiparticle energy levels due to the bare magnetic moments, $\frac{1}{2}\hbar\gamma_0 = \mu_3$, of the ^{3}He nuclei. The second term is a 'molecular field' resulting from a change in the distribution function; it is related to the molecular field encountered in the Bloch–Stoner theory of the exchange enhancement of the magnetic susceptibility of electrons in metals.

In a thermal equilibrium ensemble, the energy matrix $\underset{\sim}{\varepsilon}$ and distribution function matrix $\underset{\sim}{n}$ are both diagonal (for our choice of the field direction $\hat{\mathbf{H}}$), so we need examine only the ↑↑ and ↓↓ components. We write the quasiparticle energy shift in the form $\delta\varepsilon_{\alpha\beta} = -\frac{1}{2}\hbar\gamma H(\sigma_z)_{\alpha\beta}$, which is valid in a linear (low field) approximation; $\frac{1}{2}\hbar\gamma$ is an effective quasiparticle magnetic moment that includes the effects of the molecular field, $f^{(\mathrm{a})}$. We may write the change in the distribution function as

$$\delta n_{\alpha\beta} \cong \frac{\partial n}{\partial\varepsilon}\delta\varepsilon_{\alpha\beta}, \tag{24.31}$$

and Eq. (24.30) becomes

$$\gamma_0 = \gamma\left(1 + \frac{m^* p_{\mathrm{F}}}{\pi^2\hbar^3}\frac{1}{4\pi}\int f^{(\mathrm{a})}(\hat{\mathbf{p}},\hat{\mathbf{p}}')\mathrm{d}\Omega'\right). \tag{24.32}$$

We expand the function $f^{(\mathrm{a})}(\hat{\mathbf{p}}\cdot\hat{\mathbf{p}}')$ in Legendre polynomials analogous to Eq. (24.22)

$$F^{(\mathrm{a})} = \sum_n F_n^{(\mathrm{a})}P_n(\cos\chi) \tag{24.33}$$

and we may then write (24.32) in the form

$$\gamma = \frac{\gamma_0}{1 + F_0^{(\mathrm{a})}}. \tag{24.34}$$

7. Lifshitz and Pitaevskii (1980) use a convention where the left-hand side of Eq. (24.29) is written $f_{\alpha\gamma,\beta\delta}(\mathbf{p},\mathbf{p}')$.

This immediately leads to a generalization of the usual expression for the Pauli paramagnetic susceptibility, $\chi_0 = \frac{1}{4}\hbar^2\gamma_0^2\mathcal{N}_F$, to the form

$$\chi = \frac{1}{4}\frac{\hbar^2\gamma_0^2\mathcal{N}_F}{1 + F_0^{(a)}}. \qquad (24.35)$$

Eq. (24.35), which we call the magnetic susceptibility sum rule, can be used to fix $F_0^{(a)}$ from the measured magnetic susceptibility. $F_0^{(a)}$ is negative in ^{3}He so the magnetic susceptibility is enhanced.

24.2 Collisionless collective modes

We may loosely define a collective mode as a time (or frequency) and space (or wave vector) dependent oscillation of the fluid involving a coordinated motion of its constituent particles. We can distinguish collective modes according to whether they are hydrodynamic or collisionless.

Hydrodynamic collective modes owe their existence to various conservation laws (momentum, energy, and mass). Ordinary sound waves are the best-known hydrodynamic mode, but the damped viscous shear waves and the thermal conduction wave are also classed as hydrodynamic modes. If the liquid has a continuously broken symmetry, as in superfluid ^{3}He and ^{4}He or liquid crystals, additional modes appear (e.g., the second sound mode of He II and superfluid ^{3}He). Hydrodynamic modes are restricted to frequencies satisfying the condition $\omega\tau << 1$, where τ is a characteristic relaxation time.[8]

A necessary (but not sufficient) condition for the existence of collisionless collective modes is $\omega\tau >> 1$. According to the argument given in Subsec. 24.1, the collisions between excited quasiparticles become more infrequent as the temperature is lowered. Hence, at sufficiently low temperatures, the fluid will enter the region $\omega\tau >> 1$. As shown by Landau (1957b) collisionless collective modes in ^{3}He arise from the restoring force associated with the molecular field which occurs when the distribution function of the quasiparticles is distorted from its equilibrium form. The time evolution of the distribution function, $n = n(\mathbf{p}, \mathbf{r}, t)$, is governed by the Boltzmann equation. Although we have argued that collisions between quasiparticles become unimportant at low temperatures, this does not mean the particles do not interact. This interaction manifests itself through a *coherent* molecular field rather than through *incoherent* quasiparticle scattering events. In fact one may neglect collisions altogether in deriving the frequency–wave-vector relation of the collective modes; collisions only lead to damping.

24.2.1 The kinetic equation

We obtain the Boltzmann or kinetic equation by equating the total time derivative of the distribution function to a collision integral, $I[n]$,

8. One may understand this condition by calculating the time needed to equilibrate the temperature oscillations associated with successive nodes and antinodes of a sound wave in a gas (from kinetic theory) and comparing it with the period $t_0 = 2\pi/\omega$. One finds that adiabatic (hydrodynamic) propagation requires $t_0 >> \tau$.

$$\frac{\partial n}{\partial t} + \frac{\partial n}{\partial \mathbf{r}} \cdot \dot{\mathbf{r}} + \frac{\partial n}{\partial \mathbf{p}} \cdot \dot{\mathbf{p}} = I[n]; \tag{24.36}$$

as discussed above we will set $I[n] = 0$. From Hamilton's equations $\dot{\mathbf{r}} = \partial\varepsilon/\partial\mathbf{p}$ and $\dot{\mathbf{p}} = -\partial\varepsilon/\partial\mathbf{r}$ we obtain

$$\frac{\partial n}{\partial t} + \frac{\partial n}{\partial \mathbf{r}} \cdot \frac{\partial \varepsilon}{\partial \mathbf{p}} - \frac{\partial n}{\partial \mathbf{p}} \cdot \frac{\partial \varepsilon}{\partial \mathbf{r}} = 0. \tag{24.37}$$

Writing the distribution function as the sum of an equilibrium part and a small deviation, $n(\mathbf{p}, \mathbf{r}, t) = n_0(\mathbf{p}) + n'(\mathbf{p}, \mathbf{r}, t)$, and noting $\varepsilon = \varepsilon[n]$ through Eq. (24.3) we obtain, on retaining only the linear terms in n',

$$\frac{\partial n'(\mathbf{p}, \mathbf{r}, t)}{\partial t} + \frac{\partial n'(\mathbf{p}, \mathbf{r}, t)}{\partial \mathbf{r}} \cdot \mathbf{v} - \frac{\partial n_0}{\partial \varepsilon} \mathbf{v} \cdot \int f(\mathbf{p}, \mathbf{p}') \frac{\partial n'}{\partial \mathbf{r}} \mathrm{d}\tau' = 0. \tag{24.38}$$

Note that in a noninteracting Fermi gas the last term, which (as we will show) is the restoring force required for the existence of a collective mode, would be absent.[9] The collective modes are the *eigensolutions* of Eq. (24.38), which is an *integro-differential* equation. We seek solutions of the form $n'(\mathbf{p}, \mathbf{r}, t) \sim n'(\mathbf{p})\mathrm{e}^{-\mathrm{i}(\omega t - \mathbf{k}\cdot\mathbf{r})}$ which on substituting in (24.38) yields

$$(\omega - \mathbf{k} \cdot \mathbf{v}) n'(\mathbf{p}) + \frac{\partial n_0}{\partial \varepsilon} \mathbf{k} \cdot \mathbf{v} \int f(\mathbf{p}, \mathbf{p}') n'(\mathbf{p}') \mathrm{d}\tau' = 0. \tag{24.39}$$

The equilibrium distribution function is a step or θ function, $n_0(\mathbf{p}) = \theta(\mathbf{p}_\mathrm{F} - \mathbf{p})$; we expect the deviation $n'(\mathbf{p})$ to be proportional to a δ function which we write in the form

$$n'(\mathbf{p}) = -\nu(\hat{\mathbf{p}})\delta(p - p_\mathrm{F})/v_\mathrm{F} = \nu(\hat{\mathbf{p}})\frac{\partial n_0}{\partial \varepsilon}. \tag{24.40}$$

Operationally, $n'(\mathbf{p})$ is equivalent to an angular-dependent shift of the Fermi radius by an amount $\delta\mathbf{p} = \nu(\hat{\mathbf{p}})\hat{\mathbf{p}}/v_\mathrm{F}$:

$$n[p_\mathrm{F} - p - \nu(\hat{\mathbf{p}})/v_\mathrm{F}] \cong n_0(\mathbf{p}) + \frac{\partial n_0}{\partial \mathbf{p}} \cdot \delta\mathbf{p} + \ldots$$

$$= \theta(p_\mathrm{F} - p) - \delta(p_\mathrm{F} - p)\frac{\nu(\hat{\mathbf{p}})}{v_\mathrm{F}} + \ldots.$$

Specifically, the eigensolutions we seek will correspond to a time-, space-, and angular-dependent (shape) change of the Fermi surface. Inserting (24.40) into (24.39) and using (24.21) we have

$$(\omega - \mathbf{k} \cdot \mathbf{v})\nu(\hat{\mathbf{p}}) - \mathbf{k} \cdot \mathbf{v} \frac{1}{4\pi} \int F(\hat{\mathbf{p}}, \hat{\mathbf{p}}')\nu(\hat{\mathbf{p}}')\mathrm{d}\Omega' = 0. \tag{24.41}$$

All possible collective modes, which involve density–velocity oscillations, follow from the solution of (24.41). We rewrite the linear homogeneous integral equation (24.41) in the standard

9. Eq. (24.38) is valid only for a neutral Fermi liquid such as ^{3}He where the self-consistent field arises from short-range forces. In a charged system one must explicitly include the added contribution of the long-range Coulomb potential which is obtained by simultaneously solving the kinetic equation and Maxwell's equations (or Poisson's equation in the quasistatic limit).

form by defining a new variable

$$\psi(\hat{\mathbf{p}}) = \frac{s - \cos\theta}{\cos\theta} \nu(\hat{\mathbf{p}}), \tag{24.42}$$

which leads to the equation

$$\psi(\hat{\mathbf{p}}) = \frac{1}{4\pi} \int F(\hat{\mathbf{p}}, \hat{\mathbf{p}}') \frac{\cos\theta'}{s - \cos\theta'} \psi(\hat{\mathbf{p}}') d\Omega', \tag{24.43}$$

where θ is the angle between $\mathbf{k}$ and $\mathbf{v}$ and we have defined a dimensionless velocity $s = \omega/kv_F$; the quantity $(1/4\pi)F(\hat{\mathbf{p}}\cdot\hat{\mathbf{p}}')\cos\theta'/(s - \cos\theta')$ is the kernel of our integral equation. We will obtain Eq. (24.43) from the microscopic theory in Subsec. 54.2.

24.2.2 *Collisionless longitudinal zero sound*

The simplest possible situation leading to the existence of longitudinal zero sound is where the expansion of the Landau interaction function consists of a single term, F_0. Integrating the right-hand side of (24.43) then yields a constant; this implies $\psi(\hat{\mathbf{p}})$ is a constant or

$$\nu(\theta) = \text{const} \cdot \frac{\cos\theta}{s - \cos\theta}. \tag{24.44}$$

The condition for a solution to exist is then

$$\frac{F_0}{2} \int_{-1}^{+1} \frac{x dx}{s - x} = 1. \tag{24.45}$$

Carrying out the integration we obtain

$$\frac{1}{F_0} = \frac{s}{2} \ell n \left(\frac{s + 1}{s - 1} \right) - 1. \tag{24.46}$$

Eq. (24.46) has solutions only for $F_0 > 0$, which in this simplified model is the criterion for the existence of zero sound; i.e., in the absence of a repulsive interaction F_0 there is no zero sound. We examine the limiting cases of $F_0 \to 0$ and $F_0 \to \infty$. For the first of these we have

$$s = 1 + 2e^{-2/F_0}; \tag{24.47}$$

in this limit, the phase velocity of the longitudinal zero sound mode, $\omega/k \equiv c_0^\ell$, approaches the Fermi velocity, v_F. In the other limit, $F_0 \to \infty, s >> 1$ and we expand the logarithm (to the third order in s^{-1}) to obtain

$$s^2 = \frac{F_0}{3} \tag{24.48}$$

or $c_0^\ell = v_F(F_0/3)^{1/2}$.

Fig. 24.2 shows the maximal distortion of the Fermi surface associated with the passage of a longitudinal zero sound wave for the case $s = 4$. The largest distortion occurs at the leading edge

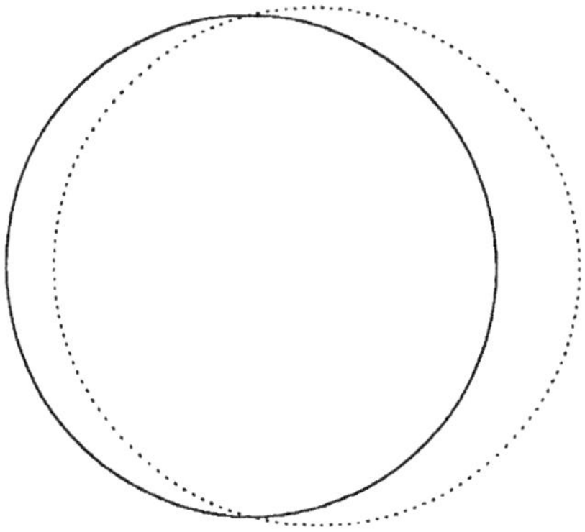

Figure 24.2 The maximal distortion of the Fermi surface associated with the passage of a longitudinal zero sound wave for the case $s = 4$.

of the Fermi surface. In the limit $s \to 1$ the distortion approaches a dimple. From Fig. 24.2 it is clear that the distortion corresponds to a longitudinal excitation of the system.

24.2.3 *Collisionless transverse zero sound*

We now consider the case where $F_1 \neq 0$ in (24.23). We may write the function $\cos\chi = \hat{\mathbf{p}}\cdot\hat{\mathbf{p}}'$ as

$$\cos\chi = \cos\theta\cos\theta' + \sin\theta\sin\theta'(\cos\phi\cos\phi' + \sin\phi\sin\phi'). \tag{24.49}$$

We seek solutions to Eq. (24.43) of the form

$$\psi(\theta,\phi) = \text{const}\cdot\sin\theta\begin{pmatrix}\cos\phi\\ \sin\phi\end{pmatrix}, \tag{24.50}$$

where the two forms correspond to the two orthogonal polarizations associated with a transverse mode; the corresponding form for $v(\hat{\mathbf{p}})$ is

$$v(\theta,\phi) = \text{const}\cdot\frac{\sin\theta\cos\theta}{s-\cos\theta}\begin{pmatrix}\cos\phi\\ \sin\phi\end{pmatrix}. \tag{24.51}$$

Inserting (24.51) into (24.43) and using (24.49) we obtain, on integrating over $d\phi$,

$$1 = \frac{F_1}{4}\int_{-1}^{+1}\frac{x(1-x^2)}{s-x}\,dx, \tag{24.52}$$

where $x = \cos\theta$ (note that for a mode having the structure of (24.51) the integral involving the F_0 term vanishes by symmetry). Carrying out the integration we obtain

$$\frac{F_1}{4}\left[s(1-s^2)\ell n\left(\frac{s+1}{s-1}\right) + 2(s^2-1) + \frac{2}{3}\right] = 1. \tag{24.53}$$

The threshold for transverse zero sound propagation corresponding to the limit $s \to 1$ is $F_1 > 6$. Since this criterion is only barely satisfied in ^{3}He we do not explore the large s limit. Fig. 24.3

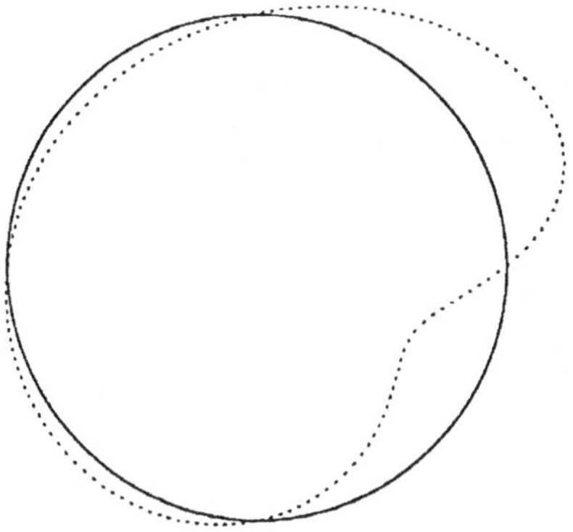

Figure 24.3 The maximal distortion of the Fermi surface associated with transverse zero sound for the case $s = 1.2$.

shows the (maximal) distortion of the Fermi surface associated with transverse zero sound for the case $s = 1.2$.

24.2.4 *Collisionless spin waves*

Spin waves in liquid ^{3}He were first considered by Silin (1958). It is appropriate to discuss the propagation of spin waves since: (i) they prepare us for more general types of collective modes occurring in superfluid ^{3}He involving the many components of the full distribution function, and (ii) they will occur (along with other modes) as eigenfunctions of the kinetic equations of the superfluid.

Our discussion of the density/velocity collective modes made use of the Boltzmann equation wherein one traces the time evolution of a distribution function, $n(\mathbf{p}, \mathbf{r}, t)$. Such a treatment is semiclassical in that we simultaneously specify the values of $\mathbf{p}$ and $\mathbf{r}$ of the quasiparticles which, since they are associated with noncommuting operators, is not strictly valid in quantum mechanics. A formal analysis, due to Wigner, makes clear the sense in which a distribution function treatment is valid (see Subsec. 55.2 for further discussion). The following arguments are not intended to be rigorous, but rather to motivate the resulting equations of motion physically for the spin-generalized distribution function, Eq. (24.56).

In general the matrices $\underset{\sim}{n}(\mathbf{p}, \mathbf{r}, t)$ and $\underset{\sim}{\varepsilon}(\mathbf{p}, \mathbf{r}, t)$ do not commute. Therefore terms in the kinetic equation like $(\partial \underset{\sim}{n}/\partial \mathbf{r}) \cdot (\partial \underset{\sim}{\varepsilon}/\partial \mathbf{p})$ are ambiguous; the ambiguity is removed by the usual quantum mechanical prescription of replacing classical expressions involving noncommuting operators by symmetrized expressions. Thus for the time dependence of $\underset{\sim}{n}$ arising from the *quasiclassical* $(\mathbf{p}, \mathbf{r})$ degrees of freedom, $\dot{\underset{\sim}{n}}^{\mathrm{c}}$, we write

$$\frac{\mathrm{d}\underset{\sim}{n}^{\mathrm{c}}}{\mathrm{d}t} = \frac{\partial \underset{\sim}{n}}{\partial t} + \frac{1}{2}\left(\frac{\partial \underset{\sim}{n}}{\partial \mathbf{r}} \cdot \frac{\partial \underset{\sim}{\varepsilon}}{\partial \mathbf{p}} + \frac{\partial \underset{\sim}{\varepsilon}}{\partial \mathbf{p}} \cdot \frac{\partial \underset{\sim}{n}}{\partial \mathbf{r}}\right) - \frac{1}{2}\left(\frac{\partial \underset{\sim}{n}}{\partial \mathbf{p}} \cdot \frac{\partial \underset{\sim}{\varepsilon}}{\partial \mathbf{r}} + \frac{\partial \underset{\sim}{\varepsilon}}{\partial \mathbf{r}} \cdot \frac{\partial \underset{\sim}{n}}{\partial \mathbf{p}}\right)$$

$$= \frac{\partial \underset{\sim}{n}}{\partial t} + \frac{1}{2}\{\underset{\sim}{n}, \underset{\sim}{\varepsilon}\}_{\mathbf{r},\mathbf{p}} - \frac{1}{2}\{\underset{\sim}{\varepsilon}, \underset{\sim}{n}\}_{\mathbf{r},\mathbf{p}}, \tag{24.54a}$$

where in the latter we have employed the Poisson bracket notation,

$$\{\underset{\sim}{n},\underset{\sim}{\varepsilon}\}_{\mathbf{r},\mathbf{p}} = \frac{\partial \underset{\sim}{n}}{\partial \mathbf{r}}\cdot\frac{\partial \underset{\sim}{\varepsilon}}{\partial \mathbf{p}} - \frac{\partial \underset{\sim}{n}}{\partial \mathbf{p}}\cdot\frac{\partial \underset{\sim}{\varepsilon}}{\partial \mathbf{r}}. \tag{24.54b}$$

When treating the time evolution of the spin degrees of freedom, which have no semiclassical limit, we must be more careful. Thus the time dependence arising from the strictly quantum mechanical character of the spin degrees of freedom, $\dot{\underset{\sim}{n}}^{\mathrm{q}}$, must be obtained from the density matrix equation of motion. Here the energy matrix, $\underset{\sim}{\varepsilon}$, plays the role of a Hamiltonian and we write

$$\frac{\mathrm{d}\underset{\sim}{n}^{\mathrm{q}}}{\mathrm{d}t} = \frac{\mathrm{i}}{\hbar}[\underset{\sim}{\varepsilon},\underset{\sim}{n}], \tag{24.55}$$

where $[\underset{\sim}{\varepsilon},\underset{\sim}{n}]$ denotes the commutator of $\underset{\sim}{\varepsilon}$ and $\underset{\sim}{n}$. The total time derivative of the distribution function is given by the sum of the terms on the right in (24.54) and (24.55) and must be equated to a spin-generalized collision integral, $\underset{\sim}{I}(\underset{\sim}{n})$:

$$\frac{\mathrm{d}\underset{\sim}{n}}{\mathrm{d}t} = \frac{\mathrm{d}\underset{\sim}{n}^{\mathrm{c}}}{\mathrm{d}t} + \frac{\mathrm{d}\underset{\sim}{n}^{\mathrm{q}}}{\mathrm{d}t} = \underset{\sim}{I}(\underset{\sim}{n}). \tag{24.56}$$

We now separate the spin (traceless) and density components of the matrices $\underset{\sim}{n}$ and $\underset{\sim}{\varepsilon}$ by writing them in the form

$$\underset{\sim}{n} = n\underset{\sim}{1} + \mathbf{m}\cdot\underset{\sim}{\boldsymbol{\sigma}} \tag{24.57a}$$

and

$$\underset{\sim}{\varepsilon} = \varepsilon\underset{\sim}{1} + \mathbf{e}\cdot\underset{\sim}{\boldsymbol{\sigma}}. \tag{24.57b}$$

To evaluate (24.56) we exploit the identity

$$(\mathbf{a}\cdot\underset{\sim}{\boldsymbol{\sigma}})(\mathbf{b}\cdot\underset{\sim}{\boldsymbol{\sigma}}) = (\mathbf{a}\cdot\mathbf{b})\underset{\sim}{1} + \mathrm{i}(\mathbf{a}\times\mathbf{b})\cdot\underset{\sim}{\boldsymbol{\sigma}}, \tag{24.58}$$

where **a** and **b** are any two vectors; we then have

$$\underset{\sim}{\varepsilon}\underset{\sim}{n} - \underset{\sim}{n}\underset{\sim}{\varepsilon} = 2\mathrm{i}(\mathbf{e}\times\mathbf{m})\cdot\underset{\sim}{\boldsymbol{\sigma}} \tag{24.59a}$$

and

$$\underset{\sim}{\varepsilon}\underset{\sim}{n} + \underset{\sim}{n}\underset{\sim}{\varepsilon} = 2(n\varepsilon + \mathbf{m}\cdot\mathbf{e})\underset{\sim}{1} + 2(n\mathbf{e} + \varepsilon\mathbf{m})\cdot\underset{\sim}{\boldsymbol{\sigma}}. \tag{24.59b}$$

The equation of motion separates into a scalar and a vector equation first derived by Silin (1958):

$$\frac{\partial n}{\partial t} + \frac{\partial n}{\partial \mathbf{r}}\cdot\frac{\partial \varepsilon}{\partial \mathbf{p}} + \frac{\partial m_i}{\partial r_j}\frac{\partial e_i}{\partial p_j} - \frac{\partial n}{\partial \mathbf{p}}\cdot\frac{\partial \varepsilon}{\partial \mathbf{r}} - \frac{\partial m_i}{\partial p_j}\frac{\partial e_i}{\partial r_j} = I(\mathbf{p}) \tag{24.60}$$

and

$$\frac{\partial \mathbf{m}}{\partial t} + \frac{\partial n}{\partial r_i}\frac{\partial \mathbf{e}}{\partial p_i} + \frac{\partial \mathbf{m}}{\partial r_i}\frac{\partial \varepsilon}{\partial p_i} - \frac{\partial n}{\partial p_i}\frac{\partial \mathbf{e}}{\partial r_i} - \frac{\partial \mathbf{m}}{\partial p_i}\frac{\partial \varepsilon}{\partial r_i} - \frac{2}{\hbar}\mathbf{e}\times\mathbf{m} = \mathbf{J}(\mathbf{p}), \tag{24.61}$$

where the repeated indices are summed. $I(\mathbf{p})$ and $\mathbf{J}(\mathbf{p})$ are collision integrals which will be neglected in the following discussion. Using Eqs. (24.29) and (24.30) together with the properties

$$\underset{\sim}{\sigma}_i^2 = \underset{\sim}{1}, \qquad \underset{\sim}{\sigma}_i\underset{\sim}{\sigma}_j = \mathrm{i}\underset{\sim}{\sigma}_k, \qquad \mathrm{Tr}\underset{\sim}{\sigma}_i = 0,$$

we have

$$\delta\varepsilon(\mathbf{p}) = \int d\tau' f^{(s)}(\mathbf{p}, \mathbf{p}')\delta n(\mathbf{p}') \tag{24.62}$$

and

$$\mathbf{e} = -\frac{1}{2}\hbar\gamma_0\mathbf{H} + \int d\tau' f^{(a)}(\mathbf{p}, \mathbf{p}')\mathbf{m}(\mathbf{p}'). \tag{24.63}$$

Use of the form (24.30) is restricted to a weakly polarized liquid; i.e., for magnetic fields such that $\hbar\gamma_0 H << \mu$. For a strongly polarized liquid the expression for $f_{\alpha\beta,\gamma\delta}(\mathbf{p}, \mathbf{p}')$ is considerably more complicated. We now examine small amplitude spin waves. We write

$$\mathbf{m} = \mathbf{m}_0 + \mathbf{m}' \tag{24.64a}$$

and

$$\mathbf{e} = \mathbf{e}^{\text{ext}} + \mathbf{e}_0 + \mathbf{e}', \tag{24.64b}$$

where $\mathbf{m}_0$ is the static component of the vector distribution function,[10] $-(\hbar/2)(\partial n_0/\partial\varepsilon)\gamma\mathbf{H}$, $\mathbf{m}'$ is the dynamic change associated with the spin wave, $\mathbf{e}^{\text{ext}} = -(\hbar/2)\gamma_0\mathbf{H}$, $\mathbf{e}_0 = \int d\tau' f^{(a)}(\mathbf{p}, \mathbf{p}')\mathbf{m}_0(\mathbf{p}')$, and $\mathbf{e}' = \int d\tau' f^{(a)}(\mathbf{p}, \mathbf{p}')\mathbf{m}'(\mathbf{p}')$.

The linearized form of Eq. (24.61) is

$$\frac{\partial\mathbf{m}'}{\partial t} + v_i\frac{\partial\mathbf{m}'}{\partial x_i} - \frac{\partial n_0}{\partial p_i}\frac{\partial\mathbf{e}'}{\partial x_i} - \frac{2}{\hbar}\mathbf{e}_0 \times \mathbf{m}' - \frac{2}{\hbar}\mathbf{e}' \times \mathbf{m}_0 + \gamma_0\mathbf{H} \times \mathbf{m}' = 0, \tag{24.65}$$

where the repeated indices are again summed. The fourth and sixth terms may be combined as $\gamma\mathbf{H} \times \mathbf{m}'$. Defining $m'_{\pm} = m'_x \pm im'_y$ our vector equation of motion becomes

$$\frac{\partial m'_z}{\partial t} + \mathbf{v}\cdot\nabla m'_z - \frac{\partial n_0}{\partial\varepsilon}\mathbf{v}\cdot\nabla\int f^{(a)}(\mathbf{p}, \mathbf{p}')m'_z(\mathbf{p}')d\tau' = 0 \tag{24.66}$$

and

$$\frac{\partial m_{\pm}}{\partial t} + (\mathbf{v}\cdot\nabla \pm i\Omega_0^a)m'_{\pm} - (\mathbf{v}\cdot\nabla \pm i\Omega_0^a)\frac{\partial n_0}{\partial\varepsilon}\int f^{(a)}(\mathbf{p}, \mathbf{p}')m'_{\pm}(\mathbf{p}')d\tau' = 0. \tag{24.67}$$

Here $\Omega_0^a = \Omega_0/(1 + F_0^a)$ and $\Omega_0 = \gamma_0 H$, the Larmor frequency.

The first of these equations is identical in structure to that for zero sound; however, since $F_0^a < 0$ it has no propagating solutions.

The second equation has propagating solutions only for long wavelengths and for $H \neq 0$. Assuming a form

$$m_{\pm}(\mathbf{p}, \mathbf{r}, t) = \frac{\partial n_0}{\partial\varepsilon}\mu_{\pm}(\hat{\mathbf{p}})e^{-i\omega t + i\mathbf{k}\cdot\mathbf{r}}$$

Eq. (24.67) becomes

$$(-\omega + \mathbf{k}\cdot\mathbf{v} \pm \Omega_0^a)\mu_{\pm}(\hat{\mathbf{p}}) = -(\mathbf{k}\cdot\mathbf{v} \pm \Omega_0^a)\frac{1}{4\pi}\int F^a(\hat{\mathbf{p}}, \hat{\mathbf{p}}')\mu_{\pm}(\hat{\mathbf{p}})d\hat{\mathbf{p}}'. \tag{24.68}$$

10. More generally it may be a uniformly processing distribution resulting from the application of a 'tipping' pulse as is commonly used in nuclear magnetic resonance.

We define a variable

$$\psi_{\pm}(\hat{\mathbf{p}}) = \frac{-\omega + \mathbf{k}\cdot\mathbf{v} \pm \Omega_0^{\mathrm{a}}}{\mathbf{k}\cdot\mathbf{v} \pm \Omega_0^{\mathrm{a}}}\mu_{\pm}(\hat{\mathbf{p}})$$

and obtain

$$\psi_{\pm}(\hat{\mathbf{p}}) = -\frac{1}{4\pi}\int F^{\mathrm{a}}(\hat{\mathbf{p}}, \hat{\mathbf{p}}')\frac{kv\cos\theta' \pm \Omega_0^{\mathrm{a}}}{-\omega + kv\cos\theta' \pm \Omega_0^{\mathrm{a}}}\psi_{\pm}(\hat{\mathbf{p}}')\mathrm{d}\hat{\mathbf{p}}'. \tag{24.69}$$

We first examine the limit $k = 0$. One of the many solutions to (24.69) is $\psi(\hat{\mathbf{p}}) = \text{const}$, yielding

$$-\frac{1}{4\pi}\frac{\pm\Omega_0^{\mathrm{a}}}{\pm\Omega_0^{\mathrm{a}} - \omega}\int F^{\mathrm{a}}(\hat{\mathbf{p}}, \hat{\mathbf{p}}')\mathrm{d}\hat{\mathbf{p}}' = 1, \tag{24.70}$$

which leads to $\omega = \pm\Omega_0$; the spins precess at the *free particle* Larmor frequency. For $k \neq 0$ we limit ourselves to a single Fermi liquid parameter. F_0^{a}; $\psi(\hat{\mathbf{p}}) = \text{const}$ is still a solution and the dispersion is governed by

$$+\frac{F_0^{\mathrm{a}}}{2}\int_0^{\pi}\frac{kv\cos\theta' \pm \Omega_0^{\mathrm{a}}}{-\omega + kv\cos\theta' \pm \Omega_0^{\mathrm{a}}}\mathrm{d}\cos\theta' = 1 \tag{24.71}$$

or

$$-\frac{F_0^{\mathrm{a}}}{2}\left[2 + \frac{\omega}{kv}\ell\mathrm{n}\left(\frac{-\omega \pm \Omega_0^{\mathrm{a}} + kv}{-\omega \pm \Omega_0^{\mathrm{a}} - kv}\right)\right] = 1. \tag{24.72}$$

Carrying out the expansion of the logarithm (again to third order) yields the long-wavelength spin-wave dispersion relation

$$\omega = \pm\Omega_0\left[1 + \frac{(1 + F_0^{\mathrm{a}})^2}{3F_0^{\mathrm{a}}}\left(\frac{kv}{\Omega_0}\right)^2\right]; \tag{24.73}$$

note that since $F_0^{\mathrm{a}} < 0, \omega(k)$ decreases with increasing wave number.

PART II THE MICROSCOPIC THEORY OF A UNIFORM SUPERCONDUCTOR

25 The Cooper problem: pairing of two electrons above a filled Fermi sea

Historically one of the important steps leading to the successful formulation of the microscopic theory of superconductivity was the model calculation performed by Cooper (1956) which showed that two electrons, interacting above a filled Fermi sea, had a bound state, in the presence of an arbitrarily weak attractive interaction. It was assumed that the only effect of the remaining electrons was to exclude Fourier components of the pair wave function with wave vectors $k \leq k_F$. We now discuss Cooper's calculation.

We write the wave equation of the two electrons selected, having coordinates $\mathbf{r}_1$ and $\mathbf{r}_2$, in the form

$$\psi(\mathbf{r}_1, \mathbf{r}_2) = \phi_{\mathbf{q}}(\boldsymbol{\rho})e^{i\mathbf{q}\cdot\mathbf{R}}\chi(\sigma_1, \sigma_2), \tag{25.1}$$

where $\mathbf{R}$ is the center of mass coordinate, $\mathbf{R} \equiv (\mathbf{r}_1 + \mathbf{r}_2)/2, \boldsymbol{\rho} \equiv \mathbf{r}_1 - \mathbf{r}_2$, and σ_1 and σ_2 denote the spins of the two electrons ($\uparrow$ or $\downarrow$); $\phi_{\mathbf{q}}(\boldsymbol{\rho})$ is the wavefunction in the center of mass and χ denotes the spin wavefunction (any spin–orbit coupling is neglected). In what follows we assume that pairing takes place in a singlet state for which

$$\chi = \frac{1}{\sqrt{2}}\left[\begin{pmatrix}1\\0\end{pmatrix}\begin{pmatrix}0\\1\end{pmatrix} - \begin{pmatrix}0\\1\end{pmatrix}\begin{pmatrix}1\\0\end{pmatrix}\right]. \tag{25.2}$$

We initially restrict ourselves to the case $\mathbf{q} = 0$ (i.e., the center of mass of the pair is at rest with respect to the Fermi sea). We first write ϕ as a Fourier expansion[1]

$$\begin{aligned}\phi(\boldsymbol{\rho}) &= \sum_{\mathbf{k}} g(\mathbf{k})e^{i\mathbf{k}\cdot\boldsymbol{\rho}}\\ &= \sum_{\mathbf{k}} g(\mathbf{k})e^{i\mathbf{k}\cdot\mathbf{r}_1}e^{-i\mathbf{k}\cdot\mathbf{r}_2};\end{aligned} \tag{25.3}$$

written in this way we see that the pair wavefunction can be regarded as being made up of plane

1. Since the singlet spin wavefunction is antisymmetric $\phi(\boldsymbol{\rho})$ must be symmetric in order to satisfy the Pauli principle; i.e., $\phi(\boldsymbol{\rho}) = \phi(-\boldsymbol{\rho})$ which in turn requires $g(\mathbf{k}) = g(-\mathbf{k})$.

wave states with equal and opposite momenta (and opposite spins for our assumed singlet pairing). The Schrödinger equation for the pair wavefunction is

$$\left[-\frac{\hbar^2}{2m}(\nabla_1^2+\nabla_2^2)+V(\mathbf{r}_1,\mathbf{r}_2)\right]\psi(\mathbf{r}_1,\mathbf{r}_2)=\left(\varepsilon+2\frac{\hbar^2k_F^2}{2m}\right)\psi(\mathbf{r}_1,\mathbf{r}_2), \tag{25.4}$$

where we have defined the energy, ε, relative to the combined energy of two electrons at the Fermi surface. Inserting (25.3) into (25.4) yields

$$\sum_{\mathbf{k}'}\left[\frac{\hbar^2}{2m}2k'^2+V(\mathbf{r}_1,\mathbf{r}_2)\right]g(\mathbf{k}')e^{i\mathbf{k}'\cdot(\mathbf{r}_1-\mathbf{r}_2)}=\left(\varepsilon+2\frac{\hbar^2k_F^2}{2m}\right)\sum_{\mathbf{k}'}g(\mathbf{k}')e^{i\mathbf{k}'\cdot(\mathbf{r}_1-\mathbf{r}_2)}. \tag{25.5}$$

We assume translational invariance and write $V(\mathbf{r}_1,\mathbf{r}_2)=V(\boldsymbol{\rho})$. We multiply by $e^{-i\mathbf{k}\cdot\boldsymbol{\rho}}$ and integrate over $d^3\rho$

$$\frac{\hbar^2}{m}\int\sum_{\mathbf{k}'}e^{i(\mathbf{k}'-\mathbf{k})\cdot\boldsymbol{\rho}}g(\mathbf{k}')k'^2d^3\rho+\int\sum_{\mathbf{k}'}V(\boldsymbol{\rho})e^{i(\mathbf{k}'-\mathbf{k})\cdot\boldsymbol{\rho}}g(\mathbf{k}')d^3\rho$$

$$=\left(\varepsilon+\frac{\hbar^2k_F^2}{m}\right)\int\sum_{\mathbf{k}'}e^{i(\mathbf{k}'-\mathbf{k})\cdot\boldsymbol{\rho}}g(\mathbf{k}')d^3\rho. \tag{25.6}$$

Now

$$\int e^{i(\mathbf{k}-\mathbf{k}')\cdot\boldsymbol{\rho}}d^3\rho=\begin{cases}0, & \mathbf{k}\neq\mathbf{k}' \quad (25.7a)\\ L^3 & \mathbf{k}=\mathbf{k}' \quad (25.7b)\end{cases}$$

$$=L^3\delta_{\mathbf{k}\mathbf{k}'}. \tag{25.7c}$$

Eq. (25.6) then becomes

$$\frac{\hbar^2}{m}k^2g(\mathbf{k})+\sum_{\mathbf{k}'}g(\mathbf{k}')V_{\mathbf{k}\mathbf{k}'}=(\varepsilon+2\varepsilon_F)g(\mathbf{k}), \tag{25.8}$$

where we defined

$$V_{\mathbf{k}\mathbf{k}'}=\frac{1}{L^3}\int V(\boldsymbol{\rho})e^{i(\mathbf{k}'-\mathbf{k})\cdot\boldsymbol{\rho}}d^3\rho. \tag{25.9}$$

Eq. (25.8) is referred to as the Bethe–Goldstone equation for the two-electron problem. The Pauli principle forbids our inclusion of any plane wave states with $k<k_F$ since, according to our assumption of a filled Fermi sea, these states are already occupied; i.e., $g(\mathbf{k})=0$ for $k<k_F$. For our model calculation we adopt the simple potential

$$V_{\mathbf{k}\mathbf{k}'}=\begin{cases}-\dfrac{V}{L^3}, & \left|\dfrac{\hbar^2k^2}{2m}-\varepsilon_F\right| \text{ and } \left|\dfrac{\hbar^2k'^2}{2m}-\varepsilon_F\right|<\hbar\omega_D, \quad (25.10a)\\[2ex] 0, & \left|\dfrac{\hbar^2k^2}{2m}-\varepsilon_F\right| \text{ or } \left|\dfrac{\hbar^2k'^2}{2m}-\varepsilon_F\right|>\hbar\omega_D, \quad (25.10b)\end{cases}$$

where $\hbar\omega_D$ is a 'cut-off frequency' which (if we assume pairing via an attractive, phonon-induced, electron–electron interaction[2]) is of order $k_B\theta_D$ where k_B is the Boltzmann constant and θ_D is the

2. That superconductivity might arise from the electron–phonon interaction was first pointed out by Fröhlich (1950).

Debye temperature. Substituting (25.10) in Eq. (25.8) and carrying out the integration we obtain

$$\left(-\frac{\hbar^2 k^2}{m} + \varepsilon + 2\varepsilon_F\right) g(\mathbf{k}) = -\frac{V}{L^3}\sum_{\mathbf{k}'} g(\mathbf{k}'). \tag{25.11}$$

The right-hand side of (25.11) being summed over all $\mathbf{k}'$ must be a constant. Dividing (25.11) by the factor $(-(\hbar^2k^2/m) + \varepsilon + 2\varepsilon_F)$ leaves $g(\mathbf{k})$ on the left-hand side of (25.11). If we sum both sides over $\mathbf{k}$, the constant $\Sigma_{\mathbf{k}} g(\mathbf{k})$ appears on both sides and may be canceled leaving the 'self-consistency' condition

$$\frac{V}{L^3}\sum_{\mathbf{k}} \frac{1}{\dfrac{\hbar^2k^2}{m} - \varepsilon - 2\varepsilon_F} = 1. \tag{25.12}$$

Writing $(\hbar^2k^2/2m) - \varepsilon_F = \xi$, converting the sum to an integration, and introducing the density of states per spin,[3] $\mathcal{N}(\xi)$, yields

$$V\int_0^{\hbar\omega_D} \frac{\mathcal{N}(\xi)\mathrm{d}\xi}{2\xi - \varepsilon} = 1.$$

$\mathcal{N}(\xi)$ may be taken as a constant over the range of integration and taken from under the integral sign; carrying out the integration we obtain

$$\frac{1}{2}\mathcal{N}(0)V\ell\mathrm{n}\left(\frac{\varepsilon - 2\hbar\omega_D}{\varepsilon}\right) = 1; \tag{25.13}$$

assuming $|\varepsilon| << \hbar\omega_D$ we obtain

$$\varepsilon = -2\hbar\omega_D \mathrm{e}^{-(2/\mathcal{N}(0)V)}. \tag{25.14}$$

We note the following properties of Eq. (25.14):

(i) The sign of ε is negative and hence (25.14) corresponds to a bound state.

(ii) A bound state suggests a gap in the energy spectrum, which is consistent with the exponentially falling heat capacity observed experimentally.

(iii) We obtain a bound state regardless of how small V is (this is in contrast to the two-body problem in the absence of a Fermi sea where $\mathcal{N}(\xi) \sim \xi^{1/2}$ and we obtain a bound state only if the attraction is greater than some minimum value).

(iv) The result (25.14) cannot be expanded in a power series in V and hence could not be obtained by a perturbation theory argument.

It is instructive to calculate the mean square radius of the pair wavefunction;

$$\overline{\rho^2} = \frac{\int |\phi(\rho)|^2 \rho^2 \mathrm{d}^3\rho}{\int |\phi(\rho)|^2 \mathrm{d}^3\rho}. \tag{25.15}$$

Now $\phi(\rho) = \Sigma_{\mathbf{k}} g(\mathbf{k})\mathrm{e}^{\mathrm{i}\mathbf{k}\cdot\rho}$ and $|\phi(\rho)|^2 = \Sigma_{\mathbf{k},\mathbf{k}'} g(\mathbf{k})g^*(\mathbf{k}')\mathrm{e}^{\mathrm{i}(\mathbf{k}-\mathbf{k}')\cdot\rho}$; we have $\int|\phi(\rho)|^2\mathrm{d}^3\rho =$

3. In Fermi liquid theory (see Sec. 24) we introduced the density of states at the Fermi energy for *both* spins, $\mathcal{N}_F$, whereas in superconductivity one generally uses the density of states per spin, $\mathcal{N}(\xi)$.

$L^3 \Sigma_{\mathbf{k}} |g(\mathbf{k})|^2$, which is the denominator of (25.15). The numerator is evaluated by noting that $\boldsymbol{\rho}$ is generated by operating on the exponential factor with $-\mathrm{i}\nabla_{\mathbf{k}}$: $\rho e^{\mathrm{i}\mathbf{k}\cdot\rho} = -\mathrm{i}\nabla_{\mathbf{k}} e^{\mathrm{i}\mathbf{k}\cdot\rho}$; we generated ρ^2 by incorporating two such terms acting with respect to **k** and **k**′. Performing two integrations by parts in the variables **k** and **k**′ yields

$$\int |\phi(\boldsymbol{\rho})|^2 \rho^2 \mathrm{d}^3\rho = \int \sum_{\mathbf{k},\mathbf{k}'} [-\mathrm{i}\nabla_{\mathbf{k}'} g^*(\mathbf{k}') \cdot \mathrm{i}\nabla_{\mathbf{k}} g(\mathbf{k}) e^{\mathrm{i}(\mathbf{k}-\mathbf{k}')\cdot\rho}] \mathrm{d}^3\rho$$

$$= L^3 \sum_{\mathbf{k}} |\nabla_{\mathbf{k}} g(\mathbf{k})|^2,$$

where we again used Eq. (25.7). Thus

$$\overline{\rho^2} = \frac{\sum_{\mathbf{k}} |\nabla_{\mathbf{k}} g(\mathbf{k})|^2}{\sum_{\mathbf{k}} |g(\mathbf{k})|^2}. \tag{25.16}$$

From Eq. (25.11) $g(\mathbf{k}) \propto \mathrm{const}/(2\xi - \varepsilon)$; noting $(\partial/\partial k) = (\partial\xi/\partial k)(\partial/\partial\xi) = \hbar v_{\mathrm{F}}(\partial/\partial\xi)$ (where v_{F} is the Fermi velocity), Eq. (25.16) then takes the form

$$\overline{\rho^2} = \frac{\hbar^2 v_{\mathrm{F}}^2 \displaystyle\int \frac{4\mathrm{d}\xi}{(2\xi-\varepsilon)^4}}{\displaystyle\int \frac{\mathrm{d}\xi}{(2\xi-\varepsilon)^2}} = \frac{-\dfrac{2}{3}(\hbar v_{\mathrm{F}})^2 \dfrac{1}{(2\xi-\varepsilon)^3}\Big|_0^\infty}{-\dfrac{1}{2}\dfrac{1}{2\xi-\varepsilon}\Big|_0^\infty} = \frac{4}{3}\frac{\hbar^2 v_{\mathrm{F}}^2}{\varepsilon^2}. \tag{25.17}$$

Assuming $\varepsilon \sim k_{\mathrm{B}} T_{\mathrm{c}}$, $T_{\mathrm{c}} \sim 10\,\mathrm{K}$ and $v_{\mathrm{F}} \sim 10^8\,\mathrm{cm/s}$ we obtain $(\overline{\rho^2})^{1/2} \sim 10^{-4}\,\mathrm{cm}$ or $10^4\,\text{Å}$; this value is of the same order as the Pippard coherence length, ξ_0, which lends an interpretation to that quantity in terms of the size of the Cooper pair wavefunction. However, since an electron typically occupies a volume $\sim (2\,\text{Å})^3$ there would be of order 10^{11} other electrons within a 'coherence volume' and hence it is certainly not reasonable to ignore these electrons in constructing a pair wavefunction. This defect is corrected by the BCS theory which treats all electrons on an equal footing.

If we were to compute the binding energy for the case when $\mathbf{q} \neq 0$ we would obtain

$$\varepsilon(q) = \varepsilon(0) + \frac{\hbar v_{\mathrm{F}} q}{2}. \tag{25.18}$$

Hence $\varepsilon(q) = 0$ for $q = 2|\varepsilon(0)|/\hbar v_{\mathrm{F}}$; we see that if q^{-1} is of order the Pippard coherence length, then the bound state (and presumably superconductivity) disappear. More important (25.18) shows that the maximum binding comes from pairing two electrons with *equal and opposite* momenta; this strongly suggests that in constructing a many-body ground state the feature of pairing electrons with no net momenta (equal and opposite **k**) should be retained.

If we convert the wave vector q into a velocity we obtain $v_{\mathrm{c}} = 2|\varepsilon(0)|/p_{\mathrm{F}}$. This suggests there is a critical velocity for the destruction of superfluidity. A similar criterion was obtained by Landau (1941) for superfluid ^{4}He. However, in a superconductor the magnetic field created by a current density complicates a simple application of this idea to estimate the critical current.

26 The BCS theory of the superconducting ground state

The success of the Cooper model calculation in producing a bound state, with its implication that a proper many-body theory would contain a gap in the excitation spectrum (as dictated by various experiments), strongly suggests that the ground-state wavefunction should be constructed from pairs of electrons; i.e., we might examine a form like

$$\Psi_N = \psi(1,2)\psi(3,4)\psi(5,6)\ldots\psi(N-1,N), \tag{26.1a}$$

where we defined the two-particle pair wavefunction

$$\psi(1,2) \equiv \phi(\mathbf{r}_1,\mathbf{r}_2)\chi(\sigma_1,\sigma_2). \tag{26.1b}$$

To insure (26.1a) satisfies the Pauli principle we must operate on it with the antisymmetrization operator, $\hat{A}$

$$\Psi_N = \hat{A}\{\psi(1,2)\psi(3,4)\ldots\psi(N-1,N)\}; \tag{26.2}$$

note the number of terms generated by $\hat{A}$ increases rapidly.[1] From our examination of the Cooper problem we take $\phi(\mathbf{r}_1,\mathbf{r}_2) = \phi(\mathbf{r}_1 - \mathbf{r}_2)$; i.e., the center of mass of each pair is at rest; and we assume translational invariance. We again perform a Fourier expansion

$$\phi(\mathbf{r}) = \sum_{\mathbf{k}} g_{\mathbf{k}} e^{i\mathbf{k}\cdot\mathbf{r}}. \tag{26.3}$$

On substituting (26.3) into (26.2) we obtain

$$\Psi_N = \sum_{\mathbf{k}_1}\ldots\sum_{\mathbf{k}_{N/2}} g_{\mathbf{k}_1}\ldots g_{\mathbf{k}_{N/2}}\hat{A}(e^{i\mathbf{k}_1\cdot(\mathbf{r}_1-\mathbf{r}_2)}\ldots e^{i\mathbf{k}_{N/2}\cdot(\mathbf{r}_{N-1}-\mathbf{r}_N)})$$
$$\times [(1\uparrow)(2\downarrow)\ldots(N-1\uparrow)(N\downarrow)], \tag{26.4}$$

where we now restrict ourselves to the pairing of spin up and spin down electrons (singlet pairing).[2] Rather than writing (26.4) in the coordinate representation, as we have, it is usually easier to perform calculations in the occupation number (Wigner–Jordan) representation; i.e., to use the language of second quantization. (This technique is reviewed in Appendix A.) Instead of

1. For four electrons the wavefunction $A\{\psi(1,2)\psi(3,4)\}$ would contain twelve terms.
2. The antisymmetrization operation may be written more explicitly using the Slater determinant notation given in Eq. (A.12). The quantum numbers $p_1, p_2, p_3, p_4, \ldots$ correspond in the present case to $\mathbf{k}_1\uparrow, -\mathbf{k}_1\downarrow, \mathbf{k}_2\uparrow, -\mathbf{k}_2\downarrow, \ldots$. We again have $g(\mathbf{k}_i) = g(-\mathbf{k}_i)$ for the singlet pairs being considered here.

writing $\hat{A}e^{i\mathbf{k}_1\cdot(\mathbf{r}_1-\mathbf{r}_2)}(1\uparrow)(2\downarrow)$ we write $\hat{c}^\dagger_{\mathbf{k}_1\uparrow}\hat{c}^\dagger_{-\mathbf{k}_1\downarrow}|\phi_0\rangle$ where $|\phi_0\rangle$ is the vacuum state and $\hat{c}^\dagger_{\mathbf{k}_i\sigma_i}$ is the Fermi creation operator for an electron i having momentum $\mathbf{k}$ and spin projection $\sigma(=\uparrow$ or $\downarrow)$. We recall the commutation relations for these operators

$$\hat{c}^\dagger_{\mathbf{k}\alpha}\hat{c}^\dagger_{\ell\beta} + \hat{c}^\dagger_{\ell\beta}\hat{c}^\dagger_{\mathbf{k}\alpha} = 0, \tag{26.5a}$$

$$\hat{c}_{\mathbf{k}\alpha}\hat{c}_{\ell\beta} + \hat{c}_{\ell\beta}\hat{c}_{\mathbf{k}\alpha} = 0, \tag{26.5b}$$

and

$$\hat{c}^\dagger_{\mathbf{k}\alpha}\hat{c}_{\ell\beta} + \hat{c}_{\ell\beta}\hat{c}^\dagger_{\mathbf{k}\alpha} = \delta_{\mathbf{k}\ell}\delta_{\alpha\beta}, \tag{26.5c}$$

where $\mathbf{k}$ and ℓ denote wave vectors and α and β spin projections. In the second quantized notation the many-body pair wavefunction (26.4) can then be written

$$\Psi_N = \sum_{\mathbf{k}_1}\cdots\sum_{\mathbf{k}_{N/2}} g_{\mathbf{k}_1}\cdots g_{\mathbf{k}_{N/2}}\hat{c}^\dagger_{\mathbf{k}_1\uparrow}\hat{c}^\dagger_{-\mathbf{k}_1\downarrow}\cdots\hat{c}^\dagger_{\mathbf{k}_{N/2}\uparrow}\hat{c}^\dagger_{-\mathbf{k}_{N/2}\downarrow}|\phi_0\rangle. \tag{26.6}$$

It is quite difficult to perform calculations with (26.6) and Bardeen, Cooper, and Schrieffer proposed an alternative wavefunction

$$\Psi_{\text{BCS}} = \prod_{\mathbf{k}}(u_{\mathbf{k}} + v_{\mathbf{k}}\hat{c}^\dagger_{\mathbf{k}\uparrow}\hat{c}^\dagger_{-\mathbf{k}\downarrow})|\phi_0\rangle, \tag{26.7}$$

where the product extends over all plane wave states. It is much easier to perform calculations with the BCS wavefunction. (We have collected all the required calculations involving the BCS wavefunction in Appendix B.) The (in general complex) quantities $u_{\mathbf{k}}$ and $v_{\mathbf{k}}$ are not independent but are fixed by the normalization condition $\langle\Psi_{\text{BCS}}|\Psi_{\text{BCS}}\rangle = 1$ which yields

$$|u_{\mathbf{k}}|^2 + |v_{\mathbf{k}}|^2 = 1. \tag{26.8}$$

The wavefunction (26.6) for the special case

$$g(\mathbf{k}) = 1, \quad |\mathbf{k}| < k_{\text{F}},$$
$$g(\mathbf{k}) = 0, \quad |\mathbf{k}| > k_{\text{F}},$$

describes the 'Fermi vacuum' or Fermi sphere: all states with $|\mathbf{k}| < k_{\text{F}}$ are filled and those with $|\mathbf{k}| > k_{\text{F}}$ are empty. The corresponding form for Eq. (26.7) is

$$u_{\mathbf{k}} = 0, \quad v_{\mathbf{k}} = 1, \quad |\mathbf{k}| < k_{\text{F}},$$
$$u_{\mathbf{k}} = 1, \quad v_{\mathbf{k}} = 0, \quad |\mathbf{k}| > k_{\text{F}}.$$

The difference between (26.6) and (26.7) for the general case is that the first defines a state with precisely $N/2$ pairs (N electrons) while the second is a superposition of pair states containing $2, 4, 6, \ldots, N, \ldots, \infty$ electrons; i.e., it does not describe a state with a fixed number of particles. However, by superposition we may relate the two wavefunctions as

$$\Psi_{\text{BCS}} = \sum_N \lambda_N \Psi_N, \tag{26.9}$$

with the normalization condition $\Sigma_N|\lambda_N|^2 = 1$. The average number of particles, $\bar{N}$, associated with the BCS wavefunction is

$$\bar{N} = \langle\Psi_{\text{BCS}}|\hat{N}|\Psi_{\text{BCS}}\rangle, \tag{26.10}$$

where we have defined the number operator

$$\hat{N} \equiv \sum_{\mathbf{k},\sigma} \hat{c}^{\dagger}_{\mathbf{k}\sigma}\hat{c}_{\mathbf{k}\sigma}. \tag{26.11}$$

Evaluating (26.10) using (26.7) gives

$$\bar{N} = \sum_{\mathbf{k}} 2|v_{\mathbf{k}}|^2. \tag{26.12}$$

The mean square fluctuation (or the variance) of the number of particles is defined as

$$\overline{\delta N^2} \equiv \overline{N^2} - \bar{N}^2 = \langle \Psi_{\mathrm{BCS}} | \hat{N}\hat{N} | \Psi_{\mathrm{BCS}} \rangle - \langle \Psi_{\mathrm{BCS}} | \hat{N} | \Psi_{\mathrm{BCS}} \rangle^2$$

$$= \sum_{\mathbf{k}} 4|v_{\mathbf{k}}|^2 |u_{\mathbf{k}}|^2 \tag{26.13}$$

(see Appendix B). Now $\Sigma_{\mathbf{k}} \to [V/(2\pi)^3] \int \mathrm{d}^3k$; therefore (26.13) is proportional to the volume or, equivalently, $\bar{N}$. Hence

$$\frac{\left(\overline{\delta N^2}\right)^{1/2}}{\bar{N}} \sim \bar{N}^{-1/2}.$$

For a metal $\bar{N} \sim 10^{22}$ and therefore $\bar{N}^{-1/2}$ is of order 10^{-11}. This says that the coefficients in the expansion (26.9) will be highly peaked around $\bar{N}$ as shown in Fig. 26.1. We therefore reach the very important conclusion that calculations performed using (26.6) or (26.7) should differ by amounts of order 10^{-11}. Since it is far easier to calculate using (26.7), this form was adopted by BCS.

The approximate ground-state energy is obtained by using (26.7) as a trial wavefunction and evaluating the expectation value of the Hamiltonian of the system of electrons; this is followed by minimization of the resultant energy with respect to the free parameter, $v_{\mathbf{k}}$, subject to the condition that the average number of particles is $\bar{N}$. We may write the Hamiltonian of our (translationally invariant) system of electrons as

$$\hat{H} = \sum_{\mathbf{k},\sigma} \frac{\hbar^2 k^2}{2m} \hat{c}^{\dagger}_{\mathbf{k}\sigma}\hat{c}_{\mathbf{k}\sigma} + \sum_{\substack{\mathbf{k},\mathbf{k}',\mathbf{q}\\ \sigma,\sigma'}} V_{\mathbf{k},\mathbf{k}',\mathbf{q};\sigma,\sigma'} \hat{c}^{\dagger}_{\mathbf{k}+\mathbf{q}\sigma}\hat{c}^{\dagger}_{\mathbf{k}'-\mathbf{q}\sigma'}\hat{c}_{\mathbf{k}'\sigma'}\hat{c}_{\mathbf{k}\sigma}, \tag{26.14}$$

where the second term accounts for all electron–electron interactions and momentum conservation (and transfer by an amount $\mathbf{q}$) has been built into the scattering potential, V. In many-body theory it is conventional to measure all excitation energies from the Fermi surface. Since our wavefunction does not fix the number of electrons we must introduce a constraint that the *average* number of particles is $\bar{N}$; i.e., $\langle \Psi_{\mathrm{BCS}} | \Sigma_{\mathbf{k},\sigma} \hat{c}^{\dagger}_{\mathbf{k},\sigma}\hat{c}_{\mathbf{k},\sigma} | \Psi_{\mathrm{BCS}} \rangle = \bar{N}$. This constraint is introduced through a Lagrange multiplier μ, which we know from statistical physics is the chemical potential. Incorporating these considerations we write the Hamiltonian as

$$\hat{H}' = \hat{H} - \mu\hat{N} = \sum_{\mathbf{k},\sigma} \xi_{\mathbf{k}} \hat{c}^{\dagger}_{\mathbf{k}\sigma}\hat{c}_{\mathbf{k}\sigma} + \sum_{\substack{\mathbf{k},\mathbf{k}',\mathbf{q}\\ \sigma,\sigma'}} V_{\mathbf{k},\mathbf{k}',\mathbf{q};\sigma,\sigma'} \hat{c}^{\dagger}_{\mathbf{k}+\mathbf{q}\sigma}\hat{c}^{\dagger}_{\mathbf{k}'-\mathbf{q}\sigma'}\hat{c}_{\mathbf{k}'\sigma'}\hat{c}_{\mathbf{k}\sigma} \tag{26.15a}$$

$$= \hat{H}'_0 + \hat{H}_{\mathrm{I}} \tag{26.15b}$$

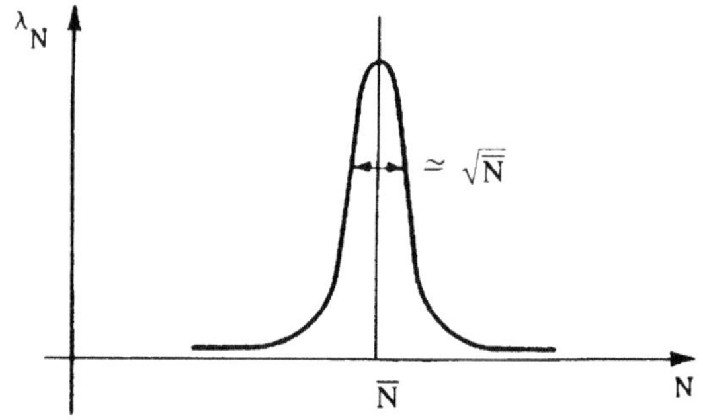

Figure 26.1 The probability amplitude λ_N of finding a configuration with N particles in the BCS wavefunction. Note λ_N is strongly peaked around $\bar{N}$.

where

$$\xi_{\mathbf{k}} \equiv \frac{\hbar^2 k^2}{2m} - \mu. \tag{26.16a}$$

The calculation outlined above is still too complex to be carried out with the 'full' Hamiltonian (26.14) and it is necessary to introduce further approximations (beyond the variational approximation). The first approximation assumes that the Landau Fermi liquid picture applies to the electrons (see Sec. 24). In this model the majority of the effects of normal-state electron–electron interactions may be incorporated by assuming the $\hat{c}^\dagger$ operators create 'quasiparticles' rather than electrons. The energy $\varepsilon(\mathbf{k})$, which for free electrons is $\hbar^2k^2/2m$, is then altered to $\hbar^2k^2/2m^*$ where m^* is a many-body corrected effective mass (not to be confused with the effective mass in semiconductors which arises from band structure effects); i.e.,

$$\begin{aligned}\xi_{\mathbf{k}} &= \frac{\hbar^2 k^2}{2m^*} - \mu \\ &= \frac{\hbar^2}{2m^*}(k^2 - k_{\mathrm{F}}^2) \\ &\simeq \hbar \mathbf{v}_{\mathrm{F}} \cdot \delta\mathbf{k},\end{aligned} \tag{26.16b}$$

where k_{F} is the Fermi wave vector, v_{F} is the Fermi velocity and $\delta\mathbf{k} \equiv \mathbf{k} - \mathbf{k}_F$. A quasiparticle is pictured as an electron together with an accompanying polarization-correlation 'cloud' which accounts for the coordinated motion of all the remaining electrons as they 'get out of the way' of the electron considered (because of either electron–electron repulsion or the Pauli principle). This idea allows us to regard the electron–electron scattering potential as weak (in some sense) since it then contains only the 'left over' component which cannot be incorporated into an effective mass.

In addition to the electron–electron contribution to the effective mass m^*, there is a contribution from the electron–phonon interaction.[3] Physically this contribution results from a polarization of the lattice of ions by the electron. In most metals this contribution is actually larger than the electron–electron contribution.[4]

3. This contribution is limited to excitation energies $\xi \lesssim \hbar\omega_{\mathrm{D}}$, where $\hbar\omega_{\mathrm{D}}$ is the Debye energy. Above this energy the lattice can no longer follow the electron's motion and this mechanism is suppressed.
4. Exceptions are the highly correlated so-called heavy fermion materials where very large electron–electron mass enhancements are encountered (>10).

The 'left over' or residual electron–electron interaction, in addition to being weaker, also has a much more complicated structure. (The bare Coulomb potential has the form $4\pi e^2/q^2$, whereas the residual potential depends on **k**, **k**′, and **q**, as indicated in Eq. (26.15).) When the effect of the indirect electron–phonon interaction is included to form the overall electron–electron interaction the interaction becomes attractive for low excitation energies, as will be discussed in more detail in Sec. 31.[5]

We now restrict ourselves to that part of the interaction Hamiltonian which contributes to superconductivity; i.e., we retain only the attractive part leading to the formation of pairs with opposite momenta and opposite spins. Replacing **k**′ by − **k**, followed by replacing **k** + **q** by **k**′ and writing $\alpha = \uparrow$ and $\beta = \downarrow$ our 'reduced' Hamiltonian becomes[6]

$$\begin{aligned}\hat{H}'_R &= \sum_{\mathbf{k},\sigma} \xi_{\mathbf{k}} \hat{c}^\dagger_{\mathbf{k}\sigma} \hat{c}_{\mathbf{k}\sigma} + \sum_{\mathbf{k},\mathbf{k}'} V_{\mathbf{k}\mathbf{k}'} \hat{c}^\dagger_{\mathbf{k}'\uparrow} \hat{c}^\dagger_{-\mathbf{k}'\downarrow} \hat{c}_{-\mathbf{k}\downarrow} \hat{c}_{\mathbf{k}\uparrow} \\ &= \hat{H}'_0 + \hat{H}_{IR}. \end{aligned} \tag{26.17}$$

The ground-state energy is the expectation value of this quantity with Ψ_{BCS}

$$E' = E - \mu \bar{N} = \langle \Psi_{BCS} | \hat{H}'_R | \Psi_{BCS} \rangle. \tag{26.18}$$

The matrix elements of $\hat{H}'_0$ and $\hat{H}_{IR}$ are evaluated in Appendix B as

$$\langle \Psi_{BCS} | \hat{H}'_0 | \Psi_{BCS} \rangle = 2 \sum_{\mathbf{k}} | v_{\mathbf{k}} |^2 \xi_{\mathbf{k}} \tag{26.19a}$$

and

$$\langle \bar{\Psi}_{BCS} | \hat{H}_{IR} | \bar{\Psi}_{BCS} \rangle = \sum_{\mathbf{k},\mathbf{k}'} V_{\mathbf{k}\mathbf{k}'} v^*_{\mathbf{k}} u_{\mathbf{k}} v_{\mathbf{k}'} u^*_{\mathbf{k}'}. \tag{26.19b}$$

To complete our evaluation of the ground-state energy using this BCS variational wavefunction we must minimize the sum of (26.18) and (26.19) subject to the condition $| u^2_{\mathbf{k}} | + | v^2_{\mathbf{k}} | = 1$. In what follows we choose the phases, so $u_{\mathbf{k}}$ and $v_{\mathbf{k}}$ are real; the normalization condition is conveniently handled by introducing the definitions

$$| u_{\mathbf{k}} | = \cos \theta_{\mathbf{k}} \tag{26.20a}$$

5. Strictly speaking the form of the quasiparticle interaction employed in (26.14) is not sufficiently general to account for the indirect effects of the electron–phonon interaction, and, to a lesser degree, the 'direct' electron–electron interaction. To account properly for such effects one must introduce into the total Hamiltonian all the dynamical variables associated with the motion of both the phonons and the electrons (including the interaction terms). Both the electron system and the phonon system require a finite time to polarize and hence all effective interactions are 'retarded' or delayed in time; this translates into a frequency dependence of this interaction. The electrons, being light, respond rapidly and it is a good approximation to regard their response as instantaneous. The ion motion (associated with the phonons) on the other hand is slow and cannot be ignored. We can qualitatively capture this effect by cutting off the effective electron–electron interaction in Eq. (26.14) for electron excitation energies $\xi > \hbar\omega_D$ where ω_D is a characteristic phonon frequency. For the electron system the cut off would be of order $\hbar\omega_p$ or μ, where ω_p is the plasmon frequency. Both of these energies are of order 10 eV.
6. Written out fully the potential $V_{\mathbf{k},\mathbf{k}',\mathbf{q}}$ coupling initial states **k**, **k**′ to final states **k** + **q**, **k**′ − **q** would be (suppressing spin indices) $V_{\mathbf{k},\mathbf{k}';\mathbf{k}+\mathbf{q},\mathbf{k}'-\mathbf{q}}$; after the substitutions we have $V_{\mathbf{k},-\mathbf{k};\mathbf{k}',-\mathbf{k}'}$ which we denote simply as $V_{\mathbf{k}\mathbf{k}'}$.

and

$$|v_{\mathbf{k}}| = \sin\theta_{\mathbf{k}}. \tag{26.20b}$$

Thus

$$\langle\Psi_{BCS}|\hat{H}'_R|\Psi_{BCS}\rangle = 2\sum_{\mathbf{k}}\xi_{\mathbf{k}}\sin^2\theta_{\mathbf{k}} + \frac{1}{4}\sum_{\mathbf{kk}'}V_{\mathbf{kk}'}\sin 2\theta_{\mathbf{k}}\sin 2\theta_{\mathbf{k}'} \tag{26.21}$$

and minimizing with respect to the parameters $\theta_{\mathbf{k}}$ yields

$$2\xi_{\mathbf{k}}\sin 2\theta_{\mathbf{k}} + \cos 2\theta_{\mathbf{k}}\sum_{\mathbf{k}'}V_{\mathbf{kk}'}\sin 2\theta_{\mathbf{k}'} = 0$$

or

$$\xi_{\mathbf{k}}\tan 2\theta_{\mathbf{k}} = -\frac{1}{2}\sum_{\mathbf{k}'}V_{\mathbf{kk}'}\sin 2\theta_{\mathbf{k}'}. \tag{26.22}$$

We define the function

$$\Delta_{\mathbf{k}} = -\sum_{\mathbf{k}'}V_{\mathbf{kk}'}u_{\mathbf{k}'}v_{\mathbf{k}'} = -\frac{1}{2}\sum_{\mathbf{k}'}V_{\mathbf{kk}'}\sin 2\theta_{\mathbf{k}'}, \tag{26.23}$$

which is called the 'gap function', and (26.22) becomes

$$\tan 2\theta_{\mathbf{k}} = \frac{\Delta_{\mathbf{k}}}{\xi_{\mathbf{k}}}$$

or

$$\sin 2\theta_{\mathbf{k}} = 2u_{\mathbf{k}}v_{\mathbf{k}} = \frac{\Delta_{\mathbf{k}}}{\varepsilon_{\mathbf{k}}}, \tag{26.24}$$

where we defined the energy

$$\varepsilon_{\mathbf{k}} = (\xi_{\mathbf{k}}^2 + \Delta_{\mathbf{k}}^2)^{1/2}. \tag{26.25}$$

In the next section we will show that $\varepsilon_{\mathbf{k}}$ is the energy required to add an electron to the system in a state **k**. We may also calculate

$$\cos 2\theta_{\mathbf{k}} = u_{\mathbf{k}}^2 - v_{\mathbf{k}}^2 = \frac{\xi_{\mathbf{k}}}{\varepsilon_{\mathbf{k}}}. \tag{26.26}$$

The combinations (26.24) and (26.26) occur regularly in evaluating various properties of superconductors; they are often referred to as 'coherence factors'. We also list for reference

$$u_{\mathbf{k}}^2 = \frac{1}{2}\left(1 + \frac{\xi_{\mathbf{k}}}{\varepsilon_{\mathbf{k}}}\right) \tag{26.27}$$

and

$$v_{\mathbf{k}}^2 = \frac{1}{2}\left(1 - \frac{\xi_{\mathbf{k}}}{\varepsilon_{\mathbf{k}}}\right). \tag{26.28}$$

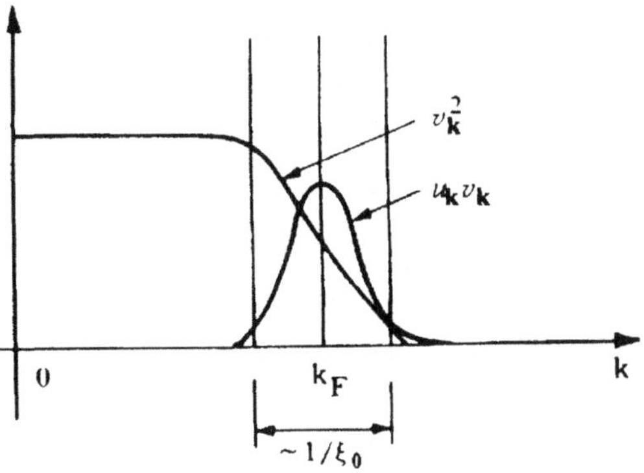

Figure 26.2 The coherence factor v_k^2 and the condensation amplitude $u_k v_k$ in the BCS ground state.

The coherence factor v_k^2 and the condensation amplitude $u_k v_k$ are shown in Fig. 26.2. Note that rather than being step functions, as was the case for the Fermi vacuum, u_k and v_k are 'smeared out' in the vicinity of k_F. The width of this smearing is of order ξ_0^{-1}, the Cooper–Pippard coherence length. The smearing is a consequence of the correlation resulting from the pairing interaction. Substituting (26.24) into (26.23) yields

$$\Delta_k = -\sum_{k'} V_{kk'} \frac{\Delta_{k'}}{2\varepsilon_{k'}} = -\sum_{k'} V_{kk'} \frac{\Delta_{k'}}{2(\xi_{k'}^2 + \Delta_{k'}^2)^{1/2}}. \tag{26.29}$$

Eq. (26.29) is a nonlinear integral equation for the gap function, Δ_k; it is called the gap equation.

Eq. (26.29) has the trivial solution $\Delta_k = 0$ and using (26.27) and (26.28) we have

$$v_k = 1, \qquad u_k = 0, \qquad \text{for } \xi_k < 0 \tag{26.30}$$

and

$$v_k = 0, \qquad u_k = 1, \qquad \text{for } \xi_k > 0. \tag{26.31}$$

The resulting BCS wavefunction is then

$$\psi_{BCS} = \prod_{k<k_F} \hat{c}_{k\uparrow}^\dagger \hat{c}_{-k\downarrow}^\dagger |\phi_0\rangle; \tag{26.32}$$

Eq. (26.32) is just the usual ground-state wavefunction of a filled Fermi sea (also called the Fermi vacuum). We will discuss this case again in the next section.

The simplest example of a nontrivial solution to Eq. (26.29) is to use the Cooper model potential,

$$V_{kk'} = \begin{cases} -V, & |\xi_k| \text{ and } |\xi_{k'}| \le \hbar\omega_D, & (26.33a) \\ 0, & |\xi_k| \ge \hbar\omega_D \text{ or } |\xi_{k'}| \ge \hbar\omega_D. & (26.33b) \end{cases}$$

The function Δ_k will then have the form

$$\Delta_k = \Delta, \quad \xi_k < \hbar\omega_D; \tag{26.34a}$$

$$\Delta_{\mathbf{k}} = 0, \quad \xi_{\mathbf{k}} > \hbar\omega_{\mathrm{D}}. \tag{26.34b}$$

Provided we cut the integral off at $\pm\hbar\omega_{\mathrm{D}}$, we may then take $\Delta_{\mathbf{k}}$ from under the integral sign in (22.29) and cancel it from both sides yielding the condition

$$\begin{aligned} 1 &= \mathcal{N}(0)V\int_{-\hbar\omega_{\mathrm{D}}}^{+\hbar\omega_{\mathrm{D}}}\frac{\mathrm{d}\xi}{2(\xi^2+\Delta^2)^{1/2}} \\ &= \mathcal{N}(0)V\sinh^{-1}\left(\frac{\hbar\omega_{\mathrm{D}}}{\Delta}\right) \end{aligned} \tag{26.35}$$

or

$$\frac{\Delta}{\hbar\omega_{\mathrm{D}}} = \left[\sinh\left(\frac{1}{\mathcal{N}(0)V}\right)\right]^{-1}, \tag{26.36}$$

where $\mathcal{N}(0)$ is the density of states at the Fermi energy. Since $\theta_{\mathrm{D}} = \hbar\omega_{\mathrm{D}}/k_{\mathrm{B}} \sim 300\,\mathrm{K}$ and $\Delta/k_{\mathrm{B}} \sim T_{\mathrm{c}} \sim 10\,\mathrm{K}$ we may approximate (26.36) as

$$\Delta \cong 2\hbar\omega_{\mathrm{D}}\mathrm{e}^{-1/\mathcal{N}(0)V}. \tag{26.37}$$

Note the close similarity between the 'gap' Δ and the bound state energy found by Cooper. Having the expressions (26.27) and (26.28) for $u_{\mathbf{k}}$ and $v_{\mathbf{k}}$ we can evaluate the ground-state energy using (26.19a) and (26.19b):

$$E - \mu\bar{N} = 2\sum_{\mathbf{k}}\frac{\xi_{\mathbf{k}}}{2}\left(1-\frac{\xi_{\mathbf{k}}}{\varepsilon_{\mathbf{k}}}\right) + \frac{1}{4}\sum_{\mathbf{k},\mathbf{k}'}V_{\mathbf{k}\mathbf{k}'}\frac{\Delta_{\mathbf{k}}\Delta_{\mathbf{k}'}}{\varepsilon_{\mathbf{k}}\varepsilon_{\mathbf{k}'}}. \tag{26.38}$$

The first (single-particle) term in (26.38) may be rewritten as[7]

$$\begin{aligned} &-\Delta^2\mathcal{N}(0)\int_{-\hbar\omega_{\mathrm{D}}/\Delta}^{\hbar\omega_{\mathrm{D}}/\Delta}\frac{x^2\mathrm{d}x}{(1+x^2)^{1/2}} + \mathcal{N}(0)\int_{-\hbar\omega_{\mathrm{D}}}^{\hbar\omega_{\mathrm{D}}}|\xi|\mathrm{d}\xi + \int_{-\mu}^{\infty}\mathcal{N}(\xi)(\xi-|\xi|)\mathrm{d}\xi \\ &= -\Delta^2\mathcal{N}(0)\frac{\hbar\omega_{\mathrm{D}}}{\Delta}\left(\frac{\hbar^2\omega_{\mathrm{D}}^2}{\Delta^2}+1\right)^{1/2} + \Delta^2\mathcal{N}(0)\sinh^{-1}\left(\frac{\hbar\omega_{\mathrm{D}}}{\Delta}\right) + \mathcal{N}(0)\hbar^2\omega_{\mathrm{D}}^2 \\ &\quad + 2\int_{-\mu}^{0}\mathcal{N}(\xi)\xi\mathrm{d}\xi, \end{aligned}$$

where we assumed the density of states was constant in the interval $\hbar\omega_{\mathrm{D}} > \xi > -\hbar\omega_{\mathrm{D}}$. Expanding the first term to first order in $(\Delta/\hbar\omega_{\mathrm{D}})^2$, using (26.35) for the second term, and noting the last term is the total normal ground-state energy, E_0, we obtain the form

$$E_0 + \frac{\Delta^2}{V} - \frac{1}{2}\mathcal{N}(0)\Delta^2. \tag{26.39}$$

Using (26.29) twice to evaluate the second (interaction) term in (26.38) we obtain $-\Delta^2/V$. The total 'condensation energy' is then

$$(E-\mu\bar{N}) - (E-\mu\bar{N})_{\Delta=0} = -\frac{1}{2}\mathcal{N}(0)\Delta^2. \tag{26.40}$$

7. $\displaystyle\int\frac{x^2}{(x^2+1)^{1/2}}\mathrm{d}x = \frac{1}{2}[x(1+x^2)^{1/2} - \sinh^{-1}x].$

Referring to Eq. (4.11) we obtain that in the BCS model

$$\frac{1}{8\pi} H_c^2(T = 0) = \frac{1}{2} \mathcal{N}(0) \Delta^2. \tag{26.41}$$

27 Elementary excitations: the Bogoliubov–Valatin transformation

We recall expressions (26.19) for the expectation value of the reduced Hamiltonian

$$\langle \Psi_{BCS} | \hat{H}'_R | \Psi_{BCS} \rangle = 2\sum_k v_k^2 \xi_k + \sum_{k,k'} u_k v_k u_{k'} v_{k'} V_{kk'}. \tag{27.1}$$

Let us calculate the energy $\tilde{\varepsilon}_k$ required to add an electron to the system in a state $|\mathbf{k}\uparrow\rangle$ assuming its companion state $|-\mathbf{k}\downarrow\rangle$ is empty. In order to do this we must: (i) account for the energy increase due to removing the amplitude of the pair[1] associated with this wave vector $\mathbf{k}$ and (ii) add the energy of the lone electron introduced into the state $|\mathbf{k}\rangle$. When we remove the bound pair from the ground state, according to (27.1) the energy of the system changes by an amount

$$-2v_k^2 \xi_k - 2u_k v_k \sum_{k'} u_{k'} v_{k'} V_{kk'} \tag{27.2}$$

(the factor 2 in the second term arises because the chosen pair state ($\mathbf{k}$, $-\mathbf{k}$) occurs twice since we have a double sum). Using Eq. (26.23) we may write the form (27.2) as

$$-2v_k^2 \xi_k + 2u_k v_k \Delta_k. \tag{27.3}$$

Adding to (27.3) the energy ξ_k of one (unbound) electron then yields the quasiparticle excitation energy, $\tilde{\varepsilon}_k$

$$\begin{aligned} \tilde{\varepsilon}_k &= \xi_k(1 - 2v_k^2) + 2u_k v_k \Delta_k \\ &= \xi_k\left[1 - \left(1 - \frac{\xi_k}{\varepsilon_k}\right)\right] + \frac{\Delta_k^2}{\varepsilon_k} \\ &= \varepsilon_k, \end{aligned} \tag{27.4}$$

where we used Eqs. (26.24) and (26.27). Thus the energy needed to add an electron in state $\mathbf{k}\uparrow$ is ε_k. If we calculate the energy required to *remove* an electron in a state $-\mathbf{k}\downarrow$ we also obtain ε_k. Note the minimum excitation energy is Δ_k; i.e., the excitation spectrum has an *energy gap.* Note also that since the pairing interaction smears out the single-particle occupation factors (u_k and v_k) it is possible to add electrons to the system for states ξ both above and below the chemical potential,

1. Note we cannot simply subtract the contribution from the interaction term only ($\hat{H}_{IR}$) because the pairing affects the single-particle energies also (H'_0).

μ. (In a normal ground state the states with $\mathbf{k} < k_F$ would be filled at $T = 0$ and hence no electrons could be added by the Pauli principle.)

Let us ask what energy is required to create a single-particle excitation from the superconducting ground state. This process combines the two processes discussed above: we must (i) *remove* an electron from the state $\mathbf{k}$, which requires a minimum energy Δ, and (ii) *add* this electron to some state $\mathbf{k}'$ which also requires a minimum energy Δ; the total minimum energy then required is 2Δ.

We next examine the wavefunctions associated with various excitations. We may write the wavefunction of a particle(electron)-like state with spin $\uparrow$ and wave vector $\mathbf{k}$ as

$$\begin{aligned}\hat{c}^\dagger_{\mathbf{k}\uparrow}|\Psi_{\mathrm{BCS}}\rangle &= (u_{\mathbf{k}}c^\dagger_{\mathbf{k}\uparrow} + v_{\mathbf{k}}\hat{c}^\dagger_{\mathbf{k}\uparrow}\hat{c}^\dagger_{\mathbf{k}\uparrow}\hat{c}^\dagger_{-\mathbf{k}\downarrow})\prod_{\mathbf{k}'\neq\mathbf{k}}(u_{\mathbf{k}'} + v_{\mathbf{k}'}\hat{c}^\dagger_{\mathbf{k}'\uparrow}\hat{c}^\dagger_{-\mathbf{k}'\downarrow})|\phi_0\rangle \\ &= u_{\mathbf{k}}\hat{c}^\dagger_{\mathbf{k}\uparrow}\prod_{\mathbf{k}'\neq\mathbf{k}}(u_{\mathbf{k}'} + v_{\mathbf{k}'}\hat{c}^\dagger_{\mathbf{k}'\uparrow}\hat{c}^\dagger_{-\mathbf{k}'\downarrow})|\phi_0\rangle \\ &= u_{\mathbf{k}}|\mathbf{k}\uparrow\rangle,\end{aligned} \tag{27.5}$$

where

$$|\mathbf{k}\sigma\rangle \equiv \hat{c}^\dagger_{\mathbf{k}\sigma}\prod_{\mathbf{k}'\neq\mathbf{k}}(u_{\mathbf{k}'} + v_{\mathbf{k}'}\hat{c}^\dagger_{\mathbf{k}'\uparrow}\hat{c}^\dagger_{-\mathbf{k}'\downarrow})|\phi_0\rangle. \tag{27.6}$$

We may write a hole state with spin $\downarrow$ and wave vector $-\mathbf{k}$ as

$$\begin{aligned}\hat{c}_{-\mathbf{k}\downarrow}|\Psi_{\mathrm{BCS}}\rangle &= (u_{\mathbf{k}}c_{-\mathbf{k}\downarrow} + v_{\mathbf{k}}\hat{c}_{-\mathbf{k}\downarrow}\hat{c}^\dagger_{\mathbf{k}\uparrow}\hat{c}^\dagger_{-\mathbf{k}\downarrow})\prod_{\mathbf{k}'\neq\mathbf{k}}(u_{\mathbf{k}'} + v_{\mathbf{k}'}\hat{c}^\dagger_{\mathbf{k}'\uparrow}\hat{c}^\dagger_{-\mathbf{k}'\downarrow})|\phi_0\rangle \\ &= -v_{\mathbf{k}}|\mathbf{k}\uparrow\rangle.\end{aligned} \tag{27.7}$$

Note we generated the same state (apart from normalization) by either adding an electron to $+\mathbf{k}\uparrow$ or removing it from $-\mathbf{k}\downarrow$. Similarly we have

$$\hat{c}_{\mathbf{k}\uparrow}|\Psi_{\mathrm{BCS}}\rangle = v_{\mathbf{k}}|-\mathbf{k}\downarrow\rangle \tag{27.8}$$

and

$$\hat{c}^\dagger_{-\mathbf{k}\downarrow}|\Psi_{\mathrm{BCS}}\rangle = u_{\mathbf{k}}|-\mathbf{k}\downarrow\rangle. \tag{27.9}$$

If we multiply (27.5) by $v_{\mathbf{k}}$ and (27.7) by $u_{\mathbf{k}}$ and add we obtain

$$(u_{\mathbf{k}}\hat{c}_{-\mathbf{k}\downarrow} + v_{\mathbf{k}}\hat{c}^\dagger_{\mathbf{k}\uparrow})|\Psi_{\mathrm{BCS}}\rangle = 0 \tag{27.10}$$

or

$$\hat{\gamma}_{-\mathbf{k}\downarrow}|\Psi_{\mathrm{BSC}}\rangle = 0, \tag{27.11}$$

where

$$\hat{\gamma}_{-\mathbf{k}\downarrow} \equiv u_{\mathbf{k}}\hat{c}_{-\mathbf{k}\downarrow} + v_{\mathbf{k}}\hat{c}^\dagger_{\mathbf{k}\uparrow}. \tag{27.12}$$

On the other hand, multiplying (27.8) by $v_{\mathbf{k}}$ and (27.9) by $u_{\mathbf{k}}$, adding, and applying the normalization condition yields

$$\hat{\gamma}^\dagger_{-\mathbf{k}\downarrow}|\Psi_{\mathrm{BSC}}\rangle = |-\mathbf{k}\downarrow\rangle. \tag{27.13}$$

Similarly using (27.5) and (27.7) yields

$$(u_{\mathbf{k}}\hat{c}^{\dagger}_{\mathbf{k}\uparrow} - v_{\mathbf{k}}\hat{c}_{-\mathbf{k}\downarrow})|\Psi_{\text{BCS}}\rangle = |\mathbf{k}\uparrow\rangle \tag{27.14}$$

or

$$\hat{\gamma}^{\dagger}_{\mathbf{k}\uparrow}|\Psi_{\text{BCS}}\rangle = |\mathbf{k}\uparrow\rangle, \tag{27.15}$$

where

$$\hat{\gamma}^{\dagger}_{\mathbf{k}\uparrow} \equiv u_{\mathbf{k}}\hat{c}^{\dagger}_{\mathbf{k}\uparrow} - v_{\mathbf{k}}\hat{c}_{-\mathbf{k}\downarrow}. \tag{27.16}$$

Using (27.8) and (27.9) we obtain

$$\hat{\gamma}_{\mathbf{k}\uparrow}|\Psi_{\text{BCS}}\rangle = 0. \tag{27.17}$$

The two pairs of operators $\hat{\gamma}_{\mathbf{k}\uparrow}, \hat{\gamma}_{-\mathbf{k}\downarrow}; \hat{\gamma}^{\dagger}_{\mathbf{k}\uparrow}, \hat{\gamma}^{\dagger}_{-\mathbf{k}\downarrow}$ play the role of quasiparticle destruction and creation operators. Since the ground state contains no quasiparticle excitations, we must have $\hat{\gamma}|\Psi_{\text{BCS}}\rangle = 0$; on the other hand $\hat{\gamma}^{\dagger}|\Psi_{\text{BCS}}\rangle = |\mathbf{k}\rangle$, an excited state with one quasiparticle. The operators $\hat{\gamma}, \hat{\gamma}^{\dagger}$ are called Bogoliubov–Valatin operators after Bogoliubov (1958) and Valatin (1958) who first introduced them. It is easy to verify these operators satisfy the Fermi anticommutation rules

$$[\hat{\gamma}_{\mathbf{k}\sigma}, \hat{\gamma}_{\mathbf{k}'\sigma'}]_{+} = 0, \tag{27.18a}$$

$$[\hat{\gamma}^{\dagger}_{\mathbf{k}\sigma}, \hat{\gamma}^{\dagger}_{\mathbf{k}'\sigma'}]_{+} = 0, \tag{27.18b}$$

and

$$[\hat{\gamma}^{\dagger}_{\mathbf{k}\sigma}, \hat{\gamma}_{\mathbf{k}'\sigma'}]_{+} = \delta_{\mathbf{k}\mathbf{k}'}\delta_{\sigma\sigma'}. \tag{27.18c}$$

Orthonormal states involving more than one quasiparticle excitation may be generated by applying successive Bogoliubov quasiparticle creation operators to the ground state

$$|\mathbf{k}_1\sigma_1, \mathbf{k}_2\sigma_2, \mathbf{k}_3\sigma_3, \ldots\rangle = (\hat{\gamma}^{\dagger}_{\mathbf{k}_1\sigma_1}\hat{\gamma}^{\dagger}_{\mathbf{k}_2\sigma_2}\hat{\gamma}^{\dagger}_{\mathbf{k}_3\sigma_3}\cdots)|\Psi_{\text{BCS}}\rangle. \tag{27.19}$$

It is important to note that the Bogoliubov quasiparticle operators incorporate two features. The first feature is apparent in the absence of a pairing correlation ($\Delta = 0$) where the operators become

$$\hat{\gamma}^{\dagger}_{\mathbf{k}\uparrow} = \begin{cases} \hat{c}^{\dagger}_{\mathbf{k}\uparrow} & k > k_{\text{F}}, \quad (27.20\text{a}) \\ -\hat{c}_{-\mathbf{k}\downarrow} & k < k_{\text{F}}, \quad (27.20\text{b}) \end{cases}$$

and

$$\hat{\gamma}^{\dagger}_{-\mathbf{k}\downarrow} = \begin{cases} \hat{c}^{\dagger}_{-\mathbf{k}\downarrow} & k > k_{\text{F}}, \quad (27.21\text{a}) \\ \hat{c}_{\mathbf{k}\uparrow} & k < k_{\text{F}}. \quad (27.21\text{b}) \end{cases}$$

Consider (27.20a) which says that to create an electron excitation $|\mathbf{k}\uparrow\rangle$ for $k > k_{\text{F}}$ we simply add an electron with quantum numbers, $\mathbf{k}, \uparrow$. However, (27.20b) says that to add an excitation with quantum numbers $\mathbf{k}, \uparrow$ when $k < k_{\text{F}}$ we must *destroy* the *existing* electron state with quantum numbers $-\mathbf{k}, \downarrow$, to *create* a *hole* state with quantum numbers $\mathbf{k}, \uparrow$. The negative sign with (27.20b) reflects both an arbitrary phase convention and the fact that our pairing has been restricted to singlet states. Eqs. (27.21) are interpreted in a similar manner.

The second feature of the Bogoliubov quasiparticle operators is more subtle. In the superconductor ($\Delta \neq 0$), the operators $\hat{c}^\dagger_{\mathbf{k}\uparrow}$ (electron-like) and $\hat{c}_{-\mathbf{k}\downarrow}$ (hole-like) become *mixed*, there being some amplitude for each to be nonzero for a region $\Delta k \sim 1/\xi_0$ about k_F. This property reflects correlations introduced by the pairing interactions.

The commutation relations (27.18) resulted from our definitions of the $\hat{\gamma}$ operators and the normalization property obeyed by $u_\mathbf{k}$ and $v_\mathbf{k}$. We could also proceed in the opposite direction by demanding the operators $\hat{\gamma}$ satisfy the Fermi commutation rules in which case the normalization condition would be a resultant property. Since the transformations $\hat{\gamma} \Leftrightarrow \hat{c}$ preserve the Fermi commutation rules they are said to be canonical (a word commonly invoked in physics when a transformation retains the overall structure of the resulting equations).

There is an alternative way (Bogoliubov's method) to obtain the ground-state energy of a superconductor in which the BCS wavefunction is not explicitly involved. To do this we (i) rewrite the Hamiltonian, $\hat{H}'_R$, in terms of the Bogoliubov operators, $\hat{\gamma}, \hat{\gamma}^\dagger$ and (ii) commute, using (27.18), all destruction operators, $\hat{\gamma}$, to the right where we assume they annihilate the ground state, which by definition contains no excited states. The resulting c-number expression is then minimized (with respect to u, v) to obtain the ground-state energy.

Implementing this strategy requires the inverses of the transformations (27.12) and (27.16) which are easily found to be

$$\hat{c}_{-\mathbf{k}\downarrow} = u_\mathbf{k}\hat{\gamma}_{-\mathbf{k}\downarrow} - v_\mathbf{k}\hat{\gamma}^\dagger_{\mathbf{k}\uparrow} \tag{27.22}$$

and

$$\hat{c}^\dagger_{\mathbf{k}\uparrow} = u_\mathbf{k}\hat{\gamma}^\dagger_{\mathbf{k}\uparrow} + v_\mathbf{k}\hat{\gamma}_{-\mathbf{k}\downarrow}. \tag{27.23}$$

We will not carry out the calculation of the ground-state energy using the Bogoliubov method because, with just a little more effort, we can obtain the free energy for any temperature, which we do in the next section. One could, in fact, dispense with the BCS wavefunction entirely and introduce the Bogoliubov transformation from the outset. However, some physical insight is lost in this approach in that a rationale for introducing the transformation is then lacking.

Eqs. (27.22) and (27.23) may be written in matrix form; introducing the notation of Eqs. (26.20) we have

$$\begin{pmatrix} \hat{c}^\dagger_{\mathbf{k}\uparrow} \\ \hat{c}_{-\mathbf{k}\downarrow} \end{pmatrix} = \begin{pmatrix} \cos\theta_\mathbf{k} & \sin\theta_\mathbf{k} \\ -\sin\theta_\mathbf{k} & \cos\theta_\mathbf{k} \end{pmatrix} \begin{pmatrix} \hat{\gamma}^\dagger_{\mathbf{k}\uparrow} \\ \hat{\gamma}_{-\mathbf{k}\downarrow} \end{pmatrix}. \tag{27.24}$$

The inverse transformations, (27.12) and (27.16), are

$$\begin{pmatrix} \hat{\gamma}^\dagger_{\mathbf{k}\uparrow} \\ \hat{\gamma}_{-\mathbf{k}\downarrow} \end{pmatrix} = \begin{pmatrix} \cos\theta_\mathbf{k} & -\sin\theta_\mathbf{k} \\ \sin\theta_\mathbf{k} & \cos\theta_\mathbf{k} \end{pmatrix} \begin{pmatrix} \hat{c}^\dagger_{\mathbf{k}\uparrow} \\ \hat{c}_{-\mathbf{k}\downarrow} \end{pmatrix}. \tag{27.25}$$

Eqs. (27.24) and (27.25) differ only in the sign of the angle of rotation in 'operator space'.

28 Calculation of the thermodynamic properties using the Bogoliubov–Valatin Method

As discussed in the previous section, excited states are easily expressed using the Bogoliubov–Valatin operators. We can also rewrite the reduced Hamiltonian of the system in terms of the Bogoliubov quasiparticle operators. In doing so we generate two types of terms: those containing products of an equal number of creation ($\hat{\gamma}^\dagger_{\mathbf{k}\sigma}$) and destruction ($\hat{\gamma}_{\mathbf{k}\sigma}$) operators and those containing an unequal number. When taking thermally averaged quantum-expectation values of such products, only the first forms yield nonzero averages. Furthermore, since the fermion–quasiparticle excitations do not interact (in our mean-field-like theory), the averages are given by

$$\langle \hat{\gamma}^\dagger_{\mathbf{k}\sigma}\hat{\gamma}_{\mathbf{k}\sigma}\rangle = f_{\mathbf{k}\sigma} \tag{28.1a}$$

and

$$\langle \hat{\gamma}_{\mathbf{k}\sigma}\hat{\gamma}^\dagger_{\mathbf{k}\sigma}\rangle = 1 - f_{\mathbf{k}\sigma}, \tag{28.1b}$$

where $f_{\mathbf{k}\sigma}$ will turn out to be the usual ideal–Fermi-gas occupation factor,

$$f_{\mathbf{k}\sigma} = \frac{1}{e^{\varepsilon_{\mathbf{k}}/k_B T} + 1}. \tag{28.2}$$

Inserting Eqs. (27.22) and (27.23) into our reduced Hamiltonian, (26.18) yields

$$\hat{H}'_R = \hat{H}'_0 + \hat{H}_{IR},$$

$$\hat{H}'_0 = \sum_{\mathbf{k}} \xi_{\mathbf{k}}[2v_{\mathbf{k}}^2 + (u_{\mathbf{k}}^2 - v_{\mathbf{k}}^2)(\hat{\gamma}^\dagger_{\mathbf{k}\uparrow}\hat{\gamma}_{\mathbf{k}\uparrow} + \hat{\gamma}^\dagger_{-\mathbf{k}\downarrow}\hat{\gamma}_{-\mathbf{k}\downarrow})]$$

$$+ \text{(terms with unequal numbers of } \hat{\gamma}^\dagger_{\mathbf{k}\sigma} \text{ and } \hat{\gamma}_{\mathbf{k}\sigma}), \tag{28.3}$$

and

$$\hat{H}_{IR} = \sum_{\mathbf{k},\mathbf{k}'} V_{\mathbf{k}\mathbf{k}'}[u_{\mathbf{k}}v_{\mathbf{k}}u_{\mathbf{k}'}v_{\mathbf{k}'}(1 - \hat{\gamma}^\dagger_{\mathbf{k}\uparrow}\hat{\gamma}_{\mathbf{k}\uparrow} - \hat{\gamma}^\dagger_{-\mathbf{k}\downarrow}\hat{\gamma}_{-\mathbf{k}\downarrow})(1 - \hat{\gamma}^\dagger_{\mathbf{k}'\uparrow}\hat{\gamma}_{\mathbf{k}'\uparrow} - \hat{\gamma}^\dagger_{-\mathbf{k}'\downarrow}\hat{\gamma}_{-\mathbf{k}'\downarrow})$$

$$+ \text{(terms with unequal numbers of } \hat{\gamma}^\dagger_{\mathbf{k}\sigma} \text{ and } \hat{\gamma}_{\mathbf{k}\sigma})]. \tag{28.4}$$

Using Eq. (28.1) the expectation value of our reduced Hamiltonian is

$$\langle \hat{H}'_R\rangle = 2\sum_{\mathbf{k}} \xi_{\mathbf{k}}[v_{\mathbf{k}}^2 + (u_{\mathbf{k}}^2 - v_{\mathbf{k}}^2)f_{\mathbf{k}}] + \sum_{\mathbf{k},\mathbf{k}'} V_{\mathbf{k}\mathbf{k}'}u_{\mathbf{k}}v_{\mathbf{k}}u_{\mathbf{k}'}v_{\mathbf{k}'}(1 - 2f_{\mathbf{k}})(1 - 2f_{\mathbf{k}'}). \tag{28.5}$$

The Helmholtz free energy in the pairing approximation is given by

$$F = \langle \hat{H}'_{\mathbf{R}} \rangle - TS, \tag{28.6}$$

where the entropy for fermions is given by the usual expression from statistical mechanics

$$\begin{aligned} S &= -k_{\mathrm{B}} \sum_{\mathbf{k},\sigma} [f_{\mathbf{k}} \ell n f_{\mathbf{k}} + (1 - f_{\mathbf{k}}) \ell n (1 - f_{\mathbf{k}})] \\ &= -2k_{\mathrm{B}} \sum_{\mathbf{k}} [f_{\mathbf{k}} \ell n f_{\mathbf{k}} + (1 - f_{\mathbf{k}}) \ell n (1 - f_{\mathbf{k}})]. \end{aligned} \tag{28.7}$$

Introducing $u_{\mathbf{k}} = \cos\theta_{\mathbf{k}}$ and $v_{\mathbf{k}} = \sin\theta_{\mathbf{k}}$ and minimizing F with respect to the parameters $\theta_{\mathbf{k}}$ yields

$$u_{\mathbf{k}}^2 = \frac{1}{2}\left(1 + \frac{\xi_{\mathbf{k}}}{\varepsilon_{\mathbf{k}}}\right) \tag{28.8}$$

and

$$v_{\mathbf{k}}^2 = \frac{1}{2}\left(1 - \frac{\xi_{\mathbf{k}}}{\varepsilon_{\mathbf{k}}}\right), \tag{28.9}$$

where we defined

$$\varepsilon_{\mathbf{k}} = (\xi_{\mathbf{k}}^2 + \Delta_{\mathbf{k}}^2)^{1/2} \tag{28.10}$$

but where now $\Delta_{\mathbf{k}}$ is defined by

$$\Delta_{\mathbf{k}} = -\sum_{\mathbf{k}'} V_{\mathbf{k}\mathbf{k}'} u_{\mathbf{k}'} v_{\mathbf{k}'} (1 - 2f_{\mathbf{k}'}). \tag{28.11}$$

This equation may also be written as

$$\Delta_{\mathbf{k}} = -\sum_{\mathbf{k}'} V_{\mathbf{k}\mathbf{k}'} u_{\mathbf{k}'} v_{\mathbf{k}'} \tanh\left(\frac{\varepsilon_{\mathbf{k}'}}{2k_{\mathrm{B}}T}\right). \tag{28.11'}$$

Note all quantities have the same form as for the $T = 0$ (ground-state) case except (28.11) which contains an extra factor $(1 - 2f_{\mathbf{k}'})$. (It is useful to point out that since $\varepsilon_{\mathbf{k}}$ is positive definite, $f_{\mathbf{k}} = 0$ for all $\mathbf{k}$ at $T = 0$.) If we minimize F with respect to $f_{\mathbf{k}}$ (i.e., $\partial F/\partial f_{\mathbf{k}} = 0$) we obtain (28.2); the Fermi occupation factor follows immediately from expression (28.7) for S which is based on the exclusion principle. Using the expressions for $u_{\mathbf{k}}$ and $v_{\mathbf{k}}$ we may rewrite the gap equation, (28.11), as

$$\Delta_{\mathbf{k}} = -\sum_{\mathbf{k}'} V_{\mathbf{k}\mathbf{k}'} \frac{(1 - 2f_{\mathbf{k}'})}{2\varepsilon_{\mathbf{k}'}} \Delta_{\mathbf{k}'}. \tag{28.12}$$

Let us first discuss behavior of the gap function in the Cooper model

$$V_{\mathbf{k}\mathbf{k}'} = \begin{cases} -V & \text{for } |\xi_{\mathbf{k}}| \text{ and } |\xi_{\mathbf{k}'}| \le \hbar\omega_{\mathrm{D}}, \\ 0 & \text{for } |\xi_{\mathbf{k}}| \text{ or } |\xi_{\mathbf{k}'}| > \hbar\omega_{\mathrm{D}}. \end{cases} \tag{28.13}$$

Eq. (28.11) may then be written as the condition

$$1 = \mathcal{N}(0) V \int_{-\hbar\omega_{\mathrm{D}}}^{+\hbar\omega_{\mathrm{D}}} \mathrm{d}\xi \frac{1 - 2f[(\xi^2 + \Delta^2(T))^{1/2}]}{2[\xi^2 + \Delta^2(T)]^{1/2}}. \tag{28.14}$$

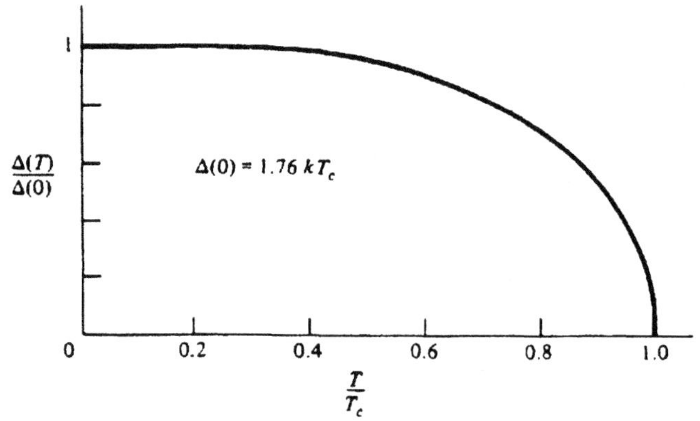

Figure 28.1 Temperature dependence of the energy gap in the BCS theory.

The solution of Eq. (28.14), which must be obtained numerically, gives $\Delta(T)$ which is shown in Fig. 28.1. The transition temperature, T_c, corresponds to $\Delta(T_c) = 0$ which from (28.14) is obtained from

$$1 = \mathcal{N}(0)V\int_{-\hbar\omega_D}^{+\hbar\omega_D} d\xi \frac{1 - 2f(\xi, T_c)}{2\xi}$$

$$= \mathcal{N}(0)V\int_{0}^{+\hbar\omega_D} \frac{d\xi}{\xi} \tanh\frac{\xi}{2k_B T_c}. \tag{28.15}$$

For large ξ, $\tanh(\xi/2k_B T_c) \to 1$ and the integral has the asymptotic form $\ell n(\hbar\omega_D/k_B T_c) + C$; a numerical calculation yields $C = \ell n\ 1.14$. Therefore

$$1 = \mathcal{N}(0)V\ell n\frac{1.14\hbar\omega_D}{k_B T_c} \tag{28.16}$$

or

$$k_B T_c = 1.14\hbar\omega_D e^{-1/\mathcal{N}(0)V}. \tag{28.17}$$

We now evaluate the free energy. The $-TS$ contribution, Eq. (28.7), can be rewritten as

$$-TS = \sum_k \left[\varepsilon_k(1 - 2f_k) - 2k_B T\ell n\left(2\cosh\frac{\varepsilon_k}{2k_B T}\right)\right]. \tag{28.18}$$

The contribution of $\langle \hat{H}_{IR}\rangle$ to the energy, E', follows from Eqs. (28.5) and (28.11):

$$\langle \hat{H}_{IR}\rangle = -\sum_k \frac{\Delta_k^2}{2\varepsilon_k}(1 - 2f_k),$$

$$= -\frac{\Delta^2(T)}{V}. \tag{28.19}$$

From Eq. (28.3) and (28.8)–(28.10) we have

$$\langle \hat{H}'_0\rangle = 2\sum_k [\xi_k v_k^2 + f_k\varepsilon_k - \frac{\Delta^2}{\varepsilon_k} f_k]. \tag{28.20}$$

From (28.19) and (28.20) we obtain the energy $E' = \langle \hat{H}'_R \rangle$ as

$$\langle \hat{H}'_R \rangle = 2\sum_k \left[\xi_k v_k^2 + f_k \varepsilon_k - \frac{\Delta^2}{2\varepsilon_k} \right] + \frac{\Delta^2}{V}. \tag{28.21a}$$

This expression can be rewritten in a more symmetrical form as

$$\langle \hat{H}'_R \rangle = \sum_k \frac{1}{2\varepsilon_k} [(\varepsilon_k + \xi_k)^2 f_k - (\varepsilon_k - \xi_k)^2 (1 - f_k)]. \tag{28.21b}$$

Using (28.2) and (28.7) we obtain

$$-TS = \sum_k \left\{ \varepsilon_k (1 - 2f_k) - 2k_B T \ell n \left[2\cosh\left(\frac{\varepsilon_k}{k_B T}\right) \right\};$$

combining this with (28.21a) yields the fee energy as:

$$F' = \sum_k \left[2\xi_k v_k + \varepsilon_k - \frac{\Delta^2}{\varepsilon_k} - 2k_B T \ell n \left(2\cosh \frac{\varepsilon_k}{2k_B T} \right) \right] - \frac{\Delta^2(T)}{V}. \tag{28.22}$$

The condensation free energy is related to the thermodynamic critical field by $F_N - F_S = (1/8\pi) H_c^2(T)$ as given by Eq. (4.11).

From thermodynamics the heat capacity is

$$\begin{aligned} C &= T\,\mathrm{d}S/\mathrm{d}T \\ &= -2k_B T \sum_k [\ell n f_k - \ell n(1 - f_k)]\, \partial f_k / \partial T \\ &= 2k_B T \sum_k \frac{\varepsilon_k}{k_B T} \frac{\mathrm{d}}{\mathrm{d}T} \left(\frac{1}{e^{\varepsilon_k / k_B T} + 1} \right). \end{aligned}$$

Noting that $\varepsilon_k = \varepsilon_k(T)$ and writing $f'(x)$ in terms of $f(x)$ we obtain

$$\begin{aligned} C &= \frac{-2}{k_B T^2} \sum_k f_k (1 - f_k) \left(-\varepsilon_k^2 + T\Delta \frac{\mathrm{d}\Delta}{\mathrm{d}T} \right) \\ &= \frac{2}{k_B T^2} \mathcal{N}(0) \int_0^\infty \mathrm{d}\xi f(\varepsilon)[1 - f(\varepsilon)] \left(\varepsilon^2 - T\Delta \frac{\mathrm{d}\Delta}{\mathrm{d}T} \right). \end{aligned} \tag{28.23}$$

At very low temperature, $f_k << 1$ and $\Delta \mathrm{d}\Delta/\mathrm{d}T \cong 0$, and the heat capacity approaches the limiting form

$$\begin{aligned} C &\cong \frac{2\mathcal{N}(0)\Delta^2(0) e^{-\Delta(0)/k_B T}}{k_B T^2} \int_0^\infty e^{-(\xi^2 / 2k_B T \Delta(0))} \mathrm{d}\xi \\ &\cong 2\mathcal{N}(0)\Delta(0)(2\pi)^{1/2} k_B (\Delta(0)/k_B T)^{3/2} e^{-(\Delta(0)/k_B T)}; \end{aligned} \tag{28.24}$$

we see that the heat capacity approaches zero asymptotically as $e^{-\Delta(0)/k_B T}$ due to the presence of the gap in the energy spectrum, as discussed in Part I Sec. 1.

29 Quasiparticle tunneling

One of the most powerful probes of the superconducting state in general (and the energy gap in particular) is quasiparticle tunneling. Measurements are usually performed using a 'tunnel junction'. One way to form a tunnel junction is to deposit two thin films separated by an insulating layer (I); the insulating layer may be a natural oxide formed by exposing the first film to air (prior to depositing the second film) or it may also be deposited. The films may involve a superconductor (S) and a normal metal (N) to form an SIN junction or two superconductors (S, S′) to form an SIS′ junction. The insulating layer must be thin enough so that a measurable current flow is observable for voltages $V \sim \Delta/|e|$. At the same time, the coupling must not be so strong that one cannot develop a voltage across the junction.

A junction is but one example of a 'superconducting weak link'. Weak links may also be formed by thinning a superconducting strip to a size of order ξ, or by bringing a sharp metal (superconductor or normal) tip with a radius $< \xi$ in contact with a bulk superconductor.

Fig. 29.1(a) depicts the situation at absolute zero for a tunnel junction formed from two normal metals biased by a voltage V polarized so as to favor charge transfer from the left-hand side (L) to the right-hand side (R). The shaded region represents occupied states and the unshaded part unoccupied states. Fig. 29.1(b) shows an SIS junction where the quasiparticle states are separated from the ground state by an energy gap Δ.

The tunneling process may be described by the introduction of a 'tunneling Hamiltonian' (Bardeen 1962, Cohen, Falicov, and Phillips 1962)

$$\hat{H}_{\mathrm{T}} = \sum_{\ell r \lambda \rho} T_{r\rho\ell\lambda}[\hat{c}^{\dagger}_{r\rho}\hat{c}_{\ell\lambda} + \text{h.c.}], \tag{29.1}$$

where (ℓ, r) and (λ, ρ) denote pairs of translational and spin quantum numbers on the left- and right-hand sides of the junction, respectively. In what follows we assume T is independent of spin and write

$$T_{r\rho\ell\lambda} = T_{r\ell}\delta_{\rho\lambda} \tag{29.2}$$

in which case our tunneling Hamiltonian becomes

$$\hat{H}_{\mathrm{T}} = \sum_{\ell r \lambda \rho} T_{r\ell}[\hat{c}^{\dagger}_{r}\hat{c}_{\ell} + \text{h.c.}]\delta_{\rho\lambda}. \tag{29.3}$$

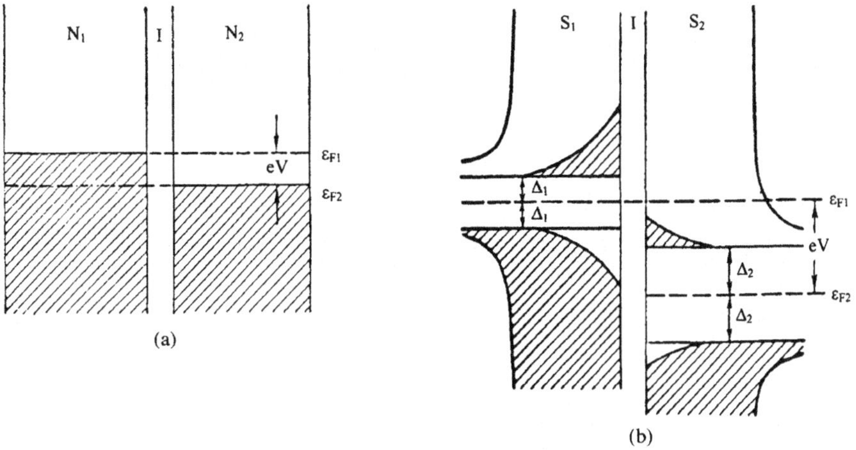

Figure 29.1 Schematic of: (a) an N1IN2 junction biased by a voltage V at absolute zero temperature; (b) an S,IS2 junction biased by a voltage $V > (\Delta_1 + \Delta_2)_2$ at a finite temperature.

Note the form chosen for $\hat{H}_T$ excludes the possibility of there being 'trapping' states within the junction. It also assumes instantaneous interactions.

We start with a discussion of (NIN') tunneling. We use first order, time-dependent perturbation theory to calculate the rate at which charge is transferred across the junction. This involves the difference of the rate of transfer from left to right, $w_{L\to R}$, and from right to left, $w_{R\to L}$:

$$w_{L\to R} = 2\frac{2\pi}{\hbar}\sum_{\ell r}|T_{r\ell}|^2(1-f_r)f_\ell\delta(\xi_\ell - eV - \xi_r) \tag{29.4a}$$

and (from the Hermitian conjugate term in (29.3))

$$w_{R\to L} = 2\frac{2\pi}{\hbar}\sum_{\ell r}|T_{r\ell}|^2(1-f_\ell)f_r\delta(\xi_\ell - eV - \xi_r). \tag{29.4b}$$

In taking the expectation value $\langle \hat{H}_T^\dagger \hat{H}_T\rangle$ we use the rules $\langle \hat{c}_n^\dagger \hat{c}_{n'}\rangle = f_n\delta_{nn'}$ and $\langle \hat{c}_n \hat{c}_{n'}^\dagger\rangle = (1-f_n)\delta_{nn'}$.

The total current, I, is given by

$$I = e(w_{L\to R} - w_{R\to L})$$

or

$$I = \frac{4\pi e}{\hbar}\sum_{\ell r}|T_{\ell r}|^2(f_\ell - f_r)\delta(\xi_\ell - eV - \xi_r). \tag{29.5}$$

Summing over ℓ is achieved by writing $\Sigma_\ell \to \int \mathcal{N}(\xi_\ell)d\xi_\ell$, where $\mathcal{N}(\xi)$ is the density of states; the δ function then results in replacing ξ_ℓ by $\xi_r + eV$. We may then write the distribution function in (29.5) as

$$f(\xi_r + eV) - f(\xi_r) = \delta(\xi_r)eV. \tag{29.6}$$

Changing Σ_ℓ to an integration and assuming $\mathcal{N}(\xi)$ and $T_{nn'}$ are independent of energy (for the range of potentials of interest) we may write the normal state junction conductance, $G_{NN} \equiv I/V$, as

$$G_{NN} = \frac{4\pi e^2}{\hbar}[\mathcal{N}(0)]^2 |\bar{T}|^2, \tag{29.7}$$

where $|\bar{T}|$ is a Fermi-surface-averaged matrix element.

Consider next the case of the SIS′ junction. Excitations in either superconductor are best treated using the Bogoliubov–Valatin operators and the same is true for the tunneling Hamiltonian so we write

$$\hat{c}_\ell^\dagger = u_\ell^* \hat{\gamma}_\ell^\dagger + v_\ell \hat{\gamma}_\ell, \tag{29.8a}$$

$$\hat{c}_\ell = u_\ell \hat{\gamma}_\ell + v_\ell^* \hat{\gamma}_\ell^\dagger, \tag{29.8b}$$

and similarly for $\hat{c}_r^\dagger$ and $\hat{c}_r$. Eq. (29.3) then becomes

$$\hat{H}_T = \sum_{\ell r \lambda \rho} [T_{r\ell}(u_r^* u_\ell \hat{\gamma}_r^\dagger \hat{\gamma}_\ell + u_r^* v_\ell^* \hat{\gamma}_r^\dagger \hat{\gamma}_\ell^\dagger + v_r u_\ell \hat{\gamma}_r \hat{\gamma}_\ell + v_r v_\ell^* \hat{\gamma}_r \hat{\gamma}_\ell^\dagger) + \text{h.c.}]\delta_{\lambda\rho} \tag{29.9}$$

(all operators in (29.9) anticommute with no resultant c numbers since the operators act on different state spaces). The initial and final states which $\hat{H}_T$ connects are each products of left and right eigenstates. We denote the superconducting ground states on the two sides by $|\phi_L\rangle$ and $|\phi_R\rangle$; excited states with quantum numbers ℓ and r are then written $|\ell\rangle = \hat{\gamma}_\ell^\dagger|\phi_L\rangle$ and $|r\rangle = \hat{\gamma}_r^\dagger|\phi_R\rangle$. An initial state consisting of a single excitation in L with quantum number ℓ and no excitations in R would be written $\hat{\gamma}_\ell^\dagger|\phi_L\rangle|\phi_R\rangle$, or $|\ell, \phi_R\rangle$ for short, and similarly for other excitations.

We will write explicitly all matrix elements which lead to the transfer of charge across the junction; we collect them into groups having the same initial and final states and, for later use, we give the energy difference $\Delta E \equiv E_{\text{initial}} - E_{\text{final}}$:

$$T_{r\ell} u_\ell^* u_r \langle \phi_L, r|\hat{\gamma}_r^\dagger \hat{\gamma}_\ell|\ell, \phi_R\rangle + T_{\ell r}^* v_r^* v_\ell \langle \phi_L, r|\hat{\gamma}_\ell \hat{\gamma}_r^\dagger|\ell, \phi_R\rangle,$$
$$\Delta E = \varepsilon_\ell - eV - \varepsilon_r; \tag{29.10a}$$

$$T_{r\ell} u_r^* v_\ell^* \langle \ell, r|\hat{\gamma}_r^\dagger \hat{\gamma}_\ell^\dagger|\phi_L, \phi_R\rangle + T_{\ell r}^* v_r^* u_\ell^* \langle \ell, r|\hat{\gamma}_\ell^\dagger \hat{\gamma}_r^\dagger|\phi_L, \phi_R\rangle,$$
$$\Delta E = eV - \varepsilon_\ell - \varepsilon_r; \tag{29.10b}$$

$$T_{\ell r}^* u_r v_\ell \langle \phi_L, \phi_R|\hat{\gamma}_\ell \hat{\gamma}_r|\ell, r\rangle + T_{r\ell} v_r u_\ell \langle \phi_L, \phi_R|\hat{\gamma}_r \hat{\gamma}_\ell|\ell, r\rangle,$$
$$\Delta E = -eV + \varepsilon_\ell + \varepsilon_r; \tag{29.10c}$$

and

$$T_{\ell r}^* u_r u_\ell^* \langle \ell, \phi_R|\hat{\gamma}_\ell^\dagger \hat{\gamma}_r|\phi_L, r\rangle + T_{r\ell} v_r v_\ell^* \langle \ell, \phi_R|\hat{\gamma}_r \hat{\gamma}_\ell^\dagger|\phi_L, r\rangle,$$
$$\Delta E = \varepsilon_r - \varepsilon_\ell + eV. \tag{29.10d}$$

Fig. 29.2 gives a representation of the four different tunneling processes corresponding to the various terms of Eqs. (29.10). In evaluating the terms that occur in $\langle \hat{H}_T^\dagger \hat{H}_T \rangle$ we use the rules $\langle \hat{\gamma}_n^\dagger \hat{\gamma}_{n'} \rangle = f_n \delta_{nn'}$, $\langle \hat{\gamma}_n \hat{\gamma}_{n'}^\dagger \rangle = (1 - f_n)\delta_{nn'}$ and $\langle \hat{\gamma}_n \hat{\gamma}_{n'} \rangle = \langle \hat{\gamma}_n^\dagger \hat{\gamma}_{n'}^\dagger \rangle = 0$. To simplify the remaining calculations we will assume both S and S′ have identical superconducting properties (i.e., they are prepared symmetrically from the same material). Furthermore, in the thermal equilibrium the coherence factors u and v drop out, as a consequence of the symmetry with regard to exchange of

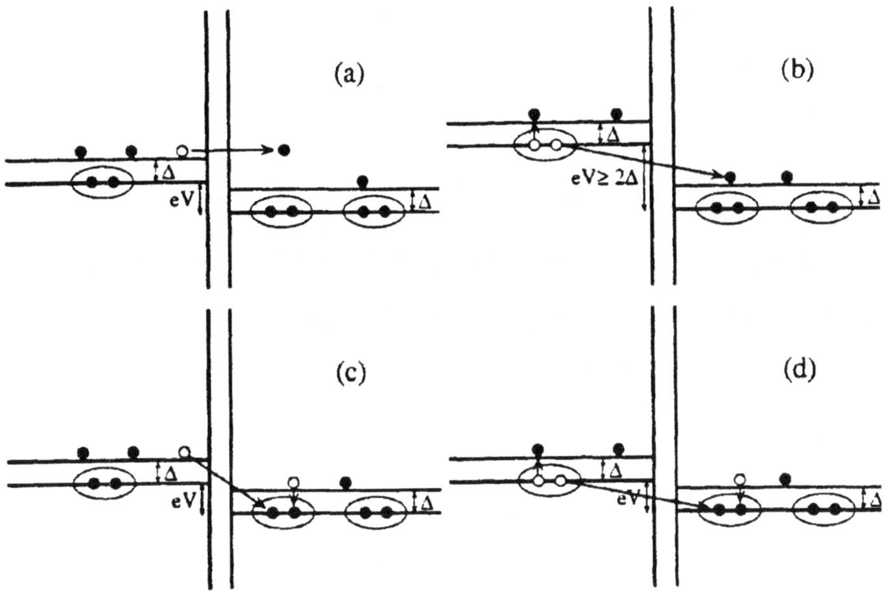

Figure 29.2 Schematic of the four different quasiparticle tunneling processes corresponding to the various terms of Eqs. (29.10): (a) simple quasiparticle tunneling; (b) Cooper pair breaking with tunneling; (c) energy violating Cooper pair formation; (d) Cooper pair mediated quasiparticle tunneling.

the electron-like branch and hole-like branch. The transition rates associated with the four processes (29.10a)–(29.10d) are then given by

$$w^{qp}_{L\to R} = 2\frac{2\pi}{\hbar}\sum_{\ell,r}|T_{\ell r}|^2 f_\ell(1-f_r)\delta(\varepsilon_\ell - \varepsilon_r - eV), \tag{29.11a}$$

$$w^{qp}_{R\to L} = 2\frac{2\pi}{\hbar}\sum_{\ell,r}|T_{\ell r}|^2(1-f_\ell)f_r\delta(\varepsilon_r - \varepsilon_\ell + eV), \tag{29.11b}$$

$$w^{pb} = 2\frac{2\pi}{\hbar}\sum_{\ell,r}|T_{\ell r}|^2(1-f_\ell)(1-f_r)\delta(-\varepsilon_r - \varepsilon_\ell + eV), \tag{29.11c}$$

and

$$w^{r} = 2\frac{2\pi}{\hbar}\sum_{\ell,r}|T_{\ell r}|^2 f_\ell f_r\delta(\varepsilon_\ell + \varepsilon_r - eV). \tag{29.11d}$$

The rates (29.11a) and (29.11b) describe quasiparticle tunneling from left to right (L → R) and right to left (R → L), respectively (with or without the assistance of a pair); (29.11c) describes pair breaking (pb) and (29.11d) describes quasiparticle recombination to form a pair (r).

The total quasiparticle current, I_{qp}, is e times the difference of (29.11a) and (29.11b)

$$I^{qp} = e[w^{qp}_{L\to R} - w^{qp}_{R\to L}]$$

or

$$I^{qp} = \frac{4\pi e}{\hbar}\sum_{\ell,r}|T_{\ell r}|^2(f_\ell - f_r)\delta(\varepsilon_\ell - \varepsilon_r - eV). \tag{29.12}$$

The total pair-breaking/recombination current I^p is e times the difference of (29.11c) and (29.11d)

$$I^p = \frac{4\pi e}{\hbar}\sum_{\ell,r}|T_{\ell r}|^2(1 - f_\ell - f_r)\delta(\varepsilon_\ell + \varepsilon_r - eV). \tag{29.13}$$

To convert the summations to integrations in (29.12) and (29.13) we need to define a quasiparticle density of states $\mathcal{N}_s(\varepsilon)$. Since the quasiparticle excitations are fermions created by the operators $\hat{\gamma}_{\mathbf{k}}^\dagger$, which are in one-to-one correspondence with the excitations created by the operators $\hat{c}_{\mathbf{k}}^\dagger$ of the normal metal, the superconducting density of states $\mathcal{N}_s(\varepsilon)$ may be obtained by equating $\mathcal{N}_s(\varepsilon)d\varepsilon = \mathcal{N}_n(\xi)d\xi$. Because we are largely interested in energies ξ only a few meV from the Fermi energy, we can take $\mathcal{N}_n(\xi) = \mathcal{N}(0)$, a constant. The quasiparticle density of states is then given by

$$\frac{\mathcal{N}_s(\varepsilon)}{\mathcal{N}(0)} = \frac{d\xi_{\mathbf{k}}}{d\varepsilon_{\mathbf{k}}} = \begin{cases} \dfrac{\varepsilon_{\mathbf{k}}}{(\varepsilon_{\mathbf{k}}^2 - \Delta^2)^{1/2}} & \varepsilon > \Delta, \\ 0, & \varepsilon < \Delta. \end{cases} \tag{29.14}$$

An important feature of the quasiparticle density of states, shown in Fig. 29.3, is the $1/(\varepsilon_{\mathbf{k}}^2 - \Delta^2)^{1/2}$ singularity at the gap edge.

Using (29.14) we may write

$$\sum_{\ell,r} = \iint \mathcal{N}_L(0)\mathcal{N}_R(0)\frac{d\xi_\ell}{d\varepsilon_\ell}\frac{d\xi_r}{d\varepsilon_r}d\varepsilon_r d\varepsilon_\ell. \tag{29.15}$$

Integrating over $d\varepsilon_r$, and assuming identical superconductors and an energy-independent matrix element, (29.12) and (29.13) become

$$I^{qp} = \frac{4\pi e}{\hbar}|T|^2[\mathcal{N}(0)]^2\int[f(\varepsilon - eV) - f(\varepsilon)]\frac{|\varepsilon - eV|}{[(\varepsilon - eV)^2 - \Delta^2]^{1/2}}\frac{|\varepsilon|}{(\varepsilon^2 - \Delta^2)^{1/2}}d\varepsilon \tag{29.16}$$

and

$$I^p = \frac{4\pi e}{\hbar}|T|^2[\mathcal{N}(0)]^2\int[1 - f(-\varepsilon + eV) - f(\varepsilon)]\frac{|eV - \varepsilon|}{[(eV - \varepsilon)^2 - \Delta^2]^{1/2}}\frac{|\varepsilon|}{(\varepsilon^2 - \Delta^2)^{1/2}}d\varepsilon. \tag{29.17}$$

Due to the behavior of the distribution functions we have no quasiparticle current at $T = 0$ for $eV < 2\Delta$. However, there is a sharp onset of current flow for $eV = 2\Delta$ and we obtain essentially normal-state behavior for $eV > 2\Delta$. Fig. 29.4 shows the experimental results for the voltage–current characteristics of an $Al/AlO_x/Al$ tunnel junction as a function of temperature. Note the curves show an identifiable increase in the slope, dI/dV, at all temperatures when $eV = 2\Delta(T)$, allowing one easily to estimate $\Delta(T)$; the quantitative temperature dependence of Δ follows from fitting the theoretical expressions.

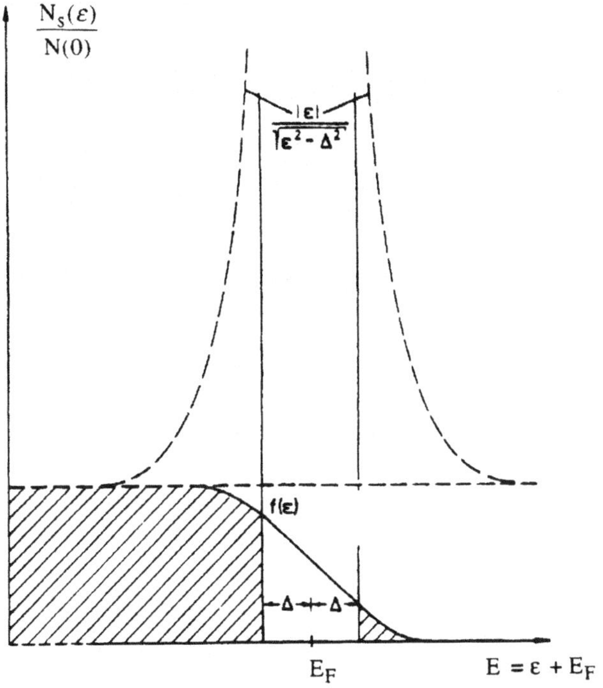

Figure 29.3 The normalized BCS quasiparticle density of states as a function of the excitation energy ε. The Fermi distribution function for a nonzero temperature is shown in the lower portion of the figure.

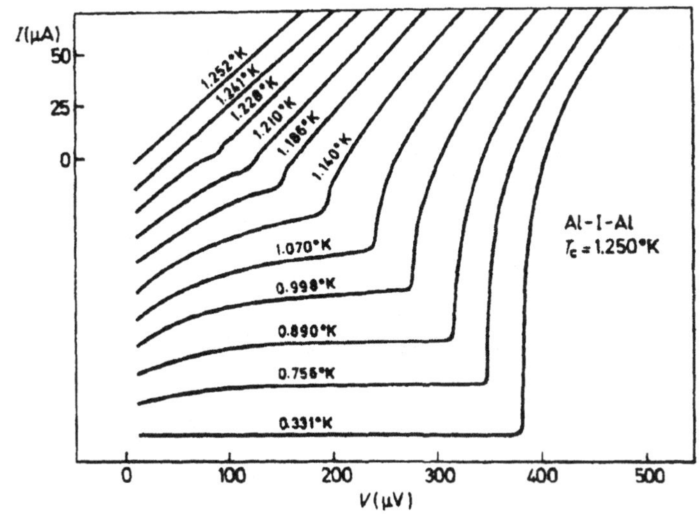

Figure 29.4 The I–V characteristics of an $Al/AlO_x/Al$ junction. Successive curves are displaced for clarity. (After Blackford and March (1968).)

30 Pair tunneling: the microscopic theory of the Josephson effects

In 1962 Josephson predicted that if the insulating barrier in an SIS junction is sufficiently thin (about 10–20 Å), Cooper pairs may tunnel across the junction. As a result, below some critical current the voltage across the junction is zero. This phenomenon is called the Josephson effect and the supercurrent arising from pair tunneling is called the Josephson supercurrent. Following the original work of Josephson (1962, 1964, 1965, 1969) we use a tunneling Hamiltonian formalism to model the basic aspects of the microscopic theory of Josephson tunneling and derive an expression for the temperature dependence of the Josephson supercurrent.

As discussed in the previous section, from a quantum mechanical point of view a tunnel junction is usually described by the following Hamiltonian (Cohen, Falicov, and Phillips 1962, Wolf 1985):

$$\hat{H} = \hat{H}_L + \hat{H}_R + \hat{H}_T = \hat{H}_0 + \hat{H}_T. \tag{30.1}$$

Here $\hat{H}_L$ and $\hat{H}_R$ are the complete Hamiltonians of the left- and right-hand superconductors, respectively, which commute with the particle number operators $\hat{N}_L$ and $\hat{N}_R$ given by

$$\hat{N}_L = \sum_{\ell\sigma} \hat{c}^\dagger_{\ell\sigma}\hat{c}_{\ell\sigma}, \qquad \hat{N}_R = \sum_{r\sigma} \hat{c}^\dagger_{r\sigma}\hat{c}_{r\sigma}. \tag{30.2}$$

$\hat{H}_T$ is the tunneling interaction term which transfers electrons from one superconductor to the other:

$$\hat{H}_T = \sum_{\ell,r,\sigma} (T_{\ell r}\hat{c}^\dagger_{\ell\sigma}\hat{c}_{r\sigma} + T^*_{\ell r}\hat{c}^\dagger_{r\sigma}\hat{c}_{\ell\sigma}), \tag{30.3}$$

where $\hat{c}^\dagger_{\ell\sigma}$ ($\hat{c}_{\ell\sigma}$) creates (destroys) one electron with translational quantum number ℓ and spin σ on the left-hand side; $\hat{c}^\dagger_{r\sigma}$ ($\hat{c}_{r\sigma}$) creates (destroys) one electron with translational quantum number r and spin σ on the right-hand side. Under the assumption of 'elastic' tunneling between states of equal energy the anticommutation relations $\{\hat{c}^\dagger_\ell, \hat{c}_r\} = \{\hat{c}_\ell, \hat{c}_r\} = 0$ may be assumed to lowest order in $\hat{H}_T$. Note that processes involving 'spin-flip' are excluded in the tunneling Hamiltonian (30.3).

The tunneling current $I(V, T)$ for $T \geq 0$ and $V \neq 0$ is obtained from the expectation value of rate change of the electron number operator $\hat{N}_R$:

$$I(V,T) = -e\langle \dot{\hat{N}}_R \rangle, \tag{30.4}$$

where $\dot{\hat{N}}_R$ is the time derivative of $\hat{N}_R$ and the current is assumed to flow from left to right. The expectation value is defined as

$$\langle \dot{\hat{N}}_R \rangle = \frac{\mathrm{Tr}\{e^{-\hat{H}/k_B T} \dot{\hat{N}}_R\}}{\mathrm{Tr}\{e^{-\hat{H}/k_B T}\}},$$

where $\hat{H}$ is the total Hamiltonian of the system, k_B is the Boltzman constant, and Tr{ } denotes the trace of the operator inside the brackets.

From the equation of motion for $\hat{N}_R$ we have

$$\dot{\hat{N}}_R = \frac{i}{\hbar}[\hat{H}, \hat{N}_R] = \frac{i}{\hbar}[\hat{H}_T, \hat{N}_R],$$

which yields by using Eq. (30.3) and the anticommutation relation for $\hat{c}_\ell, c_r$:

$$\dot{\hat{N}}_R = \frac{i}{\hbar}\sum_{\ell r\sigma}(T_{\ell r}\hat{c}^\dagger_{\ell\sigma}\hat{c}_{r\sigma} - T^*_{\ell r}\hat{c}^\dagger_{r\sigma}\hat{c}_{\ell\sigma}). \tag{30.5}$$

Substituting (30.5) in (30.4) we obtain for the tunneling current

$$I(V,T) = \frac{2e}{\hbar}\mathrm{Im}\left\{\sum_{\ell r\sigma} T_{\ell r}\langle \hat{c}^\dagger_{\ell\sigma}\hat{c}_{r\sigma}\rangle\right\}, \tag{30.6}$$

where we used the property

$$T^*_{\ell r}\langle \hat{c}^\dagger_{r\sigma}\hat{c}_{\ell\sigma}\rangle = \{T_{\ell r}\langle \hat{c}^\dagger_{\ell\sigma}\hat{c}_{r\sigma}\rangle\}^\dagger.$$

We use time-dependent perturbation theory to evaluate (30.6) which involves transitions between different states. We assume the unperturbed problem solved in the sense that the eigenfunctions $|n\rangle$ and the energy eigenvalues E_n defined by

$$\hat{H}_0|n\rangle = E_n|n\rangle \tag{30.7}$$

are known. Assuming the initial state is

$$\psi_0 = |k\rangle, \tag{30.8}$$

if at $t \geq 0$, the external perturbation $\hat{H}_T$ is turned on, the time evolution of the wavefunction obeys the Schrödinger equation

$$i\hbar\frac{\partial}{\partial t}\psi(t) = (\hat{H}_0 + \hat{H}_T)\psi(t). \tag{30.9}$$

As usual we expand $\psi(t)$ as

$$\psi(t) = \sum_n a_{nk}(t)e^{-iE_n t/\hbar}|n\rangle. \tag{30.10}$$

The time evolution of $a_{nk}(t)$ is due solely to $\hat{H}_T(t)$, and the probability of finding the system in a state $|n\rangle$ is given by $|a_{nk}(t)|^2$. We introduce a new representation, called the interaction representation, where the time evolution of the wavefunction is governed by

$$i\hbar\frac{\partial}{\partial t}\psi_I(t) = \hat{H}_T\psi_I(t), \tag{30.11}$$

which is a Schrödinger-like equation with the total $\hat{H}$ replaced by $\hat{H}_T$. We can show that for an observable $\hat{A}$

$$\frac{d\hat{A}_I}{dt} = \frac{1}{i\hbar}[\hat{A}_I, \hat{H}_0], \tag{30.12}$$

which is a Heisenberg-like equation with $\hat{H}$ replaced by $\hat{H}_0$. From (30.12) it is clear that

$$\hat{A}_I = e^{i\hat{H}_0 t/\hbar}\hat{A}_S e^{-i\hat{H}_0 t/\hbar}, \tag{30.13}$$

where the subscripts I and S stand for the interaction and Schrödinger representations, respectively. To first order in $\hat{H}_T$, the solution to Eq. (30.11) can be written as

$$|\psi_I(t)\rangle \approx \left(1 - \frac{i}{\hbar}\int_{-\infty}^{t} e^{\eta\tau}\hat{H}_{TI}(t)d\tau\right)|\psi_0\rangle, \tag{30.14}$$

where $\eta \to 0^+$; i.e., η tends to zero through positive values. Here $|\psi_0\rangle$ is an eigenfunction of the unperturbed Hamiltonian $\hat{H}_0$. The factor $e^{\eta\tau}$ implies that the perturbation is turned on adiabatically starting from $t = -\infty$. For simplicity, in the following calculation we omit the subscript I and understand that all wavefunctions and operators are taken in the interaction representation. Using (30.14), Eq. (30.6) becomes

$$I = \frac{2e}{\hbar}\text{Im}\sum_{\ell r\sigma} T_{\ell r}\left\{\langle\hat{c}^\dagger_{\ell\sigma}\hat{c}_{r\sigma}\rangle_0 - \frac{i}{\hbar}\int_{-\infty}^{t} d\tau e^{\eta\tau}\langle[\hat{c}^\dagger_{\ell\sigma}(t)\hat{c}_{r\sigma}(t),\hat{H}_T(\tau)]\rangle_0\right\}, \tag{30.15}$$

where the symbol $\langle\ \rangle_0$ denotes an expectation value with respect to the unperturbed Hamiltonian $\hat{H}_0$ and Im (Re) refers to the imaginary (real) part. The first term of this expression is zero since it represents the current when $\hat{H}_T$ is zero. Hence

$$I = -\frac{2e}{\hbar^2}\text{Re}\sum_{\ell r\sigma} T_{\ell r}\int_{-\infty}^{t} d\tau e^{\eta\tau}\langle[\hat{c}^\dagger_{\ell\sigma}(t)\hat{c}_{r\sigma}(t), \hat{H}_T(\tau)]\rangle_0. \tag{30.16}$$

Substituting (30.3) into (30.16) we obtain

$$\begin{aligned} I = &-\frac{2e}{\hbar^2}\text{Re}\sum_{\substack{\ell r\sigma \\ \ell' r'\sigma'}} T_{\ell r}\int_{-\infty}^{t} d\tau e^{\eta\tau}\{T_{\ell' r'}\langle[\hat{c}^\dagger_{\ell\sigma}(t)\hat{c}_{r\sigma}(t), \hat{c}^\dagger_{\ell'\sigma'}(\tau)\hat{c}_{r'\sigma'}(\tau)]\rangle_0 \\ &+ T^*_{\ell' r'}\langle[\hat{c}^\dagger_{\ell\sigma}(t)\hat{c}_{r\sigma}(t), \hat{c}^\dagger_{r'\sigma'}(\tau)\hat{c}_{\ell'\sigma'}(\tau)]\rangle_0\} \\ = &\int_{-\infty}^{t} d\tau e^{\eta\tau}[S(t-\tau) + R(t-\tau)]. \end{aligned} \tag{30.17}$$

Here we have defined

$$\begin{aligned} S(t-\tau) = &-\frac{2e}{\hbar^2}\text{Re}\sum_{\substack{\ell r\sigma \\ \ell' r'\sigma'}} T_{\ell r}T^*_{\ell' r'}[\langle\hat{c}^\dagger_{\ell\sigma}(t)\hat{c}_{\ell'\sigma'}(\tau)\rangle_0\langle\hat{c}_{r\sigma}(t)\hat{c}^\dagger_{r'\sigma'}(\tau)\rangle_0 \\ &- \langle\hat{c}_{\ell'\sigma'}(\tau)\hat{c}^\dagger_{\ell\sigma}(t)\rangle_0\langle\hat{c}^\dagger_{r'\sigma'}(\tau)\hat{c}_{r\sigma}(t)\rangle_0] \end{aligned} \tag{30.18a}$$

and

$$R(t-\tau) = \frac{2e}{\hbar^2} \mathrm{Re} \sum_{\substack{\ell r\sigma \\ \ell' r'\sigma'}} T_{\ell r} T_{\ell' r'} [\langle \hat{c}^\dagger_{\ell\sigma}(t) \hat{c}^\dagger_{\ell'\sigma'}(\tau) \rangle_0 \langle \hat{c}_{r\sigma}(t) \hat{c}_{r'\sigma'}(\tau) \rangle_0$$

$$- \langle \hat{c}^\dagger_{\ell'\sigma'}(\tau) \hat{c}^\dagger_{\ell\sigma}(t) \rangle_0 \langle \hat{c}_{r'\sigma'}(\tau) \hat{c}_{r\sigma}(t) \rangle_0]. \quad (30.18b)$$

In writing the above expressions we have used the anticommutation relations for $\hat{c}$s. Later we will see that $S(t-\tau)$ is related to quasiparticle tunneling and $R(t-\tau)$ describes processes in which phase coherent tunneling of Cooper pairs occurs.

We use the Bogoliubov–Valatin transformation described in Sec. 27 to calculate the tunneling current. For convenience, we rewrite Eqs. (27.12) and (27.16):

$$\hat{\gamma}_{-\mathbf{k}\downarrow} = u_{\mathbf{k}} \hat{c}_{-\mathbf{k}\downarrow} + v_{\mathbf{k}} \hat{c}^\dagger_{\mathbf{k}\uparrow} \quad (30.19a)$$

and

$$\hat{\gamma}^\dagger_{\mathbf{k}\uparrow} = u_{\mathbf{k}} \hat{c}^\dagger_{\mathbf{k}\uparrow} - v_{\mathbf{k}} \hat{c}_{-\mathbf{k}\downarrow}. \quad (30.19b)$$

From Eq. (27.13) (or (27.15)) we know that the operator $\gamma^\dagger_{\mathbf{k}\uparrow}$ (or $\gamma^\dagger_{-\mathbf{k}\downarrow}$) creates a quasiparticle excitation $|\mathbf{k}\uparrow\rangle$ (or $|-\mathbf{k}\downarrow\rangle$) from the BCS ground state. If the pair state ($\mathbf{k}\uparrow, -\mathbf{k}\downarrow$) with probability amplitude $u_{\mathbf{k}}$ is not occupied, then creation of the $\mathbf{k}\uparrow$ electron does not change the number of pairs in the condensate. However, if the pair state ($\mathbf{k}\uparrow, -\mathbf{k}\downarrow$) is occupied, generation of the quasiparticle excitation $|\mathbf{k}\uparrow\rangle$ proceeds by destroying the $|-\mathbf{k}\downarrow\rangle$ member of the $\mathbf{k}$ pair, thus reducing the number of pairs by one. To account for the pair tunneling, we must allow the pair number to change, assuming a large reservoir of pairs. In fact, this assumption is the central point of Josephson's theoretical discovery (Josephson 1962) that the pair current is of the same order in $|T|^2$ as the quasiparticle tunneling current. To allow the pair number to change and to conserve the charge, Josephson (1962) and, independently, Bardeen (1962) modified the Bogoliubov–Valatin operators to specify the creation (annihilation) of precisely one electron (hole). One can define a complete set of operators as

$$\hat{\gamma}^\dagger_{\mathrm{ek}\uparrow} = u_{\mathbf{k}} \hat{c}^\dagger_{\mathbf{k}\uparrow} - v_{\mathbf{k}} S^* \hat{c}_{-\mathbf{k}\downarrow}, \quad (30.20a)$$

$$\hat{\gamma}^\dagger_{\mathrm{hk}\uparrow} = u_{\mathbf{k}} S \hat{c}^\dagger_{\mathbf{k}\uparrow} - v_{\mathbf{k}} \hat{c}_{-\mathbf{k}\downarrow} = S \gamma^\dagger_{\mathrm{ek}\uparrow}, \quad (30.20b)$$

$$\hat{\gamma}^\dagger_{\mathrm{ek}\downarrow} = u_{\mathbf{k}} \hat{c}^\dagger_{-\mathbf{k}\downarrow} + v_{\mathbf{k}} S^* \hat{c}_{\mathbf{k}\uparrow}, \quad (30.20c)$$

$$\hat{\gamma}^\dagger_{\mathrm{hk}\downarrow} = u_{\mathbf{k}} S \hat{c}^\dagger_{-\mathbf{k}\downarrow} + v_{\mathbf{k}} \hat{c}_{\mathbf{k}\uparrow} = S \hat{\gamma}^\dagger_{\mathrm{ek}\downarrow}, \quad (30.20d)$$

where S^* adds a pair to the condensate and S destroys one. These operators create excitations while changing the charge by exactly $\mp e$, corresponding to the subscripts e and h for electron and hole, respectively. Accordingly, they are better adapted to describe the energy levels analogous to those of ordinary electrons. Note that Eqs. (30.20a) and (30.20b) differ only by one condensed pair from the reservoir at the chemical potential. The inverse operators are

$$\hat{c}^\dagger_{\mathbf{k}\uparrow} = u_{\mathbf{k}} \hat{\gamma}^\dagger_{\mathrm{ek}\uparrow} + v_{\mathbf{k}} \hat{\gamma}_{\mathrm{hk}\downarrow}, \quad (30.21a)$$

$$\hat{c}^\dagger_{-\mathbf{k}\uparrow} = u_{\mathbf{k}} \hat{\gamma}^\dagger_{\mathrm{ek}\downarrow} - v_{\mathbf{k}} \hat{\gamma}_{\mathrm{hk}\uparrow}. \quad (30.21b)$$

In this kind of modified superconductor representation, the total energy for a system with fermion excitations given by Eq. (30.20) can be written as

$$E = \sum_{\mathbf{k}} \varepsilon_{\mathbf{k}} \hat{\gamma}^{\dagger}_{\mathbf{k}} \gamma_{\mathbf{k}} + \mu N, \tag{30.22}$$

where the sum runs over all the excitations, which have $\varepsilon_{\mathbf{k}} \geq 0$. The electrochemical potential μ is introduced explicitly since we are dealing with processes which change the number N of electrons in a subsystem at a given μ by $\pm$ 1. The energy required to make an electron-like excitation is

$$\varepsilon_{\mathrm{ek}} = \varepsilon_{\mathbf{k}} + \mu, \tag{30.23a}$$

while that to make a hole-like excitation differs by 2μ and is

$$\varepsilon_{\mathrm{hk}} = -\mu + \varepsilon_{\mathbf{k}}. \tag{30.23b}$$

We first calculate the term involving $S(t - \tau)$ by substituting Eq. (30.21a) and (30.21b) into (30.17). The time dependence of the quasiparticle operators in the interaction representation may be obtained by using Eq. (30.12) and (30.22). Therefore

$$\hat{\gamma}_{\mathrm{e}\ell}(t) = \hat{\gamma}_{\mathrm{e}\ell}(0) \mathrm{e}^{-\mathrm{i}(\varepsilon_\ell + \mu_\ell)t/\hbar} \tag{30.24a}$$

and

$$\hat{\gamma}_{\mathrm{h}\ell}(t) = \hat{\gamma}_{\mathrm{h}\ell}(0) \mathrm{e}^{-\mathrm{i}(\varepsilon_\ell - \mu_\ell)t/\hbar}. \tag{30.24b}$$

For example, in the expansion of $S(t - \tau)$ in the quasiparticle operators, the first term involves

$$|u_\ell|^2 |u_r|^2 \mathrm{e}^{\mathrm{i}(\varepsilon_\ell - \varepsilon_r + \mu_\ell - \mu_r)(t-\tau)/\hbar} \langle \hat{\gamma}^{\dagger}_{\mathrm{e}\ell}(0) \hat{\gamma}_{\mathrm{e}\ell}(0) \rangle_0 \langle \hat{\gamma}_{\mathrm{e}r}(0) \hat{\gamma}^{\dagger}_{\mathrm{e}r}(0) \rangle_0.$$

Assuming the bias polarity is such that

$$\mu_\ell - \mu_r = eV, \tag{30.25}$$

in the calculation of the contribution of this term to the tunneling current via (30.17), the integral

$$\int_{-\infty}^{t} \mathrm{d}\tau \mathrm{e}^{\eta\tau} \mathrm{e}^{\mathrm{i}(\varepsilon_\ell - \varepsilon_r + eV)(t-\tau)/\hbar} |u_\ell|^2 |u_r|^2 \langle \hat{\gamma}^{\dagger}_{\mathrm{e}\ell}(0) \hat{\gamma}_{\mathrm{e}\ell}(0) \rangle_0 \langle \hat{\gamma}_{\mathrm{e}r}(0) \hat{\gamma}^{\dagger}_{\mathrm{e}r}(0) \rangle_0$$

occurs. Carrying out the integration with respect to τ yields

$$\frac{\mathrm{i}\hbar |u_\ell|^2 |u_r|^2 f_\ell (1 - f_r)}{\varepsilon_\ell - \varepsilon_r + eV + \mathrm{i}\hbar\eta},$$

which has a characteristic energy denominator $\Delta E = \varepsilon_\ell - \varepsilon_r + eV$, where the occupation numbers f are defined as

$$f_{\mathrm{e}\ell} = \langle \hat{\gamma}^{\dagger}_{\mathrm{e}\ell}(0) \hat{\gamma}_{\mathrm{e}\ell}(0) \rangle_0. \tag{30.26}$$

Altogether, there are four distinct denominators ΔE; they represent four different quasiparticle tunneling processes which conserve charge and energy.

The resulting quasiparticle tunneling current is given by

$$\begin{aligned} I_{\mathrm{qp}} &= \int_{-\infty}^{t} \mathrm{d}\tau \mathrm{e}^{\eta\tau} S(t - \tau) \\ &= -\frac{2e}{\hbar} \mathrm{Im} \sum_{\ell r} |T_{\ell r}|^2 \left[\frac{|u_\ell|^2 |u_r|^2 (f_\ell - f_r)}{\varepsilon_\ell - \varepsilon_r + eV + \mathrm{i}\hbar\eta} + \frac{|v_\ell|^2 |u_r|^2 (1 - f_\ell - f_r)}{-\varepsilon_\ell - \varepsilon_r + eV + \mathrm{i}\hbar\eta} \right. \end{aligned}$$

$$+\frac{|u_\ell|^2|v_r|^2(f_\ell+f_r-1)}{\varepsilon_\ell+\varepsilon_r+eV+i\hbar\eta}+\frac{|v_\ell|^2|v_r|^2(f_\ell-f_r)}{\varepsilon_\ell-\varepsilon_r+eV+i\hbar\eta}\Bigg]. \tag{30.27}$$

The pair tunneling current involving $R(t-\tau)$ may be calculated following the procedures outlined above. For example, in the calculation we will encounter a term containing

$$u_\ell v_\ell u_r v_r[\langle \hat\gamma_{h\ell\downarrow}(t)\hat\gamma^\dagger_{e\ell\downarrow}(\tau)\rangle_0\langle \hat\gamma^\dagger_{hr\downarrow}(t)\hat\gamma_{er\downarrow}(\tau)\rangle_0$$
$$-\langle \hat\gamma^\dagger_{e\ell\downarrow}(\tau)\hat\gamma_{h\ell\downarrow}(t)\rangle_0\langle \hat\gamma^\dagger_{er\downarrow}(\tau)\hat\gamma_{hr\downarrow}(t)\rangle_0].$$

By using Eqs. (30.20) and (30.24), the above expression can be written

$$u_\ell v_\ell u_r v_r S^*_\ell S_r e^{-i(\varepsilon_\ell-\varepsilon_r+eV)(t-\tau)/\hbar}[\langle \hat\gamma_{e\ell\downarrow}(0)\hat\gamma^\dagger_{e\ell\downarrow}(0)\rangle_0\langle \hat\gamma^\dagger_{er\downarrow}(0)\hat\gamma_{er\downarrow}(0)\rangle_0$$
$$-\langle \hat\gamma^\dagger_{e\ell\downarrow}(0)\hat\gamma_{e\ell\downarrow}(0)\rangle_0\langle \hat\gamma_{er\downarrow}(0)\hat\gamma^\dagger_{er\downarrow}(0)\rangle_0].$$

Thermal averaging brings in the factor f_r-f_ℓ. The particular significance of the $S^*_\ell S_r$ combination is that it depends on the phase difference, Φ, between the pairs of the left- and right-hand electrodes, hence we may write

$$S^*_\ell S_r = e^{-i\Phi}. \tag{30.28}$$

By collecting all terms which conserve charge, energy, and spin and completing the integration with respect to τ, we find the pair tunneling current I_J is given by

$$I_J = -\frac{e}{\hbar}\mathrm{Im}\left(\sum_{\ell_>r_>}\frac{4|T_{\ell r}|^2 u_\ell v_\ell u_r v_r e^{-i\Phi}(f_r-f_\ell)}{\varepsilon_r-\varepsilon_\ell-eV+i\hbar\eta}+\text{c.c.}\right). \tag{30.29}$$

Note that in Eq. (30.29), the summation over wave number is restricted to electron-like branches $\ell_>$ and $r_>$ only.

Eqs. (30.27) and (30.29) can be further simplified by changing the sum on wave number to integrals on energy and by using the relation

$$\frac{1}{\Delta E\pm i0}=P\left(\frac{1}{\Delta E}\right)\mp i\pi\delta(\Delta E), \tag{30.30}$$

where P indicates the principal part in taking the integral and $i\pi$ is the residue at the pole. In all terms in I_{qp}, the principal parts make no contribution. The sum in (30.27) has to be taken over the hole and electron branches on both the left- and right-hand electrodes of the junction, and noting that u and v are interchanged for the hole-like branch (and that $u^2+v^2=1$), we see that on summing over the two branches the coherence factors u and v drop out. In the four terms remaining, the first two combine if ε is allowed to take both positive and negative values, while the third and fourth cancel, leaving the simple result

$$I_{qp}=\frac{2\pi e|T|^2}{\hbar}\int_{-\infty}^{\infty}[f(\varepsilon+eV)-f(\varepsilon)]\mathcal{N}_R(\varepsilon+eV)\mathcal{N}_L(\varepsilon)d\varepsilon, \tag{30.31}$$

where $\mathcal{N}_L$ and $\mathcal{N}_R$ are quasiparticle densities of states for the left- and right-hand electrodes, respectively.

Due to the $e^{-i\Phi}$ factor, the pair tunneling current of (30.29) includes a principal part contribution as well as a contribution from the residue; we write these contributions as

$$I_J = I_1(V)\sin\Phi + I_2(V)\cos\Phi. \tag{30.32}$$

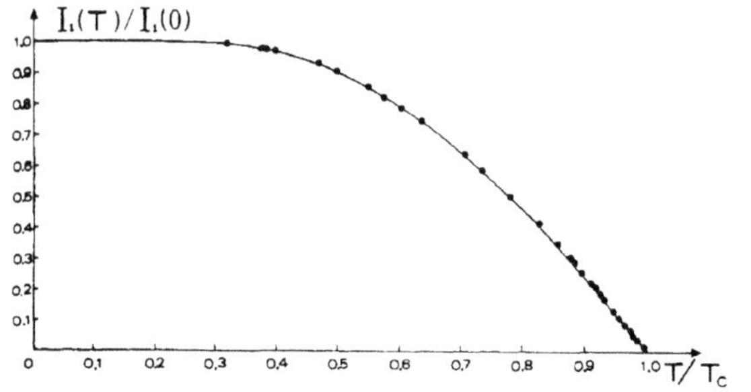

Figure 30.1 Temperature dependence of the maximum dc Josephson current for an Sn/SnO$_x$/Sn junction. The experimental data (solid circles) are compared with the theoretical curve (solid line) calculated using Eq. (30.36). (After Balsamo *et al.* (1974).)

Here

$$I_1(V) = \frac{2e|T|^2}{\hbar}\mathcal{N}_L(0)\mathcal{N}_R(0)P\int_{-\infty}^{\infty} d\xi_1 \int_{-\infty}^{\infty} d\xi_2 \frac{\Delta_1\Delta_2}{\varepsilon_1\varepsilon_2}\frac{f(\varepsilon_1)-f(\varepsilon_2)}{\varepsilon_1-\varepsilon_2+eV}, \tag{30.33}$$

arising from the principal parts, and

$$I_2(V) = \frac{2\pi e|T|^2}{\hbar}\mathcal{N}_L(0)\mathcal{N}_R(0)\int_{-\infty}^{\infty} d\xi \frac{\Delta_1\Delta_2}{(\varepsilon-eV)\varepsilon}[f(\varepsilon-eV)-f(\varepsilon)], \tag{30.34}$$

arising from the residues. The total tunneling current is thus

$$I = I_{qp} + I_J = I_{qp} + I_1(V)\sin\Phi + I_2(V)\cos\Phi. \tag{30.35}$$

The term $I_1\sin\Phi$ is usually regarded as the Josephson supercurrent. I_1 is finite at zero bias voltage. When $\Delta_1 = \Delta_2$, the temperature dependence of the Josephson supercurrent is given by (Ambegaokar and Baratoff 1963)

$$I_1(T) = \frac{\pi\Delta(T)}{2eR_N}\tanh\left(\frac{\Delta(T)}{2k_BT}\right), \tag{30.36}$$

where R_N is the normal state junction resistance. The temperature dependence of the maximum supercurrent is shown in Fig. 30.1. From the above expression we see that the maximum supercurrent at $T = 0$ corresponds to the current in the normal-state junction at an applied voltage of $\pi\Delta/2e$; i.e.,

$$I_1(0) = \frac{\pi\Delta}{2eR_N}. \tag{30.37}$$

This expression is very useful in practice. It allows a ready estimation of the expected maximum Josephson supercurrent just by looking at the I–V curve of the junction. If the quasiparticle tunneling current at the sum-gap voltage $2\Delta/e$ is I_N, the theoretically predicted maximum Josephson supercurrent is $I_1/I_N = \pi/4$ (see Fig. 30.2).

For an unsymmetric junction with $\Delta_1 \neq \Delta_2$, the maximum Josephson supercurrent at $T = 0$ is (Anderson 1964)

$$I_1(0) = \frac{2}{eR_N}\frac{\Delta_1\Delta_2}{\Delta_1+\Delta_2}K\left(\frac{|\Delta_1-\Delta_2|}{\Delta_1+\Delta_2}\right), \tag{30.38}$$

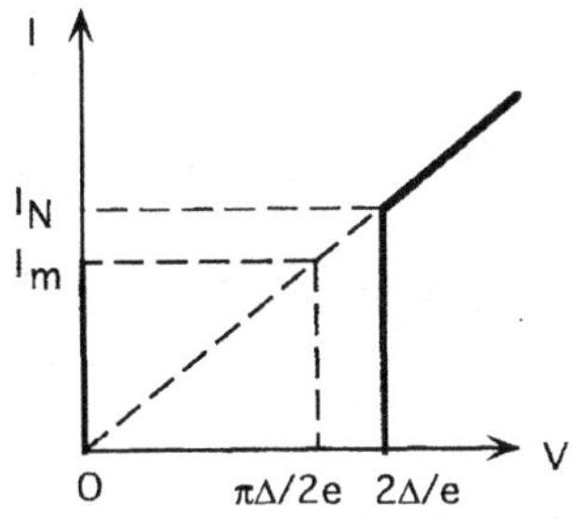

Figure 30.2 Schematic showing the ratio between the maximum Josephson current and the quasiparticle current at the sum-gap voltage; $I_m = I_1(0)$.

where $K(x)$ is the complete elliptic integral of the first kind. If $\Delta_1 \simeq \Delta_2$, using the asymptotic form $K(x) \simeq \pi/2$ for $x \to 0$, (30.38) becomes

$$I_1(0) \simeq \frac{\pi}{eR_N} \frac{\Delta_1 \Delta_2}{\Delta_1 + \Delta_2}, \tag{30.39}$$

in agreement with Eq. (30.37).

31 Simplified discussion of pairing mechanisms

31.1 The electron–phonon interaction

For the vast majority of conventional superconductors, the electron–phonon interaction is believed to be the source of the attractive component of the effective electron–electron interaction. (Other mechanisms have been proposed for the high T_c superconductors and for the heavy-fermion superconductors.) A rigorous discussion of the electron–phonon interaction is best accomplished using Green's function methods, which will not be discussed in this book. We will content ourselves with a more simple-minded approach.

The bare Coulomb interaction between electrons is

$$V(r) = \frac{e^2}{r} \tag{31.1}$$

and its Fourier transform is

$$V(q) = \frac{4\pi e^2}{q^2}. \tag{31.2}$$

The effect of electron–electron interactions may be incorporated by introducing a dielectric function $\varepsilon(\mathbf{q}, \omega)$. Although the bare Coulomb interaction is frequency-independent (reflecting the fact that it is instantaneous for most solid state time scales), a frequency dependence is introduced when account is taken of the background of positive ions and the electron gas which have response times of order ω_{ip} and ω_{ep}, respectively; here $\omega_{ip} = (4\pi n_i^2 z^2 e^2/M)^{1/2}$ is the ion-plasma frequency ($\sim 10^{13}\,\mathrm{s}^{-1}$), $\omega_{ep} = (4\pi n_e^2 e^2/m)^{1/2}$ is the electron-plasma frequency ($\sim 10^{16}\,\mathrm{s}^{-1}$), $z|e|$ is the effective nuclear charge, n_i and n_e are ion and electron densities and M and m are the ion and electron masses, respectively. Our *effective* electron–electron interaction will be written

$$V(\mathbf{q}, \omega) = \frac{4\pi e^2}{\varepsilon(\mathbf{q}, \omega) q^2}, \tag{31.3}$$

where $\varepsilon(\mathbf{q}, \omega)$ contains the effects due to both the ions and the background electrons.

We will use a self-consistent potential approach to calculating $\varepsilon(\mathbf{q}, \omega)$. Assume we have a plane wave electric field

$$\mathbf{E} = \mathbf{E}_0 \mathrm{e}^{-i\omega t + i\mathbf{q}\cdot\mathbf{r}} \tag{31.4}$$

propagating in our electron–ion system; accompanying this wave will be an electric displacement wave

$$\mathbf{D} = \mathbf{D}_0 e^{-i\omega t + i\mathbf{q}\cdot\mathbf{r}}. \tag{31.5}$$

The dielectric function is defined as

$$\varepsilon(\mathbf{q}, \omega) = |\mathbf{D}|/|\mathbf{E}|. \tag{31.6}$$

In order to allow us to consider a mode with arbitrary ω and $\mathbf{q}$ (as opposed to some natural mode of the system having a specific dispersion behavior $\omega = \omega(\mathbf{q})$) we imagine that an external charge density $\rho_{ext}(\omega, \mathbf{q})$ (not involving the metals' own electrons and ions) is introduced into the system.

From electrostatics

$$\nabla \cdot \mathbf{E} = 4\pi\rho_{total}, \tag{31.7}$$

where

$$\rho_{total} = \rho_e + \rho_i + \rho_{ext}, \tag{31.8}$$

where ρ_e and ρ_i are the electron-gas and background-ion charge densities. We regard ρ_e and ρ_i as 'bound' charge and ρ_{ext} as 'free' charge; thus

$$\nabla \cdot \mathbf{D} = 4\pi\rho_{ext} \tag{31.9}$$

and combining Eqs. (31.6)–(31.8) we obtain

$$\varepsilon = \frac{\rho_{ext}}{\rho_e + \rho_i + \rho_{ext}}. \tag{31.10}$$

The ions will be regarded as classical; i.e., their motion is governed by Newton's law

$$M\dot{\mathbf{V}}_i = z|e|\mathbf{E}, \tag{31.11}$$

where $\mathbf{V}_i$ is the ion's velocity and $z|e|$ is its effective charge. We rewrite (31.11) in terms of the ion current density $\mathbf{j} = n_i z|e|\mathbf{V}_i$

$$\frac{d\mathbf{j}_i}{dt} = \frac{n_i z^2 e^2}{M}\mathbf{E}. \tag{31.12}$$

The accompanying ionic charge density is calculated using the equation of continuity

$$\frac{\partial \rho_i}{\partial t} + \nabla \cdot \mathbf{j}_i = 0; \tag{31.13}$$

combining (31.12) and (31.13) yields

$$\frac{\partial^2 \rho_i}{\partial t^2} + \frac{n_i z^2 e^2}{M}\nabla \cdot \mathbf{E} = 0. \tag{31.14}$$

Using Eq. (31.7) we have

$$\frac{\partial^2 \rho_i}{\partial t^2} = -\omega_{ip}^2 \rho_{total} \tag{31.15}$$

or

$$\omega^2 \rho_i = \omega_{ip}^2 \rho_{total}. \tag{31.16}$$

We now turn to the response of the electron system. Since $\omega_{ep} >> \omega_{ip}$ the electrons will follow the ions adiabatically ($\mu/\hbar$ is also a characteristic frequency of the electron gas but it too greatly exceeds ω_{ip}). We will calculate the electron response in the Thomas–Fermi model in which one assumes that the local electron density is related to the local Fermi level through the free-Fermi-gas expression

$$\mu - |e|\phi = \frac{\hbar^2}{2m}(3\pi^2 n_e)^{2/3}, \tag{31.17}$$

where μ is the global chemical potential and $-|e|\phi$ is a shift due to the local electrostatic potential, ϕ. Inverting (31.17) we have

$$n_e = \frac{1}{3\pi^2}\left[\frac{2m}{\hbar^2}(\mu - |e|\phi)\right]^{3/2}. \tag{31.18}$$

Now the change in the electron density, δn_e, is much less than n_e and hence we can expand (31.18) as

$$n_e^{(0)} + \delta n_e \cong \frac{1}{3\pi^2}\left(\frac{2m\mu}{\hbar^2}\right)^{3/2}\left[1 - \frac{3}{2}\frac{|e|\phi}{\mu}\right];$$

thus

$$\delta n_e = -n_e^{(0)}\frac{3}{2}\frac{|e|\phi}{\mu},$$

where $n_e^{(0)} \equiv (1/3\pi^2)(2m\mu/\hbar^2)^{3/2}$. Noting that $\rho_e = -|e|\delta n_e$ we have

$$\rho_e = -\frac{3}{2}\frac{n_e^{(0)}e^2\phi}{\mu}. \tag{31.19}$$

From Poisson's equation

$$\nabla^2\phi = -4\pi\rho_{total}, \tag{31.20}$$

and the potential associated with our plane-wave distortion is therefore

$$\phi = \frac{4\pi\rho_{total}}{q^2}, \tag{31.21}$$

which on inserting into (31.19) gives

$$\rho_e = -\frac{6\pi n_e^{(0)}e^2}{\mu q^2}\rho_{total}$$

$$= -\frac{k_{TF}^2}{q^2}\rho_{total}, \tag{31.22}$$

where k_{TF} is the Thomas–Fermi wave vector (which is the reciprocal of the Thomas–Fermi screening length) given by $k_{TF} = (6\pi^2 n_e^{(0)} e^2/\mu)^{1/2}$. Writing $\rho_{ext} = \rho_{total} - \rho_i - \rho_e$ and using (31.16) and (31.22) we have

$$\varepsilon(\mathbf{q}, \omega) = \frac{\rho_{\text{total}} - \rho_{\text{i}} - \rho_{\text{e}}}{\rho_{\text{total}}}$$

$$= 1 - \frac{\omega_{\text{ip}}^2}{\omega^2} + \frac{k_{\text{TF}}^2}{q^2}$$

$$= \frac{\omega^2(q^2 + k_{\text{TF}}^2) - \omega_{\text{ip}}^2 q^2}{\omega^2 q^2} \tag{31.23}$$

or

$$\varepsilon(\mathbf{q}, \omega)^{-1} = \frac{\omega^2 q^2}{\omega^2(q^2 + k_{\text{TF}}^2) - \omega_{\text{ip}}^2 q^2}. \tag{31.24}$$

Inserting (31.23) into (31.3) we obtain an electron–electron interaction of the form

$$V(\mathbf{q}, \omega) = \frac{4\pi e^2}{q^2 + k_{\text{TF}}^2 - \frac{\omega_{\text{ip}}^2}{\omega^2} q^2}. \tag{31.25}$$

For $\omega >> \omega_{\text{ip}}$

$$V(\mathbf{q}, \omega) = \frac{4\pi e^2}{q^2 + k_{\text{TF}}^2} \tag{31.26}$$

and on Fourier transforming to $\mathbf{r}$ space we have

$$V(r) = \frac{e^2}{r} \mathrm{e}^{-k_{\text{TF}} r}; \tag{31.27}$$

we see that for large r, $V(r)$ falls off exponentially, an effect referred to as screening. Thus in the high frequency limit, $\omega >> \omega_{\text{ip}}$, the electron–electron interaction is repulsive. We next consider frequencies $\omega << \omega_{\text{ip}}$ but for wave vectors of order a^{-1} where a is an interatomic spacing (or equivalently wave vectors of order the Brillouin zone dimension). If $\omega_{\text{ip}}^2/\omega^2 > (q^2 + k_{\text{TF}}^2)/q^2$ we see that (31.25) is *attractive* and this behavior explicitly involves the ions through ω_{ip} which we may call an electron–phonon (electron–ion) interaction.

Let us now examine the low frequency collective modes of our electron–ion system. The collective modes follow from the equation[1]

$$\varepsilon(\mathbf{q}, \omega) = 0. \tag{31.28}$$

From Eq. (31.23) we obtain

$$\omega_q^2 = \frac{\omega_{\text{ip}}^2 q^2}{q^2 + k_{\text{TF}}^2}. \tag{31.29}$$

The behavior of this expression is shown in Fig. 31.1.

For $q << k_{\text{TF}}$ (which is reasonably well satisfied throughout the Brillouin zone since

1. We can then sustain an internal electric field while $\mathbf{D} = 0$. Since the normal component of $\mathbf{D}$ must be continuous across a boundary, a longitudinal mode can then propagate perpendicular to the boundary and produce no external field.

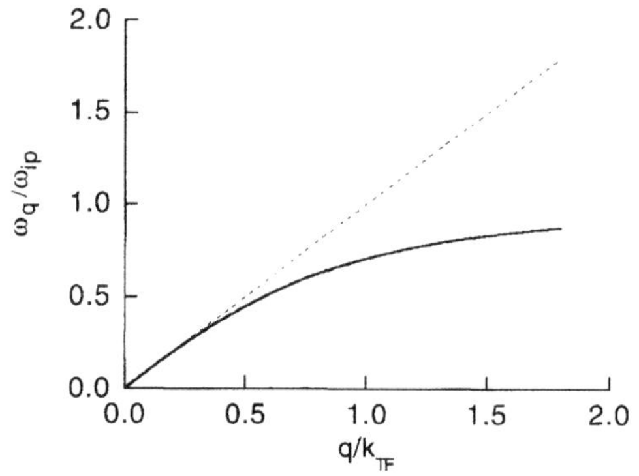

Figure 31.1 The solid line is calculated from Eq. (31.29); the dashed line shows the slope at $q = 0$.

$k_{TF} > k_{BZ} \sim \pi/a_0$, where a_0 is the lattice constant) we obtain the linear relation

$$\omega_q = \frac{\omega_{ip}}{k_{TF}} q. \tag{31.30}$$

A linear ω vs q relation is expected for the (longitudinal) phonons at small q ($<< k_{BZ}$). One can readily calculate that $\omega_{ip}/k_{TF} \sim 10^5$ cm/s, which is typical of the sound velocity in solids. We can rewrite our reciprocal dielectric function (31.24) in terms of ω_q as

$$\varepsilon^{-1}(\mathbf{q}, \omega) = \frac{q^2}{q^2 + k_{TF}^2}\left[1 + \frac{\omega_q^2}{\omega^2 - \omega_q^2}\right]; \tag{31.31}$$

for $\omega >> \omega_q$ we recover the Thomas Fermi dielectric function; for ω near, but smaller than, ω_q the interaction is attractive. Our potential may then be rewritten as

$$\begin{aligned} V(\mathbf{q}, \omega) &= \frac{4\pi e^2}{q^2 + k_{TF}^2} + \frac{4\pi e^2}{q^2 + k_{TF}^2} \frac{\omega_q^2}{\omega^2 - \omega_q^2} \\ &= V^{\text{coulomb}} + V^{(\text{el}-\text{ph})}. \end{aligned} \tag{31.32}$$

31.2 The spin fluctuation mechanism

In order to get pairing we must have an attractive interaction at low energies. For the electron–phonon pairing mechanism just discussed, this comes about through a combination of a local lattice polarization and retardation (in that the local polarization lives on for a time after the initial polarizing electron has departed). The electrons and phonons represent different degrees of freedom; i.e., we have a 'two-component' system. Now the particles of a one-component system (e.g., the electron liquid in a metal, the atoms in liquid ^{3}He, and the nucleons in a nucleus or neutron star) may locally polarize a region of the surrounding media, and this polarization may live on after this particle leaves that region; i.e., under appropriate conditions we may have the two key parts of the electron–phonon coupling mechanism, but in a one-component system. The question then arises as to whether pairing in a one-component system occurs in Nature. The experimental answer is yes: superfluid ^{3}He, consisting only of condensed ^{3}He atoms.

Superfluid ^{3}He is an example of pairing in a spin triplet ($s = 1$) state. Now the pair wavefunctions must be antisymmetric on exchanging the particles and the three $s = 1$ spin wavefunctions are symmetric. Therefore the orbital wavefunction, which we might write as a sum over products of a radial wavefunction and a spherical harmonic, must be antisymmetric. But spherical harmonics have a parity $(-1)^{\ell}$ (parity being identical to exchange symmetry in this case) and therefore only odd ℓ spatial wavefunctions are consistent with triplet pairing. To a high degree of accuracy ^{3}He is in an $s = 1, \ell = 1$ state. Therefore we are led to discuss spin- (and later ℓ-) dependent particle–particle (^{3}He–^{3}He or electron–electron) effective interaction potentials.

For an isotropic, translationally invariant system (like ^{3}He) the most general static, momentum-conserving, two-body potential acting between incoming plane-wave states $\mathbf{k}$, β and $\mathbf{k}'$, δ and outgoing plane-wave states $\mathbf{k}+\mathbf{q}$, α and $\mathbf{k}'-\mathbf{q}$, γ is

$$\hat{V}(\mathbf{q}) = \frac{1}{2}[\Gamma_{\rho}(\mathbf{q})\delta_{\alpha\beta}\delta_{\gamma\delta} + \Gamma_{s}(\mathbf{q})\boldsymbol{\sigma}_{\alpha\beta}\cdot\boldsymbol{\sigma}_{\gamma\delta}]; \tag{31.33}$$

here $\mathbf{k}$ and $\mathbf{k}'$ denote particle wave vectors, $\mathbf{q}$ is the momentum exchange (transfer) and $\alpha, \beta, \gamma, \delta$ denote spin states ($\downarrow$ or $\uparrow$).[2] We now examine the effect of (31.33) on pair states $|\mathbf{k}', -\mathbf{k}'; s, s_z\rangle$; s and s_z denote the total and the z axis projection of the pair spin. For triplet spin eigen states, $|\chi_{1s_z}\rangle$, we have the usual forms

$$\begin{aligned} |\chi_{11}\rangle &= \begin{pmatrix}1\\0\end{pmatrix}\begin{pmatrix}1\\0\end{pmatrix}; \quad |\chi_{1-1}\rangle = \begin{pmatrix}0\\1\end{pmatrix}\begin{pmatrix}0\\1\end{pmatrix}; \\ |\chi_{10}\rangle &= \frac{1}{\sqrt{2}}\left[\begin{pmatrix}1\\0\end{pmatrix}\begin{pmatrix}0\\1\end{pmatrix} + \begin{pmatrix}0\\1\end{pmatrix}\begin{pmatrix}1\\0\end{pmatrix}\right] \end{aligned} \tag{31.34}$$

while for the singlet state we have

$$|\chi_{00}\rangle = \frac{1}{\sqrt{2}}\left[\begin{pmatrix}1\\0\end{pmatrix}\begin{pmatrix}0\\1\end{pmatrix} - \begin{pmatrix}0\\1\end{pmatrix}\begin{pmatrix}1\\0\end{pmatrix}\right]. \tag{31.35}$$

For the symmetric ($+$ or $\ell = $ even) and antisymmetric ($-$ or $\ell = $ odd) orbital states we have

$$|\mathbf{k}, -\mathbf{k}\rangle_{\pm} = [e^{i\mathbf{k}\cdot(\mathbf{r}-\mathbf{r}')} \pm e^{-i\mathbf{k}\cdot(\mathbf{r}-\mathbf{r}')}]. \tag{31.36}$$

The full pair wavefunction can now be written as

$$|\mathbf{k}, -\mathbf{k}; s, s_z\rangle = \begin{cases} |\mathbf{k}, -\mathbf{k}\rangle_{+}|\chi_{00}\rangle & \text{(singlet)} \\ |\mathbf{k}, -\mathbf{k}\rangle_{-}|\chi_{1z}\rangle & \text{(triplet).} \end{cases} \tag{31.37}$$

By writing out the form $\boldsymbol{\sigma}\cdot\boldsymbol{\sigma}'$ in terms of Pauli matrices (see Part III Eq. (52.2)) it is easy to verify that the spin-dependent part of (31.33) has the properties

$$\Gamma_s\boldsymbol{\sigma}\cdot\boldsymbol{\sigma}'|\chi_{00}\rangle = -3\Gamma_s|\chi_{00}\rangle \tag{31.38a}$$

and

$$\Gamma_s\boldsymbol{\sigma}\cdot\boldsymbol{\sigma}'|\chi_{10}\rangle = +\Gamma_s|\chi_{10}\rangle, \tag{31.38b}$$

2. The indices ρ and s denote the density- and spin-dependent parts of our effective many-body interaction. The notation Γ formally denotes the many-body vertex function.

the result for $\chi_{1\pm1}$ being the same as (31.37). We may now immediately write the spin triplet, 1, and spin singlet, 0, pair matrix elements of (31.33) as

$$V_1(\mathbf{k},\mathbf{k}') = \frac{1}{2}[(\Gamma_\rho(\mathbf{k}-\mathbf{k}') + \Gamma_s(\mathbf{k}-\mathbf{k}')) - (\mathbf{k}\to-\mathbf{k})] \tag{31.39}$$

and

$$V_0(\mathbf{k},\mathbf{k}') = \frac{1}{2}[(\Gamma_\rho(\mathbf{k}-\mathbf{k}') - 3\Gamma_s(\mathbf{k}-\mathbf{k}')) + (\mathbf{k}\to-\mathbf{k})], \tag{31.40}$$

where the last term in these expressions follows from (31.36). A very important property follows from (31.39) and (31.40); interpreting Γ_s as an exchange interaction, ferromagnetic coupling ($\Gamma_s < 0$) favors $s = 1$ pairing while antiferromagnetic coupling ($\Gamma_s > 0$) favors $s = 0$ pairing. Since ^{3}He has ferromagnetic coupling ($F_0^a < 0$) we immediately have a mechanism for spin triplet pairing. One can go on to expand Γ_ρ and Γ_s in Legendre polynomials. For Γ_s we would have

$$\Gamma_s(\mathbf{q}) = \sum_\ell \Gamma_{s\ell} P_\ell(\cos\chi) = \Gamma_{s0} + \Gamma_{s1}\cos\theta + \ldots; \tag{31.41}$$

here $\mathbf{q}$ is a vector connecting two points $\mathbf{k}$ and $\mathbf{k}'$ lying near k_F and χ measures the angle between them. Therefore a large positive Γ_{s2} could result in spin-driven, d-wave pairing (a symmetry proposed by Pines and coworkers for the high T_c oxides and by Joynt and coworkers for UPt_3).

It is interesting to examine how a spin-dependent effective interaction can arise in a material like ^{3}He given that the bare potential (repulsive hard core at short distances and attractive van der Waals at large distances) is essentially spin-independent. As a model of the repulsive bare potential in ^{3}He we adopt a repulsive contact form

$$V(\mathbf{r}-\mathbf{r}') = I\delta^{(3)}(\mathbf{r}-\mathbf{r}'),$$

the Fourier transform of which is simply the $\mathbf{q}$-independent constant I. The second-quantized Hamiltonian would then be (see Eq. (26.15))

$$\hat{H}' = \sum_{\mathbf{k},\sigma} \xi_{\mathbf{k}} \hat{c}^\dagger_{\mathbf{k}\sigma}\hat{c}_{\mathbf{k}\sigma} + \frac{I}{2}\sum_{\substack{\mathbf{k},\mathbf{k}',\mathbf{q}\\ \sigma,\sigma'}} \hat{c}^\dagger_{\mathbf{k}-\mathbf{q}\sigma}\hat{c}^\dagger_{\mathbf{k}'+\mathbf{q}\sigma'}\hat{c}_{\mathbf{k}'\sigma'}\hat{c}_{\mathbf{k}\sigma}$$

$$- \mu_3 \mathbf{H}\cdot\sum_{\mathbf{k},\sigma,\sigma'} \boldsymbol{\sigma}_{\sigma\sigma'}\hat{c}^\dagger_{\mathbf{k}\sigma'}\hat{c}_{\mathbf{k}\sigma}, \tag{31.42}$$

where we have included the interaction with an external magnetic field, $\mathbf{H}$. We seek a mean field description where we pair a creation operator with a destruction operator and take an ensemble average: $n_{\mathbf{k}\sigma\sigma'} = \langle \hat{c}^\dagger_{\mathbf{k}\sigma}\hat{c}_{\mathbf{k}\sigma'}\rangle$. There are two such pairings: $\mathbf{q} = 0$ (the direct Hartree term) and $\mathbf{k} = \mathbf{k}' + \mathbf{q}$ (the exchange or Fock term). Choosing the magnetic field and the quantization axis parallel to $\hat{\mathbf{z}}$, $\underline{n}_{\mathbf{k}}$ will be diagonal (in equilibrium); i.e.,

$$\underline{n}_{\mathbf{k}} = \begin{pmatrix} n_{\mathbf{k}\uparrow} & 0 \\ 0 & n_{\mathbf{k}\downarrow}\end{pmatrix}. \tag{31.43}$$

The expectation value of $\hat{H}'$ is then

$$\langle \hat{H}' \rangle = \sum_{\mathbf{k},\sigma} \xi_{\mathbf{k}} n_{\mathbf{k}\sigma} + \frac{I}{2} \sum_{\substack{\mathbf{k},\mathbf{k}' \\ \sigma,\sigma'}} (1 - \delta_{\sigma\sigma'}) n_{\mathbf{k}'\sigma'} n_{\mathbf{k}\sigma}$$

$$- \mu_3 H \sum_{\mathbf{k},\sigma\sigma'} (\sigma_z)_{\sigma\sigma'} n_{\mathbf{k}\sigma'}. \tag{31.44}$$

If we identify the total energy density E' as $\langle \hat{H}' \rangle$, then using the Landau Fermi liquid definition of the quasiparticle energies we have

$$\frac{\delta E'}{\delta n_{\mathbf{k}\sigma}} = \varepsilon_{\mathbf{k}\sigma} = \xi_{\mathbf{k}} + I \sum_{\mathbf{k}'\sigma'} (1 - \delta_{\sigma\sigma'}) n_{\mathbf{k}'\sigma'} \mp \mu_3 H, \tag{31.45}$$

where the $-$ or $+$ in the last term is associated with $\uparrow$ or $\downarrow$ spin states and μ_3 is the magnetic moment of a ^{3}He atom. We will ignore the direct term at this point since it simply introduces a constant shift to the quasiparticle energies which may be absorbed into the definition of $\xi_{\mathbf{k}}$. (Later in this section, when discussing the effect of I on pairing, we will refer to this Hartree term.) The sum over $\mathbf{k}'$ of $n_{\mathbf{k}'\sigma'}$ is simply the total number density, $n_{\sigma'}$, associated with spin σ' ($\uparrow$ or $\downarrow$) (which we denote by dropping the subscript $\mathbf{k}$ in (31.43)). Our quasiparticle energies then become

$$\varepsilon_{\mathbf{k}\uparrow} = \xi_{\mathbf{k}} + I n_{\downarrow} - \mu_3 H \tag{31.46a}$$

and

$$\varepsilon_{\mathbf{k}\downarrow} = \xi_{\mathbf{k}} + I n_{\uparrow} + \mu_3 H. \tag{31.46b}$$

Let us define an effective magnetic moment by

$$\mu^{\text{eff}} H = -\frac{1}{2}(\varepsilon_{\mathbf{k}\uparrow} - \varepsilon_{\mathbf{k}\downarrow}).$$

Expanding the number densities as

$$n_{\uparrow}(H) = n(0) + \mathcal{N}_0 \mu^{\text{eff}} H$$

and

$$n_{\downarrow}(H) = n(0) - \mathcal{N}_0 \mu^{\text{eff}} H,$$

where $\mathcal{N}_0$ is the density of states, we obtain (Stoner 1938)

$$\mu^{\text{eff}} = \frac{\mu_3}{1 - \bar{I}}, \tag{31.47}$$

where $\bar{I} \equiv \mathcal{N}_0 I$. Hence a repulsive interaction ($I > 0$) favors parallel spin alignment. The condition $\bar{I} = 1$ can be regarded as the ferromagnetic instability threshold. For $\bar{I} > 1$ the liquid would have a net spin polarization and a straightforward extension of the above discussion allows a calculation of the spontaneous moment as well as the onset temperature (Curie temperature) of this ferromagnetic transition (Wohlforth 1953). In the present model $-\bar{I}$ is equivalent to the parameter F_0^{a} of the Landau Fermi liquid theory.

The above discussion was limited to the static response of the liquid. However, it is relatively easy to generalize our model to obtain the 'screened' low frequency dynamical response relevant to the pairing interaction using some hand-waving arguments. We consider three frequency (ω), wave vector ($\mathbf{q}$), and momentum ($\mathbf{p}$) dependent interaction-energy variations (fluctuations): $\delta e^{\text{tot}}(\omega, \mathbf{q}, \mathbf{p})$, $\delta e^{\text{ext}}(\omega, \mathbf{q}, \mathbf{p})$ and $\delta e^{\text{ind}}(\omega, \mathbf{q}, \mathbf{p})$ corresponding to the total, external, and induced contributions respectively; they are related by $\delta e^{\text{tot}}(\omega, \mathbf{q}, \mathbf{p}) = \delta e^{\text{ext}}(\omega, \mathbf{q}, \mathbf{p}) + \delta e^{\text{ind}}(\omega, \mathbf{q}, \mathbf{p})$. From Landau Fermi liquid theory we have

$$\delta e^{\text{ind}}(\omega, \mathbf{q}, \mathbf{p}) = \int \mathrm{d}\tau' f^{(a)}(\mathbf{p}, \mathbf{p}')\delta m(\omega, \mathbf{q}, \mathbf{p}'), \tag{31.48}$$

where δm and δe^{ind} denote the spin density and the associated induced spin energy fluctuations, respectively, and $\mathrm{d}\tau' = [2/(2\pi\hbar)^3]\mathrm{d}^3p$. From the Landau transport equation:

$$(\omega - \mathbf{q}\cdot\mathbf{v})\delta m(\omega, \mathbf{q}, \mathbf{p}) - \frac{\partial n_0}{\partial \varepsilon}\mathbf{q}\cdot\mathbf{v}[\delta e^{\text{ext}}(\omega, \mathbf{q}, \mathbf{p}) + \delta e^{\text{ind}}(\omega, \mathbf{q}, \mathbf{p})] = 0, \tag{31.49}$$

we have

$$\delta m(\omega, \mathbf{q}, \mathbf{p}) = \frac{\mathbf{q}\cdot\mathbf{v}}{\omega - \mathbf{q}\cdot\mathbf{v}}\frac{\partial n_0}{\partial \varepsilon}\delta e^{\text{tot}}(\omega, \mathbf{q}, \mathbf{p}),$$

which on substitution into (31.48) gives

$$\begin{aligned}\delta e^{\text{ind}}(\omega, \mathbf{q}, \mathbf{p}) &= \int \mathrm{d}\tau' f^{(a)}(\mathbf{p}, \mathbf{p}')\frac{\mathbf{q}\cdot\mathbf{v}'}{\omega - \mathbf{q}\cdot\mathbf{v}'}\frac{\partial n_0}{\partial \varepsilon'}\delta e^{\text{tot}}(\omega, \mathbf{q}, \mathbf{p}') \\ &= \delta e^{\text{tot}}(\omega, \mathbf{q}, \mathbf{p}) - \delta e^{\text{ext}}(\omega, \mathbf{q}, \mathbf{p}).\end{aligned} \tag{31.50}$$

If we interpret $-I$ as the 'external' or bare interaction, write $F^{a}(\hat{\mathbf{p}}\cdot\hat{\mathbf{p}}') = -\bar{I}$ and $\delta e^{\text{tot}}(\omega, \mathbf{q}, \mathbf{p})$ as a screened interaction $\Gamma_s(\omega, \mathbf{q}, \mathbf{p})$, then (31.50) yields

$$\Gamma_s(\omega, \mathbf{q}, \mathbf{p}) = -\frac{I}{1 - \bar{I}\chi(\omega, \mathbf{q})/\mathcal{N}_{\text{F}}}; \tag{31.51}$$

here

$$\chi(\omega, \mathbf{q}) \equiv -\frac{2}{(2\pi\hbar)^3}\int \mathrm{d}^3p\frac{\mathbf{q}\cdot\mathbf{v}}{\omega - \mathbf{q}\cdot\mathbf{v}}\frac{\partial n_0}{\partial \varepsilon} \tag{31.52}$$

(which is the long wavelength limit of the noninteracting Fermi gas susceptibility[3]). If we repeat the calculations retaining the Hartree term, which was discarded above but which leads to $F_0^s = +\bar{I}$, then we obtain

$$\Gamma_\rho(\omega, \mathbf{q}, \mathbf{p}) = \frac{I}{1 + \bar{I}\chi(\omega, \mathbf{q})/\mathcal{N}_{\text{F}}}. \tag{31.53}$$

Since $I > 0$ in ^{3}He, pairing is produced by Γ_s in the spin triplet 'channel' in this model.

We now turn to an evaluation of $\chi(\omega, \mathbf{q})$ which we rewrite as

$$\chi(\omega, \mathbf{q}) = \frac{1}{2}\mathcal{N}_{\text{F}}\int_0^\pi \frac{\cos\theta \mathrm{d}(\cos(\theta)}{s - \cos\theta + \mathrm{i}0_+}, \tag{31.54}$$

where the infinitesimal in the denominator causes the oscillating quantities to vanish as $t \to -\infty$

3. For arbitrary $\mathbf{q}$ we substitute $n_0(\mathbf{p} + \hbar\mathbf{q}) - n_0(\mathbf{p})$ for $\hbar\mathbf{q}\cdot\mathbf{v}(\partial n_0/\partial\varepsilon)$ in the numerator and $\varepsilon(\mathbf{p} + \hbar\mathbf{q}) - \varepsilon(\mathbf{p})$ for $\hbar\mathbf{q}\cdot\mathbf{v}$ in the denominator of Eq. (31.52).

(necessary for causality) and we defined $s \equiv \omega/qv$. This integral was encountered earlier in connection with longitudinal zero sound in ^{3}He (Eq. (24.45)). In that case we had $s > 1$ and no singularity was encountered in the interval of integration. However, we will be interested in the low frequency response ($s << 1$), where a singularity does occur. We split the integration into two parts

$$\chi(\omega, \mathbf{q}) = -\frac{\mathcal{N}_F}{2}\int_{-1}^{+1}\frac{x\mathrm{d}x}{s - x + \mathrm{i}0_+}$$
$$= -\lim\frac{\mathcal{N}_F}{2}\left(\int_{-1}^{s-\varepsilon}\frac{x\mathrm{d}x}{s - x + \mathrm{i}0_+} + \int_{s+\varepsilon}^{1}\frac{x\mathrm{d}x}{s - x + \mathrm{i}0_+}\right)$$
$$= \frac{\mathcal{N}_F}{2}[2 - s\ell\mathrm{n}(1 + s) + s\ell\mathrm{n}(1 - s) + \mathrm{i}\pi s]$$
$$= \mathcal{N}_F\left[1 + \frac{\mathrm{i}\pi}{2}s + \ldots\right] \tag{31.55}$$

(in the above the imaginary term comes from the usual integration around the semicircle in the complex plane associated with the pole and the sign of x was changed for the first integral in the second step). Inserting (31.55) into (31.51) yields the long wavelength form of Γ_s (which enters the triplet matrix element (31.39)) as

$$\Gamma_s(s) = \frac{-I}{1 - \bar{I}\left(1 + \frac{\mathrm{i}\pi}{2}s\right)}$$
$$= -I\frac{(1 - \bar{I}) + \mathrm{i}\frac{\pi}{2}\bar{I}s}{(1 - \bar{I})^2 + \frac{\pi^2}{4}\bar{I}^2 s^2}. \tag{31.56}$$

The real part of this expression is largest at $s = 0$: $\Gamma_s^R(0) = -I/(1 - \bar{I})$. The imaginary part on the other hand is largest at $s_{max} = (1 - \bar{I})/\bar{I}$: $\Gamma_s^I(s_{max}) = -I/(1 - \bar{I})$. This peak at low frequencies in $\Gamma_s^I(s)$ can be physically interpreted as due to long-lived (low frequency) spin fluctuations (Berk and Schrieffer 1966, Doniach and Engelsberg 1966). It is apparent then that the model of a repulsive contact interaction yields the two properties required for superfluidity: attraction and retardation.

32 The effect of Coulomb repulsion on T_c

In this section we develop an extension of the Cooper model potential which qualitatively accounts for the effects of Coulomb repulsion. If we assume the **k** dependence of Δ and V may be approximated by an energy dependence only, then we may rewrite the BCS gap equation as

$$\Delta(\xi) = -\int V(\xi - \xi')\frac{\Delta(\xi')\mathcal{N}(\xi')}{2\varepsilon'}[1 - 2f(\varepsilon')]d\xi'. \tag{32.1}$$

We next introduce the Bogoliubov model potential shown in Fig. 32.1 which we may write as

$$V(\xi - \xi') = \begin{cases} -V_p + V_c, & |\xi - \xi'| < \hbar\omega_D; \\ +V_c, & |\xi - \xi'| < \hbar\omega_c; \\ 0, & |\xi - \xi'| > \hbar\omega_c. \end{cases} \tag{32.2}$$

It differs from the Cooper potential (which contains only the attractive component $-V_p$) by the addition of a constant repulsive potential, $+V_c$, in the interval $|\xi - \xi'| < \hbar\omega_c$, where ω_c is a Coulomb cut-off frequency of order $\mu/\hbar$ or ω_{ep}.

The solution of (32.1) with the model potential (32.2) is complicated and we will resort to an approximate solution. The function $\Delta(\xi)$ will be described in terms of two average values Δ_1 and Δ_2 as follows:

$$\Delta(\xi) = \begin{cases} \Delta_1, & \xi \lesssim \hbar\omega_D; \\ \Delta_2, & \hbar\omega_D \lesssim \xi < \hbar\omega_c; \\ 0, & \xi > \hbar\omega_c. \end{cases} \tag{32.3}$$

In terms of these averages we will approximate the homogeneous integral equation (32.1) as a 2×2 set of homogeneous algebraic equations of the form

$$\begin{pmatrix} \Delta_1 \\ \Delta_2 \end{pmatrix} = \begin{pmatrix} I_{11} & I_{12} \\ I_{21} & I_{22} \end{pmatrix}\begin{pmatrix} \Delta_1 \\ \Delta_2 \end{pmatrix}, \tag{32.4}$$

here

$$I_{11} = \mathcal{N}(0)(V_p - V_c)I_1, \tag{32.5a}$$
$$I_{12} = \mathcal{N}(0)V_cI_2, \tag{32.5b}$$
$$I_{21} = \mathcal{N}(0)V_cI_1, \tag{32.5c}$$

and

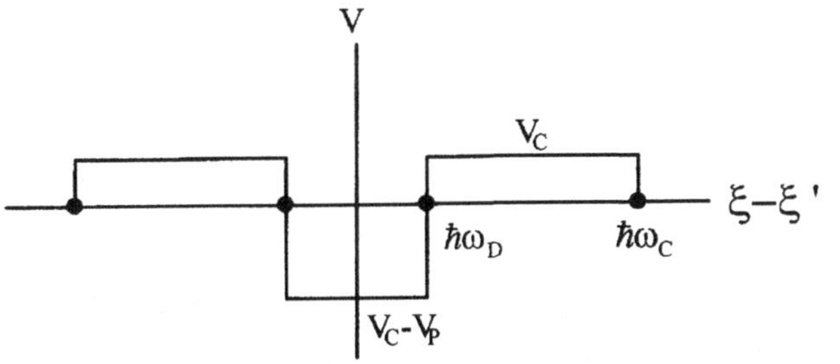

Figure 32.1 Schematic diagram of the Bogoliubov model potential.

$$I_{22} = \mathcal{N}(0)V_c I_2, \tag{32.5d}$$

where

$$I_1 = \int_0^{\omega_D} \frac{d\xi}{(\xi^2 + \Delta_1^2)^{1/2}}\{1 - 2f[(\xi^2 + \Delta_1^2)^{1/2}]\}, \tag{32.6a}$$

$$I_2 = \int_{\omega_D}^{\omega_c} \frac{d\xi}{(\xi^2 + \Delta_2^2)^{1/2}}\{1 - 2f[(\xi^2 + \Delta_2^2)^{1/2}]\}, \tag{32.6b}$$

and $\mathcal{N}(\xi)$ was again taken as a constant. The transition temperature involves the limit $\Delta_1, \Delta_2 \to 0$. We may then evaluate (32.6a) as earlier.

$$I_1 = \ell n\left(1.13\frac{\hbar\omega_D}{k_B T_c}\right). \tag{32.7a}$$

I_2 is essentially logarithmic in the relevant energy range and will be written

$$I_2 = \ell n\left(\frac{\omega_c}{\omega_D}\right). \tag{32.7b}$$

Introducing the parameters $\lambda = \mathcal{N}(0)V_p$ and $\mu = \mathcal{N}(0)V_c$ the transition temperature follows from the solution of

$$\begin{vmatrix} (\lambda - \mu)I_1 - 1 & -\mu I_2 \\ -\mu I_1 & -\mu I_2 - 1 \end{vmatrix} = 0. \tag{32.8}$$

Setting the determinant to zero yields the condition

$$-\lambda\mu I_1 I_2 + \mu I_2 - (\lambda - \mu)I_1 + 1 = 0. \tag{32.9}$$

Introducing a quantity $\mu^* = \mu/(1 + \mu I_2)$ (corresponding to $\mu = \mu^*/(1 - \mu^* I_2)$) Eq. (32.9) simplifies to $(\lambda - \mu^*)I_1 = 1$ or

$$1' = \ell n\left(\frac{1.13\,\hbar\omega_D}{k_B T_c}\right)(\lambda - \mu^*), \tag{32.10}$$

with

$$\mu^* = \frac{\mu}{1 + \mu\ell n(\omega_c/\omega_D)} \tag{32.11}$$

corresponding to

$$k_B T_c = 1.13\,\hbar\omega_D e^{\frac{1}{(\lambda - \mu^*)}}, \tag{32.12}$$

The striking property of this equation is that the effect of the Coulomb repulsion is greatly reduced in that $\mu^* \ll \mu$. In fact we may even have $\lambda - \mu < 0$, corresponding to an entirely repulsive interaction, provided $\lambda - \mu^* > 0$.

Let us examine the effect of a change in ω_D on the transition temperature, T_c. From Eq. (32.12) it follows that

$$\frac{\delta T_c}{T_c} = \frac{\delta\omega_D}{\omega_D}\left[1 - \frac{\mu^2}{1 + \mu \ell n(\omega_c/\omega_D)}\right]. \tag{32.13}$$

Now $\omega_D \propto M^{-1/2}$, where M is the ionic mass corresponding to $\delta\omega_D/\omega_D = -(1/2)\,\delta M/M$. For a pure electron–phonon interaction we expect $\delta T_C/T_c = -(1/2)\,\delta M/M$ for different isotopes of a given superconductor, which is in reasonable agreement with the experiment for 'simple' metals (those containing only s and p electrons), implying the Coulomb repulsion effects are relatively weak. Significant departures are observed for transition, rare earth, and actinide systems (containing d and f electrons) implying a stronger Coulomb repulsions (Garland 1963).

33 The two-band superconductor

In Secs. 26–28 we assumed a superconductor may be approximated by an interacting Fermi gas model. Real superconductors all have a multisheeted Fermi surface.[1] Most of the polyvalent metals having s and p derived conduction electrons have a Fermi surface which is approximately spherical except where it intersects the Brillouin zone (in the extended zone scheme); the Fermi gas model provides a good representation of these metals. However, the Fermi surfaces of superconductors involving d or f derived conduction electrons bear little resemblance to a sphere, and furthermore the amount of d or f admixture with the s–p contribution in the conduction electron wavefunction can vary considerably between sheets (and on a given sheet). The variation in the wavefunction character can, in turn, lead to a variation in the (renormalized) electron–electron interaction potential.

As a model to examine the effects of a sheet-dependent interaction potential we introduce a two-band BCS-like model with the reduced Hamiltonian (Suhl, Matthias, and Walker, 1959)

$$\hat{H}_{\mathrm{R}} = \sum_{\mathbf{k}_1,\sigma} \xi_{\mathbf{k}_1}\hat{c}^{\dagger}_{\mathbf{k}_1\sigma}\hat{c}_{\mathbf{k}_1\sigma} + \sum_{\mathbf{k}_2,\sigma} \xi_{\mathbf{k}_2}\hat{c}^{\dagger}_{\mathbf{k}_2\sigma}\hat{c}_{\mathbf{k}_2\sigma} + \sum_{\mathbf{k}_1,\mathbf{k}_1'} V_{\mathbf{k}_1\mathbf{k}_1'}\hat{c}^{\dagger}_{\mathbf{k}_1\uparrow}\hat{c}^{\dagger}_{-\mathbf{k}_1\downarrow}\hat{c}_{-\mathbf{k}_1'\downarrow}\hat{c}_{\mathbf{k}_1'\uparrow}$$
$$+ \sum_{\mathbf{k}_2,\mathbf{k}_2'} V_{\mathbf{k}_2\mathbf{k}_2'}\hat{c}^{\dagger}_{\mathbf{k}_2\uparrow}\hat{c}^{\dagger}_{-\mathbf{k}_2\downarrow}\hat{c}_{-\mathbf{k}_2'\downarrow}\hat{c}_{\mathbf{k}_2'\uparrow} + \sum_{\mathbf{k}_1,\mathbf{k}_2} [V_{\mathbf{k}_1\mathbf{k}_2}\hat{c}^{\dagger}_{\mathbf{k}_1\uparrow}\hat{c}^{\dagger}_{-\mathbf{k}_1\downarrow}\hat{c}_{-\mathbf{k}_2\downarrow}\hat{c}_{\mathbf{k}_2\uparrow}$$
$$+ \text{h.c.}]. \qquad (33.1)$$

Here $\mathbf{k}_1$ and $\mathbf{k}_2$ number the wave vectors on the two Fermi surface sheets, 1 and 2; the electron–electron potentials $V_{\mathbf{k}_1\mathbf{k}_1'}$ and $V_{\mathbf{k}_2\mathbf{k}_2'}$ are the intrasheet contributions while $V_{\mathbf{k}_1\mathbf{k}_2}$ is an intersheet contribution. The Hamiltonian (33.1) is similar to the case of the two coupled superconductors discussed in Sec. 29 if 1 and 2 are replaced by ℓ and r; the last two terms are then interpreted as an interband 'Josephson' tunneling.

If we divide the **k** space associated with Eq. (26.17) into two parts labeled as $\mathbf{k}_1$ and $\mathbf{k}_2$ it may be written as

$$\hat{H}_{\mathrm{R}} = \sum_{i=1}^{2} \sum_{\mathbf{k}_i,\sigma} \xi_{\mathbf{k}_i}\hat{c}^{\dagger}_{\mathbf{k}_i\sigma}\hat{c}_{\mathbf{k}_i\sigma} + \sum_{i,j} \sum_{\mathbf{k}_i,\mathbf{k}_j'} V_{\mathbf{k}_i\mathbf{k}_j'}\hat{c}^{\dagger}_{\mathbf{k}_i\uparrow}\hat{c}^{\dagger}_{-\mathbf{k}_i\downarrow}\hat{c}_{-\mathbf{k}_j'\downarrow}\hat{c}_{\mathbf{k}_j'\uparrow}; \qquad (33.2)$$

1. The only single-sheeted Fermi surfaces are those of the alkali metals (Li, Na, K, ...) and the noble metals (Cu, Ag, Au). To date none of these has been found to be superconducting.

this equation is formally identical to Eq. (33.1). The finite temperature gap equation generalizing Eq. (28.12) is then

$$\Delta_{\mathbf{k}_i} = -\sum_{j=1}^{2}\sum_{\mathbf{k}'_j} V_{\mathbf{k}_i\mathbf{k}'_j}\frac{1-2f_{\mathbf{k}'_j}}{2\varepsilon_{\mathbf{k}'_j}}\Delta_{\mathbf{k}'_j}. \tag{33.3}$$

As a model we will assume that the intrasheet and the intersheet pairing potentials are approximated by three constants, $-V_{11}$, $-V_{22}$, and $-V_{12}$. Eq. (33.3) then becomes

$$\Delta_i = \sum_{j=1}^{2}\mathcal{N}_j(0)V_{ij}\int_{-\hbar\omega_D}^{\hbar\omega_D}\frac{1-2f(\varepsilon_j)}{2\varepsilon_j}\Delta_j d\xi_j, \tag{33.4}$$

which is the generalization of Eq. (28.14) (for simplicity we assume the cut-off frequency is identical on both sheets of the Fermi surface). Note the Δ_i are independent of $\mathbf{k}_i$ and depend only on the sheet index i. The transition temperature is obtained from the $\Delta_i \to 0$ limit where (33.4) becomes a linear equation. The integrals are evaluated as discussed following Eq. (28.15). The condition that Eq. (33.4) has a nontrivial solution is that the determinant of the matrix of coefficients multiplying the vector Δ_i vanishes; i.e.,

$$\begin{vmatrix} \mathcal{N}_1 V_{11}\ell n\left(\dfrac{1.13\hbar\omega_D}{k_B T_c}\right) - 1 & \mathcal{N}_2 V_{12}\ell n\left(\dfrac{1.13\hbar\omega_D}{k_B T_c}\right) \\ \mathcal{N}_1 V_{12}\ell n\left(\dfrac{1.13\hbar\omega_D}{k_B T_c}\right) & \mathcal{N}_2 V_{22}\ell n\left(\dfrac{1.13\hbar\omega_D}{k_B T_c}\right) - 1 \end{vmatrix} = 0. \tag{33.5}$$

Eq. (33.5) has the solution[2]

$$\ell n\frac{1.13\hbar\omega_D}{k_B T_c} = \frac{\mathcal{N}_1 V_{11} + \mathcal{N}_2 V_{22} \pm [(\mathcal{N}_1 V_{11} - \mathcal{N}_2 V_{22})^2 + 4\mathcal{N}_1\mathcal{N}_2 V_{12}^2]^{1/2}}{2\mathcal{N}_1\mathcal{N}_2(V_{11}V_{22} - V_{12}^2)}. \tag{33.6}$$

In the limit $\mathcal{N}_1 = \mathcal{N}_2$ and $V_{11} = V_{22} = V_{12}$ we recover Eq. (28.17) provided we define $\mathcal{N} = \mathcal{N}_1 + \mathcal{N}_2$.

An interesting limit of Eq. (33.6) is $V_{11} >> V_{12}, V_{22}$. Although both Δ_1 and Δ_2 will be finite in the superconducting state, this condition leads to $\Delta_1 >> \Delta_2$. For the physically unlikely case $V_{12} = 0, V_{11} > V_{22}$ both solutions would be significant and the metal would have two transition temperatures, T_1 and T_2. To find $\Delta_1(T)$ and $\Delta_2(T)$ we must numerically solve Eq. (33.4).

The V_{12} term in the model may be regarded as an internal Josephson coupling. This leads to the possibility of an internal 'collective mode' in which the two bands exchange carriers (at constant total charge density) in an oscillatory manner (Leggett 1966).

2. Only the solution leading to the highest transition temperature is physically significant.

34 Time-dependent perturbations

In this section we give a simplified treatment of the dynamic response of a BCS superconductor to a time-dependent perturbation. Two examples which will be discussed here are (i) high frequency sound waves and (ii) a precessing nuclear spin magnetization. Two additional examples are (iii) the response following carrier injection (via a tunnel junction) or creation (via photons with $\hbar\omega >> 2\Delta$) and (iv) the response to microwaves (where $\hbar\omega \sim \Delta$) which will be treated separately in Secs. 35 and 50, respectively.

An external perturbation induces transitions between the electronic states which leads to an attenuation or decay of the perturbation. By evaluating the transition probabilities associated with the perturbation in both the normal and superconducting states and taking the ratio we can deduce the relative change in the decay rates brought on by superconductivity.

To allow for a wide class of perturbations we will write the perturbing Hamiltonian as

$$\hat{H}_1 = \sum_{\substack{\mathbf{k},\alpha \\ \mathbf{k}',\alpha'}} V_{\mathbf{k}\alpha;\mathbf{k}'\alpha'} \hat{c}^\dagger_{\mathbf{k}\alpha} \hat{c}_{\mathbf{k}'\alpha'}, \tag{34.1}$$

where V denotes a general spin- and momentum-dependent potential. We now replace the operators $\hat{c}_{\mathbf{k}\alpha}$ and $\hat{c}^\dagger_{\mathbf{k}\alpha}$ with the forms involving the Bogoliubov–Valatin operators $\hat{\gamma}_{\mathbf{k}\alpha}$ and $\hat{\gamma}^\dagger_{\mathbf{k}\alpha}$. To do this it is convenient to rewrite Eqs. (27.22) and (27.23) as

$$\begin{pmatrix} \hat{c}_{-\mathbf{k}\downarrow} \\ \hat{c}^\dagger_{\mathbf{k}\uparrow} \end{pmatrix} = \left[u_\mathbf{k} \begin{pmatrix} 1 & 0 \\ 0 & 1 \end{pmatrix} + v_\mathbf{k} \begin{pmatrix} 0 & -1 \\ 1 & 0 \end{pmatrix} \right] \begin{pmatrix} \hat{\gamma}_{-\mathbf{k}\downarrow} \\ \hat{\gamma}^\dagger_{\mathbf{k}\uparrow} \end{pmatrix}. \tag{34.2}$$

We define the spin operator $\underset{\sim}{\rho} \equiv \begin{pmatrix} 0 & -1 \\ 1 & 0 \end{pmatrix}$ which may also be written as $i\underset{\sim}{\sigma}_2$, where $\underset{\sim}{\sigma}_2$ is the Pauli matrix $\begin{pmatrix} 0 & i \\ -i & 0 \end{pmatrix}$. Note that the operator $\underset{\sim}{\rho}$ enters the definition of the Wigner time reversal operator $\mathscr{C}\underset{\sim}{\rho}$, where $\mathscr{C}$ is the complex conjugation operator. In terms of $\underset{\sim}{\rho}$ Eq. (34.2) can be written

$$\hat{c}^\dagger_{\mathbf{k}\alpha} = \sum_\beta (u_\mathbf{k} \delta_{\alpha\beta} \hat{\gamma}^\dagger_{\mathbf{k}\beta} + v_\mathbf{k} \rho_{\alpha\beta} \hat{\gamma}_{-\mathbf{k}\beta}) \tag{34.3a}$$

and

$$\hat{c}_{\mathbf{k}\alpha} = \sum_\beta (u_\mathbf{k} \delta_{\alpha\beta} \hat{\gamma}_{\mathbf{k}\beta} + v_\mathbf{k} \rho_{\alpha\beta} \hat{\gamma}^\dagger_{-\mathbf{k}\beta}). \tag{34.3b}$$

Substituting Eqs. (34.3) and their Hermitian conjugates into Eq. (34.1) we obtain

$$\hat{H}_1 = \sum_{\substack{\mathbf{k},\alpha \\ \mathbf{k}',\alpha'}} V_{\mathbf{k}\alpha;\mathbf{k}'\alpha'}\left(u_{\mathbf{k}}u_{\mathbf{k}'}\hat{\gamma}^{\dagger}_{\mathbf{k}'\alpha'}\hat{\gamma}_{\mathbf{k}'\alpha'} + v_{\mathbf{k}}v_{\mathbf{k}'}\sum_{\beta,\beta'}\rho_{\alpha\beta}\rho_{\alpha'\beta'}\hat{\gamma}_{-\mathbf{k}\beta}\hat{\gamma}^{\dagger}_{-\mathbf{k}'\beta'} \right.$$

$$\left. + u_{\mathbf{k}}v_{\mathbf{k}'}\sum_{\beta'}\rho_{\alpha'\beta'}\hat{\gamma}^{\dagger}_{\mathbf{k}\alpha}\hat{\gamma}^{\dagger}_{-\mathbf{k}'\beta'} + u_{\mathbf{k}'}v_{\mathbf{k}}\sum_{\beta}\rho_{\alpha\beta}\hat{\gamma}_{-\mathbf{k}\beta}\hat{\gamma}_{\mathbf{k}'\alpha'}\right). \quad (34.4)$$

The term $\hat{\gamma}^{\dagger}_{\mathbf{k}\alpha}\hat{\gamma}_{\mathbf{k}'\alpha'}$ describes a process where a Bogoliubov quasiparticle is scattered from state $|\mathbf{k}'\alpha'\rangle$ to $|\mathbf{k}\alpha\rangle$ and similarly for the term $\hat{\gamma}_{-\mathbf{k}\beta}\hat{\gamma}^{\dagger}_{-\mathbf{k}'\beta'}$ (see below); the term $\hat{\gamma}^{\dagger}_{\mathbf{k}\alpha}\hat{\gamma}^{\dagger}_{-\mathbf{k}'\beta'}(\hat{\gamma}_{-\mathbf{k}\beta}\hat{\gamma}_{\mathbf{k}'\alpha'})$ creates (destroys) two quasiparticles.[1]

Let us first examine the scattering processes which involve the first two terms in Eq. (34.4). The matrix element of $\hat{H}_1$ resulting in a transition from an initial state $|\mathbf{k}_i\alpha_i\rangle$ to a (different) final state $|\mathbf{k}_f\alpha_f\rangle$ is

$$\langle \mathbf{k}_f\alpha_f|\hat{H}_1|\mathbf{k}_i\alpha_i\rangle = V_{\mathbf{k}_f\alpha_f;\mathbf{k}_i\alpha_i}u_{\mathbf{k}_f}u_{\mathbf{k}_i} - v_{\mathbf{k}_i}v_{\mathbf{k}_f}\sum_{\gamma\gamma'}V_{-\mathbf{k}_i\gamma;-\mathbf{k}_f\gamma'}\rho_{\gamma\alpha_i}\rho_{\gamma'\alpha_f} \quad (34.5)$$

(where we used $\gamma_i\gamma_j^{\dagger} = -\gamma_j^{\dagger}\gamma_i$ in the second term). The second term in Eq. (34.5) is effectively the same as the first but with the spins and momenta of the initial and final states reversed. Noting that complex conjugation is equivalent to $\mathbf{k} \to -\mathbf{k}$ for our plane-wave states, we can interpret this second term as the matrix element connecting the *time-reversed states.* We will restrict ourselves to perturbations for which the time-reversed matrix element is either even (+) or odd (−) under time reversal. We may then write Eq. (34.5) as

$$\langle \mathbf{k}_f\alpha_f|\hat{H}_1|\mathbf{k}_i\alpha_i\rangle = V_{\mathbf{k}_f\alpha_f;\mathbf{k}_i\alpha_i}(u_{\mathbf{k}_f}u_{\mathbf{k}_i} - \eta v_{\mathbf{k}_f}v_{\mathbf{k}_i}), \quad (34.6)$$

where $\eta = +1$ or -1 for processes which are even or odd under time reversal. The form $u_{\mathbf{k}}u_{\mathbf{k}'} - \eta v_{\mathbf{k}}v_{\mathbf{k}'}$ is referred to as a *coherence factor.*

Using the 'golden rule' expression of time-dependent perturbation theory the transition rate $w_{i\to f}$ is

$$w_{i\to f} = \frac{2\pi}{\hbar}|\langle \mathbf{k}_f\alpha_f|\hat{H}_1|\mathbf{k}_i\alpha_i\rangle|^2\{f(\varepsilon_i)[1-f(\varepsilon_f)] - f(\varepsilon_f)[1-f(\varepsilon_i)]\}\delta(\varepsilon_f - \varepsilon_i - \hbar\omega). \quad (34.7)$$

The rate at which energy is absorbed from the perturbing potential is

$$\frac{dE}{dt} = \dot{E} = \frac{1}{(2\pi)^6}\sum_{\alpha_f=1}^{2}\sum_{\alpha_i=1}^{2}\int d^3k_f d^3k_i \hbar\omega w_{i\to f}. \quad (34.8)$$

We sum over initial and final spin states, write $d^3k = 4\pi k^2(dk/d\xi)d\xi(1/4\pi)d\Omega$, and define

$$\overline{V^2} = \frac{4}{(4\pi)^2}\int V^2(\mathbf{k}_f, \varepsilon_f; \mathbf{k}_i, \varepsilon_i)d\Omega_f d\Omega_i \quad (34.9)$$

to obtain

$$\dot{E} = 8\pi\omega\overline{V^2}\int_{\Delta}^{\infty}\mathcal{N}_s(\varepsilon_f)\mathcal{N}_s(\varepsilon_i)[u(\varepsilon_f)u(\varepsilon_i) - \eta v(\varepsilon_f)v(\varepsilon_i)]^2[f(\varepsilon_i) - f(\varepsilon_f)]\delta(\varepsilon_f - \varepsilon_i - \hbar\omega)d\varepsilon_f d\varepsilon_i, \quad (34.10)$$

1. These four terms are analogous to those encountered in our earlier discussion of quasiparticle tunneling in Sec. 29. There are also similarities in the calculations which follow.

where $\mathcal{N}_s(\varepsilon) = \mathcal{N}(0)\,|\,d\xi/d\varepsilon\,| = \mathcal{N}(0)\varepsilon/(\varepsilon^2 - \Delta^2)^{1/2}$ is the density of states in the superconductor. Using Eqs. (26.27) and (26.28) we obtain

$$(u_{\mathbf{k}_f}u_{\mathbf{k}_i} - \eta v_{\mathbf{k}_f}v_{\mathbf{k}_i})^2 = \frac{1}{2}\left[1 + \frac{\xi_i\xi_f}{\varepsilon_i\varepsilon_f} - \eta\frac{\Delta^2}{\varepsilon_i\varepsilon_f}\right]. \tag{34.11}$$

Noting that the terms containing ξ are odd (and therefore vanish on integrating over ξ) Eq. (34.10) becomes

$$\dot{E} = 4\pi\omega\overline{V^2}[\mathcal{N}(0)]^2\int_{\Delta}^{\infty}d\varepsilon_f\int_{\Delta}^{\infty}d\varepsilon_i\frac{\varepsilon_f\varepsilon_i - \eta\Delta^2}{(\varepsilon_f^2 - \Delta^2)^{1/2}(\varepsilon_i^2 - \Delta^2)^{1/2}}$$
$$\times\,[f(\varepsilon_i) - f(\varepsilon_f)]\delta(\varepsilon_f - \varepsilon_i - \hbar\omega). \tag{34.12}$$

To complete the calculation we must evaluate the contribution of the last two terms in (34.4) which involve pair breaking and pair recombination events induced by the external potential; these events are nonzero only for $\hbar\omega > 2\Delta$. The calculation of these processes is straightforward and when the result is added to (34.12) the total energy absorbed by all four processes is

$$\dot{E} = 2\pi\omega\overline{V^2}[\mathcal{N}(0)]^2\int_{-\infty}^{\infty}d\varepsilon_f\int_{-\infty}^{\infty}d\varepsilon_i\frac{\varepsilon_f\varepsilon_i - \eta\Delta^2}{(\varepsilon_f^2 - \Delta^2)^{1/2}(\varepsilon_i^2 - \Delta^2)^{1/2}}$$
$$\times\,[f(\varepsilon_i) - f(\varepsilon_f)]\delta(\varepsilon_f - \varepsilon_i - \hbar\omega);$$

here ε_i and ε_f have arbitrary signs but $|\varepsilon_{i,f}| > \Delta$. (34.13)

In the normal state this expression reduces to

$$\dot{E}_N = 2\pi\omega\overline{V^2}[\mathcal{N}(0)]^2\int_{-\infty}^{\infty}d\xi_f\int_{-\infty}^{+\infty}d\xi_i[f(\xi_i) - f(\xi_f)]\delta(\xi_f - \xi_i - \hbar\omega). \tag{34.14}$$

For temperatures for which $\hbar\omega << k_BT$ (and where we continue to assume $\mathcal{N}(0)$ is constant) we may expand the occupation functions of $f(\xi_f) = f(\xi_i) + (\partial f/d\xi)d\xi$ or

$$\dot{E}_N = -\,2\pi\hbar\omega^2[\mathcal{N}(0)]^2\overline{V^2}\int_{-\infty}^{+\infty}d\xi_i\frac{\partial f}{\partial\xi_i} = 2\pi\hbar\omega^2[\mathcal{N}(0)]^2\overline{V^2}. \tag{34.15}$$

The value of the constant $\overline{V^2}$ cannot be accurately calculated. However, by measuring the *ratio* of the energy absorbed in the superconducting, (34.13), and normal, (34.14), states we see that this factor (along with the density of states) cancels out; i.e.,

$$\frac{\dot{E}_S}{\dot{E}_N} = \frac{\displaystyle\int_{-\infty}^{+\infty}d\varepsilon_f\int_{-\infty}^{+\infty}d\varepsilon_i\frac{\varepsilon_f\varepsilon_i - \eta\Delta^2}{(\varepsilon_f^2 - \Delta^2)^{1/2}(\varepsilon_i^2 - \Delta^2)^{1/2}}[f(\varepsilon_i) - f(\varepsilon_f)]\delta(\varepsilon_f - \varepsilon_i - \hbar\omega)}{\displaystyle\int_{-\infty}^{+\infty}d\xi_i\int_{-\infty}^{+\infty}d\xi_f[f(\xi_i) - f(\xi_f)]\delta(\xi_f - \xi_i - \hbar\omega)}$$

and on using Eq. (34.15) we have

$$\frac{\dot{E}_S}{\dot{E}_N} = \frac{1}{\hbar\omega}\int_{-\infty}^{+\infty}d\varepsilon_f\int_{-\infty}^{+\infty}d\varepsilon_i\frac{\varepsilon_i\varepsilon_f - \eta\Delta^2}{(\varepsilon_f^2 - \Delta^2)^{1/2}(\varepsilon_i^2 - \Delta^2)^{1/2}}[f(\varepsilon_i) - f(\varepsilon_f)]\delta(\varepsilon_f - \varepsilon_i - \hbar\omega). \tag{34.16}$$

34.1 Ultrasonic attenuation

Using resonant piezoelectric plates one can generate sound waves in solids; the techniques are relatively straightforward in the frequency range 10^7–10^8 Hz and such ultrasonic waves have been extensively used to study normal and superconducting metals. The sound waves couple to the electrons via the self-consistent electromagnetic fields generated by the moving (positive) ions and by a shift in the electronic energy levels associated with a deformation of the lattice. However, here we consider only a simplified version of the latter mechanism and write the perturbing Hamiltonian as

$$\hat{H}_1 = \lambda q u_0 \mathrm{e}^{\mathrm{i}qx - \mathrm{i}\omega t} \sum_{\mathbf{k},\alpha} \hat{c}^{\dagger}_{\mathbf{k},\alpha} \hat{c}_{\mathbf{k}',\alpha}, \tag{34.17}$$

where $u = u_0 \mathrm{e}^{\mathrm{i}qx - \mathrm{i}\omega t}$ is the (time- and space-dependent) displacement, λ is the 'deformation potential' (a constant measuring the strength of the electron–acoustic wave coupling), and q is the sound wave vector (the coupling is linear in q since it involves the deformation of the lattice; i.e., the gradient of u). This perturbation is even under time reversal and hence $\eta = 1$ in Eq. (34.16). Furthermore, experiments are usually done in the regime $\hbar\omega << k_B T$, where we may use the expansion

$$f(\varepsilon_i) - f(\varepsilon_f) \cong \frac{\partial f}{\partial \varepsilon}(\varepsilon_i - \varepsilon_f) \cong -\hbar\omega \frac{\partial f}{\partial \varepsilon}$$

in (34.16) leading to

$$\frac{\dot{E}_S}{\dot{E}_N} = -\int_{|\varepsilon| > \Delta} \mathrm{d}\varepsilon \frac{\varepsilon^2 - \Delta^2}{\varepsilon^2 - \Delta^2} \frac{\partial f(\varepsilon)}{\partial \varepsilon} = \frac{2}{1 + \mathrm{e}^{\Delta/k_B T}}. \tag{34.18}$$

The factor of 2 comes from the fact that the integration in (34.13) extends over positive and negative ε.

Fig. 34.1 shows a fit of this expression to the original ultrasonic attenuation data of Morse and Bohm (1957) for the metal tin.

34.2 Nuclear spin relaxation

Let us consider a metal whose nuclei have a spin and hence a nuclear magnetic moment. When this system is subjected to a magnetic field, and the spins are allowed to come into equilibrium, the nuclear spin system acquires a magnetic moment, $m(H, T)$ (the magnitude of which is given by Curie's law with a Curie constant involving the number of spins and their magnetic moment). If the magnetic field is suddenly changed, the instantaneous nuclear moment, $m(t)$, evolves toward the equilibrium value appropriate to the new field according to the Bloch relaxation equation

$$\frac{\mathrm{d}m(t)}{\mathrm{d}t} = -\frac{m(t) - m(H, T)}{T_1}. \tag{34.19}$$

Since a magnetic field is expelled from a Type I superconductor, experiments must be

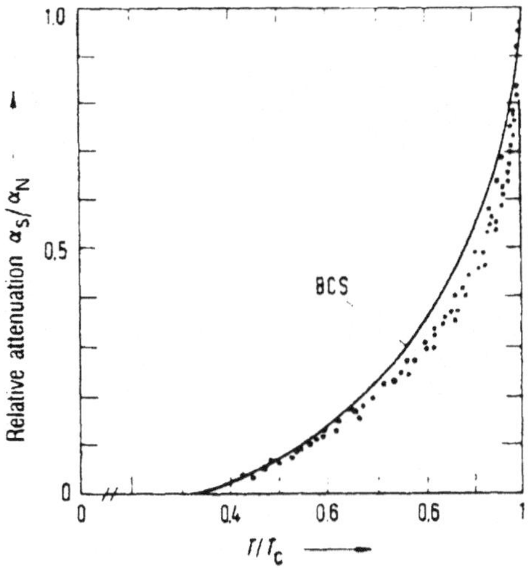

Figure 34.1 Ultrasonic absorption in superconducting tin. The solid line is calculated from Eq. (34.18) using $2\Delta = 3.5k_BT_c$. (After Morse and Bohm (1957).)

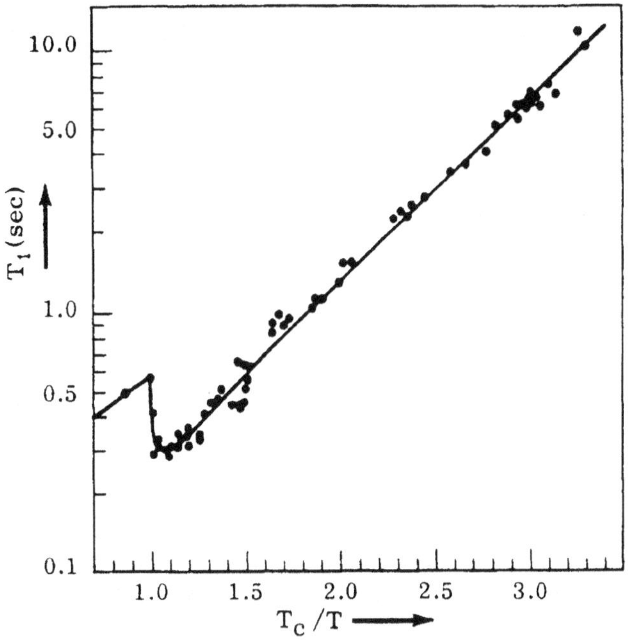

Figure 34.2 Nuclear relaxation rates in aluminum. Note the dip of T_1 at temperatures close to the transition temperature. The theoretical curve is calculated on the assumption that the peak in the BCS density of states is smeared out by 20%.

performed by (i) polarizing the nuclei in the normal state at a field $H > H_c$, (ii) passing to the superconducting state by removing the field (in a time short compared to T_1), (iii) allowing the magnetization to evolve for some period, and (iv) reapplying the field and measuring the present value of $m(t)$. Fitting the data for a series of such experiments to Eq. (34.19) one obtains T_1. The experiment is then repeated at various temperatures to obtain $T_1(T)$. Nuclei in metals relax through their interaction with a fluctuating internal magnetic field arising from the motion of thermally excited electrons (which themselves have a magnetic moment μ_B). We will not discuss this mechanism but merely note that (like the magnetization itself) it is odd under time reversal; i.e., $\eta = -1$ in Eq. (34.16). From the relation $\mathrm{d}E = 4\pi H \mathrm{d}M$ we have $\dot{E} = 4\pi \dot{m} H$ and (34.16) becomes

$$\frac{(T_1)_N}{(T_1)_S} = -\int_{|\varepsilon|\Delta} \mathrm{d}\varepsilon \frac{\partial f}{\partial \varepsilon} \frac{\varepsilon^2 + \Delta^2}{\varepsilon^2 - \Delta^2}. \tag{34.20}$$

This integral diverges logarithmically in the superconducting state in the BCS model. However, in a real metal $\Delta = \Delta(\hat{\mathbf{k}})$ and the presence of any anisotropy removes the divergence. Fig. 34.2 shows the temperature dependence of the nuclear relaxation rate in aluminum measured by Redfield (1962) and a fit to the data assuming a 20% 'smearing' of the energy gap. Note the data have the unusual property that T_1 is *shorter* just below T_c (although it is much longer for $T << T_c$). This phenomena was first observed by Hebel and Slichter (Hebel and Slichter 1957, Hebel 1959).

The very different behavior of the ultrasonic attenuation (where the numerator and denominator of the coherence factor cancel) and the nuclear relaxation (where no cancellation occurs) shows the profound effect that the coherence factor has on dynamic processes in superconductors.

35 Nonequilibrium superconductivity

In this section we examine some nonequilibrium properties of superconductors which can be studied using tunnel junctions. We will limit the treatment to a system which is spatially homogeneous, which is usually a good approximation for a thin film where the carriers are confined to the film thickness, d. Three ways in which a thin-film superconductor can be driven out of equilibrium are: (i) injecting carriers via a tunnel junction, (ii) creating excess carriers by irradiating the film with photons[1] with a frequency $\hbar\omega >> 2\Delta$, and (iii) injecting phonons via, e.g., a heat pulse. The time dependence of the decay of the nonequilibrium quasiparticle distribution can be monitored with a tunnel junction (or a second tunnel junction for case (i) above). We begin with a discussion of the various scattering processes tending to restore equilibrium.

35.1 Elastic and inelastic scattering processes

We will limit ourselves to the following scattering processes: electron–phonon, electron–electron, and electron–impurity; we start our discussion with the last process. We will assume there are no magnetic impurities which would result in spin-flip processes. In our previous discussion of the BCS ground state we formed pairs from the states $|+\mathbf{k}\uparrow\rangle$ and $|-\mathbf{k}\downarrow\rangle$. We recall the Wigner time reversal operator, $\mathscr{C}i\underset{\sim}{\sigma}_2$; when this operator acts on the single particles' states (making up a pair) it converts one into the other ($i\underset{\sim}{\sigma}_2$ flips the spin and $\mathscr{C}$ converts $\mathbf{k}$ to $-\mathbf{k}$). In the presence of elastic electron–impurity scattering, which does not destroy time reversal invariance, we may imagine that as the scattering potential is 'turned on' we create a new set of eigenstates which replace the momentum eigenstates (and could be represented by linear combinations of them). The superconducting ground state may now be formed by pairing each of the new eigenstates with its time-reversed partner. The above arguments, due to Anderson (1961), suggest that many of the properties of an s-paired superconductor, such as T_c, may be insensitive to nonmagnetic impurity scattering, as is observed experimentally. Rather than trying to construct the time-reversed pairs explicitly, it is easier simply to average quantities over the Fermi surface (such as $\Delta(\hat{\mathbf{k}})$, where $\hat{\mathbf{k}}$

1. It has been suggested that a tunnel junction might be used as a high energy resolution photon detector. The number of quasiparticles produced is $n \cong \hbar\omega/\Delta$. For a material with a gap $\Delta \sim 10^{-3}$ eV we would have 10^7 quasiparticles for a 10 keV X-ray photon. The Poisson statistical uncertainty $1/n^{1/2} \sim 0.3 \times 10^{-3}$ would then lead to an energy uncertainty of 3 eV.

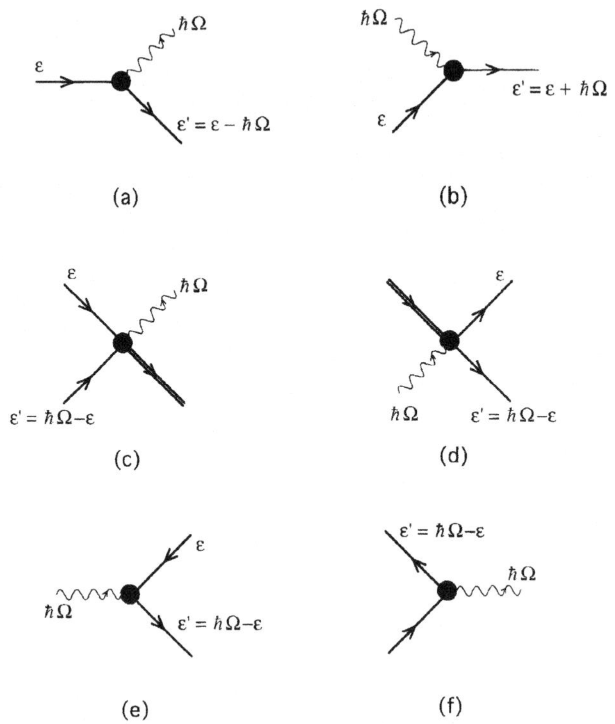

Figure 35.1 Scattering processes involving phonons (〰), quasiparticles (——>——), and the condensate (═══>═══).

measures the angle on the Fermi surface). Therefore in what follows we will ignore impurity scattering but assume all quantities are averaged over an associated constant-energy surface.

Electron–electron scattering involves that part of the electron–electron interaction which is 'left over' after the formation of quasiparticles. It tends to be rather weak in metals (relative to electron–impurity and electron–phonon processes) and will be ignored. (In highly disordered films electron–electron scattering plays an important role.)

There are several electron–phonon scattering processes, the most important of which are shown in Fig. 35.1. Fig. 35.1(a) and (b) correspond to (inelastic) processes in which a phonon of energy $\hbar\Omega$ is emitted or absorbed, respectively, but the number of quasiparticles is unchanged. Process (a) is permitted when $\varepsilon' - \varepsilon - \hbar\Omega \geq \Delta$ whereas process (b) is allowed for all Ω. Process (c) corresponds to quasiparticle recombination, resulting in the addition of a Cooper pair to the condensate and the creation of a phonon with an energy $\hbar\Omega \geq 2\Delta$. Numerical calculations show that this process dominates over process (a) when it is allowed. The inverse process of (c) is (d) and corresponds to the absorption of a phonon and the breaking of a condensate pair to produce two quasiparticles; it requires a phonon energy $\hbar\Omega \geq 2\Delta$.

The relaxation rates (transition rates) associated with these processes, which are calculated from the 'golden rule', involve the average of the square of the electron–phonon matrix element over initial and final electron states separated by an energy $\hbar\Omega$, which is denoted as $|\alpha(\Omega)|^2$. When calculating an electron lifetime the density of phonon states, which is denoted by $F(\Omega)$, will also enter. The quantity $|\alpha(\Omega)|^2F(\Omega)$ can be obtained[2] from structure that is observed in the tunneling spectra at energies $2\Delta < eV < \hbar\Omega_D$ arising from inelastic tunneling events. For low frequencies, $\Omega << \Omega_D$, $|\alpha|^2 \to$ const and $F(\Omega) \propto \Omega^2$ (where the proportionality constant is that entering the density of states in the Debye theory of the phonon specific heat); therefore $|\alpha|^2 F(\Omega) = b\Omega^2$, where b is obtained from the tunneling spectra.

In the normal state we expect from the golden rule that the energy-dependent electron scattering rate arising from the electron–phonon interaction, $w_e(= 1/\tau_e$, where τ_e is the associated lifetime), will be qualitatively given by

$$\frac{1}{\tau_e(\varepsilon)} = w_e(\varepsilon) = \frac{2\pi}{\hbar}|\alpha|^2 \mathcal{N}(0)\left(\frac{\varepsilon}{\hbar\Omega_D}\right)^3, \tag{35.1}$$

where the last two factors are the joint density of states for the electrons and the phonons, respectively. Among the superconducting elements, τ_e for an electron with energy $k_B T_c$ ranges from about 10^{-6} s (for Al) to 10^{-10} s (for Pb, a 'strong coupling' superconductor). The normal-state phonon lifetime, τ_{ph}, is dominated by the graph shown in Fig. 35.1(e) which describes a process in which a phonon decays into an electron and a hole (the latter represented by an arrow directed to the left). The joint electron–hole density of states for an energy $\hbar\Omega$ is of order $\mathcal{N}(0)\hbar\Omega/\mu$, where μ is the Fermi energy. Hence the phonon lifetime is given by

$$\frac{1}{\tau_{ph}(\Omega)} = w_{ph}(\Omega) = \frac{2\pi}{\hbar}|\alpha|^2 \mathcal{N}(0)\frac{\hbar\Omega}{\mu}. \tag{35.2}$$

The variation among the elemental metals of τ_{ph} is much smaller than τ_e and for $\hbar\Omega = k_B T_c$ is of order 10^{-10} s. Fig. 35.1(f) shows the reverse normal-state process in which an electron and hole annihilate[3] to form a phonon.

The modification of the density of states in the superconducting state together with the presence of the pair-breaking/recombination processes and the effects of the coherence factor, (34.11), considerably complicates the temperature–energy dependence of the electron and phonon lifetimes. Extensive calculations have been carried out by Kaplan *et al.* (1976) under the assumption $\alpha^2 F(\Omega) = b\Omega^2$ which will not be reviewed here.

In a thin film one must also consider changes in the phonon population resulting from transmission into and out of the (insulating) substrate. In a film of thickness d, a phonon traveling at the velocity of sound, v_s, would arrive at a boundary in a time of order d/v_s. The transmission probability, η, is governed by the mismatch in the acoustic impedance between the film, Z_f, and the substrate, Z_s, and is given by

$$\eta = \frac{4Z_f Z_s}{(Z_f + Z_s)^2}, \tag{35.3}$$

where Z is the product of the mass density and the sound velocity (ρv_s). The phonon lifetime arising from this escape process, τ_{es}, is given by

$$\tau_{es} = \frac{4}{\eta}\frac{d}{v_s}, \tag{35.4}$$

where 4 is a geometrical factor.

2. A discussion of this topic requires solving the coupled equations of motion for the electron and phonon Green's functions which is beyond the scope of this book. (See Eliashberg (1960).)
3. In the normal state, electrons and holes represent distinct excitation branches. In the superconductor, however, these states are mixed, especially near the Fermi surface.

35.2 Quasiparticle and phonon populations in a nonequilibrium superconductor

We now develop a set of coupled, Boltzmann-like equations for the nonequilibrium distribution functions for the quasiparticles, $f(\varepsilon)$, and the phonons, $n(\Omega)$. Based on the processes discussed in Subsec. 35.1 we may write the equation of motion for the quasiparticle distribution function as (Kaplan *et al.* 1976, Chang and Scalapino 1978)

$$\frac{df^{(\mathrm{tot})}(\varepsilon)}{dt} = I_{\mathrm{qp}}(\varepsilon) + \frac{df^{(\mathrm{a})}(\varepsilon)}{dt} + \frac{df^{(\mathrm{b})}(\varepsilon)}{dt} + \frac{df^{(\mathrm{cd})}(\varepsilon)}{dt}, \tag{35.5}$$

where the first term on the right results from quasiparticle injection and the superscripts on the remaining terms denote the inelastic scattering contributions associated with the graphs in Fig. 35.1. From the discussion in Sec. 34 it was clear that we can account for the effects of superconductivity on a golden rule scattering rate (resulting from a normal-state matrix element) by: (i) replacing the normal density of states $\mathcal{N}(0)$ by $\mathcal{N}(0)d\xi/d\varepsilon \equiv \mathcal{N}(0)\rho(\varepsilon)$, and (ii) including the coherence factor (34.11). We may then immediately write down the three scattering rates entering Eq. (35.5) as

$$\frac{df^{(\mathrm{a})}(\varepsilon)}{dt} = -\frac{2\pi}{\hbar}\int_0^{\varepsilon-\Delta} d\Omega\,|\alpha(\Omega)|^2 F(\Omega)\rho(\varepsilon-\hbar\Omega)\left[1-\frac{\Delta^2}{\varepsilon(\varepsilon-\hbar\Omega)}\right]$$
$$\times\,\{f(\varepsilon)[1-f(\varepsilon-\hbar\Omega)][n(\Omega)+1]-[1-f(\varepsilon)]f(\varepsilon-\hbar\Omega)n(\Omega)\}, \tag{35.6a}$$

$$\frac{\partial f^{(\mathrm{b})}(\varepsilon)}{dt} = -\frac{2\pi}{\hbar}\int_0^{\infty} d\Omega\,|\alpha(\Omega)|^2 F(\Omega)\rho(\varepsilon+\hbar\Omega)\left[1-\frac{\Delta^2}{\varepsilon(\varepsilon+\hbar\Omega)}\right]$$
$$\times\,\{f(\varepsilon)[1-f(\varepsilon+\hbar\Omega)]n(\Omega)-f(\varepsilon+\hbar\Omega)[1-f(\varepsilon)][n(\Omega)+1]\}, \tag{35.6b}$$

and

$$\frac{df^{(\mathrm{c,d})}(\varepsilon)}{dt} = -\frac{2\pi}{\hbar}\int_{\varepsilon+\Delta}^{\infty} d\Omega\,|\alpha(\Omega)|^2 F(\Omega)\rho(\hbar\Omega-\varepsilon)\left[1+\frac{\Delta^2}{\varepsilon(\hbar\Omega-\varepsilon)}\right]$$
$$\times\,\{f(\varepsilon)f(\hbar\Omega-\varepsilon)[n(\Omega)+1]-[1-f(\varepsilon)][1-f(\hbar\Omega-\varepsilon)]n(\Omega)\}. \tag{35.6c}$$

In a similar way we can describe the time evolution of the phonon distribution by

$$\frac{dn(\Omega)}{dt} = I_{\mathrm{ph}}(\Omega) - \frac{dn^{(\mathrm{a,b})}(\Omega)}{dt} - \frac{dn^{(\mathrm{c,d})}(\Omega)}{dt} - \frac{n(\Omega)-n(\Omega,T)}{\tau_{\mathrm{es}}}. \tag{35.7}$$

The first and last terms on the right account for phonon injection (e.g., via a heat pulse) and exchange of phonons with the substrate, respectively; hence $n(\Omega, T)$ is the equilibrium phonon distribution and its inclusion in the last term insures there is no net phonon exchange with the substrate in thermal equilibrium. The term $dn^{(\mathrm{a,b})}(\Omega)/dt$ describes those phonon absorption and emission processes in which the number of quasiparticles remains constant and is given by

$$\frac{dn^{(\mathrm{a,b})}(\Omega)}{dt} = \frac{8\pi}{\hbar}\frac{\mathcal{N}(0)}{n}\int_{\Delta}^{\infty} d\varepsilon\int_{\Delta}^{\infty} d\varepsilon'\,|\alpha(\Omega)|^2\rho(\varepsilon)\rho(\varepsilon')\left(1-\frac{\Delta^2}{\varepsilon\varepsilon'}\right)\delta(\varepsilon-\varepsilon'+\hbar\Omega)$$

$$\times \{f(\varepsilon)[1-f(\varepsilon')]n(\Omega) - f(\varepsilon')[1-f(\varepsilon)][n(\Omega)+1]\}, \tag{35.8a}$$

where n is the ion density. The extra factor of 4 occurring here accounts for the two initial state 'branches' with $k < k_F$ and $k > k_F$ and the corresponding two final state branches for a total of four processes $k^< \to k^<, k^< \to k^>, k^> \to k^<, k^> \to k^>$, where $k^>$ denotes $k > k_F$, etc. The term $dn^{(c,d)}(\Omega)/dt$ describes phonon absorption and emission processes which are accompanied by pair production and pair destruction, respectively, and is given by

$$\frac{dn^{(c,d)}(\Omega)}{dt} = -\frac{4\pi}{\hbar}\int_\Delta^\infty d\varepsilon \int_\Delta^\infty d\varepsilon' \,|\alpha(\Omega)|^2 \rho(\varepsilon)\rho(\varepsilon')\left(1+\frac{\Delta^2}{\varepsilon\varepsilon'}\right)\delta(\varepsilon+\varepsilon'-\hbar\Omega)$$

$$\times \{[1-f(\varepsilon)][1-f(\varepsilon')]n(\Omega) - f(\varepsilon)f(\varepsilon')[n(\Omega)+1]\}. \tag{35.8b}$$

If the deviation of the electron distribution function from its equilibrium value is large then the magnitude of the BCS gap will shift. Under the assumption that the gap responds instantaneously to the distribution function we may use the BCS integral equation to compute Δ; i.e., Δ is that quantity obtained from solving

$$1 = \mathcal{N}(0)V\int_\Delta^\infty d\varepsilon \frac{1-2f(\varepsilon)}{(\varepsilon^2-\Delta^2)^{1/2}}, \tag{35.9}$$

where $f(\varepsilon)$ is the *nonequilibrium* distribution function. In practice one would have to solve (35.5), (35.7), and (35.9) self-consistently.

A self-consistent solution for $n(\Omega)$, $f(\varepsilon)$, and Δ is computationally demanding and in applications it is common to use a set of phenomenological equations which relate the quasiparticle number density, N_{qp}, and phonon number density, N_{ph}; i.e., the energy–frequency dependence is integrated over. Clearly these equations (which can be obtained from (35.5) and (35.7) under certain conditions) have the form (Rothwarf and Taylor 1967)

$$\frac{dN_{qp}}{dt} = I_{qp} - 2RN_{qp}^2 + 2\beta N_{ph} \tag{35.10}$$

and

$$\frac{dN_{ph}}{dt} = I_{ph} + RN_{qp}^2 - \beta N_{ph} - \frac{N_{ph}-N_{ph}^{(0)}}{\tau_{es}}, \tag{35.11}$$

where I_{qp} and I_{ph} are the energy-integrated quasiparticle and phonon injection-currents, respectively, and only that part of the phonon density or current with $\hbar\Omega > 2\Delta$ is included in N_{ph} (or I_{ph}) (the remainder playing a minimal role). The second term in both these equations arises from quasiparticle recombination, which is proportional to the square of the quasiparticle density and occurs at a rate determined by the parameter R. The third term is the pair-breaking contribution occurring at a rate determined by β. The factors of 2 in (35.10) account for the fact that two quasiparticles disappear for each phonon created (and conversely two quasiparticles appear for each phonon pair-breaking event). In equilibrium where $\dot{N}_{qp} = \dot{N}_{ph} = I_{qp} = I_{ph} = 0$ and $N_{ph} = N_{ph}^{(0)}, N_{qp} = N_{qp}^{(0)}$, we have $RN_{qp}^{(0)2} = \beta N_{ph}^{(0)}$. If N_{qp} differs slightly from its equilibrium value as $N_{qp} = N_{qp}^{(0)} + \delta N_{qp}$ then from Eq. (35.10) we would have

$$\delta\dot{N}_{qp} \cong -4RN_{qp}^{(0)}\delta N_{qp},$$

leading to exponential decay with a (recombination) time constant $\tau_R = (4RN_{qp}^{(0)})^{-1}$. Similarly, if N_{ph} deviates from its equilibrium value as $N_{ph} = N_{ph}^{(0)} + \delta N_{ph}$ it would return to equilibrium with a (pair-breaking) time constant $\tau_B = \beta^{-1}$. Under steady state conditions where $\dot{N}_{qp} = \dot{N}_{ph} = 0$ but I_{qp} and I_{ph} differ from zero, the linearized forms of Eqs. (35.10) and (35.11) yield (on eliminating δN_{ph})

$$\delta N_{qp} = I_{qp}\tau_R\left(1 + \frac{\tau_{es}}{\tau_B}\right) + 2I_{ph}\tau_R\frac{\tau_{es}}{\tau_B}. \tag{35.12}$$

When $\tau_{es} = 0$, the excess quasiparticle density is determined only by τ_R; however, if $\tau_{es} \neq 0$ all relaxation times are involved.

With the above theory one can go on to examine a number of interesting nonequilibrium phenomena. One field of study involves energy gap enhancement in the presence of electromagnetic radiation. Another involves the use of tunnel junctions as generators and detectors of phonons. Thus if we subject an SIS tunnel junction to a voltage $eV > 2\Delta$ we inject a large population of quasiparticles with $\varepsilon \geq \Delta$ which on recombining create a large population of phonons with energy $\hbar\Omega = 2\Delta$. If the junction is deposited on a dielectric single crystal, one can study the propagation of phonons that escape the film. Similarly a second (appropriately biased) SIS junction can serve as a detector by converting the 2Δ phonons to quasiparticles which, in turn, tunnel through the junction to produce a current.

PART III NONUNIFORM SUPERCONDUCTIVITY

36 Bogoliubov's self-consistent potential equations

In the presence of boundaries, scattering centers, and other nonuniformities, the superfluid Fermi-gas model we have used in Part II is not sufficiently general. A system in a magnetic field, with its associated position-dependent vector potential, can be regarded as inhomogeneous. To treat such problems we must therefore generalize the microscopic pairing theory to deal with nonuniform systems. The most powerful approach to such problems is to use the many-body Green's function methods of quantum field theory which will be discussed only briefly in Sec. 55. The approach we use here, which is adequate for many purposes, is Bogoliubov's self-consistent field method (Bogoliubov, Tolmachev, and Shirkov 1959). Our treatment will be similar to that of deGennes (1966).

We write the Hamiltonian of the system in second quantized form as

$$\hat{H} = \hat{H}^{(1)} + \hat{V}^{(2)}, \tag{36.1}$$

where $\hat{H}^{(1)}$ is the one-electron Hamiltonian

$$\hat{H}^{(1)} = \int d^3r \hat{\psi}_\alpha^\dagger(\mathbf{r}) \left[-\frac{\hbar^2}{2m}\left(\nabla - \frac{ie}{\hbar c}\mathbf{A}\right)^2 \delta_{\alpha\beta} + U_{\alpha\beta}^{(1)}(\mathbf{r}) \right] \hat{\psi}_\beta(\mathbf{r}) \tag{36.2}$$

and

$$\hat{V}^{(2)} = \frac{1}{2}\int d^3r \int d^3r' \hat{\psi}_\delta^\dagger(\mathbf{r})\hat{\psi}_\gamma^\dagger(\mathbf{r}') U_{\delta\gamma,\alpha\beta}^{(2)}(\mathbf{r},\mathbf{r}')\hat{\psi}_\alpha(\mathbf{r}')\hat{\psi}_\beta(\mathbf{r}); \tag{36.3}$$

Greek letters denote spin indices and the summation convention is implied.

The term involving the vector potential accounts for the interaction of an external magnetic field with the orbital motion of the electron. If we ignore the interaction of an external magnetic field with the electron spin (and various relativistic effects) we may write the one-electron potential $U^{(1)}$, which in the absence of an external potential involves only the periodic potential associated with the lattice, as

$$U_{\alpha\beta}^{(1)} = U^{(1)}\delta_{\alpha\beta}. \tag{36.4a}$$

We will assume that the two-particle interaction is spin-independent; i.e.,

$$U^{(2)}_{\delta\gamma,\alpha\beta} = U^{(2)}\delta_{\alpha\gamma}\delta_{\delta\beta}. \tag{36.4b}$$

We further assume that $U^{(2)}$ is time-independent (i.e., nonretarded).[1] $\hat{\psi}_\alpha(\mathbf{r})\,(\hat{\psi}^\dagger_\alpha(\mathbf{r}))$ are the usual field operators which destroy (create) a particle of spin α at the point $\mathbf{r}$ (see Appendix A). They obey the Fermi commutation relations

$$\hat{\psi}_\alpha(\mathbf{r})\hat{\psi}_\beta(\mathbf{r}') + \hat{\psi}_\beta(\mathbf{r}')\hat{\psi}_\alpha(\mathbf{r}) = 0 \tag{36.5a}$$

and

$$\hat{\psi}^\dagger_\alpha(\mathbf{r})\hat{\psi}_\beta(\mathbf{r}') + \hat{\psi}_\beta(\mathbf{r}')\hat{\psi}^\dagger_\alpha(\mathbf{r}) = \delta_{\alpha\beta}\delta^{(3)}(\mathbf{r}-\mathbf{r}'). \tag{36.5b}$$

We will assume a contact form for the electron–electron interaction; i.e.,

$$U^{(2)}(\mathbf{r},\mathbf{r}') = -\,V\delta^{(3)}(\mathbf{r}-\mathbf{r}'), \tag{36.6}$$

where we have chosen the sign such that a positive V corresponds to an attractive potential and, in an inhomogeneous system, $V = V(\mathbf{r})$.

Since energies in many-body theory are conventionally measured relative to the Fermi energy, μ, we define

$$\hat{H}'^{(1)} = \hat{H}^{(1)} - \mu\hat{N}, \tag{36.7}$$

where $\hat{N} = \int d^3r\,\hat{\psi}^\dagger_\alpha(\mathbf{r})\hat{\psi}_\alpha(\mathbf{r})$ is the number operator.

In the self-consistent field (or potential) approach one seeks an approximate solution of the many-body problem in terms of an effective one-electron Hamiltonian; the best such Hamiltonian of a given form is taken to be the one which minimizes the total free energy. We write this effective Hamiltonian as[2]

$$\begin{aligned}\hat{H}^{(e)} = \int d^3r[&\hat{\psi}^\dagger_\alpha(\mathbf{r})\mathcal{H}'_0\hat{\psi}_\alpha(\mathbf{r}) + \hat{\psi}^\dagger_\alpha(\mathbf{r})U(\mathbf{r})\hat{\psi}_\alpha(\mathbf{r})\\ &+ \Delta(\mathbf{r})\hat{\psi}^\dagger_\uparrow(\mathbf{r})\hat{\psi}^\dagger_\downarrow(\mathbf{r}) + \Delta^*(\mathbf{r})\hat{\psi}_\downarrow(\mathbf{r})\hat{\psi}_\downarrow(\mathbf{r})],\end{aligned} \tag{36.8}$$

where

$$\mathcal{H}'_0 = -\frac{\hbar^2}{2m}\left(\nabla - \frac{ie}{\hbar c}\mathbf{A}\right)^2 - \mu. \tag{36.9}$$

$U(\mathbf{r})$ is a mean (one-electron) potential and $\Delta(\mathbf{r})$ is often referred to as the pairing potential.

The key feature of the mean field theory of superconductivity is that we admit nonvanishing expectation values for the spin-paired operators $\hat{\psi}^\dagger_\uparrow\hat{\psi}^\dagger_\downarrow$ and $\hat{\psi}_\downarrow\hat{\psi}_\uparrow$; such expectation values originate from the nonnumber-conserving BCS variational wavefunction and related ensembles (see Part II), and are referred to as *pair amplitudes*. In a nonsuperconductor, the $\hat{\psi}$ operators can

1. As discussed in Sec. 32, the electron–phonon interaction leads to a position (or equivalently $\mathbf{k}$) and time (frequency) dependent interaction. The Cooper model involves a $\xi(k)$-independent potential up to some cut-off, $\hbar\omega_D$. The introduction of a cut-off is equivalent to a retardation on a time scale of order ω_D.
2. The form (36.4b), in addition to 'singlet' pairing of the form $\hat{\psi}_\downarrow(\mathbf{r})\hat{\psi}_\uparrow(\mathbf{r})$ and its hermitian conjugate will also generate pairing in the 'triplet channel' leading to terms in Eq. (36.8) of the form $\hat{\psi}_\uparrow(\mathbf{r})\hat{\psi}_\uparrow(\mathbf{r})$, $\hat{\psi}_\downarrow(\mathbf{r})\hat{\psi}_\downarrow(\mathbf{r})$ as well as $\hat{\psi}_\downarrow(\mathbf{r})\hat{\psi}_\uparrow(\mathbf{r})$ and their hermitian conjugates. A potential which distinguishes between their pairings is introduced in Sec. 52.

be expanded in terms of a complete set of wavefunctions, such as the eigenfunctions w_n of $\mathscr{H}'_0$ (with $\mathbf{A} = 0$); i.e., $\mathscr{H}'_0 w_n = \xi_n w_n$, where ξ_n are the associated energies (again measured from the Fermi surface). Thus we would write

$$\hat{\psi}_\alpha(\mathbf{r}) = \sum_n w_n(\mathbf{r})\hat{c}_{n\alpha} \tag{36.10}$$

where $\hat{c}_{n\alpha}$ destroys a particle of spin α in state n. In a uniform superconductor, the Hamiltonian (36.8) is diagonalized by a Bogoliubov transformation (Bogoliubov 1958), as discussed in Sec. 27

$$\hat{c}_{n\uparrow} = \mathscr{u}_n\hat{\gamma}_{n\uparrow} - \mathscr{v}_n^*\hat{\gamma}_{n\downarrow}^\dagger \tag{36.11a}$$

and

$$\hat{c}_{n\downarrow} = \mathscr{u}_n\hat{\gamma}_{n\downarrow} + \mathscr{v}_n^*\hat{\gamma}_{n\uparrow}^\dagger, \tag{36.11b}$$

where $|\mathscr{u}_n|^2 + |\mathscr{v}_n|^2 = 1$ insures that if the $\hat{c}_{n\alpha}$ obey the Fermi commutation rules, then the $\hat{\gamma}_{n\alpha}$ also will; these commutation rules are:

$$\hat{c}_{n\alpha}\hat{c}_{n'\beta} + \hat{c}_{n'\beta}\hat{c}_{n\alpha} = 0, \tag{36.12a}$$

$$\hat{c}_{n\alpha}\hat{c}_{n'\beta}^\dagger + \hat{c}_{n'\beta}^\dagger\hat{c}_{n\alpha} = \delta_{nn'}\delta_{\alpha\beta}, \tag{36.12b}$$

and

$$\hat{\gamma}_{n\alpha}\hat{\gamma}_{n'\beta} + \hat{\gamma}_{n'\beta}\hat{\gamma}_{n\alpha} = 0, \tag{36.12c}$$

$$\hat{\gamma}_{n\alpha}\hat{\gamma}_{n'\beta}^\dagger + \hat{\gamma}_{n'\beta}^\dagger\hat{\gamma}_{n\alpha} = \delta_{nn'}\delta_{\alpha\beta}. \tag{36.12d}$$

For a nonuniform superconductor, rather than simply inserting (36.11a) and (36.11b) into (36.10) for the *two* spin states, we seek a more general transformation where two sets of eigenfunctions, $u_n(\mathbf{r})$ and $v_n(\mathbf{r})$, are introduced and we write $\hat{\psi}_\alpha$ as

$$\hat{\psi}_\uparrow(\mathbf{r}) = \sum_n [u_n(\mathbf{r})\hat{\gamma}_{n\uparrow} - v_n^*(\mathbf{r})\hat{\gamma}_{n\downarrow}^\dagger] \tag{36.13a}$$

and

$$\hat{\psi}_\downarrow(\mathbf{r}) = \sum_n [u_n(\mathbf{r})\hat{\gamma}_{n\downarrow} + v_n^*(\mathbf{r})\hat{\gamma}_{n\uparrow}^\dagger]. \tag{36.13b}$$

The greater flexibility afforded by Eqs. (36.13a) and (36.13b) can, in some cases, lead to a lower free energy, insuring a more accurate solution. We now develop a set of equations for the new functions u_n and v_n. We demand that Eqs. (36.13) diagonalize $\hat{H}_e$; i.e.,

$$\hat{H}_e = E_{0S} + \sum_{n,\alpha} \varepsilon_n\hat{\gamma}_{n\alpha}^\dagger\hat{\gamma}_{n\alpha}, \tag{36.14}$$

where E_{0S} is the superconducting ground-state energy. From Eqs. (36.12c), (36.12d), and (36.14) we have

$$[\hat{H}_e, \hat{\gamma}_{n\alpha}] = -\varepsilon_n\hat{\gamma}_{n\alpha} \tag{36.15a}$$

and

$$[\hat{H}_e, \hat{\gamma}^\dagger_{n\alpha}] = \varepsilon_n \hat{\gamma}^\dagger_{n\alpha}. \tag{36.15b}$$

Our strategy will be to evaluate the commutator of $\hat{H}_e$ with $\hat{\psi}_\uparrow$ and $\hat{\psi}_\downarrow$ in two nonequivalent ways. For the first of these we use (36.5a) and (36.5b) and Eq. (36.8) to obtain

$$[\hat{H}_e, \hat{\psi}_\uparrow(\mathbf{r})] = -[\mathscr{H}'_0 + U(\mathbf{r})]\hat{\psi}_\uparrow(\mathbf{r}) - \Delta(\mathbf{r})\hat{\psi}^\dagger_\downarrow(\mathbf{r}) \tag{36.16a}$$

and

$$[\hat{H}_e, \hat{\psi}_\downarrow(\mathbf{r})] = -[\mathscr{H}'_0 + U(\mathbf{r})]\hat{\psi}_\downarrow(\mathbf{r}) + \Delta(\mathbf{r})\hat{\psi}^\dagger_\uparrow(\mathbf{r}), \tag{36.16b}$$

after which we insert Eqs. (36.13) into the right-hand side. The second way to evaluate the commutator is to substitute Eqs. (36.13) directly, evaluating the resulting commutators of the $\hat{\gamma}$ operators using Eqs. (36.15). Comparing the coefficients of the $\hat{\gamma}$ operators in the two sets of expressions so formed we obtain Bogoliubov's equations

$$\varepsilon_n u_n(\mathbf{r}) = [\mathscr{H}'_0 + U(\mathbf{r})]u_n(\mathbf{r}) + \Delta(\mathbf{r})v_n(\mathbf{r}) \tag{36.17a}$$

and

$$\varepsilon_n v_n(\mathbf{r}) = -[\mathscr{H}'^*_0 + U(\mathbf{r})]v_n(\mathbf{r}) + \Delta^*(\mathbf{r})u_n(\mathbf{r}). \tag{36.17b}$$

Note that we have two coupled partial differential equations for our eigenfunctions u_n and v_n describing the superconducting state; however, the mean field potentials $U(\mathbf{r})$ and $\Delta(\mathbf{r})$ are still undetermined.

Eqs. (36.17) may be written in matrix form as[3]

$$\begin{bmatrix} [\mathscr{H}'_0 + U(\mathbf{r})] & \Delta(\mathbf{r}) \\ \Delta^*(\mathbf{r}) & -[\mathscr{H}'^*_0 + U(\mathbf{r})] \end{bmatrix} \begin{pmatrix} u_n(\mathbf{r}) \\ v_n(\mathbf{r}) \end{pmatrix} = \varepsilon_n \begin{pmatrix} u_n(\mathbf{r}) \\ v_n(\mathbf{r}) \end{pmatrix}. \tag{36.18}$$

Note that in a magnetic field the kinetic energy operator, $-(\hbar^2/2m)[\nabla - (ie/\hbar c)\mathbf{A}]^2$, contains an imaginary cross term which changes sign on taking the complex conjugate; hence $\mathscr{H}'^*_0 \neq \mathscr{H}'_0$.

By introducing a spinor representation $\psi = \begin{pmatrix} u(\mathbf{r}) \\ v(\mathbf{r}) \end{pmatrix}$, defining a Hamiltonian

$$\underset{\sim}{\mathscr{H}'} = \left[\frac{1}{2m}\left(\mathbf{p} - \frac{e}{c}\mathbf{A}\underset{\sim}{\tau}_z\right)^2 + U(\mathbf{r}) - \mu\right], \tag{36.19}$$

and writing the complex gap function as $\Delta(\mathbf{r}) = \Delta_1(\mathbf{r}) + i\Delta_2(\mathbf{r})$, we may rewrite Eq. (36.18) in the form

$$\varepsilon\psi = [\underset{\sim}{\mathscr{H}'}\underset{\sim}{\tau}_z + \Delta_1\underset{\sim}{\tau}_x + \Delta_2\underset{\sim}{\tau}_y]\psi \tag{36.20}$$

which is useful for some applications; here $\underset{\sim}{\tau}_i$ are the usual Pauli spin matrices.[4] Writing $\Delta_\pm = \Delta_1 + i\Delta_2$ and $\underset{\sim}{\tau}_\pm = \underset{\sim}{\tau}_x \pm i\underset{\sim}{\tau}_y$, Eq. (36.19) may also be written as

3. For the case of a uniform superconducting electron gas we can seek a solution of Eq. (36.18) in the form $u_\mathbf{k}(\mathbf{r}) = (1/L)^{3/2}\mathscr{u}_\mathbf{k}e^{i\mathbf{k}\cdot\mathbf{r}}$ and $v_\mathbf{k}(\mathbf{r}) = (1/L)^{3/2}\mathscr{v}_\mathbf{k}e^{i\mathbf{k}\cdot\mathbf{r}}$ with Δ constant and $U = 0$; L^3 is the volume of the system. The eigenvalue of Eq. (36.18) is then $\varepsilon^2 = \xi^2 + \Delta^2$ where $\xi = \hbar^2k^2/2m - \mu$. The normalization condition, (36.22a), reduces to $|\mathscr{u}_\mathbf{k}|^2 + |\mathscr{v}_\mathbf{k}|^2 = 1$ with which we can obtain the eigenvectors, $u_\mathbf{k}$ and $v_\mathbf{k}$. When these are inserted in Eq. (37.11) we obtain the usual BCS integral equations for Δ discussed in Sec. 26.
4. Note the $\underset{\sim}{\tau}_i$ introduced here act in *particle–hole space* and not in spin space.

$$\varepsilon\psi = [\mathscr{H}'\underset{\sim}{\tau}_z + \tfrac{1}{2}(\Delta_-\underset{\sim}{\tau}_+ + \Delta_+\underset{\sim}{\tau}_-)]\psi. \tag{36.21}$$

The functions u_n and v_n obey orthogonality relations which can be obtained from Eqs. (36.17). We multiply Eq. (36.17a) for u_n on left by u_m^*, multiply the complex conjugate of Eq. (36.17a) for u_m^* on the left by u_n and subtract to obtain

$$(\varepsilon_n - \varepsilon_m)\int d^3r u_m^* u_n = \int d^3r(u_m^*\mathscr{H}'u_n - u_n\mathscr{H}'^*u_m) + \int d^3r(u_n^*\Delta v_n - u_n\Delta^* v_m^*)$$

Performing similar steps with (36.17b) yields

$$(\varepsilon_n - \varepsilon_m)\int d^3r u_m^* v_n = -\int d^3r(v_m^*\mathscr{H}'v_n - v_n\mathscr{H}'v_m^*) + \int d^3r(v_n^*\Delta^* u_n - v_n\Delta u_m).$$

Performing integrations by parts on those terms on the right involving ∇ and adding these expressions we have

$$(\varepsilon_n - \varepsilon_m)\int d^3r(u_m^* u_n + v_m^* v_n) = 0.$$

The integral must therefore vanish for $n \neq m$. For $n = m$ the integral is undetermined and adopting the usual normalization condition we have

$$\int d^3r[u_m^*(\mathbf{r})u_n(\mathbf{r}) + v_m^*(\mathbf{r})v_n(\mathbf{r})] = \delta_{mn}. \tag{36.22a}$$

We next multiply Eq. (36.17a) for u_n on the left by v_m and Eq. (36.17b) for v_n on the left by u_m and subtract; performing the same operations with n and m interchanged and subtracting the resulting expressions yields a second orthogonality expression

$$\int d^3r[v_m(\mathbf{r})u_n(\mathbf{r}) - u_m(\mathbf{r})v_n(\mathbf{r})] = 0. \tag{36.22b}$$

We can also obtain expressions analogous to the completeness relations. Substituting Eqs. (36.13) into (36.5b) and using (36.12c) and (36.12d) we obtain

$$\sum_n [u_n^*(\mathbf{r})u_n(\mathbf{r}') + v_n(\mathbf{r})v_n^*(\mathbf{r}')] = \delta^{(3)}(\mathbf{r} - \mathbf{r}'). \tag{36.23a}$$

Similarly, substituting Eqs. (36.13) into (36.5a) and using (36.12c) and (36.12d) yields

$$\sum [u_n(\mathbf{r})v_n^*(\mathbf{r}') - v_n^*(\mathbf{r})u_n(\mathbf{r}')] = 0. \tag{36.23b}$$

37 Self-consistency conditions and the free energy

We must now determine $U(\mathbf{r})$ and $\Delta(\mathbf{r})$ which enter $\hat{H}_e$ from the requirement that the free energy evaluated with the full Hamiltonian (defined by Eq. (36.1)) be stationary with respect to our chosen expansion of the $\hat{\psi}$ operators. The free energy is given by

$$F = \langle \hat{H} \rangle - TS, \tag{37.1}$$

where the brackets imply a combined quantum mechanical and statistical average, to be performed with a Gibbs ensemble involving the eigenvalues and eigenfunctions of $\hat{H}_e$.

In calculating $\langle \hat{H} \rangle$ we must take the expectation value of a term involving four $\hat{\psi}$ operators arising from $\hat{V}^{(2)}$. In the mean field approximation this is accomplished by taking the product of the expectation values of all nonequivalent pairs of Fermi $\hat{\psi}$ operators, being careful to change the sign every time a pair of Fermi operators is commuted. In this approximation we have

$$\begin{aligned}\langle \hat{\psi}_\alpha^\dagger(\mathbf{r}_1)\hat{\psi}_\beta^\dagger(\mathbf{r}_2)\hat{\psi}_\gamma(\mathbf{r}_3)\hat{\psi}_\delta(\mathbf{r}_4)\rangle &= \langle \hat{\psi}_\alpha^\dagger(\mathbf{r}_1)\hat{\psi}_\beta^\dagger(\mathbf{r}_2)\rangle\langle \hat{\psi}_\gamma(\mathbf{r}_3)\hat{\psi}_\delta(\mathbf{r}_4)\rangle \\ &- \langle \hat{\psi}_\alpha^\dagger(\mathbf{r}_1)\hat{\psi}_\gamma(\mathbf{r}_3)\rangle\langle \hat{\psi}_\beta^\dagger(\mathbf{r}_2)\hat{\psi}_\delta(\mathbf{r}_4)\rangle + \langle \hat{\psi}_\alpha^\dagger(\mathbf{r}_1)\hat{\psi}_\delta(\mathbf{r}_4)\rangle\langle \hat{\psi}_\beta^\dagger(\mathbf{r}_2)\hat{\psi}_\gamma(\mathbf{r}_3)\rangle.\end{aligned} \tag{37.2}$$

For the case of the spin-independent contact interaction described by Eqs. (36.4) and (36.6), we can write the variation of the free energy as

$$\begin{aligned}\delta F = \int \mathrm{d}^3r\{&\delta\langle \hat{\psi}_\alpha^\dagger(\mathbf{r})\mathscr{H}'_0\hat{\psi}_\alpha(\mathbf{r})\rangle - V\langle \hat{\psi}_\alpha^\dagger(\mathbf{r})\hat{\psi}_\alpha(\mathbf{r})\rangle\delta\langle \hat{\psi}_\beta^\dagger(\mathbf{r})\hat{\psi}_\beta(\mathbf{r})\rangle \\ &+ V\langle \hat{\psi}_\alpha^\dagger(\mathbf{r})\hat{\psi}_\alpha(\mathbf{r})\rangle\delta\langle \hat{\psi}_\alpha^\dagger(\mathbf{r})\hat{\psi}_\alpha(\mathbf{r})\rangle - V[\langle \hat{\psi}_\uparrow^\dagger(\mathbf{r})\hat{\psi}_\downarrow^\dagger(\mathbf{r})\rangle\delta\langle \hat{\psi}_\downarrow(\mathbf{r})\hat{\psi}_\uparrow(\mathbf{r})\rangle \\ &+ \langle \hat{\psi}_\downarrow(\mathbf{r})\hat{\psi}_\uparrow(\mathbf{r})\rangle\delta\langle \hat{\psi}_\uparrow^\dagger(\mathbf{r})\hat{\psi}_\downarrow^\dagger(\mathbf{r})\rangle]\} - T\delta S = 0.\end{aligned} \tag{37.3}$$

Terms of the form $\langle \hat{\psi}_\uparrow^\dagger(\mathbf{r})\hat{\psi}_\downarrow(\mathbf{r})\rangle$ have been neglected in Eq. (37.3) since they are nonzero only in a magnetic system.

Now the free energy formed from our effective Hamiltonian,

$$F_e = \langle \hat{H}_e \rangle - TS, \tag{37.4}$$

is automatically stationary since our expectation value can be directly expressed (assuming we know the mean potentials) in terms of the eigenvalues and eigenfunctions of $\hat{H}_e$; i.e., $\delta\langle \hat{H}_e \rangle - T\delta S = 0$. Using Eq. (36.8) we have

$$\begin{aligned}\int \mathrm{d}^3r\{&\delta\langle \hat{\psi}_\alpha^\dagger(\mathbf{r})[\mathscr{H}'_0 + U(\mathbf{r})]\hat{\psi}_\alpha(\mathbf{r})\rangle \\ &+ [\Delta(\mathbf{r})\delta\langle \hat{\psi}_\uparrow^\dagger(\mathbf{r})\hat{\psi}_\downarrow^\dagger(\mathbf{r})\rangle + \Delta^*(\mathbf{r})\delta\langle \hat{\psi}_\downarrow(\mathbf{r})\hat{\psi}_\uparrow(\mathbf{r})\rangle]\} - T\delta S = 0.\end{aligned} \tag{37.5}$$

Comparing Eqs. (37.3) and (37.5) we see they are the same *provided* we define the self-consistent potentials $U(\mathbf{r})$ and $\Delta(\mathbf{r})$ according to

$$U(\mathbf{r}) = -V\langle\hat{\psi}_\uparrow^\dagger(\mathbf{r})\hat{\psi}_\uparrow(\mathbf{r})\rangle = -V\langle\hat{\psi}_\downarrow^\dagger(\mathbf{r})\hat{\psi}_\downarrow(\mathbf{r})\rangle \tag{37.6}$$

and

$$\Delta(\mathbf{r}) = -V\langle\hat{\psi}_\downarrow(\mathbf{r})\hat{\psi}_\uparrow(\mathbf{r})\rangle = V\langle\hat{\psi}_\uparrow(\mathbf{r})\hat{\psi}_\downarrow(\mathbf{r})\rangle. \tag{37.7}$$

At this point we substitute Eqs. (36.13a) and (36.13b) into (37.6) and (37.7). The mean values of the $\hat{\gamma}$ operators that then occur are evaluated using the standard rules

$$\langle\hat{\gamma}_{n\alpha}^\dagger\hat{\gamma}_{m\beta}\rangle = \delta_{\alpha\beta}\delta_{nm}f_n, \tag{37.8a}$$

$$\langle\hat{\gamma}_{m\beta}\hat{\gamma}_{n\alpha}^\dagger\rangle = \delta_{\alpha\beta}\delta_{nm}(1-f_n), \tag{37.8b}$$

and

$$\langle\hat{\gamma}_{n\alpha}\hat{\gamma}_{m\beta}\rangle = 0; \tag{37.8c}$$

here

$$f_n = \frac{1}{e^{\varepsilon_n/k_B T}+1} \tag{37.9}$$

is the usual Fermi occupation factor, which follows from a minimization of the $-TS$ contribution to F, where S is given by (28.7). Performing these calculations we have

$$U(\mathbf{r}) = -V\sum_n [|u_n(\mathbf{r})|^2 f_n + |v_n(\mathbf{r})|^2(1-f_n)] \tag{37.10}$$

and

$$\Delta(\mathbf{r}) = +V\sum_n v_n^*(\mathbf{r})u_n(\mathbf{r})(1-2f_n). \tag{37.11}$$

We see that superconductivity in an inhomogeneous system is governed by a system of equations involving the two Bogoliubov equations, Eqs. (36.17a) and (36.17b), and two self-consistency conditions, Eqs. (37.10) and (37.11).

Having obtained the equations which minimize the free energy, we now construct the corresponding expression for F_e itself. To do this we express the ψ operators entering $\langle H_e\rangle$ in Eq. (37.4) through Eq. (36.8) using Eqs. (36.13a) and (36.13b). On summing over the two spin states and defining $\langle\hat{H}_e'\rangle$ as $\langle\hat{H}_e - \mu\hat{N}\rangle$ we obtain

$$\begin{aligned}\langle H_e'\rangle = \Big\langle \int d^3r \sum_{n,n'} \{&[u_n^*(\mathbf{r})\hat{\gamma}_{n\uparrow}^\dagger - v_n(\mathbf{r})\hat{\gamma}_{n\downarrow}][\mathscr{H}_0' + U(\mathbf{r})][u_{n'}(\mathbf{r})\hat{\gamma}_{n'\uparrow} - v_{n'}^*(\mathbf{r})\hat{\gamma}_{n'\downarrow}^\dagger]\\ &+ [u_n^*(\mathbf{r})\hat{\gamma}_{n\downarrow}^\dagger + v_n(\mathbf{r})\hat{\gamma}_{n\uparrow}][\mathscr{H}_0' + U(\mathbf{r})][u_{n'}(\mathbf{r})\hat{\gamma}_{n'\downarrow} + v_{n'}^*(\mathbf{r})\hat{\gamma}_{n'\uparrow}^\dagger]\\ &+ \Delta(\mathbf{r})[u_n^*(\mathbf{r})\hat{\gamma}_{n\uparrow}^\dagger - v_n(\mathbf{r})\hat{\gamma}_{n\downarrow}][u_{n'}^*(\mathbf{r})\hat{\gamma}_{n'\downarrow}^\dagger + v_{n'}(\mathbf{r})\hat{\gamma}_{n'\uparrow}]\\ &+ \Delta^*(\mathbf{r})[u_n(\mathbf{r})\hat{\gamma}_{n\downarrow} + v_n^*(\mathbf{r})\hat{\gamma}_{n\uparrow}^\dagger][u_{n'}(\mathbf{r})\hat{\gamma}_{n'\uparrow} - v_{n'}^*(\mathbf{r})\hat{\gamma}_{n'\downarrow}^\dagger]\}\Big\rangle.\end{aligned}$$

Using (37.8a) and (37.8b) we have

$$\langle H_e'\rangle = \int d^3r \sum_n \{2u_n^*(\mathbf{r})[\mathscr{H}_0' + U(\mathbf{r})]u_n(\mathbf{r})f_n + 2v_n(\mathbf{r})[\mathscr{H}_0' + U(\mathbf{r})]v_n^*(\mathbf{r})(1-f_n)$$

$$- \Delta(\mathbf{r})u_n^*(\mathbf{r})v_n(\mathbf{r})(1 - 2f_n) - \Delta^*(\mathbf{r})u_n(\mathbf{r})v_n^*(\mathbf{r})(1 - 2f_n)\}.$$

Inserting the Bogoliubov equations, (36.17a) and (36.17b), and using the orthogonality condition (36.22) and Eq. (37.11) for $\Delta(\mathbf{r})$ we obtain

$$\langle H_e \rangle = \sum_n 2\varepsilon_n f_n - \int d^3r \left[\frac{|\Delta(\mathbf{r})|^2}{V} + 2\sum_n |v_n(\mathbf{r})|^2 \varepsilon_n \right]. \tag{37.12}$$

Comparing this equation with Eq. (36.14) we see that the term involving the integral must correspond to the ground state energy, E_{0S}

$$E_{0S} = - \int d^3r \left[\frac{|\Delta(\mathbf{r})|^2}{V} + 2\sum_n |v_n(\mathbf{r})|^2 \varepsilon_n \right]. \tag{37.13}$$

The term $-TS$ is obtained using (28.7) which can be rewritten as

$$-TS = k_B T \sum_n \left[\frac{\varepsilon_n}{k_B T}(1 - 2f_n) - 2\ell n\left(2\cosh\frac{\varepsilon_n}{2k_B T}\right)\right]. \tag{37.14}$$

The expression for the effective free energy is then

$$F_e = \sum_n \left[-2k_B T \ell n\left(2\cosh\frac{\varepsilon_n}{2k_B T}\right) + \varepsilon_n \right] + E_{0S}, \tag{37.15}$$

which is to be compared with the uniform superconductor result, (28.22).

In the presence of a magnetic field diamagnetic currents flow. The current operator in a magnetic field follows from the usual prescription of substituting $\nabla - (ie/\hbar c)\mathbf{A}(\mathbf{r})$ in the expression for the quantum mechanical current:

$$\hat{\mathbf{j}}(\mathbf{r}) = \frac{e\hbar}{2im}[\hat{\psi}_\alpha^\dagger(\mathbf{r})\nabla\hat{\psi}_\alpha(\mathbf{r}) - (\nabla\hat{\psi}_\alpha^\dagger(\mathbf{r}))\hat{\psi}_\alpha(\mathbf{r})]$$
$$- \frac{e^2}{mc}\mathbf{A}(\mathbf{r})\hat{\psi}_\alpha^\dagger(\mathbf{r})\hat{\psi}_\alpha(\mathbf{r}). \tag{37.16}$$

Inserting (36.13a) and (36.13b) and taking expectation values according to (37.8), the current density becomes

$$\langle\hat{\mathbf{j}}(\mathbf{r})\rangle \equiv \mathbf{j}(\mathbf{r}) = -\frac{ie\hbar}{2m}\sum_n \{[u_n^*(\mathbf{r})\nabla u_n(\mathbf{r}) - \text{c.c.}]f_n$$
$$+ [v_n(\mathbf{r})\nabla v_n^*(\mathbf{r}) - \text{c.c.}](1 - f_n)\}$$
$$- \frac{e^2}{mc}\mathbf{A}(\mathbf{r})\sum_n [|u_n(\mathbf{r})|^2 f_n + |v_n(\mathbf{r})|^2(1 - f_n)]. \tag{37.17}$$

We may write this as the sum of two contributions

$$\mathbf{j}(\mathbf{r}) = \mathbf{j}_1(\mathbf{r}) + \mathbf{j}_2(\mathbf{r}), \tag{37.18}$$

where $\mathbf{j}_1$ is the term in curly brackets and $\mathbf{j}_2$ is the remainder; the latter is referred to as the 'gauge current'.

38 Linearized self-consistency condition and the correlation function

38.1 Treating the gap function as a perturbation

If we restrict ourselves to a determination of the superconducting transition temperature we may apply a linear approximation to the self-consistency condition (analogous to our linearization of the G–L equation in Part I) to determine T_c. In later sections we will return to the full nonlinear theory.

In the normal state the eigenfunctions follow from the solution of

$$\mathscr{H}'_0\phi_n = \xi_n\phi_n$$

or

$$\left[-\frac{\hbar^2}{2m}\left(\nabla - \frac{ie}{\hbar c}\mathbf{A}\right)^2 + U(\mathbf{r}) - \mu\right]\phi_n = \xi_n\phi_n \tag{38.1}$$

(in what follows we will combine $U^{(1)}(\mathbf{r})$ and $U(\mathbf{r})$ and write the sum simply as $U(\mathbf{r})$). Near the superconducting transition we expect u_n and v_n to differ only slightly from ϕ_n and it is natural to seek an approximate solution using perturbation theory. We introduce the expansions

$$u_n = u_n^{(0)} + u_n^{(1)} + \ldots, \tag{38.2a}$$

$$v_n = v_n^{(0)} + v_n^{(1)} + \ldots, \tag{38.2b}$$

and

$$\varepsilon_n = \varepsilon_n^{(0)} + \varepsilon_n^{(1)} + \ldots. \tag{38.2c}$$

From the structure of Eqs. (36.17a) and (36.17b) it is apparent that when $\Delta = 0$, where $\varepsilon_n^{(0)} = |\xi_n|$, the solutions are restricted in energy and have the form

$$u_n^{(0)} = \phi_n, \quad v_n^{(0)} = 0, \quad \xi_n > 0$$

and

$$u_n^{(0)} = 0, \quad v_n^{(0)} = \phi_n^*, \quad \xi_n < 0$$

or, more compactly,

$$u_n^{(0)} = \phi_n\theta(\xi_n) \tag{38.3a}$$

and

$$v_n^{(0)} = \phi_n^* \theta(-\xi_n), \tag{38.3b}$$

where $\theta(x)$ is the Heaviside (or step) function. Note that from Eq. (36.17b), the v_n are eigenfunctions of $\mathscr{H}_0'^* + U$ when $\Delta = 0$ and hence the appropriate eigenfunctions are ϕ_n^*, as we have written in Eq. (38.3b).

It is useful to note at this point that a slight modification of Eqs. (38.3) forms the basis of an approximate solution to the Bogoliubov equations at all temperatures. Rather than writing Eqs. (38.3) in terms of the functions $\theta(\xi_n)$ and $\theta(-\xi_n)$, we use the corresponding functions of BCS theory and write $u_n(\mathbf{r}) = \mathscr{u}_n\phi_n(\mathbf{r})$ and $v_n(\mathbf{r}) = \mathscr{v}_n\phi_n^*(\mathbf{r})$ where $\mathscr{u}_n$ and $\mathscr{v}_n$ are given by Eqs. (26.27) and (26.28). Physically this approximation relies on the strong similarity between the superconducting and normal-state wavefunctions.

Since $u_n^{(0)}v_n^{(0)} = 0$ in Eq. (37.11), Δ vanishes in the zeroth order. We expand $u_n^{(1)}$ and $v_n^{(1)}$ in terms of the complete and orthonormal set ϕ_m

$$u_n^{(1)} = \sum_m a_{nm}\phi_m \tag{38.4a}$$

and

$$v_n^{(1)} = \sum_m b_{nm}\phi_m^*. \tag{38.4b}$$

If we insert these expansions into (36.17a) and (36.17b), multiply the first equation by ϕ_n^* and the second by ϕ_m, and integrate over d^3r we obtain (for $m \neq n$)

$$a_{nm} = \left(\frac{\theta(-\xi_n)}{|\xi_n| - \xi_m}\right)\int \Delta(\mathbf{r})\phi_n^*(\mathbf{r})\phi_m^*(\mathbf{r})\mathrm{d}^3r \tag{38.5a}$$

and

$$b_{nm} = \left(\frac{\theta(+\xi_n)}{|\xi_n| + \xi_m}\right)\int \Delta^*(\mathbf{r})\phi_n(\mathbf{r})\phi_m(\mathbf{r})\mathrm{d}^3r. \tag{38.5b}$$

(The diagonal elements of a_{nn} ($\xi_n > 0$) and b_{nn} ($\xi_n < 0$) must vanish if our expanded wavefunctions are to remain normalized; see Appendix C.) Inserting the values of a_{mn} and b_{mn} obtained from (38.5) into (38.4), and using the resulting $u_n^{(1)}$ and $v_n^{(1)}$ in Eq. (37.11), we obtain the first order correction to the self-consistency condition for $\Delta(\mathbf{r})$ which is an integral equation of the form

$$\Delta(\mathbf{r}) = \int K(\mathbf{r}, \mathbf{r}')\Delta(\mathbf{r}')\mathrm{d}^3r', \tag{38.6}$$

where the kernel is defined as

$$K(\mathbf{r}', \mathbf{r}) = V\sum_{m,n}[1 - 2f(|\xi_n|)]\left[\frac{\theta(\xi_n)}{|\xi_n| + \xi_m} + \frac{\theta(-\xi_n)}{|\xi_n| - \xi_m}\right]\phi_n^*(\mathbf{r}')\phi_m^*(\mathbf{r}')\phi_n(\mathbf{r})\phi_m(\mathbf{r}). \tag{38.7}$$

Noting $\theta(-\xi_n)/(|\xi_n| - \xi_m) = -\theta(-\xi_n)/(\xi_n + \xi_m)$, using the identity $1 - 2f(\xi_n) = \tanh(\xi_n/2k_BT)$

(and the fact that this function is odd) and symmetrizing the expression with respect to m and n, we may rewrite Eq. (38.7) as

$$K(\mathbf{r}',\mathbf{r}) = \frac{V(\mathbf{r})}{2}\sum_{m,n}\frac{\tanh\left(\dfrac{\xi_n}{2k_BT}\right)+\tanh\left(\dfrac{\xi_m}{2k_BT}\right)}{\xi_n+\xi_m}\phi_n^*(\mathbf{r}')\phi_m^*(\mathbf{r}')\phi_n(\mathbf{r})\phi_m(\mathbf{r}). \quad (38.8)$$

38.2 Relation to a correlation function

Although Eq. (38.8) is the formal solution to the posed problem of expressing the gap function perturbatively, it is not very useful as it stands. We now show that K can be related to a correlation function. We start by writing Eq. (38.8) in another form by noting that the function $\tanh x$ has poles with a residue of unity at $\xi = 2\pi i k_B T(\nu + \frac{1}{2})$, where ν is the set of all positive and negative integers. Using a theorem from the theory of complex functions, we may regard $\tanh x$ as the analytic continuation of the function $\Sigma_\nu 1/[x \pm (\nu + \frac{1}{2})i\pi]$ or

$$\tanh\frac{\xi}{2k_BT} = 2k_BT\sum_\omega\frac{1}{\xi \pm i\hbar\omega}, \quad (38.9)$$

where $\hbar\omega \equiv 2\pi k_BT(\nu + \frac{1}{2})$. We may then write

$$\frac{\tanh\dfrac{\xi_n}{2k_BT}+\tanh\dfrac{\xi_m}{2k_BT}}{\xi_n+\xi_m} = 2k_BT\sum_\omega\frac{1}{\xi_n+\xi_m}\left(\frac{1}{\xi_n - i\hbar\omega}+\frac{1}{\xi_m+i\hbar\omega}\right)$$

$$= 2k_BT\sum_\omega\frac{1}{(\xi_n - i\hbar\omega)(\xi_m + i\hbar\omega)}.$$

Thus,[1]

$$K(\mathbf{r}',\mathbf{r}) = V(\mathbf{r})k_BT\sum_\omega\sum_{m,n}\frac{\phi_m^*(\mathbf{r}')\phi_n^*(\mathbf{r}')\phi_m(\mathbf{r})\phi_n(\mathbf{r})}{(\xi_m + i\hbar\omega)(\xi_n - i\hbar\omega)}. \quad (38.10)$$

It will turn out to be important to write the energy denominator in Eq. (38.10) as

$$\frac{1}{(\xi_n - i\hbar\omega)(\xi_m + i\hbar\omega)} = \int d\xi d\xi'\frac{\delta(\xi - \xi_n)\delta(\xi' - \xi_m)}{(\xi - i\hbar\omega)(\xi' + i\hbar\omega)}. \quad (38.11)$$

By introducing a complex conjugation operator, $\mathscr{C}$, having the property $\mathscr{C}\phi_m = \phi_m^*$ we may write the wavefunction products occurring in (38.10) as[2]

1. It may be shown that K is given by $k_BTV\Sigma_\omega\mathscr{G}(\omega;\mathbf{r},\mathbf{r}')\mathscr{G}(-\omega;\mathbf{r},\mathbf{r}')$ where $\mathscr{G}(\omega;\mathbf{r},\mathbf{r}')$ is the one-particle thermal Green's function discussed in Sec. 55.
2. Formally, we have
$\langle m|\delta(\mathbf{R}-\mathbf{r}_1)|n\rangle = \int d^3r'd^3r''\langle m|\mathbf{r}'\rangle\langle\mathbf{r}'|\delta(\mathbf{R}-\mathbf{r}_1)|\mathbf{r}''\rangle\langle\mathbf{r}''|n\rangle = \int d^3r'd^3r''\langle m|\mathbf{r}'\rangle\delta(\mathbf{r}'-\mathbf{r}_1)$
$\times\langle\mathbf{r}'|\mathbf{r}''\rangle\langle\mathbf{r}''|n\rangle = \int d^3r'd^3r''\langle m|\mathbf{r}'\rangle\delta(\mathbf{r}'-\mathbf{r}_1)\delta(\mathbf{r}'-\mathbf{r}'')\langle\mathbf{r}'|n\rangle = \langle m|\mathbf{r}_1\rangle\langle\mathbf{r}_1|n\rangle = \phi_m^*(\mathbf{r}_1)\phi_n(\mathbf{r}_1)$.

$$\phi_n^*(\mathbf{r})\phi_m^*(\mathbf{r}) = \langle n\,|\,\delta(\mathbf{R}-\mathbf{r})\mathscr{C}\,|\,m\rangle \tag{38.12a}$$

and

$$\phi_n(\mathbf{r}')\phi_m(\mathbf{r}') = \langle m\,|\,\mathscr{C}^\dagger\delta(\mathbf{R}-\mathbf{r}')\,|\,n\rangle, \tag{38.12b}$$

where $\mathbf{R}$ is the position operator, $\mathbf{R}\,|\,\mathbf{r}\rangle = \mathbf{r}\,|\,\mathbf{r}\rangle$. Using (38.11) and (38.12) together with the identity

$$\delta(\xi-\xi_n)\delta(\xi'-\xi_m) = \delta(\xi-\xi_n-\xi'+\xi_m)\delta(\xi-\xi_n) \tag{38.13}$$

we may write Eq. (38.10) as

$$K(\mathbf{r},\mathbf{r}') = L^3 k_{\mathrm{B}} T \sum_\omega \int \frac{\mathscr{N}(\xi) V \mathrm{d}\xi \mathrm{d}\xi'}{(\xi - \mathrm{i}\hbar\omega)(\xi' + \mathrm{i}\hbar\omega)} g_\xi(\mathbf{r},\mathbf{r}',\hbar\Omega) \tag{38.14}$$

by noting $\Sigma_n \delta(\xi-\xi_n) = \mathscr{N}(\xi)L^3$ (where L^3 is the volume and $\mathscr{N}(\xi)$ is the density of states) and identifying (defining) a correlation function, g_ξ, as

$$g_\xi(\mathbf{r},\mathbf{r}',\hbar\Omega) \equiv \frac{\sum_{m,n} \langle n\,|\,\delta(\mathbf{R}-\mathbf{r})\mathscr{C}\,|\,m\rangle\langle m\,|\,\mathscr{C}^\dagger\delta(\mathbf{R}-\mathbf{r}')\,|\,n\rangle\delta(\xi-\xi_n)\delta(\xi_m-\xi_n-\hbar\Omega)}{\sum_n \delta(\xi-\xi_n)}, \tag{38.15}$$

where $\hbar\Omega \equiv \xi' - \xi$. In what follows we will drop the subscript on g; i.e., we will assume that g_ξ is independent of ξ for the energy range relevant to superconductivity. It is instructive to examine the Fourier transform of Eq. (38.15)[3]

$$g(\mathbf{r},\mathbf{r}',t) = \int \mathrm{e}^{-\mathrm{i}\Omega t} g(\mathbf{r},\mathbf{r}',\Omega)\mathrm{d}\Omega. \tag{38.16}$$

We can obtain an alternative form of $\mathbf{K}(\mathbf{r},\mathbf{r}')$ by inserting (38.16) into (38.14) and carrying out the two integrations over $\mathrm{d}\xi$ and $\mathrm{d}\xi'$ (using Cauchy's theorem). For each integral there are four cases: $t<0, \omega<0$; $t>0, \omega>0$; $t<0, \omega>0$; and $t>0, \omega<0$. The result is

$$K(\mathbf{r},\mathbf{r}') = \frac{2\pi}{\hbar} L^3 k_{\mathrm{B}} T \mathscr{N}(0) V \sum_\omega \int_0^\infty \mathrm{d}t \mathrm{e}^{-2|\omega|t} g(\mathbf{r},\mathbf{r}',t), \tag{38.14'}$$

where we assumed a constant density of states evaluated at the Fermi surface.

We now turn to a critical examination of $g(\mathbf{r},\mathbf{r}',t)$. The δ function involving Ω in $g(\Omega)$ produces a term $\mathrm{e}^{+\mathrm{i}(\xi_n-\xi_m)t/\hbar}$ in $g(t)$; we may then write this factor and the first matrix element in (38.15) as

$$\begin{aligned}\langle n\,|\,\delta(\mathbf{R}-\mathbf{r})\mathscr{C}\,|\,m\rangle \mathrm{e}^{+\mathrm{i}(\xi_n-\xi_m)t/\hbar} &= \langle n\,|\,\mathrm{e}^{\mathrm{i}\xi_n t/\hbar}\delta(\mathbf{R}-\mathbf{r})\mathrm{e}^{-\mathrm{i}\xi_m t/\hbar}\mathscr{C}\,|\,m\rangle \\ &= \langle n\,|\,\mathrm{e}^{+\mathrm{i}\xi_n t/\hbar}\delta(\mathbf{R}-\mathbf{r})\mathrm{e}^{-\mathrm{i}\xi_m t/\hbar}\mathrm{e}^{-\mathrm{i}\xi_m t/\hbar}\mathscr{C}\mathrm{e}^{-\mathrm{i}\xi_m t/\hbar}\,|\,m\rangle,\end{aligned} \tag{38.17}$$

where we exploited the properties of the complex conjugation operator in the second step. Defining the Heisenberg operator

3. There are four transforms involving the variables $\mathbf{r}$, $\mathbf{q}$, t, and Ω which occur regularly; $g(\mathbf{r},\mathbf{r}',t)$, $g(\mathbf{q},\mathbf{q}',t)$, $g(\mathbf{r},\mathbf{r}',\Omega)$, and $g(\mathbf{q},\mathbf{q}',\Omega)$. For translationally invariant systems only the difference of the space or wave vector variables enters: $\mathbf{r}-\mathbf{r}' \to \mathbf{r}$; $\mathbf{q}-\mathbf{q}' \to \mathbf{q}$.

$$\delta[\mathbf{R}(t) - \mathbf{r}] = e^{i\mathscr{H}'t/\hbar}\delta(\mathbf{R} - \mathbf{r})e^{-i\mathscr{H}'t/\hbar}, \tag{38.18a}$$

noting that the corresponding time-dependent $\mathscr{C}$ operator is defined as

$$\mathscr{C}(t) = e^{-i\mathscr{H}'t/\hbar}\mathscr{C}e^{-i\mathscr{H}'t/\hbar}, \tag{38.18b}$$

and that $\Sigma_m |m\rangle\langle m| = 1$, we have[4]

$$g(\mathbf{r}, \mathbf{r}', t) = \frac{\sum_n \langle n|\delta[\mathbf{R}(t) - \mathbf{r}]\mathscr{C}(t)\mathscr{C}^\dagger(0)\delta[\mathbf{R}(0) - \mathbf{r}']|n\rangle\delta(\xi - \xi_n)}{\sum_n \delta(\xi - \xi_n)}. \tag{38.19}$$

The time dependence of $\mathscr{C}$ is given by the usual Heisenberg equation of motion

$$\frac{d\mathscr{C}}{dt} = \frac{i}{\hbar}[\mathscr{H}', \mathscr{C}] = -\frac{e}{mc}[\nabla\cdot\mathbf{A} + \mathbf{A}\cdot\nabla]\mathscr{C}. \tag{38.20a}$$

When $\mathbf{A} = 0$, $\mathscr{C}(t)$ is time-independent, $\mathscr{C}\mathscr{C}^\dagger = 1$, and $g(\mathbf{r}, \mathbf{r}', t)$ has a simple interpretation as a correlation function: it measures the amplitude that a particle of energy ξ located at $\mathbf{r}'$ at $t = 0$ will be at the point $\mathbf{r}$ at time t. In calculating this amplitude we sum over all states n with energy ξ; the denominator in (38.19) normalizes the amplitude by the total density of such states.

When $\mathbf{A} \neq 0$, we may (in some situations) approximate the velocity operator $\mathbf{v} = (\hbar/im)(\nabla - (ie/\hbar c)\mathbf{A})$ by $\hbar\nabla/im$; this is valid provided we choose a gauge where the Fermi momentum $p_F >> (e/c)|\mathbf{A}|$. If we locally neglect the position dependence of $\mathbf{A}$ (or work in the gauge $\nabla\cdot\mathbf{A} = 0$), the previous equation of motion becomes

$$\begin{aligned}\frac{d\mathscr{C}}{dt} &\cong -\frac{2ie}{c\hbar}\mathbf{A}\cdot\mathbf{v}\mathscr{C} \\ &\equiv -i\frac{d\Phi}{dt}\mathscr{C};\end{aligned} \tag{38.20b}$$

then

$$\mathscr{C}(t) = e^{-i\Phi(t)}\mathscr{C}(0)$$

where the integral of Φ is given by

$$\Phi(t) = \frac{2e}{\hbar c}\int_0^t \mathbf{A}\cdot\mathbf{v}\,dt.$$

When account is taken of the two $\mathbf{r}$-dependent δ functions in (38.20), we may write the phase as

$$\Phi(t) = \frac{2e}{\hbar c}\mathbf{A}\cdot(\mathbf{r} - \mathbf{r}'). \tag{38.21}$$

4. Had we not earlier written our energy denominators in the form given in Eq. (38.11), we would not be able to apply the closure property here. This important trick is due to van Hove. Expressing $K(\mathbf{r}, \mathbf{r}')$ in terms of the correlation function (38.19) is due to deGennes.

Our earlier neglect of the position dependence of $\mathbf{A}$ is partially rectified by now writing[5] $\mathbf{A} = \mathbf{A}(\mathbf{r})$. One justifies this by noting that the dominant contribution to $K(\mathbf{r}, \mathbf{r}')$ arises from a small region with a radius of the order of a coherence length about the point $\mathbf{r} = \mathbf{r}'$; $\mathbf{A}(\mathbf{r})$ is then assumed to vary only slowly through this region. With these approximations the kernel of our linearized self-consistency condition is

$$K(\mathbf{r}, \mathbf{r}') = L^3 k_B T \sum_{\omega} \int \frac{\mathcal{N}(\xi) V \mathrm{d}\xi \mathrm{d}\xi'}{(\xi - \mathrm{i}\hbar\omega)(\xi' + \mathrm{i}\hbar\omega)} g(\mathbf{r}, \mathbf{r}', \Omega), \tag{38.22a}$$

where[6]

$$\begin{aligned} g(\mathbf{r}, \mathbf{r}', \Omega) &= \frac{1}{2\pi} \int \mathrm{e}^{\mathrm{i}\Omega t} g(\mathbf{r}, \mathbf{r}', t) \mathrm{d}t \\ &\cong \frac{1}{2\pi} \int \mathrm{e}^{\mathrm{i}\Omega t} g_0(\mathbf{r}, \mathbf{r}', t) \mathrm{e}^{-(2\mathrm{i}e/\hbar c)\mathbf{A}(\mathbf{r}')\cdot(\mathbf{r} - \mathbf{r}')} \mathrm{d}t \end{aligned} \tag{38.22b}$$

and g_0 is the zero-field correlation function (we likewise denote our zero field kernel as $K_0(\mathbf{r}, \mathbf{r}')$). We will discuss a more accurate prescription for obtaining g in the presence of a vector potential in Sec. 40 (see discussion leading up to Eq. (40.13)). In a homogeneous system, Eq. (38.6) may be written

$$\Delta(\mathbf{r}) = \int K(\mathbf{r} - \mathbf{r}') \Delta(\mathbf{r}') \mathrm{d}^3 r'. \tag{38.23}$$

A perfect crystal is invariant only under discrete translations and an alloy has no translational symmetry at all; i.e., neither is homogeneous. However, for most situations the distance scales of interest in superconductivity are of the order of a coherence length or larger, i.e., much larger than an interatomic separation. We will assume that we may take an 'average' of Eq. (38.6) in the form

$$\overline{\Delta(\mathbf{r})} = \int \overline{K(\mathbf{r}, \mathbf{r}') \Delta(\mathbf{r}')} \mathrm{d}^3 r$$

and that $\overline{K(\mathbf{r}, \mathbf{r}')} \to K(\mathbf{r} - \mathbf{r}')$ and $\overline{\Delta(\mathbf{r})} \to \Delta(\mathbf{r})$ to yield Eq. (38.23).[7]

Eq. (38.23) is easily solved if we take the Fourier transform

$$\begin{aligned} \Delta(\mathbf{q}) &= \int \mathrm{e}^{\mathrm{i}\mathbf{q}\cdot\mathbf{r}} K(\mathbf{r} - \mathbf{r}') \Delta(\mathbf{r}') \mathrm{d}^3 r' \mathrm{d}^3 r \\ &= \int \mathrm{e}^{\mathrm{i}\mathbf{q}\cdot(\mathbf{r} - \mathbf{r}')} K(\mathbf{r} - \mathbf{r}') \mathrm{d}^3 |\mathbf{r} - \mathbf{r}'| \mathrm{e}^{\mathrm{i}\mathbf{q}\cdot\mathbf{r}'} \Delta(\mathbf{r}') \mathrm{d}^3 r' \\ &= K(\mathbf{q}) \Delta(\mathbf{q}), \end{aligned} \tag{38.24}$$

5. A slightly more accurate expression for the phase is

$$\Phi(t) = \frac{2e}{\hbar c} \int_{\mathbf{r}(0)}^{\mathbf{r}(t)} \mathbf{A}(\mathbf{s}) \cdot \mathrm{d}\mathbf{s}.$$

6. In this section we follow deGennes convention of incorporating the $(2\pi)^{-1}$ factor in the Fourier time transform. In Secs. 54 and 55 we will follow the more standard convention of incorporating the $(2\pi)^{-1}$ factor in the Fourier frequency transform.
7. For a system having a symmetry lower than cubic, the anisotropy of the correlation function entering K would be characterized by a tensor diffusion constant in the dirty limit (see Subsec. 40.2).

which has the solutions $K(\mathbf{q}) = 1$ or $\Delta(\mathbf{q}) = 0$. In a uniform system, Δ is a constant corresponding to $q = 0$; the temperature at which $K(\mathbf{q} = 0) = 1$ defines the bulk transition temperature, T_c. Qualitatively boundary conditions on Δ can constrain $\mathbf{q}$ to some characteristic nonzero value $\mathbf{q}_0$, corresponding to an inhomogeneous gap;[8] the condition $K(\mathbf{q}_0) = 1$ then occurs for $T < T_c$.

We write the zero field Fourier transform of K as

$$K_0(\mathbf{q}) = \int d^3|\mathbf{r} - \mathbf{r}'|\,K_0(\mathbf{r} - \mathbf{r}')e^{i\mathbf{q}\cdot(\mathbf{r}-\mathbf{r}')}$$

$$= \frac{1}{L^3}\int d^3r d^3r' K_0(\mathbf{r} - \mathbf{r}')e^{i\mathbf{q}\cdot(\mathbf{r}-\mathbf{r}')}, \tag{38.25}$$

which, using Eq. (38.22), may be written

$$K_0(\mathbf{q}) = k_B T\sum_{\omega}\int\frac{\mathcal{N}(\xi)V d\xi d\xi'}{(\xi - i\hbar\omega)(\xi' + i\hbar\omega)}\frac{1}{2\pi}\int e^{i\Omega t}g_0(\mathbf{q}, t)dt, \tag{38.26a}$$

where

$$g_0(\mathbf{q}, t) = \frac{\sum_n \langle n|e^{i\mathbf{q}\cdot\mathbf{R}(t) - i\mathbf{q}\cdot\mathbf{R}(0)}|n\rangle\delta(\xi - \xi_n)}{\sum_n \delta(\xi - \xi_n)}. \tag{38.26b}$$

We can denote the average defined in Eq. (38.26b) as

$$g_0(\mathbf{q}, t) \equiv \langle e^{i\mathbf{q}\cdot\mathbf{R}(t)}e^{-i\mathbf{q}\cdot\mathbf{R}(0)}\rangle_{\xi}. \tag{38.27}$$

8. We emphasize again that describing inhomogenities by a characteristic wave vector, $\mathbf{q}_0$, gives only a qualitatively correct picture of a boundary-induced, position-dependent gap function. A more quantitative theory is given in Sec. 43.

39 Behavior of the correlation function in the clean and dirty limits

39.1 A simple model for the clean limit

The evaluation of $g_0(\mathbf{q}, t)$ in a real material is extremely difficult and we will invoke model solutions. We first consider the limit of a very pure metal with a free-electron-like band structure. In this case the average involved in (38.27) is over a spherical energy shell; the shell of interest is essentially the Fermi surface. In a semiclassical model we assume that $\mathbf{R}$ is not an operator. At $t = 0$ the particle is at $\mathbf{R}(0) = \mathbf{R}_0$; at time t it is at $\mathbf{R}_0 + \mathbf{V}t$ where $|\mathbf{V}| \cong V_F$. We may refer to this as the ballistic limit. We require

$$g_0(\mathbf{q}, t) = \langle e^{i\mathbf{q}\cdot[\mathbf{R}(t) - \mathbf{R}(0)]} \rangle = \frac{1}{4\pi}\int \sin\theta \mathrm{d}\theta \mathrm{d}\phi e^{iqV_F t\cos\theta}$$

$$= \frac{1}{2}\int \sin\theta \mathrm{d}\theta e^{iqV_F t\cos\theta}, \tag{39.1a}$$

where θ is the angle between $\mathbf{V}$ and $\mathbf{q}$. Carrying out the integral over $\mathrm{d}\theta$ we get

$$g_0(\mathbf{q}, t) = \frac{1}{qV_F t}\sin(qV_F t) \cong 1 - \frac{1}{6}q^2 V_F^2 t^2. \tag{39.1b}$$

Taking the time Fourier transform of (39.1a) yields

$$g_0(\mathbf{q}, \Omega) = \frac{1}{4\pi}\int\int \mathrm{d}t \sin\theta \mathrm{d}\theta e^{i(qV_F\cos\theta + \Omega)t}$$

$$= -\frac{1}{2}\int_0^\pi \mathrm{d}\cos\theta\, \delta(qV_F\cos\theta + \Omega), \tag{39.2}$$

where we have used $\int e^{i\omega t}\mathrm{d}t = 2\pi\delta(\omega)$. Carrying out the angular integration we obtain

$$g_0(\mathbf{q}, \Omega) = \begin{cases} \dfrac{1}{2qV_F}, & |\Omega| < qV_F, \\ 0, & |\Omega| > qV_F. \end{cases} \tag{39.3}$$

39.2 The dirty limit

To discuss the effect of collisions, we must examine Eq. (38.27) in greater depth. Our discussion is modeled after that of Singwi (1964). We return to the Schrödinger representation and write Eq. (38.27) as

$$g_0(\mathbf{q}, t) = \langle e^{i\mathscr{H}t/\hbar} e^{i\mathbf{q}\cdot\mathbf{R}(0)} e^{-i\mathscr{H}t/\hbar} e^{-i\mathbf{q}\cdot\mathbf{R}(0)} \rangle_\xi. \tag{39.4}$$

We may regard $e^{-i\mathbf{q}\cdot\mathbf{R}(0)}$ as the eigenfunction of the momentum operator with momentum $\mathbf{q}$. Alternatively we may interpret the operator product $e^{i\mathbf{q}\cdot\mathbf{R}(0)} e^{-i\mathscr{H}t/\hbar} e^{-i\mathbf{q}\cdot\mathbf{R}(0)}$ as $e^{-i\mathscr{H}_q t/\hbar}$, where $\mathscr{H}_q$ is the Hamiltonian in a frame moving with momentum $\mathbf{q}$, i.e.,

$$\mathscr{H}_q = \mathscr{H} + \frac{\mathbf{p}\cdot\hbar\mathbf{q}}{m} + \frac{\hbar^2 q^2}{2m}, \tag{39.5}$$

yielding

$$g_0(\mathbf{q}, t) = e^{-i\hbar q^2 t/2m} \langle e^{i\mathscr{H}t/\hbar} e^{-i(\mathscr{H} - \mathbf{p}\cdot\hbar\mathbf{q}/m)t/\hbar} \rangle. \tag{39.6}$$

We write $\mathscr{H} - \mathbf{p}\cdot\hbar\mathbf{q}/m$ as $\mathscr{H} + \mathscr{H}^{(1)}$ and calculate the effect of $\mathscr{H}'^{(1)}$ using time-dependent perturbation theory.

The time-dependent wavefunction which evolves from the action of $\mathscr{H} + \mathscr{H}^{(1)}$ may be written as $\psi(t) = \Sigma_n[a_n^{(0)} + a_n^{(1)}(t) + a_n^{(2)}(t) + \ldots] \,|\, n\rangle e^{i\omega_n t}$, where $|\, n\rangle$ are the eigenstates of $\mathscr{H}$ with eigenvalues $\varepsilon_n (= \hbar\omega_n)$.

The coefficients $a_n^{(i)}(t)$ are related by the usual expression of time-dependent perturbation theory (when the perturbation is turned on at $t = 0$)

$$i\hbar a_n^{(i+1)} = \sum_m a_m^{(i)} \int_0^t e^{i\omega_n t} \langle n \,|\, \mathscr{H}^{(1)} \,|\, m \rangle e^{-i\omega_m t} \tag{39.7}$$

and we assume $a_n^{(i)}(0) = 0$ for $i > 0$. We then obtain the following operator expression

$$e^{-i(\mathscr{H} + \mathscr{H}^{(1)})t/\hbar} = e^{-i\mathscr{H}t/\hbar}\left[1 + \left(\frac{-i}{\hbar}\right)\int_0^t dt_1 e^{i\mathscr{H}t_1/\hbar}\mathscr{H}^{(1)} e^{-i\mathscr{H}t_1/\hbar}\right.$$

$$\left. + \left(\frac{-i}{\hbar}\right)^2 \int_0^t dt_1 \int_0^{t_1} dt_2 (e^{i\mathscr{H}t_1/\hbar}\mathscr{H}^{(1)} e^{-i\mathscr{H}t_1/\hbar})(e^{i\mathscr{H}t_2/\hbar}\mathscr{H}^{(1)} e^{-i\mathscr{H}t_2/\hbar}) + \ldots \right. \tag{39.8}$$

Using (39.8), writing the velocity operator $\mathbf{V}$ as $\mathbf{p}/m$, and returning to the Heisenberg representation, Eq. (39.4) becomes

$$g_0(\mathbf{q}, t) = e^{-i\hbar q^2 t/2m}\left[1 - i\left\langle \int_0^t dt_1 \mathbf{q}\cdot\mathbf{V}(t_1) \right\rangle_\xi \right.$$

$$\left. - \left\langle \int_0^t dt_1 \mathbf{q}\cdot\mathbf{V}(t_1) \int_0^{t_1} dt_2 \mathbf{q}\cdot\mathbf{V}(t_2) \right\rangle_\xi + \ldots\right]. \tag{39.9}$$

The second term in (39.9) is zero since the velocity, averaged over an energy shell, vanishes. In the third term we identify the *velocity autocorrelation function* as $\langle V_i(t_1) V_j(t_2) \rangle$ (note that the limits on the integration require $t_1 \geq t_2$). For an isotropic system we average over the angle between $\mathbf{q}$ and $\mathbf{V}$

$$\frac{1}{2}\int_0^{\pi} \cos^2\theta \sin\theta \mathrm{d}\theta = \frac{1}{3}.$$

Performing the integral over $\mathrm{d}t_2$ in the last term in (39.9) gives

$$-\frac{q^2}{3}\left\langle \int_0^t \mathrm{d}t_1 V(t_1) \int_0^{t_1} \mathrm{d}t_2 V(t_2) \right\rangle_\xi = -\frac{q^2}{3}\left\langle \int_0^t \mathrm{d}t_1 V(t_1)[R(t_1) - R(0)] \right\rangle_\xi,$$

where we have written $R(t)$ as the indefinite integral of $V(t)$. Integrating a second time fixing the origin such that $R(0) = 0$ we have (on integrating by parts)

$$\left\langle \int_0^t \mathrm{d}t_1 V(t_1)R(t_1) \right\rangle = \frac{1}{2}\langle R^2(t)\rangle.$$

Our expression for g_0 then becomes[1]

$$g_0(\mathbf{q}, t) = \mathrm{e}^{-i\hbar q^2 t/2m}\left[1 - \frac{1}{6}q^2\langle R^2(t)\rangle\right]. \qquad (39.10)$$

Obtaining the long-time behavior requires our summing (39.9) to infinite order in some approximation. Higher order terms involve, successively, averages of 4, 6, 8,... velocity operators. The usual approximation adopted is to assume such averages are dominated by products of averages of two velocity operators. When the detailed calculations are carried out it turns out that the successive terms correspond to the power series expansion of a Gaussian correlation function of the form

$$g_0(\mathbf{q}, t) = \exp\left[-\frac{i\hbar q^2}{2m}t - \frac{q^2}{3}\left\langle \int_0^t \mathrm{d}t_1 V(t_1) \int_0^{t_1} \mathrm{d}t_2 V(t_2) \right\rangle\right]. \qquad (39.11a)$$

In the long-time, semiclassical approximation, represented by (39.10), we would have

$$g_0(\mathbf{q}, t) = \exp\left[-\frac{i\hbar q^2}{2m}t - \frac{1}{6}q^2\langle R^2(t)\rangle_\xi\right]. \qquad (39.11b)$$

Let us examine the behavior of the system for times which are long compared to a

1. We may formally rewrite the last term in Eq. (39.9) in a different way by integrating by parts where, in the usual expression $\int u\mathrm{d}v = uv - \int v\mathrm{d}u$, we set $\mathrm{d}v = V(t)$ and $u = \int_0^{t_1} V(t_2)\mathrm{d}t_2$.

$$\int_0^t V(t_1)\mathrm{d}t_1 \int_0^{t_1} v(t_2)\mathrm{d}t_2 = \left[\int_0^{t_1} V(t_1)\mathrm{d}t_1 \int_0^{t_1} V(t_2)\mathrm{d}t_2\right]_0^t - \int_0^t V(t_1)\mathrm{d}t_1 \int_0^{t_1} V(t_2)\mathrm{d}t_2$$

or, on taking an ensemble average,

$$\left\langle \int_0^t V(t_1)\mathrm{d}t_1 \int_0^{t_1} v(t_2)\mathrm{d}t_2 \right\rangle = \frac{1}{2}\int_0^t \mathrm{d}t_1 \int_0^t \mathrm{d}t_2 \langle \hat{T}V(t_1)V(t_2)\rangle.$$

The time ordering operator $\hat{T}$, which places earlier times to the right, has been inserted since the V are operators which do not commute for different times.

microscopic scattering time, τ. The random-walk expression connecting the displacement and time is

$$\frac{1}{6}\langle R^2(t)\rangle = Dt, \tag{39.12}$$

where D is the diffusion coefficient given by $D = \frac{1}{3}V_F^2\tau$. Provided $k_F\ell >> 1$, where $\ell = v_F\tau$ and k_F is the Fermi wave vector, the imaginary term in the exponent may be neglected and we have

$$g_0(\mathbf{q}, t) = e^{-Dq^2|t|} \tag{39.13}$$

(where the absolute value sign on t is required for the behavior to be causal). This form for g, which is referred to as the dirty limit, will be used regularly in subsequent calculations. On taking the Fourier transform of (39.13) we obtain

$$\begin{aligned} g_0(\mathbf{q}, \Omega) &= \frac{1}{2\pi}\int_{-\infty}^{+\infty} dt e^{-Dq^2|t| + i\Omega t} \\ &= \frac{1}{2\pi}\frac{2Dq^2}{D^2q^4 + \Omega^2}. \end{aligned} \tag{39.14}$$

39.3 The general case

As a simple model for interpolating between free (ballistic) electrons and the diffusion limit we introduce an ansatz for the velocity autocorrelation function as

$$\langle V(t)V(0)\rangle_{\xi=0} = V_F^2 e^{-t/\tau}. \tag{39.15}$$

To evaluate the integral in (39.11a) we require $\langle V(t_1)V(t_2)\rangle$. From the homogeneity of time this function depends only on $t_1 - t_2$. We may therefore write (39.15) as

$$\langle V(t_1)V(t_2)\rangle_{\xi=0} = V_F^2 e^{-(t_1-t_2)/\tau}. \tag{39.16}$$

The required integral is then

$$\begin{aligned} \int_0^t dt_1 \int_0^{t_1} dt_2 V_F^2 e^{-(t_1-t_2)/\tau} &= V_F^2\tau \int_0^t (1 - e^{-t_1/\tau})dt_1 \\ &= V_F^2[\tau t - \tau^2(1 - e^{-t/\tau})] \end{aligned} \tag{39.17}$$

or

$$\langle R^2(t)\rangle_{\xi=0} = 2V_F^2[\tau t - \tau^2(1 - e^{-t/\tau})]. \tag{39.18}$$

This expression has the limiting forms

$$\langle R^2(t \to 0)\rangle_{\xi=0} = V_F^2 t^2 \tag{39.19a}$$

and

$$\langle R^2(t \to \infty)\rangle_{\xi=0} = 2V_F^2\tau t. \tag{39.19b}$$

The resulting correlation function $g_\xi(\mathbf{q}, t)$ associated with (39.19b) was given as (39.13); the correlation function associated with (39.19a) is

$$g_0(\mathrm{q}, \mathrm{t}) = \exp\left(-\frac{i\hbar q^2}{2m} t - \frac{1}{6} q^2 V_{\mathrm{F}}^2 t^2 \right). \tag{39.20}$$

If we disregard the time-dependent quantum phase factor and expand the remainder of the exponential we obtain

$$g_0(\mathrm{q}, \mathrm{t}) = 1 - \frac{1}{6} q^2 V_{\mathrm{F}}^2 t^2, \tag{39.21}$$

in agreement with the semiclassical model calculation (39.1b).

40 The self-consistency condition

40.1 The dirty limit at zero magnetic field

If we insert Eq. (39.14) into (38.26) we obtain the dirty limit kernel

$$K_0(\mathbf{q}) = \frac{k_B T}{\hbar\pi}\sum_\omega \int \frac{\mathcal{N}(\xi) V \mathrm{d}\xi \mathrm{d}\xi'}{(\xi - \mathrm{i}\hbar\omega)(\xi' + \mathrm{i}\hbar\omega)} \frac{Dq^2}{(\Omega - \mathrm{i}Dq^2)(\Omega + \mathrm{i}Dq^2)}. \tag{40.1}$$

We replace ξ' by $\xi - \hbar\Omega$ and $\mathrm{d}\xi'$ by $\hbar\mathrm{d}\Omega$. The resulting expression has poles in the complex ξ plane at $\xi = \mathrm{i}\hbar\omega$ and $\xi = \hbar\Omega - \mathrm{i}\hbar\omega$. We write (40.1) in the form

$$K_0(\mathbf{q}) = K_0^{>}(\mathbf{q}) + K_0^{<}(\mathbf{q}),$$

where the two quantities on the right arise from terms with $\omega > 0$ and $\omega < 0$, respectively. Using Cauchy's theorem and completing the contour in either half plane yields

$$K_0^{>}(\mathbf{q}) = \frac{k_B T}{\hbar\pi}\mathcal{N}(0)V2\pi\mathrm{i}\sum_{\omega>0}\int \mathrm{d}\Omega \frac{Dq^2}{(2\mathrm{i}\omega - \Omega)(\Omega - \mathrm{i}Dq^2)(\Omega + \mathrm{i}Dq^2)}.$$

Here we have assumed both $\mathcal{N}(0)$ and V are independent of energy. The above expression has a single pole in the lower half of the complex Ω plane and the resulting integration yields

$$K_0^{>}(\mathbf{q}) = \frac{k_B T}{\pi\hbar}\mathcal{N}(0)V2\pi^2\sum_{\omega>0}\frac{1}{2\omega + Dq^2}.$$

Carrying out similar procedures for $\omega < 0$ yields the same expression, provided ω is replaced by $|\omega|$. Thus our complete kernel becomes[1]

$$K_0(\mathbf{q}) = \frac{2\pi}{\hbar}k_B T\mathcal{N}(0)V\sum_\omega \frac{1}{2|\omega| + Dq^2}. \tag{40.2}$$

Alternatively, we can obtain (40.2) by inserting (39.13) into the Fourier transform in ($\mathbf{r}$) and ($\mathbf{r}'$) of

1. For the case of isotropic scattering and a spherical Fermi surface one can show that (Abrikosov, Gorkov, and Dzyaloshinski 1963)

$$K_0(\mathbf{q}) = \frac{2\pi\mathcal{N}(0)}{\hbar V_F}\sum_\omega\left[\frac{q}{\tan^{-1}(\zeta\omega q)} - \frac{1}{\ell}\right],$$

where $\zeta_\omega^{-1} = \ell^{-1} + 2|\omega|/V_F$. This expression passes continuously from (40.2) to the clean limit expression (which is easily obtained from (39.3)).

(38.14′) and carry out the integration over t. As it stands, the sum in (40.2) is divergent; this behavior arises from our ignoring the BCS cut-off in V. We rewrite (40.2) as

$$K_0(\mathbf{q}) = K_0(0) + [K_0(\mathbf{q}) - K_0(0)]. \tag{40.3}$$

Returning to Eq. (40.1), we note that

$$\lim_{q\to 0}\frac{1}{\pi}\frac{Dq^2}{\Omega^2 + D^2q^4} = \delta(\Omega);$$

carrying out the integration over $d\xi'$ and using Eq. (38.9) we obtain

$$K_0(0) = \int \frac{d\xi \mathcal{N}(\xi)V}{2\xi}\tanh\frac{\xi}{2k_BT}.$$

Since $V = \text{const}$ for $|\xi| < \hbar\omega_D$ (where $\hbar\omega_D$ is a phonon cut-off energy) and zero otherwise, and $\mathcal{N}(\xi) \cong \mathcal{N}(0)$, we have

$$K_0(0) = \mathcal{N}(0)V\int_{-\hbar\omega_D}^{+\hbar\omega_D}\frac{d\xi}{2\xi}\tanh\left(\frac{\xi}{2k_BT}\right);$$

the right-hand side is the integral that occurs in the BCS theory for the superconducting transition temperature which can be written as

$$K_0(0) \cong \mathcal{N}(0)V\ell n\left(\frac{1.14\hbar\omega_D}{k_BT}\right). \tag{40.4}$$

We expect a uniform gap ($\Delta = \text{const}$) in a bulk sample, corresponding to $K(0) = 1$, and this condition yields the bulk transition temperature, $k_BT_c = 1.14\hbar\omega_D e^{-1/\mathcal{N}(0)V}$ which is the usual BCS expression.

Using (40.2) and (40.4), (40.3) becomes

$$K_0(\mathbf{q}) = \mathcal{N}(0)V\left[\ell n\left(\frac{1.14\hbar\omega_D}{k_BT}\right) + \frac{2\pi k_BT}{\hbar}\sum_{\omega}\left(\frac{1}{2|\omega| + Dq^2} - \frac{1}{2|\omega|}\right)\right]$$

$$= \mathcal{N}(0)V\left[\ell n\left(\frac{1.14\hbar\omega_D}{k_BT}\right) + \sum_{\nu}\left(\frac{1}{\xi_d^2q^2 + 2|\nu + \tfrac{1}{2}|} - \frac{1}{2|\nu + \tfrac{1}{2}|}\right)\right], \tag{40.5}$$

where we have introduced a (dirty limit) coherence length $\xi_d^2 \equiv \hbar D/2\pi k_BT = \hbar V_F\ell/6\pi k_BT$. The two summations in (40.5) are still separately divergent, but their difference is finite. We introduce the digamma function, or ψ function, defined by $\psi(z) = \Gamma'(z)/\Gamma(z)$, where $\Gamma(z)$ is the Γ function. We give, without proof (see Abramowitz and Stegun (1970), p. 259), the following identity

$$\psi\left(\frac{z}{2} + \frac{1}{2}\right) - \psi\left(\frac{1}{2}\right) = -\sum_{\nu=-\infty}^{+\infty}\left(\frac{1}{z + 2|\nu + \tfrac{1}{2}|} - \frac{1}{2|\nu + \tfrac{1}{2}|}\right) \equiv \mathcal{U}(z). \tag{40.6}$$

(Note that this would be a slowly converging expansion for large z.) Since this function occurs so often in the theory of superconductivity, we have denoted it by a single function $\mathcal{U}(z)$. Thus the final expression for our zero-field kernel is then

$$K_0(\mathbf{q}) = \mathcal{N}(0)V\left[\ell n \frac{1.14\theta_D}{T} - \mathcal{U}(\xi_d^2 q^2)\right]. \tag{40.7}$$

In a bulk superconductor, where homogeneous nucleation can occur (corresponding to $q = 0$), the transition temperature is independent of D; i.e., potential scattering has no effect on the transition temperature.

40.2 The dirty limit at finite magnetic field

Eq. (40.7) applies to a dirty homogeneous superconductor at zero field. We now extend our discussion to the case of nonzero magnetic fields and systems with arbitrary boundaries. We begin by noting that Eq. (39.13) is the Fourier transform of the function

$$g_0(\mathbf{r}, t) = \frac{1}{(4\pi D|t|)^{3/2}} e^{-r^2/4D|t|}; \tag{40.8}$$

this function is a solution of the diffusion equation,

$$D\nabla^2 g_0(\mathbf{r}, t) - \frac{\partial g_0(\mathbf{r}, t)}{\partial |t|} = 0, \tag{40.9}$$

for $\mathbf{r} \neq 0$ and satisfies the condition $g(\mathbf{r}, 0) = \delta(\mathbf{r})$. We expect the function $g_0(\mathbf{r}, \mathbf{r}')$ entering Eq. (38.22) to obey the diffusion equation for either $\mathbf{r}$ or $\mathbf{r}'$,

$$\left(D\nabla^2 - \frac{\partial}{\partial |t|}\right) g_0(\mathbf{r}, \mathbf{r}', t) = 0 \tag{40.10a}$$

and

$$\left(D\nabla'^2 - \frac{\partial}{\partial |t|}\right) g_0(\mathbf{r}, \mathbf{r}', t) = 0, \tag{40.10b}$$

where ∇ and ∇' operate with respect to the coordinates $\mathbf{r}$ and $\mathbf{r}'$ and we require $g_0(\mathbf{r}, \mathbf{r}', t) = \delta(\mathbf{r} - \mathbf{r}')$; g_0 is also presumed to satisfy whatever boundary conditions are imposed at the relevant interfaces. The effect of the vector potential in Eq. (38.22) was included in a quasiclassical manner and we now develop a more accurate description. We assert that the correlation function, in the presence of a magnetic field described by a vector potential $\mathbf{A}$, and at a time $t + \varepsilon$, may be regarded as having evolved from its form at time t under the combined influence of a phase change induced by $\mathbf{A}$, and diffusion occurring during the interval ε governed by $g_0(\varepsilon)$, according to the integral equation

$$g(\mathbf{r}, \mathbf{r}', t + \varepsilon) = \int d^3r'' g(\mathbf{r}, \mathbf{r}'', t) g_0(\mathbf{r}'', \mathbf{r}', \varepsilon) e^{-(2ie/\hbar c)\mathbf{A}(\mathbf{r}')\cdot(\mathbf{r}' - \mathbf{r}'')}. \tag{40.11}$$

We expand g on the left-hand side to first order in ε and on the right-hand side to second order in $(\mathbf{r}'' - \mathbf{r}')$ (which we will see is equivalent to first order in ε); the phase factor is also expanded to second order. Thus

$$g(\mathbf{r},\mathbf{r}',t)+\frac{\partial g(\mathbf{r},\mathbf{r}',t)}{\partial t}\varepsilon=\int d^3r''[g(\mathbf{r},\mathbf{r}',t)+\nabla' g(\mathbf{r},\mathbf{r}',t)\cdot(\mathbf{r}''-\mathbf{r}')$$
$$+\frac{1}{2}(\mathbf{r}''-\mathbf{r}')\cdot\nabla'\nabla' g(\mathbf{r},\mathbf{r}',t)\cdot(\mathbf{r}''-\mathbf{r}')]\left\{1-\frac{2ie}{\hbar c}\mathbf{A}(\mathbf{r}')\cdot(\mathbf{r}''-\mathbf{r}')\right.$$
$$\left.-\frac{2e^2}{\hbar^2c^2}[\mathbf{A}(\mathbf{r}')\cdot(\mathbf{r}''-\mathbf{r}')]^2\right\}g_0(\mathbf{r}'',\mathbf{r}',\varepsilon).$$

The zero order terms must cancel since g_0 becomes $\delta(\mathbf{r}''-\mathbf{r})$ as $\varepsilon\to 0$. The odd terms vanish since $g_0(\mathbf{r},\mathbf{r}'')$ is symmetric in its arguments. We then have

$$\frac{\partial g(\mathbf{r},\mathbf{r}',t)}{\partial|t|}\varepsilon=\left[\nabla'-\frac{2ie}{\hbar c}\mathbf{A}(\mathbf{r}')\right]^2 g(\mathbf{r},\mathbf{r}',t)\frac{1}{2}\int d^3r'' g_0(\mathbf{r},\mathbf{r}'',\varepsilon)(\mathbf{r}''-\mathbf{r}')^2. \tag{40.12}$$

For small ε, g_0 is given by (40.8) with $\mathbf{r}\to\mathbf{r}'-\mathbf{r}''$. The integral in (40.12) can then be evaluated as $2D\varepsilon$ and we see that the behavior of $g(\mathbf{r},\mathbf{r}',t)$ is governed by the modified diffusion equation[2]

$$\frac{\partial}{\partial|t|}g(\mathbf{r},\mathbf{r}',t)=D\left[\nabla'-\frac{2ie}{\hbar c}\mathbf{A}(\mathbf{r}')\right]^2 g(\mathbf{r},\mathbf{r}',t), \tag{40.13}$$

again subject to the condition $g(\mathbf{r},\mathbf{r}',0)=\delta(\mathbf{r}-\mathbf{r}')$. When operating on $\mathbf{r}$ we would use the complex conjugate of the operator in the square brackets in (40.13).

The solution of the integral equation (38.6) for Δ is most easily accomplished by finding the normalized eigenfunctions, $g_n(\mathbf{r})$, and eigenvalues, Ω_n, of the operator $D[\nabla-(2ie/\hbar c)\mathbf{A}(\mathbf{r})]^2$ subject to the prescribed boundary conditions; i.e.,

$$D\left[\nabla-\frac{2ie}{\hbar c}\mathbf{A}(\mathbf{r})\right]^2 g_n(\mathbf{r})=-\Omega_n g_n(\mathbf{r}). \tag{40.14}$$

We observe that the function

$$g(\mathbf{r},\mathbf{r}',t)=L^{-3}\sum_n g_n^*(\mathbf{r}')g_n(\mathbf{r})e^{-\Omega_n|t|} \tag{40.15}$$

satisfies the diffusion equation (with respect to $\mathbf{r}$ or $\mathbf{r}'$) and, from the completeness relation, reduces to $\delta(\mathbf{r}-\mathbf{r}')$ at $t=0$. Fourier transforming with respect to the time variable (again with $t\to|t|$) yields

$$g(\mathbf{r},\mathbf{r}',\Omega)=\frac{1}{2\pi L^3}\sum_n g_n^*(\mathbf{r}')g_n(\mathbf{r})\frac{2\Omega_n}{\Omega^2+\Omega_n^2}. \tag{40.16}$$

Inserting (40.16) into Eq. (38.22) we obtain a form similar to (40.1) which, on using (40.2), yields immediately

2. Alternatively we may show that g satisfies the inhomogeneous equation

$$\frac{\partial}{\partial|t|}g(\mathbf{r},\mathbf{r}',t)-D\left[\nabla-\frac{2ie}{\hbar c}\mathbf{A}(\mathbf{r})\right]^2 g(\mathbf{r},\mathbf{r}',t)=L^{-3}\delta(t)\delta^{(3)}(\mathbf{r}-\mathbf{r}').$$

The solutions to this equation, as well as (40.13), are complex when $\mathbf{A}\neq 0$; thus g is, strictly speaking, not a correlation function for particles, which must be positive definite. We may still think of g as an 'information correlation function'.

$$\Delta(\mathbf{r}) = \frac{2\pi}{\hbar} k_B T \mathcal{N}(0) \sum_{\omega} \sum_{n} \frac{g_n(\mathbf{r})}{2|\omega| + \Omega_n} \int d^3 r' g_n^*(\mathbf{r}')\Delta(\mathbf{r}'). \tag{40.17}$$

Multiplying (40.17) by $g_m^*(\mathbf{r})$, integrating over all $\mathbf{r}$, and using orthonormality of the g_n we see that the eigenfunctions of the integral equation (38.6) are the g_ns themselves; since the eigenvalue of (38.6) is unity we must have

$$1 = \frac{2\pi}{\hbar} k_B T \mathcal{N}(0) V \sum_{\omega} \frac{1}{2|\omega| + \Omega_n}. \tag{40.18}$$

Note the smallest eigenvalue, Ω_0, yields the highest transition temperature. From the discussion surrounding Eq. (40.7), we have immediately

$$\ell n \frac{T(0)}{T(H)} = \mathcal{U}\left(\frac{\hbar \Omega_0}{2\pi k_B T}\right), \tag{40.19}$$

where $T(H)$ is the field-dependent transition temperature. For a bulk superconductor (40.14) is equivalent to the linearized G–L equation of Part I provided we identify D with $\hbar^2/2m^*$. The lowest eigenvalue is

$$\Omega_0 = \frac{eH}{mc} = \frac{2eDH}{\hbar c}$$

or

$$\ell n \frac{T(0)}{T(H)} = \mathcal{U}\left(\frac{eDH}{\pi c k_B T}\right) = \mathcal{U}\left(\frac{\hbar D H}{\phi_0 k_B T}\right), \tag{40.20}$$

which is the well known Maki–deGennes–Werthamer result for the upper critical field of a bulk Type II superconductor.

We now treat the case of an anisotropic superconductor for which Eq. (40.14) must be replaced by

$$\left[\nabla - \frac{2ie}{\hbar c}\mathbf{A}(\mathbf{r})\right] \cdot \mathbf{D} \cdot \left[\nabla - \frac{2ie}{\hbar c}\mathbf{A}(\mathbf{r})\right] g_n(\mathbf{r}) = -\Omega_n g_n(\mathbf{r}), \tag{40.21}$$

where

$$\mathbf{D} = \begin{bmatrix} D_{\parallel} & 0 & 0 \\ 0 & D_{\parallel} & 0 \\ 0 & 0 & D_{\perp} \end{bmatrix}. \tag{40.22}$$

Eq. (40.21) is formally identical to Eq. (11.1) and, introducing an angle-dependent diffusion constant (analogous to the anisotropic cyclotron frequency of Eq. (11.5)) given by

$$D(\theta) = [D_{\parallel} D_{\perp} \sin^2\theta + D_{\parallel}^2 \cos^2\theta]^{1/2}, \tag{40.23}$$

we may write

$$\Omega_0(\theta) = \frac{2eD(\theta)H}{\hbar c}, \tag{40.24}$$

which when substituted in (40.19) yields (40.20) with $D = D(\theta)$.

We can also treat the case of a superconducting slab with a thickness which is much smaller than a coherence length. The required eigenvalue has already been obtained in Eq. (12.9); writing $m = \hbar/4D$ we have immediately

$$\Omega_0 = \frac{2eDH}{\hbar c}\cos\theta + \frac{d^2e^2}{3\hbar^2c^2}DH^2\sin^2\theta. \tag{40.25}$$

Substituting (40.25) into (40.19) yields the angular dependence of the upper critical field of a thin superconducting slab.

Finally we note that if the diffusion coefficient is inhomogeneous, but the pairing potential V is constant, the lowest eigenvalue of the equation (which is a slight generalization of (40.14))

$$\left[\nabla - \frac{2ie}{\hbar c}\mathbf{A}(\mathbf{r})\right]\cdot\mathbf{D}(\mathbf{r})\cdot\left[\nabla - \frac{2ie}{\hbar c}\mathbf{A}(\mathbf{r})\right]g_n(\mathbf{r}) = -\Omega_n g_n(\mathbf{r}) \tag{40.26}$$

would yield the field dependence of the transition temperature.

40.3 The clean limit at zero magnetic field

Since a free electron metal is translationally invariant, $K(\mathbf{r},\mathbf{r}') = K(|\mathbf{r}-\mathbf{r}'|)$. The Fourier transform of (38.14′) is then

$$K(\mathbf{q}) = \frac{2\pi}{\hbar}k_BT\mathcal{N}(0)V\sum_\omega\int_0^\infty \mathrm{d}t e^{-2|\omega|t}g(\mathbf{q},t). \tag{40.27}$$

Inserting (39.1a) for our clean limit correlation function, $g(\mathbf{q},t)$, Eq. (40.27) becomes

$$K(\mathbf{q}) = \frac{2\pi}{\hbar}k_BT\mathcal{N}(0)V\sum_\omega\int_0^\infty \mathrm{d}t e^{-2|\omega|t}\frac{1}{2}\int_0^\pi \sin\theta \mathrm{d}\theta e^{iqV_Ft\cos\theta}.$$

Carrying out the integration over $\mathrm{d}t$ gives

$$K(\mathbf{q}) = \frac{2\pi}{\hbar}k_BT\mathcal{N}(0)V\sum_\omega\frac{1}{2}\int_0^\pi \frac{-\sin\theta\mathrm{d}\theta}{-2|\omega| + iqV_F\cos\theta}$$

and integrating over $\mathrm{d}\theta$ yields

$$K(\mathbf{q}) = \frac{2\pi}{\hbar}k_BT\mathcal{N}(0)V\sum_\omega\frac{1}{2iqV_F}\ell n\left(\frac{-2|\omega| - iqV_F}{-2|\omega| + iqV_F}\right)$$

or

$$K(\mathbf{q}) = \frac{2\pi}{qV_F\hbar}k_BT\mathcal{N}(0)V\sum_\omega\tan^{-1}\frac{qV_F}{2|\omega|}. \tag{40.28}$$

In the limit $q \to 0$ Eq. (40.28) becomes

$$K(0) = \frac{2\pi}{\hbar}k_BT\mathcal{N}(0)V\sum\frac{1}{2|\omega|}. \tag{40.29}$$

This is the same as obtained for the $q = 0$ limit of the dirty kernel, Eq. (40.2), and by the same arguments will be given by Eq. (40.4). The result for finite q can then be written in the form (40.3) as

$$K(\mathbf{q}) = \mathcal{N}(0)V\ell n\left(\frac{1.14\hbar\omega_D}{k_B T}\right) + \frac{2\pi}{\hbar}k_B T\mathcal{N}(0)V\sum_{\omega}\left[\frac{1}{qV_F}\tan^{-1}\frac{qV_F}{2|\omega|} - \frac{1}{2|\omega|}\right], \tag{40.30}$$

which can be evaluated numerically.

Using the form (39.3) for $g(\mathbf{r}, t)$, which interpolates between the free electron and the dirty limits, we may obtain the expression in footnote 1 of this section.

41 Effects involving electron spin

41.1 Spin generalized Bogoliubov equations

We now generalize the Bogoliubov equations to allow for spin-dependent potentials. The more general form is required if we are to include the effects of spin paramagnetism, spin–orbit coupling, and magnetic impurities; it is also required when discussing unconventional (non-s-wave) superconductivity.

We return to the general form of the Hamiltonian given by Eqs. (36.2) and (36.3). We do not restrict the one-body potential $U^{(1)}_{\alpha\beta}$ to be diagonal in the spin indices, as we did in Eq. (36.4a), but we continue to assume a diagonal two-body term, Eq. (36.4b), of the contact form given in (36.6). The Hamiltonian then becomes (where the summation convention is again implied for the spin indices)

$$\hat{H} = \int d^3r \hat{\psi}^\dagger_\alpha(\mathbf{r}) \left\{ \left[-\frac{\hbar}{2m}\nabla - \frac{ie}{\hbar c}\mathbf{A}(\mathbf{r}) \right]^2 \delta_{\alpha\beta} + U^{(1)}_{\alpha\beta}(\mathbf{r}) \right\} \hat{\psi}_\beta(\mathbf{r})$$

$$- \frac{1}{2} V \int d^3r \hat{\psi}^\dagger_\alpha(\mathbf{r}) \hat{\psi}^\dagger_\beta(\mathbf{r}) \hat{\psi}_\beta(\mathbf{r}) \hat{\psi}_\alpha(\mathbf{r}). \tag{41.1}$$

The pair potential is still given by (37.7), which can be written in the (singlet) form

$$\Delta(\mathbf{r}) = \frac{V}{2} \rho_{\alpha\beta} \langle \hat{\psi}_\alpha(\mathbf{r}) \hat{\psi}_\beta(\mathbf{r}) \rangle, \tag{41.2}$$

where we have defined the operator

$$\underset{\sim}{\rho} = \begin{pmatrix} 0 & 1 \\ -1 & 0 \end{pmatrix} = i\underset{\sim}{\sigma}_2; \tag{41.3}$$

$\underset{\sim}{\sigma}_2$ is a Pauli spin matrix. We proceed in the same manner as in Sec. 36. In place of (36.8) we have

$$\hat{H}_e = \int d^3r [\hat{\psi}^\dagger_\alpha(\mathbf{r}) \mathscr{H}'_{\alpha\beta} \hat{\psi}_\beta(\mathbf{r})]$$

$$+ \int d^3r [\Delta(\mathbf{r}) \rho_{\alpha\beta} \hat{\psi}^\dagger_\alpha(\mathbf{r}) \hat{\psi}^\dagger_\beta(\mathbf{r}) + \Delta^*(\mathbf{r}) \rho_{\alpha\beta} \hat{\psi}_\alpha(\mathbf{r}) \hat{\psi}_\beta(\mathbf{r})]. \tag{41.4}$$

This equation is diagonalized by a generalized Bogoliubov transformation of the form

$$\hat{\psi}_\alpha(\mathbf{r}) = \sum_N [u_{N\alpha}(\mathbf{r})\hat{\gamma}_N + v^*_{N\alpha}(\mathbf{r})\hat{\gamma}^\dagger_N], \tag{41.5}$$

where the sum on N includes both the *translational and spin quantum numbers*; when we wish to distinguish these we write $N \to (n, \nu)$, where $\nu = 1, 2$ for spin up and spin down, respectively.[1] Following steps similar to those leading to Eqs. (36.22) and Eqs. (36.23) we obtain the orthonormalization and completeness relations

$$\sum_{\alpha=1}^{2} \int [u^*_{M\alpha}(\mathbf{r})u_{N\alpha}(\mathbf{r}) + v^*_{M\alpha}(\mathbf{r})v_{N\alpha}(\mathbf{r})]\mathrm{d}^3r = \delta_{MN}, \tag{41.6a}$$

$$\sum_{\alpha=1}^{2} \int [u_{N\alpha}(\mathbf{r})v_{M\alpha}(\mathbf{r}) + v_{N\alpha}(\mathbf{r})u_{M\alpha}(\mathbf{r})]\mathrm{d}^3r = 0, \tag{41.6b}$$

$$\sum_N [u_{N\alpha}(\mathbf{r})u^*_{N\beta}(\mathbf{r}') + v^*_{N\alpha}(\mathbf{r})v_{N\beta}(\mathbf{r}')] = \delta_{\alpha\beta}\delta^{(3)}(\mathbf{r} - \mathbf{r}'), \tag{41.6c}$$

and

$$\sum_N [u_{N\alpha}(\mathbf{r})v^*_{N\beta}(\mathbf{r}') + v^*_{N\alpha}(\mathbf{r})u_{N\beta}(\mathbf{r}')] = 0. \tag{41.6d}$$

We require the commutator of $\hat{H}_\mathrm{e}$ with the $\hat{\psi}$ operators which is given by

$$[\hat{H}_\mathrm{e}, \hat{\psi}_\alpha(\mathbf{r})] = \left[\frac{\hbar^2}{2m}\left(\nabla - \frac{\mathrm{i}e}{\hbar c}\mathbf{A}(\mathbf{r})\right)^2 - \mu\right]\hat{\psi}_\alpha(\mathbf{r}) + U_{\alpha\beta}(\mathbf{r})\hat{\psi}_\beta(\mathbf{r}) + \Delta(\mathbf{r})\rho_{\alpha\beta}\hat{\psi}^\dagger_\beta(\mathbf{r}). \tag{41.7a}$$

The equation of motion analogous to (36.15a) is

$$[\hat{H}_\mathrm{e}, \hat{\gamma}_N] = -\varepsilon_N\hat{\gamma}_N. \tag{41.7b}$$

We next substitute Eq. (41.5) (and its hermitian conjugate) into the right-hand side of the commutator (41.7a). As in Sec. 36, the commutator can be independently evaluated using Eqs. (36.14) and (41.7b); comparison of the two yields the 'spin-generalized' Bogoliubov equations,

$$\varepsilon u_\alpha(\mathbf{r}) = \left[\frac{1}{2m}\left(\mathbf{p} - \frac{e}{c}\mathbf{A}\right)^2 - \mu\right]u_\alpha(\mathbf{r}) + U_{\alpha\beta}(\mathbf{r})u_\beta(\mathbf{r}) + \Delta(\mathbf{r})\rho_{\alpha\beta}v_\beta(\mathbf{r}) \tag{41.8}$$

$$-\varepsilon v_\beta(\mathbf{r}) = \left[\frac{1}{2m}\left(\mathbf{p} + \frac{e}{c}\mathbf{A}\right)^2 - \mu\right]v_\alpha(\mathbf{r}) + U^*_{\alpha\beta}(\mathbf{r})v_\beta(\mathbf{r}) + \Delta^*(\mathbf{r})\rho_{\alpha\beta}u_\beta(\mathbf{r}), \tag{41.9}$$

where the quantum numbers ($N = n, \nu$) have been suppressed (note we now have four coupled Bogoliubov equations involving the wavefunctions u_1, u_2, v_1, and v_2).

1. One must be careful to distinguish spin coordinates (or components) of a wavefunction and spin quantum numbers. In particular we write $u_N = (u_{N_1}, u_{N_2})$. The upper and lower components correspond to $\alpha = 1, 2$ (or $\uparrow, \downarrow$). With $N = (n, \nu)$, there will be two two-component wavefunctions corresponding to $\nu = 1, 2$ for each value of n. In what follows we will use the letters ν, μ for spin quantum numbers and $\alpha, \beta, \gamma, \delta$ for spin wavefunction components.

41.2 The density matrix

The one-particle density matrix, $\rho_{\alpha\beta}(\mathbf{r},\mathbf{r}')$, is defined by[2]

$$\rho_{\alpha\beta}(\mathbf{r},\mathbf{r}') = \langle \hat{\psi}^\dagger_\beta(\mathbf{r}')\hat{\psi}_\alpha(\mathbf{r})\rangle, \tag{41.10}$$

where the brackets implies both quantum mechanical and statistical averaging (note that for $\mathbf{r} = \mathbf{r}'$ and $\alpha = \beta$, the term in the angle brackets is the number density operator, the integral of which is the total number of particles).

Inserting Eq. (41.5) and its transpose into (41.10) and using Eqs. (37.8) we obtain for the equilibrium density matrix

$$\rho_{\alpha\beta}(\mathbf{r},\mathbf{r}') = \sum_N [u^*_{N\beta}(\mathbf{r}')u_{N\alpha}(\mathbf{r})f_N + v_{N\beta}(\mathbf{r}')v^*_{N\alpha}(\mathbf{r})(1 - f_N)]. \tag{41.11}$$

We also define an *anomalous* density matrix as

$$\rho^{\mathrm{A}}_{\alpha\beta}(\mathbf{r},\mathbf{r}') = \langle \hat{\psi}_\beta(\mathbf{r}')\hat{\psi}_\alpha(\mathbf{r})\rangle; \tag{41.12}$$

$\rho^{\mathrm{A}}_{\alpha\beta}(\mathbf{r},\mathbf{r}')$ is sometimes written as $F_{\alpha\beta}(\mathbf{r},\mathbf{r}')$. The expression analogous to (41.11) is

$$\rho^{\mathrm{A}}_{\alpha\beta}(\mathbf{r},\mathbf{r}') = \sum_N [u_{N\alpha}(\mathbf{r})v^*_{N\beta}(\mathbf{r}')f_N + u_{N\beta}(\mathbf{r}')v^*_{N\alpha}(\mathbf{r})(1 - f_N)]. \tag{41.13}$$

For the uniform case we may write the wavefunctions as

$$u_N(\mathbf{r}) = \frac{1}{L^{3/2}}\begin{pmatrix} u_{N\uparrow} \\ u_{N\downarrow} \end{pmatrix} \mathrm{e}^{\mathrm{i}\mathbf{k}\cdot\mathbf{r}}, \tag{41.14a}$$

$$v_N(\mathbf{r}) = \frac{1}{L^{3/2}}\begin{pmatrix} v_{N\uparrow} \\ v_{N\downarrow} \end{pmatrix} \mathrm{e}^{\mathrm{i}\mathbf{k}\cdot\mathbf{r}}, \tag{41.14b}$$

where L_3 is the volume of the system. Substituting (41.14) into (41.11) we have (for the uniform case)

$$\rho_{\alpha\beta}(\mathbf{r},\mathbf{r}') = \frac{1}{L^3}\sum_{\mathbf{k}'\nu} [u^*_{\mathbf{k}'\nu\beta}u_{\mathbf{k}\nu\alpha}f_{\mathbf{k}'\nu} + v_{-\mathbf{k}'\nu\beta}v^*_{-\mathbf{k}'\nu\alpha}(1 - f_{\mathbf{k}'\nu})]\mathrm{e}^{\mathrm{i}\mathbf{k}'\cdot(\mathbf{r}-\mathbf{r}')}, \tag{41.15}$$

and taking the Fourier transform in $\mathbf{r} - \mathbf{r}'$ we obtain a function

$$n_{\alpha\beta}(\mathbf{k}) = \sum_\nu [u^*_{\mathbf{k}\nu\beta}u_{\mathbf{k}\nu\alpha}f_{\mathbf{k}\nu} + v_{-\mathbf{k}\nu\beta}v^*_{-\mathbf{k}\nu\alpha}(1 - f_{\mathbf{k}\nu})]. \tag{41.16}$$

The expression corresponding to (41.15) for ρ^{A} is

$$\rho^{\mathrm{A}}_{\alpha\beta}(\mathbf{r},\mathbf{r}') = \frac{1}{2L^3}\sum_{\mathbf{k}'\nu} (u_{-\mathbf{k}'\nu\alpha}v^*_{-\mathbf{k}'\nu\beta} - u_{\mathbf{k}'\nu\beta}v^*_{\mathbf{k}'\nu\alpha})(1 - 2f_{\mathbf{k}'\nu})\mathrm{e}^{\mathrm{i}\mathbf{k}'\cdot(\mathbf{r}-\mathbf{r}')}, \tag{41.17}$$

which on taking the Fourier transform gives

$$n^{\mathrm{A}}_{\alpha\beta}(\mathbf{k}) = \frac{1}{2}\sum_\nu (u_{-\mathbf{k}\nu\alpha}v^*_{-\mathbf{k}\nu\beta} - u_{\mathbf{k}\nu\beta}v^*_{\mathbf{k}\nu\alpha})(1 - 2f_{\mathbf{k}\nu}), \tag{41.18}$$

2. Some authors define ρ with an additional factor of N^{-1}.

Instead of performing a Bogoliubov transformation we may use the standard expansion

$$\hat{\psi}_\alpha(\mathbf{r}) = \frac{1}{L^{3/2}}\sum_{\mathbf{k}} e^{-i\mathbf{k}\cdot\mathbf{r}}\chi_\alpha \hat{c}_{\mathbf{k}\alpha}, \tag{41.19}$$

where χ is a spin wavefunction. The resulting expansion corresponding to (41.16) is

$$n_{\alpha\beta}(\mathbf{k}) = \langle \hat{c}^\dagger_{\mathbf{k}\beta}\hat{c}_{\mathbf{k}\alpha}\rangle, \tag{41.20}$$

which is the usual expression for the Fourier transform of the density matrix; $n_{\alpha\beta}(\mathbf{k})$ is thus the distribution function for the occupation of states in momentum space.

The corresponding expression for n^{A} is

$$n^{\mathrm{A}}_{\alpha\beta}(\mathbf{k}) = \langle \hat{c}_{\mathbf{k}\beta}\hat{c}_{-\mathbf{k}\alpha}\rangle. \tag{41.21}$$

By convention we call $n^{\mathrm{A}}_{\alpha\beta}$ the pair distribution function. It has the property

$$n^{\mathrm{A}}_{\alpha\beta}(\mathbf{k}) = -\, n^{\mathrm{A}}_{\beta\alpha}(-\mathbf{k}). \tag{41.22}$$

It is sometimes convenient to write the distribution function as the sum of a scalar and a (two-component) spinor

$$n_{\alpha\beta}(\mathbf{k}) = n(\mathbf{k})\delta_{\alpha\beta} + (\mathbf{m}(\mathbf{k})\cdot \underset{\sim}{\sigma})_{\alpha\beta}, \tag{41.23}$$

where

$$n(\mathbf{k}) = \frac{1}{2}[n_{\uparrow\uparrow}(\mathbf{k}) + n_{\downarrow\downarrow}(\mathbf{k})], \tag{41.24a}$$

$$m_z(\mathbf{k}) = \frac{1}{2}[n_{\uparrow\uparrow}(\mathbf{k}) - n_{\downarrow\downarrow}(\mathbf{k})], \tag{41.24b}$$

$$m_x(\mathbf{k}) + i m_y(\mathbf{k}) = n_{\uparrow\downarrow}(\mathbf{k}), \tag{41.24c}$$

and

$$m_x(\mathbf{k}) - i m_y(\mathbf{k}) = n_{\downarrow\uparrow}(\mathbf{k}); \tag{41.24d}$$

m, which transforms as a vector, plays the role of a **k**-dependent spin polarization vector.

In a similar manner we may write the pair distribution function as a sum of 'singlet', n^{s}, and 'triplet', $\mathbf{n}^{\mathrm{t}}$, terms satisfying[3]

$$n^{\mathrm{s}}_{\alpha\beta}(\mathbf{k}) = n^{\mathrm{s}}_{\alpha\beta}(-\mathbf{k}) = -\, n^{\mathrm{s}}_{\beta\alpha}(\mathbf{k}) \tag{41.25a}$$

and

$$n^{\mathrm{t}}_{\alpha\beta}(\mathbf{k}) = -\, n^{\mathrm{t}}_{\alpha\beta}(-\mathbf{k}) = -\, n^{\mathrm{t}}_{\beta\alpha}(\mathbf{k}), \tag{41.25b}$$

which may be combined in the form

$$n^{\mathrm{A}}_{\alpha\beta}(\mathbf{k}) = n^{\mathrm{s}}(\mathbf{k})(\mathrm{i}\underset{\sim}{\sigma}_2)_{\alpha\beta} + \mathbf{n}^{\mathrm{t}}(\mathbf{k})\cdot(\underset{\sim}{\sigma}\mathrm{i}\underset{\sim}{\sigma}_2)_{\alpha\beta}, \tag{41.26}$$

where $\mathbf{n}^{\mathrm{t}}(\mathbf{k})$ measures the three independent spin components of the triplet pair distribution function; $n^{\mathrm{s}}(\mathbf{k})$ and $\mathbf{n}^{\mathrm{t}}(\mathbf{k})$ have the properties

3. Triplet order parameters will be discussed in Secs. 52 and 53.

$$n_s(\mathbf{k}) = n_s(-\mathbf{k}) \tag{41.27a}$$

and

$$\mathbf{n}^t(\mathbf{k}) = -\mathbf{n}^t(-\mathbf{k}). \tag{41.26b}$$

Conventional superconductivity involves singlet pairing; the anomalous distribution function can then be written as (see Eq. (41.3))

$$\underset{\sim}{n}^{A} = F(\mathbf{k})\begin{pmatrix} 0 & 1 \\ -1 & 0 \end{pmatrix} \tag{41.28a}$$

and the anomalous density matrix as

$$\underset{\sim}{\rho}^{A} = F(\mathbf{r},\mathbf{r}')\begin{pmatrix} 0 & 1 \\ -1 & 0 \end{pmatrix}. \tag{41.28b}$$

The gap function $\underset{\sim}{\Delta}(\mathbf{r})$ is then (see Eq. (37.7))

$$\underset{\sim}{\Delta}(\mathbf{r}) = -VF(\mathbf{r},\mathbf{r})\begin{pmatrix} 0 & 1 \\ -1 & 0 \end{pmatrix} \tag{41.29a}$$

$$= \Delta(\mathbf{r})\begin{pmatrix} 0 & 1 \\ -1 & 0 \end{pmatrix}. \tag{41.29b}$$

41.3 The linearized gap equation

We now develop the linearized gap function and the associated kernel for the spin generalized case. The steps are essentially identical to those already carried out in Sec. 38 leading up to Eq. (38.10) with the exception that we must carry the $\underset{\sim}{\rho}$ operator and the associated spin component indices along through the calculations. The result is

$$K(\mathbf{r},\mathbf{r}') = \frac{1}{2} V k_B T \sum_{M,N} \sum_{\omega} \frac{\phi^*_{N\alpha}(\mathbf{r})\underset{\sim}{\mathscr{K}}^{\dagger}_{\alpha\beta}\phi_{M\beta}(\mathbf{r})\phi^*_{M\gamma}(\mathbf{r}')\underset{\sim}{\mathscr{K}}_{\gamma\delta}\phi_{N\delta}(\mathbf{r}')}{(\xi_M + i\hbar\omega)(\xi_N - i\hbar\omega)}. \tag{41.30}$$

Here $\mathscr{K}$ is the Wigner time reversal operator

$$\underset{\sim}{\mathscr{K}} \equiv \underset{\sim}{\rho}\mathscr{C} = i\underset{\sim}{\sigma}_2\mathscr{C} \tag{41.31}$$

with $\mathscr{C}$ again being the charge conjugation operator. The $\phi_{N\alpha}$ are the eigenfunctions of the normal-state, mean field Hamiltonian

$$\xi_N\phi_{N\alpha} = \left[\frac{1}{2m}\left(\mathbf{p} - \frac{e}{c}\mathbf{A}\right)^2 - \mu\right]\phi_{N\alpha} + U_{\alpha\beta}\phi_{N\beta}. \tag{41.32}$$

We introduce a correlation function analogous to Eq. (38.19)

$$g^{(\nu)}(\mathbf{r},\mathbf{r}',t) = -\frac{1}{2}\frac{\sum_n \delta(\xi - \xi_{n,\nu})\langle n,\nu\,|\,\delta[\mathbf{r} - \mathbf{R}(t)]\underset{\sim}{\mathscr{K}}^{\dagger}(t)\underset{\sim}{\mathscr{K}}(0)\delta[\mathbf{r}' - \mathbf{R}(0)]\,|\,n,\nu\rangle}{\sum \delta(\xi - \xi_{n,\nu})} \tag{41.33}$$

where we have again suppressed the ξ subscript on g. In terms of this correlation function the kernel entering the gap equation is

$$K(\mathbf{r},\mathbf{r}') = L^3 k_B T \sum_{\omega} \sum_{\nu} \int \frac{\mathcal{N}(0) V \mathrm{d}\xi \mathrm{d}\xi'}{(\xi - i\hbar\omega)(\xi' + i\hbar\omega)} g^{(\nu)}(\mathbf{r},\mathbf{r}',(\xi' - \xi)/\hbar), \tag{41.34}$$

or, in form analogous to (38.14′),

$$K(\mathbf{r},\mathbf{r}') = \frac{2\pi}{\hbar} L^3 k_B T \mathcal{N}(0) V \sum_{\omega} \sum_{\nu} \int_0^{\infty} \mathrm{d}t e^{-2|\omega|t} g^{(\nu)}(\mathbf{r},\mathbf{r}',t). \tag{41.35}$$

41.4 Spin-dependent potentials

For our purposes here we will consider the potential $U_{\alpha\beta}$ as being made up of four terms. The dominant term is the nonrelativistic crystal potential which we can write as $U(\mathbf{r})\delta_{\alpha\beta}$. The Pauli spin-paramagnetism arising from the interaction of the electron magnetic moments with an external magnetic field has the form

$$U^{(P)}_{\alpha\beta} = I(\sigma_3)_{\alpha\beta}; \tag{41.36}$$

here $I = \mu_B H \chi/\chi^{(P)}$, where $\chi/\chi^{(P)}$ is the ratio of true susceptibility to the band susceptibility (and phenomenologically allows for an exchange enhanced susceptibility) and μ_B and H are the Bohr magneton and magnetic field, respectively. We will show later that this term puts an upper limit on the upper critical field which would otherwise be divergent in the limit $D \to 0$; this phenomenon is referred to as Pauli paramagnetic limiting (or the Clogston–Chandrasekhar limit).

Dilute magnetic impurities, such as Fe and Gd, have an especially damaging effect on superconductivity. Their effect is often modeled by including a 'spin Hamiltonian' term of the form

$$\underset{\sim}{\mathcal{H}}^{(s)} = \sum_i \Gamma(\mathbf{r},\mathbf{r}_i') \underset{\sim}{\sigma} \cdot \mathbf{s}_i, \tag{41.37}$$

where σ, Γ, and $\mathbf{s}$ are the conduction electron Pauli spin operator, exchange potential (having a range of the order of an atomic diameter), and impurity spin (treated as a classical spin vector), respectively. By flipping the spins of individual conduction electrons this term acts to break the Cooper pairs.

Finally, when heavy atoms are involved it may be necessary to include the effects of spin–orbit coupling. Although we will not make use of the explicit form of the spin–orbit Hamiltonian, it has the form

$$\underset{\sim}{\mathcal{H}}^{(so)} = \frac{e\hbar}{4m^2c^2} \nabla U(\mathbf{r}) \cdot \underset{\sim}{\sigma} \times \mathbf{p}. \tag{41.38}$$

Qualitatively $H^{(so)}$ mixes spin up and spin down states and tends to counteract the effect of Pauli spin paramagnetism.

41.5 Paramagnetic impurities, electron paramagnetism, and spin–orbit coupling

We will treat the effect of various 'pair-breaking' terms entering $U_{\alpha\beta}$, by considering their effect on the time-dependent operator $\underset{\sim}{\mathscr{K}}(t)$ which is defined as (see Eq. (38.18b))

$$\underset{\sim}{\mathscr{K}}(t) = \mathrm{e}^{-\mathrm{i}\mathscr{H}'t/\hbar}\underset{\sim}{\mathscr{K}}(0)\mathrm{e}^{-\mathrm{i}\mathscr{H}'t/\hbar}. \tag{41.39}$$

Formally, $\mathscr{H}^{(\mathrm{so})}$ commutes with $\mathscr{H}$. It nonetheless has an indirect effect which we will neglect for the moment. The time derivative of $\mathscr{K}(t)$ follows from the Heisenberg equation of motion and is given by

$$\dot{\underset{\sim}{\mathscr{K}}}(t) = \left[-\mathrm{i}\frac{\mathrm{d}\Phi}{dt} + \frac{2\mathrm{i}}{\hbar}I\underset{\sim}{\sigma}_z - \frac{1}{\tau_{\mathrm{s}}}\right]\underset{\sim}{\mathscr{K}}(t). \tag{41.40}$$

The first term, arising from the vector potential, was obtained earlier in Eq. (38.21) or (38.21′) and the second term follows directly on evaluating the commutator of $\mathscr{K}$ with (41.36). The last term, which accounts for uncorrelated (or ergodic) spin-flip scattering caused by (41.37), has been incorporated in a phenomenological manner. The term in τ_s leads to an exponential decay of $\mathscr{K}(t)$, a behavior which is referred to as ergodic. The term in I on the other hand leads to a periodic behavior of $\mathscr{K}(t)$; i.e., $\mathscr{K}(t \to \infty) \neq 0$ and this behavior is termed nonergodic.

We have already discussed the effect of the vector potential on the upper critical magnetic field, leading to Eq. (40.20). We now discuss, separately, the effect of electron paramagnetism and impurity spins. Writing the equation of motion for each of the components of the matrix differential equation arising from the second term on the right of (41.40), and integrating subject to the initial condition $\underset{\sim}{\mathscr{K}}(t=0) = \underset{\sim}{\rho}\mathscr{C}$ we obtain

$$\underset{\sim}{\mathscr{K}}(t) = \begin{pmatrix} 0 & \mathrm{e}^{\mathrm{i}\Omega_{\mathrm{L}}t} \\ -\mathrm{e}^{-\mathrm{i}\Omega_{\mathrm{L}}t} & 0 \end{pmatrix}\mathscr{C}, \tag{41.41}$$

where Ω_{L} is the spin Larmor frequency, $2I/\hbar$ (the $\underset{\sim}{\mathscr{K}}(t)$ operator remains antiunitary for all times, as it should). Thus the quantity $\underset{\sim}{\mathscr{K}}^{\dagger}(t)\underset{\sim}{\mathscr{K}}(0)$ entering Eq. (41.33) is given by the unitary operator

$$\underset{\sim}{\mathscr{K}}^{\dagger}(t)\underset{\sim}{\mathscr{K}}(0) = \begin{pmatrix} \mathrm{e}^{-\mathrm{i}\Omega_{\mathrm{L}}t} & 0 \\ 0 & \mathrm{e}^{\mathrm{i}\Omega_{\mathrm{L}}t} \end{pmatrix}. \tag{41.42}$$

Inserting (41.39) into (41.33) and summing over the spin indices associated with the eigenstates N yields

$$\begin{aligned} g^{\nu}(\mathbf{r},\mathbf{r}') &= \frac{1}{2}\frac{\sum\limits_n \delta(\xi - \xi_{n,\nu})\langle n|\delta[\mathbf{r} - \mathbf{R}(t)]\delta[\mathbf{r}' - \mathbf{R}(0)]|n\rangle}{\sum\limits_n \delta(\xi - \xi_{n,\nu})} \\ &\quad \times [\mathrm{e}^{-\mathrm{i}\Omega_{\mathrm{L}}t}\delta_{\nu 1} + \mathrm{e}^{+\mathrm{i}\Omega_{\mathrm{L}}t}\delta_{\nu 2}] \\ &\cong \frac{1}{2}g(\mathbf{r},\mathbf{r}',t)[\mathrm{e}^{-\mathrm{i}\Omega_{\mathrm{L}}t}\delta_{\nu 1} + \mathrm{e}^{+\mathrm{i}\Omega_{\mathrm{L}}t}\delta_{\nu 2}], \end{aligned} \tag{41.43}$$

where $g(\mathbf{r},\mathbf{r}',t)$ is the correlation function in the absence of spin polarization. Since we are presently considering the isolated effect of spin paramagnetism we set $\mathbf{A} = 0$. Then Δ is a constant and we may cancel it from both sides of the gap equation. Integrating $K(\mathbf{r}, \mathbf{r}')$ over d^3r' and noting that $L^3 \int g(\mathbf{r}, \mathbf{r}', t)\mathrm{d}^3r' = 1$, we have using (41.43) and (41.35) the following implicit equation for the transition temperature

$$1 = \frac{2\pi}{\hbar}\mathcal{N}(0)Vk_BT\sum_{\omega}\int_0^{\infty} \mathrm{d}t e^{-2|\omega|t} \cdot \frac{1}{2}(e^{-i\Omega_L t} + e^{+\Omega_L t})$$

or

$$1 = \frac{\pi}{\hbar}\mathcal{N}(0)Vk_BT\sum_{\omega}\left(\frac{1}{2|\omega| + i\Omega_L} + \frac{1}{2|\omega| - i\Omega_L}\right). \tag{41.44}$$

Using Eq. (40.6) we have

$$1 = \mathcal{N}(0)V\left\{\ell n\frac{1.14\hbar\omega_D}{k_BT} - \frac{1}{2}\left[\mathcal{U}\left(\frac{i\hbar\Omega_L}{2\pi k_BT}\right) + \mathcal{U}\left(\frac{-i\hbar\Omega_L}{2\pi k_BT}\right)\right]\right\}$$

or, using the property $\mathrm{Im}\psi(\frac{1}{2} + iy) = \frac{1}{2}\pi\tanh(\pi y)$, we have

$$\ell n\left(\frac{T_c}{T}\right) = \mathrm{Re}\,\mathcal{U}\left(\frac{i\hbar\Omega_L}{2\pi k_BT}\right). \tag{41.45}$$

The reduction of the transition temperature with magnetic field due to the effect of Pauli paramagnetism was first examined by Chandrasekhar (1962) and Clogston (1962).

Next we consider the isolated effect of dilute magnetic impurities.[4] The time dependence of $\underset{\sim}{\mathcal{K}}(t)$ arising from the last term on the right of Eq. (41.40) is

$$\underset{\sim}{\mathcal{K}}(t) = e^{-t/\tau_s}\underset{\sim}{\mathcal{K}}(0) \tag{41.46}$$

or $\underset{\sim}{\mathcal{K}}^{\dagger}(t)\underset{\sim}{\mathcal{K}}(0) = e^{-t/\tau_s}$. Following arguments similar to those leading to Eq. (41.44) yields

$$1 = \frac{2\pi}{\hbar}\mathcal{N}(0)Vk_BT\sum_{\omega}\frac{1}{2|\omega| + \dfrac{1}{\tau_s}} \tag{41.47}$$

or

$$\ell n\left(\frac{T_c}{T}\right) = \mathcal{U}\left(\frac{\hbar}{2\pi k_BT\tau_s}\right). \tag{41.48}$$

We defer further discussion of the effects of paramagnetic impurities to Subsec. 41.6.

We now return to the question of spin–orbit coupling.[5] Our approach here will be phenomenological. By introducing an *ad hoc* spin–orbit relaxation term in Eq. (41.40), we can

4. As impurities become more concentrated they begin to precess due to their mutual interaction. Alternatively, in the presence of an external magnetic field, the classical spin vectors undergo Larmor precession. Such dynamic effects greatly complicate the theory.
5. The effect of spin–orbit coupling was introduced by Ferrell (1959) to account for the result of Knight shift experiments and was further developed by Anderson (1959) and Abrikosov and Gorkov (1962). Its effect on reducing the pair breaking arising from Pauli paramagnetism has been studied by Fulde and Maki (1966), Werthamer, Helfand, and Hohenberg (1966), and Maki (1966) using the Green's function approach.

reproduce the Fulde–Maki result (1966). The term we add must have the property that, in the absence of a time reversal breaking perturbation, it results in no evolution of the time reversal operator, $\mathscr{K}(t)$, from its form at $t = 0$. A relaxation contribution proportional to the sum of $\mathscr{K}$ and its transpose $\tilde{\mathscr{K}}$ satisfies this requirement. In what immediately follows we limit ourselves to only the Larmour and spin–orbit terms and Eq. (41.40) becomes

$$\dot{\mathscr{K}} = i\Omega_L \sigma_z \cdot \mathscr{K}(t) - \frac{1}{\tau'}(\mathscr{K} + \tilde{\mathscr{K}}), \tag{41.49}$$

where τ' is a phenomenological spin–orbit scattering time. We must solve (41.49) subject to the initial condition $\mathscr{K}(0) = \begin{pmatrix} 0 & 1 \\ -1 & 0 \end{pmatrix}\mathscr{C}$. The calculations are straightforward, though tedious, and we only give the final result for the matrix $\mathscr{K}(t)$

$$\begin{bmatrix} 0 & \dfrac{[(1-\Omega_L^2\tau'^2)^{1/2} + 1 - i\Omega_L\tau']e^{-a_+t} + [(1-\Omega_L^2\tau'^2)^{1/2} - 1 + i\Omega_L\tau']e^{-a_-t}}{2(1-\Omega_L^2\tau'^2)^{1/2}} \\ \dfrac{[(1-\Omega_L^2\tau'^2)^{1/2} + 1 + i\Omega_L\tau']e^{-a_+t} + [-(1-\Omega_L^2\tau'^2)^{1/2} + 1 + i\Omega_L\tau']e^{-a_-t}}{2(1-\Omega_L^2\tau'^2)^{1/2}} & 0 \end{bmatrix}\mathscr{C} \tag{41.50}$$

where

$$a_\pm = -\frac{1}{\tau'}[\pm(1-\Omega_L^2\tau'^2)^{1/2} - 1]. \tag{41.51}$$

The following quantity determines the transition temperature through Eq. (41.35)

$$\mathrm{Tr}\,\mathscr{K}^\dagger(t)\mathscr{K}(0) = \left(1 + \frac{1}{(1-\Omega_L^2\tau'^2)^{1/2}}\right)e^{-a_+t} + \left(1 - \frac{1}{(1-\Omega_L^2\tau'^2)^{1/2}}\right)e^{-a_-t}. \tag{41.52}$$

Carrying out the time integration and evaluating the ω sum in (41.35) as we have in previous calculations we obtain the Fulde–Maki (1966) expression for the transition temperature

$$\ell n\left(\frac{T_c}{T}\right) = \frac{1}{2}\left[1 + \frac{1}{(1-\Omega_L^2\tau'^2)^{1/2}}\right]\mathscr{U}\left(\frac{\hbar a_+}{\pi k_B T}\right) + \frac{1}{2}\left[1 - \frac{1}{(1-\Omega_L^2\tau'^2)^{1/2}}\right]\mathscr{U}\left(\frac{\hbar a_-}{\pi k_B T}\right). \tag{41.53}$$

In order to make contact with the more microscopic approach we must define $\tau' = 3\tau_{so}$ where τ_{so} is the spin–orbit scattering time introduced by Abrikosov and Gorkov (1962). Eq. (41.53) reduces to (41.45) in the limit $1/\tau' \to 0$ as it must. In the limit $\Omega_L\tau' \to \infty$, Eq. (41.53) is commonly approximated by

$$\ell n\frac{T_c}{T} \cong \mathscr{U}\left(\frac{\hbar\Omega_L^2\tau'}{2\pi k_B T}\right). \tag{41.54}$$

In the limit $\tau' \to 0$ the transition temperature is not depressed.

We may easily include the effects of the vector potential. It turns out the behavior is still given by Eq. (41.53), however, with

$$a_{\pm} = \frac{1}{\tau'}[1 \mp (1 - \Omega_L^2\tau'^2)^{1/2}] + \frac{\Omega_0}{2}, \tag{41.55}$$

where Ω_0 is given by (40.24) or (40.25) for an anisotropic bulk or thin film sample, respectively.

41.6 The Fulde–Ferrell state

Let us examine the effect of a magnetic field on the spin of the electrons in more detail. The decrease in the normal-state energy due to this effect is $\frac{1}{2}\chi_n H^2$, where χ_n is the normal-state Pauli spin susceptibility

$$\chi_n = \frac{1}{2} g^2 \mu_B^2 \mathcal{N}(0); \tag{41.56}$$

here g is the conduction electron g factor, which we take to be 2. We must now examine the susceptibility, χ_s, in the superconducting state. In order to polarize ground-state (condensed) electrons in the superconducting state one must break pairs (i.e., destroy the superconducting state). On the other hand the number of excited-state electrons decreases exponentially as $e^{-\Delta/k_B T}$ at low temperatures. Hence we expect the paramagnetic contribution to the susceptibility (i.e., the spin susceptibility) of the superconducting state to be small at low temperatures. In addition there is the diamagnetic contribution to the susceptibility[6] in the superconductor arising from London screening currents. In a perfect Type I material the Meissner effect eliminates the paramagnetism and $\chi_s = -1/4\pi$. In a Type II material where flux penetrates the sample above H_{c1}, it plays a rapidly decreasing role as the field increases, and vanishes at H_{c2}.

Qualitatively, a transition between the normal and superconducting states is expected at a field

$$\frac{1}{2}(\chi_n - \chi_s)H_p^2 = \frac{H_c^2(T)}{8\pi}, \tag{41.57}$$

where H_c is the thermodynamic critical field and H_p is the paramagnetic upper critical field (Clogston 1962, Chandrasekhar 1962). If $\chi_n = 0$ and $\chi_s = -1/4\pi, H = H_c$ as expected for the Meissner state. Solving (41.57) for H_p we obtain

$$H_p(T) = \frac{H_c(T)}{[4\pi(\chi_n - \chi_s)]^{1/2}}. \tag{41.58}$$

Whether the transition at $H = H_p$ is first or second order is a subtle question which we now examine in more detail. If we assume that the transition at H_{c2} in a Type II material is second order for some range of parameters, the condition that the second order line terminate at a critical point is that the kernel of our linearized theory satisfy the condition

6. The Meissner susceptibility is sometimes called the orbital contribution to the total susceptibility, since it arises from the motion (or 'orbits') of the electrons.

$$K(H, T) = 1 \tag{41.59a}$$

and that the coefficient of the $|\Delta|^2\Delta$ term in the nonlinear G–L theory vanish. This coefficient is calculated in Sec. 45 (see Eq. (45.3)) and involves the sum $\Sigma(1/|\omega|^3)$; in the presence of a magnetic field our condition is therefore (see Eq. (41.44))

$$\frac{1}{2}\sum_{\omega}\left[\frac{1}{\left(|\omega| + \mathrm{i}\frac{\Omega_L}{2}\right)^3} + \frac{1}{\left(|\omega| - \mathrm{i}\frac{\Omega_L}{2}\right)^3}\right] = 0, \tag{41.59b}$$

which occurs for

$$\frac{\gamma H_c}{k_B T_c(H_c)} = 1.911, \tag{41.60}$$

where $\Omega_L = 2\gamma H$ and γ is the gyromagnetic ratio. Sarma (1963) finds that the solution to Eqs. (41.59) is

$$\frac{\gamma H_c}{\Delta_0} = 0.61 \tag{41.61a}$$

and

$$\frac{k_B T_c(H_c)}{\Delta_0} = 0.31, \tag{41.61b}$$

where $\Delta_0 = 1.75 k_B T_c(H = 0)$ is the BCS energy gap at $T = 0$. For $T < T_c(H_c)$ the transition between the normal and superconducting states becomes first order.

The above analysis was carried out under the assumption that the energy gap was constant. Fulde and Ferrell (1964) suggested that a state with an energy gap function of the form

$$\Delta(\mathbf{r}) = \Delta(\mathbf{q}_0)e^{\mathrm{i}\mathbf{q}_0\cdot\mathbf{r}} \tag{41.62}$$

might be more stable. The equation analogous to (40.5) describing such a situation is

$$K(\mathbf{q}, H, T) - K(0, 0, T) = \mathcal{N}(0)V\ell n\frac{T}{T_c(H = 0, \mathbf{q} = 0)}. \tag{41.63}$$

The left-hand side of (41.63) is a function of $\mathbf{q}$, and that value of $\mathbf{q}$ leading to the highest transition temperature corresponds to the most stable wave vector, which we designated as $\mathbf{q}_0$; thus we must seek a simultaneous solution of (41.63) and the equation

$$\left.\frac{\partial K(\mathbf{q}, H, T)}{\partial \mathbf{q}}\right|_{\mathbf{q}=\mathbf{q}_0} = 0. \tag{41.64}$$

An analysis by Sarma and St James (1969) shows that a simultaneous solution to (41.63) and (41.64) can only be obtained for the case of a clean superconductor. For the free-electron-like case we have (see (40.28))

$$K(\mathbf{q}, H, T) - K(0, H, T) = \frac{\pi\mathcal{N}(0)Vk_B T}{\hbar q V_F}\mathrm{Re}\sum_{\omega}\left\{\tan^{-1}\frac{qV_F - 2\gamma H}{2|\omega|}\right.$$

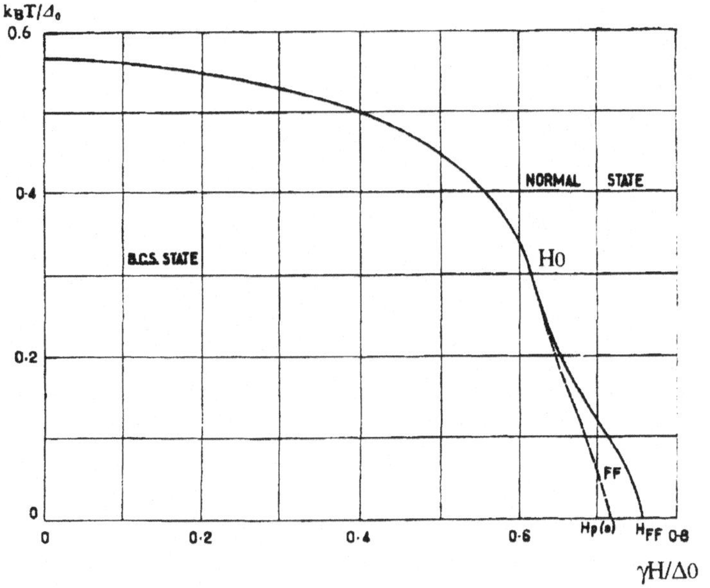

Figure 41.1 Transition curve from the BCS state and from the Fulde and Ferrell depaired state to the normal state (full curve). The domain of existence of the Fulde–Farrell state is labeled FF. The dashed line shows an approximate boundary between the BCS and the FF state. (After Sarma and Saint James (1969).)

$$+ \tan^{-1}\frac{qV_F + 2\gamma H}{2|\omega|}\Big\} - \pi\mathcal{N}(0)Vk_BT\mathrm{Re}\sum_\omega \frac{1}{|\omega| + \mathrm{i}\gamma H}. \tag{41.65}$$

A solution with $\mathbf{q} \neq 0$ exists for $H > H_F$, where H_F is the solution to

$$\left.\frac{\partial^2 K(\mathbf{q}, H, T)}{\partial q^2}\right|_{\mathbf{q}=0} = 0. \tag{41.66}$$

The condition (41.66) can be seen to be identical to (41.59b) and hence H_F is identical to H_c given by (41.60). The phase diagram as calculated by Sarma and Saint-James is shown in Fig. 41.1. The transition from the normal state to the Fulde–Ferrell state is second order while that between the Fulde–Ferrell state and the BCS state is first order.

41.7 Gapless superconductivity

We now examine the effect of paramagnetic impurities in more detail. For temperatures just below T_c, as given in Eq. (41.48), a new superconducting state appears in which there is an order parameter, Δ, but there is no gap in the excitation spectrum! The approach we take here, which uses ordinary perturbation theory, is somewhat heuristic (see deGennes (1966)). A formal treatment requires many-body techniques (Abrikosov and Gorkov 1961).

To examine the excitation spectrum we calculate the corrections to the quasiparticle energy ε using perturbation theory where we write

$$\varepsilon = \varepsilon^{(0)} + \varepsilon^{(1)} + \varepsilon^{(2)} \ldots; \tag{41.67}$$

here $\varepsilon^{(0)} = |\xi|, \varepsilon^{(1)} = 0$, and $\varepsilon^{(2)}$ is calculated in Appendix C (Eqs. (C.14a) and (C.14b) as

$$\varepsilon_n^{(2)} = \sum_{\ell}' \frac{\left| \int \phi_n^*(\mathbf{r})\Delta(\mathbf{r})\phi_\ell^*(\mathbf{r})\mathrm{d}^3r \right|^2}{\xi_n + \xi_\ell} \mathrm{sgn}(\xi_n). \tag{41.68}$$

For the case of a uniform superconductor, $\Delta = \text{const}$, we may write the energy to second order as

$$\varepsilon_n = |\xi_n| + |\Delta|^2 \sum_{\ell}' \frac{|\langle n|\mathscr{K}|\ell\rangle|^2}{\xi_n + \xi_\ell} \mathrm{sgn}(\xi_n). \tag{41.69}$$

In the absence of a magnetic perturbation (paramagnetic impurities in the present case) the normal state eigenfunctions ϕ_n and ϕ_ℓ may be chosen as real (and we need not introduce $\mathscr{K}$). In this latter case we obtain $\varepsilon_n = |\xi_n| + (\Delta^2/2|\xi_n|)$ which coincides with the first two terms in the expansion of the BCS energy spectrum, $\varepsilon_n = (\xi_n^2 + |\Delta|^2)^{1/2}$. Clearly this perturbation theory breaks down for $\xi \lesssim \Delta$ (our expansion parameter being Δ/ξ).

Let us now examine (41.69) in the presence of paramagnetic impurities. The integrands of the matrix elements involving $\mathscr{K}$, ϕ_n^*, and ϕ_ℓ occurred earlier in our study of the linearized gap equation (see Eq. (41.30)), where we showed they could be related to a correlation function $g(\mathbf{r}, \mathbf{r}', \Omega)$. For the uniform case considered here the $\mathbf{r}$ dependence of the correlation function is not involved and we need only consider the energy (frequency) dependence, $g(\Omega)$. Now $g(t)$ was given prior to Eq. (41.47) as $\mathrm{e}^{-|t|/\tau_s}$, the Fourier transform of which is

$$g(\Omega) = \frac{1}{2\pi} \int_{-\infty}^{+\infty} \mathrm{d}t \mathrm{e}^{\mathrm{i}\Omega t - |t|/\tau_s} = \frac{1}{\pi} \frac{\tau_s}{1 + \Omega^2 \tau_s^2}. \tag{41.70}$$

We may then write the energy as

$$\varepsilon = |\xi| + |\Delta|^2 P \int \mathrm{d}\xi' g\left(\frac{\xi - \xi'}{\hbar}\right) \frac{1}{\xi + \xi'} \mathrm{sgn}(\xi), \tag{41.71}$$

where we wrote the summation Σ_ℓ' as a principal value integral. Carrying out the integration we obtain

$$\varepsilon = |\xi| + \frac{2|\Delta|^2|\xi|}{(2\xi)^2 + (\hbar/\tau_s)^2}. \tag{41.72}$$

Note that ε is now well behaved in the limit $\xi \to 0$. Provided $\Delta\tau_s/\hbar << 1$, the second term on the right in (41.72) will always be much less than the first, implying our perturbation theory is valid. Fig. 41.2 shows a plot of (41.72). The most striking feature is that there is *no gap* in the excitation spectrum even though Δ is nonzero. One could go on to calculate diamagnetic screening currents; the result is that they do not vanish. We therefore conclude that one can have superconductivity in the absence of a gap: gapless superconductivity. We can use the excitation spectrum to calculate the superconducting density of states with the result

$$\mathscr{N}(\varepsilon) = \mathscr{N}(0)\frac{\partial \xi}{\partial \varepsilon} \cong \mathscr{N}(0)\left\{1 + 2\left(\frac{\Delta\tau_s}{\hbar}\right)^2 \frac{(2\varepsilon)^2 - (\hbar/\tau_s)^2}{[(2\varepsilon)^2 + (\hbar/\tau_s)^2]^2}\right\}. \tag{41.73}$$

Fig. 41.3 shows a plot of this function.

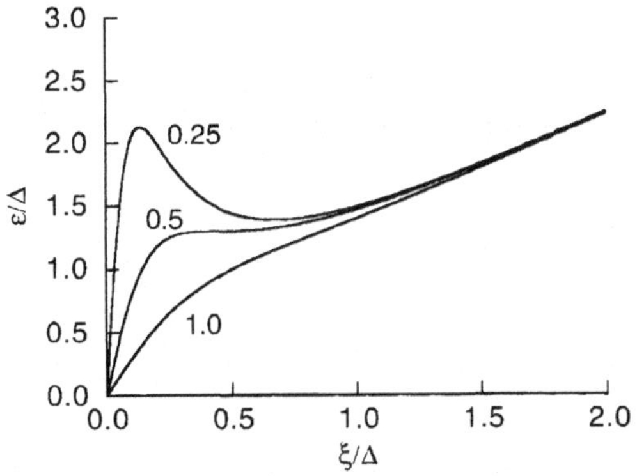

Figure 41.2 Theoretical curve calculated from Eq. (41.72) for three different values of $\hbar\tau_s/\Delta$.

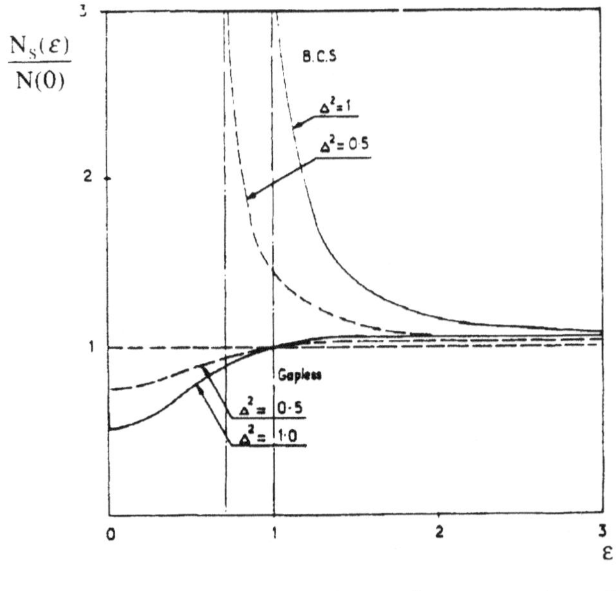

Figure 41.3 The density of states for a gapless superconductor for two different values of the pair potentials $|\Delta(\mathbf{r})|$ and for $\hbar/2\tau(T) = 1$. The corresponding BCS case with $\Delta = |\Delta(\mathbf{r})|$ is also shown. (After Saint-James, Thomas, and Sarma (1969).)

At lower temperatures, where $\Delta\tau_s/\hbar \sim 1$, our perturbation theory breaks down. A more sophisticated treatment (in which all orders of Δ are summed) shows that a gap appears. This marks a phase transition to a *second superconducting state*.

42 Boundary conditions

In this section we discuss boundary conditions that apply between insulators, normal metals, and superconductors. These boundary conditions will apply to dirty superconductors and were first derived by deGennes (1964), and later generalized to the case of finite magnetic fields by Takahashi and Tachiki (1986a).

We first derive a sum rule which we will require later. We rewrite Gorkov's linearized self-consistency condition, Eq. (38.6), in the form (deGennes 1964)

$$\Delta(\mathbf{r}) = V(\mathbf{r})k_{\mathrm{B}}T\sum_{\omega}\int \mathrm{d}^3r' Q_{\omega}(\mathbf{r},\mathbf{r}')\Delta(\mathbf{r}'), \tag{42.1}$$

where, from Eq. (38.10), $Q_{\omega}(\mathbf{r},\mathbf{r}')$ is defined by

$$Q_{\omega}(\mathbf{r},\mathbf{r}') = \sum_{m,n}\frac{\phi_m^*(\mathbf{r}')\phi_n^*(\mathbf{r}')\phi_m(\mathbf{r})\phi_n(\mathbf{r})}{(\xi_m + \mathrm{i}\hbar\omega)(\xi_n - \mathrm{i}\hbar\omega)}, \tag{42.2}$$

and

$$k_{\mathrm{B}}TV(\mathbf{r})\sum_{\omega}Q_{\omega}(\mathbf{r},\mathbf{r}') = K(\mathbf{r},\mathbf{r}'). \tag{42.3}$$

The sum rule we seek is the integral of $Q_{\omega}(\mathbf{r},\mathbf{r}')$ over either $\mathbf{r}$ or $\mathbf{r}'$, and we now limit the discussion to zero field ($H = 0$), where we may choose the eigenfunctions $\phi_n(\mathbf{r})$ to be real and write them as $w_n(\mathbf{r})$. From the orthonormality of the w_n we have

$$\int Q_{\omega}(\mathbf{r},\mathbf{r}')\mathrm{d}^3r' = \sum_n \frac{[w_n(\mathbf{r})]^2}{\xi_n^2 + (\hbar\omega)^2}.$$

We define a *local* density of states by

$$\mathcal{N}(\xi,\mathbf{r}) = \sum_n [w_n(\mathbf{r})]^2\delta(\xi - \xi_n). \tag{42.4}$$

We then have

$$\int Q_{\omega}(\mathbf{r},\mathbf{r}')\mathrm{d}^3r' = \int \frac{\mathrm{d}\xi}{\xi^2 + (\hbar\omega)^2}\sum_n [w_n(\mathbf{r})]^2\delta(\xi - \xi_n).$$

The ω_n of interest are of order $k_{\mathrm{B}}T_{\mathrm{c}}/\hbar$, and over this energy range $\mathcal{N}(\xi,\mathbf{r})$ is independent of energy;

it may then be taken outside the integral and approximated by its value at $\xi = 0$, denoted as $\mathcal{N}(\mathbf{r})$, yielding

$$\int Q_\omega(\mathbf{r},\mathbf{r}')\mathrm{d}^3r' = \mathcal{N}(\mathbf{r})\int \frac{\mathrm{d}\xi}{\xi^2 + (\hbar\omega)^2} = \frac{\pi}{\hbar|\omega|}\mathcal{N}(\mathbf{r}). \tag{42.5}$$

The following discussion will be limited to the case of dirty superconductors. We do this simply because this is the only case where the boundary conditions may be written down in a general form; it will turn out that they involve only the bulk density of states and the diffusion constants of the constituents. (The clean limit requires a much more complete description of the electronic structure than is generally available; see Alexander, Orlando, Rainer, and Tedrow (1985).)

As discussed in Sec. 38 the behavior of a dirty (or clean) superconductor is governed by a correlation function. However, the correlation function defined by Eq. (38.15) is normalized by the total density of states which, for a macroscopically inhomogeneous system, is not appropriate. We define an *unnormalized* correlation function by

$$\mathcal{g}_\xi(\mathbf{r},\mathbf{r}',\Omega) \equiv \sum_{n,m} w_n(\mathbf{r})w_m(\mathbf{r})w_n(\mathbf{r}')w_m(\mathbf{r}')\delta(\xi - \xi_n)\delta\left(\frac{\xi_n - \xi_m}{\hbar} + \Omega\right). \tag{42.6}$$

The subscript ξ signifies that $\mathcal{g}$ is a function of the energy. However, the range of energies relevant to superconductivity is restricted to the immediate vicinity of the Fermi energy and we will therefore assume $\mathcal{g}$ is independent of ξ and omit the subscript in what follows. The Fourier transform of (42.6) can, by the manipulations leading up to Eq. (38.19), be written as

$$\mathcal{g}(\mathbf{r},\mathbf{r}',t) = \sum_n \langle n|\delta[\mathbf{R}(t) - \mathbf{r}]\delta[\mathbf{R}(0) - \mathbf{r}']|n\rangle\delta(\xi - \xi_n), \tag{42.7}$$

where $\mathbf{R}(t)$ is the Heisenberg position operator; $\mathcal{g}$ measures the probability of finding a particle at the point $\mathbf{R}(t)$ that was initially at the point $\mathbf{R}(0)$.

We require the relation between $\mathcal{g}$ and Q_ω. Using Eq. (38.11) and Eq. (38.13) we can relate (42.2) and (42.6) as

$$Q_\omega(\mathbf{r},\mathbf{r}') = \int \frac{\mathrm{d}\xi\mathrm{d}\xi'}{(\xi - i\hbar\omega)(\xi' + i\hbar\omega)}\mathcal{g}(\mathbf{r},\mathbf{r}',(\xi' - \xi)/\hbar). \tag{42.8}$$

From the discussion leading up to Eq. (38.14′) we may write Q_ω in the form

$$Q_\omega(\mathbf{r},\mathbf{r}') = \frac{2\pi}{\hbar}\int_0^\infty \mathrm{d}t e^{-2|\omega|t}\mathcal{g}(\mathbf{r},\mathbf{r}',t). \tag{42.9}$$

In the diffusion approximation the behavior of $\mathcal{g}$ is governed by the analogue of Eq. (40.10)[1]

$$\frac{\partial\mathcal{g}(\mathbf{r},\mathbf{r}',t)}{\partial t} - D\nabla^2\mathcal{g}(\mathbf{r},\mathbf{r}',t) = 0; \quad t > 0. \tag{42.10}$$

The function $\mathcal{g}(\mathbf{r},\mathbf{r}',t)$ is singular at $t = 0$. The form and strength of the singularity follow from

1. In an inhomogeneous anistropic media we would replace $D\nabla^2$ by $\nabla\cdot\mathbf{D}(\mathbf{r})\cdot\nabla$, where $\mathbf{D}(\mathbf{r})$ is a position-dependent tensor.

taking the Fourier time transform of Eq. (42.6) (using the convention defined in Eq. (38.16)), setting $t = 0$, and using the closure property

$$\lim_{\mathbf{r}\to\mathbf{r}'} g(\mathbf{r},\mathbf{r}',t=0) = \delta(\mathbf{r}-\mathbf{r}')\sum_n |w_n(\mathbf{r})|^2\delta(\xi-\xi_n) = \mathcal{N}(\mathbf{r})\delta(\mathbf{r}-\mathbf{r}'). \tag{42.11}$$

We now derive a differential equation governing the spatial behavior of $Q_\omega(\mathbf{r},\mathbf{r}')$. We multiply both sides of Eq. (42.10) by $(2\pi/\hbar)\mathrm{e}^{-2|\omega|t}$ and integrate from 0 to ∞ to obtain

$$\frac{2\pi}{\hbar}\int_0^\infty \mathrm{d}t\mathrm{e}^{-2|\omega|t}\frac{\partial g(\mathbf{r},\mathbf{r}',t)}{\partial t} - D\nabla^2\frac{2\pi}{\hbar}\int_0^\infty \mathrm{d}t\mathrm{e}^{-2|\omega|t}g(\mathbf{r},\mathbf{r}',t) = 0.$$

Integrating the first term by parts and using Eq. (42.11) we have

$$2|\omega|Q_\omega(\mathbf{r},\mathbf{r}') - D\nabla^2 Q_\omega(\mathbf{r},\mathbf{r}') = \frac{2\pi}{\hbar}\mathcal{N}(\mathbf{r})\delta(\mathbf{r}-\mathbf{r}'); \tag{42.12}$$

Eq. (42.12) will form the basis for much of our subsequent analysis.

Our present interest is in plane interfaces in the dirty limit, which will be regarded as homogeneous in the two directions which are perpendicular to the plane normal and we choose the associated coordinates as x and y; we will use z for the coordinate along the plane normal. We may formally perform an in-plane average over x and y by Fourier transforming the variables $x - x'$ and $y - y'$ into q_x and q_y and then setting both q_x and q_y to zero. This transformed version of Eq. (42.12) is

$$2|\omega|Q_\omega(z,z') - D\frac{\mathrm{d}^2}{\mathrm{d}z^2}Q_\omega(z,z') = \frac{2\pi}{\hbar}\mathcal{N}(z)\delta(z-z'), \tag{42.13}$$

where $\mathcal{N}(z) = (1/L^2)\int\mathcal{N}(\mathbf{r})\mathrm{d}x\mathrm{d}y$ with L^2 the area of the interface.

By definition, the function Q_ω satisfying Eq. (42.13) must be symmetric with respect to the variables z and z', and the differential operator may be applied to either variable. $Q_\omega(z,z')$ is a shorthand for $Q_\omega(z,z',q_x=0,q_y=0)$.

Eq. (42.13) is an inhomogeneous equation and the general solution is a superposition of solutions to the homogeneous part (where the right-hand side is set equal to zero) and an explicit inhomogeneous (particular) solution. The latter has the form

$$Q_\omega(z-z') = \frac{\pi\mathcal{N}}{2\hbar|\omega|\xi_\omega}\mathrm{e}^{-|z-z'|/\xi_\omega}, \tag{42.14}$$

which can be directly verified by substituting in Eq. (42.13) (one differentiation with respect to z generates a factor $\mathrm{sgn}(z-z')$ which on differentiating a second time yields $2\delta(z-z')$); here we defined $\xi_\omega = (D/2|\omega|)^{1/2}$ and this length sets the range of $Q_\omega(z,z')$. The long-range behavior of Q will be governed by the $\omega_0 = (\pi k_B T/\hbar)$ term which leads us to define an 'alloy coherence length' $\xi \equiv (D\hbar/2\pi k_B T)^{1/2}$.

We first consider the case of a metal–vacuum interface where the metal occupies the half space $z > 0$. Since there can be no flux of particles through the boundary, we require $D[\partial g(z,z')/\partial z]|_{z=0} = 0$ which, from the definition of $Q_\omega(\mathbf{r},\mathbf{r}')$, requires $[\partial Q_\omega(z,z')/\partial z]|_{z=0} = 0$; from Eq. (42.1) this immediately leads to the boundary condition on $\Delta(z)$

$$\left.\frac{\partial\Delta(z)}{\partial z}\right|_{z=0} = 0 \quad \text{(metal–vacuum interface).} \tag{42.15}$$

We satisfy the boundary condition on $Q_\omega(z, z')$ by adding the appropriate solution of the homogeneous equation to Eq. (42.14); this corresponds to locating an image at the point $-z'$ for each source point z. By inspection we write

$$Q_\omega(z, z') = \frac{\pi \mathcal{N}}{2\hbar|\omega|\xi_\omega}(e^{-|z-z'|/\xi_\omega} + e^{-(z+z')/\xi_\omega}), \quad z, z' > 0. \tag{42.16}$$

On passing from a metal to an insulator the wavefunctions of conduction electrons typically decay in an interatomic distance (which is small relative to ξ_ω), and thus we may use this same metal–vacuum boundary condition at a plane, bulk metal–bulk insulator interface.

We now discuss the case of two metals, A and B, with densities of states $\mathcal{N}_A$ and $\mathcal{N}_B$ and diffusion constants D_A and D_B, which are separated by an interface which we again locate at $z = 0$; materials A and B will be positioned at $z > 0$ and $z < 0$, respectively. We integrate Eq. (42.13) over all z' to obtain a boundary condition

$$2|\omega| \int Q_\omega(z, z')dz' - \left[D(z')\frac{dQ_\omega(z, z')}{dz'}\right]\Bigg|_{z'=0^+}^{z'=\infty} - \left[D(z')\frac{dQ_\omega(z, z')}{dz'}\right]\Bigg|_{z'=-\infty}^{z'=0^-}$$
$$= \frac{2\pi\mathcal{N}(z)}{\hbar}. \tag{42.17}$$

Using the sum rule expressed by Eq. (42.5), we see that the first term on the left cancels the one on the right. Qualitatively, Q_ω falls off for large z' according to Eq. (42.14) and we are then left with our first boundary condition

$$\left[D(z')\frac{dQ_\omega(z, z')}{dz'}\right]_{z'=0^+} - \left[D(z')\frac{dQ_\omega(z, z')}{dz'}\right]_{z'=0^-} = 0 \tag{42.18}$$

(this condition is equivalent to conserving the flux of particles through the boundary).

The integration of Eq. (42.13), which is a second order differential equation, requires two boundary conditions. Unlike Eq. (42.18), which is general, the second boundary condition depends on the atomic details of the interface. Two limiting cases involve: (1) a highly reflecting interface (as one might have if a thin oxide layer were present at the interface) and (2) 'good' metallic contact (where good will be defined later). If we neglect any possible retardation effects at the interface (involving the trapping and later release of an electron at an interface state) we may write the second boundary condition in the (Cauchy) form

$$[Q_\omega(z, z')]_{z'=0^+} = \left[\alpha Q_\omega(z, z') + \beta\frac{dQ_\omega(z, z')}{dz'}\right]_{z'=0^-}. \tag{42.19}$$

At this point we note that our classical diffusion description of the correlation function $\mathcal{g}$ (which governs Q_ω) can be obtained as the limiting case of a semiclassical description based on the Boltzman transport equation. In this picture one studies the evolution of electrons distributed over some convenient set of one-particle states[2] (e.g., plane waves) which scatter off impurity

2. Note these states are not the exact one-particle states which define $\mathcal{g}$ in Eq. (42.6).

centers and, of interest to us here, interfaces.[3] A similar problem is encountered in the 'one-speed' theory of diffusing neutrons near the surface of a moderator. From the solution of the transport equation it is known that, in the absence of an insulating (reflecting) barrier, β/α is of the order of a mean free path, ℓ. However, from Eq. (42.14) we expect $\partial Q_\omega/\partial z'$ to be of order Q_ω/ξ_ω and thus the second term on the right of Eq. (42.19) is smaller than the first by a factor of order ℓ/ξ_ω; in the dirty limit where $\ell << \xi$ (or more accurately when the transmission coefficient of the boundary is much greater than ℓ/ξ, which defines 'good' contact), we may safely neglect this term and simplify Eq. (42.19) to

$$Q_\omega(z, 0^+) = \alpha Q_\omega(z, 0^-). \tag{42.20}$$

The constant α in Eq. (42.20) can be determined by the symmetry requirements of $Q_\omega(z, z')$, and to do this we need to construct the explicit form of the solutions. The correctly normalized solution to Eq. (42.13) for z and z' both in metal A has a form similar to Eq. (42.16), but with the strength of the image term modified by a parameter λ

$$Q_\omega(z, z') = \frac{\pi \mathcal{N}_\mathrm{A}}{2\hbar|\omega|\xi^\mathrm{A}_\omega}[\mathrm{e}^{-|z-z'|/\xi^\mathrm{A}_\omega} + \lambda \mathrm{e}^{-(z+z')/\xi^\mathrm{A}_\omega}], \; z, z' > 0 \tag{42.21}$$

(we define ξ^A_ω and ξ^B_ω as $(D_\mathrm{A}/2|\omega|)^{1/2}$ and $(D_\mathrm{B}/2|\omega|)^{1/2}$, respectively).

A solution satisfying Eq. (42.13) with the differential operator applied to either z or z' in the region $z > 0, z' < 0$ is

$$Q_\omega(z, z') = \frac{\mathcal{N}_\mathrm{B}\pi}{2\hbar|\omega|\xi^\mathrm{B}_\omega}\mu \mathrm{e}^{-z/\xi^\mathrm{A}_\omega + z'/\xi^\mathrm{B}_\omega}, \; z > 0, z' < 0; \tag{42.22}$$

here μ is another parameter and the factor $(\mathcal{N}_\mathrm{B}\pi/(2\hbar|\omega|\xi^\mathrm{B}_\omega)$ has been chosen for later convenience (the normalization cannot be established from Eq. (42.13) alone when z and z' are on opposite sides of the interface). Applying the two boundary conditions, Eqs. (42.18) and (42.20), to Eqs. (42.21) and (42.22) yields

$$(1 - \lambda)\mathcal{N}_\mathrm{A} = \mu \mathcal{N}_\mathrm{B} \tag{42.23}$$

and

$$(1 + \lambda)\frac{\mathcal{N}_\mathrm{A}}{\xi^\mathrm{A}_\omega} = \alpha\mu\frac{\mathcal{N}_\mathrm{B}}{\xi^\mathrm{B}_\omega}; \tag{42.24}$$

solving this pair of equations for μ yields

$$\mu = \frac{2\xi^\mathrm{B}_\omega}{\xi^\mathrm{B}_\omega + \alpha\xi^\mathrm{A}_\omega}\frac{\mathcal{N}_\mathrm{A}}{\mathcal{N}_\mathrm{B}}, \tag{42.25}$$

with which we can rewrite Eq. (42.22) as

$$Q_\omega(z, z') = \frac{\pi}{\hbar|\omega|}\frac{\mathcal{N}_\mathrm{A}}{\xi^\mathrm{B}_\omega + \alpha\xi^\mathrm{A}_\omega}\mathrm{e}^{-z/\xi^\mathrm{A}_\omega + z'/\xi^\mathrm{B}_\omega}; \quad z > 0, z' < 0. \tag{42.26}$$

If we repeat the calculation for the cases $z, z' < 0$ and $z < 0, z' > 0$ we obtain as the analogue of

3. This point of view has been extensively developed by Luders and Usadel (1971) in their interesting book.

Eq. (42.26)

$$Q_\omega(z, z') = \frac{\pi}{\hbar|\omega|} \frac{\mathcal{N}_B}{\xi_\omega^A + \alpha^{-1}\xi_\omega^B} e^{+z/\xi_\omega^B - z'/\xi_\omega^A}; \quad z < 0, \quad z' > 0. \tag{42.27}$$

Requiring Q_ω be symmetric in z and z' yields, on comparing Eqs. (42.26) and (42.27),

$$\frac{\mathcal{N}_A}{\xi_B + \alpha\xi_A} = \frac{\mathcal{N}_B}{\xi_A + \alpha^{-1}\xi_B}, \tag{42.28}$$

which has the solution $\alpha = \mathcal{N}_A/\mathcal{N}_B$. Returning to Eq. (42.20) we see that this implies

$$\frac{Q_\omega(z, 0^-)}{\mathcal{N}_B} = \frac{Q_\omega(z, 0^+)}{\mathcal{N}_A} \tag{42.29a}$$

for all z; by symmetry,

$$\frac{Q_\omega(0^-, z')}{\mathcal{N}_B} = \frac{Q_\omega(0^+, z')}{\mathcal{N}_A} \tag{42.29b}$$

for all z'. On examining Eq. (42.1) we see that (42.29b) implies that at the interface

$$\frac{\Delta_A}{\mathcal{N}_A V_A} = \frac{\Delta_B}{\mathcal{N}_B V_B}. \tag{42.30}$$

Eq. (42.18) applied to (42.1) yields

$$\frac{D_A}{V_A} \frac{d\Delta_A}{dz} = \frac{D_B}{V_B} \frac{d\Delta_B}{dz}. \tag{42.31}$$

The above are deGennes' (1964) dirty limit boundary conditions.

Although these boundary conditions have been established for a planar boundary only, they should provide a reasonable description of a more general boundary if we replace d/dz by $\hat{\mathbf{n}} \cdot \nabla$ where $\hat{\mathbf{n}}$ is a unit vector normal to the boundary.

43 The proximity effect at zero field

43.1 Governing equations

When a superconducting slab is placed in contact with a normal metal, the transition temperature of the superconductor is reduced. As the thickness of the slab increases the transition temperature approaches the bulk value, but the gap function will be suppressed in the vicinity of the boundary. These effects are referred to as proximity effects. Proximity effects play a role in more complex geometries. An example is an array of thin superconducting wires (e.g., NbTi) embedded in a metal host (Cu) as one has with a multifilimentary conductor. Another example would be a phase-separated alloy involving normal and superconducting phases.

In this section we use the differential equation for Q_ω, Eq. (42.13), and the boundary conditions determined in the previous section to solve the gap equation for the zero-field transition temperature of a sandwich or an artificial metallic superlattice; we will assume there is no spin-flip scattering. Our approach is to use a variant of the eigenfunction technique developed in Sec. 40. Assume we have a system consisting of alternate layers of two metals, A and B, which, respectively, are characterized by the diffusion constants and pairing potentials D_A, V_A and D_B, V_B. The particular approach used here is due to Takahashi and Tachiki (1985, 1986a,b). For other work see Auvil, Ketterson, and Song (1989), Radović, Ledvij, and Dobrosavljević-Grujić (1991) and Koperdraad (1995).

We seek a solution to (42.13) for $Q_\omega(z, z')$ in the 'symmetrical' form

$$Q_\omega(z, z') = [\mathcal{N}(z)\mathcal{N}(z')]^{1/2} \sum_{n,m} a_{nm}\, \psi_n^*(z')\psi_m(z), \tag{43.1}$$

where the $\psi_n(z)$ are eigenfunctions of the equation

$$D(z)\frac{d^2}{dz^2}\psi_n(z) + \Omega_n\psi_n(z) = 0; \tag{43.2}$$

here $D(z) = D_A$ or D_B and $\mathcal{N}(z) = \mathcal{N}_A$ or $\mathcal{N}_B$ in medium A or B respectively, and the a_{nm} are expansion coefficients yet to be determined.

The $\psi_n(z)$ will be normalized according to the condition

$$\int_0^d \psi_n^*(z)\psi_m(z)dz = \delta_{nm}, \tag{43.3}$$

where d is the repeat distance.

Note the ψ_n could be chosen to be real in which case Q_ω is completely symmetrical in z and z' (as assumed in the previous section). Our writing ψ_n^* in (43.1) and (43.3) is in anticipation of the field-dependent case where Q_ω (and hence Δ) are no longer real. Inserting (43.1) into (42.13) we obtain

$$[\mathcal{N}(z)\mathcal{N}(z')]^{1/2}\sum_{n,m}[2|\omega|+\Omega_n]a_{nm}\psi_n^*(z')\psi_m(z)=\frac{2\pi}{\hbar}\mathcal{N}(z)\delta(z-z'). \tag{43.4}$$

Dividing through by $[\mathcal{N}(z)\mathcal{N}(z')]^{1/2}$, multiplying both sides by $\psi_{n'}(z')$ and $\psi_{m'}^*(z)$, and integrating over z and z' yields

$$(2|\omega|+\Omega_n)a_{nm}=\frac{2\pi}{\hbar}\delta_{mn} \tag{43.5}$$

and from (43.1) we have

$$Q_\omega(z,z')=\frac{2\pi}{\hbar}[\mathcal{N}(z)\mathcal{N}(z')]^{1/2}\sum_n\frac{\psi_n^*(z')\psi_n(z)}{2|\omega|+\Omega_n}. \tag{43.6}$$

Inserting (43.6) into the gap equation, Eq. (42.1), we have

$$\Delta(z)=\frac{2\pi}{\hbar}k_BTV(z)[\mathcal{N}(z)]^{1/2}\sum_\omega\sum_n\int\frac{[\mathcal{N}(z')]^{1/2}\psi_n^*(z')\psi_n(z)}{2|\omega|+\Omega_n}\Delta(z')dz'. \tag{43.7}$$

We define the quantity $F(z)=\Delta(z)/V(z)$, which is referred to as the pair amplitude (as can be seen from Eq. (37.7)). We expand $F(z)$ in terms of the $\psi_n(z)$ according to

$$F(z)=[\mathcal{N}(z)]^{1/2}\sum_{n'}c_{n'}\psi_{n'}(z). \tag{43.8}$$

We substitute (43.8) into (43.7) and multiply both sides of the resulting equation by $\psi_m^*(z)$ followed by integration over z to obtain the equation

$$\sum_n\left[\delta_{nm}-\frac{2\pi}{\hbar}k_BT\sum_\omega\frac{1}{2|\omega|+\Omega_m}\int\mathcal{N}(z')V(z')\psi_m^*(z')\psi_n(z')dz'\right]c_n=0. \tag{43.9}$$

This equation has a nontrivial solution provided the determinant of the coefficients vanishes:

$$\left|\delta_{nm}-\frac{2\pi}{\hbar}k_BT\sum_\omega\frac{\langle m|\mathcal{N}V|n\rangle}{2|\omega|+\Omega_m}\right|=0, \tag{43.10a}$$

where we have defined the matrix element

$$\langle m|\mathcal{N}V|n\rangle\equiv\int\psi_m^*(z)\mathcal{N}(z)V(z)\psi_n(z)dz. \tag{43.10b}$$

Eq. (43.10) is the formal solution to the problem we have posed and the highest temperature, T_c, for which it has a solution corresponds to the physical transition temperature. The uniform superconductor is a special case of Eq. (43.10b) corresponding to $\langle m|\mathcal{N}V|n\rangle=\mathcal{N}V\delta_{mn}$; i.e., the determinant is diagonal and every eigenvalue Ω_n leads directly to a zero of the determinant. Since the smallest eigenvalue $\Omega_n=0$ leads to the highest transition temperature, we have

$$1 = \frac{2\pi}{\hbar} k_B T \mathcal{N} V \sum_{\omega} \frac{1}{2|\omega|}, \tag{43.11}$$

which, from the discussion of Sec. 40, is equivalent to the BCS expression for the transition temperature.[1]

43.2 The thin-film (Cooper–deGennes) limit

The simplest nontrivial application of the above formalism is the so-called Cooper limit (as corrected by deGennes (1964)) where the thicknesses a and b are much smaller than ξ_A and ξ_B. Then $Q_\omega(z, z')$ is essentially constant when z and z' lie within a given sublayer and from Eq. (42.29) must satisfy the requirements

$$Q_\omega(\mathrm{A}, \mathrm{A})/\mathcal{N}_\mathrm{A} = Q_\omega(\mathrm{A}, \mathrm{B})/\mathcal{N}_\mathrm{B} \tag{43.12a}$$

and

$$Q_\omega(\mathrm{B}, \mathrm{B})/\mathcal{N}_\mathrm{B} = Q_\omega(\mathrm{B}, \mathrm{A})/\mathcal{N}_\mathrm{A}, \tag{43.12b}$$

where A and B denote the layers in which the coordinates z or z' lie. From the symmetry requirement on $Q_\omega(z, z')$ we have

$$Q_\omega(\mathrm{B}, \mathrm{A}) = Q_\omega(\mathrm{A}, \mathrm{B}). \tag{43.13}$$

From Eq. (42.5) we obtain

$$Q_\omega(\mathrm{A}, \mathrm{A})a + Q_\omega(\mathrm{A}, \mathrm{B})b = \frac{\pi}{\hbar|\omega|} \mathcal{N}_A \tag{43.14a}$$

and

$$Q_\omega(\mathrm{B}, \mathrm{A})a + Q_\omega(\mathrm{B}, \mathrm{B})b = \frac{\pi}{\hbar|\omega|} \mathcal{N}_B. \tag{43.14b}$$

Solving (43.12) and (43.13) we have

$$Q_\omega(\mathrm{A}, \mathrm{A}) = \frac{\pi}{\hbar|\omega|} \frac{\mathcal{N}_\mathrm{A}^2}{a\mathcal{N}_\mathrm{A} + b\mathcal{N}_\mathrm{B}}, \tag{43.14c}$$

$$Q_\omega(\mathrm{B}, \mathrm{B}) = \frac{\pi}{\hbar|\omega|} \frac{\mathcal{N}_\mathrm{B}^2}{a\mathcal{N}_\mathrm{A} + b\mathcal{N}_\mathrm{B}}, \tag{43.14d}$$

and

$$Q_\omega(\mathrm{A}, \mathrm{B}) = \frac{\pi}{\hbar|\omega|} \frac{\mathcal{N}_\mathrm{A}\mathcal{N}_\mathrm{B}}{a\mathcal{N}_\mathrm{A} + b\mathcal{N}_\mathrm{B}}. \tag{43.14e}$$

From Eq. (42.1) we have

1. A superconductor which has an inhomogeneous diffusion constant, $D(z)$, but a constant $\mathcal{N}(z)V(z)$ also has an unaltered transition temperature (at zero field only).

$$\Delta_A = V_A k_B T \sum_\omega [aQ_\omega(A, A)\Delta_A + bQ_\omega(A, B)\Delta_B] \tag{43.15a}$$

and

$$\Delta_B = V_B k_B T \sum_\omega [aQ_\omega(B, A)\Delta_A + bQ_\omega(B, B)\Delta_B], \tag{43.15b}$$

which on substituting Eqs. (43.14) yields

$$\Delta_A = V_A k_B T \sum_\omega \frac{\pi}{\hbar|\omega|} \frac{a\mathcal{N}_A^2 \Delta_A + b\mathcal{N}_A \mathcal{N}_B \Delta_B}{a\mathcal{N}_A + b\mathcal{N}_B} \tag{43.16a}$$

and

$$\Delta_B = V_B k_B T \sum_\omega \frac{\pi}{\hbar|\omega|} \frac{a\mathcal{N}_A \mathcal{N}_B \Delta_A + b\mathcal{N}_B^2 \Delta_B}{a\mathcal{N}_A + b\mathcal{N}_B}. \tag{43.16b}$$

To find the transition temperature, we set the determinant of the coefficients of Δ_A and Δ_B in Eq. (43.16) to zero. The algebra is simplest if we assume identical cut-off frequencies in A and B. We then obtain

$$\frac{k_B T}{\hbar} \sum_\omega \frac{\pi}{|\omega|} \to \ell n \frac{1.14\hbar\omega_D}{k_B T} = \frac{1}{\rho}, \tag{43.17a}$$

where

$$\rho \equiv \frac{V_A \mathcal{N}_A^2 a + V_B \mathcal{N}_B^2 b}{\mathcal{N}_A a + \mathcal{N}_B b}, \tag{43.17b}$$

yielding

$$k_B T_c = 1.14\hbar\omega_D e^{-1/\rho}. \tag{43.18}$$

43.3 The general 1D case

To proceed to the case where $\xi_A < a$ or $\xi_B < b$ we must determine the eigenfunctions of Eq. (43.2). These eigenfunctions are constructed by joining together the solutions in the two separate media, A and B, subject the boundary conditions, Eqs. (42.29)–(42.31). From Eq. (43.1) it is clear that $Q_\omega(z, z')$ will satisfy these boundary conditions if $\psi(z)/[\mathcal{N}(z)]^{1/2}$ and $[\mathcal{N}(z)]^{1/2} D(z) d\psi(z)/dz$ are both continuous at the interfaces.[2]

We assume medium A occupies $0 < z < a$, and media B occupies $-b < z < 0$, after which the media periodically reproduces itself; we seek solutions of the form $e^{\pm ik_A z}$ and $e^{\pm ik_B z}$ in these two regions, respectively, where $k_A = (\Omega/D_A)^{1/2}$ and $k_B = (\Omega/D_B)^{1/2}$. Thus

$$\psi_A = A_+ e^{ik_A z} + A_- e^{-ik_A z} \tag{43.19a}$$

and

2. We may equivalently require that the logarithmic derivative $\mathcal{N}(z)D(z)(d/dz)\ell n[\psi(z)]$ be continuous.

$$\psi_B = B_+ e^{ik_B z} + B_- e^{-ik_B z}. \tag{43.19b}$$

Applying the above boundary conditions on ψ at $z = 0$ yields

$$\mathcal{N}_B^{-1/2}(B_+ + B_-) = \mathcal{N}_A^{-1/2}(A_+ + A_-) \tag{43.20a}$$

and

$$\mathcal{N}_B^{1/2} D_B k_B (B_+ - B_-) = \mathcal{N}_A^{1/2} D_A k_A (A_+ - A_-). \tag{43.20b}$$

Since we have a periodic structure we must also satisfy Bloch conditions of the form $f(-b)e^{ikd} = f(+a)$ on the slope and magnitude of ψ (our model being a modified Kronig–Penney form); thus

$$\mathcal{N}_B^{1/2}(B_+ e^{-ik_B b} + B_- e^{+ik_B b})e^{ikd} = \mathcal{N}_A^{-1/2}(A_+ e^{ik_A a} + A_- e^{-ik_A a}) \tag{43.21a}$$

and

$$\mathcal{N}_B^{-1/2} D_B k_B (B_+ e^{-ik_B b} - B_- e^{+ik_B b})e^{ikd} = \mathcal{N}_A^{1/2} D_A k_A (A_+ e^{ik_A a} - A_- e^{-ik_A a}). \tag{43.21b}$$

We solve Eqs. (43.20) for A_+ and A_- and insert them in (43.21). We then set the determinant of the coefficients of B_+ and B_- equal to zero.

The ensuing algebra is somewhat tedious and we give the final result:

$$\cos kd = \cos k_A a \cos k_B b - \gamma \sin k_A a \sin k_B b, \tag{43.22}$$

where

$$\gamma \equiv \frac{D_A^2 \mathcal{N}_A^2 k_A^2 + D_B^2 \mathcal{N}_B^2 k_B^2}{2 D_A D_B \mathcal{N}_A \mathcal{N}_B k_A k_B}; \tag{43.23}$$

substituting our definitions at k_A and k_B we have

$$\gamma \equiv \frac{D_A \mathcal{N}_A^2 + D_B \mathcal{N}_B^2}{2 \mathcal{N}_A \mathcal{N}_B (D_A D_B)^{1/2}}, \tag{43.24}$$

which is independent of Ω. Note $\gamma = 1$ if $\mathcal{N}(z)[D(z)]^{1/2}$ is continuous.

If $D_A = D_B$ (implying $k_A = k_B$) and $\mathcal{N}_A = \mathcal{N}_B$, then, as far as the diffusive motion of the electrons is concerned, the medium is homogeneous; we still assume $V_A \neq V_B$, however. These restrictions define the mathematically simplest model of an artificial metallic superlattice and we will treat this case first.

Our eigenfunctions (for this special case) have (in both media) the form

$$\psi_k(z) = A_+(k)e^{ikz} + A_-(k)e^{-ikz}; \tag{43.25}$$

here the allowed values of k form a continuum (and hence the index n in Eq. (43.9) is continuous). We seek the highest temperature T at which (43.10) has a solution which constitutes the ground-state eigenfunction of our integral equation for Δ which (on general quantum mechanical grounds since Δ is proportional to a pair wavefunction) we expect to be symmetric about the midpoint of any superconducting layer (which we take as the A layers). Thus our basis states for expanding $\Delta(z)$ in Eq. (43.8) must have the form

$$\psi_n(z) = \left(\frac{2}{d}\right)^{1/2} \cos\left[k_n\left(z - \frac{a}{2}\right)\right]; \quad k_n = \frac{2\pi n}{d}. \tag{43.26}$$

We require the matrix elements (43.10b) which are given by

$$\begin{aligned}
\langle m|\mathcal{N}V|n\rangle &= \frac{2}{d}\mathcal{N}\left\{V_A \int_0^a \cos\left[k_m\left(z - \frac{a}{2}\right)\right]\cos\left[k_n\left(z - \frac{a}{2}\right)\right]dz\right. \\
&\quad \left. + V_B \int_{-b}^{0} \cos\left[k_m\left(z - \frac{a}{2}\right)\right]\cos\left[k_n\left(z - \frac{a}{2}\right)dz\right]\right\} \\
&= \frac{\mathcal{N}}{\pi}(V_A - V_B)\left[\frac{1}{m-n}\sin\frac{\pi(m-n)a}{d} + \frac{1}{m+n}\sin\frac{\pi(m+n)a}{d}\right]; \quad m \neq n \\
&= \frac{\mathcal{N}}{d}(V_A a + V_B b) + \frac{\mathcal{N}}{2\pi n}[V_A - V_B]\sin\frac{2\pi n a}{d}; m = n.
\end{aligned} \tag{43.27}$$

For the special case $a = b = d/2$ we have

$$\langle m|\mathcal{N}V|n\rangle = \begin{cases} -\frac{\mathcal{N}}{\pi}(V_A - V_B)\left[\frac{(-1)^{(m-n+1)/2}}{m-n} + \frac{(-1)^{(m+n+1)/2}}{m+n}\right]; & m+n = \text{odd} \\ 0; \quad m+n = \text{even} & \\ \frac{\mathcal{N}}{2}(V_A + V_B); \quad m = n. & \end{cases} \tag{43.28}$$

Noting that $\Omega_m = D(2\pi m/d)^2$, all quantities necessary to evaluate the determinant (43.10a) for our special case $D_A = D_B = D, \mathcal{N}_A = \mathcal{N}_B = \mathcal{N}$ are now available. However, we must perform an ω sum on every element of the determinant. In general the cut-off frequencies ω_D associated with V_A and V_B will be different; since our goal here continues to be an examination of the simplest nontrivial artificial metallic superlattice, we will assume $\omega_{DA} = \omega_{DB} = \omega_D$. We can then write[3]

$$\frac{2\pi}{\hbar}k_B T \sum_{\omega} \frac{\langle m|\mathcal{N}V|n\rangle}{2|\omega| + \Omega_m} = \langle m|\mathcal{N}V|n\rangle\left[\ell n\frac{1.14\hbar\omega_D}{k_B T} - \mathcal{U}\left(\frac{\hbar\Omega_m}{2\pi k_B T}\right)\right]; \tag{43.29}$$

we must keep in mind that (43.29) is divergent for large Ω_m.

We now return to the general case where the product $\mathcal{N}D^{1/2}$ is discontinuous at the interfaces; i.e., $\gamma \neq 1$ in Eq. (43.24). We again require that the eigenfunctions be symmetric and that $k = 2n\pi/d$. We must then solve the equation

$$\cos k_A a \cos k_B b - \gamma \sin k_A a \sin k_B b = 1 \tag{43.30}$$

for the eigenfrequencies Ω (where $k_A^2 = \Omega/D_A$ and $k_B^2 = \Omega/D_B$). With some trigonometric manipulations Eq. (43.30) may be rewritten in the form

$$\mathcal{N}_A D_A^{1/2} \tan\left(\frac{k_A a}{2}\right) = -\mathcal{N}_B D_B^{1/2} \tan\left(\frac{k_B b}{2}\right). \tag{43.31}$$

Note there is a special case where $\mathcal{N}D^{1/2}$ is continuous but $\mathcal{N}$ and D are not separately

3. The symbol k_B is used as both a wavevector and Boltzmann's constant; the distinction between these two uses is, however, obvious from the structure of the equations.

continuous; from (43.30) this leads to the eigenfrequencies

$$\Omega_n^{(0)} = 4\pi^2 n^2 \left[\frac{a}{D_A^{1/2}} + \frac{b}{D_B^{1/2}} \right]^2 . \tag{43.32}$$

For the case $\gamma \cong 1$ we may expand Ω in terms of $1-\gamma$ in the form $\Omega_n^{(0)} + \Delta\Omega_n$. After some calculations we obtain

$$\Delta\Omega_n^2 = \frac{8(1-\gamma)}{\left[\dfrac{a}{(\Omega_n^{(0)} D_A)^{1/2}} + \dfrac{b}{(\Omega_n^{(0)} D_B)^{1/2}} \right]^2} \sin\left(\frac{\Omega_n^{(0)} a^2}{D_A} \right)^{1/2} \sin\left(\frac{\Omega_n^{(0)} b^2}{D_B} \right)^{1/2} . \tag{43.33}$$

If γ differs significantly from one we must solve Eq. (43.31) numerically.

Rather than writing our eigenfunctions in the general form (43.21) we pick the alternative form (valid for $k = 0$)

$$\psi_{nA} = A_n \cos\left[k_{nA} \left(z - \frac{a}{2} \right) \right] \tag{43.34a}$$

and

$$\psi_{nB} = B_n \cos\left[k_{nB} \left(z + \frac{b}{2} \right) \right], \tag{43.34b}$$

where k_{nA} and k_{nB} are the solutions to Eq. (43.19) corresponding to the eigenvalue Ω_n. We normalize the ψ_n according to Eq. (43.3)

$$A_n^2 I_{nn}^{(A)} + B_n^2 I_{nn}^{(B)} = 1, \tag{43.35}$$

where

$$I_{nn}^{(A)} = \int_0^a \cos^2\left[k_{nA} \left(z - \frac{a}{2} \right) \right] dz = \frac{a}{2} + \frac{1}{2k_{nA}} \sin(k_{nA} a) \tag{43.36a}$$

and

$$I_{nn}^{(B)} = \int_{-b}^0 \cos^2\left[k_{nB} \left(z + \frac{b}{2} \right) \right] dz = \frac{b}{2} + \frac{1}{2k_{nB}} \sin(k_{nB} b). \tag{43.36b}$$

To solve for A_n and B_n we must combine (43.19) with either of the boundary conditions; applying the condition on the magnitude yields

$$\mathcal{N}_A^{-1/2} A_n \cos\left(\frac{k_{nA} a}{2} \right) = \mathcal{N}_B^{-1/2} B_n \cos\left(\frac{k_{nB} b}{2} \right)$$

or

$$B_n = \left(\frac{\mathcal{N}_B}{\mathcal{N}_A} \right)^{1/2} \frac{\cos\left(\dfrac{k_{nA} a}{2} \right)}{\cos\left(\dfrac{k_{nB} b}{2} \right)} A_n,$$

which yields

$$A_n^2 = \left[I_{nn}^{(\mathrm{A})} + \frac{\mathcal{N}_{\mathrm{B}}}{\mathcal{N}_{\mathrm{A}}} \frac{\cos^2\left(\dfrac{k_{n\mathrm{A}}a}{2}\right)}{\cos^2\left(\dfrac{k_{n\mathrm{B}}b}{2}\right)} I_{nn}^{(\mathrm{B})} \right]^{-1} \tag{43.37a}$$

and

$$B_n^2 = \left[I_{nn}^{(\mathrm{B})} + \frac{\mathcal{N}_{\mathrm{A}}}{\mathcal{N}_{\mathrm{B}}} \frac{\cos^2\left(\dfrac{k_{n\mathrm{B}}b}{2}\right)}{\cos^2\left(\dfrac{k_{n\mathrm{A}}a}{2}\right)} I_{nn}^{(\mathrm{A})} \right]^{-1}. \tag{43.37b}$$

We require the matrix element $\langle m | \mathcal{N} V | n \rangle$ which is given by

$$\langle m | \mathcal{N} V | n \rangle = \mathcal{N}_{\mathrm{A}} V_{\mathrm{A}} A_n A_m I_{mn}^{(\mathrm{A})} + \mathcal{N}_{\mathrm{B}} V_{\mathrm{B}} B_n B_m I_{mn}^{(\mathrm{B})}, \tag{43.38}$$

where

$$I_{mn}^{(\mathrm{A})} = \int_0^a \cos\left[k_{m\mathrm{A}}\left(z - \frac{a}{2}\right)\right] \cos\left[k_{n\mathrm{A}}\left(z - \frac{a}{2}\right)\right] \mathrm{d}z$$

$$= \frac{\sin\left[(k_{m\mathrm{A}} - k_{n\mathrm{A}})\dfrac{a}{2}\right]}{k_{m\mathrm{A}} - k_{n\mathrm{A}}} + \frac{\sin\left[(k_{m\mathrm{A}} + k_{n\mathrm{A}})\dfrac{a}{2}\right]}{k_{m\mathrm{A}} + k_{n\mathrm{A}}}, \quad m \neq n; \tag{43.39a}$$

$$I_{mn}^{(\mathrm{B})} = \int_{-b}^0 \cos\left[k_{m\mathrm{B}}\left(z + \frac{b}{z}\right)\right] \cos\left[k_{n\mathrm{B}}\left(z + \frac{b}{2}\right)\right] \mathrm{d}z$$

$$= \frac{\sin\left[(k_{m\mathrm{B}} - k_{n\mathrm{B}})\dfrac{b}{2}\right]}{k_{m\mathrm{B}} - k_{n\mathrm{B}}} + \frac{\sin\left[(k_{m\mathrm{B}} + k_{n\mathrm{B}})\dfrac{b}{2}\right]}{k_{m\mathrm{B}} + k_{n\mathrm{B}}}, \quad m \neq n; \tag{43.39b}$$

and the diagonal elements, $I_{nn}^{(\mathrm{A})}$ and $I_{nn}^{(\mathrm{B})}$, were given previously. In place of Eq. (43.29) we have the following for the general case of nonidentical Debye temperatures in media A and B

$$\frac{2\pi}{\hbar} k_{\mathrm{B}} T \sum_{\omega} \frac{\langle m | \mathcal{N} V | n \rangle}{2|\omega| + \Omega_m} = \mathcal{N}_{\mathrm{A}} V_{\mathrm{A}} A_m A_n I_{mn}^{(\mathrm{A})} \left[\ell n\left(\frac{1.14\hbar\omega_{\mathrm{DA}}}{k_{\mathrm{B}}T}\right) + \mathcal{U}\left(\frac{\hbar\Omega_m}{2\pi k_{\mathrm{B}}T}\right) \right]$$
$$+ \mathcal{N}_{\mathrm{B}} V_{\mathrm{B}} B_m B_n I_{mn}^{(\mathrm{B})} \left[\ell n\left(\frac{1.14\hbar\omega_{\mathrm{DB}}}{k_{\mathrm{B}}T}\right) + \mathcal{U}\left(\frac{\hbar\Omega_m}{2\pi k_{\mathrm{B}}T}\right) \right]. \tag{43.40}$$

This quantity is substituted in Eq. (43.10a) and we seek the highest temperature for which the determinant vanishes. In the limit where the layers become very thin, the Ω_n for $n \neq 0$ become large; alternatively if we use a very large basis set the Ω_n diverge as n^2. The approximation involved in using the digamma function form then breaks down. We discuss here an alternative technique to the subtraction procedure (which diverges as $\Omega_n \to \infty$). The procedure is to simply cut off the ω sums at $\omega = \omega_{\mathrm{D}}$. When $\Omega_n >> 2\pi k_{\mathrm{B}} T/\hbar$, we may accurately replace the sum by an integral

$$\frac{2\pi}{\hbar}k_BT\sum_{\omega}^{\omega=\omega_D}\frac{1}{2|\omega|+\Omega_n}\to\int_0^{\omega_D}\frac{d\omega}{2\omega+\Omega_n}=\frac{1}{2}\ell n\left(\frac{2\omega_D+\Omega_n}{\Omega_n}\right); \tag{43.41a}$$

for small Ω_n we may continue to use the digamma form

$$\frac{2\pi}{\hbar}k_BT\sum_{\omega}^{\omega=\omega_D}\frac{1}{2|\omega|+\Omega_n}\to\ell n\frac{1.14\hbar\omega_D}{k_BT}+\mathscr{U}\left(\frac{\hbar\Omega_n}{k_BT}\right). \tag{43.41b}$$

We now show that the Cooper–deGennes limit is obtained from our general formalism in the limit $d\to 0$. In this limit $\Omega_n\to\infty$ for $n>0$ (Ω_0 is always zero); from the above discussion all the terms in our determinant vanish except the 00 term. From Eqs. (43.37a) and (43.37b) we have $A_0^2=\mathscr{N}_A/(\mathscr{N}_Aa+\mathscr{N}_Bb)$ and $B_0^2=\mathscr{N}_B/(\mathscr{N}_Aa+\mathscr{N}_Bb)$ which in turn leads to the matrix element

$$\langle 0|\mathscr{N}V|0\rangle=\frac{\mathscr{N}_A^2V_Aa+\mathscr{N}_B^2V_Bb}{\mathscr{N}_Aa+\mathscr{N}_Bb}\equiv\rho. \tag{43.42}$$

Referring to Eq. (43.10a) we obtain the Cooper–deGennes thin-film limit.

In addition to the Cooper limit, there is another physically interesting situation (first examined by deGennes) in which we can obtain an exact solution; this is the limit of a very thick normal (B) layer ($b>>\xi_\omega^{(B)}$) and a thin superconducting (A) layer ($a<<\xi_\omega^{(A)}$). From symmetry this superlattice is equivalent to a single superconducting layer of thickness $a/2$ on an infinite B layer. The normal metal and superconductor will occupy the regions $-\infty<z<0$ and $0<z<a/2$, respectively.

From our discussion of the boundary conditions we can immediately write down the form of $Q_\omega(z,z')$ in the various regions

$$Q_\omega(A,A)=\text{const};\quad 0<z,z'<\frac{a}{2}, \tag{43.43a}$$

$$Q_\omega(z,A)=\frac{\mathscr{N}_B\pi}{2\hbar|\omega|\xi_\omega}\mu e^{z/\xi_\omega};\quad z<0,\ 0<z'<\frac{a}{2}, \tag{43.43b}$$

$$Q_\omega(A,z')=\frac{\mathscr{N}_B\pi}{2\hbar|\omega|\xi_\omega}\mu e^{z'/\xi_\omega};\quad 0<z<\frac{a}{2},z'<0, \tag{43.43c}$$

$$Q_\omega(z,z')=\frac{\mathscr{N}_B\pi}{2\hbar|\omega|\xi_\omega}[e^{-|z-z'|/\xi_\omega}+\lambda e^{(z+z')/\xi_\omega}];\quad z,z'<0; \tag{43.43d}$$

here we have written $\xi_\omega=\xi_\omega^{(B)}$ and the notation A in the argument of Q_ω denotes a coordinate that is within the superconductor. Applying the boundary condition (42.29) we have

$$\frac{Q_\omega(A,A)}{\mathscr{N}_A}=\frac{Q_\omega(A,z'=0)}{\mathscr{N}_B} \tag{43.44a}$$

and

$$\frac{Q_\omega(z,A)}{\mathscr{N}_A}=\frac{Q_\omega(z,z'=0)}{\mathscr{N}_B}, \tag{43.44b}$$

which lead to

$$Q_\omega(A, A) = \frac{\pi}{2\hbar|\omega|}\frac{\mathcal{N}_A}{\xi_\omega}\mu \tag{43.45}$$

and

$$\mu\frac{\mathcal{N}_B}{\mathcal{N}_A} = 1 + \lambda. \tag{43.46}$$

We next apply the sum rule (42.5) for the case of z in A

$$\int_0^{a/2} Q_\omega(A, A)\mathrm{d}z' + \int_{-\infty}^{0} Q_\omega(A, z')\mathrm{d}z' = \frac{\pi}{\hbar|\omega|}\mathcal{N}_A$$

or

$$Q_\omega(A, A') = \frac{2}{a\hbar}\frac{\pi}{|\omega|}\left(\mathcal{N}_A - \frac{\mathcal{N}_B}{2}\mu\right) \tag{43.47}$$

(the sum rule for z in B leads to an equation which is linearly dependent on the above equations). Solving these equations yields

$$\lambda_\omega = \left(1 - \frac{a}{2\xi_\omega}\frac{\mathcal{N}_A}{\mathcal{N}_B}\right)\Big/\left(1 + \frac{a}{2\xi_\omega}\frac{\mathcal{N}_A}{\mathcal{N}_B}\right) \tag{43.48}$$

and

$$\mu_\omega = 2\frac{\mathcal{N}_A}{\mathcal{N}_B}\Big/\left(1 + \frac{a}{2\xi_\omega}\frac{\mathcal{N}_A}{\mathcal{N}_B}\right). \tag{43.49}$$

For the case where z lies in A the gap equation takes the form

$$\frac{\Delta_A}{V_A\mathcal{N}_A} = \frac{2\pi}{\hbar}k_BT\sum_\omega \frac{1}{2|\omega|}\left[\frac{\left(\frac{\mathcal{N}_A}{\mathcal{N}_B}\right)\left(\frac{a}{2\xi_\omega}\right)}{1 + \frac{a}{2\xi_\omega}\frac{\mathcal{N}_A}{\mathcal{N}_B}}\Delta_A + \frac{\xi_\omega^{-1}}{1 + \frac{a}{2\xi_\omega}\frac{\mathcal{N}_A}{\mathcal{N}_B}}\int_{-\infty}^{0} e^{z'/\xi_\omega}\Delta_B(z')\mathrm{d}z'\right]; \tag{43.50}$$

for $z < 0$ we have

$$\frac{\Delta_B(z)}{V_B\mathcal{N}_B} = \frac{2\pi}{\hbar}k_BT\sum_\omega \frac{1}{2|\omega|}\left[\frac{\frac{a}{2\xi_\omega}\frac{\mathcal{N}_A}{\mathcal{N}_B}e^{z/\xi_\omega}}{1 + \frac{a}{2\xi_\omega}\frac{\mathcal{N}_A}{\mathcal{N}_B}}\Delta_A + \frac{1}{2\xi_\omega}\int_{-\infty}^{0}\left(e^{-|z-z'|/\xi_\omega} + \frac{1 - \frac{a}{2\xi_\omega}\frac{\mathcal{N}_A}{\mathcal{N}_B}}{1 + \frac{a}{2\xi_\omega}\frac{\mathcal{N}_A}{\mathcal{N}_B}}e^{(z+z')/\xi_\omega}\right)\Delta_A(z')\mathrm{d}z'\right]. \tag{43.51}$$

This coupled set of equations has a simple solution only for the case in which $V_B = 0$ in which case we have

$$1 = \frac{2\pi}{\hbar} k_B T \mathcal{N}_A V_A \sum_\omega \frac{1}{2|\omega|} \frac{\left(\frac{\mathcal{N}_A}{\mathcal{N}_B}\right)\left(\frac{a}{2\xi_\omega}\right)}{1 + \frac{a}{2\xi_\omega}\frac{\mathcal{N}_A}{\mathcal{N}_B}}, \tag{43.52}$$

where $\xi_\omega = (D_B/2|\omega|)^{1/2}$. As discussed by Auvil and Ketterson (1988), correctly evaluating the ω sum leads to an answer slightly different from that of deGennes.

43.4 A microscopic theory of the 1D Josephson superlattice

In Sec. 16 we discussed the superconductor/insulator superlattice in the G–L model. We may carry over most of this discussion to the microscopic case, provided that we restrict ourselves to the diffusion-dominated (dirty) limit. For a uniform system (or a stepwise discontinuous media) the correlation function $g(\mathbf{r},\mathbf{r}',t)$ is governed by a diffusion equation. As a model for the superconductor/insulator (Josephson) superlattice, rather than the 3D diffusion equation, we can use the corresponding form of the Lawrence–Doniach equation, Eq. (16.3), which we may regard as an infinite set of coupled 2D diffusion equations of the form (with $\bar{A}_z$ defined after Eq. (16.2))

$$-D\left(\nabla - \frac{2ie}{\hbar c}\mathbf{A}\right)^2 g^{(n)} - \frac{J}{2}\left[g^{(n+1)}e^{-(2ie/\hbar c)\bar{A}_z s} - 2g^{(n)} + g^{(n-1)}e^{(2ie/\hbar c)\bar{A}_z s}\right] = \frac{\partial g^{(n)}}{\partial t}; \tag{43.53}$$

here $g^{(n)} = g^{(n)}(\mathbf{r},t)$ where $\mathbf{r}$ is a 2D position vector restricted to the conducting planes of the superlattice. Analogous to Eq. (40.14), we introduce the eigenvalue equation

$$-D\left(\nabla - \frac{2ie}{\hbar c}\mathbf{A}\right)^2 g_m^{(n)}(\mathbf{r}) - \frac{J}{2}\left[g_m^{(n+1)}(\mathbf{r})e^{-(2ie/\hbar c)\bar{A}_z s} - 2g_m^{(n)}(\mathbf{r}) + g_m^{(n-1)}(\mathbf{r})e^{(2ie/\hbar c)\bar{A}_z s}\right] = \Omega_m^{(n)} g_m^{(n)}(\mathbf{r}), \tag{43.54}$$

where the eigenfunctions $g_m^{(n)}(\mathbf{r})$ satisfy the orthonormality condition

$$\sum_n \int d^2 r\, g_m^{(n)}(\mathbf{r})^* g_{m'}^{(n)}(\mathbf{r}) = \delta_{mm'}. \tag{43.55}$$

We introduce a correlation $g^{(n,n')}(\mathbf{r},\mathbf{r}',t)$, which measures the gauge-invariant 'probability' that a particle initially in layer n' with position $\mathbf{r}'$ will later be found in layer n with position $\mathbf{r}$. We expand $g^{(n,n')}(\mathbf{r},\mathbf{r}',t)$ in the form

$$g^{(n,n')}(\mathbf{r},\mathbf{r}',t) = L^{-2}\sum_m g_m^{(n')*}(\mathbf{r}')g_m^{(n)}(\mathbf{r})e^{-\Omega_m|t|}. \tag{43.56}$$

In place of Eq. (38.6), the gap equation in this model takes the form

$$\Delta^{(n)}(\mathbf{r}) = \sum_{n'} \int d^2r' K^{(n,n')}(\mathbf{r},\mathbf{r}')\Delta^{(n')}(\mathbf{r}). \tag{43.57}$$

The analogue of Eq. (38.14′) is

$$K^{(n,n')}(\mathbf{r},\mathbf{r}') = \frac{2\pi}{\hbar} L^2 k_B T \mathcal{N}(0) V \sum_{\omega} \int dt e^{-2|\omega|t} g^{(n,n')}(\mathbf{r},\mathbf{r}',t). \tag{43.58}$$

Expanding the gap function in the form $\Delta^{(n)}(\mathbf{r}) = \Sigma_m a_m g_m^{(n)}(\mathbf{r})$, inserting Eq. (43.56) into Eq. (43.58), multiplying both sides by $g_m^{(n)*}(\mathbf{r})$ and integrating and summing over $\mathbf{r}$ and n, respectively, we obtain on using (43.55)

$$\frac{2\pi}{\hbar} k_B T \mathcal{N}(0) V \sum_{\omega} \frac{1}{2|\omega| + \Omega_m} = 1. \tag{43.59}$$

Thus the problem of finding the transition temperature for a superconductor/insulator superlattice reduces to the same form as for a uniform superconductor; Ω_m is just the lowest eigenvalue of Eq. (16.7). For a more complete discussion of the solutions of this equation see Klemm (1974) and Klemm *et al.* (1975). Pair-breaking effects associated with paramagnetic impurities and Larmor precession may be included by using Eq.(41.55) (see Klemm (1983) for a more complete discussion). The above model has implicitly assumed that the pairing potential is restricted to the metallic layers, i.e., there is no added tendency to pair during the tunneling process, which is in any case instantaneous in the Lawrence–Doniach model. An interesting extension would be to include retardation (and pairing) effects in the tunneling process. A different microscopic approach has been employed by Menon and Arnold (1985).

44 The proximity effect in a magnetic field

44.1 Governing equations in the presence of magnetic fields, spin susceptibility, paramagnetic impurities, and spin–orbit coupling

In this section we discuss the behavior of an artificial metallic superlattice in a magnetic field along the lines developed by Takahashi and Tachiki (1986a). We must generalize our expression for $Q_\omega(\mathbf{r}, \mathbf{r}')$ for the case of a finite magnetic field; for later required generality we also include the effects of Pauli paramagnetism, spin–orbit coupling and spin-flip scattering. From Eq. (41.30) we have

$$Q_\omega(\mathbf{r}, \mathbf{r}') = \frac{1}{2} \sum_{MN} \frac{\phi^*_{N\alpha}(\mathbf{r}) \mathscr{K}^\dagger_{\alpha\beta} \phi_{M\beta}(\mathbf{r}) \phi^*_{M\gamma}(\mathbf{r}') \mathscr{K}_{\gamma\delta} \phi_{N\delta}(\mathbf{r}')}{(\xi_M + i\hbar\omega)(\xi_N - i\hbar\omega)}. \tag{44.1}$$

In terms of a correlation function, we have from (41.35)

$$Q_\omega(\mathbf{r}, \mathbf{r}') = \frac{2\pi}{\hbar} \int_0^\infty dt e^{-2|\omega|t} [g^{(1)}(\mathbf{r}, \mathbf{r}', t) + g^{(2)}(\mathbf{r}, \mathbf{r}', t)], \tag{44.2}$$

where

$$g^{(\nu)}(\mathbf{r}, \mathbf{r}', t) = -\frac{1}{2} \sum_n \delta(\xi - \xi_{n\nu}) \langle n, \nu | \delta(\mathbf{r} - \mathbf{R}(t)) \underset{\sim}{\mathscr{K}}^\dagger(t) \underset{\sim}{\mathscr{K}}(0) \delta(\mathbf{r}' - \mathbf{R}(0)) | n, \nu \rangle. \tag{44.3}$$

The time evolution of $\underset{\sim}{\mathscr{K}}(t)$ is governed by Eq. (41.40) or (41.49)

$$\dot{\underset{\sim}{\mathscr{K}}} = \left[-i\frac{d\Phi}{dt} + \frac{2i}{\hbar} I \underset{\sim}{\sigma}_z - \frac{1}{\tau_s} \right] \underset{\sim}{\mathscr{K}}(t) - \frac{1}{\tau'} (\mathscr{K} + \tilde{\underset{\sim}{\mathscr{K}}}) \tag{44.4}$$

and, referring to the discussion of Subsec. 41.6, we may write (neglecting the spin–orbit term)

$$\underset{\sim}{\mathscr{K}}^\dagger(t) \underset{\sim}{\mathscr{K}}(0) = \exp\left(-\frac{2ie}{\hbar c} \int_{\mathbf{r}(0)}^{\mathbf{r}(t)} \mathbf{A}(\mathbf{s}) \cdot d\mathbf{s} - \frac{t}{\tau_s} \right) \begin{bmatrix} e^{-i\Omega_L t} & 0 \\ 0 & e^{i\Omega_L t} \end{bmatrix},$$

where $\Omega_L = 2I/\hbar$. From Eq. (41.43) we have (in this semiclassical approximation)

$$g^{(\nu)}(\mathbf{r}, \mathbf{r}', t) = \frac{1}{2} g(\mathbf{r}, \mathbf{r}', t) \exp\left(-\frac{2ie}{\hbar c} \int_{\mathbf{r}(0)}^{\mathbf{r}(t)} \mathbf{A}(\mathbf{s}) \cdot d\mathbf{s} - \frac{t}{\tau_s} \right) [e^{-i\Omega_L t} \delta_{\nu 1} + e^{+i\Omega_L t} \delta_{\nu 2}], \tag{44.5}$$

where $g(\mathbf{r}, \mathbf{r}', t)$ is given by Eq. (42.7). According to the discussion leading up to Eq. (40.13), rather than including the effect of the vector potential as a global phase factor, it is more accurate to modify the differential operator ∇ to the (gauge-invariant) form $[\nabla - (2ie/\hbar c)\mathbf{A}]$. Including the other contributions to the time evolution of $\underset{\sim}{\mathscr{K}}$, we must then solve the following pair of equations of motion for $g^{(\nu)}(\mathbf{r}, \mathbf{r}', t)$

$$\left[\frac{\partial}{\partial t} + i\Omega_L(\mathbf{r}) + \frac{1}{\tau_s(\mathbf{r})} + \frac{1}{\tau'(\mathbf{r})}\right] g^{(1)} - \frac{1}{\tau'(\mathbf{r})} g^{(2)} - D(\mathbf{r})\left[\nabla - \frac{2ie}{\hbar c}\mathbf{A}(\mathbf{r})\right]^2 g^{(1)} = 0, \tag{44.6a}$$

$$\left[\frac{\partial}{\partial t} - i\Omega_L(\mathbf{r}) + \frac{1}{\tau_s(\mathbf{r})} + \frac{1}{\tau'(\mathbf{r})}\right] g^{(2)} - \frac{1}{\tau'(\mathbf{r})} g^{(1)} - D(\mathbf{r})\left[\nabla - \frac{2ie}{\hbar c}\mathbf{A}(\mathbf{r})\right]^2 g^{(2)} = 0, \tag{44.6b}$$

where we take the complex conjugate of the operator in the square brackets containing ∇ when acting on $\mathbf{r}'$. Eqs. (44.6a) and (44.6b) are solved subject to the initial condition (see Eq. (42.11))

$$g^{(\nu)}(\mathbf{r}, \mathbf{r}', 0) = \frac{1}{2}\mathcal{N}(\mathbf{r})\delta(\mathbf{r} - \mathbf{r}'). \tag{44.7}$$

In a manner analogous to Eq. (44.2) we introduce an auxiliary function

$$R_\omega(\mathbf{r}, \mathbf{r}') = \frac{2\pi}{\hbar}\int_0^\infty dt e^{-2|\omega|t}[g^{(1)}(\mathbf{r}, \mathbf{r}', t) - g^{(2)}(\mathbf{r}, \mathbf{r}', t)]. \tag{44.8}$$

Combining Eqs. (44.6a), (44.6b), (44.2), and (44.8), we obtain the following coupled set of differential equations for Q_ω and R_ω

$$\left[2|\omega| + \frac{1}{\tau_s(\mathbf{r})} + \mathscr{L}\right] Q_\omega(\mathbf{r}, \mathbf{r}') + i\Omega_L(\mathbf{r}) R_\omega(\mathbf{r}, \mathbf{r}') = \frac{2\pi}{\hbar}\mathcal{N}(\mathbf{r})\delta(\mathbf{r} - \mathbf{r}') \tag{44.9a}$$

and

$$\left[2|\omega| + \frac{1}{\tau_s(\mathbf{r})} + \frac{2}{\tau'(\mathbf{r})} + \mathscr{L}\right] R_\omega(\mathbf{r}, \mathbf{r}') + i\Omega_L(\mathbf{r}) Q_\omega(\mathbf{r}, \mathbf{r}') = 0, \tag{44.9b}$$

where we have defined the operator

$$\mathscr{L} = -D(\mathbf{r})\left[\nabla - \frac{2ie}{\hbar c}\mathbf{A}(\mathbf{r})\right]^2. \tag{44.10}$$

To avoid excessive complexity we will drop the spin–orbit term for the remaining calculations. In a treatment analogous to that in Subsec. 40.2 we expand Q_ω and R_ω in terms of the eigenfunctions of the (gauge-invariant) differential equation

$$\left[\mathscr{L} + \frac{1}{\tau_s(\mathbf{r})}\right]\psi_n(\mathbf{r}) = \Omega_n \psi_n(\mathbf{r}); \tag{44.11}$$

here the Ω_n are the eigenvalues[1] and we observe that $1/\tau_s(\mathbf{r})$ behaves as a potential in the Schrödinger equation analogy. Q_ω and R_ω take the form

$$Q_\omega(\mathbf{r},\mathbf{r}') = [\mathcal{N}(\mathbf{r})\mathcal{N}(\mathbf{r}')]^{1/2} \sum_{mn} a_{mn}\psi_n^*(\mathbf{r}')\psi_m(\mathbf{r}) \tag{44.12a}$$

and

$$R_\omega(\mathbf{r},\mathbf{r}') = [\mathcal{N}(\mathbf{r})\mathcal{N}(\mathbf{r}')]^{1/2} \sum_{mn} b_{mn}\psi_n^*(\mathbf{r}')\psi_m(\mathbf{r}). \tag{44.12b}$$

Inserting (44.12a) and (44.12b) into (44.9a) and (44.9b), multiplying by $\psi_m^*(\mathbf{r})\psi_n(\mathbf{r}')$, and integrating over all $\mathbf{r}$ and $\mathbf{r}'$ yields (on relabeling some of the indices for convenience in subsequent calculations)

$$(2|\omega| + \Omega_m)a_{mn} + \mathrm{i}\sum_k \langle m|\Omega_\mathrm{L}|k\rangle b_{kn} = \frac{2\pi}{\hbar}\delta_{mn} \tag{44.13a}$$

and

$$(2|\omega| + \Omega_k)b_{kn} + \mathrm{i}\sum_\ell \langle k|\Omega_\mathrm{L}|\ell\rangle a_{\ell n} = 0, \tag{44.13b}$$

where the matrix element $\langle m|\Omega_\mathrm{L}|k\rangle$ is defined as

$$\langle m|\Omega_\mathrm{L}|k\rangle = \int \mathrm{d}^3r\psi_m^*(\mathbf{r})\Omega_\mathrm{L}(\mathbf{r})\psi_k(\mathbf{r}). \tag{44.14}$$

To solve this set of equations we first find b_{kn} from (44.13b)

$$b_{kn} = -\mathrm{i}\sum_\ell \frac{\langle k|\Omega_\mathrm{L}|\ell\rangle}{2|\omega| + \Omega_k} a_{\ell n};$$

substituting this expression in Eq. (44.13a) yields

$$\sum_\ell \left\{(2|\omega| + \Omega_m)\delta_{m\ell} + \sum_k \frac{\langle m|\Omega_\mathrm{L}|k\rangle\langle k|\Omega_\mathrm{L}|\ell\rangle}{2|\omega| + \Omega_k}\right\} a_{\ell n} = \frac{2\pi}{\hbar}\delta_{mn}. \tag{44.15}$$

From the structure of Eq. (44.15) it is clear that the quantity in curly brackets must be $2\pi/\hbar$ times the inverse of the matrix formed from the quantities $a_{\ell m}$. We may then write immediately

$$a_{mn} = \frac{2\pi}{\hbar}(\Gamma^{-1})_{mn}, \tag{44.16}$$

where the (unsymmetric) matrix Γ has the elements

$$(\Gamma)_{m\ell} = (2|\omega| + \Omega_m)\delta_{m\ell} + \sum_k \frac{\langle m|\Omega_\mathrm{L}|k\rangle\langle k|\Omega_\mathrm{L}|\ell\rangle}{2|\omega| + \Omega_k}. \tag{44.17}$$

This yields

1. We must be careful to distinguish eigenvalues, Ω_n, Ω_k, and a position-dependent Larmor frequency, $\Omega_\mathrm{L}(\mathbf{r})$, in what follows.

$$Q_\omega(\mathbf{r},\mathbf{r}') = \frac{2\pi}{\hbar}[\mathcal{N}(\mathbf{r})\mathcal{N}(\mathbf{r}')]^{1/2}\sum_{m,n}(\Gamma^{-1})_{mn}\psi_n^*(\mathbf{r}')\psi_m(\mathbf{r}), \tag{44.18}$$

from which we may immediately write the integral equation for the gap function as

$$\Delta(\mathbf{r}) = \frac{2\pi k_B T}{\hbar} V(\mathbf{r})[\mathcal{N}(\mathbf{r})]^{1/2}\sum_\omega\sum_{m,n}\psi_m(\mathbf{r})(\Gamma^{-1})_{mn} \times \int d^3r'\psi_n^*(\mathbf{r}')[\mathcal{N}(\mathbf{r}')]^{1/2}\Delta(\mathbf{r}'). \tag{44.19}$$

Introducing the pair wavefunction $F(\mathbf{r}) = \Delta(\mathbf{r})/V(\mathbf{r})$, using the expansion (43.8) for $F(\mathbf{r})$, multiplying by $\psi_n^*(\mathbf{r})$, and integrating over $\mathbf{r}$ we obtain a secular equation for the expansion coefficients c_n of $F(\mathbf{r})$; the determinant of the coefficients of this equation must vanish yielding the condition

$$\left|\delta_{mn} - \frac{2\pi}{\hbar}k_B T\sum_\omega\sum_\ell(\Gamma^{-1})_{m\ell}\langle\ell\,|\,V\mathcal{N}\,|\,n\rangle\right| = 0, \tag{44.20}$$

which is a generalization of (43.10a) to the case of finite magnetic fields, spin paramagnetism and spin-flip scattering.

We now discuss the question of the appropriate boundary conditions. The boundary condition on the gradient of Q_ω, Eq. (42.18), was established with the aid of a sum rule, which, in turn, relied on the fact that the wavefunctions could be chosen to be real. In the presence of a field the wavefunctions are not necessarily real. In what follows we will *adopt* a gauge-invariant form of the boundary conditions (42.30) and (42.31); specifically we will assume (with Takahashi and Tachiki) the continuity of the quantities[2]

$$\frac{\Delta(\mathbf{r})}{\mathcal{N}(\mathbf{r})V(\mathbf{r})} = \frac{F(\mathbf{r})}{\mathcal{N}(\mathbf{r})} \tag{44.21}$$

and

$$\frac{D(\mathbf{r})}{V(\mathbf{r})}\left[\nabla - \frac{2ie}{\hbar c}\mathbf{A}(\mathbf{r})\right]\Delta(\mathbf{r}) = D(\mathbf{r})\left[\nabla - \frac{2ie}{\hbar c}\mathbf{A}(\mathbf{r})\right]F(\mathbf{r}); \tag{44.22}$$

these conditions will be satisfied provided $[\mathcal{N}(\mathbf{r})]^{1/2}\psi(\mathbf{r})$ satisfies the same boundary conditions as $F(\mathbf{r})$. The critical magnetic field will correspond to the highest field for a given temperature for which Eq. (44.20) has a solution.

We require the eigenfunctions, $\psi_\mathbf{n}(\mathbf{r})$ of Eq. (44.11); in what follows $\mathbf{n}$ stands for n_x, n_y, n_z. We will treat only the cases $\mathbf{H}\parallel\hat{\mathbf{z}}$ and $\mathbf{H}\parallel\hat{\mathbf{x}}$ and we will use the gauges $\mathbf{A} = Hx\hat{\mathbf{y}}$ and $\mathbf{A} = -Hz\hat{\mathbf{y}}$, respectively. The equations are then separable and we write the eigenfunctions in the form

$$\psi_\mathbf{n}(\mathbf{r}) = e^{ik_y y + ik_z z}u_{n_x}(x), \quad \mathbf{H}\parallel\hat{\mathbf{z}}; \tag{44.23}$$

and

$$\psi_\mathbf{n}(\mathbf{r}) = e^{ik_x x + ik_y y}w_{n_z}(z), \quad \mathbf{H}\parallel\hat{\mathbf{x}}; \tag{44.24}$$

which lead to the differential equations

2. It is plausible that arguments similar to those in Subsec. 41.5 could establish these boundary conditions rigorously.

$$\frac{d^2u_{n_x}(x)}{dx^2} - \left[\frac{4e^2H^2}{\hbar^2c^2}(x - x_0)^2 + k_z^2 - \frac{\Omega_n - \tau_s^{-1}}{D}\right]u_{n_x}(x) = 0, \quad \mathbf{H} \parallel \hat{\mathbf{z}}; \tag{44.25}$$

and

$$\frac{d^2w_{n_z}(z)}{dz^2} - \left[\frac{4e^2H^2}{\hbar^2c^2}(z - z_0)^2 + k_x^2 - \frac{\Omega_n - \tau_s^{-1}}{D}\right]w_{n_z}(z) = 0, \quad \mathbf{H} \parallel \hat{\mathbf{x}}, \tag{44.26}$$

where $x_0, z_0 = \hbar c k_y/2eH$. For the case of $\mathbf{H} \parallel \hat{\mathbf{z}}$ we may put $x_0 = 0$ (corresponding to $k_y = 0$ or equivalently $n_y = 0$). From symmetry considerations in (44.25), the only relevant x-dependent eigenfunction will be the Gaussian ground-state harmonic oscillator wavefunction ($n_x = 0$) which has the eigenvalue $2DeH/\hbar c$ or D/ξ^2, where we define $\xi^2 \equiv \phi_0/2\pi H$. The values of k_z in media A and B take the values k_{n_zA} and k_{n_zB}, with frequencies $D_A k_{n_zA}$ and $D_B k_{n_zB}$, which are determined from Eq. (43.31). Since only the n_z eigenfunctions are relevant we may drop the suffix z in what follows. The total eigenfrequency is given by

$$\Omega_n = \tau_{sA}^{-1} + \frac{D_A}{\xi^2} + D_A k_{nA}^2 = \tau_{sB}^{-1} + \frac{D_B}{\xi^2} + D_B k_{nB}^2. \tag{44.27}$$

The z-dependent parts of the eigenfunction are given by (43.34a) and (43.34b), respectively. Note that if $\tau_{sB}^{-1} + 2DeH/\hbar^2 c > \Omega_n$, k_{nB} will be imaginary, which will require the use of the appropriate hyperbolic functions in Eqs. (43.31), (43.34), (43.36b), (43.37), and (43.39b).

Eq. (44.26) for the $\mathbf{H} \parallel \hat{\mathbf{y}}$ case is much more problematic in that the translational invariance is now lost in the z direction and we must use the boundary conditions to match the eigenfunctions w_{n_z} at each interface. We may set $k_x = 0$ with no loss of generality. By moving the origin to z_0, defining a variable $x = \sqrt{2}z/\xi$, and defining a dimensionless eigenvalue $\lambda(n_z)$ as $\lambda + \frac{1}{2} \equiv (\Omega - \tau_s^{-1})/(2D/\xi^2)$, we may rewrite Eq. (44.26) in canonical form as

$$\left(-\frac{d^2}{dx^2} + \frac{1}{4}x^2\right)w_\lambda(x) = \left(\lambda + \frac{1}{2}\right)w_\lambda(x). \tag{44.28}$$

Although this equation is identical to the (dimensionless) equation for the harmonic oscillator, the solutions differ because of the boundary conditions (harmonic oscillator functions vanish as $x \to \infty$). The eigenvalues λ, though discrete (for given boundary conditions), are no longer (necessarily) integers and furthermore (unlike the harmonic oscillator) there are two linearly independent solution, $w_\lambda(+x)$ and $w_\lambda(-x)$, for each λ; these are the Weber functions. The total eigenfrequency is given by (where we write n for n_z)

$$\Omega_n = \tau_{sA}^{-1} + \left(\lambda_A + \frac{1}{2}\right)\frac{2D_A}{\xi^2} = \tau_{sB}^{-1} + \left(\lambda_B + \frac{1}{2}\right)\frac{2D_B}{\xi^2}. \tag{44.29}$$

The eigenfunctions in a given layer i have the form

$$\psi_n^{(i)}(z) = \begin{cases} e^{ik_y y}\left\{A_{+n}^{(i)} w_\lambda\left[\frac{\sqrt{2}}{\xi}(z - z_0)\right] + A_{-n}^{(i)} w_\lambda\left[-\frac{\sqrt{2}}{\xi}(z - z_0)\right]\right\}; & \text{A layers} \\ e^{ik_y y}\left\{B_{+n}^{(i)} w_\lambda\left[\frac{\sqrt{2}}{\xi}(z - z_0)\right] + B_{-n}^{(i)} w_\lambda\left[-\frac{\sqrt{2}}{\xi}(z - z_0)\right]\right\}; & \text{B layers.} \end{cases} \tag{44.30}$$

The coefficients in successive layers follow from the boundary condition on the logarithmic derivative at each interface:

$$\mathcal{N}_A D_A \frac{d}{dz} \ell n[\psi_{nA}^{(i)}(z)] = \mathcal{N}_B D_B \frac{d}{dz} \ell n[\psi_{nB}^{(i)}(z)];\ z = 0, \pm d, \pm 2d, \ldots \tag{44.31a}$$

$$\mathcal{N}_A D_A \frac{d}{dz} \ell n[\psi_{nA}^{(i)}(z)] = \mathcal{N}_B D_B \frac{d}{dz} \ell n[\psi_{nB}^{(i)}(z)];\ z = a, a \pm d, a \pm 2d, \ldots \tag{44.31b}$$

On physical grounds we expect the eigenfunctions to be symmetric about the center of a superconducting (A) layer which, with the appropriate choice of z_0 (or equivalently k_y) can be fixed at $z = a/2$. One of the constants is fixed by normalization; all others follow on applying (44.31).

The Weber functions are related to the hypergeometric function as follows (Morse and Feshbach 1953):

$$w_\lambda(x) = 2^{\lambda/2} e^{-1/4x^2} \left[\frac{\pi^{1/2}}{\Gamma\left(\frac{1}{2} - \frac{\lambda}{2}\right)} F\left(-\frac{\lambda}{2} \middle| \frac{1}{2} \middle| \frac{x^2}{2} \right) - \left(\frac{\pi}{2}\right)^{1/2} \frac{2x}{\Gamma\left(-\frac{\lambda}{2}\right)} F\left(\frac{1}{2} - \frac{\lambda}{2} \middle| \frac{3}{2} \middle| \frac{x^2}{2} \right) \right]. \tag{44.32}$$

The derivative of $w_\lambda(x)$ may be calculated from the relations

$$\frac{dw_\lambda(x)}{dz} + \frac{x}{2} w_\lambda(x) - \lambda w_{\lambda-1}(x) = 0 \tag{44.33a}$$

or

$$\frac{dw_\lambda(x)}{dz} - \frac{x}{2} w_\lambda(x) + \lambda w_{\lambda+1}(x) = 0. \tag{44.33b}$$

44.2 Representative numerical solutions

As can be seen from the above discussion the application of the Takahashi–Tachiki formalism in the case of finite magnetic fields is relatively complex requiring extensive numerical analysis. Under the simplifying assumption that only *one* of the three quantities, $\mathcal{N}(z)$, $D(z)$, or $V(z)$ varies (in a stepwise fashion), Takahashi and Tachiki (1986a) have carried out numerical calculations (in some cases using a variational approach); their analysis is complete enough to bring out most of the expected qualitative features.

We first discuss the case where the density of states is discontinuous at the interface. Fig. 44.1 shows the parallel, $H_{c2\parallel}$ (solid), and the perpendicular, $H_{c2\perp}$ (dashed), upper critical fields (normalized to the zero-temperature upper critical field of the bulk superconductor, $H_{c2}^S(0)$) as a function of T/T_c (where T_c is the zero-field transition temperature of the superlattice); here we have $\mathcal{N}_N/\mathcal{N}_S = 0.15$, $V_S = V_N$, $D_S = D_N$, and $a = b$, where a is the thickness of the superconducting

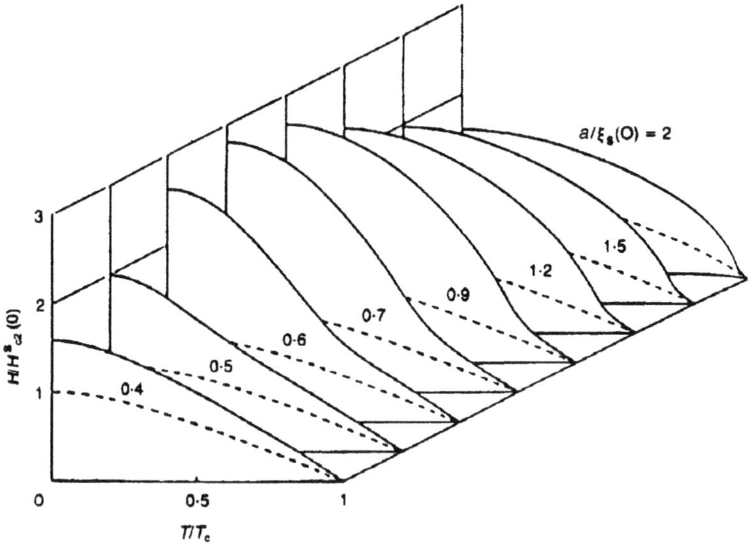

Figure 44.1 Temperature dependence of the parallel (solid lines) and perpendicular (dashed lines) upper critical fields for $a = b$ and varying values of $a/\xi_s(0)$ for a case where the density of states is discontinuous (see text). (After Takahashi and Tachiki (1986a).)

layer. The numbers beside the different sets of curves denote the value of $a/\xi_S(0)$ where $\xi_S(0) \equiv [\phi_0/2\pi H^S_{c2}(0)]^{1/2}$. We note the following features. For T close to T_c the perpendicular critical field always behaves linearly in $T_c - T$. However, the behavior of the parallel critical field depends on the magnitude of $a/\xi_S(0)$: when this quantity is small we obtain linear, or 3D-like, behavior of the upper critical field, while for large values we have square-root-like, or 2D, behavior (except for a small region near T_c); for values near unity we have a 2D–3D crossover analogous to that obtained in our analysis of the superconductor/insulator superlattice in the Lawrence–Doniach model.

Fig. 44.2(a) displays the results for $H_{c2\parallel}$ when the diffusion constants are unequal. The results are for the case $a = b = 0.75\xi_S(0)$, $V_S = V_N$, and $\mathcal{N}_S = \mathcal{N}_N$. The critical fields are displayed in reduced units, $H/H^N_{c2}(0)$, where $H^N_{c2}(0)$ is the upper critical field of the bulk metal with diffusion constant D_N.

Fig. 44.2(b) shows the temperature dependence of the perpendicular upper critical field for the same conditions as Fig. 44.2(a). Note the remarkable upturn in the temperature dependence of $H_{c2\perp}$ for large D_N/D_S. Takahashi and Tachiki explain this by noting that for T close to T_c the pair wavefunction is relatively uniform along the field direction, and thus the diffusion constant in the N layers dominates $H_{c2\perp}$. As T decreases the pair wavefunction becomes restricted to the S layers which, with their smaller diffusion constant, have a higher upper critical field; when this occurs we have an upturn in the upper critical field. Fig. 44.2(c) shows the behavior of the perpendicular upper critical field, for $V_N = V_S$, $\mathcal{N}_N = \mathcal{N}_S$, and a large ratio $D_N/D_S = 20$; here we plot several values of $a_S/\xi_N(0)$ with $b/\xi_N(0) = 2$.

In a later study Takahashi and Tachiki (1987) reexamined the behavior of the parallel upper critical field for the case of widely differing diffusion constants. They noted that by symmetry there are two possible solutions, one where nucleation occurs in the N layers and one where it occurs in the S layers. It turns out that these two nucleation temperatures (i.e., the temperatures satisfying our linearized integral equation for Δ) have different field dependences, as

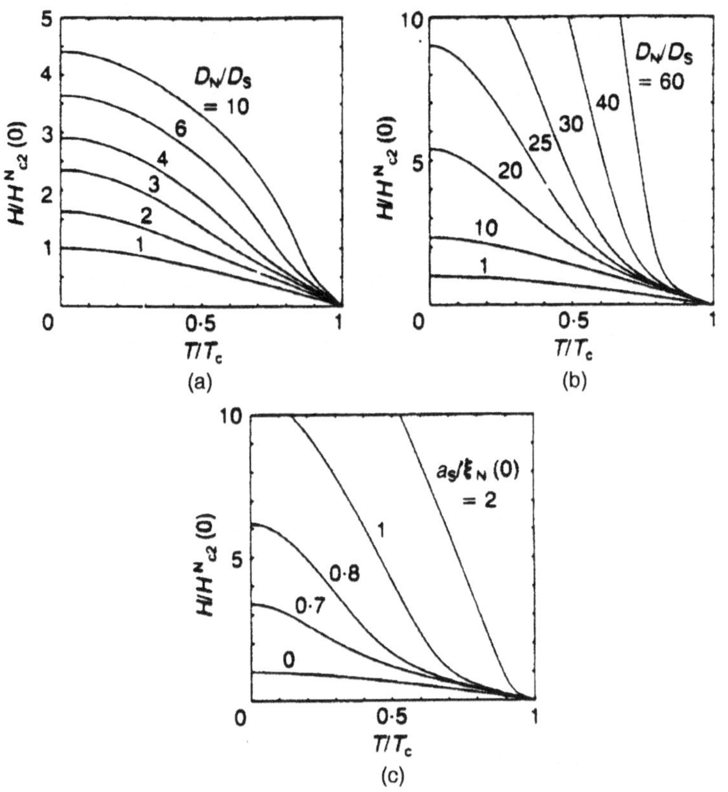

Figure 44.2 Temperature dependence of the normalized upper critical field: (a) $H_{c2\parallel}$ calculated using different diffusion constants; (b) $H_{c2\perp}$ calculated using different diffusion constants; and (c) $H_{c2\perp}$ calculated using $D_N/D_S = 20$, $b/\xi_N(0) = 2$, and different values of $a/\xi_N(0)$. (After Takahashi and Tachiki (1986a).)

shown in Fig. 44.3 for the case of $D_N/D_S = 12.5$, $a = b = \xi_N(0)$ and $V_N = V_S$. The dashed and dash-dot curves refer to solutions when $\Delta(\mathbf{r})$ is centered in the N (clean) and S (dirty) layers, respectively. For T close to T_c nucleation first occurs in the superconducting layer (since nucleation will always occur at the highest field satisfying the linearized gap equation). However, on passing the point T^*, H^* the solution centered in the N (clean) layers has a smaller field. In the absence of pinning a second order phase transition will occur at this point. Fig. 44.4 shows the parallel upper critical field (for $a = b = \xi_N(0)$ and $V_N = V_S$) as a function of D_N/D_S. Note the abrupt change in slope for the three curves for which the phase transition occurs.

Lastly, we examine the effect of a change in the BCS pairing potential. Fig. 44.5 shows the parallel and perpendicular upper critical fields for the case $a = b = d$, $V_S\mathcal{N}_S = 0.3$ and $V_N\mathcal{N}_N = 0$, $\mathcal{N}_N = \mathcal{N}_S$ and $D_N = D_S$. Crossover behavior in the parallel upper critical field is again observed. Taskahashi and Tachiki also present graphs displaying the effect of spin paramagnetism on the upper critical field that will not be reproduced here.

Extensive calculations of the magnetic-field-dependent transition temperature of superconductor/normal multilayers have been performed by Koperdraad (1995); his thesis contains an in-depth review of the theory. Although we have restricted all our discussion to the reversible

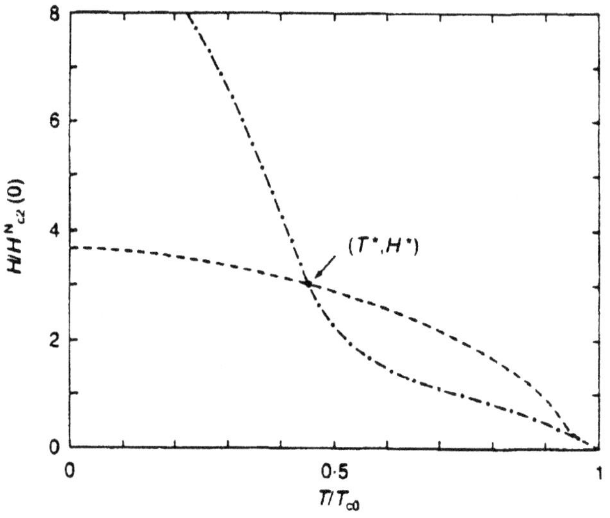

Figure 44.3 Temperature dependence for the nucleation field in the N (dashed line) and S (dotted line) layers. Note that at the point (T^*, H^*), the solutions are degenerate, implying a phase transition. (After Takahashi and Tachiki (1987).)

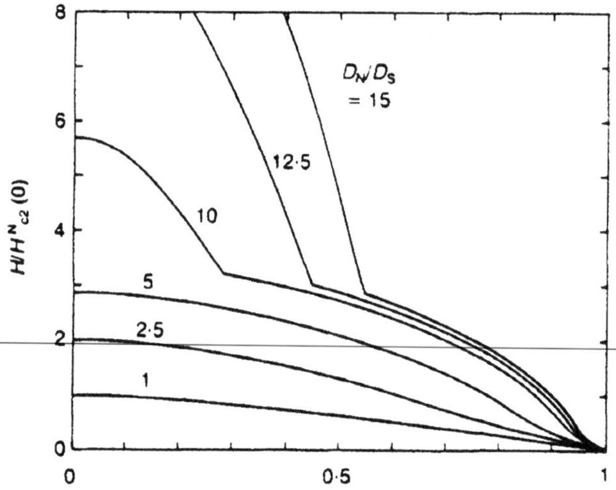

Figure 44.4 Temperature dependence of the upper critical field as a function of D_N/D_S. (After Takahashi and Tachiki (1987).)

behavior, we note in passing the work of Takahashi and Tachiki (1986b) on vortex pinning in superconductor/ferromagnetic superlattices.

Before concluding the discussion of the behavior of artificial superlattices in the presence of a field, we note that the occurrence of surface superconductivity has not been addressed. This interesting and complex problem could presumably be treated with an extension of the aforementioned techniques.

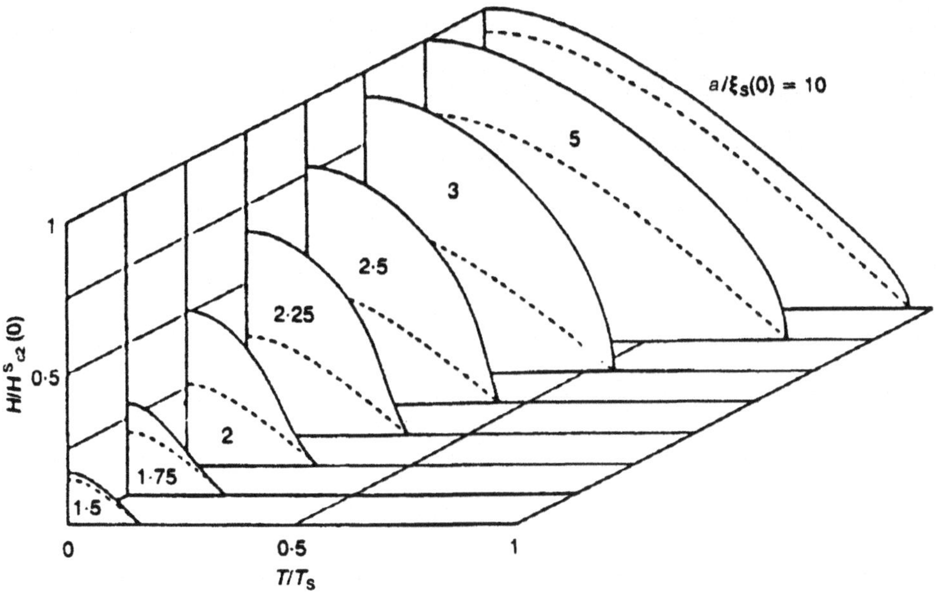

Figure 44.5 Temperature dependence of the normalized parallel (solid lines) and the perpendicular (dashed lines) upper critical fields for $a \doteq b$ and varying values of $a/\xi_s(0)$ for a case where the BCS pairing potential is discontinuous (see text). (After Takahashi and Tachiki (1986a).)

45 Derivation of the G–L theory

45.1 The first G–L equation

The G–L theory was introduced earlier (see Sec. 9) on phenomenological grounds. In this section we will derive the equations from the microscopic theory of an inhomogeneous superconductor (Gorkov 1959). However, for orientation purposes it is better first to derive the Landau theory of a second order phase transition for a homogeneous superconductor. Our starting point is the BCS gap equation

$$\Delta(T) = \mathcal{N}(0)V\Delta(T)\int_{-\hbar\omega_D}^{\hbar\omega_D} \frac{\tanh\left(\dfrac{\varepsilon}{2k_BT}\right)}{2\varepsilon}\,\mathrm{d}\xi, \tag{28.14}$$

where $\varepsilon = [\xi^2 + \Delta^2(T)]^{1/2}$ and we have restricted ourselves to the case of an energy-independent pairing potential and gap function. To facilitate the subsequent calculations we add and subtract the value of the integral at T_c and employ Eq. (38.9)

$$\Delta(T) = \mathcal{N}(0)V\Delta(T)\left\{\int_{-\hbar\omega_D}^{\hbar\omega_D} \mathrm{d}\xi\frac{\tanh(\xi/2k_BT)}{2\xi}\right.$$

$$\left. + k_BT\sum_\omega \int_{-\infty}^{+\infty}\mathrm{d}\xi\left[\frac{1}{(\varepsilon + \mathrm{i}\hbar\omega)(\varepsilon - \mathrm{i}\hbar\omega)} - \frac{1}{(\xi + \mathrm{i}\hbar\omega)(\xi - \mathrm{i}\hbar\omega)}\right]\right\}, \tag{45.1}$$

where the second term is convergent and we may extend the limits of the integral to $\pm\infty$ with no significant loss in accuracy. The first term in the square brackets has poles at $\xi = \pm\mathrm{i}(\hbar^2\omega^2 + \Delta^2)^{1/2}$ with residues of $\pm\pi/(\hbar^2\omega^2 + \Delta^2)^{1/2} \cong \pm\pi/\hbar\omega(1 + \Delta^2/2\hbar^2\omega^2)$; the corresponding poles of the second term are at $\xi = \pm\mathrm{i}\hbar\omega$ with residues of $\pm\pi/\hbar\omega$. Using the property

$$\int_{-\hbar\omega_D}^{\hbar\omega_D}\frac{\mathrm{d}\xi}{2\xi}\tanh\frac{\xi}{2k_BT} = \ell n\left(\frac{1.14\hbar\omega_D}{k_BT}\right)$$

discussed prior to Eq. (40.4), and completing the contour in either half plane, we may write (45.1) as

$$\Delta(T)\left[\mathcal{N}(0)V\ell n\left(\frac{1.14\hbar\omega_D}{k_BT}\right) - 1\right]$$

$$+ 2\pi\mathcal{N}(0)V\Delta(T)k_BT\sum_\omega\left[\frac{1}{\hbar|\omega|\left(1+\dfrac{\Delta^2(T)}{2\hbar^2\omega^2}\right)}-\frac{1}{\hbar|\omega|}\right]=0. \tag{45.2}$$

Expanding the denominator of the first term in the second square bracket to order $\Delta^2(T)/(\hbar^2\omega^2)$ yields

$$\Delta(T)\left[1-\mathcal{N}(0)V\ell n\left(\frac{1.14\hbar\omega_D}{k_BT}\right)\right]$$

$$+\pi k_BT\mathcal{N}(0)V\Delta^3(T)\sum_{|\omega|}\frac{1}{\hbar^3|\omega|^3}=0. \tag{45.3}$$

The sum[1] $\Sigma_{|\omega|}1/\hbar^3|\omega|^3=(\pi k_BT_c)^{-3}\Sigma_{n=0}^{\infty}1/(2n+1)^3=(\pi k_BT_c)^{-3}\,\frac{7}{8}\zeta(3)$, where T has been set equal to T_c. Our final form for the self-consistency condition through third order in $\Delta(T)$ is then

$$\left[1-\mathcal{N}(0)V\ell n\left(\frac{1.14\hbar\omega_D}{k_BT}\right)\right]\Delta(T)+\frac{\mathcal{N}(0)V}{\pi^2k_B^2T_c^2}\frac{7}{8}\zeta(3)\Delta^3(T)=0. \tag{45.4}$$

Near T_c the coefficient of $\Delta(T)$ may be expanded as $-\mathcal{N}(0)V(T_c-T)/T_c$ and we see that the structure of Eq. (45.4) becomes identical to Eq. (9.15) (in the uniform limit with $\mathbf{A}=0$) and hence we can say that ψ is proportional to Δ.

45.2 The gradient term in the clean limit

Still missing from Eq. (45.4) is the gradient term of the first G–L equation. To examine this term we return to the linearized self-consistency condition, Eq. (38.6). We first examine the case of a clean, free-electron-like metal where we may employ plane waves in Eq. (38.10) giving

$$K(\mathbf{r},\mathbf{r}')=V(\mathbf{r})k_BT_c\sum_\omega\int\frac{d^3kd^3k'}{(2\pi)^6}\frac{e^{i(\mathbf{k}+\mathbf{k}')\cdot(\mathbf{r}-\mathbf{r}')}}{(\xi_\mathbf{k}+i\hbar\omega)(\xi_{\mathbf{k}'}-i\hbar\omega)}e^{(2ie/\hbar c)\mathbf{A}(\mathbf{r})\cdot(\mathbf{r}-\mathbf{r}')}; \tag{45.5}$$

here we included the semiclassical phase factor given by Eq. (38.21) to account for the effects of a vector potential. The kernel (45.5) acts on the gap function $\Delta(\mathbf{r}')$ and we expand the latter as

$$\Delta(\mathbf{r}')=\Delta(\mathbf{r})-\left(\frac{\partial\Delta}{\partial\mathbf{r}'}\right)_{(\mathbf{r}'=\mathbf{r})}\cdot(\mathbf{r}'-\mathbf{r})+\frac{1}{2}(\mathbf{r}'-\mathbf{r})\cdot\left(\frac{\partial^2\Delta}{\partial\mathbf{r}'\partial\mathbf{r}'}\right)_{(\mathbf{r}'=\mathbf{r})}\cdot(\mathbf{r}'-\mathbf{r})+\ldots. \tag{45.6}$$

According to the discussion surrounding Eq. (38.21), the semiclassical approximation for including the effects of the vector potential assumes we choose a gauge where $(2e/c)|\mathbf{A}(\mathbf{r})|<<p_F$; we may then expand the exponential involving $\mathbf{A}$ as

$$e^{(2ie/\hbar c)\mathbf{A}(\mathbf{r})\cdot(\mathbf{r}-\mathbf{r}')}=1+\frac{2ie}{\hbar c}\mathbf{A}(\mathbf{r})\cdot(\mathbf{r}-\mathbf{r}')+\frac{1}{2}\left(\frac{2ie}{\hbar c}\right)^2[\mathbf{A}(\mathbf{r})\cdot(\mathbf{r}-\mathbf{r}')]^2+\ldots. \tag{45.7}$$

Writing $\mathbf{r}-\mathbf{r}'=\mathbf{R}$, and noting $\int RK(R)d^3R=0$, we may write Eq. (38.6) as

1. Abramowitz and Stegun (1970), Eq. (23.2.20).

$$\Delta(\mathbf{r}) = \left[\int K(R)\mathrm{d}^3R\right]\Delta(\mathbf{r}) + \left[\frac{1}{3}\int R^2K(R)\mathrm{d}^3R\right]\frac{1}{2}\left[\nabla - \frac{2ie}{\hbar c}\mathbf{A}(\mathbf{r})\right]^2\Delta(\mathbf{r}). \qquad (45.8)$$

Now $\int K(R)\mathrm{d}^3R = K(\mathbf{q}=0) = \mathcal{N}(0)V\ell n(1.14\hbar\omega_D/k_BT)$, where we used Eq. (40.4) in the last step; therefore the term on the left and the first term on the right of (45.8) combine to give the first term in our homogeneous G–L equation (45.4). The second term on the right of (45.8) corresponds to the gradient term of the G–L equation (Eq. (9.15)). We first perform the calculation for free-electron-like wavefunctions in the clean limit. Inserting plane wave forms $\phi_\mathbf{k}(\mathbf{r}) = (1/L^{3/2})e^{i\mathbf{k}\cdot\mathbf{r}}$ into Eq. (38.10) we obtain the integral as

$$\begin{aligned}\int R^2K(R)\mathrm{d}^3R &= -Vk_BT_c\frac{1}{(2\pi)^6}\sum_\omega\int\frac{\mathrm{d}^3k\mathrm{d}^3k'\mathrm{d}^3R}{(\xi_\mathbf{k}+i\hbar\omega)(\xi_{\mathbf{k}'}-i\hbar\omega)}\frac{\partial^2}{\partial\mathbf{k}\cdot\partial\mathbf{k}'}e^{i(\mathbf{k}+\mathbf{k}')\cdot\mathbf{R}} \\ &= -Vk_BT_c\frac{1}{(2\pi)^6}\sum_\omega\int\mathrm{d}^3k\mathrm{d}^3k'\mathrm{d}^3Re^{i(\mathbf{k}+\mathbf{k}')\cdot\mathbf{R}}\frac{\partial^2}{\partial\mathbf{k}\cdot\partial\mathbf{k}'}\frac{1}{(\xi_k+i\hbar\omega)(\xi_{k'}-i\hbar\omega)} \\ &= -\frac{1}{(2\pi)^3}Vk_BT_c\sum_\omega\int\mathrm{d}^3k\mathrm{d}^3k'\delta^{(3)}(\mathbf{k}+\mathbf{k}')\frac{\hbar^2\mathbf{v}_F(\mathbf{k})\cdot\mathbf{v}_F(\mathbf{k}')}{(\xi_\mathbf{k}+i\hbar\omega)^2(\xi_{\mathbf{k}'}-i\hbar\omega)^2} \\ &= \frac{1}{(2\pi)^3}Vk_BT_c\sum_\omega\int\mathrm{d}^3k\frac{\hbar^2v_F^2}{(\xi_\mathbf{k}+i\hbar\omega)^2(\xi_k-i\hbar\omega)^2} \qquad (45.9)\end{aligned}$$

where we used $\xi_\mathbf{k} = \hbar v_F(k-k_F)$ and $\xi_\mathbf{k} = \xi_{-\mathbf{k}}$. Evaluating the angular part of the integral and introducing the density of states per spin $\mathcal{N}(0) = [4\pi/(2\pi)^3]k_F^2/\hbar v_F$, Eq. (45.9) can be written

$$\int R^2K(R)\mathrm{d}^3R = \hbar^2v_F^2k_BT\mathcal{N}(0)V\sum_\omega\frac{\mathrm{d}\xi}{(\xi+i\hbar\omega)^2(\xi-i\hbar\omega)^2}. \qquad (45.10)$$

Eq. (45.10) has second order poles at $\xi_\mathbf{k} = \pm i\hbar\omega$ which we evaluate with $\int[f(z)/(z-a)^2]\mathrm{d}z = 2\pi i f'(a)$ to obtain for the last term in Eq. (45.8)

$$\int R^2K(R)\mathrm{d}^3R = \pi\hbar^2v_F^2k_BT_c\mathcal{N}(0)V\sum_{|\omega|}\frac{1}{|\hbar\omega|^3} = \frac{\hbar^2v_F^2\mathcal{N}(0)V}{(\pi k_BT_c)^2}\frac{7}{8}\zeta(3). \qquad (45.11)$$

The resulting linearized, clean-limit G–L equation then takes the form

$$\begin{aligned}&\left[\mathcal{N}(0)V\ell n\left(\frac{1.14\hbar\omega_D}{k_BT}\right)-1\right]\Delta(\mathbf{r}) \\ &+\frac{7}{8}\zeta(3)\frac{\hbar^2v_F^2\mathcal{N}(0)V}{6(\pi k_BT_c)^2}\left[\nabla-\frac{2ie}{\hbar c}\mathbf{A}(\mathbf{r})\right]^2\Delta(\mathbf{r}) = 0. \qquad (45.12)\end{aligned}$$

45.3 The gradient term in the dirty limit

Let us next examine the gradient term for the opposite limit of a dirty superconductor. For this case the Fourier transform of the kernel $K(\mathbf{q})$ was given by Eq. (40.5) which in R space becomes the operator expression

$$K(R) = \mathcal{N}(0)V\left[\ell n\left(\frac{1.14\hbar\omega_D}{k_B T}\right) + \sum_n\left(\frac{1}{2n+1-\xi_d^2\nabla^2} - \frac{1}{2n+1}\right)\right] \tag{45.13}$$

with $\xi_d^2 \equiv \hbar v_F^2\tau/6\pi k_B T_c$. Assuming $\Delta(\mathbf{r})$ is slowly varying we may expand (45.13) in powers of ∇^2 and restrict to the leading term to obtain[2]

$$\begin{aligned} K(R) &= \mathcal{N}(0)V\left[\ell n\left(\frac{1.14\hbar\omega_D}{k_B T}\right) + \left(\sum_{n=-\infty}^{\infty}\frac{1}{(2n+1)^2}\right)\xi_d^2\nabla^2 + \ldots\right] \\ &= \mathcal{N}(0)V\left[\ell n\left(\frac{1.14\hbar\omega_D}{k_B T}\right) + \frac{\pi^2}{4}\xi_d^2\nabla^2\right]. \end{aligned} \tag{45.14}$$

On including the semiclassical phase factor $e^{(2ie/\hbar c)\mathbf{A}(\mathbf{r})\cdot\mathbf{R}}$ and expanding to second order as we did above, the gradient term again adopts a gauge-invariant form and the linearized, dirty-limit G–L equation becomes

$$\left(\mathcal{N}(0)V\ell n\frac{1.14\hbar\omega_D}{k_B T} - 1\right)\Delta(\mathbf{r}) + \frac{\pi^2}{4}\mathcal{N}(0)V\xi_d^2\left(\nabla - \frac{2ie}{\hbar c}\mathbf{A}\right)^2\Delta(\mathbf{r}) = 0. \tag{45.15}$$

45.4 The gradient term in the general case

We may treat cases which lie between the dirty and clean limits by substituting the interpolation form (39.18) into (39.10); if we again suppress the quantum phase factor we have

$$g_0(\mathbf{q}, t) = 1 - 2v_F^2 q^2\,[\tau t - \tau^2(1 - e^{-t/\tau})]. \tag{45.16}$$

On Fourier transforming to real space the first term becomes $\delta^{(3)}(\mathbf{R})$ and, in gauge invariant form, the second term becomes

$$2v_F^2[\tau t - \tau^2(1 - e^{-t/\tau}]\left(\nabla - \frac{2ie}{\hbar c}\mathbf{A}\right)^2. \tag{45.17}$$

Using Eq. (38.14′), the coefficient entering the gradient term in Eq. (45.8) becomes

$$\begin{aligned} \frac{1}{6}\int R^2 K(R)\mathrm{d}^3R &= \frac{1}{6}\frac{2\pi}{\hbar}k_B T_c\mathcal{N}(0)V\sum_\omega\int_0^\infty \mathrm{d}t e^{-2|\omega|t}2v_F^2[\tau t - \tau^2(1 - e^{-t/\tau})] \\ &= \frac{8}{6}\frac{\pi}{\hbar}k_B T_c\mathcal{N}(0)V\sum_{|\omega|}\frac{v_F^2}{|2\omega|^2(2|\omega| + 1/\tau)} \\ &= \frac{\hbar^2}{6}\mathcal{N}(0)V\left(\frac{v_F}{\pi k_B T_c}\right)^2\sum_{n=0}^{\infty}\frac{1}{(2n+1)^2(2n+1+\rho)}, \end{aligned} \tag{45.18}$$

where $\rho \equiv \hbar/2\pi k_B T_c\tau$. Gorkov introduced the function

2. The sum is evaluated by writing $\tanh z = \Sigma_{n=-\infty}^{+\infty} 1/[z - i(\pi/2)(2n+1)]$ (see Eq. (38.9)), differentiating with respect to z to obtain $\mathrm{sech}^2 z = \Sigma_n 1/[z - (i\pi/2)(2n+1)]^2$, and evaluating the latter at $z = 0$ to give $\Sigma_{n=-\infty}^{+\infty} 1/(2n+1)^2 = \pi^2/4$.

$$\chi(\rho) = \frac{8}{7\zeta(3)} \sum_{n=0}^{\infty} \frac{1}{(2n+1)^2(2n+1+\rho)} \tag{45.19}$$

and in terms of this function we may write the coefficient of the gradient term as

$$\frac{1}{6}\int R^2 K(R) \mathrm{d}^3R = \frac{1}{6}\frac{7}{8}\zeta(3)\mathcal{N}(0)V\left(\frac{\hbar v_F}{\pi k_B T_c}\right)^2 \chi(\rho). \tag{45.20}$$

In the clean limit ($\tau \to \infty$), $\chi(\rho) = \chi(0) = 1$; in the dirty limit ($\rho \gtrsim \hbar\omega_D/k_B T_c$),

$$\chi(\rho) = \frac{\pi^2}{7\zeta(3)}\frac{1}{\rho} = \frac{2\pi^3}{7\zeta(3)}\frac{k_B T_c \tau}{\hbar}.$$

In these limits the gradient term reduces to the earlier clean limit (45.12) and dirty limit (45.15) expressions, respectively.

In leading order the inclusion of impurity scattering does not affect the coefficient of the cubic term in the G–L equation and we may simply take this term from Eq. (45.4) and add it to any of the linearized forms to obtain the first G–L equation. By combining Eqs. (45.4), (45.8), and (45.20), we obtain

$$\left[\left(1 - \frac{T}{T_c}\right) + \frac{1}{6}\left(\frac{\hbar v_F}{\pi k_B T_c}\right)^2 \frac{7}{8}\zeta(3)\chi(\rho)\left(\nabla - \frac{2ie}{\hbar c}\mathbf{A}\right)^2 - \frac{7\zeta(3)}{8(\pi k_B T_c)^2}|\Delta(\mathbf{r})|^2\right]\Delta(\mathbf{r}) = 0. \tag{45.21}$$

Eq. (45.21) is the final result for the first G–L equation derived from the microscopic theory in the limit $T \to T_c$. To complete this identification of the G–L theory with the microscopic theory we must fix the proportionality constant between the order parameter, $\psi(\mathbf{r})$ and the gap function, $\Delta(\mathbf{r})$. Although the microscopic theory fixes the magnitude of $\Delta(\mathbf{r})$ unambiguously, the magnitude of $\psi(\mathbf{r})$ is arbitrary: a change in the magnitude of $\psi(\mathbf{r})$ may be absorbed into the parameters α, β, and m^* of the theory. We will adopt a definition of $\psi(\mathbf{r})$ in which coefficient of the gradient term of the G–L expansion is given by $\hbar^2/2m^* = \hbar^2/4m$ (in the clean limit). This results in the following:

$$\psi(\mathbf{r}) = \left[\frac{7\zeta(3)n}{8(\pi k_B T_c)^2}\right]^{1/2} \Delta(\mathbf{r}); \tag{45.22a}$$

$$\alpha = -\frac{6\pi^2(k_B T_c)^2}{7\zeta(3)\varepsilon_F}\left(1 - \frac{T}{T_c}\right); \tag{45.22b}$$

$$\beta = \frac{12\pi^2(k_B T_c)^2}{7\zeta(3)\varepsilon_F n}; \tag{45.22c}$$

here n is the total electron density and ε_F is the Fermi energy.

45.5 The second G–L equation

We now turn to a discussion of the second G–L equation, Eq. (9.17). Here we will sketch how the calculation could be carried out rigorously; however, to obtain the required equation we will resort to a device. The microscopic expression for the current density is given by Eq. (37.17). We start by substituting the expansions (38.2a) and (38.2b) into (37.17). The wavefunctions $u_n^{(i)}(\mathbf{r})$ and $v_n^{(i)}(\mathbf{r})$ are expanded as

$$u_n^{(i)}(\mathbf{r}) = \sum_m a_{nm}^{(i)} \phi_m(\mathbf{r}) \tag{45.23a}$$

and

$$v_n^{(i)}(\mathbf{r}) = \sum_m b_{nm}^{(i)} \phi_m^*(\mathbf{r}); \tag{45.23b}$$

$a_{nm}^{(1)}$ and $b_{nm}^{(1)}$ are given by (38.5a) and (38.5b) and $a_{nm}^{(2)}$ and $b_{nm}^{(2)}$ are given in Appendix C by (C.13a) and (C.13b). Due to the action of the θ functions in these expansions the surviving contributions to $\mathbf{j}_1(\mathbf{r})$ through second order are

$$\begin{aligned} \mathbf{j}_1(\mathbf{r}) = \frac{ie\hbar}{2m}\Bigg\{ & \Bigg[\sum_{nm\ell} b_{nm}^{(1)*}\phi_m(\mathbf{r})\frac{\partial}{\partial \mathbf{r}} b_{n\ell}^{(1)}\phi_\ell^*(\mathbf{r}) + \sum_{nm} b_{nm}^{(2)*}\phi_m(\mathbf{r})\frac{\partial}{\partial \mathbf{r}}\phi_n^*(\mathbf{r})\theta(-\xi_n) \\ & + \sum_{nm\ell} \phi_n^*(\mathbf{r})\theta(-\xi_n)\frac{\partial}{\partial \mathbf{r}} b_{nm}^{(2)}\phi_m(\mathbf{r}) - \text{c.c.}\Bigg](1-f_n) \\ & + \Bigg[\sum_{nm\ell} a_{nm}^{(1)*}\phi_m^*(\mathbf{r})\frac{\partial}{\partial \mathbf{r}} a_{n\ell}^{(1)}\phi_\ell(\mathbf{r}) + \sum_{nm} a_{nm}^{(2)*}\phi_m^*(\mathbf{r})\frac{\partial}{\partial \mathbf{r}}\phi_n(\mathbf{r})\theta(\xi_n) \\ & + \sum_{nm} \phi_n^*(\mathbf{r})\theta(\xi_n)\frac{\partial}{\partial \mathbf{r}} a_{nm}^{(2)}\phi_m(\mathbf{r}) - \text{c.c.}\Bigg] f_n \Bigg\}. \end{aligned} \tag{45.24}$$

As mentioned above we will not actually perform the calculations leading to the final form for $\mathbf{j}$; they are somewhat tedious. Instead, and with the hindsight knowledge that the Gorkov–deGennes calculations lead to the microscopic G–L equations, we argue that the parameter m^* entering the first phenomenological equation, (9.15), must be the same as that entering the second phenomenological equation, (9.17), provided $e^* = 2e$.

46 Gauge invariance: diamagnetism in the low field limit

46.1 Gauge invariance

We first examine the case of the 'pure' gauge field (one not having an accompanying magnetic field)

$$\mathbf{A}(\mathbf{r}) = \nabla\chi(\mathbf{r}). \tag{46.1}$$

It is easy to show that the effect of introducing this form into the Bogoliubov equations is to alter the wavefunctions and the energy gap according to

$$u_n(\mathbf{r}) = u_n^{(0)}(\mathbf{r})e^{(ie/\hbar c)\chi(\mathbf{r})}, \tag{46.2a}$$

$$v_n(\mathbf{r}) = v_n^{(0)}(\mathbf{r})e^{-(ie/\hbar c)\chi(\mathbf{r})}, \tag{46.2b}$$

and

$$\Delta(\mathbf{r}) = \Delta^{(0)}(\mathbf{r})e^{(2ie/\hbar c)\chi(\mathbf{r})}, \tag{46.3}$$

where $u_n^{(0)}$, $v_n^{(0)}$, and $\Delta^{(0)}$ are the solutions with $\chi = 0$. Therefore in the presence of a pure gauge field the wavefunctions take on a phase $\frac{1}{2}\Phi(\mathbf{r}) = (e/\hbar c)\chi(\mathbf{r})$ and the gap function acquires a phase $\Phi(\mathbf{r})$ (however, the magnitude of the gap, $|\Delta(\mathbf{r})|$, is not affected). This phase invariance of $|\Delta|$ is the analog for the microscopic theory of the phase invariance of $|\psi|$ discussed earlier for the G–L theory.

46.2 The magnetic field as a perturbation

For magnetic fields less than H_{c1} (or H_c in a Type I material) a superconductor expels flux. From our earlier discussion of the Meissner effect we know that the tangential component of the magnetic field decays exponentially with the distance from the surface. For a planar surface located at $z = 0$ with a tangential field parallel to $\hat{\mathbf{x}}$ we may (in the London model) write the field inside the superconductor ($z > 0$) as $H_x(z) = H_x(z = 0)e^{-z/\lambda}$, where λ is a penetration depth; we may then choose a vector potential of the form $A_y(z) = \lambda H_x(z = 0)e^{-z/\lambda}$ (outside the superconductor ($z < 0$), where we suppose the field is uniform, we would write $A_y(z) = -H_x(z = 0)z$, which is unbounded).

Since $\mathbf{A}$ is bounded for $z > 0$ we may treat the vector potential as a perturbation in the Meissner regime; if the field at the surface is sufficiently small we may use first order perturbation theory to solve the Bogoliubov equations. We expand our wavefunctions in the form

$$u_n(\mathbf{r}) = u_n^{(0)}(\mathbf{r}) + u_n^{(1)}(\mathbf{r}) + \ldots \tag{46.4a}$$

and

$$v_n(\mathbf{r}) = v_n^{(0)}(\mathbf{r}) + v_n^{(1)}(\mathbf{r}) + \ldots, \tag{46.4b}$$

where $u_n^{(0)}$ and $v_n^{(0)}$ are the solutions to Bogoliubov's equations for $\mathbf{A} = 0$; we also expand the pair potential, $\Delta(\mathbf{r})$, as

$$\Delta(\mathbf{r}) = \Delta^{(0)}(\mathbf{r}) + \Delta^{(1)}(\mathbf{r}) + \ldots. \tag{46.5}$$

Expanding the (exact) results, (46.2) and (46.3), to first order in χ we have

$$u_n^{(1)}(\mathbf{r}) = \frac{ie}{\hbar c}\chi(\mathbf{r})u_n^{(0)}(\mathbf{r}), \tag{46.6a}$$

$$v_n^{(1)}(\mathbf{r}) = -\frac{ie}{\hbar c}\chi(\mathbf{r})v_n^{(0)}(\mathbf{r}), \tag{46.6b}$$

and

$$\Delta^{(1)}(\mathbf{r}) = \frac{2ie}{\hbar c}\chi(\mathbf{r})\Delta^{(0)}(\mathbf{r}). \tag{46.7}$$

We see that for the case of a pure gauge field the first order corrections are all *linear* in the perturbation. Recalling that the relation between the current and the vector potential was nonlocal in the Pippard model (see Eq. (3.8)), we assume the relation between $\Delta^{(1)}$ and $\mathbf{A}$ in the general case has the linear, nonlocal form

$$\Delta^{(1)}(\mathbf{r}) = \int \mathbf{P}(\mathbf{r},\mathbf{r}')\cdot\mathbf{A}(\mathbf{r}')\mathrm{d}^3r', \tag{46.8}$$

where we must establish the form of the vector function $\mathbf{P}$. We restrict ourselves to a uniform isotropic material where

$$\mathbf{P}(\mathbf{r},\mathbf{r}') = P(R)\frac{\mathbf{R}}{R} \tag{46.9}$$

with $\mathbf{R} = \mathbf{r}' - \mathbf{r}$. In lowest order the kernel $\mathbf{P}$ of Eq. (46.8) is independent of $\mathbf{A}$; its form can therefore be established from our special case of a *pure* gauge field; combining (46.8) and (46.7) yields

$$\frac{2ie}{\hbar c}\Delta^{(0)}\chi(\mathbf{r}) = \int \mathbf{P}(|\mathbf{r}-\mathbf{r}'|)\cdot\nabla'\chi(\mathbf{r}')\mathrm{d}^3r'. \tag{46.10}$$

Integrating (46.8) by parts and limiting ourselves to gauge functions, $\chi(\mathbf{r})$, which vanish as $r \to \infty$ (allowing us to discard the surface term) we have

$$\frac{2ie}{\hbar c}\Delta^{(0)}\chi(\mathbf{r}) = -\int \chi(\mathbf{r}')\nabla'\cdot\left[|\mathbf{P}(R)|\frac{\mathbf{R}}{R}\right]\mathrm{d}^3r'. \tag{46.11}$$

Assuming $\Delta^{(0)}$ is independent of $\mathbf{r}$, the solution to (46.11) is

$$\nabla \cdot \mathbf{P}(R) = -\frac{2ie}{\hbar c}\Delta^{(0)}\delta^{(3)}(\mathbf{r}-\mathbf{r}'). \tag{46.12}$$

Integrating both sides of (46.12) over d^3R and using Green's theorem yields

$$P(R) = \frac{-2ie\Delta^{(0)}}{4\pi\hbar c R^2}; \tag{46.13a}$$

or, in vector form,

$$\mathbf{P}(R) = \frac{ie\Delta^{(0)}}{2\pi\hbar c}\nabla\left(\frac{1}{R}\right). \tag{46.13b}$$

If we insert (46.13b) into (46.8) and integrate by parts, placing the surface term at the superconductor's surface, we obtain

$$\Delta^{(1)}(\mathbf{r}) = \frac{ie\Delta^{(0)}}{2\pi\hbar c}\left[-\int\frac{1}{R}(\nabla\cdot\mathbf{A})d^3R + \int\frac{\mathbf{A}\cdot\hat{\mathbf{n}}}{R}d^2R\right] \tag{46.14}$$

where $\hat{\mathbf{n}}$ is a unit vector normal to the superconductor surface. Hence if we use the London gauge ($\nabla\cdot\mathbf{A}=0$), and if $\mathbf{A}$ has no component perpendicular to the surface, then $\Delta^{(1)}=0$. It is therefore simplest to work in the London gauge, but at the expense of losing explicit gauge invariance.

Treating $\mathbf{A}$ as a perturbation and collecting the first order terms in Bogoliubov's equations we have

$$\left[\varepsilon + \frac{\hbar^2}{2m}\nabla^2 + \mu - U(\mathbf{r})\right]u_n^{(1)}(\mathbf{r}) - \Delta^{(0)}(\mathbf{r})v_n^{(1)}(\mathbf{r})$$
$$= \frac{ie\hbar}{2mc}[\mathbf{A}(\mathbf{r})\cdot\nabla + \nabla\cdot\mathbf{A}(\mathbf{r})]u_n^{(0)}(\mathbf{r}) \tag{46.15a}$$

and

$$\left[\varepsilon - \frac{\hbar^2}{2m}\nabla^2 - \mu + U(\mathbf{r})\right]v_n^{(1)}(\mathbf{r}) - \Delta^{(0)*}(\mathbf{r})u_n^{(1)}(\mathbf{r})$$
$$= \frac{ie\hbar}{2mc}[\mathbf{A}(\mathbf{r})\cdot\nabla + \nabla\cdot\mathbf{A}(\mathbf{r})]v_n^{(0)}(\mathbf{r}) \tag{46.15b}$$

(note there is no correction to ε to lowest order in $\mathbf{A}$ as shown in Subsec. 41.6). We expand $u_n^{(1)}$ and $v_n^{(1)}$ in terms of the normal-state eigenfunctions, ϕ_n (which form a complete set),

$$u_n^{(1)}(\mathbf{r}) = \sum_m a_{nm}\phi_m(\mathbf{r}) \tag{46.16a}$$

and

$$v_n^{(1)}(\mathbf{r}) = \sum_m b_{nm}\phi_m(\mathbf{r}). \tag{46.16b}$$

At this point we introduce two approximations: (i) we assume $\Delta^{(0)}$ (which we now write as

Δ) is independent of $\mathbf{r}$ (even in an alloy as discussed following Eq. (38.23)) and (ii) according to the discussion following Eq. (38.3) we assume the zeroth order superconducting wavefunctions may be written

$$u_n^{(0)}(\mathbf{r}) = u(\xi_n)\phi_n(\mathbf{r}) \tag{46.17a}$$

and

$$v_n^{(0)}(\mathbf{r}) = v(\xi_n)\phi_n(\mathbf{r}) \tag{46.17b}$$

with u_n and v_n given by Eqs. (26.27) and (26.28). Inserting (46.16) and (46.17) into (46.15), multiplying by $\phi_m^*(\mathbf{r})$, integrating over all $\mathbf{r}$, and using the orthonormality of the ϕ_n we obtain

$$(\varepsilon_n - \xi_m)a_{nm} - \Delta b_{nm} = M_{nm}u_n \tag{46.18a}$$

and

$$-\Delta a_{nm} + (\varepsilon_n + \xi_m)b_{nm} = M_{nm}v_n \tag{46.18b}$$

where the matrix element M_{nm} is defined by

$$M_{nm} = \frac{ie\hbar}{2mc}\int \phi_m^*(\mathbf{r})[\mathbf{A}(\mathbf{r})\cdot\nabla + \nabla\cdot\mathbf{A}(\mathbf{r})]\phi_n(\mathbf{r})d^3r = -M_{nm}^* = M_{mn}^*, \tag{46.19a}$$

where in the second term the gradient operates on both $\mathbf{A}(\mathbf{r})$ and $\phi_n(\mathbf{r})$. In a London gauge (46.19a) may be written

$$M_{nm} = \frac{ie\hbar}{mc}\int d^3r(\phi_m^*(\mathbf{r})\nabla\phi_n(\mathbf{r}))\cdot\mathbf{A}(\mathbf{r}). \tag{46.19b}$$

Integrating the second term in (46.19a) by parts (again with $\nabla\cdot\mathbf{A} = 0$) we may write it as

$$M_{nm} = \frac{ie\hbar}{2mc}\int d^3r[\phi_m^*(\mathbf{r})\nabla\phi_n(\mathbf{r}) - \phi_n(\mathbf{r})\nabla\phi_m^*(\mathbf{r})]\cdot\mathbf{A}(\mathbf{r}). \tag{46.19c}$$

Solving (46.18) we obtain

$$a_{nm} = \frac{M_{nm}}{\xi_n^2 - \xi_m^2}[(\varepsilon_n + \xi_m)u_n + \Delta v_n], \tag{46.20a}$$

$$b_{nm} = \frac{M_{nm}}{\xi_n^2 - \xi_m^2}[(\varepsilon_n - \xi_m)v_n + \Delta u_n]. \tag{46.20b}$$

46.3 The diamagnetic current

The expression for the diamagnetic current operator arising from a vector potential $\mathbf{A}(\mathbf{r})$ was given earlier in Eq. (37.16). Inserting our wavefunctions $u_n(\mathbf{r})$ and $v_n(\mathbf{r})$, corrected to first order, and noting that the ground state carries no current[1] we obtain for the expectation value $\langle\hat{\mathbf{j}}_1(\mathbf{r})\rangle$:

1. This property was proved by F. Bloch before the BCS theory of superconductivity.

$$\langle \hat{\mathbf{j}}_1(\mathbf{r})\rangle = -\frac{ie\hbar}{2m}\sum_n \{[v_n^{(0)*}(\mathbf{r})\nabla v_n^{(1)}(\mathbf{r}) + v_n^{(1)*}(\mathbf{r})\nabla v_n^{(0)}(\mathbf{r}) - \text{c.c.}](1-f_n)$$

$$+ [u_n^{(0)}(\mathbf{r})\nabla u_n^{(1)}(\mathbf{r}) + u_n^{(1)}(\mathbf{r})\nabla u_n^{(0)*}(\mathbf{r}) - \text{c.c.}]f_n\}, \tag{46.21}$$

where $f_n = f(\varepsilon_n)$. On inserting (46.16a), (46.16b), (46.17a), and (46.17b) this becomes

$$\langle \hat{\mathbf{j}}_1(\mathbf{r})\rangle = -\frac{ie\hbar}{2m}\sum_{n,m}\{v_n[\phi_n^*(\mathbf{r})\nabla\phi_m(\mathbf{r})b_{nm} + \phi_m^*(\mathbf{r})\nabla\phi_n(\mathbf{r})b_{nm}^* - \text{c.c.}](1-f_n)$$

$$+ u_n[\phi_n(\mathbf{r})\nabla\phi_m^*(\mathbf{r})a_{nm}^* + \phi_m(\mathbf{r})\nabla\phi_n^*(\mathbf{r})a_{nm} - \text{c.c.}]f_n\}. \tag{46.22}$$

If we restrict to $T = 0$, where $f_n = 0$, we have

$$\langle \hat{\mathbf{j}}_1(\mathbf{r})\rangle = -\frac{ie\hbar}{2m}\sum_{n,m} v_n[(\phi_m^*(\mathbf{r})\nabla\phi_n(\mathbf{r}) - \phi_n(\mathbf{r})\nabla\phi_m^*(\mathbf{r}))b_{nm}^* - \text{c.c.}]. \tag{46.23}$$

Using (46.20) and (46.19c), this equation becomes

$$\mathbf{j}_1(\mathbf{r}) = -\frac{e^2\hbar^2}{4m^2c}\sum_{m,n}\int d^3r'\left[2v_n\frac{(\varepsilon_n-\xi_m)v_n + \Delta u_n}{\xi_n^2-\xi_m^2}\right]\{[\phi_m^*(\mathbf{r})\nabla\phi_n(\mathbf{r}) - \phi_n(\mathbf{r})\nabla\phi_m^*(\mathbf{r})]$$

$$\times [\phi_m(\mathbf{r}')\nabla'\phi_n^*(\mathbf{r}') - \phi_n^*(\mathbf{r}')\nabla'\phi_m(\mathbf{r}')] + \text{c.c.}\}\cdot\mathbf{A}(\mathbf{r}'), \tag{46.24}$$

where we have written $\mathbf{j}_1(\mathbf{r})$ for $\langle\hat{\mathbf{j}}_1(\mathbf{r})\rangle$. Substituting the definitions of u_n and v_n and using $\varepsilon_n^2 = \xi_n^2 + \Delta^2$, the multiplicative factor in large square brackets in (46.24) may be written as

$$L(\xi_n, \xi_m) \equiv \frac{(\varepsilon_n - \xi_n)(\varepsilon_n - \xi_m) + \Delta^2}{\varepsilon_n(\xi_n^2 - \xi_m^2)} = \frac{(\varepsilon_n - \xi_n)(\varepsilon_n - \xi_m) + \Delta^2}{\varepsilon_n(\varepsilon_n^2 - \varepsilon_m^2)}.$$

Since the remainder of (46.24) is symmetric in n and m (and we sum over both indices), $L(\xi_n, \xi_m)$ may be replaced by its symmetric part:

$$L(\xi_n, \xi_m) \to \frac{1}{2}[L(\xi_m, \xi_n) + L(\xi_n, \xi_m)]$$

yielding

$$L(\xi_n, \xi_m) = -\frac{\Delta^2 + \xi_n\xi_m - \varepsilon_n\varepsilon_m}{2\varepsilon_n\varepsilon_m(\varepsilon_n + \varepsilon_m)}. \tag{46.25}$$

Eq. (46.24) then becomes

$$\mathbf{j}_1(\mathbf{r}) = \frac{\hbar^2e^2}{4m^2c}\sum_{n,m}\int d^3r' L(\xi_n, \xi_m)\{[\phi_m^*(\mathbf{r})\nabla\phi_n(\mathbf{r}) - \phi_n(\mathbf{r})\nabla\phi_m^*(\mathbf{r})]$$

$$\times [\phi_m(\mathbf{r}')\nabla'\phi_n^*(\mathbf{r}') - \phi_n^*(\mathbf{r}')\nabla'\phi_m(\mathbf{r}')] + \text{c.c.}\}\cdot\mathbf{A}(\mathbf{r}'). \tag{46.26}$$

If we carry out the calculations for finite temperature the function $L(\varepsilon_n, \varepsilon_m)$ entering (46.26) becomes

$$L(\varepsilon_n, \varepsilon_m) = \frac{1}{2}\left(1 + \frac{\xi_n\xi_m + \Delta^2}{\varepsilon_n\varepsilon_m}\right)\left(\frac{f_m - f_n}{\varepsilon_n - \varepsilon_m}\right)$$

$$+\frac{1}{2}\left(1-\frac{\xi_n\xi_m+\Delta^2}{\varepsilon_n\varepsilon_m}\right)\left(\frac{1-f_n-f_m}{\varepsilon_n+\varepsilon_m}\right) \tag{46.27a}$$

$$=\frac{1}{2}\frac{(1-2f_n)\varepsilon_n-(1-2f_m)\varepsilon_m}{\varepsilon_n^2-\varepsilon_m^2}$$

$$+\frac{1}{2}\frac{\xi_n\xi_m+\Delta^2}{\varepsilon_n\varepsilon_m}\frac{(1-2f_n)\varepsilon_m-(1-2f_m)\varepsilon_n}{\varepsilon_n^2-\varepsilon_m^2}. \tag{46.27b}$$

The gauge current $\langle\hat{\mathbf{j}}_2\rangle$ is explicitly first order in $\mathbf{A}$ and hence we only require the zeroth order contribution of u_n and v_n to $\hat{\psi}_\alpha^\dagger(\mathbf{r})\hat{\psi}_\alpha(\mathbf{r})$:

$$\langle\hat{\mathbf{j}}_2(\mathbf{r})\rangle=-\frac{e^2}{mc}\mathbf{A}(\mathbf{r})\sum_n[|u_n^{(0)}(\mathbf{r})|^2f_n+|v_n^{(0)}(\mathbf{r})|^2(1-f_n)]. \tag{46.28}$$

Note $\Sigma_n|v_n^{(0)}|^2=n$, the conduction electron number density. The remaining contribution, $\Sigma_n(|u_n^{(0)}|^2-|v_n^{(0)}|^2)f_n=\Sigma_n(\xi_n/\varepsilon_n)f_n=0$ since it is an odd function. Our expression for $\langle\hat{\mathbf{j}}_2(\mathbf{r})\rangle$ is therefore

$$\langle\hat{\mathbf{j}}_2(\mathbf{r})\rangle=-\frac{e^2n}{mc}\mathbf{A}(\mathbf{r}). \tag{46.29}$$

Collecting the above we may write the total current as

$$\langle j_i(\mathbf{r})\rangle=\int d^3r' S_{ij}(\mathbf{r},\mathbf{r}')A_j(\mathbf{r}'), \tag{46.30}$$

where the nonlocal susceptibility tensor S_{ij} is given by

$$S_{ij}(\mathbf{r},\mathbf{r}')=\frac{e^2\hbar^2}{4m^2c}\sum_{m,n}L(\xi_n,\xi_m)\{[\phi_m^*(\mathbf{r})\nabla_i\phi_n(\mathbf{r})-\phi_n(\mathbf{r})\nabla_i\phi_m^*(\mathbf{r})]$$

$$\times[\phi_m(\mathbf{r}')\nabla_j'\phi_n^*(\mathbf{r}')-\phi_n^*(\mathbf{r}')\nabla_j'\phi_m(\mathbf{r}')]+\text{c.c.}\}$$

$$-\frac{ne^2}{mc}\delta^{(3)}(\mathbf{r}-\mathbf{r}')\delta_{ij}. \tag{46.31}$$

Since the Meissner screening currents represent a thermodynamic response of the system, they must follow from performing the appropriate differentiations of the free energy. As argued in Sec. 21 (see Eq. (21.5)), we may write

$$\mathbf{j}(\mathbf{r})=\frac{\delta F}{\delta\mathbf{A}(\mathbf{r})} \tag{46.32a}$$

from which it follows that

$$S_{ij}(\mathbf{r},\mathbf{r}')=\frac{\delta^2F}{\delta A_i(\mathbf{r})\delta A_j(\mathbf{r}')}. \tag{46.32b}$$

From the equality of the cross functional derivatives we have

$$S_{ij}(\mathbf{r},\mathbf{r}')=S_{ji}(\mathbf{r}',\mathbf{r}). \tag{46.33}$$

Eq. (46.31) satisfies (46.33) explicitly. However, if we start with Eq. (46.19b) (rather than (46.19c)), in place of (46.31) we obtain the alternative form

$$S_{ij}(\mathbf{r}, \mathbf{r}') = \frac{e^2\hbar^2}{2m^2c} \sum_{m,n} L(\xi_m, \xi_n)\{[\phi_m^*(\mathbf{r})\nabla_i\phi_n(\mathbf{r}) - \phi_n(\mathbf{r})\nabla_i\phi_m^*(\mathbf{r})]\phi_m(\mathbf{r}')\nabla_j'\phi_n^*(\mathbf{r}') + \text{c.c.}\}$$

$$- \frac{ne^2}{mc}\delta^{(3)}(\mathbf{r} - \mathbf{r}')\delta_{ij}, \tag{46.34}$$

which is not explicitly symmetric as it stands. However, in doing the integration by parts that produced the form $\phi_m(\mathbf{r}')\nabla_j'\phi_n^*(\mathbf{r}')$ we could equally well have generated the form $-\phi_n^*(\mathbf{r}')\nabla_j'\phi_m(\mathbf{r}')$. A symmetric form results if we pair the first form with the second term in the square bracket and pair the second form with the first term and then relabel m and n in the latter pair. The resulting expression for the susceptibility tensor is

$$S_{ij}(\mathbf{r}, \mathbf{r}') = -\frac{e^2\hbar^2}{2m^2c} \sum_{m,n} L(\xi_m, \xi_n)\{[\phi_n(\mathbf{r})\nabla_i\phi_m^*(\mathbf{r})][\phi_m(\mathbf{r}')\nabla_j'\phi_n^*(\mathbf{r}')] + \text{c.c.}\}$$

$$- \frac{ne^2}{mc}\delta^{(3)}(\mathbf{r} - \mathbf{r}')\delta_{ij}. \tag{46.35}$$

46.4 Diamagnetism of the superconducting Fermi gas

As our first application of the above formalism we consider the case of a superconducting Fermi gas. The orbital part of the wavefunctions has the form r $_{\mathbf{k}}(\mathbf{r}) = L^{-3/2}e^{i\mathbf{k}\cdot\mathbf{r}}$ and (46.35) becomes

$$S_{ij}(\mathbf{r} - \mathbf{r}') = \frac{e^2\hbar^2}{m^2c} \int \frac{d^3k}{(2\pi)^3} \frac{d^3k'}{(2\pi)^3} L(\xi_{\mathbf{k}}, \xi_{\mathbf{k}'})k_ik_j'$$

$$\times (e^{i(\mathbf{k}' - \mathbf{k})\cdot(\mathbf{r} - \mathbf{r}')} + e^{-i(\mathbf{k}' - \mathbf{k})\cdot(\mathbf{r} - \mathbf{r}')}) - \frac{ne^2}{mc}\delta^{(3)}(\mathbf{r} - \mathbf{r}')\delta_{ij}. \tag{46.36}$$

Multiplying by $e^{-i\mathbf{q}\cdot(\mathbf{r} - \mathbf{r}')}$ and integrating over $d^3|\mathbf{r} - \mathbf{r}'|$ we obtain the Fourier transform of (46.36) as

$$S_{ij}(\mathbf{q}) = \frac{2e^2\hbar^2}{m^2c} \int \frac{d^3k d^3k'}{(2\pi)^3} L(\xi_{\mathbf{k}}, \xi_{\mathbf{k}'})k_ik_j'\delta^{(3)}(\mathbf{k}' - \mathbf{k} - \mathbf{q}) - \frac{ne^2}{mc}\delta_{ij}; \tag{46.37}$$

in obtaining (46.37) we changed the sign of $\mathbf{k}$ and $\mathbf{k}'$ in the second exponential (which is permitted since we are integrating over both variables) and used the property $\xi_{\mathbf{k}} = \xi_{-\mathbf{k}}$. Integrating over d^3k' we have immediately

$$S_{ij}(\mathbf{q}) = \frac{2e^2\hbar^2}{m^2c} \int \frac{d^3k}{(2\pi)^3} L(\xi_{\mathbf{k}}, \xi_{\mathbf{k}+\mathbf{q}})k_i(k_j + q_j) - \frac{ne^2}{mc}\delta_{ij} \tag{46.38a}$$

$$= \frac{2e^2\hbar^2}{m^2c} \int \frac{d^3k}{(2\pi)^3} L(\xi_-, \xi_+)\left(k_i - \frac{q_i}{2}\right)\left(k_j + \frac{q_j}{2}\right) - \frac{ne^2}{mc}\delta_{ij}, \tag{46.38b}$$

where we moved the origin $\mathbf{k} \to \mathbf{k} - \mathbf{q}/2$ in the second step and we defined $\xi_\pm = \xi(\mathbf{k} \pm \mathbf{q}/2)$. In our gauge the requirement $\nabla \cdot \mathbf{A} = 0$ is equivalent to $\mathbf{q} \cdot \mathbf{A} = 0$. The vector $\mathbf{A}$, and with it $\mathbf{j}$, are therefore perpendicular (transverse) to the vector $\hat{\mathbf{q}}$, where the latter is chosen as the polar axis for the integrations over d^3k. According to the discussion in Subsec. 46.1 $\hat{\mathbf{q}}$ would be the surface normal. The function $L(\xi_+, \xi_-)$ is large only for $k \sim k_\mathrm{F}$, whereas the important values of q will turn out to be less than or of the order of the reciprocal of the effective field penetration depth; i.e., we assume $q << k_\mathrm{F}$. With the understanding that $\mathbf{A}, \mathbf{j} \perp \hat{\mathbf{q}}$, and averaging over the azimuthal angle $(\overline{\sin^2 \phi} = \overline{\cos^2 \phi} = \frac{1}{2})$ we may write the relation between $\mathbf{j}$ and $\mathbf{A}$ as

$$\mathbf{j}(\mathbf{q}) = S(\mathbf{q})\mathbf{A}(\mathbf{q}), \tag{46.39a}$$

where

$$S(\mathbf{q}) = \frac{e^2\hbar^2}{(2\pi)^3 m^2 c} \int k^2 \sin^2 \theta \, L(\xi_-, \xi_+) \mathrm{d}^3k - \frac{ne^2}{mc}. \tag{46.39b}$$

The integration in (46.39b) is rather complicated and the calculations may be simplified by expanding the function

$$1 - 2f(\varepsilon) = 2k_\mathrm{B}T \sum_{n=-\infty}^{+\infty} \frac{1}{\varepsilon - \mathrm{i}\hbar\omega_n} \tag{46.40}$$

as we did earlier in Eq. (38.9); here $\hbar\omega_n \equiv 2\pi k_\mathrm{B}T(n + 1/2)$. Inserting (46.40) into Eq. (46.27b), and noting that sums involving an odd power of ω_n vanish, we may write $L(\xi_+, \xi_-)$ as

$$L(\xi_+, \xi_-) = -k_\mathrm{B}T \sum_n \frac{(\xi_+ + \mathrm{i}\hbar\omega_n)(\xi_- + \mathrm{i}\hbar\omega_n) + \Delta^2}{(\varepsilon_+^2 + \hbar^2\omega_n^2)(\varepsilon_-^2 + \hbar^2\omega_n^2)}. \tag{46.41}$$

The expression for $S(\mathbf{q})$ obtained by substituting (46.41) into (46.39b) is divergent, depending on the order the integrations and summations are performed. To avoid this problem we subtract the function $S(\mathbf{q}, \Delta = 0)$ since the latter must vanish, corresponding to the normal state. Our final expression then becomes

$$S(\mathbf{q}) = -\frac{e^2\hbar^2 k_\mathrm{B}T}{m^2 c} \sum_{n=-\infty}^{n=\infty} \int k^2 \sin^2 \theta \left[\frac{(\mathrm{i}\hbar\omega_n + \xi_+)(\mathrm{i}\hbar\omega_n + \xi_-) + \Delta^2}{(\hbar^2\omega_n^2 + \varepsilon_+^2)(\hbar^2\omega_n^2 + \varepsilon_-^2)} \right.$$
$$\left. - \frac{1}{(\mathrm{i}\omega_n - \xi_+)(\mathrm{i}\omega_n - \xi_-)} \right] \frac{\mathrm{d}^3k}{(2\pi)^3}. \tag{46.42}$$

Now the relevant values of q and $k - k_\mathrm{F}$ are such that $\xi_\pm$ is less than or of the order of Δ, otherwise the two integrands tend to cancel each other. In this regime we can approximate $\xi_\pm = \xi \pm \frac{1}{2}\hbar v_\mathrm{F} q \cos\theta = \hbar v_\mathrm{F}(k - k_\mathrm{F}) \pm \frac{1}{2}\hbar v_\mathrm{F} q \cos\theta$; we may also approximate k^2 by k_F^2 in the integrand and write $\mathrm{d}^3k/(2\pi)^3 = [1/(2\pi\hbar)^2] m k_\mathrm{F} \mathrm{d}\xi \mathrm{d}(\cos\theta)$ (the resulting factor $k_\mathrm{F}^3 = 3\pi^2 n$, where n is again the number density). The poles of the second term in the integrand are at $\xi = \mp \frac{1}{2}\hbar v_\mathrm{F} q \cos\theta + \mathrm{i}\hbar\omega_n$; i.e., they are either both in the upper or both in the lower half of the complex ξ plane. By closing the contour in the other half plane we see that the integral vanishes. We now examine the ξ-dependent part of the first integrand which, on factoring the denominator, becomes

$$\frac{(\mathrm{i}\hbar\omega_n + \xi_+)(\mathrm{i}\hbar\omega_n + \xi_-) + \Delta^2}{[\xi_+ + \mathrm{i}(\hbar^2\omega_n^2 + \Delta^2)^{1/2}][\xi_+ - \mathrm{i}(\hbar^2\omega_n^2 + \Delta^2)^{1/2}][\xi_- + \mathrm{i}(\hbar^2\omega_n^2 + \Delta^2)^{1/2}][\xi_- - \mathrm{i}(\hbar^2\omega_n^2 + \Delta^2)^{1/2}]}.$$

This expression has two poles in the upper and two poles in the lower half plane; by closing the contour in the upper half plane we need consider only the two poles occurring when $\xi_\pm = i(\hbar^2\omega_n^2 + \Delta^2)^{1/2}$. Noting $\xi_+ = \xi_- + \hbar v_F q \cos\theta$, the residue of the two poles is

$$\left\{\frac{[i\hbar\omega_n + i(\hbar^2\omega_n^2+\Delta^2)^{1/2}][i\hbar\omega_n - \hbar v_F q x + i(\hbar^2\omega_n^2+\Delta^2)^{1/2}] + \Delta^2}{2i(\hbar^2\omega_n^2+\Delta^2)^{1/2}[-\hbar v_F q x + 2i(\hbar^2\omega_n^2+\Delta^2)^{1/2}](-\hbar v_F q x)}\right.$$

$$\left. + \frac{[i\hbar\omega_n + \hbar v_F q x + i(\hbar^2\omega_n^2+\Delta^2)^{1/2}][i\hbar\omega_n + i(\hbar^2\omega_n^2+\Delta^2)^{1/2}] + \Delta^2}{[\hbar v_F q x + 2i(\hbar^2\omega_n^2+\Delta^2)^{1/2}](\hbar v_F q x)[2i(\hbar^2\omega_n^2+\Delta^2)^{1/2}]}\right\}$$

where we have written $x \equiv \cos\theta$. Combining the two terms yields

$$\frac{2\Delta^2}{i(\hbar^2\omega_n^2+\Delta^2)^{1/2}[\hbar^2 v_F^2 q^2 x^2 + 4(\hbar^2\omega_n^2+\Delta^2)]}.$$

Multiplying the residue by $2\pi i$, and collecting all the terms our final expression for $S(\mathbf{q})$ is:

$$S(\mathbf{q}) = -\frac{3\pi k_B T n e^2}{4mc}\sum_{n=-\infty}^{n=+\infty}\int_{-1}^{+1}\frac{\Delta^2(1-x^2)\mathrm{d}x}{[\hbar^2\omega_n^2 + \Delta^2 + (\tfrac{1}{2}\hbar v_F q x)^2](\hbar^2\omega_n^2+\Delta^2)^{1/2}}. \tag{46.43}$$

In the limit $\hbar q << \Delta/v_F$ we can neglect the term containing x in the denominator, and then (46.43) can be integrated to yield the London form:

$$S(\mathbf{q}) = -\frac{e^2 n_s}{mc} \tag{46.44}$$

where we have written[2]

$$n_s = \frac{3\pi n}{8}\sum_{n=-\infty}^{n=+\infty}\frac{k_B T\Delta^2(T)}{[\hbar^2\omega_n^2 + \Delta^2(T)]^{3/2}}. \tag{46.45}$$

Since $\xi_0^{-1} \equiv \pi\Delta(T=0)/\hbar v_F$, this regime corresponds to $q\xi_0 << 1$. In the opposite limit where $q\xi_0 >> 1$ the important range of the integration is where $x \simeq 0$ and we can neglect the x^2 term in the numerator; we then extend the limits of x to $\pm\infty$ and again use Cauchy's theorem; there are two poles on the imaginary x axis at $x = (2i/v_F\hbar q)(\hbar^2\omega_n^2+\Delta^2)^{1/2}$ and thus

$$\int_{-\infty}^{+\infty}\frac{\mathrm{d}x}{\hbar^2\omega_n^2 + \Delta^2 + \left(\frac{\hbar}{2}v_F q x\right)^2}$$

$$= \frac{1}{\left(\frac{\hbar}{2}v_F q\right)^2}\oint\frac{\mathrm{d}x}{\left[x + 2i\frac{(\hbar^2\omega_n^2+\Delta^2)^{1/2}}{\hbar v_F q}\right]\left[x - 2i\frac{(\hbar^2\omega_n^2+\Delta^2)^{1/2}}{\hbar v_F q}\right]}$$

$$= \frac{2\pi}{\hbar v_F q(\hbar^2\omega_n^2+\Delta^2)^{1/2}}$$

or

2. The sum, which we will not evaluate, is proportional to $\Delta(T)^2/(k_B T)^2$ as $T \to T_c$; thus the London depth $\lambda_L \equiv (mc^2/4\pi n e^2)^{1/2} \propto 1/\Delta \propto (1 - T/T_c)^{-1/2}$. As $T \to 0$, $n_s \to n$, apart from a constant of order unity.

$$S(\mathbf{q}) = \frac{3\pi^2 k_B T n e^2}{2mc\hbar v_F q} \sum_{n=-\infty}^{n=+\infty} \frac{\Delta^2}{\hbar^2\omega_n^2 + \Delta^2}. \tag{46.46}$$

$$S(\mathbf{q}) = -\frac{ne^2}{mc}\frac{3\pi^2}{4\hbar q v_F}\Delta \tanh\left(\frac{\Delta}{2k_B T}\right). \tag{46.47}$$

For later use we write (46.47) in the form

$$S(\mathbf{q}) = -\frac{c\beta}{4\pi q}, \tag{46.48a}$$

where

$$\beta \equiv \frac{4\pi n e^2}{mc^2}\frac{3\pi^2}{4\hbar v_F}\Delta \tanh\left(\frac{\Delta}{2k_B T}\right). \tag{46.48b}$$

As $T \to 0$, $\Delta \to \Delta(T=0) = 1.7k_B T_c$ and $\tanh(\Delta/2k_B T) \to 1$. As $T \to T_c$, $\Delta \to 3.06k_B T_c(1 - T/T_c)^{1/2}$ and $\tanh(\Delta/2k_B T_c) \to \Delta/2k_B T_c$. From footnote 2 and recalling $\xi_0^{-1} \equiv \pi\Delta(T=0)/\hbar v_F$, we see that $qS \propto 1/\lambda_L^2 \xi_0$ (in both limits).

Summarizing the above we see that $S(\mathbf{q})$ is approximately constant for $q \lesseqqgtr \xi_0^{-1}$, and falls off as q^{-1} for $q >> \xi_0^{-1}$; the corresponding behavior in position space is $S(R) \propto 1/R^2$, for $R \lesseqqgtr \xi_0$, and $S(R) \sim e^{-R/\xi_0}$, for $R >> \xi_0$, where $\mathbf{R} = \mathbf{r} - \mathbf{r}'$. Therefore the correlation, or the range of nonlocality between $\mathbf{A}(\mathbf{r})$ and $\mathbf{j}(\mathbf{r}')$, extends over distance $R \cong \xi_0$. Furthermore, the general behavior of $S(R)$ is consistent with Pippard's phenomenological form introduced in Part I.

46.5 Magnetic field behavior near a vacuum–superconductor interface

The results of the previous subsection apply, strictly speaking, only to an infinite medium (the Fourier transform was extended over all space). Near a boundary $S(\mathbf{q})$ will depend on the nature of the scattering of the electrons from the surface. There are two limiting cases of this scattering; specular and diffuse. In the specular case the electrons reflect in mirror-like fashion from the surface; i.e., the component of momentum in the plane of the surface is conserved while the component perpendicular to the surface changes sign. In the diffuse limit there is some probability of the electron leaving the surface at all angles (although the magnitude of $\mathbf{k}$ is conserved for an isotropic system). Here we discuss only the specular case since it may be treated simply by imposing reflection symmetry about a plane on an otherwise translationally invariant (homogeneous) problem.

We consider the case where $\mathbf{H}$ is parallel to the surface and in the z direction. The vector potential must be consistent with the Maxwell equation $\nabla \times \mathbf{H} = (4\pi/c)\mathbf{j}$; substituting $\mathbf{H} = \nabla \times \mathbf{A}$ yields, for our gauge where $\nabla \cdot \mathbf{A} = 0$,

$$\nabla^2 \mathbf{A} = -\frac{4\pi}{c}\mathbf{j}. \tag{46.49}$$

For our gauge $\mathbf{A} \parallel \mathbf{j}$ and both depend only on x; i.e., $\mathbf{A} = A(x)\hat{\mathbf{y}}$, and $\mathbf{j} = j(x)\hat{\mathbf{y}}$. In order to convert our half-space problem into a symmetric homogeneous problem we impose the condition $\mathbf{A}(x) = \mathbf{A}(-x)$; if $\mathbf{A}'$ were continuous the field $H_z = \partial A_y/\partial x = \partial A/\partial x$ would be zero at $x = 0$ and hence $A'(x)$ must be *discontinuous* such that

$$\lim_{\delta \to 0} A'(0 + \delta) = H_0 = -A'(0 - \delta). \tag{46.50}$$

Thus our model field will have a *cusp* at $x = 0$. To apply (46.39) and (46.43) we require the Fourier transform of (46.49), subject to (46.50); thus we write

$$\int_{-\infty}^{+\infty} \mathrm{d}x A''(x)\mathrm{e}^{-\mathrm{i}qx} = \int_{-\infty}^{0-\delta} \frac{\mathrm{d}}{\mathrm{d}x}[A'(x)\mathrm{e}^{-\mathrm{i}qx}]\mathrm{d}x + \int_{0+\delta}^{+\infty} \frac{\mathrm{d}}{\mathrm{d}x}[A'(x)\mathrm{e}^{-\mathrm{i}qx}]\mathrm{d}x$$
$$-(-\mathrm{i}q)\int_{-\infty}^{+\infty} A'(x)\mathrm{e}^{-\mathrm{i}qx}\mathrm{d}x = -\frac{4\pi}{c}j(q).$$

The first two terms, being integrals of total differentials, yield

$$-[A'(x)\mathrm{e}^{-\mathrm{i}qx}]_{-\infty} + A'(-\delta) - A'(\delta) + [A'(x)\mathrm{e}^{\mathrm{i}qx}]_{+\infty}.$$

Since $A'(\pm\infty) = 0$ we have, from (46.50), that the first two terms yield $-2H_0$. We integrate the third term by parts which, since $A(x)$ is continuous at $x = 0$, yields $-q^2A(q)$; thus the transform of (46.49) reads:

$$2H_0 + q^2A(q) = \frac{4\pi}{c}j(q).$$

Since $j(q) = S(|\mathbf{q}|)A(q)$ we have

$$A(q) = -\frac{2H_0}{q^2 - \dfrac{4\pi}{c}S(|\mathbf{q}|)}. \tag{46.51}$$

We will define the penetration depth λ by

$$\lambda = \frac{1}{H_0}\int_0^{\infty} H(x)\mathrm{d}x = \frac{1}{H_0}\int_0^{\infty}\frac{\mathrm{d}A}{\mathrm{d}x} = -\frac{A(x=0)}{H_0}. \tag{46.52}$$

Fourier transforming (46.51) for $x = 0$, (46.52) becomes

$$\lambda = \frac{2}{2\pi}\int_{-\infty}^{+\infty}\frac{\mathrm{d}q}{q^2 - \dfrac{4\pi}{c}S(|\mathbf{q}|)}. \tag{46.53}$$

For the London case where $S = -e^2n_s/mc = -c/4\pi\lambda_L^2$, (46.53) becomes

$$\lambda = \frac{2}{2\pi}\int_{-\infty}^{+\infty}\frac{\mathrm{d}_q}{q^2 + \lambda_L^{-2}} = \lambda_L \tag{46.54}$$

In the Pippard limit, where $S(|\mathbf{q}|)$ is given by (46.48), we have

$$\lambda_P = \frac{2}{2\pi}\int_{-\infty}^{+\infty}\frac{|q|\mathrm{d}k}{|q|q^2 + \beta} = \frac{2}{\pi}\int_0^{\infty}\frac{q\mathrm{d}q}{q^3 + \beta}$$

$$= \frac{2}{\pi}\left[\frac{1}{6\beta^{1/3}}\ln\frac{\beta+q^3}{(\beta^{1/3}+q)^3} + \frac{1}{\sqrt{3}\beta^{1/3}}\tan^{-1}\left(\frac{2q-\beta^{1/3}}{3^{1/2}\beta^{1/3}}\right)\right]_0^\infty$$

$$= \frac{2}{\pi 3^{1/2}\beta^{1/3}}\left[\frac{\pi}{2} + \tan 1^{-1}\left(\frac{1}{3^{1/2}}\right)\right]. \tag{46.55}$$

Eq. (46.55) is, apart from a constant of order unity, identical to Eq. (3.12) which was obtained in Part I from Pippard's equation; i.e., $\lambda_P \cong (\lambda_L^2\xi_0)^{1/3}$.

46.6 Relation between the normal-state conductivity and the superconducting diamagnetic response

Although Eqs. (46.30) and (46.35) are a formal solution to the low field diamagnetic response of a superconductor, the physical interpretation of the tensor S_{ij} given in Eq. (46.35) is not immediately obvious. DeGennes (1966) showed that S_{ij} can be related to the real part of the nonlocal, frequency-dependent, electrical conductivity tensor, $\sigma_{ij}(\mathbf{r},\mathbf{r}',\Omega)$, in the normal phase which is defined by the expression

$$\langle \hat{j}_i(\mathbf{r},\Omega)\rangle = \int d^3r'\sigma_{ij}(\mathbf{r},\mathbf{r}',\Omega)E_j(\mathbf{r}',\Omega). \tag{46.56}$$

The electric field is given in the time domain by $\mathbf{E} = -\nabla\phi - (1/c)(\partial\mathbf{A}/\partial t)$, where ϕ is the electrostatic potential; for the frequencies of interest here the electrostatic potential is negligible since it would have to be produced by a nonequilibrium charge density and the latter is screened out for frequencies below the electron plasma frequency. Hence the electric field enters the Hamiltonian only via the vector potential. We may then write the Hamiltonian in the form

$$\mathscr{H} = \mathscr{H}^{(0)'} + \mathscr{H}^{(1)}, \tag{46.57}$$

where we write

$$\mathscr{H}^{(0)'}(\mathbf{r}) = \frac{p^2}{2m} + U(\mathbf{r}) - \mu \tag{46.58}$$

and

$$\mathscr{H}^{(1)}(\mathbf{r},t) = \frac{ie\hbar}{2mc}[\nabla\cdot\mathbf{A}(\mathbf{r},t) + \mathbf{A}(\mathbf{r},t)\cdot\nabla]. \tag{46.59}$$

We write the real quantity $\mathbf{A}(\mathbf{r},t)$ in the form

$$\mathbf{A}(\mathbf{r},t) = \frac{1}{2}[\mathbf{A}(\mathbf{r})e^{-i\Omega t} + \mathbf{A}^*(\mathbf{r})e^{i\Omega t}] \tag{46.60}$$

$$= \mathrm{Re}\mathbf{A}(\mathbf{r})e^{-i\Omega t}, \tag{46.61}$$

where $\mathbf{A}(\mathbf{r})$ may be complex. The associated electric field is

$$\mathbf{E}(\mathbf{r}, t) = \mathrm{Re}\mathbf{E}(\mathbf{r})\mathrm{e}^{-\mathrm{i}\Omega t}, \tag{46.62}$$

where

$$\mathbf{E}(\mathbf{r}) = \frac{\mathrm{i}\Omega}{c}\mathbf{A}(\mathbf{r}). \tag{46.63}$$

It will turn out that we only need the real part of the conductivity, which governs the power dissipation, W, through the usual Joule form

$$\begin{aligned} W &= \frac{1}{2}\mathrm{Re}\int E_i^*(\mathbf{r})j_i(\mathbf{r})\mathrm{d}^3r \\ &= \frac{1}{2}\mathrm{Re}\int E_i^*(\mathbf{r})\sigma_{ij}(\mathbf{r}, \mathbf{r}')E_j(\mathbf{r}')\mathrm{d}^3r\mathrm{d}^3r' \\ &= \frac{1}{4}\int \mathrm{d}^3r\mathrm{d}^3r'[E_i^*(\mathbf{r})E_j(\mathbf{r}') + \mathrm{c.c.}]\mathrm{Re}\sigma_{ij}(\mathbf{r}, \mathbf{r}'), \end{aligned} \tag{46.64}$$

where we exploited the symmetry $\sigma_{ij}(\mathbf{r}, \mathbf{r}') = \sigma_{ji}(\mathbf{r}', \mathbf{r})$ in the last step.

We may calculate W quantum mechanically by examining the total rate of exchange of energy among all levels in the system which we may write as

$$W = \sum_{nm} r_{nm}(\xi_m - \xi_n), \tag{46.65}$$

where r_{nm} is the total transition rate between states n and m. We recall the usual expression from first order, time-dependent perturbation theory for the probability per unit time for a transition between an initial (occupied) state n and a final (unoccupied) state m under the influence of a harmonic perturbation $\mathscr{H}^{(1)}(\mathbf{r}, t)$:

$$w_{nm}^{(\mp)} = \frac{2\pi}{\hbar}|\mathscr{H}_{nm}^{(1)}|^2\delta(\xi_m - \xi_n \mp \hbar\Omega), \tag{46.66}$$

where the matrix elements $\mathscr{H}_{nm}^{(1)}$ of the time-dependent amplitudes of $\mathscr{H}^{(1)}$ will be given below and the + and − signs correspond to the emission and absorption of a photon, respectively. Taking into account the Fermi occupation factors f_n and $1 - f_m$ for the initial and final states we may write

$$r_{nm} = f_n(1 - f_m)[w_{nm}^{(-)} + w_{nm}^{(+)}] \tag{46.67}$$

from which the power dissipated follows immediately as

$$\begin{aligned} W &= \frac{2\pi}{\hbar}\sum_{n,m}(\xi_m - \xi_n)f_n(1 - f_m)|\mathscr{H}_{nm}^{(1)}|^2[\delta(\xi_m - \xi_n - \hbar\Omega) + \delta(\xi_m - \xi_n + \hbar\Omega)] \\ &= \frac{\pi}{\hbar}\sum_{n,m}(f_n - f_m)(\xi_m - \xi_n)|\mathscr{H}_{nm}^{(1)}|^2[\delta(\xi_m - \xi_n - \hbar\Omega) + \delta(\xi_m - \xi_n + \hbar\Omega)] \\ &\cong \frac{2\pi}{\hbar}\sum_{n,m}(f_n - f_m)(\xi_m - \xi_n)|\mathscr{H}_{nm}^{(1)}|^2\delta(\xi_m - \xi_n - \hbar\Omega), \end{aligned} \tag{46.68}$$

where we combined the expression with the form with the dummy labels m and interchanged to obtain the second line, and assumed symmetry of the matrix element with respect to positive and negative frequencies in the third line. The appropriate matrix elements of $\mathscr{H}^{(1)}(\mathbf{r}, t)$ involving the position-dependent amplitude of the first term in (46.60) are

$$\mathscr{H}_{nm}[\mathbf{A}(\mathbf{r})] = \frac{\mathrm{i}e\hbar}{2mc}\int \mathrm{d}^3r\phi_m^*(\mathbf{r})[\nabla\cdot\mathbf{A}(\mathbf{r}) + \mathbf{A}(\mathbf{r})\cdot\nabla]\phi_n(\mathbf{r})$$

$$= \frac{\mathrm{i}e\hbar}{mc}\int \mathrm{d}^3r(\phi_m^*(\mathbf{r})\nabla\phi_n(\mathbf{r}))\cdot\mathbf{A}(\mathbf{r}), \tag{46.69}$$

with a similar form for $\mathbf{A}^*(\mathbf{r})$; here we again assumed a London gauge in the second step. Combining the above and comparing the result with (46.64) we have

$$\mathrm{Re}\sigma_{ij}(\mathbf{r}, \mathbf{r}', \Omega) = \left(\frac{e\hbar}{mc}\right)^2 \frac{2\pi}{\hbar} c^2\hbar^2 \sum_{mn} \frac{f_n - f_m}{\hbar\Omega}(\phi_n^*(\mathbf{r})\nabla\phi_m(\mathbf{r}))$$

$$\times (\phi_m^*(\mathbf{r}')\nabla'\phi_n(\mathbf{r}'))\delta(\xi_n - \xi_m - \hbar\Omega). \tag{46.70}$$

Following Mattis and Bardeen (1958) we now introduce a quantity

$$\rho_\xi(\mathbf{r}, \mathbf{r}') = L^3\langle\phi_n^*(\mathbf{r})\phi_n(\mathbf{r}')\rangle_\xi, \tag{46.71}$$

where the angle brackets denote an average over all quantum numbers n having an energy ξ. For the case where the $\mathbf{k}$ vectors are good quantum numbers we write

$$\sum_n = \frac{L^3}{(2\pi)^3}\int \mathrm{d}^3k = L^3\int \mathscr{N}(\xi)\mathrm{d}\xi\cdot\frac{1}{4\pi}\int \mathrm{d}\Omega_{\hat{\mathbf{k}}} \tag{46.72}$$

and we would write

$$\rho_\xi(\mathbf{r}, \mathbf{r}') = \frac{L^3}{4\pi}\int \phi_{\mathbf{k}}^*(\mathbf{r})\phi_{\mathbf{k}}(\mathbf{r}')\mathrm{d}\Omega_{\hat{\mathbf{k}}}. \tag{46.73}$$

For the case of free electrons, (46.71) is easily evaluated as

$$\rho_{\xi_k}(\mathbf{r} - \mathbf{r}') = \frac{1}{2}\int \mathrm{e}^{\mathrm{i}kR\cos\theta}\sin\theta\mathrm{d}\theta = \frac{\sin kR}{kR}, \tag{46.74}$$

where we defined $R = |\mathbf{r} - \mathbf{r}'|$ and $k = |\mathbf{k}|$. In the presence of elastic scattering characterized by a mean free path ℓ the proper evaluation of (46.71), even for the case of 'free' electrons, is much more complex. The most rigorous derivation is due to Abrikosov and Gorkov (1961) who find[3]

$$\rho_{\xi_k}(r) = \frac{\sin kR}{kR}\mathrm{e}^{-R/2\ell}; \tag{46.75}$$

for a metal the significant values of k lie close to k_F. The above form can be rationalized by assuming the propagation constant k governing the amplitude–phase relationship for an electron

3. This expression is valid only in the limit of weak scattering where $k_F\ell >> 1$; very dirty alloys in which $k_F\ell \sim 1$ require significant modifications of the theory and lead to so-called 'localization' effects. The expression (46.75) arises from a configuration average and hence is not a solution to the Schrödinger equation for some model potential; it, furthermore, assumes isotropic scattering and is not gauge-invariant. Finally it assumes $\rho_\xi(R)$ and $\rho_{\xi'}(R)$ are uncorrelated.

at two points in space acquires an imaginary part such that $ik \rightarrow ik - 1/2\ell$, corresponding to a decay of the correlation. In a so-called 'cloudy crystal ball' model in which one introduces a phenomenological square well potential of the form $V = V_1 - iV_2$, the mean free path ℓ would be given by $\ell = \hbar^2 k/2mV_2$; this model is also referred to as the optical model.

In terms of Eq. (46.71) we can rewrite (46.70) as

$$\mathrm{Re}\sigma_{ij}(\mathbf{r},\mathbf{r}',\Omega) = \left(\frac{e\hbar}{m}\right)^2 2\pi\hbar \int \mathrm{d}\xi\mathrm{d}\xi' \mathcal{N}(\xi)\mathcal{N}(\xi')\frac{f_\xi - f_{\xi'}}{\hbar\Omega} \times [\nabla'_i\rho_\xi(\mathbf{r}',\mathbf{r})][\nabla_j\rho_{\xi'}(\mathbf{r},\mathbf{r}')]\delta(\xi - \xi' - \hbar\Omega). \tag{46.76}$$

We may simplify (46.76) by noting that when $\hbar\Omega << k_B T$

$$\frac{f_\xi - f_{\xi'}}{\hbar\Omega} = \frac{\partial f}{\partial \xi} = -\delta(\xi), \tag{46.77}$$

where the energy variable may be taken as either ξ or ξ'. The explicit energy dependence of (46.76) may then be written as $\delta(\xi)\delta(\xi - \xi' - \hbar\Omega)$. Eq. (46.76) then becomes

$$\mathrm{Re}\sigma_{ij}(\mathbf{r},\mathbf{r}',\Omega) = \left(\frac{e\hbar}{m}\right)^2 2\pi\hbar\mathcal{N}^2 \int \mathrm{d}\xi\mathrm{d}\xi' M^2_{ij}(\mathbf{r},\mathbf{r}',\xi,\xi')\delta(\xi)\delta(\xi - \xi' - \hbar\Omega), \tag{46.78}$$

where we defined the (energy) shell-averaged matrix element squared as

$$M^2_{ij}(\mathbf{r},\mathbf{r}';\xi,\xi') = \nabla'_i\rho_\xi(\mathbf{r}',\mathbf{r})\nabla_j\rho_{\xi'}(\mathbf{r},\mathbf{r}'). \tag{46.79}$$

Let us now make the connection between the real part of the conductivity tensor, σ_{ij}, and the diamagnetic susceptibility, S_{ij}. We note in passing that in the limit $\Delta \rightarrow 0$ we have

$$L(\xi,\xi') \rightarrow (f_{\xi'} - f_\xi)/(\xi - \xi') \tag{46.80}$$

which we may associate with a corresponding term in (46.70). We rewrite Eq. (46.31) for the diamagnetic susceptibility as

$$\begin{aligned} S_{ij}(\mathbf{r},\mathbf{r}') &= \left(\frac{e\hbar}{m}\right)^2 \frac{\mathcal{N}^2}{c} \int \mathrm{d}\xi\mathrm{d}\xi' L(\xi,\xi') M^2_{ij}(\mathbf{r},\mathbf{r}';\xi,\xi') - \frac{ne^2}{mc}\delta^{(3)}(\mathbf{r}-\mathbf{r}')\delta_{ij} \\ &= \left(\frac{e\hbar}{m}\right)^2 \frac{\mathcal{N}^2}{c} \int \mathrm{d}\xi\mathrm{d}\xi' L(\xi,\xi') \\ &\quad \times Z\,\mathrm{d}\xi_1 d\xi_2 M^2_{ij}(\mathbf{r},\mathbf{r}';\xi_1,\xi_2)\delta(\xi-\xi_1)\delta(\xi-\xi'-\xi_1+\xi_2) \\ &\quad - \frac{ne^2}{mc}\delta^{(3)}(\mathbf{r}-\mathbf{r}')\delta_{ij} \\ &= \frac{1}{2\pi\hbar c}\int \mathrm{d}\xi\mathrm{d}\xi' L(\xi,\xi')\sigma_{ij}\left(\mathbf{r},\mathbf{r}',\frac{\xi-\xi'}{\hbar}\right) - \frac{ne^2}{mc}\delta^{(3)}(\mathbf{r}-\mathbf{r}')\delta_{ij}, \end{aligned} \tag{46.81}$$

where we utilized the fact that σ is relatively insensitive to the zero of energy in the form (46.78). Eq. (46.81) formally establishes the connection between σ_{ij} and S_{ij}.

46.7 Calculations of the diamagnetic response using Chambers' method

We now obtain the expression for the real part of the electrical conductivity for the case of free electrons in the presence of scattering characterized by a mean free path, ℓ. Combining Eqs. (46.75), (46.76), and (46.77) and carrying out the integrations over ξ and ξ' we obtain

$$\mathrm{Re}\,\sigma_{ij}(\mathbf{r},\mathbf{r}',\Omega) = 2\pi\hbar\left(\frac{e\hbar}{m}\right)^2 \mathcal{N}^2 \frac{R_i R_j}{R^2}\frac{\partial\rho_\xi}{\partial R}\frac{\partial\rho_{\xi'}}{\partial R} \tag{46.82}$$

where $\xi = \xi' + \hbar\Omega$ and

$$\frac{\partial\rho_\xi}{\partial R}\frac{\partial\rho_{\xi'}}{\partial R} = \frac{\mathrm{e}^{-R/\ell}}{kk'R^4}\left[kR\cos kR - \left(1+\frac{R}{2\ell}\right)\sin kR\right] \times \left[k'R\cos k'R - \left(1+\frac{R}{2\ell}\right)\sin k'R\right]. \tag{46.83}$$

Only the first term in the square brackets is significant for large k and we may rewrite the product of these two terms as

$$\begin{aligned}\frac{\mathrm{e}^{-R/\ell}}{R^2}\cos kR\cos k'R &= \frac{\mathrm{e}^{-R/\ell}}{2R^2}\{\cos[(k-k')R] + \cos[(k+k')R]\}\\ &\cong \frac{1}{2R^2}\mathrm{e}^{-R/\ell}\cos[(k-k')R]\\ &= \frac{1}{2R^2}\mathrm{e}^{-R/\ell}\cos\left(\frac{\xi-\xi'}{\hbar v_\mathrm{F}}R\right),\end{aligned} \tag{46.84}$$

where we neglected a rapidly oscillating term and we used $\xi = \hbar v_\mathrm{F}(k - k_\mathrm{F})$ in the second line. Our expression for the conductivity is therefore

$$\begin{aligned}\mathrm{Re}\,\sigma_{ij}(\mathbf{r},\mathbf{r}',\Omega) &= \pi\hbar\left(\frac{e\hbar}{m}\right)^2 \mathcal{N}^2 \frac{R_i R_j}{R^4}\cos\frac{\Omega R}{v_\mathrm{F}}\mathrm{e}^{-R/\ell}\\ &= \frac{e^2\mathcal{N}v_\mathrm{F}}{2\pi}\frac{R_i R_j}{R^4}\cos\frac{\Omega R}{v_\mathrm{F}}\mathrm{e}^{-R/\ell},\end{aligned} \tag{46.85}$$

where we used $\mathcal{N} = k_\mathrm{F}^2/2\pi^2\hbar v_\mathrm{F}$ and $v_\mathrm{F} = \hbar k_\mathrm{F}/m$ in the second step. Had we used time-dependent perturbation theory (see Mattis and Bardeen (1958) and the discussion in Sec. 50) rather than our energy dissipation argument, to calculate σ_{ij} we would have obtained in place of (46.85)

$$\sigma_{ij}(\mathbf{r},\mathbf{r}',\Omega) = \frac{e^2\mathcal{N}}{2\pi}v_\mathrm{F}\frac{R_i R_j}{R^4}\mathrm{e}^{-\mathrm{i}\Omega R/v_\mathrm{F}}\mathrm{e}^{-R/\ell}. \tag{46.86}$$

Eq. (46.86) is Chambers' semiclassical expression for the electrical conductivity of a nearly free electron gas. It may also be obtained from a Boltzmann equation argument (see Ziman (1964)). Strictly speaking the Chambers model applies only in the bulk of an isotropic medium. It can be used near a surface provided: (i) the scattering of electrons from the surface is diffuse; (ii) we set

$\sigma_{ij}(r) = 0$ if either **r** or **r**′ lies outside the surface; and (iii) **E** has no component perpendicular to the surface (which is automatically satisfied in the London gauge).

Combining (46.85) and (46.81) we have

$$S_{ij}(R) = \frac{3ne^2}{4\pi mc\xi_0}\frac{R_iR_j}{R^4}e^{-R/\ell}J(R) - \frac{ne^2}{mc}\delta(R)\delta_{ij}, \tag{46.87}$$

where we defined the integral

$$J(R,T) = \frac{1}{\pi^2\Delta(0)}\int_{-\infty}^{+\infty}\int_{-\infty}^{+\infty} d\xi d\xi' L(\xi,\xi')\cos\left(\frac{\xi-\xi'}{\hbar v_F}R\right), \tag{46.88}$$

introduced the coherence length $\xi_0 = \hbar v_F/\pi\Delta(0)$, and wrote the density of states per spin in its free electron form as $\mathcal{N} = 3n/2\pi^2 m v_F^2$ in which n is the number density of electrons. We now evaluate $J(R, T = 0)$ (for a more general discussion see Appendix C of BCS (1957)). An examination of $J(R)$ shows that it has a range of order $\hbar v_F/\Delta_0 \sim \xi_0$. Following deGennes (1966) we introduce the variables $\xi = \Delta\sinh\theta$ and $\xi' = \Delta\sinh\theta'$ and then set $\alpha = (\theta - \theta')/2$ and $\beta = (\theta + \theta')/2$ in the resulting expression to obtain

$$J(R) = \frac{1}{\pi^2}\int d\alpha d\beta \frac{\sinh^2\alpha}{\cosh\alpha\cosh\beta}\cos\left[\frac{2\Delta(0)R}{\hbar v_F}\sinh\alpha\cosh\beta\right];$$

setting $u = \sinh\alpha$

$$\begin{aligned} J(R) &= \frac{1}{\pi^2}\int_{-\infty}^{+\infty}\frac{d\beta}{\cosh\beta}\int_{-\infty}^{+\infty}du\left(1 - \frac{1}{1+u^2}\right)\cos\left[\frac{2\Delta(0)R}{\hbar v_F}u\cosh\beta\right] \\ &= \frac{1}{\pi}\frac{\hbar v_F}{\Delta(0)}\delta(R) - \frac{1}{\pi}\int_{-\infty}^{+\infty}\frac{d\beta}{\cosh\beta}e^{-2\Delta(0)R\cosh\beta/\hbar v_F}. \end{aligned} \tag{46.89}$$

The first term in (46.89) cancels the term involving $\delta(R)$ in (46.87); the final result for S_{ij} is then

$$S_{ij}(R) = -\frac{3ne^2}{4\pi mc\xi_0}\frac{R_iR_j}{R^4}e^{-R/\ell}I(R), \tag{46.90}$$

where

$$I(R) = \frac{2}{\pi}\int_{-\infty}^{+\infty}\frac{d\beta}{\cosh\beta}e^{-(2R/\pi\xi_0)\cosh\beta}. \tag{46.91}$$

$I(R)$ has the properties

$$I(0) = 1 \tag{46.92a}$$

and

$$\int_0^\infty I(R)dR = \xi_0. \tag{46.92b}$$

Eqs. (46.90) and (46.91) completely define the low field magnetic response of a superconductor within the Chambers model for the conductivity at $T = 0$. The Pippard model discussed in Sec. 3 is equivalent to replacing $I(R)$ by e^{-R/ξ_0}. Note the Pippard form satisfies Eqs. (46.92). Further analysis (deGennes 1966) shows that the various limiting forms of the Pippard equation given earlier in Sec. 3 also follow from (46.91). By using the finite temperature form of $L(\xi, \xi')$ we may calculate (numerically) the response at all temperatures (BCS 1957). Of particular interest is that the kernel $S_{ij}(R)$ has a range of order ξ_0 (in a pure material) at all temperatures; i.e., the spatial form of S is nearly independent of temperature; however, the current becomes weaker as $T \to T_c$. Recalling London's equation in the form

$$\mathbf{j} = -\frac{n_s e^2}{mc}\mathbf{A}, \tag{3.5}$$

it is natural to define an effective density of superconducting electrons n_s through the equation

$$\int S(\mathbf{r}, \mathbf{r}', T)\mathrm{d}^3 r' = -\frac{n_s(T)e^2}{mc}. \tag{46.93}$$

Numerical calculations (BCS 1957) show that $n_s(T)$ approaches zero linearly as $T \to T_c$, in agreement with the G–L prediction in which $\mathbf{j} \propto |\psi(T)|^2$ with $\psi(T) \propto (T_c - T)^{1/2}$. As $T \to 0, n_s \to n$, apart from a numerical factor (see Eq. (46.45)).

47 The quasiclassical case

47.1 Quasiclassical limit of the Schrödinger equation

When the quantum numbers associated with the states of a system are large we expect its motion to be nearly classical. We may then use the so-called quasiclassical or WKBJ method which will be developed in this section. We begin by seeking a solution of the time-independent Schrödinger equation,

$$-\frac{\hbar^2}{2m}\nabla^2\psi + U\psi = E\psi, \tag{47.1}$$

in the form

$$\psi = e^{(i/\hbar)S}. \tag{47.2}$$

Substituting (47.2) into (47.1) yields

$$\frac{1}{2m}[(\nabla S)^2 - i\hbar\nabla^2 S] + U = E. \tag{47.3}$$

Eq. (47.3) is still exact since it is simply an alternative way of writing the Schrödinger equation. The quasiclassical approximation involves first expanding S as a power series in $\hbar$ and then calculating the leading term and, in some cases, the next leading term. Hence we write

$$S = S^{(0)} + \frac{\hbar}{i}S^{(1)} + \left(\frac{\hbar}{i}\right)^2 S^{(2)}\ldots \tag{47.4}$$

(the factor i is included for later convenience). Inserting (47.4) into (47.3) and collecting terms with equal powers of $\hbar$ in the resulting hierarchy of equations yields

$$\frac{1}{2m}(\nabla S^{(0)})^2 - E + U = 0 \tag{47.5a}$$

$$\nabla S^{(0)}\cdot\nabla S^{(1)} + \frac{1}{2}\nabla^2 S^{(0)} = 0 \tag{47.5b}$$

$$\vdots$$

The integration of this coupled set of equations cannot be carried out in a general form and to proceed further we must make assumptions about the nature of the system. For the case of one-particle, one-dimensional motion we may integrate (47.5a) in the form

$$S^{(0)}(x) = \pm \int \{2m[E - U(x)]\}^{1/2} \mathrm{d}x + \text{const} \tag{47.6a}$$

or

$$S^{(0)}(x) = \pm \int p(x) \mathrm{d}x + \text{const}, \tag{47.6b}$$

where we have introduced the position-dependent momentum, $p(x)$. Eq. (47.5b) may now be written as

$$p\frac{\mathrm{d}S^{(1)}}{\mathrm{d}x} + \frac{1}{2}\frac{\mathrm{d}p}{\mathrm{d}x} = 0;$$

integrating we have

$$S^{(1)}(x) = \ell \mathrm{n} p^{-1/2} + \text{const}. \tag{47.7}$$

Substituting (47.6b) and (47.7) into (47.4) and the resulting S into (47.2) we obtain the quasiclassical wavefunction as

$$\psi(x) = \frac{A_+}{[p(x)]^{1/2}} \exp\left[+ \frac{\mathrm{i}}{\hbar} \int p(x) \mathrm{d}x \right] + \frac{A_-}{[p(x)]^{1/2}} \exp\left[- \frac{\mathrm{i}}{\hbar} \int p(x) \mathrm{d}x \right], \tag{47.8}$$

where the constants of integration have been absorbed into complex amplitude factors, and both of the solutions in (47.6b) were included. Note that the resulting probability density is $|\psi(x)|^2 \propto 1/p$ which is consistent with our intuition that the likelihood of finding a particle is proportional to the reciprocal of its velocity (the slower it moves the longer the 'residence' time).

The quantity $S^{(0)}$ is precisely the classical action.[1] Comparing Eq. (47.3) with Eq. (47.5a) we see that the criterion for the applicability of classical mechanics is

$$\left(\frac{\mathrm{d}S}{\mathrm{d}x}\right)^2 >> \hbar \frac{\mathrm{d}^2 S}{\mathrm{d}x^2}; \tag{47.9}$$

in the classical approximation $\mathrm{d}S/\mathrm{d}x = p(x)$ and, using the de Broglie relation $p = \hbar/\lambda\!\!\!^{-}$, Eq. (47.9) becomes

$$\frac{\mathrm{d}\lambda\!\!\!^{-}}{\mathrm{d}x} << 1. \tag{47.10}$$

At a classical turning point (where a particle reverses its direction of motion) the momentum goes to zero ($\lambda\!\!\!^{-} \rightarrow \infty$) and Eq. (47.10) is no longer satisfied. The quasiclassical approximation always breaks down at such points. A detailed analysis (see Landau and Lifshitz (1977)) shows that we may treat such cases by superimposing incoming and outgoing quasiclassical wavefunctions in the form

1. More accurately it is the abbreviated action. The total action, S^{total}, is time-dependent and for our case where the energy, E, is constant it is given by $S^{\text{total}} = S - Et$.

$$\psi^{\text{total}} = \psi^{\text{in}} + \psi^{\text{out}}$$
$$= e^{(i/\hbar)S^{\text{in}}} + e^{(i/\hbar)S^{\text{out}}},$$

where far from the boundary

$$S^{\text{out}} = S^{\text{in}} + \frac{n\pi}{4}$$

and n is a positive or negative integer.

We now consider cases where the number of particles and/or dimensions is larger than 1. Classically, if there exists a canonical transformation which results in separate Hamiltonian equations involving single pairs of canonically conjugate coordinates, the classical action may be written as a sum

$$S = \sum_{i=1}^{3N} S_i, \tag{47.11}$$

where N is the number of particles (moving in three dimensions). This immediately leads to a total wavefunction which has the form of a product

$$\psi = \prod_{i=1}^{3N} \psi_i, \tag{47.12}$$

where $\psi_i = e^{(i/\hbar)S_i}$.

The applicability of the quasiclassical approximation in higher dimensions is complicated by the fact that the motion is characterized by multiple quantum numbers, not all of which may be large. In addition 'turning points' may occur which involve only a restricted set of coordinates or dimensions (e.g., a particle deflecting from a wall).

For the case of a free electron in three dimensions the unperturbed wavefunction has the form $\psi \propto e^{i\mathbf{k}\cdot\mathbf{r}}$ and hence $S(\mathbf{r}) = \hbar\mathbf{k}\cdot\mathbf{r} = \mathbf{p}\cdot\mathbf{r}$. In the presence of perturbing scalar, $U(\mathbf{r})$, and vector, $\mathbf{A}(\mathbf{r})$, potentials Eq. (47.3) takes the form

$$E = \frac{1}{2m}\left[\left(\nabla S - \frac{e}{c}\mathbf{A}\right)^2 - i\hbar\nabla^2 S + \frac{e}{c}\nabla\cdot\mathbf{A}\right] + U. \tag{47.13}$$

In the leading approximation

$$E = \frac{1}{2m}\left(\nabla S^{(0)} - \frac{e}{c}\mathbf{A}\right)^2 + U(\mathbf{r}); \tag{47.14}$$

inserting $\nabla S = \mathbf{p}$, Eq. (47.14) becomes the Hamiltonian of a charged particle in a magnetic field. For the case of an electron in a *uniform* magnetic field, which classically leads to spiral orbits, the requirement that the wavefunction be single-valued requires that S change by a multiple of $2\pi\hbar$ in completing an orbit; i.e.,

$$\oint\left(\hbar\mathbf{k}_\perp + \frac{e}{c}\mathbf{A}\right)\cdot d\mathbf{r} = nh, \tag{47.15}$$

where $k_\perp$ and $k_\parallel$ are the components of the wave vector perpendicular and parallel to the magnetic field and $(\hbar^2/2m)(k_\perp^2 + k_\parallel^2) = E$; $\mathbf{A}$ may be chosen in any gauge leading to a constant magnetic field.

We next consider the case of a degenerate free electron gas. For applications involving low energy excitations we may write the wavefunction of an electron near the Fermi surface as

$$\psi = e^{(i/\hbar)(S_F + s)} \tag{47.16}$$

where $S_F = \mathbf{p}_F \cdot \mathbf{r}$ is the action function of an electron on the Fermi surface and s represents a small correction. The equation analogous to (47.13) is

$$E = \frac{1}{2m}\left[\left(\mathbf{p}_F + \nabla s - \frac{e}{c}\mathbf{A}\right)^2 - i\hbar\nabla^2 s + \frac{e}{c}\nabla\cdot\mathbf{A}\right] + U. \tag{47.17}$$

If we define the excitation energies in the usual way as $\varepsilon = |E - \mu|$ where $\mu \equiv p_F^2/2m$, neglect the small quadratic terms $(\nabla s)^2$ and $-(2e/c)\nabla s\cdot\mathbf{A}$ as well as the term $-i\hbar\nabla^2 s$ and work in the London gauge we obtain

$$\varepsilon = \mathbf{v}_F\cdot\left(\nabla s - \frac{e}{c}\mathbf{A}\right) + \frac{e^2}{2mc^2}A^2 + U. \tag{47.18}$$

We may rewrite (47.18) as a Schrödinger-like equation

$$\mathscr{H}'\psi' = \varepsilon\psi' \tag{47.19}$$

where

$$\mathscr{H}' = \frac{\hbar}{i}\mathbf{v}_F\cdot\left(\nabla - \frac{ie}{\hbar c}\mathbf{A}\right) + \frac{e^2}{2mc^2}A^2 + U; \tag{47.20}$$

the wavefunction $\psi' \sim e^{(i/\hbar)s}$ in (47.19) is now an *envelope wavefunction* since the rapid oscillations, associated with the factor $e^{i\mathbf{k}_F\cdot\mathbf{r}}$ in the original wavefunction, ψ, have been removed.

The case of an electron moving in a periodic lattice is more complex. In the absence of external potentials the wavefunctions have the usual Bloch form $\psi \sim e^{i\mathbf{k}\cdot\mathbf{r}}u_{\mathbf{k}n}(\mathbf{r})$. The behavior of a Bloch electron in a magnetic field can involve a number of subtle behaviors including the Onsager–Lifshitz quantization of the orbit areas (leading to the de Haas van Alphen effect) and magnetic breakdown. We will not discuss this problem but instead assume that we can obtain a reasonably accurate representation by writing $\mathbf{v}_F = (1/m)\nabla S_F$ in the above expression (note $\mathbf{v}_F$ is not, in general, parallel to $\mathbf{p}_F$ for a real Fermi surface). S_F follows from the solution of

$$\frac{1}{2m}[(\nabla S_F)^2 - i\hbar\nabla^2 S_F] + U = \mu. \tag{47.21}$$

47.2 Quasiclassical limit of the Bogoliubov equations

We now turn to the case of a superconductor. In place of the Schrödinger equation we have the two Bogoliubov equations. We first seek a quasiclassical solution in the form

$$u(\mathbf{r}) = \mathrm{e}^{(\mathrm{i}/\hbar)(S_\mathrm{F} + s_u)} \tag{47.22a}$$

and

$$v(\mathbf{r}) = \mathrm{e}^{(\mathrm{i}/\hbar)(S_\mathrm{F} + s_v)}. \tag{47.22b}$$

Substituting (47.22a) and (47.22b) into (36.17a) and (36.17b) we obtain

$$\varepsilon = \frac{1}{2m}\left[2\nabla S_\mathrm{F}\cdot\nabla s_u - \frac{2e}{c}\nabla S_\mathrm{F}\cdot\mathbf{A} + \frac{e^2}{c^2}A^2\right] + \Delta(\mathbf{r})\mathrm{e}^{(\mathrm{i}/\hbar)(s_v - s_u)} \tag{47.23}$$

and

$$\varepsilon = -\frac{1}{2m}\left[2\nabla S_\mathrm{F}\cdot\nabla s_v + \frac{2e}{c}\nabla S_\mathrm{F}\cdot\mathbf{A} + \frac{e^2}{c^2}A^2\right] + \Delta(\mathbf{r})\mathrm{e}^{(\mathrm{i}/\hbar)(s_u - s_v)}, \tag{47.24}$$

where we neglected terms of order $\nabla^2 s_{u,v}$, $(\nabla s_{u,v})^2$, and $\mathbf{A}\cdot\nabla s_{u,v}$ under the assumptions that $s_{u,v}$ are slowly varying and $s_{u,v} << S_\mathrm{F}$; we also assumed a London gauge.

We introduce the variables

$$\bar{s} = \frac{1}{2}(s_u + s_v) \tag{47.25a}$$

and

$$s = \frac{1}{2}(s_u - s_v). \tag{47.25b}$$

Inverting these equations we have

$$s_u = \bar{s} + s \tag{47.26a}$$

and

$$s_v = \bar{s} - s. \tag{47.26b}$$

Taking the sum and difference of (47.23) and (47.24) and using our definition $\mathbf{v}_\mathrm{F} = (1/m)\nabla S_\mathrm{F}$ we have

$$\varepsilon = -\frac{e}{c}\mathbf{v}_\mathrm{F}\cdot\mathbf{A} + \mathbf{v}_\mathrm{F}\cdot\nabla s + \Delta(\mathbf{r})\cos\left(\frac{2s}{\hbar}\right) \tag{47.27}$$

and

$$0 = \mathbf{v}_\mathrm{F}\cdot\nabla\bar{s} + \frac{e^2}{2mc^2}A^2 - \mathrm{i}\Delta(\mathbf{r})\sin\left(\frac{2s}{\hbar}\right). \tag{47.28}$$

These equations constitute the quasiclassical approximation to Bogoliubov's equations. They are subject to the normalization condition (36.22a) which becomes

$$\int \mathrm{d}^3r[\mathrm{e}^{(\mathrm{i}/\hbar)(s_u - s_u^*)} + \mathrm{e}^{(\mathrm{i}/\hbar)(s_v - s_v^*)}] = 1 \tag{47.29a}$$

or

$$\int d^3r[e^{(i/\hbar)(\bar{s}+s-\bar{s}^*-s^*)} + e^{(i/\hbar)(\bar{s}-s-\bar{s}^*+s^*)}]d^3r = 1. \tag{47.29b}$$

Through a proper choice of gauge it is usually possible to neglect the term in A^2 in (47.28). In a uniform Fermi gas ($\mathbf{A} = 0$), $\nabla s = 0$, and $\nabla\bar{s} = \mathbf{p} - \mathbf{p}_F$; writing $\xi = \mathbf{v}_F \cdot (\mathbf{p} - \mathbf{p}_F)$, Eqs. (47.27) and (47.28) reduce to

$$\varepsilon^2 = \xi^2 + \Delta^2,$$

as they should.

One can work with an alternate set of wavefunctions (the Andreev wavefunctions) having the form

$$u(\mathbf{r}) = u'(\mathbf{r})e^{(i/\hbar)S_F} \tag{47.30a}$$

and

$$v(\mathbf{r}) = v'(\mathbf{r})e^{(i/\hbar)S_F}, \tag{47.30b}$$

where $u'(\mathbf{r})$ and $v'(\mathbf{r})$ are envelope wavefunctions. In terms of the wavefunctions (47.30a) and (47.30b) we may write the Bogoliubov equations as

$$\varepsilon u'(\mathbf{r}) = \mathscr{H}'u'(\mathbf{r}) + \Delta(\mathbf{r})v'(\mathbf{r}) \tag{47.31a}$$

and

$$\varepsilon v'(\mathbf{r}) = -\bar{\mathscr{H}}'v'(\mathbf{r}) + \Delta^*(\mathbf{r})u'(\mathbf{r}), \tag{47.31b}$$

where $\mathscr{H}'$ is given by Eq. (47.20) and $\bar{\mathscr{H}}'(\mathbf{A}) = \mathscr{H}'(-\mathbf{A})$ (or equivalently $\mathscr{H}'^*(\mathbf{A})$).[2]

47.3 Andreev scattering

As an application of the quasiclassical Bogoliubov theory we discuss the behavior of the Andreev wavefunctions resulting from a normal/superconductor planar interface (Andreev 1964). We have in mind the interfaces which form spontaneously in a Type I superconductor in the intermediate state. Assume initially we have a single N–S interface centered on the plane $z = 0$ with the superconductor occupying the region $z > 0$. The Andreev–Bogoliubov equations, (47.31a) and (47.31b), then take the form (we neglect the vector potential for simplicity)

$$\varepsilon u'(\mathbf{r}) = \frac{\hbar v_F}{i}\hat{\mathbf{n}}\cdot\nabla u'(\mathbf{r}) + \Delta(z)v'(\mathbf{r}) \tag{47.32a}$$

and

$$\varepsilon v'(\mathbf{r}) = -\frac{\hbar v_F}{i}\hat{\mathbf{n}}\cdot\nabla v'(\mathbf{r}) + \Delta(z)u'(\mathbf{r}); \tag{47.32b}$$

here v_F is the Fermi velocity and $\hat{\mathbf{n}}$ is a unit vector.

2. Complex conjugation corresponds to time reversal which changes the sign of the magnetic field and, with it, the vector potential, **A**.

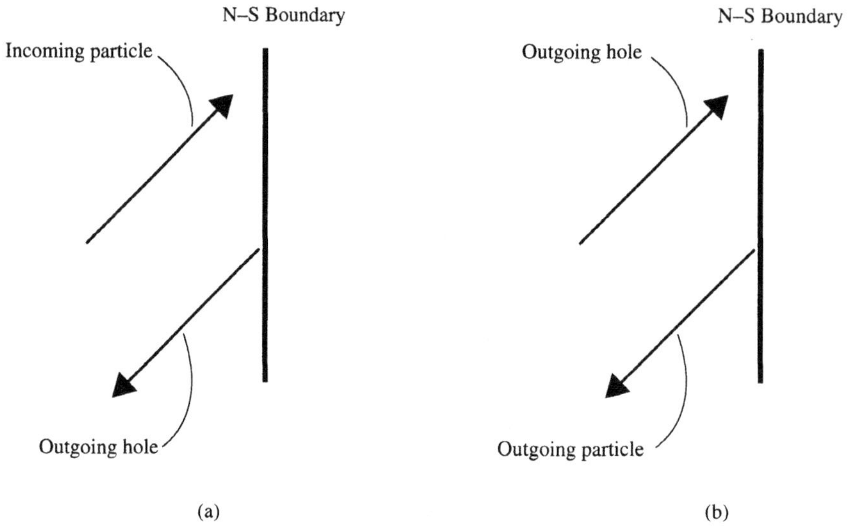

Figure 47.1 Andreev reflection: (a) $n_z > 0$; (b) $n_z < 0$.

$\Delta(z)$ can only be obtained numerically (or variationally) by performing a self-consistent calculation. However, here we only want to establish the qualitative behavior of $u'(\mathbf{r})$ and $v'(\mathbf{r})$ for which we will not need the explicit behavior of $\Delta(z)$. Qualitatively $\Delta(z)$ rises from 0 (its value for $z \to -\infty$) to the equilibrium value Δ (its value for $z \to +\infty$) over a characteristic length of order ξ.

For $z \to -\infty$ (deep in the normal metal) we set $\Delta = 0$ and the solution to Eqs. (47.32) has the form[3]

$$\begin{pmatrix} u'(\mathbf{r}) \\ v'(\mathbf{r}) \end{pmatrix} = A \begin{pmatrix} 1 \\ 0 \end{pmatrix} e^{i\mathbf{k}^{(n)}\cdot\mathbf{r}} + B \begin{pmatrix} 0 \\ 1 \end{pmatrix} e^{-i\mathbf{k}^{(n)}\cdot\mathbf{r}} \tag{47.33}$$

where the normal state wave vector, $\mathbf{k}^{(n)}$, satisfies $\hat{\mathbf{n}}\cdot\mathbf{k}^{(n)} = \varepsilon/\hbar v_F$ and the coefficients A and B are arbitrary at this point. If the z component of $\hat{\mathbf{n}}, \hat{n}_z$, is greater than 0 the first term in (47.33) corresponds to a particle incoming to the boundary (travelling parallel to $\hat{\mathbf{n}}$) and a hole outgoing from the boundary (traveling antiparallel to $\hat{\mathbf{n}}$). This situation is depicted in Fig. 47.1(a). On the other hand if $n_z < 0$ the second term in (47.33) corresponds to an incoming hole while the first corresponds to an outgoing particle.

We next examine the solutions of Eqs. (47.32) in the limit $z \to -\infty$ (deep in the superconductor) where we may set $\Delta(z) = \Delta$. For the case when $\varepsilon > \Delta$ the solution is

$$\begin{pmatrix} u'(\mathbf{r}) \\ v'(\mathbf{r}) \end{pmatrix} = \frac{C}{\sqrt{2}} \begin{bmatrix} \left(1 + \hbar\frac{v_F}{\varepsilon}\hat{\mathbf{n}}\cdot\mathbf{k}^{(s)}\right)^{1/2} \\ \left(1 - \hbar\frac{v_F}{\varepsilon}\hat{\mathbf{n}}\cdot\mathbf{k}^{(s)}\right)^{1/2} \end{bmatrix} e^{i\mathbf{k}^{(s)}\cdot\mathbf{r}} \tag{47.34}$$

where $C/\sqrt{2}$ is a constant and the wave vector in the superconductor, $\mathbf{k}^{(s)}$, satisfies

$$\mathbf{k}^{(s)}\cdot\hat{\mathbf{n}} = \frac{1}{\hbar v_F}(\varepsilon^2 - \Delta^2)^{1/2}, \quad n_z > 0 \tag{47.35a}$$

3. Note $\mathbf{k}^{(n)}$ and $\mathbf{k}^{(s)}$ are the wave vectors of the *envelope* wavefunctions $u'(\mathbf{r})$ and $v'(\mathbf{r})$ and therefore have magnitudes much smaller than k_F. In a uniform system $\xi \cong \hbar(k - k_F)v_F = \hbar k^{(n),(s)}v_F$.

and

$$\mathbf{k}^{(s)} \cdot \hat{\mathbf{n}} = -\frac{1}{\hbar v_F}(\varepsilon^2 - \Delta^2)^{1/2}, \quad n_z < 0. \tag{47.35b}$$

When $\varepsilon < \Delta$ we have only exponentially decaying solutions in the superconductor. Note that since our superconductor is completely homogeneous $k_x^{(s)} = k_x^{(n)}$; $k_y^{(s)} = k_y^{(n)}$.

As a simple model, Andreev considered a step function gap:

$$\Delta(z) = \begin{cases} \Delta, & z > 0, \\ 0, & z < 0. \end{cases} \tag{47.36}$$

Requiring continuity of the envelope wavefunctions at $z = 0$ yields

$$\begin{pmatrix} A \\ B \end{pmatrix} = \frac{C}{\sqrt{2}} \begin{bmatrix} \left(1 + \dfrac{(\varepsilon^2 - \Delta^2)^{1/2}}{\varepsilon}\right)^{1/2} \\ \left(1 - \dfrac{(\varepsilon^2 - \Delta^2)^{1/2}}{\varepsilon}\right)^{1/2} \end{bmatrix}; \quad n_z > 0 \tag{47.37a}$$

or

$$\begin{pmatrix} A \\ B \end{pmatrix} = \frac{C}{\sqrt{2}} \begin{bmatrix} \left(1 - \dfrac{(\varepsilon^2 - \Delta^2)^{1/2}}{\varepsilon}\right)^{1/2} \\ \left(1 + \dfrac{(\varepsilon^2 - \Delta^2)^{1/2}}{\varepsilon}\right)^{1/2} \end{bmatrix}; \quad n_z < 0. \tag{47.37b}$$

The transmission coefficient, T, is defined as

$$T \equiv \frac{(S_z)^{\text{transmitted}}}{(S_z)^{\text{incoming}}}, \tag{47.38}$$

where S_z is the z component of the probability current density. To obtain an expression for the probability density we start with the time-dependent form of the Bogoliubov equations

$$i\hbar \frac{\partial u(\mathbf{r})}{\partial t} = \mathscr{H} u(\mathbf{r}) + \Delta(\mathbf{r}) v(\mathbf{r}) \tag{47.39a}$$

and

$$i\hbar \frac{\partial v(\mathbf{r})}{\partial t} = -\mathscr{H}^* v(\mathbf{r}) + \Delta^*(\mathbf{r}) u(\mathbf{r}). \tag{47.39b}$$

Multiplying the first equation by $u^*(\mathbf{r})$ and the complex conjugate of the second by $v(\mathbf{r})$ and subtracting yields

$$i\hbar \left[u^*(\mathbf{r}, t) \frac{\partial u(\mathbf{r}, t)}{\partial t} + v(\mathbf{r}, t) \frac{\partial v^*(\mathbf{r}, t)}{\partial t} \right] = u^*(\mathbf{r}, t) \mathscr{H} u(\mathbf{r}, t) + v(\mathbf{r}, t) \mathscr{H} v^*(\mathbf{r}, t).$$

Subtracting the complex conjugate of this equation and integrating over all space yields

$$i\hbar \frac{\partial}{\partial t} \int [u^*(\mathbf{r}, t) u(\mathbf{r}, t) + v^*(\mathbf{r}, t) v(\mathbf{r}, t)] \mathrm{d}^3 r =$$

$$\int [u^*(\mathbf{r}, t) \mathscr{H} u(\mathbf{r}, t) - u(\mathbf{r}, t) \mathscr{H}^* u^*(\mathbf{r}, t) + v(\mathbf{r}, t) \mathscr{H} v^*(\mathbf{r}, t) - v^*(\mathbf{r}, t) \mathscr{H}^* v(\mathbf{r}, t)] \mathrm{d}^3 r.$$

Writing $\mathscr{H} = -(\hbar^2/2m)[\nabla - (\mathrm{i}e/\hbar c)\mathbf{A}]^2 + U(\mathbf{r})$ and using Green's identity gives

$$\frac{\partial}{\partial t}\rho(\mathbf{r}, t) + \nabla \cdot \mathbf{S}(\mathbf{r}, t) = 0, \tag{47.40}$$

where

$$\rho(\mathbf{r}, t) = u^*(\mathbf{r}, t)u(\mathbf{r}, t) + v^*(\mathbf{r}, t)v(\mathbf{r}, t) \tag{47.41a}$$

and[4]

$$\mathbf{S}(\mathbf{r}, t) = -\frac{\mathrm{i}\hbar}{2m}[u^*(\mathbf{r}, t)\nabla u(\mathbf{r}, t) - u(\mathbf{r}, t)\nabla u^*(\mathbf{r}, t) - v^*(\mathbf{r}, t)\nabla v(\mathbf{r}, t) + v(\mathbf{r}, t)\nabla v^*(\mathbf{r}, t)]$$
$$-\frac{e}{mc}\mathbf{A}(\mathbf{r})[u^*(\mathbf{r}, t)u(\mathbf{r}, t) + v^*(\mathbf{r}, t)v(\mathbf{r}, t)]. \tag{47.41b}$$

Recalling the full wavefunctions have the form $u, v \propto \mathrm{e}^{\mathrm{i}(\mathbf{k}_\mathrm{F} + \mathbf{k}^{(n,s)})\cdot\mathbf{r}}$ we can use the above to calculate the incoming current using (47.33) and (47.41b) as

$$S_z^{\text{incoming}} \cong |A|^2 \frac{\hbar\mathbf{k}_\mathrm{F} \cdot \hat{\mathbf{z}}}{m}, \quad n_z > 0 \tag{47.42a}$$

or

$$S_z^{\text{incoming}} = |B|^2 \frac{\hbar\mathbf{k}_\mathrm{F} \cdot \hat{\mathbf{z}}}{m}, \quad n_z < 0 \tag{47.42b}$$

and the transmitted current using (47.34) and (47.41b) as

$$S_z^{\text{transmitted}} \cong \begin{cases} |C|^2 \dfrac{\hbar v_\mathrm{F}}{\varepsilon} \hat{\mathbf{n}} \cdot \mathbf{k}^{(s)} \left(\dfrac{\hbar \mathbf{k}_\mathrm{F} \cdot \hat{\mathbf{z}}}{m}\right), & \varepsilon > \Delta, \\ 0, & \varepsilon < \Delta. \end{cases} \tag{47.43}$$

We then obtain T as

$$T = \frac{(\varepsilon^2 - \Delta^2)^{1/2}}{\varepsilon}\left|\frac{C}{A}\right|^2, \quad n_z > 0 \tag{47.44a}$$

or

$$T = \frac{(\varepsilon^2 - \Delta^2)^{1/2}}{\varepsilon}\left|\frac{C}{B}\right|^2, \quad n_z < 0. \tag{47.44b}$$

For our step function model we have

$$T = \frac{\hbar v_\mathrm{F} \hat{\mathbf{n}} \cdot \mathbf{k}^{(s)}}{\varepsilon + (\varepsilon^2 - \Delta^2)^{1/2}} = \frac{2(\varepsilon^2 - \Delta^2)^{1/2}}{\varepsilon + (\varepsilon^2 - \Delta^2)^{1/2}}, \quad n_z > 0, \tag{47.45}$$

which for small $\varepsilon - \Delta > 0$ (i.e., states with energies slightly larger than the gap) becomes

$$T = 2^{3/2}\left(\frac{\varepsilon - \Delta}{\Delta}\right)^{1/2}. \tag{47.46}$$

4. We must not confuse $\mathbf{S}$, which is the probability current associated with a single Bogoliubov particle–hole wavefunction, with the total equilibrium current density $\mathbf{j}$ given by Eq. (37.18).

In this same limit the transmission coefficient for the exact gap function $\Delta(z)$ would have the form

$$T = f(n_z)\left(\frac{\varepsilon - \Delta}{\Delta}\right)^{1/2}, \tag{47.47}$$

where $f(n_z)$ is a function of the incident angle and is of order unity. Physically, when $\varepsilon - \Delta \gtrsim \Delta$ the wavelengths of the incoming and transmitted waves become much shorter than the barrier width and the wavefunctions 'adiabatically' deform into one another.

The Andreev reflection process is unusual in the following sense. When electrons are specularly reflected from a planar normal metal–normal metal boundary, only the component of the velocity normal to the boundary changes sign; the in-plane component is unchanged. However, at a normal metal–superconductor boundary all three components of the velocity change sign, i.e., the particle direction is completely reversed.

The Andreev reflection process results in a significant increase in the thermal resistance. Here we will be content with an order of magnitude estimate of the effect in the limit $T << T_c$. (We will neglect the thermal conduction due to the phonons.) In this limit it is only the quasiparticles with an energy of order Δ which contribute to the thermal flux. These excitations lie in the rapidly decreasing tail of the Boltzmann-like distribution and hence the total number of particles passing through the boundary is small. The thermal flux, w, from the normal metal to the superconductor is given by

$$w = \int_{v_z > 0} 2n(\varepsilon)\varepsilon v_z T(\varepsilon, v_z)\frac{d^3p}{(2\pi\hbar)^3}$$

where $v_z = \partial\varepsilon/\partial p_z$ and n is the distribution function. Substituting Eq. (47.47) we have

$$w = \frac{p_F^2}{\pi^2\hbar^3}\int \varepsilon n(\varepsilon)\left(\frac{\varepsilon - \Delta}{\Delta}\right)^{1/2} d\xi \int \cos\theta f(\cos\theta) d(\cos\theta). \tag{47.48}$$

In our limit, $k_B T << \Delta$ and we can set the distribution function equal to its limiting (Boltzmann) form, $n = e^{-\varepsilon/k_B T}$; carrying out the integral we obtain

$$w = \frac{1}{2}\pi^{1/2} f_1 \frac{p_F^2}{\pi^2\hbar^3} k_B T (k_B T\Delta)^{1/2} e^{-\Delta/k_B T}, \tag{47.49}$$

where f_1 is the first moment of the function $f(\cos\theta)$:

$$f_1 \equiv \int_0^\pi \cos\theta f(\cos\theta) d(\cos\theta). \tag{47.50}$$

In thermal equilibrium there would be a flux of equal magnitude flowing from the superconductor to the normal metal. In a thermal steady state (i.e., in the presence of a net thermal flux) there would be a temperature drop, δT, and the resulting thermal flux would be $(dw/dT)\delta T$. Performing the differentiation and recalling $\Delta >> k_B T$ yields

$$Q = \frac{\pi^{1/2}}{2} f_1 \frac{p_F^2}{\pi^2\hbar^3}\Delta\left(\frac{\Delta}{k_B T}\right)^{1/2} e^{-\Delta/k_B T}\delta T. \tag{47.51}$$

In the intermediate state numerous N–S boundaries are present and a temperature drop δT occurs at each. The number and form of these boundaries depend, of course, on the magnetic field and the geometry of the sample. The cylindrical case was studied by Andreev (1964), but will not be discussed here.

48 The isolated vortex line

48.1 Bogoliubov's equations for the isolated vortex line

Isolated vortex lines do not exist in a microscopically homogeneous system since they are not stable for $H < H_{c1}$ and are only present in the form of a lattice for $H > H_{c1}$. (Pinning effects do permit isolated vortices, but then the effect of the pinning potential needs to be considered.) In spite of these difficulties the isolated vortex is a valuable model problem, a description of which is important in attempting to understand the nature of more complex flux structures in superconductors. The properties of an isolated vortex line were discussed in Sec. 7; this model neglects the effect of the position dependence of the order parameter. For temperatures close to T_c one may use the G–L theory to calculate the properties of vortex lines, as discussed briefly in Sec. 14. However, if we are interested in problems such as the spectrum of the electronic excitations in vortex cores or the position dependence of the field, current, and order parameter far from T_c, we must use the microscopic theory.

To keep our discussion within reasonable bounds we will limit ourselves to an electron gas model with no impurities at $T = 0$. Even with these simplifications an exact solution is not possible and further assumptions, or detailed numerical calculations, are required. We will follow the approach used by Caroli *et al.* (1964) and later employed by Bardeen, Kummel, Jacobs, and Tewordt (1969), which uses the Bogoliubov method.

To carry out the calculations we must choose a gauge. In the London model (discussed following (7.10)), a vortex line has a circulating density $j_\theta = ne\hbar/2mr$ in the limit $r \to 0$. In the microscopic theory the current density is related to the wavefunctions u_n and v_n and the vector potential, $\mathbf{A}$, by Eq. (37.17). The effect of a gauge function, $\chi(\mathbf{r})$, on the vector potential, the Bogoliubov wavefunctions, and the gap function is governed by Eqs. (46.1), (46.2), and (46.3), respectively. In the London gauge, Eq. (7.10) suggests that the form of the vector potential in the limit $r \to 0$ is $A_\theta \underset{r\to 0}{=} (\hbar c/2er)$; this implies that all of the current arises from the last term in Eq. (37.17) (i.e., the azimuthal component of the gradient operator, $(1/r)(\partial/\partial\theta)\hat{\boldsymbol{\theta}}$, entering the first two terms produces no contribution). As one proceeds outward from the vortex core $A_\theta(r)$ deviates from the form given and we may write it as

$$A_\theta(r) = \frac{\hbar c}{2er} + A'_\theta(r), \tag{48.1}$$

where $A'_\theta(r) \to 0$ as $r \to 0$.

We can perform a gauge transformation $A_\theta \to A_\theta + (1/r)(\partial/\partial\theta)\chi(\theta)$ where $\chi(\theta) = -(\hbar c/2e)\theta$; in this gauge the vector potential *vanishes* in the small r limit (rather than diverging as in the London gauge). However, from Eqs. (46.2a) and (46.2b) the Bogoliubov wavefunctions will acquire an explicit θ dependence

$$u_n \to u_n e^{-i(\theta/2)}$$

and

$$v_n \to v_n e^{+i(\theta/2)},$$

and j_θ now arises from the first two terms in (37.17); from (46.3) the gap function also acquires the θ dependence

$$\Delta(\mathbf{r}) \to \Delta(\mathbf{r}) e^{-i\theta}.$$

On the basis of the G–L theory we expect $|\Delta(\mathbf{r})| \propto r$ as $r \to 0$ and $|\Delta(\mathbf{r})| = \Delta_\infty$ as $r \to \infty$ where Δ_∞ is the equilibrium gap. To determine the structure of a vortex line we must solve Bogoliubov's equations (36.17a) and (36.17b). We write $u(\mathbf{r})$ and $v(\mathbf{r})$ in the two-component spinor form $\psi = \begin{pmatrix} u(\mathbf{r}) \\ v(\mathbf{r}) \end{pmatrix}$ as in Eq. (36.19). In a gauge where $\Delta(r) = |\Delta(r)| e^{-i\theta}$ the solution to Eq. (36.19) can be written as a product of three wavefunctions involving the cylindrical coordinates, z, θ, and r

$$\psi(z, \theta, r) = e^{ik_z z} e^{i\mu\theta} e^{-\frac{1}{2}i\tau_z\theta} \begin{pmatrix} R_+(r) \\ R_-(r) \end{pmatrix}, \tag{48.2}$$

where τ_i denote the Pauli matrices (acting in particle–hole space) and 2μ is an odd integer.[1] The term in $e^{-\frac{1}{2}i\tau_z\theta}$ has the effect of removing the $e^{i\theta}$ phase factor associated with the gap. Expressing the Laplacian in cylindrical coordinates, substituting the product wavefunction (48.2), and expressing the gap function as $\Delta(r)e^{-i\theta}$, we find that the variables (r, θ, z) in Eq. (36.20) separate; the resulting total differential equation in r is

$$\left\{ \frac{\hbar^2}{2m} \left[-\frac{d^2}{dr^2} - \frac{1}{r}\frac{d}{dr} + \frac{1}{r^2}\left(\mu - \frac{\tau_z e r}{\hbar c} A_\theta(r) \right)^2 - k_\perp^2 \right] \tau_z \right.$$
$$\left. + |\Delta(\mathbf{r})| \tau_x \right\} \begin{pmatrix} R_+(r) \\ R_-(r) \end{pmatrix} = \varepsilon \begin{pmatrix} R_+(r) \\ R_-(r) \end{pmatrix}, \tag{48.3}$$

where $k_\perp$ is a separation constant given by $k_z^2 + k_\perp^2 = k_F^2$ and $A_\theta(r)$ is defined by Eq. (48.1).

Eq. (48.3) can be solved exactly for the case $|\Delta(r)| = \Delta_\infty$ and $A_\theta = 0$ which corresponds to the $r \to \infty$ limit of a vortex line; the result is:

$$\begin{pmatrix} R_+ \\ R_- \end{pmatrix} = \frac{\text{const}}{\sqrt{2}} \begin{pmatrix} \left[1 \pm \dfrac{(\varepsilon^2 - \Delta_\infty^2)^{1/2}}{\varepsilon} \right]^{1/2} \\ \left[1 \mp \dfrac{(\varepsilon^2 - \Delta_\infty^2)^{1/2}}{\varepsilon} \right]^{1/2} \end{pmatrix} \quad H_\mu^{(1),(2)} \left\{ \left[k_\perp^2 \pm \frac{2m}{\hbar^2} (\varepsilon^2 - \Delta_\infty^2)^{1/2} \right]^{1/2} r \right\}, \tag{48.4}$$

1. The paired states u_n and v_n^* have angular momenta $\mu - 1/2$ and $-\mu - 1/2$ for a net angular momenta of -1; the net angular momentum is associated with the circulating current of the vortex.

where $H_\mu^{(1),(2)}$ are Hankel functions of the first and second kind. Eq. (48.4) describes four independent solutions. For $\varepsilon < \Delta_\infty$ there are two exponentially decaying and two exponentially growing forms (the latter being excluded). For $\varepsilon > \Delta_\infty$ we have four independent propagating cylindrical waves. In the limit $r \to \infty$ one has the asymptotic form of the Hankel functions[2]

$$H_\mu^{(1),(2)}(x) = \left(\frac{2}{\pi x}\right)^{1/2} \exp\left[\pm \mathrm{i}\left(x + \frac{\mu^2}{2x} - \frac{\mu\pi}{2} - \frac{\pi}{4}\right)\right]. \tag{48.5a}$$

For later use we also give the asymptotic form for the Bessel functions

$$J_\mu(x) = \left(\frac{2}{\pi x}\right)^{1/2} \sin\left[x + \frac{\mu^2}{2x} - \frac{\mu\pi}{2} + \frac{\pi}{4}\right]. \tag{48.5b}$$

Eq. (48.3) may also be solved exactly when $\Delta = 0, A'_\theta = 0$ (see Eq, (48.1)), which applies at the core of a vortex line. Eq. (48.3) then becomes

$$\frac{\hbar^2}{2m}\left[-\frac{\mathrm{d}^2}{\mathrm{d}r^2} - \frac{1}{r}\frac{\mathrm{d}}{\mathrm{d}r} + \frac{1}{r^2}\left(\mu - \frac{\underset{\sim}{\tau}_z}{2}\right)^2 - k_\perp^2\right]\underset{\sim}{\tau}_z\begin{pmatrix}R_+\\R_-\end{pmatrix} = \varepsilon\begin{pmatrix}R_+\\R_-\end{pmatrix}. \tag{48.6}$$

Since we are here interested in the small r region, we must retain only the solution which is regular at $r = 0$, which is

$$R_\pm(r) = A_\pm J_{\mu\mp\frac{1}{2}}\left[\left(k_\perp^2 \pm \frac{2m}{\hbar^2}\varepsilon\right)^{1/2} r\right] \cong A_\pm J_{\mu\mp\frac{1}{2}}[(k_\perp \pm q)r] \tag{48.7}$$

where $q = m\varepsilon/\hbar^2 k_\perp$ and is positive or negative depending on whether ε is greater or less than zero; $\mu \mp \frac{1}{2}$ is an integer.

Actually to solve for the structure of the vortex line we would have to integrate Eq. (48.3) numerically; however, this equation contains the unknown functions $A_\theta(r)$ and $\Delta(r)$. $A_\theta(r)$ can be obtained by simultaneously integrating Maxwell's fourth equation, which in the present case is

$$-\frac{\partial H_z}{\partial r} = \frac{4\pi}{c} j_\theta(r), \tag{48.8a}$$

where

$$H_z = \frac{\partial}{\partial r} A_\theta(r); \tag{48.8b}$$

thus we must integrate the equation

$$j_\theta(r) = -\frac{c}{4\pi}\frac{\mathrm{d}^2 A_\theta(r)}{\mathrm{d}r^2}. \tag{48.8c}$$

The current $j_\theta(r)$ must be obtained from Eq. (37.17), the $\hat{\boldsymbol{\theta}}$ component of which is

$$j_\theta(r) = \sum_n \left\{\frac{e\hbar}{mr}\left[\left(-\mu - \frac{1}{2}\right)|R_{n-}(r)|^2(1 - f_n) + \left(\mu - \frac{1}{2}\right)|R_{n+}(r)|^2 f_n\right] - \frac{e^2}{mc}[|R_{n-}(r)|^2(1 - f_n) + |R_{n+}(r)|^2 f_n]A_\theta(r)\right\}, \tag{48.9}$$

2. Abramowitz and Stegun (1965) p. 364; the + and − signs in the exponential correspond to $H^{(1)}$ and $H^{(2)}$, respectively. See also footnote 3 of this section.

where n stands for the three quantum numbers k_z, ε, and μ; $\Delta(r)$ is evaluated using the self-consistency condition, Eq. (37.11), which in the present case is

$$\Delta(r) = V \sum_n R^*_{n-}(r) R_{n+}(r)(1 - 2f_n). \tag{48.10}$$

Starting with an approximate form for $\Delta(r)$ and $A_\theta(r)$ one would numerically integrate (48.3) and then construct new $j_\theta(r)$ and $\Delta(r)$. This process would be continued until R, j_θ, and Δ stabilized for successive iterations.

48.2 The quasiclassical equations for a vortex line

In order to proceed further with the analytical discussion we will apply the quasiclassical approximation to the Bogoliubov equations. The actual solutions to Eq. (48.3) must smoothly join the limiting forms represented by (48.4) and (48.7). We write the full solution to (48.3) as

$$\begin{pmatrix} R_+(r) \\ R_-(r) \end{pmatrix} = \begin{pmatrix} \rho_+(r) \\ \rho_-(r) \end{pmatrix} H^{(1)}_\mu(k_\perp r) + \text{c.c.} \tag{48.11}$$

This form has a structure, for cylindrical coordinates, which is analogous to Eqs. (47.22) if we make the identifications:

$$e^{(i/\hbar)S_F(r)} = H^{(1)}_\mu(k_\perp r), \rho_+(r) = e^{(i/\hbar)s_u(r)}, \rho_-(r) = e^{(i/\hbar)s_v(r)};$$

i.e., the Hankel function contains the rapidly oscillating part of the radial wavefunction while the functions $\rho_\pm(r)$ are 'envelope functions' which account for slow variations (modulations) in the amplitude and phase caused by the slowly varying functions $A'_\theta(r)$ and $|\Delta(r)|$. Inserting Eq. (48.11) into Eq. (48.3) and noting that $H_\mu(k_\perp r)$ satisfies the equation

$$\left[-\frac{d^2}{dr^2} - \frac{1}{r}\frac{d}{dr} + \frac{\mu^2}{r^2} - k_\perp^2 \right] H^{(1)}_\mu(k_\perp r) = 0 \tag{48.12}$$

we obtain

$$\begin{aligned} \Bigg\{ \frac{\hbar^2}{2m} \Bigg[&- H^{(1)}_\mu \frac{d^2}{dr^2} - \left(2\frac{dH^{(1)}_\mu}{dr} + \frac{H^{(1)}_\mu}{r} \right) \frac{d}{dr} - \frac{2e}{\hbar c}\frac{\mu}{r} A_\theta H^{(1)}_\mu \underset{\sim}{\tau}_z \\ &+ \frac{e^2}{\hbar^2 c^2} A_\theta^2 H^{(1)}_\mu \Bigg] \underset{\sim}{\tau}_z + H^{(1)}_\mu \Delta \underset{\sim}{\tau}_x \Bigg\} \begin{pmatrix} \rho_+ \\ \rho_- \end{pmatrix} = \varepsilon \begin{pmatrix} \rho_+ \\ \rho_- \end{pmatrix}. \end{aligned} \tag{48.13}$$

Since $\rho_\pm(r)$ is expected to vary significantly over a distance of order ξ_0 (where ξ_0 is a coherence length), whereas $H^{(1)}_\mu$ varies significantly over a distance of order $k_\perp^{-1}$, the first term in Eq. (48.13) may be neglected since it is smaller by a factor $(k_\perp \xi_0)^{-1}$. To the same order one may also neglect the terms in A_θ^2.

Since we are working in a quasiclassical approximation we will use the WKB solutions to (48.12):[3]

$$H_\mu^{(1),(2)}(k_\perp r) = \exp\left[\pm \mathrm{i}\int_{r_t}^{r} \beta(r')\mathrm{d}r'\right]/(r^2 - r_t^2)^{1/4}, \tag{48.14}$$

in (48.13); here $r_t = \mu/k_\perp$ is the radius of the turning point and

$$\beta(r) \equiv \frac{k_\perp}{r}(r^2 - r_t^2)^{1/2}. \tag{48.15}$$

Since we will not require the integral of β we do not list it. (This form is somewhat more accurate, except for $r \lesssim r_t$, than the asymptotic form (48.5a).) Inserting (48.14) into (48.13), and making the approximations discussed above, we obtain[4]

$$\left(-\frac{\mathrm{i}\hbar^2}{m}\beta(r)\underset{\sim}{\tau}_z\frac{\mathrm{d}}{\mathrm{d}r} + |\Delta(r)|\underset{\sim}{\tau}_x\right)\begin{pmatrix}\rho_+(x)\\ \rho_-(x)\end{pmatrix} = \left(\varepsilon + \frac{\mu e\hbar}{mcr}A_\theta(r)\right)\begin{pmatrix}\rho_+(x)\\ \rho_-(x)\end{pmatrix}. \tag{48.16}$$

In order to be well behaved at the origin the two solutions $\rho_\pm$ must be real and equal at the turning point so that the Hankel functions add to make a Bessel function. We now rewrite (48.16) in dimensionless form as

$$\left(-\mathrm{i}\underset{\sim}{\tau}_z\frac{\mathrm{d}}{\mathrm{d}x} + \delta(x)\underset{\sim}{\tau}_x\right)\begin{pmatrix}\rho_+(x)\\ \rho_-(x)\end{pmatrix} = [\lambda + \alpha(x)]\begin{pmatrix}\rho_+(x)\\ \rho_-(x)\end{pmatrix}, \tag{48.17}$$

where we defined

$$x = \frac{m\Delta_\infty}{\hbar^2 k_\perp}(r^2 - r_t^2)^{1/2}, \tag{48.18a}$$

$$\lambda = \frac{\varepsilon}{\Delta_\infty}, \tag{48.18b}$$

$$\alpha(x) = \frac{\mu e\hbar}{mcr\Delta_\infty}A_\theta(r), \tag{48.18c}$$

and

$$\delta(x) = \frac{\Delta(r)}{\Delta_\infty}, \tag{48.18d}$$

3. The WKB method assumes a solution of the form $e^{\mathrm{i}\sigma(r)}$, where $\sigma(r)$ plays the role of a dimensionless 'action'. Inserting this form into Eq. (48.12) yields the equation $-\mathrm{i}\sigma'' + \sigma'^2 - (\mathrm{i}\sigma'/r) + (\mu^2/r^2) - k^2 = 0$. Now σ' is of order k which is assumed large in WKB. We assume a small expansion parameter $\varepsilon \propto k^{-1}$ and collecting equal powers of ε we obtain the first two equations of the hierarchy so generated as:

$$\sigma_0'^2 = \frac{k^2r^2 - \mu^2}{r^2} \equiv \beta^2;\ \sigma_1' = \frac{\mathrm{i}}{2}\left(\frac{\sigma_0''}{\sigma_0'} - \frac{1}{r}\right).$$

Inserting σ_0'' and σ_0' computed from the solution of the first of these equations into the second and integrating the resulting expression yields $\sigma_1' = (\mathrm{i}/4)\ell\mathrm{n}[(k^2r^2/\mu^2) - 1]$ from which the r-dependent amplitude of (48.14) follows immediately.

4. In obtaining this equation we assumed

$$\beta >> \frac{1}{r} - \frac{r}{(r^2 - r_t^2)^{1/2}}.$$

where $\Delta_\infty = |\Delta(r \to \infty)|$ and λ is the eigenvalue of (48.17). Identifying our quasiclassical form of H_μ, Eq. (48.14), as $e^{(i/\hbar)S_F}$ in Eq. (47.22) we may write $\rho_\pm$ as

$$\begin{pmatrix} \rho_+(x) \\ \rho_-(x) \end{pmatrix} = \begin{pmatrix} e^{(i/\hbar)s_u(x)} \\ e^{(i/\hbar)s_v(x)} \end{pmatrix}. \tag{48.19}$$

Using Eqs. (47.25a) and (47.25b) we may rewrite (48.17) in a form analogous to Eqs. (47.27) and (47.28) as

$$\frac{d\theta(x)}{dx} + \delta(x)\cos[2\theta(x)] = \lambda + \alpha(x) \tag{48.20a}$$

and

$$\frac{d\bar{\theta}(x)}{dx} = i\delta(x)\sin[(2\theta(x)], \tag{48.20b}$$

where we wrote $s/\hbar = \theta$ and $\bar{s}/\hbar = \bar{\theta}$. These equations are to be solved subject to the boundary conditions that as $r \to \infty$, $\alpha(x) \to 0$, $\delta(x) \to 1$, and $s(x) \to s_\infty$, where $\cos(2s_\infty) = \lambda = \varepsilon/\Delta_\infty$; the solution then goes to the asymptotic form of (48.4). For bound core states $|\varepsilon| < \Delta_\infty$, θ is real and $\bar{\theta}$ is pure imaginary. If we are to have decaying solutions as $x \to \infty$, $2\theta_\infty$ must be between 0 and π so that $\sin 2s_\infty > 0$. For $\rho_\pm$ to be real at the turning point $x = 0$ we must have $2\theta_0 = 0$ or π (mod 2π). These boundary conditions determine the eigenvalues, $\lambda = \varepsilon/\Delta_\infty$, for a given μ and $k_\perp$. For $\varepsilon > \Delta_\infty$ scattering states are present, which will not be discussed here (see Bardeen *et al.* (1969)).

We must keep in mind that $\delta(x)\,(\Delta(r))$ and $\alpha(x)\,(A_\theta(r))$ are required input data for solving Eqs. (48.20); as mentioned earlier $\Delta(r)$ is determined from the self-consistency condition (48.10) and $A_\theta(r)$ from the fourth Maxwell equation with $j_\theta(r)$ calculated with (48.9). Because of the complexities of a complete numerical calculation,[5] Bardeen *et al.* (1969) employed model forms for $\delta(x)$ and $\alpha(x)$:

$$\delta(x) = \begin{cases} 0, & x < x_c \\ 1, & x > x_c \end{cases} \qquad \alpha(x) = \frac{b}{x^2 + b^2}.$$

Here x_c denotes a core radius and b is a penetration depth (in reduced units); we refer the reader to the original work for further details. The general conclusion is that there is one bound state for each μ and spin orientation; we have $\varepsilon > 0$ for $\mu > 0$ and, conversely, $\varepsilon < 0$ for $\mu < 0$. In the ground state ($T = 0$) the negative energy levels are filled and the positive levels empty. The occupied states contribute to the current with the same sign as the free states and they also contribute to the pairing potential.

48.3 A model calculation for the bound core states

We conclude this section by presenting a simple model calculation for the energy levels of the core

5. A numerical calculation was performed in which the free energy was minimized with respect to the parameters in the functions modeling $H(r)$ and $\Delta(r)$.

states which is useful in the extreme Type II limit ($\lambda >> \xi_0$) (Caroli *et al.* 1964). In this limit we may neglect A'_θ; and Eq. (48.3) becomes

$$\left\{\frac{\hbar^2}{2m}\left[-\frac{d^2}{dr^2}-\frac{1}{r}\frac{d}{dr}+\frac{1}{r^2}\left(\mu-\frac{1}{2}\underset{\sim}{\tau}_z\right)^2-k_\perp^2\right]\underset{\sim}{\tau}_z + \Delta(r)\underset{\sim}{\tau}_x\right\}\begin{pmatrix}R_+(r)\\ R_-(r)\end{pmatrix}=\varepsilon\begin{pmatrix}R_+(r)\\ R_-(r)\end{pmatrix}; \tag{48.21a}$$

writing $(\mu-\frac{1}{2}\underset{\sim}{\tau}_z)^2=\mu^2+\frac{1}{4}-\mu\underset{\sim}{\tau}_z$ we may rewrite Eq. (48.21a) as

$$\left\{\frac{\hbar^2}{2m}\left[-\frac{d^2}{dr^2}-\frac{1}{r}\frac{d}{dr}+\frac{1}{r^2}\left(\mu^2+\frac{1}{4}\right)-k_\perp^2\right]\underset{\sim}{\tau}_z + \Delta(r)\underset{\sim}{\tau}_x\right\}\begin{pmatrix}R_+(r)\\ R_-(r)\end{pmatrix}=\left(\varepsilon+\frac{\mu\hbar^2}{2mr^2}\right)\begin{bmatrix}R_+(r)\\ R_-(r)\end{bmatrix}. \tag{48.21b}$$

These equations can be accurately solved in the region $0<\mu<<k_F\xi_0$. We consider a radius r_c such that

$$\left(\mu+\frac{1}{2}\right)k_F^{-1}<<r_c<<\xi_0.$$

For $r<<r_c$ we may neglect $\Delta(r)$ in which case the solution is given by (48.7). For $r>r_c$ we seek a solution in the form[6]

$$R_\pm(r)=\rho_\pm(r)H_{\mu'}(k_\perp r)+\text{c.c.}, \tag{48.22}$$

where $\mu'=(\mu^2+\frac{1}{4})^{1/2}$ and $\rho_\pm(r)$ are again slowly varying functions.

Now $H_\mu(k_\perp r)$ satisfies the equation $-H''-(i/r)H'+[(\mu'^2/r^2)-k_\perp^2]H=0$ and substituting (48.22) into (48.21b) we obtain

$$\left\{\frac{\hbar^2}{2m}\left[-H_{\mu'}\frac{d^2}{dr^2}-\left(2H'_{\mu'}+\frac{1}{r}H_{\mu'}\right)\frac{d}{dr}\right]\underset{\sim}{\tau}_z + \Delta(r)H_{\mu'}\underset{\sim}{\tau}_x\right\}\begin{pmatrix}\rho_+(r)\\ \rho_-(r)\end{pmatrix}=\left(\varepsilon+\frac{\mu\hbar^2}{2mr^2}\right)H_{\mu'}\begin{pmatrix}\rho_+(r)\\ \rho_-(r)\end{pmatrix}.$$

In the limit in which we are working $k_\perp r>>1$ and we may neglect the terms in $H\rho''$ and $H'\rho'$ in this equation; furthermore, H may be approximated by the form (48.5a) where we have $H'\cong ik_\perp H$. Our envelope equation then becomes

$$\left[-i\underset{\sim}{\tau}_z\frac{\hbar^2k_\perp}{m}\frac{d}{dr}+\Delta(r)\underset{\sim}{\tau}_x\right]\begin{pmatrix}\rho_+(r)\\ \rho_-(r)\end{pmatrix}=\left(\varepsilon+\frac{\mu\hbar^2}{2mr^2}\right)\begin{pmatrix}\rho_+\\ \rho_-\end{pmatrix}. \tag{48.23}$$

6. Note the Hankel function index, $\mu'=(\mu^2+\frac{1}{4})^{1/2}$, differs from that in Eq. (48.11). Likewise Eq. (48.23) differs from (48.13).

For $\varepsilon << \Delta_\infty$ and $k_F r >> \mu$ the right-hand side of (48.23) may be treated as a perturbation; for the moment we assume $\Delta(x)$ is given. We seek a quasiclassical solution to $\rho_\pm$ of the form

$$\begin{pmatrix}\rho_+(r)\\ \rho_-(r)\end{pmatrix} = \begin{pmatrix}e^{is/\hbar}\\ e^{-is/\hbar}\end{pmatrix} e^{-\bar{\sigma}/\hbar}. \tag{48.24}$$

The real qualities s and $\bar{\sigma}(= -i\bar{s})$ entering (48.24) are the appropriate bound state forms of the actions defined in Eqs. (47.25). Inserting (48.24) and (48.23) we have

$$v_\perp\left(\frac{ds}{dr} + i\frac{d\bar{\sigma}}{dr}\right) + \Delta e^{-2is/\hbar} = \varepsilon + \frac{\mu\hbar^2}{2mr^2} \tag{48.25a}$$

and

$$v_\perp\left(\frac{ds}{dr} - i\frac{d\bar{\sigma}}{dr}\right) + \Delta e^{2is/\hbar} = \varepsilon + \frac{\mu\hbar^2}{2mr^2}, \tag{48.25b}$$

where $v_\perp \equiv \hbar k_\perp/m$. Adding and subtracting Eqs. (48.25a,b) yields

$$v_\perp\frac{ds}{dr} + \Delta\cos\frac{2s}{\hbar} = \varepsilon + \frac{\mu\hbar^2}{2mr^2} \tag{48.26a}$$

and

$$v_\perp\frac{d\bar{\sigma}}{dr} - \Delta\sin\frac{2s}{\hbar} = 0, \tag{48.26b}$$

which are the bound-state forms corresponding to Eqs. (47.27) and (47.28). When the quantity $\varepsilon + \mu\hbar^2/2mr^2$ is set to zero these equations have the zeroth order solution

$$s_0 = \frac{\pi\hbar}{4}; \quad \bar{\sigma}_0 = v_\perp^{-1}\int_0^r \Delta(r)dr. \tag{48.27}$$

We seek an iterative solution in the form

$$s(r) = s_0 + s_1(r) + \ldots; \quad \bar{\sigma}(r) = \bar{\sigma}_0 + \bar{\sigma}_1(r)\ldots.$$

Inserting these forms into Eqs. (48.26a) and (48.26b) and expanding the sine and cosine we find that to leading order we obtain $\bar{\sigma}_1(r) = 0$ and $s_1(r)$ obeys the equation

$$v_\perp\frac{ds_1}{dr} - \Delta(r)\frac{2s_1(r)}{\hbar} = \varepsilon + \frac{\mu\hbar^2}{2mr^2}, \tag{48.28}$$

which has the solution

$$s_1(r) = -\hbar\int_r^\infty \exp\left[\frac{2}{\hbar}\bar{\sigma}_0(r) - \frac{2}{\hbar}\bar{\sigma}_0(r')\right]\left(q + \frac{\mu}{2k_\perp r'^2}\right)dr', \tag{48.29}$$

where $q \equiv \varepsilon/\hbar v_\perp$. We will later require the quantity $s_1(r_c)$. We may let the lower limit of the integral in (48.29) approach zero if we simultaneously subtract $\int_0^r(ds_1/dr)dr$; by expanding $s_1(r) = (a_{-1}/r) + a_0 + a_1 r + \ldots$ and $\Delta = \Delta_1 r + \ldots$ and using (48.28) we find

$$\frac{ds_1}{dr} = \frac{\mu\hbar^2}{2mv_\perp r^2} - \frac{\Delta\mu\hbar}{mv_\perp^2 r} + \frac{\varepsilon}{v_\perp}, \quad r < r_c. \tag{48.30}$$

Noting that $e^{2\bar{\sigma}_0(r)} \cong 1$ for $r < r_c$, and that the divergent terms then cancel as $r \to 0$, we may then approximate $s_1(r_c)$ as

$$s_1(r_c) = -\frac{\mu}{2k_\perp r_c} + qr_c - \int_0^\infty e^{-2\bar{\sigma}_0(r')/\hbar}\left(q - \frac{\mu\Delta(r')}{\hbar k_\perp v_\perp r'}\right)dr'. \tag{48.31}$$

Our final form for the bound state wavefunction at r_c is

$$\begin{pmatrix} R_+(r_c) \\ R_-(r_c) \end{pmatrix} = \begin{pmatrix} \rho_+(r_c) \\ \rho_-(r_c) \end{pmatrix} H_m(k_\perp r_c)$$
$$= A' \begin{pmatrix} \exp\left[\frac{i\pi}{4} + \frac{is_1(r_c)}{\hbar}\right] \\ \exp\left[-\frac{i\pi}{4} - \frac{is_1(r_c)}{\hbar}\right] \end{pmatrix} e^{-\bar{\sigma}_0(r_c)/\hbar} H_{\mu'}(k_\perp r_c) \tag{48.32}$$

with $\bar{\sigma}_0$ and s_1 given by Eqs. (48.27) and (48.31) and A' a constant. Irrespective of the form we have obtained above for $s_1(r_c)$, Eq. (48.32) must join smoothly to the $r << \xi_0$ form given by Eq. (48.7). For $k_\perp^{-1} < r_c$ we may use the asymptotic forms of the Bessel and Hankel functions given in (48.5a) and (48.5b). Comparing these forms we obtain the condition

$$s_1(r_c) = qr_c - \frac{\mu}{2k_\perp r_c}. \tag{48.33}$$

Comparing Eqs. (48.33) and (48.31) we obtain the condition

$$\int_0^\infty e^{-2\bar{\sigma}_0(k_\perp, r)/\hbar}\left[q - \frac{\mu\Delta(r)}{\hbar k_\perp v_\perp r}\right]dr = 0$$

or

$$\varepsilon_\mu(k_\perp) = \frac{\mu}{k_\perp}\int_0^\infty \frac{\Delta(r)}{r} e^{-2\bar{\sigma}_0(k_\perp, r)/\hbar} dr \Big/ \int_0^\infty e^{-2\bar{\sigma}_0(k_\perp gr)/\hbar} dr; \tag{48.34a}$$

we rewrite the right-hand side as

$$\frac{\mu}{k_\perp}\frac{d\Delta(r)}{dr}\bigg|_{r=0} g(k_\perp), \tag{48.34b}$$

where $g(k_\perp)$ is a dimensionless function which depends on the precise shape of $\Delta(r)$, but which is close to unity. Now $d\Delta/dr \cong \Delta/\xi_0$, hence the bound-state energies of the core states are approximately

$$\varepsilon_\mu(k_\perp) = \frac{\mu\Delta}{k_\perp \xi_0} \sim \frac{\mu\Delta^2}{E_F},$$

where E_F is the Fermi energy; the lowest level corresponds to $\mu = \frac{1}{2}$. These levels would lead to a linear term in the specific heat of a Type II superconductor as $T \to 0$. They may be studied directly using scanning tunneling microscopy.

49 Time-dependent Bogoliubov equations

49.1 Basic equations

Our previous discussions have largely been limited to time-independent problems. In this section we discuss the effect of time-dependent perturbations.[1] The generalization of the gap function, Eq. (37.7), to the time-dependent case is

$$\Delta(\mathbf{r}, t) = V\langle\hat{\psi}(\mathbf{r}, t)\hat{\psi}(\mathbf{r}, t)\rangle \tag{49.1}$$

and the time-dependent versions of Eqs. (36.13) are

$$\hat{\psi}_{\uparrow}(\mathbf{r}, t) = \sum_n [u_n(\mathbf{r}, t)\hat{\gamma}_{n\uparrow} - v_n^*(\mathbf{r}, t)\hat{\gamma}_{n\downarrow}^{\dagger}] \tag{49.2a}$$

and

$$\hat{\psi}_{\downarrow}(\mathbf{r}, t) = \sum_n [u_n(\mathbf{r}, t)\hat{\gamma}_{n\downarrow} + v_n^*(\mathbf{r}, t)\hat{\gamma}_{n\uparrow}^{\dagger}]. \tag{49.2b}$$

Inserting these expansions in the operator form of the Schrödinger equation

$$i\hbar\frac{\partial\hat{\psi}_{\alpha}(\mathbf{r}, t)}{\partial t} = \hat{H}^{(e)}\hat{\psi}_{\alpha}(\mathbf{r}, t), \tag{49.3}$$

with $\hat{H}^{(e)}$ given by (36.8), and using the commutation relations (36.12c) and (36.12d) we obtain the time-dependent Bogoliubov equations

$$i\hbar\frac{\partial u_n(\mathbf{r}, t)}{\partial t} = [\mathscr{H}'_0 + U(\mathbf{r}, t)]u_n(\mathbf{r}, t) + \Delta(\mathbf{r}, t)v_n(\mathbf{r}, t) \tag{49.4}$$

and

$$i\hbar\frac{\partial v_n(\mathbf{r}, t)}{\partial t} = -[\mathscr{H}'^*_0 + U(\mathbf{r}, t)]v_n(\mathbf{r}, t) + \Delta^*(\mathbf{r}, t)u_n(\mathbf{r}, t). \tag{49.5}$$

For the time-independent case the mean fields are given by Eqs. (37.10) and (37.11). However, the

1. A rigorous discussion of time-dependent phenomena in many-body systems requires the introduction of rather formal techniques (see Lifshitz and Pitaevskii (1981), Ch. X). Here we give a heuristic treatment.

meaning of the equilibrium static Fermi occupation functions, $f(\varepsilon_n)$, which occur in these expressions is problematic in the time-dependent case. To the extent that the time-dependent disturbances of the system are small we may interpret the $f(\varepsilon_n)$ as the occupation numbers of the *equilibrium* system[2]

$$U(\mathbf{r},t) = -V\sum_n [\,|u_n(\mathbf{r},t)|^2 f_n + |v_n(\mathbf{r},t)|^2(1-f_n)] \tag{49.6}$$

and

$$\Delta(\mathbf{r},t) = +V\sum_n v_n^*(\mathbf{r},t)u_n(\mathbf{r},t)(1-2f_n). \tag{49.7}$$

49.2 The time-dependent, linearized, self-consistency condition

To obtain the time-dependent, self-consistency condition, analogous to Eqs. (38.6) and (38.7), we again treat the gap as a perturbation and calculate the first order shift in the wavefunctions $u_n(\mathbf{r},t)$ and $v_n(\mathbf{r},t)$, but now using time-dependent perturbation theory. We expand these wavefunctions as we did in Eqs. (38.2a) and (38.2b). The time-dependent forms generalizing (38.3a) and (38.3b) are[3]

$$u_n^{(0)}(\mathbf{r},t) = \theta(\xi_n)\phi_n(\mathbf{r})e^{-(i/\hbar)|\xi_n|t} \tag{49.8a}$$

and

$$v_n^{(0)}(\mathbf{r},t) = \theta(-\xi_n)\phi_n^*(\mathbf{r})e^{-(i/\hbar)|\xi_n|t} \tag{49.8b}$$

while those generalizing (38.4a) and (38.4b) are

$$u_n^{(1)}(\mathbf{r},t) = \sum_m a_{nm}(t)\phi_m(\mathbf{r})e^{-(i/\hbar)|\xi_m|t} \tag{49.9a}$$

and

$$v_n^{(1)}(\mathbf{r},t) = \sum_m b_{nm}(t)\phi_m^*(\mathbf{r})e^{-(i/\hbar)|\xi_m|t}. \tag{49.9b}$$

We insert Eqs. (49.8) and Eqs. (49.9) into Eq. (49.4) and Eq. (49.5), linearize, multiply by ϕ_m^* and ϕ_m, respectively, and integrate over d^3r (exploiting orthonormality of the ϕ_m) to obtain the first order differential equations

$$i\hbar\dot{a}_{nm} + |\xi_m|a_{nm} = \xi_m a_{nm} + \theta(-\xi_n)e^{-(i/\hbar)(|\xi_n|-|\xi_m|)t}\int \phi_n^*(\mathbf{r}')\phi_m^*(\mathbf{r}')\Delta(\mathbf{r}',t)d^3r' \tag{49.10a}$$

2. Formally if we regarded the time-dependent mean fields as being adiabatically evolved from their equilibrium forms, then the f_n are the occupation numbers at $t = -\infty$.
3. Eqs. (49.8) are the time-dependent solutions to Eqs. (49.4) and (49.5) for the case $\Delta = U = 0$; u_n and v_n are the wavefunctions of particle-like and hole-like excitations, respectively.

and

$$i\hbar\dot{b}_{nm} + |\xi_m| b_{nm} = -\xi_m b_{nm} + \theta(\xi_n)e^{-(i/\hbar)(|\xi_n|-|\xi_m|)t}\int \phi_n(\mathbf{r}')\phi_m(\mathbf{r}')\Delta^*(\mathbf{r}',t)d^3r', \qquad (49.10b)$$

which have the solutions

$$a_{nm}(t) = \frac{-i}{\hbar}\theta(-\xi_n)e^{(i/\hbar)(|\xi_m|-\xi_m)t}\int_{-\infty}^{t} dt' \int d^3r' e^{-(i/\hbar)(|\xi_n|-\xi_m)t'}\Delta(\mathbf{r}',t')\phi_n^*(\mathbf{r}')\phi_m^*(\mathbf{r}') \qquad (49.11a)$$

and

$$b_{nm}(t) = \frac{-i}{\hbar}\theta(\xi_n)e^{(i/\hbar)(|\xi_m|+\xi_m)t}\int_{-\infty}^{t} dt \int d^3r' e^{-(i/\hbar)(|\xi_n|+\xi_m)t'}\Delta^*(\mathbf{r}',t')\phi_n(\mathbf{r}')\phi_m(\mathbf{r}') \qquad (49.11b)$$

(which reduce to Eqs. (38.5) for the case of a static gap function). Inserting the first order, corrected wavefunctions into the time-dependent, self-consistency condition, (49.7), we obtain

$$\Delta(\mathbf{r},t) = \int_{-\infty}^{t} dt' \int d^3r' K(\mathbf{r},t;\mathbf{r}',t')\Delta(\mathbf{r}',t'), \qquad (49.12)$$

where we defined

$$K(\mathbf{r},\mathrm{t};\mathbf{r}',t') = \frac{iV}{\hbar}\sum_{n,m}(1-2f(|\xi_n|))[\theta(\xi_n)e^{-(i/\hbar)(|\xi_n|+\xi_m)(t-t')} - \theta(-\xi_n)e^{(i/\hbar)(|\xi_n|-\xi_m)(t-t')}]$$
$$\times \phi_n(\mathbf{r})\phi_m(\mathbf{r})\phi_n^*(\mathbf{r}')\phi_m^*(\mathbf{r}'); \qquad (49.13)$$

note these expressions depend only on the time difference $t - t' \equiv \tau$. Eq. (49.13) may be rearranged as

$$K(\mathbf{r},t;\mathbf{r}',t') = \frac{iV}{\hbar}\sum_{n,m} e^{-(i/\hbar)(\xi_n+\xi_m)(t-t')}\tanh\left(\frac{\xi_n}{2k_BT}\right)\phi_n(\mathbf{r})\phi_m(\mathbf{r})\phi_n^*(\mathbf{r}')\phi_m^*(\mathbf{r}')$$
$$= \frac{iV}{2\hbar}\sum_{n,m} e^{-(i/\hbar)(\xi_n+\xi_m)(t-t')}\left[\tanh\left(\frac{\xi_n}{2k_BT}\right) + \tanh\left(\frac{\xi_m}{2k_BT}\right)\right]\phi_n(\mathbf{r})\phi_m(\mathbf{r})\phi_n^*(\mathbf{r}')\phi_m^*(\mathbf{r}')$$
$$= \frac{iV}{\hbar}k_BT\sum_{\omega}\sum_{n,m} e^{-(i/\hbar)(\xi_n+\xi_m)(t-t')}\frac{\xi_n+\xi_m}{(\xi_n - i\hbar\omega)(\xi_m + i\hbar\omega)}\phi_n(\mathbf{r})\phi_m(\mathbf{r})\phi_n^*(\mathbf{r}')\phi_m^*(\mathbf{r}'), \qquad (49.14)$$

where in the second step we symmetrized with respect to m and n and in the third step we used Eq. (38.9). Eq. (49.14) is the formal solution for the kernel of the time-dependent, linearized, self-consistency condition.

For the case where $\Delta \neq \Delta(t)$ we may integrate (49.14) from $-\infty$, to t (assuming $\Delta = \lim_{s\to 0}\Delta e^{+st}$ and hence vanishes at $t = -\infty$), which yields

$$K(\mathbf{r},\mathbf{r}') = k_BTV\sum_{\omega}\sum_{n,m}\frac{\phi_m^*(\mathbf{r}')\phi_n^*(\mathbf{r}')\phi_m(\mathbf{r})\phi_n(\mathbf{r})}{(\xi_m + i\hbar\omega)(\xi_n - i\hbar\omega)}, \qquad (38.10)$$

as obtained for the static case in Sec. 38.

As with the time-dependent case we can relate the kernel to a correlation function. Using Eq. (38.11) we write (49.14) as

$$K(\mathbf{r},t;\mathbf{r}',t') = \frac{ik_BTV}{\hbar}\sum_\omega \int d\xi d\xi' e^{-(i/\hbar)(\xi+\xi')(t-t')} \frac{\xi+\xi'}{(\xi - i\hbar\omega)(\xi' + i\hbar\omega)} \sum_{n,m} \delta(\xi - \xi_n)\delta(\xi' - \xi_m)$$

$$\times \langle n|\delta(\mathbf{R}-\mathbf{r})\mathscr{K}|m\rangle\langle m|\mathscr{K}^\dagger\delta(\mathbf{R}-\mathbf{r}')|n\rangle, \tag{49.15}$$

where we included the full time reversal operator $\mathscr{K}$ (rather than $\mathscr{C}$) since this is the only change necessary for the spin-generalized case. Following the discussion leading to Eq. (38.14) we may write (49.15) as

$$K(\mathbf{r},t;\mathbf{r}',t') = \frac{iL^3k_BTV}{\hbar}\sum_\omega \int \frac{d\xi d\xi' \mathscr{N}(\xi)(\xi+\xi')}{(\xi - i\hbar\omega)(\xi' + i\hbar\omega)} e^{-(i/\hbar)(\xi+\xi')(t-t')} g_\xi(\mathbf{r},\mathbf{r}';\hbar\Omega), \tag{49.16}$$

where

$$g_\xi(\mathbf{r},\mathbf{r}';\hbar\Omega) \equiv \frac{\sum_{n,m} \langle n|\delta(\mathbf{R}-\mathbf{r})\mathscr{K}|m\rangle\langle m|\mathscr{K}^\dagger\delta(\mathbf{R}-\mathbf{r}')|n\rangle\delta(\xi-\xi_n)\delta(\xi_m - \xi_n - \hbar\Omega)}{\sum_n \delta(\xi - \xi_n)}, \tag{49.17}$$

with $\hbar\Omega \equiv \xi' - \xi$. We replace ξ' by $\hbar\Omega + \xi$ and $d\xi'$ by $\hbar d\Omega$; integrating over $d\xi$ (assuming g is independent of ξ and using Cauchy's theorem) followed by an integration over $d\Omega$ (using the definition of the Fourier transform of $g(\Omega)$ given by Eq. (38.16)) we obtain

$$K(\mathbf{r},t;\mathbf{r}',t') = \frac{2\pi}{\hbar} k_BTL^3\mathscr{N}(0)V\sum_\omega g(\mathbf{r},\mathbf{r}';t-t')e^{-2|\omega|(t-t')}. \tag{49.18}$$

As for the time-independent case, we find that the behavior of the linearized gap equation is completely governed by the correlation function $g(t)$ (see Eq. (38.14′)).

49.3 The linearized, time-dependent, G–L equation

It would be desirable to have a derivation of the time-dependent G–L theory discussed phenomenologically in Part I. The problem was first discussed by Schmid (1966) and later by Caroli and Maki (1967a,b) and by Gorkov and Eliashberg (1968). We refer the reader to Cyrot (1973) for an excellent review. To obtain a G–L-like equation we expand the gap function entering Eq. (49.12) as (see Eq. (45.6))

$$\Delta(\mathbf{r}',t') = \Delta(\mathbf{r},t) + \left(\frac{\partial\Delta}{\partial\mathbf{r}}\right)\cdot(\mathbf{r}-\mathbf{r}') + \frac{1}{2}(\mathbf{r}-\mathbf{r}')\cdot\left(\frac{\partial^2\Delta}{\partial\mathbf{r}\partial\mathbf{r}}\right)\cdot(\mathbf{r}-\mathbf{r}') + \ldots$$

$$+ \left(\frac{\partial\Delta}{\partial t}\right)(t-t') + \ldots. \tag{49.19}$$

As discussed in Sec. 45, the linear term in $\mathbf{r}-\mathbf{r}'$ vanishes; the linearized gap equation then takes the form (see Eq. (45.8))

$$\Delta(\mathbf{r},t) = \int_{-\infty}^{0} \mathrm{d}\tau \int \mathrm{d}^3R K(R,\tau)\Delta(\mathbf{r},t) + \frac{1}{3}\int_{-\infty}^{0} \mathrm{d}\tau \int \mathrm{d}^3R R^2 K(R,\tau)\frac{1}{2}\left[\nabla - \frac{2ie}{\hbar c}\mathbf{A}(\mathbf{r},t)\right]^2 \Delta(\mathbf{r},t)$$

$$+ \int_{-\infty}^{0} \mathrm{d}\tau\tau \int \mathrm{d}^3R K(R,\tau)\left[\frac{\partial}{\partial t} + \frac{2ie}{\hbar}\phi(\mathbf{r},t)\right]\Delta(\mathbf{r},t)\ldots, \qquad (49.20)$$

where we wrote $\tau = t - t'$. Note we have written terms involving the space and time derivatives in a gauge invariant manner. The integral in the first term on the right was obtained earlier as $\mathcal{N}(0)V\ell n(1.14\hbar\omega_D/k_BT)$ (see Eq. (40.4)) while the integral entering the second term was obtained for the clean limit in Eq. (45.11) and for the dirty limit in Eq. (45.14). This leaves only the evaluation of the integral

$$\int_{-\infty}^{0} \mathrm{d}\tau\tau \int \mathrm{d}^3R K(R,\tau) = \frac{iV}{\hbar}\int \mathrm{d}^3R \int_{-\infty}^{0} \mathrm{d}\tau\tau \left(\frac{1}{(2\pi)^3}\right)^2 L^6 \int \mathrm{d}^3k \mathrm{d}^3k' \mathrm{e}^{-(i/\hbar)(\xi+\xi')\tau}$$

$$\times \tanh\left(\frac{\xi}{2k_BT}\right)\left[\frac{1}{L^{3/2}}\right]^4 \mathrm{e}^{i(\mathbf{k}+\mathbf{k}')\cdot(\mathbf{r}-\mathbf{r}')} \qquad (49.21)$$

which we have written for plane waves using the first of the forms on the right of Eq. (49.14). Carrying out the integral over $\mathrm{d}\tau$ we have

$$\int_{-\infty}^{0} \tau \mathrm{d}\tau \int \mathrm{d}^3R K(R,\tau) = i\left(\frac{1}{(2\pi)^3}\right)^2 \hbar V \int \frac{\mathrm{d}^3k \mathrm{d}^3k' \mathrm{d}^3R}{(\xi + \xi' + 2i\delta)^2} \tanh\left(\frac{\xi}{2k_BT}\right) \mathrm{e}^{i(\mathbf{k}+\mathbf{k}')\cdot\mathbf{R}}; \qquad (49.22)$$

here we have assumed an infinitesimal contribution to the energy, $\xi \to \xi + i\delta$ (which is equivalent to adiabatically turning on the gap perturbation) so that the integral vanishes at $t = -\infty$. The integral over d^3R gives $(2\pi)^3\delta^{(3)}(\mathbf{k}+\mathbf{k}')$; integrating over d^3k', (49.22) becomes[4]

$$\int_{-\infty}^{0} \tau \mathrm{d}\tau \int \mathrm{d}^3R K(R,\tau) = \frac{i\hbar}{(2\pi)^3}\frac{V}{4}\int \frac{\mathrm{d}^3k}{(\xi + i\delta)^2} \tan\mathrm{h}\left(\frac{\xi}{2k_BT}\right)$$

$$= \frac{i\mathcal{N}(0)V\hbar}{4}\int \frac{\mathrm{d}\xi}{(\xi + i\delta)^2}\tanh\left(\frac{\xi}{2k_BT}\right)$$

$$= -\frac{\pi\mathcal{N}(0)V\hbar}{8k_BT_c}. \qquad (49.23)$$

The validity of including a cubic term in the time-dependent, G–L equation is the subject of some controversy. The problems have been reviewed by Cyrot (1973). A calculation of the current density using time-dependent perturbation theory leads to the same equation as for the static case discussed in Sec. 45, i.e., the second G–L equation.

4. If the contour is closed in the upper half plane we only need to consider the poles of $\tanh x$ occurring at $x = (2n+1)\frac{i\pi}{2}$; the residue then involves the sum $\Sigma_{n=0}^{100} \frac{1}{(2n+1)^2} = \frac{\pi^2}{8}$.

50 The response of a superconductor to an electromagnetic field

50.1 The vector potential as a time-dependent perturbation

In this section we generalize the discussion of Sec. 46 to apply to the case of time-dependent electric and magnetic fields. The vector potential continues to be the fundamental field and it is related to the electric and magnetic fields, by the usual relations:

$$\mathbf{E}(\mathbf{r},t) = -\frac{1}{c}\frac{\partial \mathbf{A}(\mathbf{r},t)}{\partial t}; \tag{50.1a}$$

and

$$\mathbf{H}(\mathbf{r},t) = \nabla \times \mathbf{A}(\mathbf{r},t). \tag{50.1b}$$

If the external fields are sufficiently weak we may regard $\mathbf{A}(\mathbf{r},t)$ as a perturbation on the time-dependent Bogoliubov equations, (49.4) and (49.5). If we work in the London gauge where $\Delta^{(1)} = 0$ (by a time-dependent generalization of the argument leading to Eq. (46.14)), the analogues of (46.15a) and (46.15b) are

$$\left[i\hbar\frac{\partial}{\partial t} + \frac{\hbar^2}{2m}\nabla^2 - U(\mathbf{r})\right]u_n^{(1)}(\mathbf{r},t) - \Delta^{(0)}(\mathbf{r},t)v_n^{(1)}(\mathbf{r},t)$$
$$= \frac{ie\hbar}{2mc}[\mathbf{A}(\mathbf{r},t)\cdot\nabla + \nabla\cdot\mathbf{A}(\mathbf{r},t)]u_n^{(0)}(\mathbf{r},t) \tag{50.2a}$$

and

$$\left[i\hbar\frac{\partial}{\partial t} - \frac{\hbar^2}{2m}\nabla^2 + U(\mathbf{r})\right]v_n^{(1)}(\mathbf{r},t) - \Delta^{(0)*}(\mathbf{r},t)u_n^{(1)}(\mathbf{r},t)$$
$$= \frac{ie\hbar}{2mc}[\mathbf{A}(\mathbf{r},t)\cdot\nabla + \nabla\cdot\mathbf{A}(\mathbf{r},t)]v_n^{(0)}(\mathbf{r},t). \tag{50.2b}$$

For the time-dependent case Eqs. (46.16) become

$$u_n^{(1)}(\mathbf{r},t) = \sum_m a_{nm}(t)\phi_m(\mathbf{r})e^{-(i/\hbar)\epsilon_m t} \tag{50.3a}$$

and

$$v_n^{(1)}(\mathbf{r},t) = \sum_m b_{nm}(t)\phi_m(\mathbf{r})e^{-(i/\hbar)\varepsilon_m t}. \tag{50.3b}$$

We continue to assume $\Delta^{(0)}$ is independent of $\mathbf{r}$ and drop the superscript and approximate the zeroth order Bogoliubov wavefunctions as

$$u_n^{(0)}(\mathbf{r},t) = \mathscr{u}_n(\varepsilon_n)\phi_n(\mathbf{r})e^{-(i/\hbar)\varepsilon_n t} \tag{50.4a}$$

and

$$v_n^{(0)}(\mathbf{r},t) = \mathscr{v}_n(\varepsilon_n)\phi_n(\mathbf{r})e^{-(i/\hbar)\varepsilon_n t}. \tag{50.4b}$$

Inserting (50.3a) and (50.3b) into (50.2a) and (50.2b), multiplying by $\phi_m^*(\mathbf{r})$, integrating over all $\mathbf{r}$, and using the orthonormality of the wavefunctions yields

$$i\hbar\dot{a}_{nm}(t) + (\varepsilon_m - \xi_m)a_{nm}(t) - \Delta b_{nm}(t) = M_{nm}(t)\mathscr{u}_n(\varepsilon_n)e^{-(i/\hbar)(\varepsilon_n - \varepsilon_m)t} \tag{50.5a}$$

and

$$i\hbar\dot{b}_{nm}(t) + (\varepsilon_m + \xi_m)b_{nm}(t) - \Delta a_{nm}(t) = M_{nm}(t)\mathscr{v}_n(\varepsilon_n)e^{-(i/\hbar)(\varepsilon_n - \varepsilon_m)t}, \tag{50.5b}$$

where the matrix elements $M_{nm}(t)$ were given earlier as

$$M_{nm}(t) = \frac{ie\hbar}{2mc}\int \phi_m^*(\mathbf{r})[\mathbf{A}(\mathbf{r},t)\cdot\nabla + \nabla\cdot\mathbf{A}(\mathbf{r},t)]\phi_n(\mathbf{r})d^3r = M_{mn}^*(t). \tag{46.19a}$$

As a first step in solving these equations we rewrite them in matrix form as

$$\left[i\hbar\frac{d}{dt} + \varepsilon_n - \underset{\sim}{h}_m\right]\begin{bmatrix} a_{nm} \\ b_{nm}\end{bmatrix} = \begin{bmatrix} \mathscr{u}_n \\ \mathscr{v}_n\end{bmatrix} M_{nm}(t)e^{-(i/\hbar)(\varepsilon_n - \varepsilon_m)t}, \tag{50.6}$$

where we introduced the matrix

$$\underset{\sim}{h}_m \equiv \begin{pmatrix} \xi_m & \Delta \\ \Delta & -\xi_m \end{pmatrix} = \xi_m\underset{\sim}{\tau}_z + \Delta\underset{\sim}{\tau}_x \tag{50.7}$$

and factored it using the Pauli matrices.

We make a change in variables according to

$$\begin{bmatrix} a_{nm}(t) \\ b_{nm}(t)\end{bmatrix} = e^{(i/\hbar)\varepsilon_m t}e^{-(i/\hbar)\underset{\sim}{h}_m t}\begin{bmatrix} a'_{nm}(t) \\ b'_{nm}(t)\end{bmatrix}, \tag{50.8}$$

which on substitution into (50.6) gives

$$i\hbar\begin{bmatrix} \dot{a}'_{nm}(t) \\ \dot{b}'_{nm}(t)\end{bmatrix} = e^{-(i/\hbar)\varepsilon_n t}e^{-(i/\hbar)\underset{\sim}{h}_m t}\begin{bmatrix} \mathscr{u}_n \\ \mathscr{v}_n\end{bmatrix} M_{nm}(t), \tag{50.9}$$

which may be directly integrated provided $M_{nm}(t)$ is given.

In order to proceed with the discussion we assume $\mathbf{A}$ involves a single frequency ω and write it in the form

$$\mathbf{A}(\mathbf{r},t) = \lim_{s\to 0}\frac{1}{2}[\mathbf{A}(\mathbf{r})e^{i\omega t + st} + \mathbf{A}^*(\mathbf{r})e^{-i\omega t + st}]. \tag{50.10}$$

The factor e^{st} insures the perturbation vanishes for $t = -\infty$. Carrying out the integration we obtain

$$\begin{bmatrix} a'_{nm} \\ b'_{nm} \end{bmatrix} = \{(\varepsilon_n - \underset{\sim}{h}_m - \hbar\omega + i\delta)^{-1} M_{nm} e^{-(i/\hbar)(\varepsilon_n - \underset{\sim}{h}_m - \hbar\omega)t}$$

$$- (\varepsilon_n - \underset{\sim}{h}_m + \hbar\omega + i\delta)^{-1} M^*_{nm} e^{-(i/\hbar)(\varepsilon_n - \underset{\sim}{h}_m + \hbar\omega)t}\} \begin{bmatrix} u_n \\ v_n \end{bmatrix}, \tag{50.11}$$

where we defined the time-independent matrix element (in a London gauge) as

$$M_{nm}[\mathbf{A}(\mathbf{r})] = \frac{ie\hbar}{mc} \int \phi^*_m(\mathbf{r})[\mathbf{A}(\mathbf{r}) \cdot \nabla]\phi_n(\mathbf{r}) d^3r \tag{50.12}$$

(with a similar form for $\mathbf{A}^*(\mathbf{r})$) and we wrote $\delta = \hbar s$. Using Eq. (50.8) and computing the inverse matrices appearing in (50.11) as

$$[\varepsilon_n - \underset{\sim}{h}_m \mp \hbar\omega]^{-1} = \frac{1}{(\varepsilon_n \mp \hbar\omega)^2 - \varepsilon_m^2} \begin{bmatrix} \varepsilon_n + \xi_m \mp \hbar\omega & \Delta \\ \Delta & \varepsilon_n - \xi_m \mp \hbar\omega \end{bmatrix} \tag{50.13}$$

we obtain our final expressions for $a_{nm}(t)$ and $b_{nm}(t)$ as

$$a_{nm}(t) = M_{nm} \frac{e^{-(i/\hbar)(\varepsilon_n - \varepsilon_m - \hbar\omega)t}}{(\varepsilon_n - \hbar\omega + i\delta)^2 - \varepsilon_m^2} [(\varepsilon_n + \xi_m - \hbar\omega)u_n + \Delta v_n]$$

$$- M^*_{nm} \frac{e^{-(i/\hbar)(\varepsilon_n - \varepsilon_m + \hbar\omega)t}}{(\varepsilon_n + \hbar\omega + i\delta)^2 - \varepsilon_m^2} [(\varepsilon_n + \xi_m + \hbar\omega)u_n + \Delta v_n] \tag{50.14a}$$

and

$$b_{nm}(t) = M_{nm} \frac{e^{-(i/\hbar)(\varepsilon_n - \varepsilon_m - \hbar\omega)t}}{(\varepsilon_n - \hbar\omega + i\delta)^2 - \varepsilon_m^2} [\Delta u_n + (\varepsilon_n - \xi_m - \hbar\omega)v_n]$$

$$- M^*_{nm} \frac{e^{-(i/\hbar)(\varepsilon_n - \varepsilon_m + \hbar\omega)t}}{(\varepsilon_n + \hbar\omega + i\delta)^2 - \varepsilon_m^2} [\Delta u_n + (\varepsilon_n - \xi_m + \hbar\omega)v_n] \tag{50.14b}$$

which are the time-dependent versions of Eqs. (46.20).

50.2 Relation between the current density and the vector potential

If the first-order-corrected, time-dependent wavefunctions are inserted in Eq. (46.21) for the current density, $\langle \mathbf{j}_1(\mathbf{r}) \rangle$, one obtains (after some algebra) an expression having the same form as Eq. (46.26), where the expression generalizing (46.27a) for $L(\xi_n, \xi_m)$ is now

$$L(\omega, \xi_n, \xi_m) = \frac{1}{4}\left(1 + \frac{\xi_n\xi_m + \Delta^2}{\varepsilon_n\varepsilon_m}\right)\left(\frac{f_m - f_n}{\varepsilon_n - \varepsilon_m - \hbar\omega + i\delta} + \frac{f_m - f_n}{\varepsilon_n - \varepsilon_m + \hbar\omega - i\delta}\right)$$

$$+ \frac{1}{4}\left(1 - \frac{\xi_n\xi_m + \Delta^2}{\varepsilon_n\varepsilon_m}\right)\left(\frac{1 - f_n - f_m}{\varepsilon_n + \varepsilon_m - \hbar\omega + i\delta} + \frac{1 - f_n - f_m}{\varepsilon_n + \varepsilon_m + \hbar\omega - i\delta}\right); \tag{50.15a}$$

we can construct the following unsymmetrized form of Eq. (50.15a), which will be more convenient for the subsequent integration over ξ_m

$$L(\omega, \xi_n, \xi_m) = -\frac{1}{2}(1 - 2f_n)\left\{\frac{\varepsilon_n + \hbar\omega - i\delta + [(\Delta^2 + \xi_n\xi_m)/\varepsilon_n]}{\varepsilon_m^2 - (\varepsilon_n + \hbar\omega - i\delta)^2}\right.$$

$$\left. + \frac{\varepsilon_n - \hbar\omega + i\delta + [(\Delta^2 + \xi_n\xi_m)/\varepsilon_n]}{\varepsilon_m^2 - (\varepsilon_n - \hbar\omega + i\delta)^2}\right\}. \tag{50.15b}$$

These expressions were first obtained by Mattis and Bardeen (1958).[1] The expression generalizing (46.79) which connects the Fourier components of $\mathbf{A}(t, \mathbf{r})$ and $\mathbf{j}(t, \mathbf{r})$ is clearly

$$\langle j_i(\omega, \mathbf{r})\rangle = \int d^3r S_{ij}(\omega, \mathbf{r}, \mathbf{r}')A_j(\omega, \mathbf{r}'), \tag{50.16}$$

where

$$S_{ij}(\omega, \mathbf{r}, \mathbf{r}') = \left(\frac{e\hbar}{m}\right)^2 \frac{1}{c}\int d\xi d\xi' L(\omega, \xi, \xi')M_{ij}^2(\mathbf{r}, \mathbf{r}', \xi, \xi') - \frac{ne^2}{mc}\delta^{(3)}(\mathbf{r} - \mathbf{r}')\delta_{ij}. \tag{50.17}$$

As discussed in Subsec. 46.6 and leading up to Eq. (46.81), the matrix elements entering S_{ij} can be obtained from the nonlocal, frequency-dependent conductivity, $\sigma_{ij}(\mathbf{r}, \mathbf{r}', \Omega)$, where $\hbar\Omega = \xi - \xi'$ and we may accurately approximate (50.17) as

$$S_{ij}(\omega, \mathbf{r}, \mathbf{r}') = \frac{1}{2\pi\hbar c}\int d\xi d\xi' L(\omega, \xi, \xi')\mathrm{Re}\left[\sigma_{ij}\left(\mathbf{r}, \mathbf{r}', \frac{\xi - \xi'}{\hbar}\right)\right]$$

$$- \frac{ne^2}{mc}\delta^{(3)}(\mathbf{r} - \mathbf{r}')\delta_{ij}. \tag{50.18}$$

Note the only difference between (50.18) and (46.81) is that the function L is now frequency-dependent. Substituting Chambers' expression (46.85), for σ_{ij}, Eq. (50.18) becomes

$$S_{ij}(\omega, \mathbf{r}, \mathbf{r}') = \frac{e^2\mathcal{N}v_F}{2\pi^2\hbar c}\frac{R_iR_j}{R^4}e^{-R/\ell}I(\omega, R, T), \tag{50.19}$$

where

$$I(\omega, R, T) = \int d\xi d\xi'\left[L(\omega, \xi, \xi') - \frac{f(\xi) - f(\xi')}{\xi - \xi'}\right]\cos\left(\frac{\xi - \xi'}{\hbar v_F}R\right); \tag{50.20}$$

in obtaining this expression we subtracted the normal-state value of S_{ij} for $\omega = 0$ to avoid convergence problems (see Eq. (46.80)). We will not discuss the subsequent integrations in detail but will only list the pertinent results as obtained by Mattis and Bardeen (1958). The first integral is performed using contour integration and convergence is facilitated by introducing a factor

$$\lim_{a\to\infty}\frac{a^2}{\xi^2 + \xi'^2 + a^2}, \tag{50.21}$$

1. These expressions were also obtained using Green's function methods by Abrikosov, Gorkov, and Khalatnikov (1959). The extension to the case of alloys has been discussed by Abrikosov and Gorkov (1959).

where the limit is taken after the integration. The result (in the limit $\delta \to 0$) is

$$I(\omega, R, T) = -\pi i \int_{\Delta-\hbar\omega}^{\Delta} [1 - 2f(\varepsilon + \hbar\omega)]\left[g(\varepsilon)\cos\left(\frac{\xi_\omega R}{\hbar v_F}\right) - i\sin\left(\frac{\xi_\omega R}{\hbar v_F}\right)\right] e^{i(\xi R/\hbar v_F)} d\varepsilon$$

$$- \pi i \int_{\Delta}^{\infty} \left\{[1 - 2f(\varepsilon + \hbar\omega)]\left[g(\varepsilon)\cos\left(\frac{\xi_\omega R}{\hbar v_F}\right) - i\sin\left(\frac{\xi_\omega R}{\hbar v_F}\right)\right] e^{i(\xi R/\hbar v_F)}\right.$$

$$\left. - [1 - 2f(\varepsilon)]\left[g(\varepsilon)\cos\left(\frac{\xi R}{\hbar v_F}\right) + i\sin\left(\frac{\xi R}{\hbar v_F}\right)\right] e^{-i(\xi_\omega R/\hbar v_F)}\right\} d\varepsilon, \qquad (50.22)$$

where

$$\xi_\omega \equiv [(\varepsilon + \hbar\omega)^2 - \Delta^2]^{1/2} \qquad (50.23)$$

and

$$g(\varepsilon) = (\varepsilon^2 + \Delta^2 + \hbar\omega\varepsilon)/\xi\xi_\omega. \qquad (50.24)$$

For a negative argument, $-x, (-x)^{1/2} = +ix^{1/2}$. The negative sign of the square root is taken when $\hbar\omega - \varepsilon > \Delta$. Eq. (50.22) reduces to $-\pi i\hbar\omega e^{-iR\omega/v_F}$ in the normal state ($\Delta \to 0$). Further integrations must be performed numerically for the general case.

In the limit where the penetration of the field is small compared with the coherence length, ξ_0 (referred to as the anomalous limit), we may set factors containing $R/\hbar v_F$ to zero. The quantity measured experimentally is usually proportional to the change in the complex conductivity, $\sigma = \sigma_1 + i\sigma_2$, relative to the normal-state conductivity, σ_N, which in the anomalous limit is given by

$$\frac{\sigma_1 - i\sigma_2}{\sigma_N} = \frac{I(\omega, 0, T)}{-\pi i\hbar\omega}; \qquad (50.25)$$

the real and imaginary parts of (50.25) are given by

$$\frac{\sigma_1}{\sigma_N} = \frac{2}{\hbar\omega}\int_{\Delta}^{\infty} [f(\varepsilon) - f(\varepsilon + \hbar\omega)]g(\varepsilon)d\varepsilon + \frac{1}{\hbar\omega}\int_{\Delta-\hbar\omega}^{-\Delta} [1 - 2f(\varepsilon + \hbar\omega)]g(\varepsilon)d\varepsilon \qquad (50.26)$$

and

$$\frac{\sigma_2}{\sigma_N} = \frac{1}{\hbar\omega}\int_{\Delta-\hbar\omega}^{\Delta} \frac{[1 - 2f(\varepsilon + \hbar\omega)](\varepsilon^2 + \Delta^2 + \hbar\omega\varepsilon)}{(\Delta^2 - \varepsilon^2)^{1/2}[(\varepsilon + \hbar\omega)^2 - \Delta^2]^{1/2}} d\varepsilon. \qquad (50.27)$$

Eq. (50.26) has the $\eta = +1$ form of interaction discussed in Sec. 34. The second term in (50.26) does not contribute unless $\hbar\omega > 2\Delta(T)$, in which case the lower limit of the integral in (50.27) is replaced by $-\Delta$ (rather than $\Delta - \hbar\omega$). The signs of the square roots are such that $g(\varepsilon)$ is positive in both integrals in (50.26).

For the $T = 0$ case ($f(\varepsilon) = 0$), the integrals for the anomalous case may be performed analytically in terms of the Euler elliptic functions $E(x)$ and $K(x)$. The first integral in (50.26) vanishes and we have absorption only when $\hbar\omega > 2\Delta(0)$; the result for σ_1 is

$$\frac{\sigma_1}{\sigma_N} = \left(1 + \frac{2\Delta}{\hbar\omega}\right)E\left(\frac{|2\Delta - \hbar\omega|}{|2\Delta + \hbar\omega|}\right) - 2\left(\frac{2\Delta}{\hbar\omega}\right)K\left(\frac{|2\Delta - \hbar\omega|}{|2\Delta + \hbar\omega|}\right); \quad \hbar\omega > 2\Delta. \qquad (50.28)$$

The expression for σ_2 following form (50.27) (valid at all ω) is

$$\frac{\sigma_2}{\sigma_N} = \frac{1}{2}\left(\left(\frac{2\Delta}{\hbar\omega} + 1\right)E\left\{\left[1 - \left(\frac{2\Delta - \hbar\omega}{2\Delta + \hbar\omega}\right)^2\right]^{1/2}\right\}\right.$$
$$\left. + \left(\frac{2\Delta}{\hbar\omega} - 1\right)K\left\{\left[1 - \left(\frac{2\Delta - \hbar\omega}{2\Delta + \hbar\omega}\right)^2\right]^{1/2}\right\}\right). \tag{50.29}$$

51 The Bogoliubov equations for an unconventional superfluid

Most superconductors are what are now called conventional superconductors, which means that the Cooper pairs form in a singlet state (with oppositely directed spins) having no net orbital angular momentum ($\ell = 0$). However, there is no fundamental reason why pairs with nonzero angular momentum cannot form. From the Pauli principle the total pair wavefunction must be antisymmetric on interchanging the particles. For the orbital wavefunction (see Eq. (25.1), $\phi(\rho) = (-1)^{\ell}\phi(-\rho)$; therefore even ℓ are associated with an antisymmetric spin wavefunction χ, the $S = 0$ singlet state of Eq. (25.2), while odd ℓ require a symmetric $S = 1$ triplet spin state. The only established unconventional superfluid is ^{3}He below 2.3×10^{-3} K (for the melting pressure), but there is growing evidence that other materials may be unconventional (e.g., the intermetallic heavy fermion compound UPt_3 and various high temperature cuprates). We will say more about ^{3}He in Sec. 53. In the next two sections we generalize the apparatus of BCS superconductivity to apply to unconventional superfluids (with particular emphasis on ^{3}He).

To begin let us consider an effective Hamiltonian which generalizes Eq. (41.1) to the case of a spin-dependent gap function:

$$\hat{H}_e = \int d^3r \Big\{ \hat{\psi}^\dagger_\alpha(\mathbf{r})\mathscr{H}'_0\hat{\psi}_\alpha(\mathbf{r}) + \int d^3r' [\hat{\psi}^\dagger_\alpha(\mathbf{r}) U_{\alpha\beta}(\mathbf{r},\mathbf{r}')\hat{\psi}_\beta(\mathbf{r}')$$

$$+ \Delta^*_{\alpha\beta}(\mathbf{r},\mathbf{r}')\hat{\psi}_\beta(\mathbf{r}')\hat{\psi}_\alpha(\mathbf{r}) + \Delta_{\alpha\beta}(\mathbf{r},\mathbf{r}')\hat{\psi}^\dagger_\alpha(\mathbf{r})\hat{\psi}^\dagger_\beta(\mathbf{r}')] \Big\}. \tag{51.1}$$

By relabeling the dummy coordinates $\alpha, \beta; \mathbf{r}, \mathbf{r}'$, the third term in this expression (which is the Hermitian conjugate of the fourth term) may be written as $\Delta^*_{\beta\alpha}(\mathbf{r}',\mathbf{r})\hat{\psi}_\alpha(\mathbf{r})\hat{\psi}_\beta(\mathbf{r}')$ or $\Delta^*_{\beta\alpha}(\mathbf{r},\mathbf{r}')\hat{\psi}_\alpha(\mathbf{r}')\hat{\psi}_\beta(\mathbf{r})$. Pairing with $\ell > 0$ cannot occur for a δ function (or contact) potential; hence we must introduce mean potentials that depend on two coordinates (as real many-body potentials do). We recall the two-particle operator in the full Hamiltonian, Eq. (36.3),

$$V^{(2)} = \frac{1}{2}\int d^3r \int d^3r' \hat{\psi}^\dagger_\delta(\mathbf{r})\hat{\psi}^\dagger_\gamma(\mathbf{r}') U^{(2)}_{\delta\gamma,\alpha\beta}(\mathbf{r},\mathbf{r}')\psi_\alpha(\mathbf{r}')\hat{\psi}_\beta(\mathbf{r}). \tag{51.2}$$

Using the Hartree–Fock–Gorkov factorization (Eq. (37.2)) we can generalize the expression (37.3) for the variation of the free energy in the mean potential approximation to:

$$\delta F = \int d^3r \Big\{ \mathscr{H}'_0 \delta\langle \hat{\psi}^\dagger_\alpha(\mathbf{r})\hat{\psi}_\alpha(\mathbf{r})\rangle + \frac{1}{2}\int d^3r' U^{(2)}_{\delta\gamma,\alpha\beta}(\mathbf{r},\mathbf{r}')$$

$$\times [-\langle \hat{\psi}^\dagger_\delta(\mathbf{r})\hat{\psi}_\alpha(\mathbf{r}')\rangle \delta\langle \hat{\psi}^\dagger_\gamma(\mathbf{r}')\hat{\psi}_\beta(\mathbf{r})\rangle - \langle \hat{\psi}^\dagger_\gamma(\mathbf{r}')\hat{\psi}_\beta(\mathbf{r})\rangle \delta\langle \hat{\psi}^\dagger_\delta(\mathbf{r})\hat{\psi}_\alpha(\mathbf{r}')\rangle$$

$$+ \langle\hat{\psi}_\delta^\dagger(\mathbf{r})\hat{\psi}_\beta(\mathbf{r})\rangle\delta\langle\hat{\psi}_\gamma^\dagger(\mathbf{r}')\hat{\psi}_\alpha(\mathbf{r}')\rangle + \langle\hat{\psi}_\gamma^\dagger(\mathbf{r}')\hat{\psi}_\alpha(\mathbf{r}')\rangle\delta\langle\hat{\psi}_\delta^\dagger(\mathbf{r})\hat{\psi}_\beta(\mathbf{r})\rangle$$

$$+ \langle\hat{\psi}_\delta^\dagger(\mathbf{r})\hat{\psi}_\gamma^\dagger(\mathbf{r}')\rangle\delta\langle\hat{\psi}_\alpha(\mathbf{r}')\hat{\psi}_\beta(\mathbf{r})\rangle + \langle\hat{\psi}_\alpha(\mathbf{r}')\hat{\psi}_\beta(\mathbf{r})\rangle\delta\langle\hat{\psi}_\delta^\dagger(\mathbf{r})\hat{\psi}_\gamma^\dagger(\mathbf{r}')\rangle]\Big\}$$

$$- T\delta S = 0. \tag{51.3}$$

We define an effective free energy from Eq. (51.1), analogous to Eq. (37.4), and perform a variation

$$\delta F_e = \int d^3r \Big\{ \mathscr{H}'_0 \delta\langle\hat{\psi}_\alpha^\dagger(\mathbf{r})\hat{\psi}_\alpha(\mathbf{r})\rangle + \int d^3r' [U_{\alpha\beta}(\mathbf{r},\mathbf{r}')\delta\langle\hat{\psi}_\alpha^\dagger(\mathbf{r})\hat{\psi}_\beta(\mathbf{r}')\rangle$$

$$+ \Delta_{\beta\alpha}^*(\mathbf{r},\mathbf{r}')\delta\langle\hat{\psi}_\alpha(\mathbf{r}')\hat{\psi}_\beta(\mathbf{r})\rangle + \Delta_{\alpha\beta}(\mathbf{r},\mathbf{r}')\delta\langle\hat{\psi}_\alpha^\dagger(\mathbf{r})\hat{\psi}_\beta^\dagger(\mathbf{r}')\rangle]\Big\}$$

$$- T\delta S = 0. \tag{51.4}$$

The first four terms in the square brackets of Eq. (51.3) contribute to the self-consistent potential U. Making the substitutions: $\gamma \to \alpha, \alpha \to \gamma$; $\alpha \to \beta, \delta \to \alpha, \beta \to \gamma, \gamma \to \delta$; $\gamma \to \alpha, \alpha \to \beta, \beta \to \gamma$; and $\delta \to \alpha, \alpha \to \gamma, \gamma \to \delta$, respectively, in these four terms, and comparing with Eq. (51.4), we obtain (on relabeling the position coordinates in the first, third, and fourth terms)

$$U_{\alpha\beta}(\mathbf{r},\mathbf{r}') = \frac{1}{2}\Big\{ - [U^{(2)}_{\delta\alpha,\gamma\beta}(\mathbf{r}',\mathbf{r}) + U^{(2)}_{\alpha\delta,\beta\gamma}(\mathbf{r},\mathbf{r}')]\langle\hat{\psi}_\delta^\dagger(\mathbf{r}')\hat{\psi}_\gamma(\mathbf{r})\rangle$$

$$+ \delta^{(3)}(\mathbf{r}-\mathbf{r}') \int d^3r'' [U^{(2)}_{\delta\alpha,\beta\gamma}(\mathbf{r}'',\mathbf{r}) + U^{(2)}_{\alpha\delta,\gamma\beta}(\mathbf{r},\mathbf{r}'')]\langle\hat{\psi}_\delta^\dagger(\mathbf{r}'')\hat{\psi}_\gamma(\mathbf{r}'')\rangle\Big\}. \tag{51.5}$$

The first two terms in Eq. (51.5) are referred to as the exchange (or Fock) potential and the last two as the Hartree potential. On comparing the last two terms in the square bracket of Eq. (51.3) with the corresponding terms in Eq. (51.4) we obtain the mean pair potentials

$$\Delta_{\beta\alpha}^*(\mathbf{r},\mathbf{r}') = \frac{1}{2} U^{(2)}_{\delta\gamma,\alpha\beta}(\mathbf{r},\mathbf{r}')\langle\hat{\psi}_\delta^\dagger(\mathbf{r})\hat{\psi}_\gamma^\dagger(\mathbf{r}')\rangle \tag{51.6}$$

and

$$\Delta_{\alpha\beta}(\mathbf{r},\mathbf{r}') = \frac{1}{2} U^{(2)}_{\alpha\beta,\delta\gamma}(\mathbf{r},\mathbf{r}')\langle\hat{\psi}_\delta(\mathbf{r}')\hat{\psi}_\gamma(\mathbf{r})\rangle \tag{51.7}$$

(where we have used the relabeling $\gamma \to \beta$, $\delta \to \alpha$, $\alpha \to \delta$, and $\beta \to \gamma$ in obtaining the last term).

The potential $U^{(2)}$ does not change if the particles are physically interchanged (or equivalently the coordinates are relabeled); hence

$$U^{(2)}_{\gamma\delta,\beta\alpha}(\mathbf{r}',\mathbf{r}) = U^{(2)}_{\delta\gamma,\alpha\beta}(\mathbf{r},\mathbf{r}'). \tag{51.8}$$

Applying relation (51.8) to Eq. (51.5) shows that the two exchange terms are identical, as are the Hartree terms. We now use (51.8) to examine the behavior of $\underset{\sim}{\Delta}$ on interchanging the spin and position coordinates; from Eq. (51.7)

$$\Delta_{\beta\alpha}(\mathbf{r}',\mathbf{r}) = \frac{1}{2} U^{(2)}_{\beta\alpha,\delta\gamma}(\mathbf{r}',\mathbf{r})\langle\hat{\psi}_\delta(\mathbf{r})\hat{\psi}_\gamma(\mathbf{r}')\rangle$$

$$= \frac{1}{2} U^{(2)}_{\alpha\beta,\gamma\delta}(\mathbf{r},\mathbf{r}')\langle\hat{\psi}_\delta(\mathbf{r})\hat{\psi}_\gamma(\mathbf{r}')\rangle$$

$$= -\frac{1}{2} U^{(2)}_{\alpha\beta,\delta\gamma}(\mathbf{r},\mathbf{r}')\langle\hat{\psi}_\delta(\mathbf{r}')\hat{\psi}_\gamma(\mathbf{r})\rangle$$

$$= -\Delta_{\alpha\beta}(\mathbf{r},\mathbf{r}') \tag{51.9}$$

(where in the third step we commuted the $\hat{\psi}$ operators and relabeled the indices γ and δ). From (51.9) we see that $\underset{\sim}{\Delta}$ has the expected behavior for a pair wavefunction; i.e., it obeys the Pauli principle.

In deriving a set of Bogoliubov equations for a spin-generalized gap we will need the commutator of $\hat{\psi}_\alpha$ with $\hat{H}_e$, which is obtained using Eqs. (36.5) together with Eq. (51.1):

$$[\hat{H}_e, \hat{\psi}_\alpha(\mathbf{r})] = -\mathscr{H}'_0\hat{\psi}_\alpha(\mathbf{r}) - \int d^3r' \Big[U_{\alpha\beta}(\mathbf{r},\mathbf{r}')\hat{\psi}_\beta(\mathbf{r}') + \frac{1}{2}\Delta_{\alpha\beta}(\mathbf{r},\mathbf{r}')\hat{\psi}^\dagger_\beta(\mathbf{r}') - \frac{1}{2}\Delta_{\beta\alpha}(\mathbf{r}',\mathbf{r})\hat{\psi}^\dagger_\beta(\mathbf{r}')\Big]$$

$$= -\mathscr{H}'_0\hat{\psi}_\alpha(\mathbf{r}) - \int d^3r' [U_{\alpha\beta}(\mathbf{r},\mathbf{r}')\hat{\psi}_\beta(\mathbf{r}') + \Delta_{\alpha\beta}(\mathbf{r},\mathbf{r}')\hat{\psi}^\dagger_\beta(\mathbf{r}')],$$

where we have used Eq. (51.9) in the second step. From this point on the discussion proceeds in a manner similar to that used in Secs. 36 and 41 and the resulting Bogoliubov equations for a spin-dependent potential and gap function are

$$\varepsilon u_\alpha(\mathbf{r}) = \mathscr{H}'_0(\mathbf{r})u_\alpha(\mathbf{r}) + \int d^3r' [U_{\alpha\beta}(\mathbf{r},\mathbf{r}')u_\beta(\mathbf{r}') + \Delta_{\alpha\beta}(\mathbf{r},\mathbf{r}')v_\beta(\mathbf{r}')] \tag{51.10a}$$

and

$$-\varepsilon v_\alpha(\mathbf{r}) = \mathscr{H}'^*_0(\mathbf{r})v_\alpha(\mathbf{r}) + \int d^3r' [U^*_{\alpha\beta}(\mathbf{r},\mathbf{r}')v_\beta(\mathbf{r}') + \Delta^*_{\alpha\beta}(\mathbf{r},\mathbf{r}')u_\beta(\mathbf{r}')], \tag{51.10b}$$

where we have suppressed the index N associated with ε, u, and v. Eqs. (51.10) consist of four coupled integro-differential equations when account is taken of the spin index α. The orthogonality and completeness relations associated with the wavefunctions u_N and v_N are given in Eqs. (41.6a)–(41.6d).

We require an expression for the mean potentials, Eqs. (51.5)–(51.7). Substituting $\hat{\psi}$ and $\hat{\psi}^\dagger$ obtained from Eq. (51.5), and using the rules

$$\langle\hat{\gamma}^\dagger_N\hat{\gamma}_M\rangle = \delta_{NM}f_N, \tag{51.11a}$$

$$\langle\hat{\gamma}_N\hat{\gamma}^\dagger_M\rangle = \delta_{NM}(1-f_N), \tag{51.11b}$$

and

$$\langle\hat{\gamma}_N\hat{\gamma}_M\rangle = 0, \tag{51.11c}$$

where, with $\beta = 1/k_B T$,

$$f_N = \frac{1}{e^{\beta\varepsilon_N}+1}, \tag{51.12}$$

we obtain

$$U_{\alpha\beta}(\mathbf{r},\mathbf{r}') = -\,U^{(2)}_{\alpha\delta,\beta\gamma}(\mathbf{r},\mathbf{r}')\sum_N [u^*_{N\delta}(\mathbf{r}')u_{N\gamma}(\mathbf{r})f_N + v_{N\delta}(\mathbf{r}')v^*_{N\gamma}(\mathbf{r})(1-f_N)]$$

$$+\,\delta^{(3)}(\mathbf{r}-\mathbf{r}')\int \mathrm{d}^3r''\,U^{(2)}_{\alpha\delta,\gamma\beta}(\mathbf{r},\mathbf{r}'')\sum_N [u^*_{N\delta}(\mathbf{r}'')u_{N\gamma}(\mathbf{r}'')f_N + v_{N\delta}(\mathbf{r}'')v^*_{N\gamma}(\mathbf{r}'')(1-f_N)] \qquad (51.13)$$

and

$$\Delta_{\alpha\beta}(\mathbf{r},\mathbf{r}') = \frac{1}{2}U^{(2)}_{\alpha\beta,\delta\gamma}(\mathbf{r},\mathbf{r}')\sum_N [v^*_{N\delta}(\mathbf{r}')u_{N\gamma}(\mathbf{r})f_N + v^*_{N\gamma}(\mathbf{r})u_{N\delta}(\mathbf{r}')(1-f_N)]. \qquad (51.14)$$

If we now use the symmetry property expressed by Eq. (51.9) to rewrite Eq. (51.14) we obtain

$$\Delta_{\alpha\beta}(\mathbf{r},\mathbf{r}') = -\frac{1}{2}U^{(2)}_{\beta\alpha,\delta\gamma}(\mathbf{r}',\mathbf{r})\sum_N [v^*_{N\delta}(\mathbf{r})u_{N\gamma}(\mathbf{r}')f_N + v^*_{N\gamma}(\mathbf{r}')u_{N\delta}(\mathbf{r})(1-f_N)].$$

Interchanging the dummy indices γ and δ and using Eq. (51.8) yields

$$\Delta_{\alpha\beta}(\mathbf{r},\mathbf{r}') = -\frac{1}{2}U_{\alpha\beta,\delta\gamma}(\mathbf{r},\mathbf{r}')\sum_N [v^*_{N\gamma}(\mathbf{r})u_{N\delta}(\mathbf{r}')f_N + v^*_{N\delta}(\mathbf{r}')u_{N\gamma}(\mathbf{r})(1-f_N)]. \qquad (51.15)$$

Finally, we write $\Delta_{\alpha\beta}$ as half the sum of (51.14) and (51.15)

$$\Delta_{\alpha\beta}(\mathbf{r},\mathbf{r}') = \frac{1}{4}U^{(2)}_{\alpha\beta,\delta\gamma}(\mathbf{r},\mathbf{r}')\sum_N (1-2f_N)[v^*_{N\gamma}(\mathbf{r})u_{N\delta}(\mathbf{r}') - v^*_{N\delta}(\mathbf{r}')u_{N\gamma}(\mathbf{r})]. \qquad (51.16)$$

Eqs. (51.13) and (51.16) are the generalizations of (37.10) and (37.11).

52 Mean potentials, density matrices, and distribution functions in a uniform superfluid

52.1 The mean potentials

To proceed beyond the equations of the previous section we must make assumptions about the nature of the interparticle potential, $U^{(2)}$. For the case of a translationally invariant Fermi liquid, $U^{(2)}(\mathbf{r}, \mathbf{r}') = U^{(2)}(|\mathbf{r} - \mathbf{r}'|)$. If we make the further assumption that there is no spin–orbit coupling we may write (Nakajima 1973)

$$\frac{1}{2} U^{(2)}_{\delta\gamma,\alpha\beta}(\mathbf{r}, \mathbf{r}') = \Gamma_\rho(\mathbf{r}, \mathbf{r}')\delta_{\delta\beta}\delta_{\gamma\alpha} + \Gamma_s(\mathbf{r}, \mathbf{r}')(\underset{\sim}{\sigma})_{\delta\beta} \cdot (\underset{\sim}{\sigma})_{\gamma\alpha}, \tag{52.1}$$

where the terms in Γ_ρ and Γ_s correspond to density (ρ) and spin (s) dependent interactions. The form of the second term in (52.1) reflects the fact that the exchange interaction must depend only on the relative orientation of the spins and not on their absolute orientation in space. In writing Eq. (52.1) we have ignored the fact that the bare potential is assumed to be given, e.g., by some experimentally determined potential between the fermions (which we assume is not spin-dependent). However, our discussion in Sec. 51 has involved the Hartree–Fock mean potential approximation which is not valid for a strongly interacting system. Thus it is essential to think of Eq. (52.1) as some kind of 'renormalized' interaction potential (or vertex), which we will parametrize from various measured properties of our system. In so doing we are assuming that the basic structure of our theory remains essentially the same in a correctly derived many-body theory of the system.

For convenience we write out the product $\underset{\sim}{\sigma} \cdot \underset{\sim}{\sigma}'$ occurring in (52.1)

$$\underset{\sim}{\sigma} \cdot \underset{\sim}{\sigma}' = \begin{pmatrix} 0 & 1 \\ 1 & 0 \end{pmatrix}\begin{pmatrix} 0 & 1 \\ 1 & 0 \end{pmatrix} + \begin{pmatrix} 0 & -1 \\ 1 & 0 \end{pmatrix}\begin{pmatrix} 0 & 1 \\ -1 & 0 \end{pmatrix} + \begin{pmatrix} 1 & 0 \\ 0 & -1 \end{pmatrix}\begin{pmatrix} 1 & 0 \\ 0 & -1 \end{pmatrix}. \tag{52.2}$$

Inserting (52.1) into (51.16) yields the pairing potential as

$$\Delta_{\uparrow\uparrow}(\mathbf{r}, \mathbf{r}') = \sum_N [\Gamma_\rho(\mathbf{r}, \mathbf{r}') + \Gamma_s(\mathbf{r}, \mathbf{r}')][v^*_{N\uparrow}(\mathbf{r})u_{N\uparrow}(\mathbf{r}') - v^*_{N\uparrow}(\mathbf{r}')u_{N\uparrow}(\mathbf{r})](1 - 2f_N), \tag{52.3a}$$

$$\begin{aligned}\Delta_{\downarrow\uparrow}(\mathbf{r}, \mathbf{r}') = \sum_N \{&[\Gamma_\rho(\mathbf{r}, \mathbf{r}') - \Gamma_s(\mathbf{r}, \mathbf{r}')]v^*_{N\uparrow}(\mathbf{r}')u_{N\downarrow}(\mathbf{r}) \\ &+ 2\Gamma_s(\mathbf{r}, \mathbf{r}')v^*_{N\downarrow}(\mathbf{r}')u_{N\uparrow}(\mathbf{r}) - [\Gamma_\rho(\mathbf{r}, \mathbf{r}') - \Gamma_s(\mathbf{r}, \mathbf{r}')]v^*_{N\downarrow}(\mathbf{r})u_{N\uparrow}(\mathbf{r}') \\ &- 2\Gamma_s(\mathbf{r}, \mathbf{r}')v^*_{N\uparrow}(\mathbf{r})u_{N\downarrow}(\mathbf{r}')\}(1 - 2f_N),\end{aligned} \tag{52.3b}$$

$$\Delta_{\uparrow\downarrow}(\mathbf{r},\mathbf{r}') = (\downarrow \to \uparrow, \uparrow \to \downarrow), \tag{52.3c}$$

and

$$\Delta_{\downarrow\downarrow}(\mathbf{r},\mathbf{r}') = (\uparrow \to \downarrow, \uparrow \to \downarrow). \tag{52.3d}$$

We can rewrite (52.3b) in a more symmetric form

$$\begin{aligned}\Delta_{\downarrow\uparrow}(\mathbf{r},\mathbf{r}') = \frac{1}{2}\sum_N \{&[\Gamma_\rho(\mathbf{r},\mathbf{r}') + \Gamma_s(\mathbf{r},\mathbf{r}')][v^*_{N\downarrow}(\mathbf{r}')u_{N\uparrow}(\mathbf{r}) \\ &- v^*_{N\downarrow}(\mathbf{r})u_{N\uparrow}(\mathbf{r}') + v^*_{N\uparrow}(\mathbf{r}')u_{N\downarrow}(\mathbf{r}) - v^*_{N\uparrow}(\mathbf{r})u_{N\downarrow}(\mathbf{r}')] \\ &+ [\Gamma_\rho(\mathbf{r},\mathbf{r}') - 3\Gamma_s(\mathbf{r},\mathbf{r}')][v^*_{N\uparrow}(\mathbf{r}')u_{N\downarrow}(\mathbf{r}) + v^*_{N\uparrow}(\mathbf{r})u_{N\downarrow}(\mathbf{r}') \\ &- v^*_{N\downarrow}(\mathbf{r}')u_{N\uparrow}(\mathbf{r}) - v^*_{N\downarrow}(\mathbf{r})u_{N\uparrow}(\mathbf{r}')]\}(1 - 2f_N).\end{aligned} \tag{52.3b$'$}$$

Expressed in this way all the elements of the pairing potential matrix consist of two contributions involving the functions $\Gamma_\rho + \Gamma_s$ and $\Gamma_\rho - 3\Gamma_s$. The combination of wavefunctions multiplying the first of these is antisymmetric under space inversion ($\mathbf{r} \to \mathbf{r}'$) and symmetric under spin exchange, while the second wavefunction combination has the opposite symmetries; these two-particle wavefunctions thus correspond, respectively, to spin triplet ($s = 1$) and spin singlet ($s = 0$) pair states.

We introduce the definitions for singlet and triplet pairing potentials as

$$\mathscr{V}^{(s)}(\mathbf{r},\mathbf{r}') = \Gamma_\rho(\mathbf{r},\mathbf{r}') - 3\Gamma_s(\mathbf{r},\mathbf{r}') \tag{52.4}$$

and

$$\mathscr{V}^{(t)}(\mathbf{r},\mathbf{r}') = \Gamma_\rho(\mathbf{r},\mathbf{r}') + \Gamma_s(\mathbf{r},\mathbf{r}'); \tag{52.5}$$

note that if Γ_s is attractive in the triplet 'channel' it will be repulsive in the singlet channel. Eqs. (52.3a)–(52.3d) then become

$$\Delta_{\uparrow\uparrow}(\mathbf{r},\mathbf{r}') = \sum_N \mathscr{V}^{(t)}(\mathbf{r},\mathbf{r}')[v^*_{N\uparrow}(\mathbf{r})u_{N\uparrow}(\mathbf{r}') - v^*_{N\uparrow}(\mathbf{r}')u_{N\uparrow}(\mathbf{r})](1 - 2f_N), \tag{52.6a}$$

$$\begin{aligned}\Delta_{\downarrow\uparrow}(\mathbf{r},\mathbf{r}') = \frac{1}{2}\sum_N \{&\mathscr{V}^{(t)}(\mathbf{r},\mathbf{r}')[v^*_{N\downarrow}(\mathbf{r}')u_{N\uparrow}(\mathbf{r}) + v^*_{N\uparrow}(\mathbf{r}')u_{N\downarrow}(\mathbf{r}) \\ &- v^*_{N\downarrow}(\mathbf{r})u_{N\uparrow}(\mathbf{r}') - v^*_{N\uparrow}(\mathbf{r})u_{N\downarrow}(\mathbf{r}')] \\ &+ \mathscr{V}^{(s)}(\mathbf{r},\mathbf{r}')[v^*_{N\uparrow}(\mathbf{r}')u_{N\downarrow}(\mathbf{r}) - v^*_{N\downarrow}(\mathbf{r}')u_{N\uparrow}(\mathbf{r}) \\ &+ v^*_{N\uparrow}(\mathbf{r})u_{N\downarrow}(\mathbf{r}') - v^*_{N\downarrow}(\mathbf{r})u_{N\uparrow}(\mathbf{r}')]\}(1 - 2f_N),\end{aligned} \tag{52.6b}$$

$$\Delta_{\uparrow\downarrow}(\mathbf{r},\mathbf{r}') = (\downarrow \to \uparrow, \uparrow \to \downarrow), \tag{52.6c}$$

and

$$\Delta_{\downarrow\downarrow}(\mathbf{r},\mathbf{r}') = (\uparrow \to \downarrow, \uparrow \to \downarrow). \tag{52.6d}$$

We now discuss the mean potentials. Inserting (52.1) into (51.13) yields

$$
\begin{aligned}
U_{\uparrow\uparrow}(\mathbf{r},\mathbf{r}') = -\sum_N \Big(& [\Gamma_\rho(\mathbf{r},\mathbf{r}') + \Gamma_s(\mathbf{r},\mathbf{r}')][u^*_{N\uparrow}(\mathbf{r}')u_{N\uparrow}(\mathbf{r})f_N \\
& + v_{N\uparrow}(\mathbf{r}')v^*_{N\uparrow}(\mathbf{r})(1 - f_N)] + 2\Gamma_s(\mathbf{r},\mathbf{r}')[u^*_{N\downarrow}(\mathbf{r}')n_{N\downarrow}(\mathbf{r})f_N \\
& + v_{N\downarrow}(\mathbf{r}')v^*_{N\downarrow}(\mathbf{r})(1 - f_N)] - \delta^{(3)}(\mathbf{r}-\mathbf{r}')\int d^3r''\{[\Gamma_\rho(\mathbf{r},\mathbf{r}'') \\
& + \Gamma_s(\mathbf{r},\mathbf{r}'')][u^*_{N\uparrow}(\mathbf{r}'')u_{N\uparrow}(\mathbf{r}'')f_N + v_{N\uparrow}(\mathbf{r}'')v^*_{N\uparrow}(\mathbf{r}'')(1 - f_N)] \\
& + [\Gamma_\rho(\mathbf{r},\mathbf{r}'') - \Gamma_s(\mathbf{r},\mathbf{r}'')][u^*_{N\downarrow}(\mathbf{r}'')u_{N\downarrow}(\mathbf{r}'')f_N \\
& + v_{N\downarrow}(\mathbf{r}'')v^*_{N\downarrow}(\mathbf{r}'')(1 - f_N)]\}\Big), \qquad (52.7a)
\end{aligned}
$$

$$
\begin{aligned}
U_{\uparrow\downarrow}(\mathbf{r},\mathbf{r}') = -\sum_N \Big\{ & (\Gamma_\rho(\mathbf{r},\mathbf{r}') - \Gamma_s(\mathbf{r},\mathbf{r}'))[u^*_{N\downarrow}(\mathbf{r}')u_{N\uparrow}(\mathbf{r})f_N + v_{N\downarrow}(\mathbf{r}')v^*_{N\uparrow}(\mathbf{r})(1 - f_N)] \\
& - \delta^{(3)}(\mathbf{r}-\mathbf{r}')\int d^3r'' 2\Gamma_s(\mathbf{r},\mathbf{r}'')[u^*_{N\downarrow}(\mathbf{r}'')u_{N\uparrow}(\mathbf{r}'')f_N + v_{N\downarrow}(\mathbf{r}'')v^*_{N\uparrow}(\mathbf{r}'')(1 - f_N)]\Big\}, \qquad (52.7b)
\end{aligned}
$$

$$U_{\downarrow\uparrow} = (\uparrow \to \downarrow, \downarrow \to \uparrow), \qquad (52.7c)$$

and

$$U_{\downarrow\downarrow} = (\uparrow \to \downarrow, \uparrow \to \downarrow). \qquad (52.7d)$$

These equations may be rewritten in a form which will later allow us to make contact with Landau Fermi liquid theory if we introduce the following integral operators[1]

$$f^{(s)}(\mathbf{r},\mathbf{r}') = -\Gamma_\rho(\mathbf{r},\mathbf{r}') - 3\Gamma_s(\mathbf{r},\mathbf{r}') + 2\delta^{(3)}(\mathbf{r}-\mathbf{r}')\int d^3r''\Gamma_\rho(\mathbf{r},\mathbf{r}'') \qquad (52.8)$$

and

$$f^{(a)}(\mathbf{r},\mathbf{r}') = -\Gamma_\rho(\mathbf{r},\mathbf{r}') + \Gamma_s(\mathbf{r},\mathbf{r}') + 2\delta^{(3)}(\mathbf{r}-\mathbf{r}')\int d^3r''\Gamma_s(\mathbf{r},\mathbf{r}''), \qquad (52.9)$$

where it is understood that the term containing the integral operates on the combinations $u^*(\mathbf{r}'')u(\mathbf{r}'')$ and $v(\mathbf{r}'')v^*(\mathbf{r}'')$ in Eqs. (52.7). Inserting (52.8) and (52.9) into Eqs. (52.7) we obtain

$$
\begin{aligned}
U_{\uparrow\uparrow}(\mathbf{r},\mathbf{r}') = \frac{1}{2}\sum_N \{ & [f^{(s)}(\mathbf{r},\mathbf{r}') + f^{(a)}(\mathbf{r},\mathbf{r}')][u^*_{N\uparrow}(\mathbf{r}')u_{N\uparrow}(\mathbf{r})f_N \\
& + v_{N\uparrow}(\mathbf{r}')v^*_{N\uparrow}(\mathbf{r})(1 - f_N)] + [f^{(s)}(\mathbf{r},\mathbf{r}') \\
& - f^{(a)}(\mathbf{r},\mathbf{r}')][u^*_{N\downarrow}(\mathbf{r}')u_{N\downarrow}(\mathbf{r})f_N + v_{N\downarrow}(\mathbf{r}')v^*_{N\downarrow}(\mathbf{r})(1 - f_N)]\}, \qquad (52.10a)
\end{aligned}
$$

$$U_{\uparrow\downarrow}(\mathbf{r},\mathbf{r}') = \frac{1}{2}\sum_N f^{(a)}(\mathbf{r},\mathbf{r}')[u^*_{N\downarrow}(\mathbf{r}')u_{N\uparrow}(\mathbf{r})f_N + v_{N\downarrow}(\mathbf{r}')v^*_{N\uparrow}(\mathbf{r})(1 - f_N)]. \qquad (52.10b)$$

We now discuss the form of u and v in the uniform Fermi liquid. In the absence of spin–orbit coupling the translational symmetry allows us to number the 'orbital' quantum numbers n of our eigenfunctions with the wave vector $\mathbf{k}$. Denoting the two states associated with ν by 1 and 2, and

1. We must be careful not to confuse these functions with the Fermi occupation factor, f_N.

the components α and β by $\uparrow$ and $\downarrow$, u and v take the form

$$u_{\mathbf{k}1}(\mathbf{r}) = \frac{1}{L^{3/2}} e^{i\mathbf{k}\cdot\mathbf{r}} \begin{pmatrix} \mathscr{u}_{\mathbf{k}1\uparrow} \\ \mathscr{u}_{\mathbf{k}1\downarrow} \end{pmatrix}, \tag{52.11a}$$

$$u_{\mathbf{k}2}(\mathbf{r}) = \frac{1}{L^{3/2}} e^{i\mathbf{k}\cdot\mathbf{r}} \begin{pmatrix} \mathscr{u}_{\mathbf{k}2\uparrow} \\ \mathscr{u}_{\mathbf{k}2\downarrow} \end{pmatrix}, \tag{52.11b}$$

$$v_{\mathbf{k}1}(\mathbf{r}) = \frac{1}{L^{3/2}} e^{i\mathbf{k}\cdot\mathbf{r}} \begin{pmatrix} \mathscr{v}_{\mathbf{k}1\uparrow} \\ \mathscr{v}_{\mathbf{k}1\downarrow} \end{pmatrix}, \tag{52.11c}$$

and

$$v_{\mathbf{k}2}(\mathbf{r}) = \frac{1}{L^{3/2}} e^{i\mathbf{k}\cdot\mathbf{r}} \begin{pmatrix} \mathscr{v}_{\mathbf{k}2\uparrow} \\ \mathscr{v}_{\mathbf{k}2\downarrow} \end{pmatrix}. \tag{52.11d}$$

At this point it is convenient to Fourier transform the nonlocal mean potentials, (52.6) and (52.10). When evaluating the nonlocal terms in Eqs. (51.10a) and (51.10b), integrals having the following structure occur

$$\int d^3r' \mathscr{V}(\mathbf{r}, \mathbf{r}') e^{\pm i\mathbf{k}\cdot(\mathbf{r}-\mathbf{r}')} e^{i\mathbf{k}'\cdot\mathbf{r}'} d^3r', \tag{52.12a}$$

where $\mathscr{V}(\mathbf{r}, \mathbf{r}')$ stands for any of the terms involving $\mathscr{V}^{(s)}$, $\mathscr{V}^{(t)}$, $f^{(s)}$, or $f^{(a)}$ entering Eqs. (52.6) and (52.10). We rewrite Eq. (52.12a) as:

$$\begin{aligned} &e^{i\mathbf{k}'\cdot\mathbf{r}} \int d^3(\mathbf{r}' - \mathbf{r}) \mathscr{V}(\mathbf{r}, \mathbf{r}') e^{i(\mathbf{k}' \mp \mathbf{k})\cdot(\mathbf{r}' - \mathbf{r})} \\ &\equiv \mathscr{V}(\mathbf{k}' \mp \mathbf{k}) e^{i\mathbf{k}'\cdot\mathbf{r}}. \end{aligned} \tag{52.12b}$$

We will also need to exploit the following property: if the Fourier transform of $f(r)$ is $f(k)$, then the Fourier transform of $f^*(r)$ is $[f(-k)]^*$, which we denote as $f^*(-k)$. Clearly if $f(r)$ is real, $f^*(-k) = f(k)$. Eqs. (51.10a) and (51.10b) may then be written

$$[(\xi_{0\mathbf{k}\nu} - \varepsilon_{\mathbf{k}\nu})\delta_{\alpha\beta} + U_{\alpha\beta}(\mathbf{k})]\mathscr{u}_{\mathbf{k}\nu\beta} + \Delta_{\alpha\beta}(\mathbf{k})\mathscr{v}_{\mathbf{k}\nu\beta} = 0 \tag{52.13a}$$

and

$$\Delta^*_{\alpha\beta}(-\mathbf{k})\mathscr{u}_{\mathbf{k}\nu\beta} + [(\xi_{0\mathbf{k}\nu} + \varepsilon_{\mathbf{k}\nu})\delta_{\alpha\beta} + U^*_{\alpha\beta}(-\mathbf{k})]\mathscr{v}_{\mathbf{k}\nu\beta} = 0, \tag{52.13b}$$

where ξ_0 are the eigenvalues of $\mathscr{H}'_0$ (and are the bare single-particle energies); here the pairing potentials are given by

$$\Delta_{\uparrow\uparrow}(\mathbf{k}) = \frac{1}{L^3} \sum_{\mathbf{k}'\nu} V^{(t)}(\mathbf{k}, \mathbf{k}') \mathscr{v}^*_{\mathbf{k}'\nu\uparrow} \mathscr{u}_{\mathbf{k}'\nu\uparrow}(1 - 2f_{\mathbf{k}'\nu}), \tag{52.14a}$$

$$\begin{aligned} \Delta_{\uparrow\downarrow}(\mathbf{k}) = \frac{1}{L^3} \sum_{\mathbf{k}'\nu} [&V^{(t)}(\mathbf{k}, \mathbf{k}')(\mathscr{v}^*_{\mathbf{k}'\nu\downarrow}\mathscr{u}_{\mathbf{k}'\nu\uparrow} + \mathscr{v}^*_{\mathbf{k}'\nu\uparrow}\mathscr{u}_{\mathbf{k}'\nu\downarrow}) \\ &+ V^{(s)}(\mathbf{k}, \mathbf{k}')(\mathscr{v}^*_{\mathbf{k}'\nu\downarrow}\mathscr{u}_{\mathbf{k}'\nu\uparrow} - \mathscr{v}^*_{\mathbf{k}'\nu\uparrow}\mathscr{u}_{\mathbf{k}'\nu\downarrow})](1 - 2f_{\mathbf{k}'\nu}), \end{aligned} \tag{52.14b}$$

$$\Delta_{\downarrow\uparrow}(\mathbf{k}) = (\uparrow \to \downarrow; \downarrow \to \uparrow), \tag{52.14c}$$

$$\Delta_{\downarrow\downarrow}(\mathbf{k}) = \frac{1}{L^3}\sum_{\mathbf{k}'\nu} V^{(t)}(\mathbf{k},\mathbf{k}') v^*_{\mathbf{k}'\nu\downarrow} u_{\mathbf{k}'\nu\downarrow}(1 - 2f_{\mathbf{k}'\nu}), \tag{52.14d}$$

and the mean fields by

$$U_{\uparrow\uparrow}(\mathbf{k}) = \frac{1}{2L^3}\sum_{\mathbf{k}',\nu} [f^{(s)}(\mathbf{k},\mathbf{k}') + f^{(a)}(\mathbf{k},\mathbf{k}')][u^*_{\mathbf{k}'\nu\uparrow} u_{\mathbf{k}'\nu\uparrow} f_{\mathbf{k}'\nu} + v_{-\mathbf{k}'\nu\uparrow} v^*_{-\mathbf{k}'\nu\uparrow}(1 - f_{-\mathbf{k}'\nu})]$$

$$+ [f^{(s)}(\mathbf{k},\mathbf{k}') - f^{(a)}(\mathbf{k},\mathbf{k}')][u^*_{\mathbf{k}'\nu\downarrow} u_{\mathbf{k}'\nu\downarrow} f_{\mathbf{k}'\nu} + v_{-\mathbf{k}'\nu\downarrow} v^*_{-\mathbf{k}'\nu\downarrow}(1 - f_{-\mathbf{k}'\nu})], \tag{52.15a}$$

$$U_{\uparrow\downarrow}(\mathbf{k}) = \frac{1}{2L^3}\sum_{\mathbf{k}'\nu} f^{(a)}(\mathbf{k},\mathbf{k}')[u^*_{\mathbf{k}'\nu\downarrow} u_{\mathbf{k}'\nu\uparrow} f_{\mathbf{k}'\nu} + v_{-\mathbf{k}'\nu\downarrow} v^*_{-\mathbf{k}'\nu\uparrow}(1 - f_{-\mathbf{k}'\nu})], \tag{52.15b}$$

$$U_{\downarrow\uparrow}(\mathbf{k}) = (\uparrow \to \downarrow, \downarrow \to \uparrow), \tag{52.15c}$$

$$U_{\downarrow\downarrow}(\mathbf{k}) = (\uparrow \to \downarrow, \uparrow \to \downarrow). \tag{52.15d}$$

For convenience we collect all the interaction potentials

$$\begin{aligned} V^{(s)}(\mathbf{k},\mathbf{k}') &= \Gamma_\rho(\mathbf{k}-\mathbf{k}') - 3\Gamma_s(\mathbf{k}-\mathbf{k}') + \Gamma_\rho(\mathbf{k}+\mathbf{k}') - 3\Gamma_s(\mathbf{k}+\mathbf{k}') \\ &= \mathscr{V}^{(s)}(\mathbf{k}-\mathbf{k}') + \mathscr{V}^{(s)}(\mathbf{k}+\mathbf{k}'), \end{aligned} \tag{52.16}$$

$$\begin{aligned} V^{(t)}(\mathbf{k},\mathbf{k}') &= \Gamma_\rho(\mathbf{k}-\mathbf{k}') + \Gamma_s(\mathbf{k}-\mathbf{k}') - \Gamma_\rho(\mathbf{k}+\mathbf{k}') - \Gamma_s(\mathbf{k}+\mathbf{k}'), \\ &= \mathscr{V}^{(t)}(\mathbf{k}-\mathbf{k}') - \mathscr{V}^{(t)}(\mathbf{k}+\mathbf{k}') \end{aligned} \tag{52.17}$$

$$f^{(s)}(\mathbf{k},\mathbf{k}') = 2\Gamma_\rho(0) - \Gamma_\rho(\mathbf{k}-\mathbf{k}') - 3\Gamma_s(\mathbf{k}-\mathbf{k}'), \tag{52.18}$$

and

$$f^{(a)}(\mathbf{k},\mathbf{k}') = 2\Gamma_s(0) - \Gamma_\rho(\mathbf{k}-\mathbf{k}') - 3\Gamma_s(\mathbf{k}-\mathbf{k}'). \tag{52.19}$$

We may combine Eqs. (52.13a) and (52.13b) into a four by four matrix equation (where we suppress the quantum numbers $\mathbf{k}$ and ν on u and v

$$\begin{pmatrix} \xi_0(\mathbf{k}) + U_{11}(\mathbf{k}) & U_{12}(\mathbf{k}) & \Delta_{11}(\mathbf{k}) & \Delta_{12}(\mathbf{k}) \\ U_{21}(\mathbf{k}) & \xi_0(\mathbf{k}) + U_{22}(\mathbf{k}) & \Delta_{21}(\mathbf{k}) & \Delta_{22}(\mathbf{k}) \\ -\Delta^*_{11}(-\mathbf{k}) & -\Delta^*_{12}(-\mathbf{k}) & -\xi_0(\mathbf{k}) - U_{11}(\mathbf{k}) & -U^*_{12}(-\mathbf{k}) \\ -\Delta^*_{21}(-\mathbf{k}) & -\Delta^*_{22}(-\mathbf{k}) & -U^*_{21}(-\mathbf{k}) & -\xi_0(\mathbf{k}) - U_{22}(\mathbf{k}) \end{pmatrix} \begin{pmatrix} u_\uparrow \\ u_\downarrow \\ v_\uparrow \\ v_\downarrow \end{pmatrix} = \varepsilon(\mathbf{k}) \begin{pmatrix} u_\uparrow \\ u_\downarrow \\ v_\uparrow \\ v_\downarrow \end{pmatrix}. \tag{52.20}$$

Defining[2]

$$\underset{\sim}{\xi}(\mathbf{k}) = \begin{pmatrix} \xi_0(\mathbf{k}) + U_{11}(\mathbf{k}) & U_{12}(\mathbf{k}) \\ U_{21}(\mathbf{k}) & \xi_0(\mathbf{k}) + U_{22}(\mathbf{k}) \end{pmatrix} \tag{52.21}$$

and

$$\underset{\sim}{\Delta}(\mathbf{k}) = \begin{pmatrix} \Delta_{11}(\mathbf{k}) & \Delta_{12}(\mathbf{k}) \\ \Delta_{21}(\mathbf{k}) & \Delta_{22}(\mathbf{k}) \end{pmatrix}, \tag{52.22}$$

2. In the absence of a pairing potential the quasiparticle energies are given by $\underset{\sim}{\xi}(\mathbf{k})$ which is then identical to $\underset{\sim}{\varepsilon}(\mathbf{k})$. We now reserve $\underset{\sim}{\varepsilon}(\mathbf{k})$ for the quasiparticle energies in the presence of pairing effects.

we may write (52.20) as

$$\begin{pmatrix} \xi(\mathbf{k}) & \underset{\sim}{\Delta}(\mathbf{k}) \\ -\underset{\sim}{\Delta}^*(-\mathbf{k}) & -\underset{\sim}{\xi}^*(-\mathbf{k}) \end{pmatrix} \begin{pmatrix} u \\ v \end{pmatrix} = \varepsilon(\mathbf{k}) \begin{pmatrix} u \\ v \end{pmatrix}, \tag{52.23}$$

where u and v are now two-component spin wavefunctions and we have suppressed the quantum numbers $\mathbf{k}$ and ν. From Eq. (51.16) it follows that

$$\underset{\sim}{\Delta}(-\mathbf{k}) = -\underset{\sim}{\Delta}(\mathbf{k})$$

and thus

$$-\underset{\sim}{\Delta}^*(-\mathbf{k}) = \underset{\sim}{\Delta}^\dagger(\mathbf{k}). \tag{52.24}$$

The eigenfunctions of Eq. (52.20) must satisfy orthonormality conditions which are obtained from Eqs. (41.6a) and (41.6b). For our plane-wave eigenfunctions these conditions may be written

$$\sum_\alpha (u^*_{\mathbf{k}\nu\alpha} u_{\mathbf{k}\mu\alpha} + v^*_{\mathbf{k}\nu\alpha} v_{\mathbf{k}\mu\alpha}) = \delta_{\mu\nu} \tag{52.25}$$

and

$$\sum_\alpha (u_{\mathbf{k}\nu\alpha} v_{-\mathbf{k}\mu\alpha} + u_{-\mathbf{k}\mu\alpha} v_{\mathbf{k}\nu\alpha}) = 0. \tag{52.26}$$

52.2 Singlet–triplet separation

The matrix $\underset{\sim}{\Delta}(\mathbf{k})$ plays the role of an order parameter in the superfluid. For a given value of $\mathbf{k}$, the four components of the matrix form a representation of the spin wavefunction, a two-component spinor. Noting that the spin wavefunction of a singlet state is antisymmetric under an interchange of the particles (spin indices), we can write the order parameter immediately as

$$\underset{\sim}{\Delta}(\mathbf{k}) = d_0(\mathbf{k}) \begin{pmatrix} 0 & 1 \\ -1 & 0 \end{pmatrix} = d_0(\mathbf{k}) \mathrm{i}\underset{\sim}{\sigma}_2 \tag{52.27a}$$

$$= d_0(\mathbf{k}) \chi_0. \tag{52.27b}$$

In terms of the usual spin up (↑) and spin down (↓) wavefunctions of individual electrons, the quantity $\chi_0 = \mathrm{i}\underset{\sim}{\sigma}_2$ corresponds to the singlet wavefunction $\frac{1}{\sqrt{2}}[(\uparrow)(\downarrow) - (\downarrow)(\uparrow)]$. The factors $\mathrm{i}\underset{\sim}{\sigma}_2$ and $d_0(\mathbf{k})$ are, respectively, the antisymmetric spin and the symmetric orbital parts of the order parameter (or pair wavefunction); $d_0(\mathbf{k}) = d_0(-\mathbf{k})$.

For triplet pairing there are three spin wavefunctions each of which may have a different, and $\mathbf{k}$-dependent, amplitude. The order parameter may be written in the form

$$\underset{\sim}{\Delta}(\mathbf{k}) = \mathbf{d}(\mathbf{k}) \cdot \underset{\sim}{\sigma} \mathrm{i}\underset{\sim}{\sigma}_2 \tag{52.28a}$$

$$= \mathbf{d}(\mathbf{k}) \cdot \boldsymbol{\chi}. \tag{52.28b}$$

For a given value of **k**, the vector **d** measures the amplitude of three independent triplet ($s = 1$) spin wavefunctions represented by

$$\chi_1 = i\underset{\sim}{\sigma}_1\underset{\sim}{\sigma}_2 = \begin{pmatrix} -1 & 0 \\ 0 & 1 \end{pmatrix}; \ \chi_2 = i\underset{\sim}{\sigma}_2^2 = \begin{pmatrix} i & 0 \\ 0 & i \end{pmatrix}; \ \chi_3 = i\underset{\sim}{\sigma}_3\underset{\sim}{\sigma}_2 = \begin{pmatrix} 0 & 1 \\ 1 & 0 \end{pmatrix}, \tag{52.29}$$

which have $s_i = 0$ along the 1, 2, and 3 axes of spin space, respectively; they correspond to the conventional wavefunctions

$$\chi_1 = \frac{-1}{\sqrt{2}}[(\uparrow)(\uparrow) - (\downarrow)(\downarrow)]; \ \chi_2 = \frac{i}{\sqrt{2}}[(\uparrow)(\uparrow) + (\downarrow)(\downarrow)];$$

$$\chi_3 = \frac{1}{\sqrt{2}}[(\uparrow)(\downarrow) + (\downarrow)(\uparrow)]. \tag{52.30}$$

Note that all three of these quantities are symmetric on interchanging the indices, as required for the symmetric triplet spin wavefunction. From Eqs. (52.28) and (52.29) we have the following representation for our triplet order parameter

$$\underset{\sim}{\Delta}(\mathbf{k}) = \begin{pmatrix} -d_x(\mathbf{k}) + id_y(\mathbf{k}) & d_z(\mathbf{k}) \\ d_z(\mathbf{k}) & d_x(\mathbf{k}) + id_y(\mathbf{k}) \end{pmatrix}. \tag{52.31}$$

Under a rotation in spin space, $\underset{\sim}{\Delta}$ transforms according to[3]

$$\Delta'_{\alpha\beta} = (e^{i\sigma\cdot\theta/2})_{\alpha\gamma}(e^{i\sigma\cdot\theta/2})_{\beta\delta}\Delta_{\gamma\delta} \tag{52.32}$$

and we leave it as an exercise to show that the vector $\mathbf{d}'$ transforms as

$$\mathbf{d}' = \mathbf{R}^{-1}(\theta)\cdot\mathbf{d}, \tag{52.33}$$

where **R** is the three-dimensional matrix for a rotation by an angle θ about an axis $\hat{\boldsymbol{\theta}}$ (the Euler matrix in general); thus **d** transforms as a vector in spin space.

From Eq. (52.24) it follows that

$$\mathbf{d}(\mathbf{k}) = -\mathbf{d}(-\mathbf{k}). \tag{52.34}$$

Noting that the parity of spherical harmonics is $(-1)^\ell$, Eq. (52.34) is in accord with our intuition for p wave (or, more generally, odd angular momentum) pairing.

The excitation energies, ε, are the eigenvalues of the matrix entering Eq. (52.20). At this

3. More generally the transformation involves the Wigner rotation matrix, $\underset{\sim}{U}(\theta, \varphi, \psi)$, where θ, φ, and ψ are the Euler angles, in the form

$$\underset{\sim}{\Delta}' = \underset{\sim}{U}\underset{\sim}{\Delta}\underset{\sim}{\tilde{U}}$$

with $\underset{\sim}{U}$ given by (see Sakurai (1985))

$$U(\theta, \varphi, \psi) = \begin{bmatrix} \cos\frac{\theta}{2}e^{-i(\varphi+\psi)/2} & \sin\frac{\theta}{2}e^{-i(\varphi-\psi)/2} \\ \sin\frac{\theta}{2}e^{i(\varphi-\psi)/2} & \cos\frac{\theta}{2}e^{i(\varphi+\psi)/2} \end{bmatrix}.$$

point we assume the ξ block has the form $\underset{\sim}{\xi} = \xi\underset{\sim}{1}$; i.e. we assume the off-diagonal elements vanish, we absorb U_{11} and U_{22} into the definition of ξ, and we assume $\xi_{11} = \xi_{22}$.

The eigenvalues in the singlet case follow from the solution of

$$\begin{vmatrix} \xi(\mathbf{k}) - \varepsilon(\mathbf{k}) & 0 & 0 & d_0(\mathbf{k}) \\ 0 & \xi(\mathbf{k}) - \varepsilon(\mathbf{k}) & -d_0(\mathbf{k}) & 0 \\ 0 & -d_0^*(\mathbf{k}) & -\xi(\mathbf{k}) - \varepsilon(\mathbf{k}) & 0 \\ d_0^*(\mathbf{k}) & 0 & 0 & -\xi(\mathbf{k}) - \varepsilon(\mathbf{k}) \end{vmatrix} = 0 \quad \text{(singlet)} \tag{52.35}$$

or

$$\xi^2 - \varepsilon^2 + |d_0(\mathbf{k})|^2 = 0 \text{ (singlet).} \tag{52.36}$$

Thus the excitation energies (which are positive definite) are given by

$$\varepsilon(\mathbf{k}) = [\xi^2(\mathbf{k}) + |d_0(\mathbf{k})|^2]^{1/2} \quad \text{(singlet).} \tag{52.37}$$

The triplet case is somewhat more complicated, the eigenvalues following from the equation

$$\begin{vmatrix} \xi(\mathbf{k}) - \varepsilon(\mathbf{k}) & 0 & -d_x(\mathbf{k}) + id_y(\mathbf{k}) & d_z(\mathbf{k}) \\ 0 & \xi(\mathbf{k}) - \varepsilon(\mathbf{k}) & d_z(\mathbf{k}) & +d_x(\mathbf{k}) + id_y(\mathbf{k}) \\ -d_x^*(\mathbf{k}) - id_y^*(\mathbf{k}) & d_z^*(\mathbf{k}) & -\xi(\mathbf{k}) - \varepsilon(\mathbf{k}) & 0 \\ d_z^*(\mathbf{k}) & +d_x^*(\mathbf{k}) - id_y^*(\mathbf{k}) & 0 & -\xi(\mathbf{k}) - \varepsilon(\mathbf{k}) \end{vmatrix} = 0 \tag{52.38}$$

or

$$\begin{aligned}(\xi^2 - \varepsilon^2)^2 &+ (\xi^2 - \varepsilon^2)[2|d_z|^2 + (d_x - id_y)(d_x^* + id_y^*) + (d_x + id_y)(d_x^* - id_y^*)] \\ &+ [|(d_x + id_y)(d_x - id_y)|^2 + d_z^{*2}(d_x + id_y)(d_x - id_y) + d_z^2(d_x^* - id_y^*)(d_x^* + id_y^*) \\ &+ |d_z|^4] = 0 \quad \text{(triplet).}\end{aligned} \tag{52.39}$$

We see that the energy spectrum is potentially much more complex in a triplet superfluid. Fortunately, the states occurring in ^{3}He (at zero magnetic field) fall into a special class which have a relatively simple structure: the vector $\mathbf{d}(\mathbf{k})$ has the form

$$\mathbf{d}(\mathbf{k}) = a(\mathbf{k})\hat{\mathbf{n}}(\mathbf{k}), \tag{52.40}$$

where $a(\mathbf{k})$ is a complex number and $\hat{\mathbf{n}}(\mathbf{k})$ is a real unit vector; we then have the property

$$\mathbf{d}(\mathbf{k}) \times \mathbf{d}^*(\mathbf{k}) = 0. \tag{52.41}$$

States where the $\mathbf{d}$ vector has this property are referred to as unitary states because $\underset{\sim}{\Delta}\underset{\sim}{\Delta}^\dagger$, which generally has the property

$$\underset{\sim}{\Delta}(\mathbf{k})\underset{\sim}{\Delta}^\dagger(\mathbf{k}) = |\mathbf{d}|^2\underset{\sim}{1} + i(\mathbf{d} \times \mathbf{d}^*)\cdot\underset{\sim}{\boldsymbol{\sigma}}, \tag{52.42}$$

is then proportional to a unit matrix. Thus, for the unitary states, Eq. (52.39) simplifies to

$$(\xi^2 - \varepsilon^2 + |\mathbf{d}(\mathbf{k})|^2)^2 = 0 \tag{52.43}$$

or

$$\varepsilon^2(\mathbf{k}) = \xi^2(\mathbf{k}) + |\mathbf{d}(\mathbf{k})|^2 \quad \text{(unitary triplet),} \tag{52.44}$$

which has the same structure as in the BCS theory. Unitary states also have the property that the gap matrix has no net 'polarization', i.e., the spin up and spin down components have equal amplitudes. Note that each of the three triplet states we have chosen for the representation of the triplet spin wavefunction, Eq. (52.29), has no net polarization and, provided the components of **d** are all proportional to a common complex constant, a, the superposition of spin states also has no net polarization.

The components of the energy gap matrix follow from Eq. (52.14), and to evaluate them we require the eigenvectors of Eq. (52.20), subject to the normalization conditions given in Eq. (52.25). There is some arbitrariness in the choice of these eigenvectors, particularly for triplet cases, which will be discussed in Sec. 53.

For the singlet case a suitable set of wavefunctions is

$$\mathscr{u}_{k1} = \begin{pmatrix} \mathscr{u}_k \\ 0 \end{pmatrix}; \ \mathscr{v}_{k1} = \begin{pmatrix} 0 \\ \mathscr{u}_{\mathbf{k}} \end{pmatrix}; \ \mathscr{u}_{k2} = \begin{pmatrix} 0 \\ \mathscr{u}_{\mathbf{k}} \end{pmatrix}; \ \mathscr{v}_{k2} = \begin{pmatrix} -\mathscr{v}_k \\ 0 \end{pmatrix}, \tag{52.45}$$

where

$$|\mathscr{u}_{\mathbf{k}}|^2 = \frac{1}{2}\left\{1 + \frac{\xi(\mathbf{k})}{[\xi^2(\mathbf{k}) + |d_0(\mathbf{k})|^2]^{1/2}}\right\} = \frac{1}{2}\left[1 + \frac{\xi(\mathbf{k})}{\varepsilon(\mathbf{k})}\right] \tag{52.46}$$

and

$$|\mathscr{v}_{\mathbf{k}}|^2 = \frac{1}{2}\left\{1 - \frac{\xi(\mathbf{k})}{[\xi^2(\mathbf{k}) + |d_0(\mathbf{k})|^2]^{1/2}}\right\} = \frac{1}{2}\left[1 - \frac{\xi(\mathbf{k})}{\varepsilon(\mathbf{k})}\right]. \tag{52.47}$$

We also list the combinations:

$$\mathscr{u}_{\mathbf{k}}\mathscr{v}_{\mathbf{k}} = \frac{\Delta(\mathbf{k})}{2\varepsilon(\mathbf{k})} \tag{52.48}$$

and

$$|\mathscr{u}_{\mathbf{k}}|^2 - |\mathscr{v}_{\mathbf{k}}|^2 = \frac{\xi(\mathbf{k})}{\varepsilon(\mathbf{k})} = \frac{\partial\varepsilon(\mathbf{k})}{\partial\xi(\mathbf{k})}. \tag{52.49}$$

The latter two combinations occur regularly in various applications of the theory. Eqs. (52.46)–(52.49) were encountered in Part II (see Eqs. (26.27) and (26.28)). When Eqs. (52.45), (52.48), and (52.49) are substituted in Eqs. (52.14a)–(52.14d) we obtain

$$\Delta_{\uparrow\downarrow}(\mathbf{k}) = -\Delta_{\downarrow\uparrow}(\mathbf{k}) = d_0(\mathbf{k}) = \frac{1}{L^3}\sum_{\mathbf{k}'}\frac{V^{(s)}(\mathbf{k}, \mathbf{k}')d_0(\mathbf{k}')}{2\varepsilon(\mathbf{k}')}\{1 - 2f[\varepsilon(\mathbf{k}')]\}, \tag{52.50a}$$

and

$$\Delta_{\uparrow\uparrow}(\mathbf{k}) = \Delta_{\downarrow\downarrow}(\mathbf{k}) = 0; \tag{52.50b}$$

Eq. (52.50a) is the usual BCS integral equation for Δ. Note that only the singlet potential, $V^{(s)}$ (Eq. 52.16)), enters the gap equation.

52.3 Density matrices and distribution functions

By introducing the one-particle density matrix we can make contact with the Landau Fermi liquid theory in the normal ($\Delta = 0$) state. The one-particle density matrix, $\rho_{\alpha\beta}(\mathbf{r}, \mathbf{r}')$, is defined by[4]

$$\rho_{\alpha\beta}(\mathbf{r}, \mathbf{r}') = \langle \hat{\psi}^\dagger_\beta(\mathbf{r}')\hat{\psi}_\alpha(\mathbf{r})\rangle, \tag{52.51}$$

where the angle brackets imply both quantum mechanical and statistical averaging (note that for $\mathbf{r} = \mathbf{r}'$ and $\alpha = \beta$ the term in the angle brackets is the number operator, the expectation value of which is the total number of particles).

Inserting Eq. (41.5) and its transpose into (52.51) and using Eqs. (51.11) we obtain

$$\rho_{\alpha\beta}(\mathbf{r}, \mathbf{r}') = \sum_N [u^*_{N\beta}(\mathbf{r}')u_{N\alpha}(\mathbf{r})f_N + v_{N\beta}(\mathbf{r}')v^*_{N\alpha}(\mathbf{r})(1 - f_N)]. \tag{52.52}$$

Substituting (52.11) into (52.52) we have (for the uniform case)

$$\rho_{\alpha\beta}(\mathbf{r}, \mathbf{r}') = \frac{1}{L^3}\sum_{\mathbf{k}'\nu} [u^*_{\mathbf{k}'\nu\beta}u_{\mathbf{k}'\nu\alpha}f_{\mathbf{k}'\nu} + v_{-\mathbf{k}'\nu\beta}v^*_{-\mathbf{k}'\nu\alpha}(1 - f_{\mathbf{k}'\nu})]\mathrm{e}^{\mathrm{i}\mathbf{k}'\cdot(\mathbf{r}-\mathbf{r}')}. \tag{52.53}$$

Taking the Fourier transform in $\mathbf{r} - \mathbf{r}'$ we obtain a function

$$n_{\alpha\beta}(\mathbf{k}) = \sum_\nu [u^*_{\mathbf{k}\nu\beta}u_{\mathbf{k}\nu\alpha}f_{\mathbf{k}\nu} + v_{-\mathbf{k}\nu\beta}v^*_{-\mathbf{k}\nu\alpha}(1 - f_{\mathbf{k}\nu})]. \tag{52.54}$$

Instead of performing a Bogoliubov transformation we may use the standard expansion

$$\hat{\psi}_\alpha(\mathbf{r}) = \frac{1}{L^{3/2}}\sum_{\mathbf{k}} \mathrm{e}^{-\mathrm{i}\mathbf{k}\cdot\mathbf{r}}\chi_\alpha\hat{c}_{\mathbf{k}\alpha},$$

where χ is a spin wavefunction. The resulting expansion corresponding to (52.54) is

$$n_{\alpha\beta}(\mathbf{k}) = \langle \hat{c}^\dagger_{\mathbf{k}\beta}\hat{c}_{\mathbf{k}\alpha}\rangle, \tag{52.55}$$

which is the usual expression for the Fourier transform of the density matrix; $n_{\alpha\beta}(\mathbf{k})$ is thus the distribution function for the occupation of states in momentum space. We may now rewrite our self-consistent potential (Eq. (52.15)) in terms of $n_{\alpha\beta}(\mathbf{k})$ (Eq. (52.54)) obtaining

$$U_{\uparrow\uparrow}(\mathbf{k}) = \frac{1}{2L^3}\sum_{\mathbf{k}'} \{[f^{(s)}(\mathbf{k}, \mathbf{k}') + f^{(a)}(\mathbf{k}, \mathbf{k}')]n_{\uparrow\uparrow}(\mathbf{k}') + [f^{(s)}(\mathbf{k}, \mathbf{k}') - f^{(a)}(\mathbf{k}, \mathbf{k}')]n_{\downarrow\downarrow}(\mathbf{k}')\} \tag{52.56a}$$

and

$$U_{\uparrow\downarrow}(\mathbf{k}) = \frac{1}{2L^3}\sum_{\mathbf{k}'} f^{(a)}(\mathbf{k}, \mathbf{k}')n_{\uparrow\downarrow}(\mathbf{k}'). \tag{52.56b}$$

As in Subsec. 41.2 we write the distribution function as the sum of a scalar and a (two-component) spinor

$$n_{\alpha\beta}(\mathbf{k}) = n(\mathbf{k})\delta_{\alpha\beta} + [\mathbf{m}(\mathbf{k})\cdot\underset{\sim}{\sigma}]_{\alpha\beta}, \tag{52.57}$$

4. Some authors define ρ with an additional factor of N^{-1}.

where

$$n(\mathbf{k}) = \frac{1}{2}[n_{\uparrow\uparrow}(\mathbf{k}) + n_{\downarrow\downarrow}(\mathbf{k})], \tag{52.58a}$$

$$m_z(\mathbf{k}) = \frac{1}{2}[n_{\uparrow\uparrow}(\mathbf{k}) - n_{\downarrow\downarrow}(\mathbf{k})], \tag{52.58b}$$

$$m_x(\mathbf{k}) + im_y(\mathbf{k}) = n_{\uparrow\downarrow}(\mathbf{k}), \tag{52.58c}$$

and

$$m_x(\mathbf{k}) - im_y(\mathbf{k}) = n_{\downarrow\uparrow}(\mathbf{k}). \tag{52.58d}$$

We may then write the complete self-consistent potential in the form

$$U_{\alpha\beta}(\mathbf{k}) = \frac{1}{L^3}\sum_{\mathbf{k}'}(f^{(s)}(\mathbf{k},\mathbf{k}')n(\mathbf{k}')\delta_{\alpha\beta} + f^{(a)}(\mathbf{k},\mathbf{k}')(\mathbf{m}\cdot\boldsymbol{\sigma})_{\alpha\beta}). \tag{52.59}$$

Alternatively $U_{\alpha\beta}$ may be written as

$$U_{\alpha\beta}(\mathbf{k}) = \frac{1}{L^3}\sum_{\mathbf{k}'} f_{\alpha\beta;\gamma\delta}(\mathbf{k},\mathbf{k}')n_{\delta\gamma}(\mathbf{k}'), \tag{52.60}$$

where

$$f_{\alpha\beta;\gamma\delta}(\mathbf{k},\mathbf{k}') = f^{(s)}(\mathbf{k},\mathbf{k}')\delta_{\alpha\beta}\delta_{\gamma\delta} + f^{(a)}(\mathbf{k},\mathbf{k}')\boldsymbol{\sigma}_{\alpha\beta}\cdot\boldsymbol{\sigma}_{\gamma\delta}. \tag{52.61}$$

We may write $\xi(\mathbf{k})$, defined in (52.21), as

$$\xi_{\alpha\beta}(\mathbf{k}) = \xi(\mathbf{k})\delta_{\alpha\beta} + [\mathbf{e}(\mathbf{k})\cdot\underset{\sim}{\sigma}]_{\alpha\beta} \tag{52.62}$$

with

$$\xi(\mathbf{k}) = \xi_0(\mathbf{k}) + \frac{1}{L^3}\sum_{\mathbf{k}'} f^{(s)}(\mathbf{k},\mathbf{k}')n(\mathbf{k}') \tag{52.63}$$

and

$$\mathbf{e}(\mathbf{k}) = -\mu_3\mathbf{H} + \frac{1}{L^3}\sum_{\mathbf{k}'} f^{(a)}(\mathbf{k},\mathbf{k}')\mathbf{m}(\mathbf{k}'); \tag{52.64}$$

to distinguish the bare single-particle energies we have used the notation $\xi_0(\mathbf{k})$ in (52.63) and we have included the effect of an external field which contributes an energy $-(\hbar/2)\gamma H = -\mu H$, where μ is the magnetic moment of an electron or a ^{3}He atom.

If we consider only the *change* in the molecular field, $\delta U_{\alpha\beta}(\mathbf{k})$, caused by a *change* in the distribution, $\delta n_{\alpha\beta}(\mathbf{k})$, and identify the former as $\delta\varepsilon_{\alpha\beta}(\mathbf{k})$, we recover the fundamental equation of the Landau Fermi liquid theory, which generalizes Eq. (24.3), as

$$\delta\varepsilon_{\alpha\beta}(\mathbf{k}) = \frac{1}{L^3}\sum_{\mathbf{k}'} f_{\alpha\beta;\gamma\delta}(\mathbf{k},\mathbf{k}')\delta n_{\delta\gamma}(\mathbf{k}'). \tag{52.65}$$

As with our earlier discussion following Eq. (52.1), we should regard $f^{(s)}(\mathbf{k},\mathbf{k}')$ and $f^{(a)}(\mathbf{k},\mathbf{k}')$ as phenomenological functions to be parametrized and determined experimentally.

Rather than the density matrix which entered our discussion of the molecular fields, $U_{\alpha\beta}$, we may rewrite the pair potential, $\Delta_{\alpha\beta}$, in terms of an *anomalous* density matrix defined as

$$\rho^{(\mathrm{A})}_{\alpha\beta}(\mathbf{r},\mathbf{r}') = \langle \hat{\psi}_\beta(\mathbf{r}')\hat{\psi}_\alpha(\mathbf{r})\rangle. \tag{52.66}$$

The expression analogous to (52.52) is

$$\rho^{(\mathrm{A})}_{\alpha\beta}(\mathbf{r},\mathbf{r}') = \sum_N [u_{N\alpha}(\mathbf{r})v^*_{N\beta}(\mathbf{r}')f_N + u_{N\beta}(\mathbf{r}')v^*_{N\alpha}(\mathbf{r})(1-f_N)]; \tag{52.67}$$

for a uniform Fermi liquid this becomes

$$\rho^{(\mathrm{A})}_{\alpha\beta}(\mathbf{r},\mathbf{r}') = \frac{1}{2L^3}\sum_{\mathbf{k}'\nu}(u_{-\mathbf{k}'\nu\alpha}v^*_{-\mathbf{k}'\nu\beta} - u_{\mathbf{k}'\nu\beta}v^*_{\mathbf{k}'\nu\alpha})(1-2f_{\mathbf{k}'\nu})\mathrm{e}^{\mathrm{i}\mathbf{k}'\cdot(\mathbf{r}'-\mathbf{r})}, \tag{52.68}$$

which on taking the Fourier transform gives

$$n^{(\mathrm{A})}_{\alpha\beta}(\mathbf{k}) = \frac{1}{2}\sum_\nu (u_{-\mathbf{k}\nu\alpha}v^*_{-\mathbf{k}\nu\beta} - u_{\mathbf{k}\nu\beta}v^*_{\mathbf{k}\nu\alpha})(1-2f_{\mathbf{k}\nu}). \tag{52.69}$$

(Compare this expression to the 'diagonal' distribution function given by Eq. (52.54).) In terms of the conventional operators (analogous to (52.55)) the expression is

$$n^{(\mathrm{A})}_{\alpha\beta}(\mathbf{k}) = \langle \hat{c}_{\mathbf{k}\beta}\hat{c}_{-\mathbf{k}\alpha}\rangle. \tag{52.70}$$

By convention we call $n^{(\mathrm{A})}_{\alpha\beta}$ the pair distribution function. It has the property

$$n^{(\mathrm{A})}_{\alpha\beta}(\mathbf{k}) = -\,n^{(\mathrm{A})}_{\beta\alpha}(-\mathbf{k}). \tag{52.71}$$

We may write the pair distribution function as a sum of singlet, $n^{(\mathrm{s})}$, and triplet, $n^{(\mathrm{t})}$, terms satisfying

$$n^{(\mathrm{s})}_{\alpha\beta}(\mathbf{k}) = n^{(\mathrm{s})}_{\alpha\beta}(-\mathbf{k}) = -\,n^{(\mathrm{s})}_{\beta\alpha}(\mathbf{k}) \tag{52.72}$$

and

$$n^{(\mathrm{t})}_{\alpha\beta}(\mathbf{k}) = -\,n^{(\mathrm{t})}_{\alpha\beta}(-\mathbf{k}) = n^{(\mathrm{t})}_{\beta\alpha}(\mathbf{k}), \tag{52.73}$$

which may be combined in the form

$$n^{(\mathrm{A})}_{\alpha\beta}(\mathbf{k}) = n^{(\mathrm{s})}(\mathbf{k})(\mathrm{i}\underset{\sim}{\sigma}_2)_{\alpha\beta} + \mathbf{n}^{(\mathrm{t})}(\mathbf{k})\cdot(\underset{\sim}{\sigma}\mathrm{i}\underset{\sim}{\sigma}_2)_{\alpha\beta}, \tag{52.74}$$

where $\mathbf{n}^{(\mathrm{t})}(\mathbf{k})$ (analogous to $\mathbf{d}$) measures the three independent spin components of the triplet pair distribution function; $n^{(\mathrm{s})}(\mathbf{k})$ and $\mathbf{n}^{(\mathrm{t})}(\mathbf{k})$ have the properties

$$n^{(\mathrm{s})}(\mathbf{k}) = n^{(\mathrm{s})}(-\mathbf{k}) \tag{52.75}$$

and

$$\mathbf{n}^{(\mathrm{t})}(\mathbf{k}) = -\,\mathbf{n}^{(\mathrm{t})}(-\mathbf{k}). \tag{52.76}$$

The pairing potential, $\Delta_{\alpha\beta}$, may then be expressed as

$$\Delta_{\alpha\beta}(\mathbf{k}) = \sum_{\mathbf{k}'}[V^{(\mathrm{s})}(\mathbf{k},\mathbf{k}')n^{(\mathrm{s})}(\mathbf{k}')(\mathrm{i}\underset{\sim}{\sigma}_2)_{\alpha\beta} + V^{(\mathrm{t})}(\mathbf{k},\mathbf{k}')\mathbf{n}^{(\mathrm{t})}(\mathbf{k}')\cdot(\underset{\sim}{\sigma}\mathrm{i}\underset{\sim}{\sigma}_2)_{\alpha\beta}]. \tag{52.77}$$

Eq. (52.77) for $\Delta_{\alpha\beta}(\mathbf{k})$ may be written in a form analogous to (52.60) as

$$\Delta_{\alpha\beta}(\mathbf{k}) = \frac{1}{L^3}\sum_{\mathbf{k}'} V_{\alpha\beta;\gamma\delta}(\mathbf{k},\mathbf{k}')(\underset{\sim}{n}^{(\mathrm{A})}(\mathbf{k})\mathrm{i}\underset{\sim}{\sigma}_2)_{\delta\gamma}, \tag{52.78}$$

where

$$V_{\alpha\beta;\gamma\delta}(\mathbf{k},\mathbf{k}') = -V^{(\mathrm{s})}(\mathbf{k},\mathbf{k}')\delta_{\alpha\beta}\delta_{\gamma\delta} - V^{(\mathrm{t})}(\mathbf{k},\mathbf{k}')\boldsymbol{\sigma}_{\alpha\beta}\cdot\boldsymbol{\sigma}_{\gamma\delta}. \tag{52.79}$$

These latter two equations may be written as

$$\Delta_{\alpha\beta}(\mathbf{k}) = \{d_0(\mathbf{k})\delta_{\alpha\gamma} + [\mathbf{d}(\mathbf{k})\cdot\underset{\sim}{\sigma}]_{\alpha\gamma}\}\mathrm{i}(\underset{\sim}{\sigma}_2)_{\gamma\beta}, \tag{52.80}$$

where

$$d_0(\mathbf{k}) = -\frac{1}{L^3}\sum_{\mathbf{k}'} V^{(\mathrm{s})}(\mathbf{k},\mathbf{k}')n^{(\mathrm{s})}(\mathbf{k}') \tag{52.81}$$

and

$$\mathbf{d}(\mathbf{k}) = -\frac{1}{L^3}\sum_{\mathbf{k}'} V^{(\mathrm{t})}(\mathbf{k},\mathbf{k}')\mathbf{n}^{(\mathrm{t})}(\mathbf{k}'). \tag{52.82}$$

Up to this point we have not discussed the $\mathbf{k}$ dependence of $\mathbf{d}(\mathbf{k})$. In the BCS model of superconductivity, the pairing potential $V^{(\mathrm{s})}(\mathbf{k},\mathbf{k}')$ is assumed independent of both the magnitude and the direction of $\mathbf{k}$ and $\mathbf{k}'$; this in turn leads to a $\mathbf{k}$-independent energy gap. For higher angular momentum pairing we must explicitly include the dependence of the potential on the angle between $\mathbf{k}$ and $\mathbf{k}'$, but we may continue to model the behavior with a potential that is independent of $|\mathbf{k}|$ and $|\mathbf{k}'|$; i.e.,

$$V^{(\mathrm{t})}(\mathbf{k},\mathbf{k}') = \sum_{\ell}(2\ell+1)V_{\ell}^{(\mathrm{t})}(\mathbf{k},\mathbf{k}') = \sum_{\ell}(2\ell+1)V_{\ell}^{(\mathrm{t})}(k,k')P_{\ell}(\hat{\mathbf{k}}\cdot\hat{\mathbf{k}}'). \tag{52.83}$$

Here the $V_{\ell}^{(\mathrm{t})}$ are the coefficients in an expansion of $V^{(\mathrm{t})}$ in Legendre polynomials and the expansion is limited to odd values of ℓ. Of course, we must still cut the potential off if either $\mathbf{k}$ or $\mathbf{k}'$ corresponds to an energy greater than a cut-off energy, ω_c. In this model $\mathbf{d}(\mathbf{k}) = \mathbf{d}(\hat{\mathbf{k}})$.

53 Superfluid ^{3}He

53.1 Some experimental properties of superfluid ^{3}He

Liquid ^{3}He is perhaps the best example of a Fermi liquid (the orbital and spin angular momenta vanish leaving only the nuclear spin, $I = 1/2$). It liquefies at 3.2 K at 1 bar and below about 50 mK behaves as a degenerate Fermi liquid, which is well described by Landau's theory discussed in Sec. 24. The heat capacity is linear in temperature and the associated many-body-mass enhancement is large (relative to most electronic systems) ranging from $m^*/m_3 = 2.8$ to 6 (where m_3 is the mass of a ^{3}He atom), as the pressure changes from 0 to 32 bar. The magnetic susceptibility obeys the Pauli law. All relaxation times have the characteristic T^{-2} behavior discussed earlier. Below 2.3×10^{-3} K (at the melting pressure) the liquid is superfluid. The phase diagram in the pressure–temperature plane is shown in Fig. 53.1. We note that there are actually two superfluid phases, which are designated as the A (or axial or ABM) phase (Anderson and Morel 1960, Anderson and Brinkman 1973) and the B (or isotropic or BW) phase (Balian and Werthamer 1963). Both phases are examples of various possible BCS superfluids which are, however, paired in an $\ell = 1, s = 1$ state, there being 18 such states in general. Extensive nuclear magnetic resonance and acoustic studies have confirmed the precise symmetry of these states. The next two sections discuss the BCS theory of the A and B phases. For an extensive discussion of the properties of normal and superfluid ^{3}He the reader is referred to the book of Vollhardt and Wölfle (1990).

53.2 Structure of the gap in an $\ell = 1, s = 1$ superfluid

In Subsec. 52.2 we discussed the structure of a general gap matrix and noted the triplet contribution has the structure

$$\Delta_{\alpha\beta} = \sum_{\nu,\gamma} [d_\nu(\mathbf{k})\underset{\sim}{\sigma}_\nu]_{\alpha\gamma}(i\sigma_2)_{\gamma\beta} \tag{52.28a}$$

or equivalently

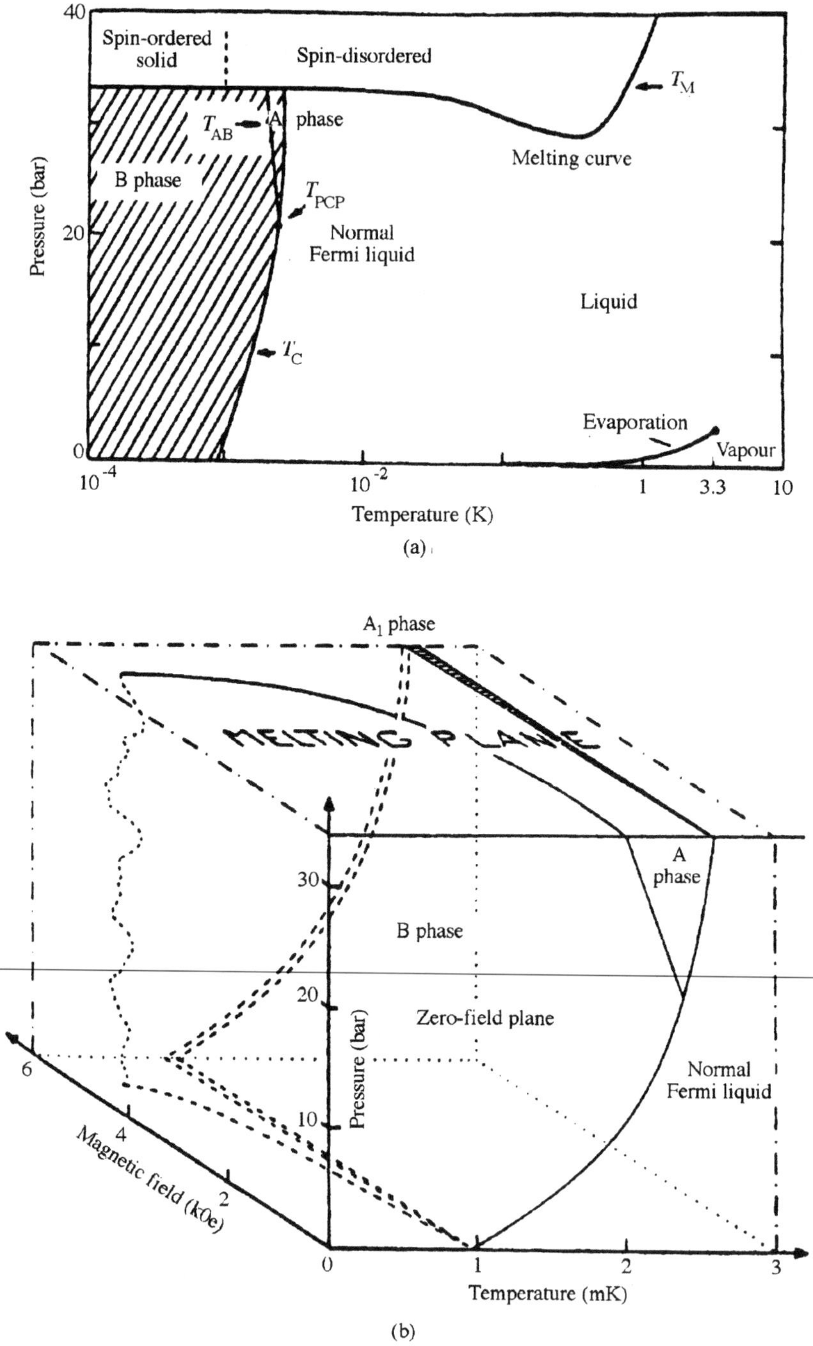

Figure 53.1 The phase diagram of superfluid ^3He.

$$\underset{\sim}{\Delta}(\mathbf{k}) = \begin{pmatrix} -d_x(\mathbf{k}) + id_y(\mathbf{k}) & d_z(\mathbf{k}) \\ d_z(\mathbf{k}) & d_x(\mathbf{k}) + id_y(\mathbf{k}) \end{pmatrix}. \tag{52.31}$$

We recall that the $d_\nu(\mathbf{k})$ are $\mathbf{k}$-dependent functions which define the relative contribution of three triplet spin wavefunctions which, when written as two-component spinors, have the form

$$\chi_1 = \underset{\sim}{\sigma}_1 i\underset{\sim}{\sigma}_2 = \begin{pmatrix} -1 & 0 \\ 0 & 1 \end{pmatrix}; \ \chi_2 = \underset{\sim}{\sigma}_2 i\underset{\sim}{\sigma}_2 = \begin{pmatrix} i & 0 \\ 0 & i \end{pmatrix}; \ \chi_3 = \underset{\sim}{\sigma}_3 i\underset{\sim}{\sigma}_2 = \begin{pmatrix} 0 & 1 \\ 1 & 0 \end{pmatrix}. \tag{52.29}$$

The corresponding wavefunctions in the conventional form were given in Eqs. (52.30).

Note that the first two forms in Eq. (52.29) do not correspond to the more familiar $s_z = \pm 1$ forms $\begin{pmatrix} 1 \\ 0 \end{pmatrix}\begin{pmatrix} 1 \\ 0 \end{pmatrix}$ and $\begin{pmatrix} 0 \\ 1 \end{pmatrix}\begin{pmatrix} 0 \\ 1 \end{pmatrix}$. Two reasons for our chosen representation are: (i) the three χ_ν transform as components of a vector under a rotation of the coordinate axes in spin space and (ii) a linear combination of the χ_ν with real coefficients has no net spin polarization (note the ground state of ^{3}He is unpolarized). The χ_ν have the further property that $\underset{\sim}{\sigma}_\nu \chi_\nu = 0$. Recall that $\underset{\sim}{\Delta}$ must satisfy the Pauli antisymmetry property

$$\Delta_{\alpha\beta}(\mathbf{k}) = -\Delta_{\beta\alpha}(-\mathbf{k}), \tag{52.24}$$

as was discussed; since our $s = 1$ spin wavefunctions are symmetric, we must have

$$\mathbf{d}(\mathbf{k}) = -\mathbf{d}(-\mathbf{k}). \tag{52.34}$$

Hence our 'orbital' wavefunctions must have odd contributions only (recall the parity of spherical harmonics is $(-1)^\ell$). The lowest value, $\ell = 1$, corresponds to the three p states which in Cartesian form (which again transform as a vector) are:

$$\hat{k}_x = \sin\theta\cos\phi; \quad \hat{k}_y = \sin\theta\sin\phi; \quad \hat{k}_z = \cos\theta. \tag{53.1}$$

The most general gap structure would involve a sum over all possible spin-orbital wavefunction products, each with its own complex amplitude, which we may write as

$$\underset{\sim}{\Delta} = \sum_{i\alpha} D_{\alpha i}\chi_\alpha \hat{k}_i \tag{53.2}$$

(here we again denoted the orbital and spin wavefunction components by i and α, respectively; see footnote 3, Sec. 23). The matrix $D_{\alpha i}$ contains 9 complex (18 real) coefficients. In principle, there could exist 18 different $\ell = 1, s = 1$ superfluids. As discussed in the previous section, at zero magnetic field only two of these are stable ground states of superfluid ^{3}He, depending on the pressure; they can be represented by the $D_{\alpha i}$ matrices:

$$\mathbf{D} = \Delta_0 \begin{pmatrix} 1 & 0 & 0 \\ 0 & 1 & 0 \\ 0 & 0 & 1 \end{pmatrix} \quad \text{(B phase)} \tag{53.3}$$

and

$$\mathbf{D} = \Delta_0 \begin{pmatrix} 0 & 0 & 0 \\ 0 & 0 & 0 \\ 1 & i & 0 \end{pmatrix}. \quad \text{(A phase).} \tag{53.4}$$

The corresponding gap matrices are

$$\underset{\sim}{\Delta} = \Delta_0 \begin{pmatrix} -k_x + ik_y & k_z \\ k_z & k_x + ik_y \end{pmatrix} \quad \text{(B phase)} \tag{53.5}$$

and (for $\mathbf{d} \parallel \hat{\mathbf{z}}$)

$$\underset{\sim}{\Delta} = \Delta_0 \begin{pmatrix} 0 & k_x + ik_y \\ k_x + ik_y & 0 \end{pmatrix} \quad \text{(A phase).} \tag{53.6}$$

The relative stability of the possible ground states can be analyzed using a generalized G–L approach in which one constructs rotationally invariant quadratic and quartic invariants from the matrices $D_{\alpha i}$ (which form a representation of the order parameter) there being one and five of these, respectively; hence five coefficients β_i replace the single coefficients β of the s-wave case. The relative size of these coefficients, which depend on pressure, determines which possible ground state will be stable as discussed in Subsec 23.3.1; for further discussion of this analysis the reader is referred to Vollhardt and Wölfle (1990).

Note the forms chosen for the B and A phases are rotationally degenerate. In both the A and B phases the free energy is unchanged if we rotate the coordinate system used to define spin space and real space[1] by some arbitrary angle θ about an arbitrary axis $\hat{\theta}$. For the B phase we may write $\mathbf{D}$ as

$$\mathbf{d} = \mathbf{R}(\theta), \tag{53.7}$$

where $\mathbf{R}$ is the Euler rotation matrix. In a similar manner, for the A phase we could orient the spin quantization axis in any direction relative to the orbital wavefunction $k_x + ik_y$, i.e., we could write $\chi \to \chi\mathbf{R}(\theta)$. If we orient $\mathbf{d}$ along the y or x axis Eq. (53.6) becomes

$$\underset{\sim}{\Delta} = i\Delta_0 \begin{pmatrix} k_x + ik_y & 0 \\ 0 & k_x + ik_y \end{pmatrix}, \quad \mathbf{d} \parallel \hat{\mathbf{y}}, \tag{53.8}$$

and

$$\underset{\sim}{\Delta} = \Delta_0 \begin{pmatrix} -k_x - ik_y & 0 \\ 0 & k_x + ik_y \end{pmatrix}, \quad \mathbf{d} \parallel \hat{\mathbf{x}}, \tag{53.9}$$

respectively. When written in terms of conventional wavefunctions the case $\mathbf{d} \parallel \hat{\mathbf{y}}$ has the form

$$\Delta(\theta, \phi) = i\Delta_0 \sin\theta e^{i\varphi} \left[\begin{pmatrix} 1 \\ 0 \end{pmatrix}\begin{pmatrix} 1 \\ 0 \end{pmatrix} + \begin{pmatrix} 0 \\ 1 \end{pmatrix}\begin{pmatrix} 0 \\ 1 \end{pmatrix} \right]. \tag{53.10}$$

For the B phase the corresponding form is

$$\Delta = \Delta_0 \left\{ -\sin\theta e^{-i\varphi} \begin{pmatrix} 1 \\ 0 \end{pmatrix}\begin{pmatrix} 1 \\ 0 \end{pmatrix} + \sin\theta e^{i\varphi} \begin{pmatrix} 0 \\ 1 \end{pmatrix}\begin{pmatrix} 0 \\ 1 \end{pmatrix} \right.$$
$$\left. + \cos\theta \left[\begin{pmatrix} 1 \\ 0 \end{pmatrix}\begin{pmatrix} 0 \\ 1 \end{pmatrix} + \begin{pmatrix} 0 \\ 1 \end{pmatrix}\begin{pmatrix} 1 \\ 0 \end{pmatrix} \right] \right\}. \tag{53.11}$$

The A phase is sometimes called the up-up/down-down phase as is apparent from the spin

1. Actually the magnetic dipole–dipole interaction is not invariant under such a rotation. Its effect is small, however, and is neglected in the present discussion.

wavefunction in (53.10); on the other hand the B phase contains all three spin wavefunctions. Note the admixture of the three spin wavefunctions varies with $\hat{\mathbf{k}}$ in the B phase whereas it is independent of $\hat{\mathbf{k}}$ in the A phase. The magnitude of the gap is determined by taking $|\Delta|^2 \equiv \frac{1}{2}\mathrm{Tr}(\underset{\sim}{\Delta}^\dagger \underset{\sim}{\Delta})$:

$$|\Delta|^2 = |\Delta_0|^2 \qquad \text{(B phase)} \tag{53.12}$$

and

$$|\Delta|^2 = |\Delta_0|^2 \sin^2\theta \quad \text{(A phase).} \tag{53.13}$$

We see that the magnitude of the B phase order parameter is isotropic whereas the A phase has a node along the polar axis ($\theta = 0$).

53.3 The Balian–Werthamer (BW) phase

The structures of the gap matrix determined in the previous section are, strictly speaking, valid only in the immediate vicinity of the transition temperature. However, for the special case of a pure $\ell = 1$ pairing potential they are valid at all temperatures as we will demonstrate for the BW and ABM phases. We begin with the BW phase which has the gap matrix given by Eq. (53.5). Since $\underset{\sim}{\Delta}\underset{\sim}{\Delta}^\dagger = \underset{\sim}{1}|\Delta_0|^2$, or equivalently $\Delta_0\Delta_0^*$, (53.5) automatically satisfies Eq. (52.44) with

$$\varepsilon^2 = \xi^2 + |\Delta_0|^2 \tag{53.14}$$

as in BCS theory. Inserting (53.5) into (52.20) yields the following set of equations (where we suppress the quantum number **k**, and take Δ_0 real)

$$(\xi - \varepsilon)u_\uparrow - \sin\theta e^{-i\varphi}\Delta_0 v_\uparrow + \cos\theta\Delta_0 v_\downarrow = 0, \tag{53.15a}$$

$$(\xi - \varepsilon)u_\downarrow + \cos\theta\Delta_0 v_\uparrow + \sin\theta e^{i\varphi}\Delta_0 v_\downarrow = 0, \tag{53.15b}$$

$$-\sin\theta e^{i\varphi}\Delta_0 u_\uparrow + \cos\theta\Delta_0 u_\downarrow - (\xi + \varepsilon)v_\uparrow = 0, \tag{53.15c}$$

$$\cos\theta\Delta_0 u_\uparrow + \sin\theta e^{-i\varphi}\Delta_0 u_\downarrow - (\xi + \varepsilon)v_\downarrow = 0. \tag{53.15d}$$

We seek a solution to (53.15a)–(53.15d) in the form $(u_{\mathbf{k}\uparrow}, u_{\mathbf{k}\downarrow}, v_{\mathbf{k}\uparrow}, v_{\mathbf{k}\downarrow}) = (\alpha_{\mathbf{k}\uparrow}u_{\mathbf{k}}, \alpha_{\mathbf{k}\downarrow}u_{\mathbf{k}}, \beta_{\mathbf{k}\uparrow}v_{\mathbf{k}}, \beta_{\mathbf{k}\downarrow}v_{\mathbf{k}})$, where $u_{\mathbf{k}}$ and $v_{\mathbf{k}}$ are the quantities occurring in singlet BCS theory which are given in Eqs. (52.45) and (52.46). Equations (53.15a) and (53.15b) may then be written as

$$\vec{\alpha}_{\hat{\mathbf{k}}} = \frac{\underset{\sim}{\Delta}}{\Delta_0}\vec{\beta}_{\hat{\mathbf{k}}} \tag{53.16a}$$

and Eqs. (53.15c) and (53.15d) as

$$\vec{\beta}_{\hat{\mathbf{k}}} = \frac{\underset{\sim}{\Delta}^\dagger}{\Delta_0}\vec{\alpha}_{\hat{\mathbf{k}}}, \tag{53.16b}$$

where $\vec{\alpha}_{\hat{\mathbf{k}}} = (\alpha_{\hat{\mathbf{k}}\uparrow}, \alpha_{\hat{\mathbf{k}}\downarrow})$, $\vec{\beta}_{\hat{\mathbf{k}}} = (\beta_{\hat{\mathbf{k}}\uparrow}, \beta_{\hat{\mathbf{k}}\downarrow})$. Substituting (53.16a) into (53.16b) yields $\underset{\sim}{\Delta}^\dagger\underset{\sim}{\Delta}/\Delta_0^2 = \underset{\sim}{1}$; i.e., our trial solution will yield a consistent solution provided $\underset{\sim}{\Delta}$ is unitary. With $\underset{\sim}{\Delta}$ unitary, it also follows that (53.16a) and (53.16b) are equivalent.

In terms of $\vec{\alpha}_{\hat{k}}$ and $\vec{\beta}_{\hat{k}}$ Eqs. (52.25) and (52.26) become

$$\vec{\alpha}^{\dagger}_{\hat{k}\mu} \cdot \vec{\alpha}_{\hat{k}\nu} = \delta_{\mu\nu}, \tag{53.17a}$$

$$\vec{\beta}^{\dagger}_{\hat{k}\mu} \cdot \vec{\beta}_{\hat{k}\nu} = \delta_{\mu\nu}, \tag{53.17b}$$

and

$$\vec{\alpha}_{\hat{k}\mu} \cdot \vec{\beta}_{-\hat{k}\nu} + \vec{\alpha}_{-\hat{k}\nu} \cdot \vec{\beta}_{\hat{k}\mu} = 0. \tag{53.17c}$$

The solution to Eqs. (53.16a) and (53.16b) is not unique. A set of eigenvectors may be generated immediately by assuming the eigenvectors $\vec{\alpha}_{k\nu}$ have the form (which satisfies (53.17a))

$$\vec{\alpha}_{\hat{k}1} = \begin{pmatrix} 1 \\ 0 \end{pmatrix}; \quad \vec{\alpha}_{\hat{k}2} = \begin{pmatrix} 0 \\ 1 \end{pmatrix}. \tag{53.18a}$$

It then follows immediately from (53.16b) that

$$\vec{\beta}_{\hat{k}1} = \begin{pmatrix} -\sin\theta\,\mathrm{e}^{\mathrm{i}\varphi} \\ \cos\theta \end{pmatrix}; \quad \vec{\beta}_{\hat{k}2} = \begin{pmatrix} \cos\theta \\ \sin\theta\,\mathrm{e}^{-\mathrm{i}\varphi} \end{pmatrix}. \tag{53.18b}$$

It is easy to verify that these eigenvectors satisfy Eqs. (53.17b) and (53.17c). This set of eigenvectors does not behave 'symmetrically' with respect to $\vec{\alpha}_{\hat{k}}$ and $\vec{\beta}_{\hat{k}}$ (particles and holes) and we will develop an alternative set. First we note that the B phase order parameter can be generated by operating (as in Eq. (52.32)) with the Wigner spin rotation matrix on the $s = 1, s_z = 0$ spin wavefunction (see Eq. (52.29))

$$\begin{aligned}
\begin{bmatrix} -\sin\theta\mathrm{e}^{-\mathrm{i}\varphi} & \cos\theta \\ \cos\theta & \sin\theta\mathrm{e}^{\mathrm{i}\varphi} \end{bmatrix}
&= \begin{bmatrix} \cos\frac{\theta}{2}\mathrm{e}^{-\mathrm{i}\varphi/2} & -\sin\frac{\theta}{2}\mathrm{e}^{-\mathrm{i}\varphi/2} \\ \sin\frac{\theta}{2}\mathrm{e}^{\mathrm{i}\varphi/2} & \cos\frac{\theta}{2}\mathrm{e}^{\mathrm{i}\varphi/2} \end{bmatrix}
\begin{bmatrix} 0 & 1 \\ 1 & 0 \end{bmatrix}
\begin{bmatrix} \cos\frac{\theta}{2}\mathrm{e}^{-\mathrm{i}\varphi/2} & \sin\frac{\theta}{2}\mathrm{e}^{\mathrm{i}\varphi/2} \\ -\sin\frac{\theta}{2}\mathrm{e}^{-\mathrm{i}\varphi/2} & \cos\frac{\theta}{2}\mathrm{e}^{\mathrm{i}\varphi/2} \end{bmatrix} \\
&= \begin{bmatrix} \cos\frac{\theta}{2}\mathrm{e}^{-\mathrm{i}\varphi/2} & -\sin\frac{\theta}{2}\mathrm{e}^{-\mathrm{i}\varphi/2} \\ \sin\frac{\theta}{2}\mathrm{e}^{\mathrm{i}\varphi/2} & \cos\frac{\theta}{2}\mathrm{e}^{\mathrm{i}\varphi/2} \end{bmatrix}
\begin{bmatrix} -\sin\frac{\theta}{2}\mathrm{e}^{-\mathrm{i}\varphi/2} & \cos\frac{\theta}{2}\mathrm{e}^{\mathrm{i}\varphi/2} \\ \cos\frac{\theta}{2}\mathrm{e}^{-\mathrm{i}\varphi/2} & \sin\frac{\theta}{2}\mathrm{e}^{\mathrm{i}\varphi/2} \end{bmatrix} \\
&\equiv \underset{\sim}{U}_{\hat{k}} \cdot (-\mathrm{i}\tilde{U}_{-\hat{k}}),
\end{aligned} \tag{53.19}$$

where we used the convention $\theta \to \pi - \theta, \varphi \to \varphi - \pi$ for the operation $\hat{k} \to -\hat{k}$ (note these matrices have determinants 1 and -1, respectively, consistent with $\|\underset{\sim}{\Delta}\| = -1$). Eq. (53.16a) then becomes

$$\underset{\sim}{U}^{\dagger\hat{k}}\vec{\alpha}_{\hat{k}} = -\mathrm{i}\tilde{\underset{\sim}{U}}_{-\hat{k}}\vec{\beta}_{k}. \tag{53.20}$$

The unitary of the matrices multiplying $\vec{\alpha}_{\hat{k}}$ and $\vec{\beta}_{\hat{k}}$ allow us immediately to write down the eigenvectors as the Hermitian conjugates of these matrices. Thus

$$\vec{\alpha}_{\hat{k}1} = \begin{bmatrix} \cos\frac{\theta}{2}\mathrm{e}^{-\mathrm{i}\varphi/2} \\ \sin\frac{\theta}{2}\mathrm{e}^{\mathrm{i}\varphi/2} \end{bmatrix}; \quad \vec{\alpha}_{\hat{k}2} = \begin{bmatrix} -\sin\frac{\theta}{2}\mathrm{e}^{-\mathrm{i}\varphi/2} \\ \cos\frac{\theta}{2}\mathrm{e}^{\mathrm{i}\varphi/2} \end{bmatrix} \tag{53.21a}$$

and

$$\vec{\beta}_{\mathbf{k}1} = \begin{bmatrix} -\sin\frac{\theta}{2}\mathrm{e}^{\mathrm{i}\varphi/2} \\ \cos\frac{\theta}{2}\mathrm{e}^{-\mathrm{i}\varphi/2} \end{bmatrix}; \vec{\beta}_{\mathbf{k}2} = \begin{bmatrix} \cos\frac{\theta}{2}\mathrm{e}^{\mathrm{i}\varphi/2} \\ \sin\frac{\theta}{2}\mathrm{e}^{-\mathrm{i}\varphi/2} \end{bmatrix}. \tag{53.21b}$$

It is again easy to verify that $\vec{\alpha}_{\mathbf{k}}$ and $\vec{\beta}_{\mathbf{k}}$ satisfy (53.15a)–(53.15d). Note that our second set of wavefunctions depends on the half angles $\theta/2$ and $\phi/2$, which is characteristic of rotated spin 1/2 wavefunctions; they are also more 'symmetric' than (53.18a) and (53.18b).

On substituting either Eqs. (53.18) or Eqs. (53.21) into Eq. (52.14), we obtain the gap equation

$$\Delta_0\begin{pmatrix} -\sin\theta\mathrm{e}^{-\mathrm{i}\varphi} & \cos\theta \\ \cos\theta & \sin\theta\mathrm{e}^{\mathrm{i}\varphi}\end{pmatrix} = \frac{1}{L^3}\sum_{\mathbf{k}'} V^{(\mathrm{t})}(\mathbf{k},\mathbf{k}')\Delta_0 \times \begin{pmatrix} -\sin\theta'\mathrm{e}^{-\mathrm{i}\varphi'} & \cos\theta' \\ \cos\theta' & \sin\theta'\mathrm{e}^{\mathrm{i}\varphi'}\end{pmatrix} u_{\mathbf{k}'}v_{\mathbf{k}'}(1-2f_{\mathbf{k}'}) \tag{53.22a}$$

or

$$\underset{\sim}{\Delta}(\mathbf{k}) = \frac{1}{L^3}\sum_{\mathbf{k}'} V^{(\mathrm{t})}(\mathbf{k},\mathbf{k}')\frac{\underset{\sim}{\Delta}(\mathbf{k}')}{2\varepsilon_{\mathbf{k}'}}(1-2f_{\mathbf{k}'}), \tag{53.22b}$$

where $\underset{\sim}{\Delta}$ is given by Eq. (53.6).

Each element of our assumed form for the gap matrix is proportional to a spherical harmonic. Furthermore, we may use the addition theorem of spherical harmonics to write the triplet pairing potential, Eq. (52.83), as

$$V_\ell^{(\mathrm{t})}(\mathbf{k},\mathbf{k}') = \frac{4\pi}{2\ell+1} V_\ell^{(\mathrm{t})}(k,k')\sum_m Y_\ell^m(\hat{\mathbf{k}})Y_\ell^{m*}(\hat{\mathbf{k}}'). \tag{53.23}$$

The angular part of the integration in Eq. (53.22) then becomes

$$\int \mathrm{d}\Omega' Y_{\ell'}^{m'}(\mathbf{k}')P_\ell(\hat{\mathbf{k}}\cdot\hat{\mathbf{k}}') = \frac{4\pi}{2\ell+1}\delta_{\ell\ell'}Y_{\ell'}^{m'}(\hat{\mathbf{k}}). \tag{53.24}$$

Provided we limit ourselves to only the $\ell = 1$ component of the pairing potential, V_1, each spherical harmonic component of the gap matrix on the right-hand side is, after the angular integration, reproduced. We may then cancel $\underset{\sim}{\Delta}$ from both sides of the equation and our matrix equation reduces to the scalar equation of BCS theory

$$1 = \mathcal{N}_0 V_1 \int_0^{\hbar\omega_c} \frac{\mathrm{d}\xi'\{1-2f[\varepsilon(k')]\}}{2\varepsilon(k')}, \tag{53.25}$$

where we have made the usual approximation of assuming the pairing potential $V(k,k') = V(\xi,\xi')$ is constant for $\xi,\xi' < \hbar\omega_c$ and zero if either ξ or $\xi' > \hbar\omega_c$; ω_c is the BCS cut-off frequency. The transition temperature follows from setting $\Delta = 0$ in the expression for ε in the denominator; it too is given by the BCS expression

$$k_B T_c = 1.14\hbar\omega_c \mathrm{e}^{-1/\mathcal{N}_0 V_1}. \tag{53.26}$$

At any temperature $T < T_c$ the gap follows from numerically solving Eq. (53.25).

The above discussion is limited to the pure $\ell = 1$ pairing case. If the pairing potential contains additional components ($\ell = 3, 5, \ldots$), the gap matrix no longer has the simple form of Eq. (53.5) at all temperatures.

The pair distribution function, $\underset{\sim}{n}^{(A)}$, and the quasiparticle distribution function, $\underset{\sim}{n}$ can be easily calculated using their definitions given in Eqs. (52.54) and (52.69), and our expressions for $u_{\mathbf{k}\nu}$ and $v_{\mathbf{k}\nu}$. One finds

$$\underset{\sim}{n}^{(A)} = -\frac{\underset{\sim}{\Delta}}{2\varepsilon}\tanh\frac{\varepsilon}{2k_B T} \tag{53.27}$$

or, equivalently, $\mathbf{n}^{(t)} = -(\mathbf{d}/2\varepsilon)\tanh(\varepsilon/2k_B T)$, and

$$\underset{\sim}{n} = n\underset{\sim}{1}, \tag{53.28}$$

where

$$n = \frac{1}{2}\left(1 - \frac{\xi}{\varepsilon}\tanh\frac{\varepsilon}{2k_B T}\right). \tag{53.29}$$

With the appropriate forms for ε and $\underset{\sim}{\Delta}$ these same expressions are valid for any unitary phase (e.g., the ABM phase).

53.5 The Anderson–Brinkman–Morel (ABM) or A phase

The gap matrix for the ABM phase for $\mathbf{d}(\hat{\mathbf{k}})$ parallel to $\hat{\mathbf{z}}$ is given by (53.6) which on inserting into (52.20) yields

$$(\xi - \varepsilon)u_\uparrow + \Delta_0 \sin\theta e^{i\varphi} v_\downarrow = 0, \tag{53.30a}$$

$$(\xi - \varepsilon)u_\downarrow + \Delta_0 \sin\theta e^{i\varphi} v_\uparrow = 0, \tag{53.30b}$$

$$\Delta_0 \sin\theta e^{-i\varphi} u_\downarrow - (\xi + \varepsilon)v_\uparrow = 0, \tag{53.30c}$$

$$\Delta_0 \sin\theta e^{-i\varphi} u_\uparrow - (\xi + \varepsilon)v_\downarrow = 0. \tag{53.30d}$$

Proceeding in the same manner as with the B phase we may rewrite (53.30a) and (53.30b) as

$$\begin{pmatrix} \alpha_{\hat{k}\uparrow} \\ \alpha_{\hat{k}\downarrow} \end{pmatrix} = e^{i\varphi}\begin{pmatrix} 0 & 1 \\ 1 & 0 \end{pmatrix}\begin{pmatrix} \beta_{\hat{k}\uparrow} \\ \beta_{\hat{k}\downarrow} \end{pmatrix}; \tag{53.31}$$

Eqs. (53.30c) and (53.30d) yield an identical equation. In obtaining (53.31) we have written

$$\varepsilon^2 = \xi^2 + \Delta_0^2 \sin^2\theta; \tag{53.32}$$

note that this implies that u and v are now angle-dependent. We again encounter in (53.31) a unitary matrix and a suitable set of eigenvectors is

$$\left.\begin{aligned} \alpha_{\hat{k}1} &= \begin{pmatrix} 1 \\ 0 \end{pmatrix} e^{-i\varphi/2}; \quad \alpha_{\hat{k}2} = \begin{pmatrix} 0 \\ 1 \end{pmatrix} e^{-i\varphi/2}, \\ \beta_{\hat{k}1} &= \begin{pmatrix} 0 \\ 1 \end{pmatrix} e^{i\varphi/2}; \quad \beta_{\hat{k}2} = \begin{pmatrix} 1 \\ 0 \end{pmatrix} e^{i\varphi/2}. \end{aligned}\right\} \quad (\mathbf{d} \parallel \hat{\mathbf{z}}) \tag{53.33}$$

For reference we also list the gap function for $\mathbf{d(k)}$ aligned parallel to the $\hat{\mathbf{x}}$ and $\hat{\mathbf{y}}$ axes, respectively,

$$\underset{\sim}{\Delta}(\hat{\mathbf{k}}) = \Delta_0 \begin{pmatrix} -\sin\theta e^{i\varphi} & 0 \\ 0 & \sin\theta e^{i\varphi} \end{pmatrix} \quad (\mathbf{d} \parallel \hat{\mathbf{x}}), \tag{53.34a}$$

and

$$\underset{\sim}{\Delta}(\hat{\mathbf{k}}) = \Delta_0 \begin{pmatrix} i\sin\theta e^{i\varphi} & 0 \\ 0 & i\sin\theta e^{i\varphi} \end{pmatrix} \quad (\mathbf{d} \parallel \hat{\mathbf{y}}). \tag{53.34b}$$

Suitable sets of associated wavefunctions for these two orientations are

$$\left.\begin{aligned} \alpha_{\hat{\mathbf{k}}1} &= \begin{pmatrix} -1 \\ 0 \end{pmatrix} e^{-i\varphi/2}, & \alpha_{\hat{\mathbf{k}}2} &= \begin{pmatrix} 0 \\ 1 \end{pmatrix} e^{-i\varphi/2}, \\ \beta_{\hat{\mathbf{k}}1} &= \begin{pmatrix} 1 \\ 0 \end{pmatrix} e^{i\varphi/2}, & \beta_{\hat{\mathbf{k}}2} &= \begin{pmatrix} 0 \\ 1 \end{pmatrix} e^{i\varphi/2}, \end{aligned}\right\} (\mathbf{d} \parallel \mathbf{x}') \tag{53.35a}$$

and

$$\left.\begin{aligned} \alpha_{\hat{\mathbf{k}}1} &= \begin{pmatrix} i \\ 0 \end{pmatrix} e^{-i\varphi/2}, & \alpha_{\hat{\mathbf{k}}2} &= \begin{pmatrix} 0 \\ i \end{pmatrix} e^{-i\varphi/2}, \\ \beta_{\hat{\mathbf{k}}1} &= \begin{pmatrix} 1 \\ 0 \end{pmatrix} e^{i\varphi/2}, & \beta_{\hat{\mathbf{k}}2} &= \begin{pmatrix} 0 \\ 1 \end{pmatrix} e^{i\varphi/2}. \end{aligned}\right\} (\mathbf{d} \parallel \mathbf{y}') \tag{53.35b}$$

Inserting (53.33) into Eqs. (52.14a)–(52.14d) we obtain the gap equation for the A phase

$$\underset{\sim}{\Delta}(\mathbf{k}) = \frac{1}{L^3} \sum_{\mathbf{k}'} V^{(t)}(\mathbf{k}, \mathbf{k}') \frac{\underset{\sim}{\Delta}(\mathbf{k}')}{2\varepsilon_{\mathbf{k}'}} (1 - 2f_{\mathbf{k}'}), \tag{53.36}$$

which has the same structure as Eq. (53.22) but with $\underset{\sim}{\Delta}$ and ε replaced by Eqs. (53.6) and (53.32), respectively. The angular dependence associated with ε considerably complicates the solution of (53.36) but it will turn out that, when $V^{(t)}$ contains only an $\ell = 1$ component, our form for Δ is a solution of (53.36) for all temperatures. We write $V^{(t)}(\mathbf{k}, \mathbf{k}')$ as

$$\begin{aligned} V^{(t)}(\mathbf{k}, \mathbf{k}') &= 3V_1(\xi, \xi')\hat{\mathbf{k}} \cdot \hat{\mathbf{k}}' \\ &= 3V_1(\xi, \xi')\left[\cos\theta\cos\theta' + \frac{1}{2}\sin\theta\sin\theta'(e^{i(\varphi-\varphi')} + e^{-i(\varphi-\varphi')})\right]. \end{aligned} \tag{53.37}$$

Carrying out the integration over $d\varphi'$ in (53.37) yields

$$1 = \mathcal{N}_0 V_1 \frac{3}{4} \int_0^{\omega_c} \int_0^{\pi} \frac{d\xi' d\theta' \sin^3\theta' \{1 - 2f[\varepsilon(\theta')]\}}{2\varepsilon(\theta')}, \tag{53.38}$$

which must be solved numerically to obtain $\Delta(T)$. Although the transition temperature, T_c, for the A phase is still given by Eq. (53.26), the value of $\Delta(T)$ at any other temperature will differ from that obtained from Eq. (53.25); i.e., $\Delta_{ABM}(T) \neq \Delta_{BW}(T)$. It is easy to verify that (53.27)–(53.29) are also valid for the A phase.

54 Collective modes in normal and superfluid Fermi systems

54.1 General formalism

As discussed in Secs. 23 and 24, superfluid and normal Fermi liquids may undergo collisionless collective oscillations. For a normal system these oscillations involve a coherent distortion of the position- and time-dependent shape of the (spin-dependent) distribution function in momentum space. The restoring force is provided by the accompanying mean field generated by the Landau quasiparticle interaction function (and in metals the self-consistent Coulomb potential). In a superfluid, in addition to a distortion of the distribution function, one may have a distortion of the anomalous distribution function; in this latter case the restoring force results from an accompanying distortion of the gap function (the off-diagonal mean field) which is generated by the pairing potential.

In this section we will develop a thoery of collective modes (in both normal and superfluid systems) based on the time-dependent Bogoliubov mean field theory. Our approach will be to calculate perturbatively the change in the Bogoliubov wavefunctions, $u^{(1)}_{n\nu\alpha}(\mathbf{r}, t)$, $v^{(1)}_{n\nu\alpha}(\mathbf{r}, t)$, resulting from a time- and space-dependent change in the mean fields, $U^{(1)}_{\alpha\beta}(\mathbf{r}, \mathbf{r}', t)$, $\Delta^{(1)}_{\alpha\beta}(\mathbf{r}, \mathbf{r}', t)$. Here n and ν denote the quantum numbers associated with the translational and spin degrees of freedom; α and β denote the components of spin operators and eigenfunctions. The theory is made self-consistent by requiring that the perturbations in the wavefunctions produce (through the self-consistency conditions) the time-dependent mean fields.

As in earlier discussions, we expand the Bogoliubov wavefunctions as

$$u_{n\nu\alpha}(\mathbf{r}, t) = u^{(0)}_{n\nu\alpha}(\mathbf{r}, t) + u^{(1)}_{n\nu\alpha}(\mathbf{r}, t) + \ldots, \tag{54.1a}$$

$$v_{n\nu\alpha}(\mathbf{r}, t) = v^{(0)}_{n\nu\alpha}(\mathbf{r}, t) + v^{(1)}_{n\nu\alpha}(\mathbf{r}, t) + \ldots \tag{54.1b}$$

and the nonlocal mean field integral operators as

$$U_{\alpha\beta}(\mathbf{r}, \mathbf{r}', t) = U^{(0)}_{\alpha\beta}(\mathbf{r}, \mathbf{r}', t) + U^{(1)}_{\alpha\beta}(\mathbf{r}, \mathbf{r}', t) + \ldots \tag{54.2a}$$

$$\Delta_{\alpha\beta}(\mathbf{r}, \mathbf{r}', t) = \Delta^{(0)}_{\alpha\beta}(\mathbf{r}, \mathbf{r}', t) + \Delta^{(1)}_{\alpha\beta}(\mathbf{r}, \mathbf{r}', t) + \ldots. \tag{54.2b}$$

As usual, $U^{(0)}_{\alpha\beta}$ is absorbed into the definition of ξ. Inserting Eqs. (54.1) into Eqs. (52.6) and (52.10) we obtain the first order corrections to the off-diagonal and the diagonal mean fields as:

$$\Delta^{(1)}_{\uparrow\uparrow}(\mathbf{r}, \mathbf{r}', t) = -\sum_{n\nu} \mathscr{V}^{(t)}(\mathbf{r}, \mathbf{r}')[v^{*(0)}_{n\nu\uparrow}(\mathbf{r}', t)u^{(1)}_{n\nu\uparrow}(\mathbf{r}, t)$$

$$+ v^{*(1)}_{n\nu\uparrow}(\mathbf{r}', t)u^{(0)}_{n\nu\uparrow}(\mathbf{r}, t) - (\mathbf{r} \to \mathbf{r}', \mathbf{r}' \to \mathbf{r})](1 - 2f_n); \tag{54.3a}$$

$$\Delta^{(1)}_{\uparrow\downarrow}(\mathbf{r},\mathbf{r}',t) = \frac{1}{2}\sum_{nv}\{\mathscr{V}^{(t)}(\mathbf{r},\mathbf{r}')[v^{*(0)}_{nv\uparrow}(\mathbf{r}',t)u^{(1)}_{nv\downarrow}(\mathbf{r},t) + v^{*(1)}_{nv\uparrow}(\mathbf{r}',t)u^{(0)}_{nv\downarrow}(\mathbf{r},t)$$

$$+ v^{*(0)}_{nv\downarrow}(\mathbf{r}',t)u^{(1)}_{nv\uparrow}(\mathbf{r},t) + v^{*(1)}_{nv\downarrow}(\mathbf{r}',t)u^{(0)}_{nv\uparrow}(\mathbf{r},t) - (\mathbf{r}'\to\mathbf{r},\mathbf{r}\to\mathbf{r}')]$$

$$+ \mathscr{V}^{(s)}(\mathbf{r},\mathbf{r}')[v^{*(1)}_{nv\downarrow}(\mathbf{r}',t)u^{(0)}_{nv\uparrow}(\mathbf{r},t) + v^{*(0)}_{nv\downarrow}(\mathbf{r}',t)u^{(1)}_{nv\uparrow}(\mathbf{r},t)$$

$$- v^{*(1)}_{nv\uparrow}(\mathbf{r}',t)u^{(0)}_{nv\downarrow}(\mathbf{r},t) - v^{*(0)}_{nv\uparrow}(\mathbf{r}',t)u^{(1)}_{nv\downarrow}(\mathbf{r},t)$$

$$+ (\mathbf{r}'\to\mathbf{r},\mathbf{r}\to\mathbf{r}')]\}(1-2f_n); \tag{54.3b}$$

$$U^{(1)}_{\uparrow\uparrow}(\mathbf{r},\mathbf{r}',t) = \frac{1}{2}\sum_{nv}\Big([f^{(s)}(\mathbf{r},\mathbf{r}') + f^{(a)}(\mathbf{r},\mathbf{r}')]\{[u^{*(0)}_{nv\uparrow}(\mathbf{r}',t)u^{(1)}_{nv\uparrow}(\mathbf{r},t)$$

$$+ u^{*(1)}_{nv\uparrow}(\mathbf{r}',t)u^{(0)}_{nv\uparrow}(\mathbf{r},t)]f_n$$

$$+ [v^{(0)}_{nv\uparrow}(\mathbf{r}',t)v^{*(1)}_{nv\uparrow}(\mathbf{r},t) + v^{(1)}_{nv\uparrow}(\mathbf{r}',t)v^{*(0)}_{nv\uparrow}(\mathbf{r},t)](1-f_n)\}$$

$$+ [f^{(s)}(\mathbf{r},\mathbf{r}') - f^{(a)}(\mathbf{r},\mathbf{r}')]$$

$$\times \{[u^{*(0)}_{nv\downarrow}(\mathbf{r}',t)u^{(1)}_{nv\downarrow}(\mathbf{r},t) + u^{*(1)}_{nv\downarrow}(\mathbf{r}',t)u^{(0)}_{nv\downarrow}(\mathbf{r},t)]f_n$$

$$+ [v^{(0)}_{nv\downarrow}(\mathbf{r}',t)v^{*(1)}_{nv\downarrow}(\mathbf{r},t) + v^{(1)}_{nv\downarrow}(\mathbf{r}',t)v^{*(0)}_{nv\downarrow}(\mathbf{r},t)](1-f_n)\}\Big); \tag{54.4a}$$

and

$$U^{(1)}_{\uparrow\downarrow}(\mathbf{r},\mathbf{r}',t) = \frac{1}{2}\sum_{nv} f^{(a)}(\mathbf{r},\mathbf{r}')\{[u^{*(0)}_{nv\downarrow}(\mathbf{r}',t)u^{(1)}_{nv\uparrow}(\mathbf{r},t) + u^{*(1)}_{nv\downarrow}(\mathbf{r}',t)u^{(0)}_{nv\uparrow}(\mathbf{r},t)]f_n$$

$$+ [v^{(0)}_{nv\downarrow}(\mathbf{r}',t)v^{*(1)}_{nv\uparrow}(\mathbf{r},t) + v^{(1)}_{nv\downarrow}(\mathbf{r}',t)v^{*(0)}_{nv\uparrow}(\mathbf{r},t)](1-f_n)\}. \tag{54.4b}$$

We now calculate $u^{(1)}_{nv\alpha}(\mathbf{r},t)$ and $v^{(1)}_{nv\alpha}(\mathbf{r},t)$. We insert the expansions (54.2a) and (54.2b) into the time-dependent generalization of the time-independent Bogoliugov equations, (51.10a) and (51.10b) (which are the counterpart of the spin singlet form (49.5)). Restricting ourselves to the first order corrections we have

$$i\hbar\dot{u}^{(1)}_{nv\alpha}(\mathbf{r},t) = \sum_{\beta}\Big\{\mathscr{H}^{(0)}\delta_{\alpha\beta}u^{(1)}_{nv\beta}(\mathbf{r},t) + \int d^3r'$$

$$\times [U^{(1)}_{\alpha\beta}(\mathbf{r},\mathbf{r}',t)u^{(0)}_{nv\beta}(\mathbf{r}',t) + \Delta^{(0)}_{\alpha\beta}(\mathbf{r},\mathbf{r}')v^{(1)}_{nv\beta}(\mathbf{r}',t) + \Delta^{(1)}_{\alpha\beta}(\mathbf{r},\mathbf{r}',t)v^{(0)}_{nv\beta}(\mathbf{r}',t)]\Big\} \tag{54.5a}$$

and

$$i\hbar\dot{v}^{(1)}_{nv\alpha}(\mathbf{r},t) = \sum_{\beta}\Big\{-\mathscr{H}^{(0)}\delta_{\alpha\beta}v^{(1)}_{nv\beta}(\mathbf{r},t) - \int d^3r'$$

$$\times [U^{(1)*}_{\alpha\beta}(\mathbf{r},\mathbf{r}',t)v^{(0)}_{nv\beta}(\mathbf{r}',t) + \Delta^{(0)*}_{\alpha\beta}(\mathbf{r},\mathbf{r}')u^{(1)}_{nv\beta}(\mathbf{r}',t)\,\Delta^{(1)*}_{\alpha\beta}(\mathbf{r},\mathbf{r}',t)u^{(0)}_{nv\beta}(\mathbf{r}',t)]\Big\}. \tag{54.5b}$$

We expand $u^{(1)}_{nv}$ and $v^{(1)}_{nv}$ in terms of plane-wave, normal-state wavefunctions as[1]

1. Disorder tends to damp collective oscillations. For this reason translationally invariant plane wavefunctions will provide an adequate model.

$$u^{(1)}_{n\nu\alpha}(\mathbf{r}, t) = \sum_{m\mu} a_{nm\nu\mu}(t) e^{i(\mathbf{k}_m \cdot \mathbf{r} - \varepsilon_m t/\hbar)} \chi_{\mu\alpha} \tag{54.6a}$$

and

$$v^{(1)}_{n\nu\alpha}(\mathbf{r}, t) = \sum_{m\mu} b_{nm\nu\mu}(t) e^{i(\mathbf{k}_m \cdot \mathbf{r} - \varepsilon_m t/\hbar)} \chi_{\mu\alpha}, \tag{54.6b}$$

where $\chi_{\mu\alpha}$ denotes the two two-component spin eigenfunctions accompanying each real space wavefunction with quantum number $\mathbf{k}_m$. In what follows we use the representation $\chi_1 = \begin{pmatrix}1\\0\end{pmatrix}, \chi_2 = \begin{pmatrix}0\\1\end{pmatrix}$. The zeroth order wavefunctions are given by

$$u^{(0)}_{n\nu\alpha}(\mathbf{r}, t) = \mathcal{u}_{n\nu\alpha} e^{i(\mathbf{k}_n \cdot \mathbf{r} - \varepsilon_n t/\hbar)} \tag{54.7a}$$

and

$$v^{(0)}_{n\nu\alpha}(\mathbf{r}, t) = \mathcal{v}_{n\nu\alpha} e^{i(\mathbf{k}_n \cdot \mathbf{r} - \varepsilon_n t/\hbar)}, \tag{54.7b}$$

where $\mathcal{u}_{n\nu}$ and $\mathcal{v}_{n\nu}$ are two-component BCS spinors. Inserting these forms into Eqs. (54.5), multiplying by $(1/V)e^{-i\mathbf{k}_m \cdot \mathbf{r}}\chi^\dagger_\mu$ (where V is the volume), taking the expectation value of the bispinor $\chi^\dagger_\mu \chi_{\nu}$, and integrating over all $\mathbf{r}$ we have

$$i\hbar\left(\dot{a}_{nm\nu\mu}(t) - \frac{i}{\hbar}\varepsilon_m a_{nm\nu\mu}(t)\right) = \xi_m a_{nm\nu\mu}(t) + \sum_{\mu'} \Delta^{(0)}_{m\mu\mu'} b_{nm\nu\mu'}(t)$$
$$+ \sum_{\nu'} (U^{(1)}_{mn\mu\nu'}(t)\mathcal{u}_{n\nu\nu'} + \Delta^{(1)}_{mn\mu\nu'}(t)\mathcal{v}_{n\nu\nu'}) e^{-(i/\hbar)(\varepsilon_n - \varepsilon_m)t} \tag{54.8a}$$

and

$$i\hbar\left[\dot{b}_{nm\nu\mu}(t) - \frac{i}{\hbar}\varepsilon_m b_{nm\nu\mu}(t)\right] = -\xi_m b_{nm\nu\mu}(t) - \sum_{\mu'} \Delta^{(0)*}_{-m\mu\mu'} a_{nm\nu\mu'}(t)$$
$$+ \sum_{\nu'} \{-(U^{(1)*})_{mn\mu\nu'}(t)\mathcal{v}_{n\nu\nu'} - (\Delta^{(1)*})_{mn\mu\nu'}(t)\mathcal{u}_{n\nu\nu'}\} e^{-(i/\hbar)(\varepsilon_n - \varepsilon_m)t}, \tag{54.8b}$$

or in the matrix form

$$\sum_{\mu'} \begin{pmatrix} \left(i\hbar\frac{\partial}{\partial t} + \varepsilon_m - \xi_m\right)\delta_{\mu\mu'} & -\Delta^{(0)}_{m\mu\mu'} \\ \Delta^{(0)*}_{-m\mu\mu'} & \left(i\hbar\frac{\partial}{\partial t} + \varepsilon_m + \xi_m\right)\delta_{\mu\mu'} \end{pmatrix} \begin{pmatrix} a_{nm\nu\mu'}(t) \\ b_{nm\nu\mu'}(t) \end{pmatrix}$$
$$= e^{-(i/\hbar)(\varepsilon_n - \varepsilon_m)t} \sum_{\nu'} \begin{pmatrix} U^{(1)}_{mn\mu\nu'}(t) & \Delta^{(1)}_{mn\mu\nu'}(t) \\ -(\Delta^{(1)*})_{mn\mu\nu'}(t) & -(U^{(1)*})_{mn\mu\nu'}(t) \end{pmatrix} \begin{pmatrix} \mathcal{u}_{n\nu\nu'} \\ \mathcal{u}_{n\nu\nu'} \end{pmatrix}. \tag{54.9}$$

Here $\Delta^{(0)}_{\pm m}$ denotes $\Delta^{(0)}(\pm \mathbf{k}_m)$ and we defined the matrix elements of the diagonal and off-diagonal mean fields as[2]

2. The notation $U^*_{mn\mu\nu}(t)$ has been used as a compact way of writing $\langle m\mu | U^*_{\alpha\beta}(\mathbf{r}, \mathbf{r}', t) | n\nu \rangle$; we emphasize that this quantity is not the same as $[U_{mn\mu\nu}(t)]^* = \langle m\mu | U_{\alpha\beta}(\mathbf{r}, \mathbf{r}', t) | n\nu \rangle^*$.

$$U^{(1)}_{mn\mu\nu}(t) = \frac{1}{V}\int\int\sum_{\alpha\beta} e^{-i\mathbf{k}_m\cdot\mathbf{r}}\chi^*_{\mu\alpha}U^{(1)}_{\alpha\beta}(\mathbf{r},\mathbf{r}',t)\chi_{\nu\beta}e^{i\mathbf{k}_n\cdot\mathbf{r}'}d^3r d^3r', \tag{54.10a}$$

$$\Delta^{(1)}_{mn\mu\nu}(t) = \frac{1}{V}\int\int\sum_{\alpha\beta} e^{-i\mathbf{k}_m\cdot\mathbf{r}}\chi^*_{\mu\alpha}\Delta^{(1)}_{\alpha\beta}(\mathbf{r},\mathbf{r}',t)\chi_{\nu\beta}e^{i\mathbf{k}_n\cdot\mathbf{r}'}d^3r d^3r', \tag{54.10b}$$

$$(U^{(1)*})_{mn\mu\nu'}(t) = \frac{1}{V}\int\int\sum_{\alpha\beta} e^{-i\mathbf{k}_m\cdot\mathbf{r}}\chi^*_{\mu\alpha}U^{(1)*}_{\alpha\beta}(\mathbf{r},\mathbf{r}',t)\chi_{\nu\beta}e^{i\mathbf{k}_n\cdot\mathbf{r}'}d^3r d^3r', \tag{54.10c}$$

and

$$(\Delta^{(1)*})_{mn\mu\nu}(t) = \frac{1}{V}\int\int\sum_{\alpha\beta} e^{-i\mathbf{k}_m\cdot\mathbf{r}}\chi^*_{\mu\alpha}\Delta^{(1)*}_{\alpha\beta}(\mathbf{r},\mathbf{r}',t)\chi_{\nu\beta}e^{i\mathbf{k}_n\cdot\mathbf{r}'}d^3r d^3r' \tag{54.10d}$$

and we introduce the spinor matrix element between $\chi^\dagger$ and $\mathscr{u}$ as

$$\mathscr{u}_{n\nu'\nu} \equiv \sum_{\gamma}\chi^*_{\nu'\gamma}\mathscr{u}_{n\nu\gamma}, \tag{54.10e}$$

and a similar form involving $\chi^\dagger$ and $\mathscr{v}$. In analogy with Eq. (50.6) we rewrite Eq. (54.9) as

$$\sum_{\mu'}\left[\left(i\hbar\frac{\partial}{\partial t} + \varepsilon_m\right)\delta_{\mu\mu'}\underset{\sim}{1} - \underset{\sim}{h}^{(0)}_{m\mu\mu'}\right]\begin{pmatrix} a_{nm\nu\mu'}(t) \\ b_{nm\nu\mu'}(t)\end{pmatrix}$$

$$= \sum_{\nu'}\underset{\sim}{h}^{(1)}_{mn\mu\nu'}(t)\begin{pmatrix}\mathscr{u}_{n\nu\nu'} \\ \mathscr{v}_{n\nu\nu'}\end{pmatrix}e^{-(i/\hbar)(\varepsilon_n - \varepsilon_m)t}, \tag{54.11}$$

where we defined

$$\underset{\sim}{h}^{(0)}_{m\mu\mu'} = \begin{bmatrix} \xi_m\delta_{\mu\mu'} & \Delta^{(0)}_{m\mu\mu'} \\ -\Delta^{(0)*}_{-m\mu\mu'} & -\xi_m\delta_{\mu\mu'}\end{bmatrix}; \tag{54.12a}$$

and

$$\underset{\sim}{h}^{(1)}_{mn\mu\nu}(t) = \begin{bmatrix} U^{(1)}_{mn\mu\nu}(t) & \Delta^{(1)}_{mn\mu\nu}(t) \\ -(\Delta^{(1)*})_{mn\mu\nu}(t) & -(U^{(1)*})_{mn\mu\nu}(t)\end{bmatrix}. \tag{54.12b}$$

We now take the Fourier transform of Eq. (54.11). In doing so we assume all time-dependent functions are multiplied by $e^{0_+t}\theta(-t)$, where 0_+ is a positive infinitesimal; this insures causal behavior. We also use the property

$$\int f(t)e^{\mp i\omega_{nm}t}e^{i\omega t}dt = f(\omega \mp \omega_{nm}). \tag{54.13}$$

Let us examine the Fourier transform of the matrix elements (54.10). Writing $\mathbf{k}'$ and $\mathbf{k}$ for $\mathbf{k}_m$ and $\mathbf{k}_n$, the transform of (54.10a) is written $V^{(1)}_{\mathbf{k}'\mathbf{k}\mu\nu}(\omega)$. From (54.10a) we have, using the hermiticity property $U_{\alpha\beta}(\mathbf{r},\mathbf{r}',t) = U^*_{\beta\alpha}(\mathbf{r}',\mathbf{r},t)$,

$$\frac{1}{V}\iiint\sum_{\alpha\beta} e^{-i\mathbf{k}'\cdot\mathbf{r}}\chi^*_{\mu\alpha}U^{(1)*}_{\beta\alpha}(\mathbf{r}',\mathbf{r},t)\chi_{\nu\beta}e^{i\mathbf{k}\cdot\mathbf{r}'+i\omega t}d^3r d^3r' dt$$

$$=\left[\frac{1}{V}\iiint\sum_{\alpha\beta} e^{i\mathbf{k}'\cdot\mathbf{r}}\chi_{\mu\alpha}U^{(1)}_{\beta\alpha}(\mathbf{r}',\mathbf{r},t)\chi^*_{\nu\beta}e^{-i\mathbf{k}\cdot\mathbf{r}'-i\omega t}d^3r d^3r' dt\right]^*$$

$$=\left[\frac{1}{V}\iiint\sum_{\alpha\beta} e^{-i\mathbf{k}'\cdot\mathbf{r}}\chi^*_{\nu\alpha}U^{(1)}_{\alpha\beta}(\mathbf{r},\mathbf{r}',t)\chi_{\mu\beta}e^{i\mathbf{k}'\cdot\mathbf{r}'-i\omega t}d^3r d^3r' dt\right]^*,$$

where we relabeled α, β and $\mathbf{r}, \mathbf{r}'$ in the second step. Hence

$$U^{(1)}_{\mathbf{k}'\mathbf{k}\mu\nu}(\omega) = [U^{(1)}_{\mathbf{k}\mathbf{k}'\nu\mu}(-\omega)]^*. \tag{54.14a}$$

Equivalently we have from (54.10c)

$$(U^{(1)*})_{\mathbf{k}\mathbf{k}'\mu\nu}(\omega) = [(U^{(1)*})_{\mathbf{k}\mathbf{k}'\nu\mu}(-\omega)]^*. \tag{54.14b}$$

The transform of Eq. (54.14b) is written $\Delta_{\mathbf{k}'\mathbf{k}\mu\nu}(\omega)$. Relabeling the coordinates in (54.10d) we have

$$\frac{1}{V}\iiint\sum_{\alpha\beta} e^{-i\mathbf{k}'\cdot\mathbf{r}}\chi^*_{\mu\alpha}\Delta^{(1)*}_{\alpha\beta}(\mathbf{r},\mathbf{r}',t)\chi_{\nu\beta}e^{i\mathbf{k}\cdot\mathbf{r}'+i\omega t}d^3r d^3r' dt$$

$$=\left[\frac{1}{V}\iiint\sum_{\alpha\beta} e^{-i\mathbf{k}\cdot\mathbf{r}}\chi^*_{\nu\alpha}\Delta^{(1)}_{\beta\alpha}(\mathbf{r}',\mathbf{r},t)\chi_{\mu\beta}e^{i\mathbf{k}'\cdot\mathbf{r}'-i\omega t}d^3r d^3r' dt\right]^*. \tag{54.15}$$

We recall the Pauli principle

$$\Delta_{\alpha\beta}(\mathbf{r},\mathbf{r}',t) = -\Delta_{\beta\alpha}(\mathbf{r}',\mathbf{r},t), \tag{54.16}$$

which allows us to write (54.15) as

$$(\Delta^{(1)*})_{\mathbf{k}'\mathbf{k}\mu\nu}(\omega) = -[\Delta^{(1)}_{\mathbf{k}\mathbf{k}'\nu\mu}(-\omega)]^*. \tag{54.17}$$

With the above results we may write the Fourier transform of (54.11) as

$$\sum_{\mu'}[(\hbar\omega+\varepsilon_m)\delta_{\mu\mu'}\underset{\sim}{1} - \underset{\sim}{h}^{(0)}_{m\mu\mu'}]\begin{pmatrix} a_{nm\nu\mu'}(\omega) \\ b_{nm\nu\mu'}(\omega)\end{pmatrix} = \sum_{\nu'}\underset{\sim}{h}^{(1)}_{mn\mu\nu'}(\omega-\omega_{nm})\begin{pmatrix} u_{n\nu\nu'} \\ v_{n\nu\nu'}\end{pmatrix}, \tag{54.18}$$

where

$$\underset{\sim}{h}^{(1)}_{mn\mu\nu}(\omega) = \begin{pmatrix} U^{(1)}_{mn\mu\nu}(\omega) & \Delta^{(1)}_{mn\mu\nu}(\omega) \\ -(\Delta^{(1)*})_{mn\mu\nu}(\omega) & -(U^{(1)*})_{mn\mu\nu}(\omega)\end{pmatrix}$$

$$= \begin{pmatrix} U^{(1)}_{mn\mu\nu}(\omega) & \Delta^{(1)}_{mn\mu\nu}(\omega) \\ (\Delta^{(1)}_{nm\nu\mu}(-\omega))^* & -(U^{(1)*}_{mn\nu\mu}(-\omega))^*\end{pmatrix}. \tag{54.19}$$

Multiplying both sides of (54.18) by the 4×4 matrix (2×2 in spin space and 2×2 in particle-hole space)

$$(\varepsilon_m+\hbar\omega)\underset{\sim}{1}\delta_{\mu\mu'} + \underset{\sim}{h}^{(0)}_{m\mu\mu'},$$

we obtain

$$\begin{pmatrix} a_{nm\nu\mu}(\omega) \\ b_{nm\nu\mu}(\omega)\end{pmatrix} = \sum_{\mu'\nu'}\frac{(\varepsilon_m+\hbar\omega)\underset{\sim}{1}\delta_{\mu\mu'} + \underset{\sim}{h}^{(0)}_{m\mu\mu'}}{(\varepsilon_m+\hbar\omega)^2-\varepsilon_m^2}\underset{\sim}{h}^{(1)}_{mn\mu'\nu'}\begin{pmatrix} u_{n\nu\nu'} \\ v_{n\nu\nu'}\end{pmatrix}, \tag{54.20}$$

where we assumed $\Delta_{\mu\mu'}$ satisfies the 'unitary' condition

$$\sum_{\nu} \Delta_{\mu\nu}^{(0)*}\Delta_{\nu\mu'}^{(0)} = \delta_{\mu\mu'}|\Delta^{(0)}|^2. \tag{54.21}$$

We note that our expansion coefficients are given in terms of the Fourier time transform of the matrix elements of the first order mean fields. For this reason it is most convenient to perform the same operations on (54.3a), (54.3b), (54.4a) and (54.4b). The calculations are somewhat tedious and we will do them in detail only for (54.3a). Substituting Eqs. (54.6) and Eqs. (54.7) into (54.3a), forming matrix elements between states $V^{-1/2}\chi_{\alpha 1}^{*}e^{-i(\mathbf{k}+\mathbf{q})\cdot\mathbf{r}}$ and $V^{-1/2}\chi_{\beta 1}e^{i\mathbf{k}\cdot\mathbf{r}'}$ and Fourier transforming in t yields the expression

$$\begin{aligned}\Delta_{\mathbf{k}+\mathbf{q},\mathbf{k}11}^{(1)}(\omega) = &-V^{-1}\sum_{nm\nu}\iiint dt d^3r d^3r'(1-2f_n)\mathscr{V}^{(t)}(\mathbf{r},\mathbf{r}')e^{i\omega t} \\ &\times [e^{-i(\mathbf{k}+\mathbf{q}-\mathbf{k}_m)\cdot\mathbf{r}+(i/\hbar)\varepsilon_n t}e^{i(\mathbf{k}-\mathbf{k}_n)\cdot\mathbf{r}'-(i/\hbar)\varepsilon_m t}v_{n\nu 1}^{*}a_{nm\nu 1}(t) \\ &+e^{-i(\mathbf{k}+\mathbf{q}-\mathbf{k}_n)\cdot\mathbf{r}+(i/\hbar)\varepsilon_m t}e^{i(\mathbf{k}-\mathbf{k}_m)\cdot\mathbf{r}'-(i/\hbar)\varepsilon_n t}u_{n\nu 1}b_{nm\nu 1}(t)^{*} \\ &-e^{-i(\mathbf{k}+\mathbf{q}+\mathbf{k}_n)\cdot\mathbf{r}+(i/\hbar)\varepsilon_n t}e^{i(\mathbf{k}+\mathbf{k}_m)\cdot\mathbf{r}'-(i/\hbar)\varepsilon_m t}v_{n\nu 1}^{*}a_{nm\nu 1}(t) \\ &-e^{-i(\mathbf{k}+\mathbf{q}+\mathbf{k}_m)\cdot\mathbf{r}+(i/\hbar)\varepsilon_m t}e^{i(\mathbf{k}+\mathbf{k}_n)\cdot\mathbf{r}'-(i/\hbar)\varepsilon_n t}u_{n\nu 1}b_{nm\nu 1}(t)^{*}].\end{aligned} \tag{54.22}$$

Since $\mathscr{V}^{(t)} = \mathscr{V}^{(t)}(|\mathbf{r}-\mathbf{r}'|)$ we rewrite the exponents in terms of the variables $\mathbf{r}$ and $\mathbf{r}-\mathbf{r}'$. The integral over d^3r vanishes unless $\mathbf{q}-\mathbf{k}_m+\mathbf{k}_n$, $\mathbf{q}-\mathbf{k}_n+\mathbf{k}_m$, $\mathbf{q}+\mathbf{k}_n-\mathbf{k}_m$, or $\mathbf{q}+\mathbf{k}_m-\mathbf{k}_n$ vanish, respectively; when any of these quantities vanish the integral yields a factor V which cancels the V^{-1} prefactor. The integral over dt is accomplished using (54.13). Eliminating $\mathbf{k}_m$, writing $\mathbf{k}'$ for $\mathbf{k}_n$, and integrating over $d^3|\mathbf{r}-\mathbf{r}'|$, (54.22) may be written

$$\begin{aligned}\Delta_{\mathbf{k}+\mathbf{q},\mathbf{k}11}^{(1)}(\omega) = &-\sum_{\mathbf{k}'\nu}(1-2f_{\mathbf{k}'})\{[\mathscr{V}^{(t)}(\mathbf{k}-\mathbf{k}')-\mathscr{V}^{(t)}(\mathbf{k}+\mathbf{k}'+\mathbf{q})]v_{\mathbf{k}'\nu 1}^{*}a_{\mathbf{k}',\mathbf{k}'+\mathbf{q}\nu 1}(\omega+\omega_{\mathbf{k}',\mathbf{k}'+\mathbf{q}}) \\ &+[(\mathscr{V}^{(t)}(\mathbf{k}-\mathbf{k}'+\mathbf{q})-\mathscr{V}^{(t)}(\mathbf{k}+\mathbf{k}')]u_{\mathbf{k}'\nu 1}b_{\mathbf{k}',\mathbf{k}'-\mathbf{q}\nu 1}(-\omega+\omega_{\mathbf{k}',\mathbf{k}'-\mathbf{q}})^{*}\},\end{aligned} \tag{54.23a}$$

where $\hbar\omega_{\mathbf{k}',\mathbf{k}'+\mathbf{q}} \equiv \varepsilon_{\mathbf{k}'}-\varepsilon_{\mathbf{k}'+\mathbf{q}}$. In a similar way we obtain the remaining mean fields as

$$\begin{aligned}\Delta_{\mathbf{k}+\mathbf{q},\mathbf{k}12}^{(1)}(\omega) = &\frac{1}{2}\sum_{\mathbf{k}'\nu}(1-2f_{\mathbf{k}'})(\{[\mathscr{V}^{(t)}(\mathbf{k}-\mathbf{k}')-\mathscr{V}^{(t)}(\mathbf{k}+\mathbf{k}'+\mathbf{q})]v_{\mathbf{k}'\nu 1}^{*}a_{\mathbf{k}',\mathbf{k}'+\mathbf{q}\nu 2}(\omega+\omega_{\mathbf{k}',\mathbf{k}'+\mathbf{q}}) \\ &+[\mathscr{V}^{(t)}(\mathbf{k}-\mathbf{k}'+\mathbf{q})-\mathscr{V}^{(t)}(\mathbf{k}+\mathbf{k}')]u_{\mathbf{k}'\nu 2}b_{\mathbf{k}',\mathbf{k}'-\mathbf{q}\nu 1}(-\omega+\omega_{\mathbf{k}',\mathbf{k}'-\mathbf{q}})^{*} \\ &+(1\to 2, 2\to 1)\} \\ &+\{[\mathscr{V}^{(s)}(\mathbf{k}-\mathbf{k}')+\mathscr{V}^{(s)}(\mathbf{k}+\mathbf{k}'+\mathbf{q})]v_{\mathbf{k}'\nu 2}^{*}a_{\mathbf{k}',\mathbf{k}'+\mathbf{q}\nu 1}(\omega+\omega_{\mathbf{k}',\mathbf{k}'+\mathbf{q}}) \\ &+[\mathscr{V}^{(s)}(\mathbf{k}-\mathbf{k}'+\mathbf{q})+\mathscr{V}^{(s)}(\mathbf{k}+\mathbf{k}')]u_{\mathbf{k}'\nu 1}b_{\mathbf{k}',\mathbf{k}'-\mathbf{q}\nu 2}(-\omega+\omega_{\mathbf{k}',\mathbf{k}'-\mathbf{q}})^{*} \\ &-(1\to 2, 2\to 1)\}),\end{aligned} \tag{54.23b}$$

$$\begin{aligned}U_{\mathbf{k}+\mathbf{q},\mathbf{k}11}^{(1)}(\omega) = &\frac{1}{2}\sum_{\mathbf{k}'\nu'}(\{[f^{(s)}(\mathbf{k}-\mathbf{k}')+f^{(a)}(\mathbf{k}-\mathbf{k}')]u_{\mathbf{k}'1\nu}^{*}a_{\mathbf{k}',\mathbf{k}'+\mathbf{q}\nu 1}(\omega+\omega_{\mathbf{k}',\mathbf{k}'+\mathbf{q}}) \\ &+[f^{(s)}(\mathbf{k}-\mathbf{k}'+\mathbf{q})+f^{(a)}(\mathbf{k}-\mathbf{k}'+\mathbf{q})]u_{\mathbf{k}'1\nu}a_{\mathbf{k}',\mathbf{k}'-\mathbf{q}\nu 1}(-\omega+\omega_{\mathbf{k}',\mathbf{k}'-\mathbf{q}})^{*}\}f_{\mathbf{k}'} \\ &+\{[f^{(s)}(\mathbf{k}+\mathbf{k}')+f^{(a)}(\mathbf{k}+\mathbf{k}')]v_{\mathbf{k}'1\nu}b_{\mathbf{k}',\mathbf{k}'-\mathbf{q}\nu 1}(-\omega+\omega_{\mathbf{k}',\mathbf{k}'-\mathbf{q}})^{*}\end{aligned}$$

$$+ [f^{(s)}(\mathbf{k}+\mathbf{k}'+\mathbf{q}) + f^{(a)}(\mathbf{k}+\mathbf{k}'+\mathbf{q})]v^*_{\mathbf{k}'1\nu}b_{\mathbf{k}',\mathbf{k}'+\mathbf{q}\nu 1}(\omega+\omega_{\mathbf{k}',\mathbf{k}'+\mathbf{q}})\}(1-f_{\mathbf{k}'})$$
$$+ \{[f^{(s)}(\mathbf{k}-\mathbf{k}') - f^{(a)}(\mathbf{k}-\mathbf{k}')]u^*_{\mathbf{k}'2\nu}a_{\mathbf{k}',\mathbf{k}'+\mathbf{q}\nu 2}(\omega+\omega_{\mathbf{k}',\mathbf{k}'+\mathbf{q}})$$
$$+ [f^{(s)}(\mathbf{k}-\mathbf{k}'+\mathbf{q}) - f^{(a)}(\mathbf{k}-\mathbf{k}'+\mathbf{q})]u_{\mathbf{k}'2\nu}a_{\mathbf{k}',\mathbf{k}'-\mathbf{q}\nu 2}(-\omega+\omega_{\mathbf{k}',\mathbf{k}'-\mathbf{q}})^*\}f_{\mathbf{k}'}$$
$$+ \{[f^{(s)}(\mathbf{k}+\mathbf{k}') - f^{(a)}(\mathbf{k}+\mathbf{k}')]v^*_{\mathbf{k}'2\nu}b_{\mathbf{k}',\mathbf{k}'-\mathbf{q}\nu 2}(-\omega+\omega_{\mathbf{k}',\mathbf{k}'-\mathbf{q}})^*$$
$$+ [f^{(s)}(\mathbf{k}+\mathbf{k}'+\mathbf{q}) - f^{(a)}(\mathbf{k}+\mathbf{k}'+\mathbf{q})]v^*_{\mathbf{k}'2\nu}b_{\mathbf{k}',\mathbf{k}'+\mathbf{q}\nu 2}(\omega+\omega_{\mathbf{k}',\mathbf{k}'+\mathbf{q}})\}(1-f_{\mathbf{k}'})), \quad (54.23c)$$

and

$$U^{(1)}_{\mathbf{k}+\mathbf{q},\mathbf{k}12}(\omega) = \frac{1}{2}\sum_{\mathbf{k}'\nu}\{[f^{(a)}(\mathbf{k}-\mathbf{k}')u^*_{\mathbf{k}'2\nu}a_{\mathbf{k}',\mathbf{k}'+\mathbf{q}\nu 1}(\omega+\omega_{\mathbf{k}',\mathbf{k}'+\mathbf{q}})$$
$$+ f^{(a)}(\mathbf{k}-\mathbf{k}'+\mathbf{q})u_{\mathbf{k}'1\nu}a_{\mathbf{k}',\mathbf{k}'-\mathbf{q}\nu 2}(-\omega+\omega_{\mathbf{k}',\mathbf{k}'-\mathbf{q}})^*]f_{\mathbf{k}'}$$
$$+ [f^{(a)}(\mathbf{k}+\mathbf{k}')v_{\mathbf{k}'1\nu}b_{\mathbf{k}',\mathbf{k}'-\mathbf{q}\nu 2}(-\omega+\omega_{\mathbf{k}',\mathbf{k}'-\mathbf{q}})^*]$$
$$+ [f^{(a)}(\mathbf{k}+\mathbf{k}'+\mathbf{q})v^*_{\mathbf{k}'2\nu}b_{\mathbf{k}',\mathbf{k}'+\mathbf{q}\nu 1}(\omega+\omega_{\mathbf{k}',\mathbf{k}'+\mathbf{q}})^*](1-f_{\mathbf{k}'})\}. \quad (54.23d)$$

Eqs. (54.23a)–(54.23d) together with Eq. (54.20) form a closed set of homogeneous integral equations in $\mathbf{k}$. They may have solutions for various frequencies $\omega_i = \omega_i(\mathbf{q})$, where the index i numbers various independent modes. Such oscillatory solutions are referred to as collective modes, and both normal fluids and superfluids may support multiple collective modes.

54.2 Zero sound in a normal Fermi liquid

The simplest example of a collisionless collective mode is zero sound, which is a density wave in a normal, neutral Fermi liquid such as ^{3}He. In addition a normal Fermi liquid may also support spin waves. Here we restrict the discussion to zero sound; we also work at $T = 0$ where $f_{\mathbf{k}} = 0$. Density variations involve the trace of the diagonal mean field which we write as

$$U^{(1)}(\mathbf{r},\mathbf{r}',t) \equiv [U^{(1)}_{\uparrow\uparrow}(\mathbf{r},\mathbf{r}',t) + U^{(1)}_{\downarrow\downarrow}(\mathbf{r},\mathbf{r}',t)]; \quad (54.24)$$

the corresponding matrix element is

$$U^{(1)}_{\mathbf{k}+\mathbf{q},\mathbf{k}}(\omega) \equiv [U^{(1)}_{\mathbf{k}+\mathbf{q},\mathbf{k}11}(\omega) + U^{(1)}_{\mathbf{k}+\mathbf{q},\mathbf{k}22}(\omega)]. \quad (54.25)$$

We work in a representation where

$$v_{\mathbf{k}1} = v_{\mathbf{k}}\begin{pmatrix}1\\0\end{pmatrix}, \; v_{\mathbf{k}2} = v_{\mathbf{k}}\begin{pmatrix}0\\1\end{pmatrix}, \quad (54.26)$$

with $v_{\mathbf{k}}$ real. Eq. (54.23c) then becomes

$$U^{(1)}_{\mathbf{k}+\mathbf{q},\mathbf{k}}(\omega) = \sum_{\mathbf{k}'}[f^{(s)}(\mathbf{k}+\mathbf{k}')v_{\mathbf{k}'}b_{\mathbf{k}',\mathbf{k}'-\mathbf{q}}(-\omega+\omega_{\mathbf{k}',\mathbf{k}'-\mathbf{q}})^*$$
$$+ f^{(s)}(\mathbf{k}+\mathbf{k}'+\mathbf{q})v_{\mathbf{k}'}b_{\mathbf{k}',\mathbf{k}'+\mathbf{q}}(\omega+\omega_{\mathbf{k}',\mathbf{k}'+\mathbf{q}})], \quad (54.27)$$

where

$$b_{\mathbf{k}',\mathbf{k}'+\mathbf{q}} \equiv \frac{1}{2}[b_{\mathbf{k}',\mathbf{k}'+\mathbf{q}11} + b_{\mathbf{k}',\mathbf{k}'+\mathbf{q}22}]. \tag{54.28}$$

For the present case (54.20) becomes

$$b_{\mathbf{k}',\mathbf{k}'+\mathbf{q}}(\omega) = -\frac{|\xi_{\mathbf{k}'+\mathbf{q}}| + \hbar\omega - \xi_{\mathbf{k}'+\mathbf{q}}}{(|\xi_{\mathbf{k}'+\mathbf{q}}| + \hbar\omega)^2 - \xi^2_{\mathbf{k}'+\mathbf{q}}} v_{\mathbf{k}'}[U^{(1)}_{\mathbf{k}',\mathbf{k}'+\mathbf{q}}(-\omega + \omega_{\mathbf{k}',\mathbf{k}'+\mathbf{q}})]^* \tag{54.29}$$

or

$$\begin{aligned} b_{\mathbf{k}',\mathbf{k}'+\mathbf{q}}(\omega + \omega_{\mathbf{k}',\mathbf{k}'+\mathbf{q}}) &= -\frac{|\xi_{\mathbf{k}'}| + \hbar\omega - \xi_{\mathbf{k}'+\mathbf{q}}}{(|\xi_{\mathbf{k}'}| + \hbar\omega)^2 - \xi^2_{\mathbf{k}'+\mathbf{q}}} v_{\mathbf{k}'}[U^{(1)}_{\mathbf{k}',\mathbf{k}'+\mathbf{q}}(-\omega)]^* \\ &= -\frac{1}{|\xi_{\mathbf{k}'}| + \hbar\omega + \xi_{\mathbf{k}'+\mathbf{q}}} v_{\mathbf{k}'}[U^{(1)}_{\mathbf{k}',\mathbf{k}'+\mathbf{q}}(-\omega)]^*, \end{aligned} \tag{54.30}$$

where we factored the denominator in the second step. Similarly

$$b_{\mathbf{k}',\mathbf{k}'-\mathbf{q}}(-\omega + \omega_{\mathbf{k}',\mathbf{k}'-\mathbf{q}})^* = -\frac{1}{|\xi_{\mathbf{k}'}| - \hbar\omega + \xi_{\mathbf{k}'-\mathbf{q}}} v^*_{\mathbf{k}'} U^{(1)}_{\mathbf{k}',\mathbf{k}'-\mathbf{q}}(\omega). \tag{54.31}$$

Inserting (54.30) and (54.31) into (54.27) and moving the origin of the first term from $\mathbf{k}'$ to $\mathbf{k}' + \mathbf{q}$ gives

$$\begin{aligned} U^{(1)}_{\mathbf{k}+\mathbf{q},\mathbf{k}}(\omega) &= \sum_{\mathbf{k}'} f^{(s)}(\mathbf{k} + \mathbf{k}' + \mathbf{q}) \left\{ -\frac{|v_{\mathbf{k}'+\mathbf{q}}|^2 U^{(1)}_{\mathbf{k}'+\mathbf{q},\mathbf{k}'}(\omega)}{|\xi_{\mathbf{k}'+\mathbf{q}}| - \hbar\omega + \xi_{\mathbf{k}'}} - \frac{|v_{\mathbf{k}'}|^2 [U^{(1)}_{\mathbf{k}',\mathbf{k}'+\mathbf{q}}(-\omega)]^*}{|\xi_{\mathbf{k}'}| + \hbar\omega + \xi_{\mathbf{k}'+\mathbf{q}}} \right\} \\ &= \sum_{\mathbf{k}'} f^{(s)}(\mathbf{k} + \mathbf{k}' + \mathbf{q}) \left\{ \frac{|v_{\mathbf{k}'}|^2 - |v_{\mathbf{k}'+\mathbf{q}}|^2}{\xi_{\mathbf{k}'} - \xi_{\mathbf{k}'+\mathbf{q}} - \hbar\omega - \mathrm{i}0_+} \right\} U^{(1)}_{\mathbf{k}'+\mathbf{q},\mathbf{k}'}(\omega), \end{aligned} \tag{54.32}$$

where we used Eq. (54.14) and the fact when $|v_{\mathbf{k}'}|^2 = \theta(-\xi_{\mathbf{k}'}) \neq 0$ we may write $|\xi_{\mathbf{k}'}| = -\xi_{\mathbf{k}'}$ in the second step. The eigenvalues, $\omega = \omega(\mathbf{q})$, of the integral equation Eq. (54.32) are the collective modes.

One usually restricts the study of collective modes to wave vectors $q << k_F$ (for large wave vectors the modes tend to be highly damped). In this limit we write $|v_{\mathbf{k}'}|^2 - |v_{\mathbf{k}'+\mathbf{q}}|^2 = -\delta(\xi_{\mathbf{k}'})\mathbf{v}\cdot\mathbf{q}$ and $\xi_{\mathbf{k}} - \xi_{\mathbf{k}+\mathbf{q}} = -(\partial\xi/\partial\mathbf{k})\cdot\mathbf{q} = -\hbar\mathbf{v}\cdot\mathbf{q}$, where $\mathbf{v}$ is the quasiparticle velocity. Due to the action of the δ function all quantities of interest are restricted to the Fermi surface; e.g., $\mathbf{v} = \mathbf{v}_F$ and $\mathbf{k} = \mathbf{k}_F$. This also applies to the mean field which we write as $U^{(1)}_{\mathbf{k}}(\mathbf{q}, \omega)$. Suppressing the variables ω and $|\mathbf{q}|$ and measuring $\hat{\mathbf{k}}$ from $\hat{\mathbf{q}}$ via the polar angles θ, ϕ we may assume the dynamic mean field is proportional to a function $\psi(\theta, \phi)$. For small $\mathbf{q}$ we may neglect the dependence of $f^{(s)}$ on $\mathbf{q}$. Since $\mathbf{k}$ and $\mathbf{k}'$ are restricted to the Fermi surface, $f^{(s)} = f^{(s)}(\mathbf{k} + \mathbf{k}')$, which can only depend on the angle χ between $\mathbf{k}$ and $\mathbf{k}'$, is written as $f^{(s)}(\cos\chi)$ where

$$\cos\chi = \cos\theta\cos\theta' - \sin\theta\sin\theta'\cos(\phi - \phi').$$

Going from a sum over $\mathbf{k}'$ to an integral and assembling the above we have

$$\psi(\theta, \phi) = \frac{2k_F^2}{(2\pi)^3 v_F} \int f^{(s)}(\theta, \phi; \theta'\phi') \frac{\cos\theta'}{s - \cos\theta'} \psi(\theta', \phi') \mathrm{d}\Omega', \tag{54.33}$$

where $s \equiv \omega/qv_F$. Writing $F = [2(4\pi)p_F^2/(2\pi\hbar^3)v_F]f = \mathcal{N}(0)f$, where $\mathcal{N}(0)$ is the density of states, we may write (54.33) as

$$\psi(\theta,\phi) = \frac{1}{4\pi}\int F(\theta,\phi;\theta',\phi')\frac{\cos\theta'}{s-\cos\theta'}\psi(\theta',\phi')\mathrm{d}\Omega'. \tag{54.34}$$

Eq. (54.34) is identical to Eq. (24.43) obtained earlier from the phenomenological Landau Fermi liquid theory which has a zero sound solution, assuming F has the appropriate structure, as discussed in Subsec. 24.22.

One could go on and include the spin structure of the mean field, $U_{\mu\nu}$, from which the equations for spin waves would emerge.

54.3 Collective modes in superfluid ^{3}He B

As an example of collective modes involving the off diagonal mean field, $\Delta_{\alpha\beta}(\mathbf{r},\mathbf{r}',t)$, we consider the case of ^{3}He B. The equilibrium order parameter, $\Delta^{(0)}$, was studied in Sec. 53. To simplify the discussion we restrict ourselves to a model where $f^{(s)} = f^{(a)} = 0$; furthermore we will only study the $\mathbf{q} = 0, T = 0$ case. Since ^{3}He B involves triplet pairing we may set $\mathscr{V}^{(s)} = 0$. With these simplifications Eqs. (54.23), when written as a single equation, become

$$\Delta^{(1)}_{\mathbf{k}\kappa\lambda}(\omega) = \frac{1}{2}\sum_{\mathbf{k}'\nu} V^{(t)}(\mathbf{k},\mathbf{k}')\{v^*_{\mathbf{k}'\nu\kappa}a_{\mathbf{k}'\nu\lambda}(\omega) + u_{\mathbf{k}'\nu\lambda}[b_{\mathbf{k}'\nu\kappa}(-\omega)]^*$$
$$+ v^*_{\mathbf{k}'\nu\lambda}a_{\mathbf{k}'\nu\kappa}(\omega) + u_{\mathbf{k}'\nu\kappa}[b_{\mathbf{k}'\nu\lambda}(-\omega)]^*\}, \tag{54.35a}$$

which in matrix form reads

$$\underset{\sim}{\Delta}^{(1)}_{\mathbf{k}}(\omega) = \frac{1}{2}\sum_{\mathbf{k}'} V^{(t)}(\mathbf{k},\mathbf{k}')\{\tilde{\underset{\sim}{v}}^*_{\mathbf{k}'}\underset{\sim}{a}_{\mathbf{k}'}(\omega) + [\tilde{\underset{\sim}{b}}_{\mathbf{k}'}(-\omega)]^*\underset{\sim}{u}_{\mathbf{k}'}$$
$$+ \tilde{\underset{\sim}{a}}_{\mathbf{k}'}(\omega)\underset{\sim}{v}^*_{\mathbf{k}'} + \tilde{\underset{\sim}{u}}_{\mathbf{k}'}[\underset{\sim}{b}_{\mathbf{k}'}(-\omega)]^*\}; \tag{54.35b}$$

here

$$V^{(t)}(\mathbf{k},\mathbf{k}') \equiv \mathscr{V}^{(t)}(\mathbf{k}-\mathbf{k}) - \mathscr{V}^{(t)}(\mathbf{k}+\mathbf{k}'). \tag{54.36}$$

From Eq. (54.20) the expansion coefficients take the form

$$\begin{pmatrix} a_{\mathbf{k}\nu\mu} \\ b_{\mathbf{k}\nu\mu} \end{pmatrix} = \frac{1}{(2\varepsilon_{\mathbf{k}}+\hbar\omega)\hbar\omega}\sum_{\mu'\nu'}\begin{bmatrix} (\varepsilon_{\mathbf{k}}+\hbar\omega+\xi_{\mathbf{k}})\delta_{\mu\mu'} & \Delta^{(0)}_{\mathbf{k}\mu\mu'} \\ -(\Delta^{(0)}_{-\mathbf{k}\mu'\mu})^* & (\varepsilon_{\mathbf{k}}+\hbar\omega-\xi_{\mathbf{k}})\delta_{\mu\mu'} \end{bmatrix}$$
$$\times\begin{bmatrix} \Delta^{(1)}_{\mathbf{k}\mu'\nu'}(\omega)v_{\mathbf{k}\nu\nu'} \\ -(\Delta^{(1)}_{-\mathbf{k}\nu'\mu'}(-\omega))^* u_{\mathbf{k}\nu\nu'} \end{bmatrix}; \tag{54.37a}$$

in matrix form we have

$$\begin{pmatrix} \tilde{\underset{\sim}{a}}_{\mathbf{k}}(\omega) \\ \tilde{\underset{\sim}{b}}_{\mathbf{k}}(\omega) \end{pmatrix} = \frac{1}{(2\varepsilon_{\mathbf{k}}+\hbar\omega)\hbar\omega}\begin{pmatrix} (\varepsilon_{\mathbf{k}}+\hbar\omega+\xi_{\mathbf{k}})\underset{\sim}{1} & \underset{\sim}{\Delta}^{(0)}_{\mathbf{k}} \\ -(\underset{\sim}{\Delta}^{(0)}_{-\mathbf{k}})^* & (\varepsilon_{\mathbf{k}}+\hbar\omega-\xi_{\mathbf{k}})\underset{\sim}{1} \end{pmatrix}$$
$$\begin{pmatrix} \underset{\sim}{\Delta}^{(1)}_{\mathbf{k}}(\omega)\tilde{\underset{\sim}{v}}_{\mathbf{k}} \\ -[\underset{\sim}{\Delta}^{(1)}_{-\mathbf{k}}(-\omega)]^*\tilde{\underset{\sim}{u}}_{\mathbf{k}} \end{pmatrix}, \tag{54.37b}$$

where, due to the ordering of the indices on the right-hand side, it was convenient to write the left-hand side as the transpose. From (54.37b) we have the four spinors $\underset{\sim}{a}_{\mathbf{k}}(\omega)$, $\underset{\sim}{b}_{\mathbf{k}}(-\omega)$, $\tilde{\underset{\sim}{a}}_{\mathbf{k}}(\omega)$, $\tilde{\underset{\sim}{b}}_{\mathbf{k}}(-\omega)$ entering (54.35b) as (where we invoked the triplet pairing properties $\underset{\sim}{v}_{\mathbf{k}} = \tilde{\underset{\sim}{v}}_{\mathbf{k}}$, $\underset{\sim}{u}_{\mathbf{k}} = \tilde{\underset{\sim}{u}}_{\mathbf{k}}$, $\underset{\sim}{\Delta}_{\mathbf{k}} = -\underset{\sim}{\Delta}_{-\mathbf{k}}$, and $\underset{\sim}{\Delta}_{\mathbf{k}} = \tilde{\underset{\sim}{\Delta}}_{\mathbf{k}}$)

$$\tilde{\underset{\sim}{a}}_{\mathbf{k}}(\omega) = \frac{1}{(2\varepsilon_{\mathbf{k}} + \hbar\omega)\hbar\omega}\{(\varepsilon_{\mathbf{k}} + \hbar\omega + \xi_{\mathbf{k}})\underset{\sim}{\Delta}^{(1)}_{\mathbf{k}}(\omega)\underset{\sim}{v}_{\mathbf{k}} + \underset{\sim}{\Delta}^{(0)}_{\mathbf{k}}[\underset{\sim}{\Delta}^{(1)}_{\mathbf{k}}(-\omega)]^*\underset{\sim}{u}_{\mathbf{k}}\}, \tag{54.38a}$$

$$[\tilde{\underset{\sim}{b}}_{\mathbf{k}}(-\omega)]^* = -\frac{1}{(2\varepsilon_{\mathbf{k}} - \hbar\omega)\hbar\omega}\{\underset{\sim}{\Delta}^{(0)}_{\mathbf{k}}[\underset{\sim}{\Delta}^{(1)}_{\mathbf{k}}(-\omega)]^*\underset{\sim}{v}^*_{\mathbf{k}} + (\varepsilon_{\mathbf{k}} - \hbar\omega - \xi_{\mathbf{k}})\underset{\sim}{\Delta}^{(1)}_{\mathbf{k}}(\omega)\underset{\sim}{u}^*_{\mathbf{k}}\}, \tag{54.38b}$$

$$\underset{\sim}{a}_{\mathbf{k}}(\omega) = \frac{1}{(2\varepsilon_{\mathbf{k}} + \hbar\omega)\hbar\omega}\{(\varepsilon_{\mathbf{k}} + \hbar\omega + \xi_{\mathbf{k}})\underset{\sim}{v}_{\mathbf{k}}\underset{\sim}{\Delta}^{(1)}_{\mathbf{k}}(\omega) + \underset{\sim}{u}_{\mathbf{k}}[\underset{\sim}{\Delta}^{(1)}_{\mathbf{k}}(-\omega)]^*\underset{\sim}{\Delta}^{(0)}_{\mathbf{k}}\}, \tag{54.38c}$$

$$[\underset{\sim}{b}_{\mathbf{k}}(-\omega)]^* = -\frac{1}{(2\varepsilon_{\mathbf{k}} - \hbar\omega)\hbar\omega}\{\underset{\sim}{v}^*_{\mathbf{k}}[\underset{\sim}{\Delta}^{(1)}_{\mathbf{k}}(-\omega)]^*\underset{\sim}{\Delta}^{(0)}_{\mathbf{k}} + (\varepsilon_{\mathbf{k}} - \hbar\omega - \xi_{\mathbf{k}})\underset{\sim}{u}^*_{\mathbf{k}}\underset{\sim}{\Delta}^{(1)}_{\mathbf{k}}(\omega)\}. \tag{54.38d}$$

Inserting Eqs. (54.38a)–(54.38d) into Eq. (54.35b) yields

$$\begin{aligned}
\underset{\sim}{\Delta}^{(1)}_{\mathbf{k}}(\omega) = \frac{1}{2\hbar\omega}\sum_{\mathbf{k}'} V^{(\mathrm{t})}(\mathbf{k},\mathbf{k}')\Bigg(&\frac{1}{2\varepsilon_{\mathbf{k}'} + \hbar\omega}\{(\varepsilon_{\mathbf{k}'} + \hbar\omega + \xi_{\mathbf{k}'})[\underset{\sim}{v}^*_{\mathbf{k}'}\underset{\sim}{v}_{\mathbf{k}'}\underset{\sim}{\Delta}^{(1)}_{\mathbf{k}'}(\omega) + \underset{\sim}{\Delta}^{(1)}_{\mathbf{k}'}(\omega)\underset{\sim}{v}_{\mathbf{k}'}\underset{\sim}{v}^*_{\mathbf{k}'}] \\
&+ \underset{\sim}{v}^*_{\mathbf{k}'}\underset{\sim}{u}_{\mathbf{k}'}[\underset{\sim}{\Delta}^{(1)}_{\mathbf{k}'}(-\omega)]^*\underset{\sim}{\Delta}^{(0)}_{\mathbf{k}'} + \underset{\sim}{\Delta}^{(0)}_{\mathbf{k}'}[\underset{\sim}{\Delta}^{(1)}_{\mathbf{k}'}(-\omega)]^*\underset{\sim}{u}_{\mathbf{k}'}\underset{\sim}{v}^*_{\mathbf{k}'}\} \\
&- \frac{1}{2\varepsilon_{\mathbf{k}'} - \hbar\omega}\{\underset{\sim}{\Delta}^{(0)}_{\mathbf{k}'}[\underset{\sim}{\Delta}^{(1)}_{\mathbf{k}'}(-\omega)]^*\underset{\sim}{v}^*_{\mathbf{k}'}\underset{\sim}{u}_{\mathbf{k}'} + \underset{\sim}{u}_{\mathbf{k}'}\underset{\sim}{v}^*_{\mathbf{k}'}[\underset{\sim}{\Delta}^{(1)}_{\mathbf{k}'}(-\omega)]^*\underset{\sim}{\Delta}^{(0)}_{\mathbf{k}'} \\
&+ (\varepsilon_{\mathbf{k}'} - \hbar\omega - \xi_{\mathbf{k}'})[\underset{\sim}{\Delta}^{(1)}_{\mathbf{k}'}(\omega)\underset{\sim}{u}^*_{\mathbf{k}'}\underset{\sim}{u}_{\mathbf{k}'} + \underset{\sim}{u}_{\mathbf{k}'}\underset{\sim}{u}^*_{\mathbf{k}'}\underset{\sim}{\Delta}^{(1)}_{\mathbf{k}'}(\omega)]\}\Bigg).
\end{aligned} \tag{54.39}$$

To evaluate (54.39) we require the matrix products $\underset{\sim}{v}^*\underset{\sim}{v}, \underset{\sim}{u}\underset{\sim}{u}^*, \underset{\sim}{v}^*\underset{\sim}{u}$, etc. In the representation (53.18)

$$\underset{\sim}{u}_{\mathbf{k}} = u_{\mathbf{k}}\begin{pmatrix} 1 & 0 \\ 0 & 1 \end{pmatrix}, \tag{54.40a}$$

$$\underset{\sim}{v}_{\mathbf{k}} = v_{\mathbf{k}}\begin{pmatrix} -\sin\theta\,\mathrm{e}^{\mathrm{i}\varphi} & \cos\theta \\ \cos\theta & \sin\theta\,\mathrm{e}^{-\mathrm{i}\varphi} \end{pmatrix}, \tag{54.40b}$$

where $u_{\mathbf{k}}$ and $v_{\mathbf{k}}$ are the usual BCS functions. We also recall the equilibrium gap function (53.5) for the B phase

$$\underset{\sim}{\Delta}^{(0)}_{\mathbf{k}} = \underset{\sim}{\Delta}^{(0)}(\theta, \varphi) = \Delta^{(0)}\begin{pmatrix} -\sin\theta\,\mathrm{e}^{-\mathrm{i}\varphi} & \cos\theta \\ \cos\theta & \sin\theta\,\mathrm{e}^{\mathrm{i}\varphi} \end{pmatrix}. \tag{54.41}$$

From Eqs. (54.40) we have the following for the matrix products

$$\underset{\sim}{u}_{\mathbf{k}}\cdot\underset{\sim}{u}^*_{\mathbf{k}} = \underset{\sim}{u}^*_{\mathbf{k}}\cdot\underset{\sim}{u}_{\mathbf{k}} = |u_{\mathbf{k}}|^2\begin{pmatrix} 1 & 0 \\ 0 & 1 \end{pmatrix}, \tag{54.42a}$$

$$\underset{\sim}{v}_{\mathbf{k}}\cdot\underset{\sim}{v}^*_{\mathbf{k}} = \underset{\sim}{v}^*_{\mathbf{k}}\cdot\underset{\sim}{v}_{\mathbf{k}} = |v_{\mathbf{k}}|^2\begin{pmatrix} 1 & 0 \\ 0 & 1 \end{pmatrix}, \tag{54.42b}$$

and

$$\underset{\sim}{u}_{\mathbf{k}}\cdot\underset{\sim}{v}_{\mathbf{k}}^{*}=\underset{\sim}{v}_{\mathbf{k}}^{*}\cdot\underset{\sim}{u}_{\mathbf{k}}=u_{\mathbf{k}}v_{\mathbf{k}}^{*}\begin{pmatrix}-\sin\theta\,\mathrm{e}^{-\mathrm{i}\varphi} & \cos\theta\\ \cos\theta & \sin\theta\,\mathrm{e}^{\mathrm{i}\varphi}\end{pmatrix}$$

$$=u_{\mathbf{k}}v_{\mathbf{k}}^{*}\frac{\underset{\sim}{\Delta}^{(0)}}{|\Delta^{(0)}|}, \tag{54.42c}$$

where $\Delta^{(0)}$ is the magnitude of the equilibrium gap. Writing (54.39) in terms of a common denominator and using the properties $|u_{\mathbf{k}}|^2+|v_{\mathbf{k}}|^2=1$, $|u_{\mathbf{k}}|^2-|v_{\mathbf{k}}|^2=\xi_{\mathbf{k}}/\varepsilon_{\mathbf{k}}$, $2u_{\mathbf{k}}v_{\mathbf{k}}^{*}=\Delta^{(0)}/\varepsilon_{\mathbf{k}}$, and $\xi_{\mathbf{k}}^2+\Delta_{\mathbf{k}}^2=\varepsilon_{\mathbf{k}}^2$ we obtain after a short calculation

$$\underset{\sim}{\Delta}_{\mathbf{k}}^{(1)}(\omega)=\frac{1}{2\hbar\omega}\sum_{\mathbf{k}'}\frac{V^{(t)}(\mathbf{k},\mathbf{k}')}{4\varepsilon_{\mathbf{k}'}-\hbar^2\omega^2}\left\{\frac{1}{\varepsilon_{\mathbf{k}'}}[2\hbar\omega\varepsilon_{\mathbf{k}'}^2+2\hbar^2\omega^2\xi_{\mathbf{k}'}+2\hbar\omega\xi_{\mathbf{k}'}^2]\underset{\sim}{\Delta}_{\mathbf{k}'}(\omega)\right.$$

$$\left.+\frac{2\hbar\omega}{\varepsilon_{\mathbf{k}'}}\underset{\sim}{\Delta}_{\mathbf{k}'}^{(0)}[\underset{\sim}{\Delta}_{\mathbf{k}'}^{(1)}(-\omega)]^{*}\underset{\sim}{\Delta}_{\mathbf{k}'}^{(0)}\right\}. \tag{54.43}$$

This equation may be greatly simplified noting that $\Delta^{(1)}$ must obey the self-consistency condition for the gap function,[3]

$$\underset{\sim}{\Delta}_{\mathbf{k}}^{(1)}(\omega)=\sum_{\mathbf{k}'}V(\mathbf{k},\mathbf{k}')\frac{\underset{\sim}{\Delta}_{\mathbf{k}'}^{(1)}(\omega)}{2\varepsilon_{\mathbf{k}'}}. \tag{54.44}$$

To exploit this requirement we write the term in the first square brackets in (54.43) as $[\hbar\omega(4\varepsilon_{\mathbf{k}'}^2-\hbar\omega)-2\hbar\omega\varepsilon_{\mathbf{k}'}^2+\hbar^3\omega^3+2\hbar^2\omega^2\xi_{\mathbf{k}'}+\hbar\omega\xi_{\mathbf{k}'}^2]$. We next substitute $\varepsilon_{\mathbf{k}'}^2=\xi_{\mathbf{k}'}^2+|\Delta_{\mathbf{k}'}^{(0)}|^2$ yielding $[\hbar\omega(4\varepsilon_{\mathbf{k}'}^2-\hbar\omega)-2\hbar\omega|\Delta_{\mathbf{k}'}^{(0)}|^2+\hbar^3\omega^3+2\hbar^2\omega^2\xi_{\mathbf{k}'}^2]$. Now the first term in this form cancels a corresponding term in the denominator; the remaining terms (associated with this factor only) reduce to the right-hand side of (54.44), which therefore cancels the term on the left-hand side of (54.43). We are then left with

$$\frac{1}{\hbar\omega}\sum_{\mathbf{k}'}\frac{V(\mathbf{k},\mathbf{k}')}{2\varepsilon_{\mathbf{k}'}(4\varepsilon_{\mathbf{k}'}^2-\hbar^2\omega^2)}\{[-2\hbar\omega|\underset{\sim}{\Delta}_{\mathbf{k}'}^{(0)}|^2+\hbar^3\omega^3+2\hbar^2\omega^2\xi_{\mathbf{k}'}]\underset{\sim}{\Delta}_{\mathbf{k}'}(\omega)$$

$$-2\hbar\omega\underset{\sim}{\Delta}_{\mathbf{k}'}^{(0)}[\underset{\sim}{\Delta}_{\mathbf{k}'}^{(1)}(-\omega)]^{*}\underset{\sim}{\Delta}_{\mathbf{k}'}^{(0)}\}=0. \tag{54.45}$$

At this point it is convenient to write $\Sigma_{\mathbf{k}'}=[L^3k_{\mathrm{F}}^2/(2\pi)^3\hbar v_{\mathrm{F}}]\int\mathrm{d}\Omega'\int\mathrm{d}\xi'$. Now $\Delta_{\mathbf{k}'}$ and $\varepsilon_{\mathbf{k}'}$ are even functions of $\xi_{\mathbf{k}'}$; therefore the contribution of the third term in the first square brackets of (54.45) vanishes since $\xi_{\mathbf{k}'}$ is very nearly an odd function with respect to the Fermi energy (any departure being referred to as particle–hole asymmetry). If we introduce the dimensionless function

$$\lambda(\omega)\equiv 4|\Delta^{(0)}|^2\int_{-\infty}^{+\infty}\frac{\mathrm{d}\xi}{2\varepsilon(4\varepsilon^2-\hbar^2\omega^2)}, \tag{54.46}$$

Eq. (54.45) becomes

$$\lambda(\omega)\frac{k_{\mathrm{F}}^2}{(2\pi)^3\hbar v_{\mathrm{F}}}\int\mathrm{d}\Omega' V(\hat{\mathbf{k}},\hat{\mathbf{k}}')\{(\hbar^2\omega^2-2|\underset{\sim}{\Delta}_{\mathbf{k}'}^{(0)}|^2)\underset{\sim}{\Delta}_{\mathbf{k}'}^{(1)}(\omega)-2\underset{\sim}{\Delta}_{\mathbf{k}'}^{(0)}[\underset{\sim}{\Delta}_{\mathbf{k}'}^{(1)}(-\omega)]^{*}\underset{\sim}{\Delta}_{\mathbf{k}'}^{(0)}\}=0; \tag{54.47}$$

3. In the BCS theory the interactions are essentially instantaneous on the time scale of collective mode frequencies ($\sim\hbar/\Delta^{(0)}$). Therefore even though $\Delta^{(1)}$ is a nonequilibrium (time-dependent) mean field it must still obey the self-consistency condition.

the prefactor can be dropped for the following discussion.[4] The zeroth and first order mean fields may be written as

$$\underset{\sim}{\Delta}_{\mathbf{k}}^{(0)} = \Delta\hat{\mathbf{k}}\cdot\underset{\sim}{\sigma}\mathrm{i}\underset{\sim}{\sigma}_2, \qquad \underset{\sim}{\Delta}_{\mathbf{k}}^{(1)}(\omega) = \mathbf{d}_{\mathbf{k}}(\omega)\cdot\underset{\sim}{\sigma}\mathrm{i}\underset{\sim}{\sigma}_2;$$

in what follows we will write Δ and $\mathbf{d}$ for $\Delta(0)$ and $\mathbf{d}_{\mathbf{k}}(\omega)$, respectively. We now examine the second term in (54.47).

$$\begin{aligned}\underset{\sim}{\Delta}_{\mathbf{k}}^{(0)}[\underset{\sim}{\Delta}_{\mathbf{k}}^{(1)}(-\omega)]^*\underset{\sim}{\Delta}_{\mathbf{k}}^{(0)} &= (\Delta\hat{\mathbf{k}}\cdot\underset{\sim}{\sigma}\mathrm{i}\underset{\sim}{\sigma}_2)(-\mathrm{i}\underset{\sim}{\sigma}_2\mathbf{d}^*\cdot\underset{\sim}{\sigma})(\Delta\hat{\mathbf{k}}\cdot\underset{\sim}{\sigma}\mathrm{i}\underset{\sim}{\sigma}_2)\\ &= \Delta[\hat{\mathbf{k}}\cdot\mathbf{d}^* + \mathrm{i}(\hat{\mathbf{k}}\times\mathbf{d}^*)\cdot\underset{\sim}{\sigma}](\hat{\mathbf{k}}\cdot\underset{\sim}{\sigma})\mathrm{i}\underset{\sim}{\sigma}_2\Delta\\ &= \Delta^2[(\hat{\mathbf{k}}\cdot\mathbf{d}^*)\hat{\mathbf{k}} + \mathrm{i}(\hat{\mathbf{k}}\times\mathbf{d}^*)\cdot\hat{\mathbf{k}} - (\hat{\mathbf{k}}\times\mathbf{d}^*)\times\hat{\mathbf{k}}]\cdot\underset{\sim}{\sigma}\mathrm{i}\underset{\sim}{\sigma}_2\\ &= \Delta^2[2(\hat{\mathbf{k}}\cdot\mathbf{d}^*)\hat{\mathbf{k}} - \mathbf{d}^*]\cdot\underset{\sim}{\sigma}\mathrm{i}\underset{\sim}{\sigma}_2,\end{aligned}$$

where we used the identity $(\mathbf{a}\cdot\underset{\sim}{\sigma})(\mathbf{b}\cdot\underset{\sim}{\sigma}) = (\mathbf{a}\cdot\mathbf{b})\underset{\sim}{1} + \mathrm{i}(\mathbf{a}\times\mathbf{b})\cdot\underset{\sim}{\sigma}$ twice along with the vector identity $(\mathbf{a}\times\mathbf{b})\times\mathbf{c} = (\mathbf{a}\cdot\mathbf{c})\mathbf{b} - (\mathbf{b}\cdot\mathbf{c})\mathbf{a}$. Eq. (54.47) then becomes (on rebeling $\hat{\mathbf{k}}$ and $\hat{\mathbf{k}}'$)

$$\int \mathrm{d}\Omega\, V(\hat{\mathbf{k}}',\hat{\mathbf{k}})\{(\hbar^2\omega^2 - 2\Delta^2)\mathbf{d} - 2\Delta^2[2(\hat{\mathbf{k}}\cdot\mathbf{d}^*)\hat{\mathbf{k}} - \mathbf{d}^*]\} = 0. \tag{54.48}$$

We introduce pure real and pure imaginary variables as

$$\mathbf{d}^+ = \mathbf{d} + \mathbf{d}^* \tag{54.49a}$$

and

$$\mathbf{d}^- = \mathbf{d} - \mathbf{d}^*. \tag{54.49b}$$

Combining (54.48) with its complex conjugate we see that the collective modes factor into two classes, real and imaginary, governed by the equations

$$\int \mathrm{d}\Omega\, V_1(\hat{\mathbf{k}}'\cdot\hat{\mathbf{k}})[\omega^2\mathbf{d}^+ - \omega_g^2(\mathbf{d}^+\cdot\hat{\mathbf{k}})\hat{\mathbf{k}}] = 0 \tag{54.50}$$

and

$$\int \mathrm{d}\Omega\, V_1(\hat{\mathbf{k}}'\cdot\hat{\mathbf{k}})[\omega^2 - \omega_g^2\mathbf{d}^- + \omega_g^2(\mathbf{d}^-\cdot\hat{\mathbf{k}})\hat{\mathbf{k}}] = 0; \tag{54.51}$$

here $V(\hat{\mathbf{k}}',\hat{\mathbf{k}}) = V_1(\hat{\mathbf{k}}\cdot\hat{\mathbf{k}}')$, where V_1 is the $\ell = 1$ pairing amplitude and we have defined a 'gap frequency' $\hbar\omega_g = 2\Delta$. From the relationship between $\mathbf{d}$ and $\hat{\mathbf{k}}$ for p-wave pairing, $d_\mu = D_{\mu i}\hat{k}_i$, we introduce the real and imaginary combinations

$$D_{\mu i}^+ = D_{\mu i} + D_{\mu i}^* \tag{54.52a}$$

and

$$D_{\mu i}^- = D_{\mu i} - D_{\mu i}^*. \tag{54.52b}$$

4. The properties (54.42a)–(54.42c) likewise hold for the A phase; therefore Eq. (54.47) also serves as the starting point for a discussion of the A phase collective modes.

We first discuss the real modes. Introducing (54.52a) into (54.50) we have

$$k'_j \int d\Omega \hat{k}_j \left(\omega^2 D^+_{\mu i} \hat{k}_i - \omega_g^2 \hat{k}_\nu D^+_{\nu i} \hat{k}_i \hat{k}_\mu \right). \tag{54.53}$$

We integrate (54.53) over dΩ and use the relations

$$\frac{1}{2\pi} \int d\Omega \hat{k}_i \cdot \hat{k}_j = \frac{2}{3} \delta_{ij} \tag{54.54a}$$

and

$$\frac{1}{2\pi} \int d\Omega k_i k_j k_\mu k_\nu = \frac{2}{15} (\delta_{ij}\delta_{\mu\nu} + \delta_{i\mu}\delta_{j\nu} + \delta_{i\nu}\delta_{j\mu}); \tag{54.54b}$$

since the variables θ' and ϕ' defining $\hat{\mathbf{k}}'$ are arbitrary, the terms with the index j in (54.53) must separately vanish and we obtain the three sets of equations:

$j = 1, \mu = 1, 2, 3$:

$$\frac{2}{3}\omega^2 D^+_{11} - \omega_g^2 \left[\frac{2}{15}(D^+_{22} + D^+_{33}) + \frac{2}{5} D^+_{11} \right] = 0, \tag{54.55a}$$

$$\frac{2}{3}\omega^2 D^+_{21} - \frac{2}{15}\omega_g^2 (D^+_{21} + D^+_{12}) = 0, \tag{54.55b}$$

$$\frac{2}{3}\omega^2 D^+_{31} - \frac{2}{15}\omega_g^2 (D^+_{31} + D^+_{13}) = 0; \tag{54.55c}$$

$j = 2, \mu = 1, 2, 3$:

$$\frac{2}{3}\omega^2 D^+_{12} - \frac{2}{15}\omega_g^2 (D^+_{12} + D^+_{21}) = 0, \tag{54.55d}$$

$$\frac{2}{3}\omega^2 D^+_{22} - \omega_g^2 \left[\frac{2}{15}(D^+_{11} + D^+_{33}) + \frac{2}{15} D^+_{22} \right] = 0, \tag{54.55e}$$

$$\frac{2}{3}\omega^2 D^+_{32} - \frac{2}{15}\omega_g^2 (D^+_{32} + D^+_{23}) = 0; \tag{54.55f}$$

$j = 3, \mu = 1, 3$:

$$\frac{2}{3}\omega^2 D^+_{13} - \frac{2}{15}\omega_g^2 (D^+_{13} + D^+_{31}) = 0, \tag{54.55g}$$

$$\frac{2}{3}\omega^2 D^+_{23} - \frac{2}{15}\omega_g^2 (D^+_{23} + D^+_{32}) = 0, \tag{54.55h}$$

$$\frac{2}{3}\omega^2 D^+_{33} - \omega_g^2 \left[\frac{2}{15}(D^+_{11} + D^+_{22}) + \frac{2}{5} D^+_{33} \right] = 0. \tag{54.55i}$$

These equations may be written in matrix form as follows:

$$\begin{pmatrix} \frac{1}{5}-\lambda & \frac{1}{5} \\ \frac{1}{5} & \frac{1}{5}-\lambda \end{pmatrix} \begin{pmatrix} D_{12}^{+} & D_{23}^{+} & D_{31}^{+} \\ D_{21}^{+} & D_{32}^{+} & D_{13}^{+} \end{pmatrix} = 0, \tag{54.56}$$

$$\begin{pmatrix} \frac{3}{5}-\lambda & \frac{1}{5} & \frac{1}{5} \\ \frac{1}{5} & \frac{3}{5}-\lambda & \frac{1}{5} \\ \frac{1}{5} & \frac{1}{5} & \frac{3}{5}-\lambda \end{pmatrix} \begin{pmatrix} D_{11}^{+} \\ D_{22}^{+} \\ D_{33}^{+} \end{pmatrix} = 0; \tag{54.57}$$

here $\lambda \equiv \omega^2/\omega_g^2$. The eigenvectors and eigenvalues of (54.56) are

$$\vec{V}_{1,2,3}^{+} = \frac{1}{\sqrt{2}}\begin{pmatrix} 1 \\ 1 \end{pmatrix}; \quad \lambda_{1,2,3}^{+} = \frac{2}{5},$$

$$\vec{V}_{4,5,6}^{+} = \frac{1}{\sqrt{2}}\begin{pmatrix} 1 \\ -1 \end{pmatrix}; \quad \lambda_{4,5,6}^{+} = 0, \tag{54.58}$$

while those of (54.57) are

$$\vec{V}_{7}^{+} = \frac{1}{\sqrt{2}}\begin{pmatrix} 1 \\ -1 \\ 1 \end{pmatrix}; \quad \lambda_{7}^{+} = \frac{2}{5},$$

$$\vec{V}_{8}^{+} = \frac{1}{\sqrt{6}}\begin{pmatrix} 1 \\ 1 \\ -2 \end{pmatrix}; \quad \lambda_{8}^{+} = \frac{2}{5},$$

and

$$\vec{V}_{9}^{+} = \frac{1}{\sqrt{3}}\begin{pmatrix} 1 \\ 1 \\ 1 \end{pmatrix}; \quad \lambda_{9}^{+} = 1. \tag{54.59}$$

Thus we have five degenerate modes with $\hbar\omega = (8/5)^{1/2}\Delta$, three degenerate modes with $\hbar\omega = 0$, and one mode with $\hbar\omega = 2\Delta$. The eigenvectors above only included the *nonzero* components of a nine component vector.

$$\vec{V} = (D_{13}^{+}, D_{31}^{+}, D_{23}^{+}, D_{32}^{+}, D_{12}^{+}, D_{21}^{+}, D_{11}^{+}, D_{22}^{+}, D_{33}^{+}).$$

We now turn to the imaginary modes involving the combination $D_{\mu i}^{-}$. Starting from Eq. (54.51), the remaining treatment is analogous to that for the real modes and leads to the matrix equations

$$\begin{pmatrix} \frac{4}{5}-\lambda & -\frac{1}{5} \\ -\frac{1}{5} & \frac{4}{5}-\lambda \end{pmatrix} \begin{pmatrix} D_{12}^{-} & D_{23}^{-} & D_{31}^{-} \\ D_{21}^{-} & D_{32}^{-} & D_{13}^{-} \end{pmatrix} = 0 \tag{54.60}$$

and

$$\begin{pmatrix} \frac{2}{5}-\lambda & -\frac{1}{5} & -\frac{1}{5} \\ -\frac{1}{5} & \frac{2}{5}-\lambda & -\frac{1}{5} \\ -\frac{1}{5} & -\frac{1}{5} & \frac{2}{5}-\lambda \end{pmatrix} \begin{pmatrix} D_{11}^- \\ D_{22}^- \\ D_{33}^- \end{pmatrix} = 0. \tag{54.61}$$

The eigenvectors of (54.60) are

$$\vec{V}_{1,2,3}^- = \frac{\mathrm{i}}{\sqrt{2}}\begin{pmatrix} 1 \\ 1 \end{pmatrix}; \quad \lambda_{1,2,3}^- = \frac{3}{5},$$
$$\vec{V}_{4,5,6}^- = \frac{\mathrm{i}}{\sqrt{2}}\begin{pmatrix} 1 \\ -1 \end{pmatrix}; \quad \lambda_{4,5,6}^- = 1, \tag{54.62}$$

while those of (54.61) are

$$\vec{V}_7^- = \frac{\mathrm{i}}{\sqrt{2}}\begin{pmatrix} 1 \\ -1 \\ 0 \end{pmatrix}; \quad \lambda_7^- = \frac{3}{5},$$
$$\vec{V}_8^- = \frac{\mathrm{i}}{\sqrt{6}}\begin{pmatrix} 1 \\ 1 \\ -2 \end{pmatrix}; \quad \lambda_8^- = \frac{3}{5},$$
$$\vec{V}_9^- = \frac{\mathrm{i}}{\sqrt{3}}\begin{pmatrix} 1 \\ 1 \\ 1 \end{pmatrix}; \quad \lambda_9^- = 0. \tag{54.63}$$

Thus we have five degenerate modes with frequency $\hbar\omega = (12/5)^{1/2}\Delta$, three degenerate modes with frequency 2Δ, and one mode with frequency 0. This completes the enumeration of the 18 B-phase collective modes for $q = 0$. Note the results are identical to those obtained earlier in Subsec. 23.5.1.

55 Green's functions

Much of the theoretical literature on superconductivity makes use of Green's function methods. The discussions usually involve rather formal techniques borrowed from quantum field theory. However, in this section we employ a more simplified (though much less rigorous) approach. Our goal here is only to give the reader a 'flavor' of the Green's function methods with the hope that some of its applications to superconductivity will then be 'recognizable'.

55.1 Green's functions for the Bogoliubov equations; the Nambu–Gorkov equations

In Sec. 49 we introduced the Bogoliubov transformation for the time-dependent $\hat{\psi}$ operators for a (singlet) superconductor as

$$\hat{\psi}_{\uparrow}(\mathbf{r}, t) = \sum_n [u_n(\mathbf{r}, t)\hat{\gamma}_{n\uparrow} - v_n^*(\mathbf{r}, t)\hat{\gamma}_{n\downarrow}^\dagger] \tag{49.2a}$$

and

$$\hat{\psi}_{\downarrow}(\mathbf{r}, t) = \sum_n [u_n(\mathbf{r}, t)\hat{\gamma}_{n\downarrow} + v_n^*(\mathbf{r}, t)\hat{\gamma}_{n\uparrow}^\dagger]; \tag{49.2b}$$

the corresponding hermitian conjugate operators are

$$\hat{\psi}_{\uparrow}^\dagger(\mathbf{r}, t) = \sum_n [u_n^*(\mathbf{r}, t)\hat{\gamma}_{n\uparrow}^\dagger - v_n(\mathbf{r}, t)\hat{\gamma}_{n\downarrow}] \tag{55.1a}$$

and

$$\hat{\psi}_{\downarrow}^\dagger(\mathbf{r}, t) = \sum_n [u_n^*(\mathbf{r}, t)\hat{\gamma}_{n\downarrow}^\dagger + v_n(\mathbf{r}, t)\hat{\gamma}_{n\uparrow}]. \tag{55.1b}$$

We initially introduce the Green's function via a definition; however, we will show later that this definition reduces to the more familiar definition of a Green's function involving a partial differential equation with a δ function source. The Green's function is defined by

$$G_{\alpha\beta}(\mathbf{r}_1, t_1; \mathbf{r}_2, t_2) = \begin{cases} -(\mathrm{i}/\hbar)\langle\hat{\psi}_\alpha(\mathbf{r}_1, t_1)\hat{\psi}^\dagger_\beta(\mathbf{r}_2, t_2)\rangle, & t_1 > t_2; \\ +(\mathrm{i}/\hbar)\langle\hat{\psi}^\dagger_\beta(\mathbf{r}_2, t_2)\hat{\psi}_\alpha(\mathbf{r}_1, t_1)\rangle, & t_1 < t_2. \end{cases} \tag{55.2}$$

Here the brackets refer to both a quantum mechanical and a statistical mechanical average, and the significance of the differing sign for $t_1 > t_2$ and $t_1 < t_2$ will emerge shortly. The definition (55.2) is quite general and is not limited to a mean field theory (although we may approximate it by a mean field form). Note that for $t_2 = t_1 + 0_+$ we have from Eq. (52.51)

$$\rho_{\alpha\beta}(\mathbf{r}, \mathbf{r}', t) = \frac{\hbar}{\mathrm{i}} G_{\alpha\beta}(\mathbf{r}, t; \mathbf{r}', t + 0_+); \tag{55.3}$$

i.e., we have a connection between the Green's function and the density matrix. However, since the Green's function is defined for two different times it is clearly a more general object than the density matrix. Inserting (49.2a) and (55.1a) into (55.2) and using the commutation relations (36.12c) and (36.12d) we obtain

$$G_{\uparrow\uparrow}(\mathbf{r}_1, t_1; \mathbf{r}_2, t_2) = \begin{cases} -\dfrac{i}{\hbar}\sum\limits_n [u_n^*(\mathbf{r}_2, t_2)u_n(\mathbf{r}_1, t_1)(1 - f_{n\uparrow}) + v_n(\mathbf{r}_2, t_2)v_n^*(\mathbf{r}_1, t_1)f_{n\downarrow}], & t_1 > t_2; \\ +\dfrac{i}{\hbar}\sum\limits_n [u_n^*(\mathbf{r}_2, t_2)u_n(\mathbf{r}_1, t_1)f_{n\uparrow} + v_n(\mathbf{r}_2, t_2)v_n^*(\mathbf{r}_1, t_1)(1 - f_{n\downarrow})], & t_1 < t_2. \end{cases} \tag{55.4}$$

The expression for $G_{\downarrow\downarrow}$ is identical with the exception that $f_\uparrow$ and $f_\downarrow$ are interchanged. For systems having no net spin polarization ($f_\uparrow = f_\downarrow$) $G_{\uparrow\uparrow} = G_{\downarrow\downarrow} \equiv G$.

It will turn out that to describe a superconductor we will require a second kind of Green's function, referred to as an *anomalous Green's function*, which is defined as

$$F_{\alpha\beta}(\mathbf{r}_1, t_1; \mathbf{r}_2, t_2) \equiv \begin{cases} -(\mathrm{i}/\hbar)\langle\hat{\psi}_\alpha(\mathbf{r}_1, t_1)\hat{\psi}_\beta(\mathbf{r}_2, t_2)\rangle, & t_1 > t_2; \\ +(\mathrm{i}/\hbar)\langle\hat{\psi}_\beta(\mathbf{r}_2, t_2)\hat{\psi}_\alpha(\mathbf{r}_1, t_1)\rangle, & t_1 < t_2. \end{cases} \tag{55.5}$$

We see that the anomalous Green's function is related to the pair amplitude associated with the Hartree–Fock–Gortov factorization of the Hamiltonian introduced in Sec. 36, although, again, it is a more general object being defined for two different space-time points. From the commutation relations for the $\hat{\psi}$ operators it follows that

$$F_{\alpha\beta}(\mathbf{r}_1, t_1; \mathbf{r}_2, t_2) = -F_{\beta\alpha}(\mathbf{r}_2, t_2; \mathbf{r}_1, t_1). \tag{55.6a}$$

For singlet superconductivity, where only the off diagonal spin components are nonzero, we have the symmetries

$$F_{\uparrow\downarrow}(\mathbf{r}_1, t_1; \mathbf{r}_2, t_2) = -F_{\downarrow\uparrow}(\mathbf{r}_1, t_1; \mathbf{r}_2, t_2) = F_{\uparrow\downarrow}(\mathbf{r}_2, t_2; \mathbf{r}_1, t_1) \equiv F(\mathbf{r}_1, t_1; \mathbf{r}_2, t_2). \tag{55.6b}$$

Substituting (49.2a) and (49.2b) into (55.5) and again using (36.12c) and (36.12d) we obtain

$$F_{\uparrow\downarrow}(\mathbf{r}_1, t_1; \mathbf{r}_2, t_2) = \begin{cases} -\dfrac{\mathrm{i}}{\hbar}\sum\limits_n [v_n^*(\mathbf{r}_2, t_2)u_n(\mathbf{r}_1, t_1)(1 - f_{n\uparrow}) - u_n(\mathbf{r}_2, t_2)v_n^*(\mathbf{r}_1, t_1)f_{n\downarrow}], & t_1 > t_2; \\ +\dfrac{\mathrm{i}}{\hbar}\sum\limits_n [v_n^*(\mathbf{r}_2, t_2)u_n(\mathbf{r}_1, t_1)f_{n\uparrow} - u_n(\mathbf{r}_2, t_2)v_n^*(\mathbf{r}_1, t_1)(1 - f_{n\downarrow})], & t_1 < t_2. \end{cases} \tag{55.7}$$

If we evaluate $F_{\downarrow\uparrow}(\mathbf{r}_2, t_2; \mathbf{r}_1, t_1)$ we may verify that (55.6a) is satisfied. Eq. (55.7) satisfies the symmetry (55.6b) with respect to the interchange of $(\mathbf{r}_1, t_1)$ and $(\mathbf{r}_2, t_2)$ (on setting $f_\uparrow = f_\downarrow = f$). We will also require a function $F^*_{\alpha\beta}$ defined by

$$F^*_{\alpha\beta}(\mathbf{r}_1,t_1;\mathbf{r}_2,t_2) \equiv \begin{cases} -\dfrac{\mathrm{i}}{\hbar}\langle\hat{\psi}^\dagger_\alpha(\mathbf{r}_1,t_1)\hat{\psi}^\dagger_\beta(\mathbf{r}_2,t_2)\rangle, & t_1 > t_2; \\ +\dfrac{\mathrm{i}}{\hbar}\langle\hat{\psi}^\dagger_\beta(\mathbf{r}_2,t_2)\hat{\psi}^\dagger_\alpha(\mathbf{r}_1,t_1)\rangle, & t_1 < t_2. \end{cases} \tag{55.8}$$

For singlet superconductivity we have the symmetry properties

$$F^*_{\uparrow\downarrow}(\mathbf{r}_1,t_1;\mathbf{r}_2,t_2) = -F^*_{\downarrow\uparrow}(\mathbf{r}_1,t_1;\mathbf{r}_2,t_2) = F^*_{\uparrow\downarrow}(\mathbf{r}_2,t_2;\mathbf{r}_1,t_1) \equiv F^*(\mathbf{r}_1,t_1;\mathbf{r}_2,t_2). \tag{55.6c}$$

Proceeding as above we find

$$F^*_{\uparrow\downarrow}(\mathbf{r}_1,t_1;\mathbf{r}_2,t_2) = \begin{cases} -\dfrac{\mathrm{i}}{\hbar}\sum_n [v_n(\mathbf{r}_2,t_2)u^*_n(\mathbf{r}_1,t_1)f_{n\uparrow} - u^*_n(\mathbf{r}_2,t_2)v_n(\mathbf{r}_1,t_1)(1-f_{n\downarrow})], & t_1 > t_2; \\ +\dfrac{\mathrm{i}}{\hbar}\sum_n [v_n(\mathbf{r}_2,t_2)u^*_n(\mathbf{r}_1,t_1)(1-f_{n\uparrow}) - u^*_n(\mathbf{r}_2,t_2)v_n(\mathbf{r}_1,t_1)f_{n\downarrow}], & t_1 < t_2. \end{cases} \tag{55.9}$$

Let us now consider only the case $T = 0$ where $f_n = 0$; Eqs. (55.4), (55.7), and (55.9) then become

$$G(\mathbf{r}_1,t_1;\mathbf{r}_2,t_2) = -\frac{\mathrm{i}}{\hbar}\sum_n [u^*_n(\mathbf{r}_2,t_2)u_n(\mathbf{r}_1,t_1)\theta(t_1-t_2) - v_n(\mathbf{r}_2,t_2)v^*_n(\mathbf{r}_1,t_1)\theta(t_2-t_1)], \tag{55.10}$$

$$F(\mathbf{r}_1,t_1;\mathbf{r}_2,t_2) = -\frac{\mathrm{i}}{\hbar}\sum_n [v^*_n(\mathbf{r}_2,t_2)u_n(\mathbf{r}_1,t_1)\theta(t_1-t_2) + u_n(\mathbf{r}_2,t_2)v^*_n(\mathbf{r}_1,t_1)\theta(t_2-t_1)], \tag{55.11}$$

and

$$F^*(\mathbf{r}_1,t_1;\mathbf{r}_2,t_2) = \frac{\mathrm{i}}{\hbar}\sum_n [u^*_n(\mathbf{r}_2,t_2)v_n(\mathbf{r}_1,t_1)\theta(t_1-t_2) + v_n(\mathbf{r}_2,t_2)u^*_n(\mathbf{r}_1,t_1)\theta(t_2-t_1)]. \tag{55.12}$$

We recall the time-dependent Bogoliubov equations obtained in Sec. 49:

$$\mathrm{i}\hbar\frac{\partial}{\partial t}u_n(\mathbf{r},t) = \mathscr{H}'u_n(\mathbf{r},t) + \Delta v_n(\mathbf{r},t) \tag{55.13a}$$

and

$$\mathrm{i}\hbar\frac{\partial}{\partial t}v_n(\mathbf{r},t) = -\mathscr{H}'^*v_n(\mathbf{r},t) + \Delta^*u_n(\mathbf{r},t). \tag{55.13b}$$

If $\mathscr{H}'$ and Δ are time-independent, u_n and v_n have the form

$$u_n(\mathbf{r},t) = u_n(\mathbf{r})\mathrm{e}^{-(\mathrm{i}/\hbar)\varepsilon_n t} \tag{55.14a}$$

and

$$v_n(\mathbf{r},t) = v_n(\mathbf{r})\mathrm{e}^{-(\mathrm{i}/\hbar)\varepsilon_n t}. \tag{55.14b}$$

If we use Eqs. (55.13) to operate on the $(\mathbf{r}_1,t_1)$ coordinates of the functions $G(\mathbf{r}_1,t_1;\mathbf{r}_2,t_2)$ and $F^*(\mathbf{r}_1,t_1;\mathbf{r}_2,t_2)$ it is straightforward to show that they satisfy the equations[1]

$$\left(\mathrm{i}\hbar\frac{\partial}{\partial t_1} - \mathscr{H}'_1\right)G(\mathbf{r}_1,t_1;\mathbf{r}_2,t_2) + \Delta(\mathbf{r}_1)F^*(\mathbf{r}_1,t_1;\mathbf{r}_2,t_2) = \delta^{(3)}(\mathbf{r}_1-\mathbf{r}_2)\delta(t_1-t_2) \tag{55.15a}$$

1. These equations can be obtained directly (without using the Bogoliubov equations) from the definitions of the Green's functions by using the equation of motion for $\hat{\psi}$, $\mathrm{i}\hbar\dot{\hat{\psi}} = [\hat{\psi}, \hat{H}']$, and the commutation relations for the $\hat{\psi}$ operators along with a Hartree–Fock–Gorkov mean field factorization of interaction term in the Hamiltonian.

and

$$\Delta^*(\mathbf{r}_1)G(\mathbf{r}_1,t_1;\mathbf{r}_2,t_2) + \left(\mathrm{i}\hbar\frac{\partial}{\partial t_1} + \mathscr{H}_1'^*\right)F^*(\mathbf{r}_1,t_1;\mathbf{r}_2,t_2) = 0; \tag{55.15b}$$

here we employed the completeness relations, Eqs. (36.23a) and (36.23b), and the property

$$\frac{\partial}{\partial t_1}\theta(t_1 - t_2) = -\frac{\partial}{\partial t_1}\theta(t_2 - t_1) = \delta(t_1 - t_2).$$

Note that our earlier inclusion of a sign change for $t_1 > t_2$ and $t_1 < t_2$ in the Green's function definitions is what generated $\delta(t_1 - t_2)$ in this equation; the completeness relation generated $\delta^{(3)}(\mathbf{r}_1 - \mathbf{r}_2)$. The pair of equations (55.15a) and (55.15b) are referred to as the Nambu–Gorkov equations (Nambu 1960, Gorkov 1959). It is at this point that our earlier definition of the Green's function makes contact with the more familiar definition; for the case $\Delta = 0$ (a normal Fermi gas) we would have

$$\left(\mathrm{i}\hbar\frac{\partial}{\partial t_1} - \mathscr{H}'_1\right)G(\mathbf{r}_1,t_1;\mathbf{r}_2,t_2) = \delta^{(3)}(\mathbf{r}_1 - \mathbf{r}_2)\delta(t_1 - t_2). \tag{55.16}$$

If we further restrict the wavefunctions $u_n(\mathbf{r},t)$ and $v_n(\mathbf{r},t)$ to plane waves (with $\mathbf{k}'$ replacing n) we have the forms

$$u_{\mathbf{k}'}(\mathbf{r},t) = \frac{\mathscr{u}_{k'}}{L^{3/2}}\mathrm{e}^{\mathrm{i}\mathbf{k}'\cdot\mathbf{r} - (\mathrm{i}/\hbar)\varepsilon_{\mathbf{k}'}t} \tag{55.17a}$$

and

$$v_{\mathbf{k}'}(\mathbf{r},t) = \frac{\mathscr{v}_{k'}}{L^{3/2}}\mathrm{e}^{\mathrm{i}\mathbf{k}'\cdot\mathbf{r} - (\mathrm{i}/\hbar)\varepsilon_{\mathbf{k}'}t}. \tag{55.17b}$$

Next we insert these forms into (55.10) and (55.12), write $\mathbf{r} = \mathbf{r}_1 - \mathbf{r}_2$ and $t = t_1 - t_2$ and take the Fourier transform according to the prescriptions

$$G(\mathbf{k},\omega) = \int G(\mathbf{r},t)\mathrm{e}^{-\mathrm{i}(\mathbf{k}\cdot\mathbf{r} - \omega t)}\mathrm{d}^3r\mathrm{d}t \tag{55.18a}$$

and

$$F^*(\mathbf{k},\omega) = \int F^*(\mathbf{r},t)\mathrm{e}^{-\mathrm{i}(\mathbf{k}\cdot\mathbf{r} - \omega t)}\mathrm{d}^3r\mathrm{d}t. \tag{55.18b}$$

Converting the sums over $\mathbf{k}'$ to integrals over d^3k' according to the prescription $\Sigma_{\mathbf{k}'} = [L^3/(2\pi)^3]\int \mathrm{d}^3k'$, performing the integration over d^3r using the property $\int \mathrm{d}^3r\mathrm{e}^{\mathrm{i}(\mathbf{k}' - \mathbf{k})\cdot\mathbf{r}} = (2\pi)^3\delta^{(3)}(\mathbf{k} - \mathbf{k}')$, and then performing the integration over d^3k' we obtain

$$G(\mathbf{k},t) = -\frac{\mathrm{i}}{\hbar}[|\mathscr{u}_k|^2\theta(t)\mathrm{e}^{-(\mathrm{i}/\hbar)\varepsilon_k t} - |\mathscr{v}_k^2|\theta(-t)\mathrm{e}^{(\mathrm{i}/\hbar)\varepsilon_k t}] \tag{55.19}$$

and

$$F^*(\mathbf{k},t) = \frac{\mathrm{i}}{\hbar}\mathscr{u}_k^*\mathscr{v}_k[\theta(t)\mathrm{e}^{-(\mathrm{i}/\hbar)\varepsilon_k t} + \theta(-t)\mathrm{e}^{(\mathrm{i}/\hbar)\varepsilon_k t}]. \tag{55.20}$$

Carrying out the Fourier time transform on (55.19) yields

$$G(\mathbf{k},\omega) = \frac{|u_k|^2}{\hbar\omega - \varepsilon_k + i0} + \frac{|v_k|^2}{\hbar\omega + \varepsilon_k - i0}, \tag{55.21a}$$

or on substituting the forms for $|u_k|^2$ and $|v_k|^2$

$$G(\mathbf{k},\omega) = \frac{\hbar\omega + \xi_k}{(\hbar\omega - \varepsilon_k + i0)(\hbar\omega + \varepsilon_k - i0)}, \tag{55.21b}$$

where the infinitesimal imaginary parts were incorporated into the frequencies to insure convergence of the integrals. Performing similar calculations on (55.20) yields

$$F^*(\mathbf{k},\omega) = \frac{-\Delta^*}{(\hbar\omega - \varepsilon_k + i0)(\hbar\omega + \varepsilon_k - i0)} \tag{55.22a}$$

and

$$F(\mathbf{k},\omega) = \frac{-\Delta}{(\hbar\omega - \varepsilon_k - i0)(\hbar\omega + \varepsilon_k + i0)}, \tag{55.22b}$$

where the latter may be calculated directly or obtained using the general rule $f^*(\omega) = [f(-\omega)]^*$, where f is some function. The infinitesimals in the denominators will also guarantee that the θ functions are properly reproduced on performing the inverse Fourier time transform (which is accomplished using Cauchy's formula).

Let us examine the behavior of a Fermi gas at $T = 0$ where $\Delta = 0$, $u_k = \theta(\xi_k)$, and $v_k = \theta(-\xi_k)$, corresponding to electron-like and hole-like excitations, respectively. The pole definition for the electron-like term in (55.21) corresponds to the presence of positive frequency excitations for $t > 0$; on the other hand the hole-like term in (55.21) corresponds to the presence of negative frequency excitations for $t < 0$. (An analogous situation arises in relativistic quantum mechanics.) Continuing our examination of the case $\Delta = 0$ we may rewrite (55.21) in a more compact form. Writing $\xi_k = +\varepsilon_k$ for $k > k_F$ and $\xi_k = -\varepsilon_k$ for $k < k_F$, Eq. (55.21) is equivalent to

$$G(\mathbf{k},\omega) = [\hbar\omega - \xi_k + i0\,\mathrm{sgn}(\omega)]^{-1}. \tag{55.23a}$$

The imaginary part in the denominator is important only near the pole where $\hbar\omega \cong \xi_k$. We may therefore replace $\mathrm{sgn}(\omega)$ by $\mathrm{sgn}(\xi_k)$; i.e.,

$$G(\mathbf{k},\omega) = [\hbar\omega - \xi_k + i0\,\mathrm{sgn}(\xi_k)]^{-1}. \tag{55.23b}$$

This form is useful when performing integrations over the variable ω.

There is a connection between G and F and the ordinary and anomalous density matrices ρ and ρ^A introduced earlier as Eqs. (52.52) and (52.67); in particular for $f_k = 0$ ($T = 0$) we have for our singlet case (generalizing slightly to allow for a time dependence)

$$\rho(\mathbf{r}_1,\mathbf{r}_2,t) = \sum_n v_n(\mathbf{r}_2,t)v_n^*(\mathbf{r}_1,t) \tag{55.24}$$

and

$$\rho^A(\mathbf{r}_1,\mathbf{r}_2,t) = \sum u_n(\mathbf{r}_2,t)v_n^*(\mathbf{r}_1,t). \tag{55.25}$$

Comparing (55.24) and (55.25) with (55.10) and (55.11) we have immediately

$$\rho(\mathbf{r}_1, \mathbf{r}_2, t) = \frac{\hbar}{\mathrm{i}} G(\mathbf{r}_1, t; \mathbf{r}_2, \mathbf{t}')_{t=t'-0_+} \tag{55.26}$$

and

$$\rho^{\mathrm{A}}(\mathbf{r}_1, \mathbf{r}_2, t) = -\frac{\hbar}{\mathrm{i}} F(\mathbf{r}_1, t; \mathbf{r}_2, \mathbf{t}')_{t=t'-0_+}. \tag{55.27}$$

The time-dependent normal and anomalous momentum distribution functions will be defined as

$$n(\mathbf{k}, t) = \int \mathrm{e}^{-\mathrm{i}\mathbf{k}\cdot(\mathbf{r}_1 - \mathbf{r}_2)} \rho(\mathbf{r}_1, \mathbf{r}_2, t) \mathrm{d}^3(\mathbf{r}_1 - \mathbf{r}_2) \tag{55.28}$$

and

$$n^{\mathrm{A}}(\mathbf{k}, t) = \int \mathrm{e}^{-\mathrm{i}\mathbf{k}\cdot(\mathbf{r}_1 - \mathbf{r}_2)} \rho^{\mathrm{A}}(\mathbf{r}_1, \mathbf{r}_2, t) \mathrm{d}^3(\mathbf{r}_1 - \mathbf{r}_2). \tag{55.29}$$

To complete the formalism we must define the gap function in terms of F. From Eq. (37.11) we have (generalizing to the time-dependent case)

$$\Delta(\mathbf{r}, t) = V \sum_n v_n^*(\mathbf{r}, t) u_n(\mathbf{r}, t)(1 - 2f_n)$$

and

$$\Delta(\mathbf{r}, t) = -\mathrm{i}\hbar V F(\mathbf{r}_1, t_1; \mathbf{r}_2, t_2)\Big|_{\substack{\mathbf{r}_1, \mathbf{r}_2 = \mathbf{r} \\ t_1 = t_2 + 0_+ = t}}. \tag{55.30}$$

In terms of $F(\mathbf{k}, \omega)$ we have for the homogeneous case

$$\Delta = -\frac{\mathrm{i}\hbar V}{(2\pi)^4} \int \mathrm{d}\omega \mathrm{d}^3 k F(\mathbf{k}, \omega). \tag{55.31}$$

As usual, a cut-off must be introduced in carrying out the k space integration.

The Green's function formalism is a very powerful one for solving problems involving an inhomogeneous superconductor. Its application to the problem of low field diamagnetism is discussed by Lifshitz and Pitaevskii (1980); numerous applications are contained in the book by Abrikosov *et al.* (1963).

55.2 Quasiclassical Green's functions in a normal system

When a normal metal or a superconductor is subjected to weak disturbances which vary sufficiently slowly in space and time, we may introduce several approximations into the Green's function equations of motion. The resulting equations are easier to work with and, in addition, resemble the familiar Boltzmann equation of transport theory.

We begin our discussion with the normal-metal case. Eq. (55.16) and its companion equation (involving $\mathbf{r}_2, t_2$) are

$$\left(i\hbar\frac{\partial}{\partial t_1} - \mathcal{H}'_1\right)G(\mathbf{r}_1, t_1; \mathbf{r}_2, t_2) = \delta^{(3)}(\mathbf{r}_1 - \mathbf{r}_2)\delta(t_1 - t_2) \tag{55.32}$$

and

$$-\left(i\hbar\frac{\partial}{\partial t_2} + \mathcal{H}'^{*}_2\right)G(\mathbf{r}_1, t_1; \mathbf{r}_2, t_2) = \delta^{(3)}(\mathbf{r}_1 - \mathbf{r}_2)\delta(t_1 - t_2), \tag{55.33}$$

where ∇^2 in $\mathcal{H}_1$ and $\mathcal{H}_2^*$ acts with respect to variables $\mathbf{r}_1$ and $\mathbf{r}_2$, respectively. We introduce sum and difference coordinates according to:[2]

$$T = \frac{1}{2}(t_1 + t_2); \quad \mathbf{R} = \frac{1}{2}(\mathbf{r}_1 + \mathbf{r}_2); \quad t = t_1 - t_2; \quad \mathbf{r} = \mathbf{r}_1 - \mathbf{r}_2. \tag{55.34}$$

In terms of these variables the differential operators entering Eqs. (55.32) and (55.33) are:

$$\frac{\partial}{\partial t_{1,2}} = \frac{1}{2}\frac{\partial}{\partial T} \pm \frac{\partial}{\partial t}; \tag{55.35a}$$

$$\nabla_{1,2} = \frac{1}{2}\nabla_{\mathbf{R}} \pm \nabla_{\mathbf{r}}; \tag{55.35b}$$

$$\nabla^2_{1,2} = \frac{1}{4}\nabla^2_{\mathbf{R}} \pm \nabla_{\mathbf{R}}\cdot\nabla_{\mathbf{r}} + \nabla^2_{\mathbf{r}}. \tag{55.35c}$$

Now the term involving $\nabla^2_{\mathbf{r}}$ acting on a free-electron-like wavefunction is approximately $-k^2$ and we will assume that for slowly varying disturbances this term dominates over $\frac{1}{4}(\partial^2/\partial\mathbf{R}^2)$ and we therefore neglect the latter. Introducing the quasiparticle velocity as $\mathbf{v} = \hbar\mathbf{k}/m$ we have

$$-\frac{\hbar^2}{2m}\frac{\partial}{\partial\mathbf{r}}\cdot\frac{\partial}{\partial\mathbf{R}} \to -\frac{i\hbar}{2}\mathbf{v}\cdot\frac{\partial}{\partial\mathbf{R}}.$$

We must also rewrite the potentials entering $\mathcal{H}'$ in terms of our sum and difference variables. As discussed in Sec. 36, the bare Hamiltonian contains both one-body,[3] $U(\mathbf{r}_{1,2})$, and two-body, $V(\mathbf{r}_2 - \mathbf{r}_1)$, terms. However, in our mean field treatment the two-body potentials enter through the self-consistency conditions. The sum and difference of Eqs. (55.32) and (55.33) may now be written (for a real $\mathcal{H}'$) as

2. Rewriting in terms of these variables, and carrying out the associated expansions in the variables, is commonly referred to as a Wigner transformation.
3. More generally the potentials entering (55.32) and (55.33) should be written as $U(\mathbf{r}_{1,2}, t_{1,2}) = U(\mathbf{R} \mp \mathbf{r}/2, T \mp t/2)$. If we assume that the bare and/or self-consistent one-body potentials are established rapidly relative to some characteristics time scale Ω^{-1} (where Ω is the characteristic frequency of some perturbation), then we may write $U = U(\mathbf{R} \mp \mathbf{r}/2, T)$.

$$\left[i\hbar\frac{\partial}{\partial t}+\frac{\hbar^2}{2m}\nabla_{\mathbf{r}}^2+\mu-\frac{1}{2}U\left(\mathbf{R}-\frac{\mathbf{r}}{2}\right)-\frac{1}{2}U\left(\mathbf{R}+\frac{\mathbf{r}}{2}\right)\right]G(\mathbf{R},T;\quad \mathbf{r},t)=\delta^{(3)}(\mathbf{r})\delta(t)(55.36)$$

and

$$\left[i\hbar\frac{\partial}{\partial T}+i\hbar\mathbf{v}\cdot\nabla_{\mathbf{R}}-U\left(\mathbf{R}-\frac{\mathbf{r}}{2}\right)+U\left(\mathbf{R}+\frac{\mathbf{r}}{2}\right)\right]G(\mathbf{R},T;\quad \mathbf{r},t)=0. \qquad (55.37)$$

We next perform a Fourier transform on Eqs. (55.36) and (55.37) in the variables $\mathbf{r}, t$. We write

$$\int d^3r dt e^{-i\mathbf{k}\cdot\mathbf{r}+i\omega t}G(\mathbf{R},T,\mathbf{r},t)\equiv G(\mathbf{R},T;\mathbf{k},\omega). \qquad (55.38)$$

Let us examine the two terms involving U in Eqs. (55.36) and (55.37):

$$\int d^3r e^{-i\mathbf{k}\cdot\mathbf{r}}\left[U\left(\mathbf{R}-\frac{\mathbf{r}}{2}\right)\pm U\left(\mathbf{R}+\frac{\mathbf{r}}{2}\right)\right]G(\mathbf{R},\mathbf{r}), \qquad (55.39)$$

where we temporarily suppress the time/frequency variables. We Taylor expand the one-body potential through linear terms about the point $\mathbf{R}$ and this form becomes

$$\int d^3r e^{-i\mathbf{k}\cdot\mathbf{r}}\left[U(\mathbf{R})+\frac{1}{2}\frac{\partial U}{\partial \mathbf{R}}\cdot\mathbf{r}\pm U(\mathbf{R})\mp\frac{1}{2}\frac{\partial U}{\partial \mathbf{R}}\cdot\mathbf{r}\right]G(\mathbf{R},\mathbf{r}); \qquad (55.40)$$

assuming $G(\mathbf{r})$ is short-range on the scale over which U varies, we may neglect the higher order terms. To leading order the form entering (55.36) is given by $U(\mathbf{R})G(\mathbf{R},\mathbf{k})$ which may then be written

$$[\hbar\omega-\xi_k-U(\mathbf{R},T)]G(\mathbf{R},T;\mathbf{k},\omega)=1$$

or (suppressing the imaginary parts defining the poles)

$$G(\mathbf{R},T;\mathbf{k},\omega)=\frac{1}{\hbar\omega-\xi_k-U(\mathbf{R},T)}. \qquad (55.41)$$

In this approximation the Green's function resembles that of a free particle; however, the 'zero of energy', as measured by $U(\mathbf{R},T)$, varies slowly in space and time. The term involving the derivative of the potential in (55.40) may be rewritten as

$$\int d^3r e^{-i\mathbf{k}\cdot\mathbf{r}}\frac{\partial U}{\partial \mathbf{R}}\cdot\mathbf{r}G(\mathbf{R},\mathbf{r})=i\frac{\partial U}{\partial \mathbf{R}}\cdot\int d^3r G(\mathbf{R},\mathbf{r})\frac{\partial}{\partial \mathbf{k}}e^{-i\mathbf{k}\cdot\mathbf{r}}$$

$$=i\frac{\partial U}{\partial \mathbf{R}}\cdot\frac{\partial}{\partial \mathbf{k}}G(\mathbf{R},\mathbf{k}). \qquad (55.42)$$

Eq. (55.37) can now be written

$$\frac{\partial G}{\partial T}+\frac{\partial G}{\partial \mathbf{R}}\cdot\dot{\mathbf{R}}+\frac{\partial G}{\partial \mathbf{k}}\cdot\dot{\mathbf{k}}=0, \qquad (55.43)$$

where we wrote $\dot{\mathbf{R}}=\mathbf{v}$, and $\hbar\dot{\mathbf{k}}=-\nabla_{\mathbf{R}}U$.

Since the variable ω enters only as an argument of G, it is customary to 'integrate it out'. To define this operation we recall the definition of the distribution function $n(\mathbf{k}, t)$ introduced in Eq. (55.28). From Eq. (55.26) it is clear that we may write[4,5]

$$n(\mathbf{R}, T, \mathbf{k}) = \frac{-\mathrm{i}\hbar}{2\pi} \lim_{t=0_-} \int_{-\infty}^{+\infty} \mathrm{e}^{-\mathrm{i}\omega t} G(\mathbf{R}, T; \mathbf{k}, \omega) \mathrm{d}\omega, \tag{55.44}$$

where T replaces t in those expressions and we have generalized the expression to include the slow variation of n with $\mathbf{R}$. In place of Eq. (55.37) we have

$$\frac{\partial n(\mathbf{R}, T, \mathbf{k})}{\partial T} + \frac{\partial n(\mathbf{R}, T, \mathbf{k})}{\partial \mathbf{R}} \cdot \dot{\mathbf{R}} + \frac{\partial n(\mathbf{R}, T, \mathbf{k})}{\partial \mathbf{k}} \cdot \dot{\mathbf{k}} = 0; \tag{55.45}$$

Eq. (55.45) has precisely the same structure as the Landau–Boltzmann kinetic equation of semiclassical transport theory discussed in Subsec. 24.2.1. By solving (55.45) for $n(\mathbf{R}, T, \mathbf{k})$, we may compute averages of semiclassical quantities, provided we normalize n according to the usual phase space filling condition $[1/(2\pi)^3] \int n(\mathbf{k}) \mathrm{d}^3 k = N/V$, where N/V is the particle number density.

At this point we replace $(\mathbf{R}, T)$ by $(\mathbf{r}, t)$, which is the more conventional notation. As we did earlier in our discussion of zero sound, Eq. (55.45) is linearized by writing $n = n^{(0)}(\xi) - \delta n(\mathbf{r}, \mathbf{k}, t)$. Writing

$$\frac{\partial n^{(0)}}{\partial \mathbf{k}} = \frac{\partial n^{(0)}}{\partial \xi} \frac{\partial \xi}{\partial \mathbf{k}} = \frac{\partial \theta(-\xi)}{\partial \xi} \hbar \mathbf{v} = -\delta(\xi) \hbar \mathbf{v}$$

and

$$\hbar \dot{\mathbf{k}} = -\frac{\partial \delta \xi}{\partial \mathbf{r}},$$

where from Fermi liquid theory $\delta\xi = [2/(2\pi)^3] \int f(\mathbf{k}, \mathbf{k}') \delta n(\mathbf{r}, \mathbf{k}', t) \mathrm{d}^3 k'$, Eq. (55.45) becomes

$$\frac{\partial \delta n(\mathbf{r}, \mathbf{k}, t)}{\partial t} + \frac{\partial \delta n(\mathbf{r}, \mathbf{k}, t)}{\partial \mathbf{r}} \cdot \mathbf{v} + \delta(\xi) \mathbf{v} \cdot \frac{\partial \delta \xi}{\partial \mathbf{r}} = 0. \tag{55.46}$$

Now for a collective (coherent) disturbance, δn corresponds to an angle-dependent shift of the Fermi radius; i.e.,

$$\delta n = \theta[k_{\mathrm{F}}(\theta, \phi, \mathbf{r}, t) - k] - \theta[k_{\mathrm{F}}^{(0)} - k], \tag{55.47}$$

where $k_{\mathrm{F}}(\theta, \phi, \mathbf{r}, t)$ is the dynamic (angle-, position-, and time-dependent) Fermi radius and $k_{\mathrm{F}}^{(0)}$ is its equilibrium value. To sufficient accuracy we may write (55.47) as

$$\delta n(\hat{\mathbf{k}}) = -\delta(\xi) \frac{\partial \xi}{\partial \mathbf{k}} \cdot \delta \mathbf{k} = -\delta(\xi) \hbar \mathbf{v} \cdot \delta \mathbf{k}, \tag{55.48}$$

4. Note that formally we must have t finite when doing integration in (55.44), since the integral is divergent for $t = 0$.
5. For the case of the equilibrium Fermi gas at $T = 0$ (55.44) and (55.23a) give

$$n(\mathbf{k}) = -\frac{\mathrm{i}\hbar}{2\pi} \lim_{t \to 0_-} \int_{-\infty}^{+\infty} \frac{\mathrm{e}^{-\mathrm{i}\omega t}}{\hbar\omega - \xi_k + \mathrm{i}0 \operatorname{sgn} \xi} \mathrm{d}\omega;$$

carrying out the integration using Cauchy's formula (with the contour closed in the upper half plane) we obtain $n = \theta(-\xi)$, the T = 0 Fermi function.

where $\delta\mathbf{k} \equiv [k_F(\theta, \phi, \mathbf{r}, t) - k_F^{(0)}]\hat{\mathbf{k}}(\theta, \phi)$ and $\hat{\mathbf{k}}(\theta, \phi)$ is a unit vector in the direction specified by the polar angles θ, ϕ. We introduce the function $\nu(\hat{\mathbf{k}}_F(\theta, \phi))$ defined by

$$\nu(\hat{\mathbf{k}}) = \int \delta n(\hat{\mathbf{k}}) \mathrm{d}\xi = \hbar \mathbf{v}_F \cdot \delta\mathbf{k}(\hat{\mathbf{k}}). \tag{55.49}$$

Our final form for the *energy integrated* quasiclassical equation of motion is then

$$\frac{\partial}{\partial t}\nu(\hat{\mathbf{k}}, \mathbf{r}, t) + \mathbf{v}_F(\hat{\mathbf{k}}) \cdot \frac{\partial}{\partial \mathbf{r}}\nu(\hat{\mathbf{k}}, \mathbf{r}, t) + \mathbf{v}_F(\hat{\mathbf{k}}) \cdot \frac{\partial}{\partial \mathbf{r}}\left[\frac{1}{4\pi}\int F(\hat{\mathbf{k}}, \hat{\mathbf{k}}') \cdot \nu(\hat{\mathbf{k}}', \mathbf{r}, t) \mathrm{d}\Omega'\right]$$
$$= -\frac{\nu(\hat{\mathbf{k}}, \mathbf{r}, t)}{\tau}, \tag{55.50}$$

where $F(\hat{\mathbf{k}}, \hat{\mathbf{k}}') \equiv \mathcal{N}(0) f(\hat{\mathbf{k}}, \hat{\mathbf{k}}')$ and we included a phenomenological collision integral on the right-hand side of our Boltzmann equation of the form

$$I(n) = -\frac{\delta n}{\tau}. \tag{55.51}$$

Note that for a normal system we were able to integrate over both ω and ξ. In the next subsection, where we treat superfluid systems, we will find that we may integrate over ξ or ω but not both due to a nontrivial behavior of the energy spectrum in the vicinity of the Fermi surface.

55.3 Quasiclassical Green's functions in a superfluid system

55.3.1 Keldysh formalism

Rather than the usual Green's functions (55.2) and (55.5) (which have awkward analytic properties at finite temperatures) there is another commonly used formalism where we introduce three matrix functions having the form (Keldysh 1965, Shelankov 1985)

$$\underset{\sim}{G}^{\mathrm{R}}(x_1, x_2) = -\frac{\mathrm{i}}{\hbar}\theta(t_1 - t_2)\begin{pmatrix} \langle \hat{\psi}_\uparrow(x_1), \hat{\psi}_\uparrow^\dagger(x_2)\rangle_+ & -\langle \hat{\psi}_\uparrow(x_1), \hat{\psi}_\downarrow(x_2)\rangle_+ \\ \langle \hat{\psi}_\downarrow^\dagger(x_1), \hat{\psi}_\uparrow^\dagger(x_2)\rangle_+ & -\langle \hat{\psi}_\downarrow^\dagger(x_1), \hat{\psi}_\downarrow(x_2)\rangle_+ \end{pmatrix}, \tag{55.52a}$$

$$\underset{\sim}{G}^{\mathrm{A}}(x_1, x_2) = \frac{\mathrm{i}}{\hbar}\theta(t_2 - t_1)\begin{pmatrix} \langle \hat{\psi}_\uparrow(x_1), \hat{\psi}_\uparrow^\dagger(x_2)\rangle_+ & -\langle \hat{\psi}_\uparrow(x_1), \hat{\psi}_\downarrow(x_2)\rangle_+ \\ \langle \hat{\psi}_\downarrow^\dagger(x_1), \hat{\psi}_\uparrow^\dagger(x_2)\rangle_+ & -\langle \hat{\psi}_\downarrow^\dagger(x_1), \hat{\psi}_\downarrow(x_2)\rangle_+ \end{pmatrix}, \tag{55.52b}$$

and

$$\underset{\sim}{G}^{\mathrm{K}}(x_1, x_2) = -\frac{\mathrm{i}}{\hbar}\begin{pmatrix} \langle \hat{\psi}_\uparrow(x_1), \hat{\psi}_\uparrow^\dagger(x_2)\rangle_- & -\langle \hat{\psi}_\uparrow(x_1), \hat{\psi}_\downarrow(x_2)\rangle_- \\ \langle \hat{\psi}_\downarrow^\dagger(x_1), \hat{\psi}_\uparrow^\dagger(x_2)\rangle_- & -\langle \hat{\psi}_\downarrow^\dagger(x_1), \hat{\psi}_\downarrow(x_2)\rangle_- \end{pmatrix}; \tag{55.52c}$$

here we have used the shorthand notation $x_1 = (\mathbf{r}_1, t_1), x_2 = (\mathbf{r}_2, t_2)$ (as is common in many-body theory), and $\langle\ \rangle_\pm$ denotes a quantum statistical average of the anticommutator (+) or commutator (−) of the corresponding $\hat{\psi}$ operators. These matrix functions are referred to as the retarded (R), advanced (A) and Keldysh (K) Green's functions although the third is, strictly speaking, not a

Green's function since it does not contain the discontinuity at $t_1 = t_2$. The anticommutators (+) in the above expressions, when expressed in terms of Bogoliubov wavefunctions via (49.2) and (55.1), are given by

$$\langle \hat{\psi}_\uparrow(x_1), \hat{\psi}_\uparrow^\dagger(x_2)\rangle_+ = \sum_n [u_n(x_1)u_n^*(x_2) + v_n^*(x_1)v_n(x_2)], \tag{55.53a}$$

$$-\langle \hat{\psi}_\uparrow(x_1), \hat{\psi}_\downarrow(x_2)\rangle_+ = -\sum_n [u_n(x_1)v_n^*(x_2) - v_n^*(x_1)u_n(x_2)], \tag{55.53b}$$

$$\langle \hat{\psi}_\downarrow^\dagger(x_1), \hat{\psi}_\uparrow^\dagger(x_2)\rangle_+ = \sum_n [v_n(x_1)u_n^*(x_2) - u_n^*(x_1)v_n(x_2)], \tag{55.53c}$$

and

$$-\langle \hat{\psi}_\downarrow^\dagger(x_1), \hat{\psi}_\downarrow(x_2)\rangle_+ = -\sum_n [v_n(x_1)v_n^*(x_2) + u_n^*(x_1)u_n(x_2)], \tag{55.53d}$$

while the commutators (−) are given by

$$\langle \hat{\psi}_\uparrow(x_1), \hat{\psi}_\uparrow^\dagger(x_2)\rangle_- = \sum_n [u_n(x_1)u_n^*(x_2)(1 - 2f_{n\uparrow}) - v_n^*(x_1)v_n(x_2)(1 - 2f_{n\downarrow})], \tag{55.54a}$$

$$-\langle \hat{\psi}_\uparrow(x_1), \hat{\psi}_\downarrow(x_2)\rangle_- = -\sum_n [u_n(x_1)v_n^*(x_2)(1 - 2f_{n\uparrow}) + v_n^*(x_1)u_n(x_2)(1 - 2f_{n\downarrow})], \tag{55.54b}$$

$$\langle \hat{\psi}_\downarrow^\dagger(x_1), \hat{\psi}_\uparrow^\dagger(x_2)\rangle_- = \sum_n [v_n(x_1)u_n^*(x_2)(1 - 2f_{n\uparrow}) + u_n^*(x_1)v_n(x_2)(1 - 2f_{n\downarrow})], \tag{55.54c}$$

and

$$-\langle \hat{\psi}_\downarrow^\dagger(x_1), \hat{\psi}_\downarrow(x_2)\rangle_- = -\sum_n [v_n(x_1)v_n^*(x_2)(1 - 2f_{n\uparrow}) - u_n^*(x_1)u_n(x_2)(1 - 2f_{n\downarrow})]. \tag{55.54d}$$

Note that neither the retarded nor the advanced Green's function contains any statistical factors; all statistical information is contained in the Keldysh functions.

Collecting the Bogoliubov equations and their Hermitian conjugates

$$\left(i\hbar\frac{\partial}{\partial t} - \mathscr{H}'\right)u_n = \Delta v_n; \quad \left(i\hbar\frac{\partial}{\partial t} + \mathscr{H}'^*\right)v_n = \Delta^* u_n;$$

$$\left(-i\hbar\frac{\partial}{\partial t} - \mathscr{H}'^*\right)u_n^* = \Delta^* v_n^*; \quad \left(-i\hbar\frac{\partial}{\partial t} + \mathscr{H}'\right)v_n^* = \Delta u_n^*,$$

introducing the operator (which we may regard as the inverse of our Green's function matrix)

$$\underset{\sim}{Q}_1 \equiv \begin{pmatrix} i\hbar\dfrac{\partial}{\partial t_1} - \mathscr{H}'_1 & -\Delta_1 \\ \Delta_1^* & -i\hbar\dfrac{\partial}{\partial t_1} - \mathscr{H}'^*_1 \end{pmatrix}, \tag{55.55}$$

and using the completeness relations (36.23a) and (36.23b), it is straightforward to show that

$$Q_1 \cdot G^{R,A}(x_1, x_2) = \delta^{(4)}(x_1 - x_2)1 \tag{55.56a,b}$$

and

$$Q_1 \cdot G^{K}(x_1, x_2) = 0, \tag{55.56c}$$

where $\delta^{(4)}(x_1 - x_2) = \delta(t_1 - t_2)\delta^{(3)}(\mathbf{r}_1 - \mathbf{r}_2)$ and 1 is a 2×2 unit matrix.[6] We see that $G^{R,A}(x_1, x_2)$ are Green's functions of the matrix operator Q_1. Performing similar calculations with the operator

$$Q_2^\dagger \equiv \begin{pmatrix} -i\hbar\dfrac{\partial}{\partial t_2} - \mathscr{H}_2'^* & -\Delta_2 \\ \Delta_2^* & i\hbar\dfrac{\partial}{\partial t_2} - \mathscr{H}'_2 \end{pmatrix}, \tag{55.57}$$

acting to the left, we find

$$G^{R,A}(x_1, x_2) \cdot Q_2^\dagger = \delta^{(4)}(x_1 - x_2)1 \tag{55.58a,b}$$

and

$$G^{K}(x_1, x_2) \cdot Q_2^\dagger = 0. \tag{55.58c}$$

Since the physically measured quantities are usually expressed in terms of G, we wish to relate this quantity to $G^{R,A,K}$. Comparing the definitions (55.2) and (55.6) with (55.52a)–(55.52c) we find that

$$G(x_1, x_2) = \frac{1}{2}[G^R(x_1, x_2) + G^A(x_1, x_2) + G^K(x_1, x_2)] \tag{55.59}$$

where

$$G(x_1, x_2) = \begin{pmatrix} G_{\uparrow\uparrow}(x_1, x_2) & -F_{\uparrow\downarrow}(x_1, x_2) \\ F^*_{\downarrow\uparrow}(x_1, x_2) & G_{\downarrow\downarrow}(x_2, x_1) \end{pmatrix} \tag{55.60a}$$

$$\equiv \begin{pmatrix} G_{11} & G_{12} \\ G_{21} & G_{22} \end{pmatrix} \tag{55.60b}$$

and $F^*_{\downarrow\uparrow}(x_1, x_2) = -F^*_{\uparrow\downarrow}(x_1, x_2)$ by (55.6c).

55.3.2 Quasiclassical equations of motion

We are now in a position to construct quasiclassical equations of motion. Subtracting (55.58a)–(55.58c) from (55.56a)–(55.56c) we obtain

$$\left[i\hbar\left(\frac{\partial}{\partial t_1} + \frac{\partial}{\partial t_2}\right) - \mathscr{H}'_1 + \mathscr{H}_2'^*\right]G_{11} - \Delta_2^* G_{12} - \Delta_1 G_{21} = 0, \tag{55.61a}$$

$$\left[i\hbar\left(\frac{\partial}{\partial t_1} - \frac{\partial}{\partial t_2}\right) - \mathscr{H}'_1 + \mathscr{H}'_2\right]G_{12} + \Delta_1 G_{22} - \Delta_2 G_{11} = 0, \tag{55.61b}$$

6. Eqs. (55.56a)–(55.56c) may be written in a more compact form as

$$\begin{pmatrix} Q_1 & 0 \\ 0 & Q_1 \end{pmatrix}\begin{pmatrix} G^R(x_1, x_2) & G^K(x_1, x_2) \\ 0 & G^A(x_1, x_2) \end{pmatrix} = \begin{pmatrix} 1 & 0 \\ 0 & 1 \end{pmatrix}\delta^{(4)}(x_1, x_2).$$

$$\left[-i\hbar\left(\frac{\partial}{\partial t_1}-\frac{\partial}{\partial t_2}\right)-\mathscr{H}_1'^{*}+\mathscr{H}_2'^{*}\right]G_{21}+\Delta_1^{*}G_{11}-\Delta_2^{*}G_{22}=0, \tag{55.61c}$$

and

$$\left[-i\hbar\left(\frac{\partial}{\partial t_1}+\frac{\partial}{\partial t_2}\right)-\mathscr{H}_1'^{*}+\mathscr{H}_2'\right]G_{22}+\Delta_2G_{21}+\Delta_1^{*}G_{12}=0. \tag{55.61d}$$

The next step is to rewrite Eqs. (55.61a)–(55.61d) in terms of the sum and difference variables introduced in Eq. (55.34) which immediately results in the substitutions $(\partial/\partial t_1)+(\partial/\partial t_2)=\partial/\partial T$ and $(\partial/\partial t_1)-(\partial/\partial t_2)=\partial/2\partial t$. Under the assumption that G_{ij} is either short-ranged or rapidly oscillating with $\mathbf{r}$, we only need to retain the leading term in the expansion of $\mathscr{H}'$ and Δ with respect to $\mathbf{r}$. Including the effects of a vector potential $\mathbf{A}$, we then have $\mathscr{H}'_1-\mathscr{H}'_2=-(\hbar^2/m)\nabla_{\mathbf{R}}\cdot\nabla_{\mathbf{r}}+(2ie\hbar/mc)\mathbf{A}(\mathbf{R})\cdot\nabla_{\mathbf{r}}$ and $\mathscr{H}'_1-\mathscr{H}_2'^{*}=-(\hbar^2/m)\nabla_{\mathbf{R}}\cdot\nabla_{\mathbf{r}}$ and we write $\Delta(R)$ for either Δ_1 or Δ_2. At this point we Fourier transform $G(\mathbf{R},T,\mathbf{r},t)$ to $G(\mathbf{R},T,\mathbf{k},\omega)$ and replace $\nabla_{\mathbf{r}}$ by $i\mathbf{k}$ (with $\hbar\mathbf{k}=m\mathbf{v}$) and $\partial/\partial t=-i\omega$ in Eqs. (55.61a)–(55.61d). For an unpolarized system (with spatial inversion symmetry) we may write $G_{11}=-G_{22}\equiv G$. Using $G_{12}=-F$ and $G_{21}=F^*$ Eqs. (55.61a)–(55.61d) now take the form[7]

$$\left[i\hbar\frac{\partial}{\partial T}+i\hbar\mathbf{v}\cdot\nabla_{\mathbf{R}}\right]G(\mathbf{R},T,\mathbf{k},\omega)+\Delta^{*}(\mathbf{R},T)F(\mathbf{R},T,\mathbf{k},\omega)-\Delta(\mathbf{R},T)F^{*}(R,T,\mathbf{k},\omega)=0, \tag{55.62a}$$

$$\left[2\hbar\omega+i\hbar\mathbf{v}\cdot\nabla_{\mathbf{R}}+\frac{2e}{c}\mathbf{v}\cdot\mathbf{A}(\mathbf{R},\mathrm{T})\right]F(\mathbf{R},T,\mathbf{k},\omega)+2\Delta(\mathbf{R},T)G(\mathbf{R},T,\mathbf{k},\omega)=0, \tag{55.62b}$$

and

$$\left[-2\hbar\omega+i\hbar\mathbf{v}\cdot\nabla_{\mathbf{R}}-\frac{2e}{c}\mathbf{v}\cdot\mathbf{A}(\mathbf{R},\mathrm{T})\right]F^{*}(\mathbf{R},T,\mathbf{k},\omega)+2\Delta^{*}(\mathbf{R},T)G(\mathbf{R},T,\mathbf{k},\omega)=0. \tag{55.62c}$$

Eqs. (55.62a)–(55.62c) must be supplemented by two 'self-consistency' conditions which fix the two potentials, $\mathbf{A}(\mathbf{R},T)$ and $\Delta(\mathbf{R},T)$; $\Delta(\mathbf{R},T)$ is determined by Eq. (55.31). The total magnetic field, $\mathbf{B}(\mathbf{R})$, is given by

$$\mathbf{B}(\mathbf{R})=\nabla_{\mathbf{R}}\times\mathbf{A}(\mathbf{R}).$$

Any shift of the total magnetic field from the external field, $\mathbf{B}_e(\mathbf{R})$, must arise from circulating supercurrents, $\mathbf{j}(\mathbf{R})$, via Ampère's law,

$$\nabla_{\mathbf{R}}\times[\mathbf{B}(\mathbf{R})-\mathbf{B}_e(\mathbf{R})]=\frac{4\pi}{c}\mathbf{j}(\mathbf{R}). \tag{55.63}$$

7. For the real time Green's functions, F and F^* are trivially related. We include Eq. (55.62) in anticipation of the quasiclassical formalism for temperature Green's functions, where $\mathscr{F}$ and $\bar{\mathscr{F}}$ may be independent.

The expression for the current operator, $\hat{\mathbf{j}}(\mathbf{r})$, is

$$\hat{\mathbf{j}}(\mathbf{r}) = \frac{e\hbar}{2im}[\hat{\psi}_\alpha^\dagger(\mathbf{r})\nabla\hat{\psi}_\alpha(\mathbf{r}) - (\nabla\hat{\psi}_\alpha^\dagger(\mathbf{r}))\hat{\psi}_\alpha(\mathbf{r})]$$
$$- \frac{e^2}{mc}\mathbf{A}(\mathbf{r})\psi_\alpha^\dagger(\mathbf{r})\psi_\alpha(\mathbf{r}). \tag{37.16}$$

The expectation value of the current can be calculated from the Green's function (55.2) as follows

$$\langle\hat{\mathbf{j}}(\mathbf{r}_1, t_1)\rangle = -\frac{e\hbar^2}{m}(\nabla_1 - \nabla_2)G(\mathbf{r}_1, t_1; \mathbf{r}_2, t_2)\Bigg|_{\substack{r_1 = r_2 \\ t_1 = t_2 - 0}} - \frac{ne^2}{m}\mathbf{A}(\mathbf{r}_1, t_1), \tag{55.64}$$

where 0 is the positive infinitesimal, n is the number density of electrons and we have included a factor 2 to account for the two-spin components. Rewriting in terms of the sum and difference variables (55.34), expressing $G(\mathbf{R}, T, \mathbf{r} = 0, t = -0)$ in terms of $G(\mathbf{R}, T, \mathbf{k}, \omega)$, and using (55.35b) we obtain the expectation value of $\hat{\mathbf{j}}$ as

$$\mathbf{j}(\mathbf{R}, \mathbf{T}) = \lim_{t \to -0} -ie\hbar\int \mathbf{v}G(\mathbf{R}, T, \mathbf{k}, \omega)\mathrm{e}^{-\mathrm{i}\omega t}\frac{\mathrm{d}\omega}{2\pi}\frac{\mathrm{d}^3k}{2\pi}, \tag{55.65}$$

where the limit taken insures the proper ordering of the $\hat{\psi}$ operators in G.

As was the case with the normal-state quasiclassical Green's function equation of motion, the variable t (or its transform ω) appears only as an argument of G suggesting again that we may suppress it, the result being a density matrix (as in Eqs. (55.26) and (55.27)). We can define the ordinary and anomalous density matrices as

$$\rho_{11}(\mathbf{R}, T, \mathbf{r}) = \frac{\hbar}{\mathrm{i}}G_{11}(\mathbf{R}, T, \mathbf{r}, t)_{t=0_-}, \tag{55.66a}$$

$$\rho_{22}(\mathbf{R}, T, \mathbf{r}) = \frac{\hbar}{\mathrm{i}}G_{22}(\mathbf{R}, T, -\mathbf{r}, -t)_{t=0_-}, \tag{55.66b}$$

$$\rho_{12}(\mathbf{R}, T, \mathbf{r}) = -\frac{\hbar}{\mathrm{i}}G_{12}(\mathbf{R}, T, \mathbf{r}, t)_{t=0_-} = -\rho_{\mathrm{A}}(\mathbf{R}, T, \mathbf{r}), \tag{55.66c}$$

and

$$\rho_{21}(\mathbf{R}, T, \mathbf{r}) = \frac{\hbar}{\mathrm{i}}G_{21}(\mathbf{R}, T, \mathbf{r}, t)_{t=0_-} = \rho_{\mathrm{A}}^*(\mathbf{R}, T, \mathbf{r}). \tag{55.66d}$$

The distribution function is the Fourier transform in $\mathbf{r}$ of the density matrix. In terms of the $(\mathbf{r}, t)$ Fourier transform of $G(\mathbf{R}, T, \mathbf{r}, t)$ we have the distribution function matrix elements (see Eq. (55.44)) as

$$n_{ij}(\mathbf{R}, T, \mathbf{k}) = \frac{-\mathrm{i}\hbar}{2\pi}\lim_{t=0_-}\int \mathrm{e}^{-\mathrm{i}\omega t}G_{ij}(\mathbf{R}, T, \mathbf{k}, \omega)\mathrm{d}\omega. \tag{55.67}$$

The resulting equations, which we will not write down explicitly, are a generalization of the Boltzmann equation to the case where there is a BCS condensate present. Note, as mentioned earlier, in a superconductor we cannot follow the integration over ω by an integration over the energy, ξ, as we did in obtaining (55.50), since the distribution functions have a nontrivial energy dependence in the vicinity of the Fermi surface; i.e., they are not simple θ functions, as in the normal state.

It is straightforward to generalize the above theory to the case of unconventionally paired superfluids. For a detailed discussion the reader is referred to the book of Vollhardt and Wölfle (1990).

55.3.3 Eilenberger equations

There is another kind of quasiclassical quantity which can be formed from $\underset{\sim}{G}(\mathbf{R}, T, \mathbf{k}, \omega)$. This is the so-called 'energy integrated' Green's function $\underset{\sim}{g}(\mathbf{R}, T, \hat{\mathbf{k}}, \omega)$, where the ω variable remains but the magnitude of $\mathbf{k}$ (actually ξ_k) is integrated over (this is the analogue of our earlier integration over ω to form the distribution functions n_{ij}); the unit vector $\hat{\mathbf{k}}$ remains since, for a nonuniform system, $\xi_{\mathbf{k}}$ depends on the direction in $\mathbf{k}$ space. From Eqs. (55.23a) and (55.23b) it is clear that ω and ξ enter the Fermi gas Green's function in a very similar way and hence it is reasonable to consider integrating out either variable. We define the energy integrated Green's function through the prescription

$$\underset{\sim}{g}(\mathbf{R}, T, \hat{\mathbf{k}}, \omega) = \frac{\mathrm{i}}{\pi} \int \mathrm{d}\xi_k \underset{\sim}{G}(\mathbf{R}, T, \mathbf{k}, \omega) \tag{55.68}$$

which is analogous to the 'ω-integrated' distribution functions; however, a convergence factor analogous to the $\mathrm{e}^{-\mathrm{i}\omega t}$ factor in (55.65) is absent. A second problem is that the 'normalization', lost when subtracting the original equations of motion to form Eqs. (55.61a)–(55.61d), must be separately imposed. To demonstrate this point we note that $G(\mathbf{r}_1, t_1, \mathbf{r}_2, t_2) = 0$ is a solution of (55.61a)–(55.61d), whereas it is not a solution to Eqs. (55.56) or Eqs. (55.58) due to the presence of the δ function.

The following discussion is not intended to be rigorous (for more rigorous discussions see Eilenberger (1968) and Shelankov (1985)). We make the following assumptions: (1) The upper and lower limits are arranged so that the large ξ divergence of (55.68) is avoided. (2) We will evaluate the integrals for the equilibrium, $T = 0$, isotropic Fermi gas Green's functions (55.21b) and (55.22a,b) and observe that the resulting expressions obey a 'normalization condition'. (3) We will then assume that this same normalization condition applies to the nonequilibrium, nonisotropic $T \neq 0$ case. The resulting equations of motion are (where we replace the variables $(\mathbf{R}, T)$ by $(\mathbf{r}, t)$)

$$\left[\mathrm{i}\hbar\frac{\partial}{\partial t} + \mathrm{i}\hbar\mathbf{v}_{\mathrm{F}} \cdot \nabla\right] g(\mathbf{r}, t, \hat{\mathbf{k}}, \omega) + \Delta^*(\mathbf{r}, t) f(\mathbf{r}, t, \hat{\mathbf{k}}, \omega) - \Delta(\mathbf{r}, t) f^*(\mathbf{r}, t, \hat{\mathbf{k}}, \omega) = 0, \tag{55.69a}$$

$$\left[2\hbar\omega + \mathrm{i}\hbar\mathbf{v}_{\mathrm{F}} \cdot \nabla + \frac{2e}{c}\mathbf{v}_{\mathrm{F}} \cdot \mathbf{A}(\mathbf{r}, t)\right] f(\mathbf{r}, t, \hat{\mathbf{k}}, \omega) + 2\Delta(\mathbf{r}, t) g(\mathbf{r}, t, \hat{\mathbf{k}}, \omega) = 0, \tag{55.69b}$$

and

$$\left[-2\hbar\omega + \mathrm{i}\hbar\mathbf{v}_{\mathrm{F}} \cdot \nabla - \frac{2e}{c}\mathbf{v}_{\mathrm{F}} \cdot \mathbf{A}(\mathbf{r}, t)\right] f^*(\mathbf{r}, t, \hat{\mathbf{k}}, \omega) + 2\Delta^*(\mathbf{r}, t) g(\mathbf{r}, t, \hat{\mathbf{k}}, \omega) = 0. \tag{55.69c}$$

These equations were first obtained by Eilenberger (1968) and, together with the self-consistency conditions on $\mathbf{A}(\mathbf{r}, t)$, Eqs. (55.63) and (55.67), and the gap equation (55.31), provide a complete quasiclassical description of a superconductor.[8]

To carry out the above program we require the integrals (where we use (55.21b), (55.22a), and (55.22b))

$$g(\omega) = \frac{i}{\pi}\int G(\mathbf{k}, \omega)d\xi_k = \frac{i}{\pi}\int \frac{(\hbar\omega + \xi_k)d\xi_k}{(\hbar\omega - \varepsilon_k + i0)(\hbar\omega + \varepsilon_k - i0)} \tag{55.70}$$

and

$$f^*(\omega) = \frac{i}{\pi}\int F^*(\mathbf{k}, \omega)d\xi_k = -\frac{i}{\pi}\int \frac{\Delta^* d\xi_k}{(\hbar\omega - \varepsilon_k + i0)(\hbar\omega + \varepsilon_k - i0)}. \tag{55.71}$$

The term involving ξ_k in the numerator of (55.70) is odd and vanishes on symmetrical integration. Writing $\varepsilon^2 = \xi^2 + \Delta^2$, (55.70) and (55.71) may be written in the form

$$g(\omega) = -\frac{i}{\pi}\int \frac{\hbar\omega d\xi}{(\xi - \xi_0)(\xi + \xi_0) - i0} \tag{55.72a}$$

and

$$f^*(\omega) = \frac{i}{\pi}\int \frac{\Delta^* d\xi}{(\xi - \xi_0)(\xi + \xi_0) - i0}, \tag{55.72b}$$

where we dropped the subscript on ξ_k and we defined $\xi_0 = (\hbar^2\omega^2 - \Delta^2)^{1/2}$. Eqs. (55.72) have identical poles located at $\xi_1 = \xi_0 + i0$ and $\xi_2 = -\xi_0 - i0$ (lying in the upper and lower half planes, respectively); the corresponding residues (R_1, R_2) for $g(\omega)$ are $R_1^g = (1/2\pi i)(\hbar\omega/\xi_0)$ and $R_2^g = -(1/2\pi i)(\hbar\omega/\xi_0)$ while those for $f^*(\omega)$ are $R_1^f = -(1/2\pi i)(\Delta^*/\xi_0)$ and $R_2^f = (1/2\pi i)(\Delta^*/\xi_0)$. When ξ_0 is real ($\hbar\omega > |\Delta|$) the poles lie infinitesimally close to the real axis. In integrating along the real axis we avoid the pole in the lower half plane with an infinitesimal clockwise semicircle in the upper half plane; likewise we avoid the pole in the upper half plane with an infinitesimal counterclockwise semicircle in the lower half plane. We take the sum of $\pm\pi i \times$ (residue) with a $(+)$ and $(-)$ sign for counterclockwise and clockwise circuits, respectively. Alternatively we displace the contour slightly into the upper half plane in which case we avoid the pole in the lower half plane entirely. The pole in the upper half plane is now completely enclosed and connected to the main contour (running parallel to the real axis) by two short self-canceling segments. Assuming symmetrical integration, all but the pole contribution vanishes. It then follows that our integrals are given by

$$g(\omega) = \pi i R_1^g - \pi i R_2^g = \frac{\hbar\omega}{(\hbar^2\omega^2 - \Delta^2)^{1/2}} \tag{55.73}$$

and

$$f^*(\omega) = \pi i R_1^f - \pi i R_2^f = \frac{-\Delta^*}{(\hbar^2\omega^2 - \Delta^2)^{1/2}}. \tag{55.74}$$

8. Eqs. (55.69a)–(55.69c) are given in the real time representation. The Matsubara time representation, utilized by Eilenberger, will be treated in Subsec. 55.7 along with the effects of impurity scattering.

Using the property $f(\omega) = -[f^*(-\omega)]^*$ (which follows from its definition and the conjugation rule for Fourier transforms) we have immediately

$$f(\omega) = \frac{\Delta}{(\hbar^2\omega^2 - \Delta^2)^{1/2}}. \tag{55.75}$$

Assembling the above we have

$$\underset{\sim}{g}(\omega) = \begin{pmatrix} g(\omega) & f(\omega) \\ f^*(\omega) & -g(\omega) \end{pmatrix} \tag{55.76a}$$

or

$$g(\omega) = \begin{pmatrix} \dfrac{\hbar\omega}{(\hbar^2\omega^2 - \Delta^2)^{1/2}} & \dfrac{\Delta}{(\hbar^2\omega^2 - \Delta^2)^{1/2}} \\ \dfrac{-\Delta^*}{(\hbar^2\omega^2 - \Delta^2)^{1/2}} & \dfrac{-\hbar\omega}{(\hbar^2\omega^2 - \Delta^2)^{1/2}} \end{pmatrix}. \tag{55.76b}$$

By multiplying these matrices we obtain

$$\underset{\sim}{g}(\omega) \cdot \underset{\sim}{g}(\omega) = \underset{\sim}{1}, \tag{55.77a}$$

or equivalently

$$[g(\omega)]^2 + f^*(\omega) f(\omega) = 1; \tag{55.77b}$$

this is our sought-after normalization condition. Performing the calculations for the $\hbar\omega < \Delta$ yields the identical normalization condition.

One can now go on to linearize the theory, or develop an appropriate perturbation expansion, to solve transport problems. For a general review of quasiclassical theory the reader is referred to Serene and Rainer (1983).

55.4 Temperature Green's functions

To describe the behavior of an inhomogeneous system, involving boundaries, static or dynamic external fields, etc., our earlier discussions have involved solving the Bogoliubov equations perturbatively. One can also construct a time-dependent perturbation theory for the Green's function equations of motion. At $T = 0$ this is relatively straightforward, although we will not discuss it here, but at finite temperatures this perturbation theory turns out to be very complex. However, having successfully calculated the finite temperature Green's function of a system one would then have information on its behavior at two different space-time points, $(\mathbf{r}_1, t_1)$ and $(\mathbf{r}_2, t_2)$, as a function of temperature, which allows one to calculate many of the quantities measured in typical experiments.

If we are less ambitious and restrict ourselves to calculating the thermodynamic response of a system (such as the equilibrium response to an external static magnetic field), where the time coordinates do not appear, one can develop a kind of thermodynamic perturbation theory which is much easier to apply in practice. This theory involves the introduction of a new kind of Green's function called a *temperature* or *Matsubara* Green's function. The 'time' coordinate in this

Green's function corresponds (in a way that we will make more precise) to imaginary time in a conventional Green's function (the latter then being referred to as a real time Green's function). Since one's intuition concerning imaginary time is initially limited, the reader must accept a certain amount of formal development.

Suppose we rewrite the Bogoliubov equations with the time coordinate replaced by $-i\tau$, resulting in $i(\partial/\partial t) \to -(\partial/\partial\tau)$. Eqs. (55.13) then become

$$-\hbar\frac{\partial u_n(\mathbf{r},\tau)}{\partial\tau} = \mathscr{H}'u_n(\mathbf{r},\tau) + \Delta(\mathbf{r})v_n(\mathbf{r},\tau) \tag{55.78a}$$

and

$$-\hbar\frac{\partial v_n(\mathbf{r},\tau)}{\partial\tau} = -\mathscr{H}'^{*}v_n(\mathbf{r},\tau) + \Delta^{*}(\mathbf{r})u_n(\mathbf{r},\tau). \tag{55.78b}$$

In place of the time-dependent eigenfunctions $u_n(\mathbf{r},t)$, $v_n(\mathbf{r},t)$ given by Eqs. (55.13) we would have

$$u_n(\mathbf{r},\tau) = u_n(\mathbf{r})e^{-(1/\hbar)\varepsilon_n\tau} \tag{55.79a}$$

and

$$v_n(\mathbf{r},\tau) = v_n(\mathbf{r})e^{-(1/\hbar)\varepsilon_n\tau}; \tag{55.79b}$$

however, in the place of the Hermitian conjugate wavefunctions $u_n^{*}(\mathbf{r},t)$, $v_n^{*}(\mathbf{r},t)$ we introduce the *adjoint* wavefunctions $\bar{u}_n(\mathbf{r},t)$, $\bar{v}(\mathbf{r},t)$ defined by

$$\bar{u}_n(\mathbf{r},\tau) = u_n^{*}(\mathbf{r},-\tau) \tag{55.80a}$$

and

$$\bar{v}_n(\mathbf{r},\tau) = v_n^{*}(\mathbf{r},-\tau) \tag{55.80b}$$

which would be the solutions to the complex conjugate of Eqs. (55.78) with τ replaced by $-\tau$. Next we introduce *temperature Green's* functions $\mathscr{G}$, $\mathscr{F}$, and $\bar{\mathscr{F}}$ defined by

$$\mathscr{G}_{\alpha\beta}(\mathbf{r}_1,\tau_1;\mathbf{r}_2,\tau_2) = \begin{cases} -\dfrac{1}{\hbar}\langle\hat{\psi}_\alpha(\mathbf{r}_1,\tau_1)\hat{\bar{\psi}}_\beta(\mathbf{r}_2,\tau_2)\rangle, & \tau_1 > \tau_2, \\ \dfrac{1}{\hbar}\langle\hat{\bar{\psi}}_\beta(\mathbf{r}_2,\tau_2)\hat{\psi}_\alpha(\mathbf{r}_1,\tau_1)\rangle, & \tau_1 < \tau_2; \end{cases} \tag{55.81}$$

$$\mathscr{F}_{\alpha\beta}(\mathbf{r}_1,\tau_1;\mathbf{r}_2,\tau_2) = \begin{cases} \dfrac{1}{\hbar}\langle\hat{\psi}_\alpha(\mathbf{r}_1,\tau_1)\hat{\psi}_\beta(\mathbf{r}_2,\tau_2)\rangle, & \tau_1 > \tau_2, \\ -\dfrac{1}{\hbar}\langle\hat{\psi}_\beta(\mathbf{r}_2,\tau_2)\hat{\psi}_\alpha(\mathbf{r}_1,\tau_1)\rangle, & \tau_1 < \tau_2; \end{cases} \tag{55.82a}$$

and

$$\bar{\mathscr{F}}_{\alpha\beta}(\mathbf{r}_1,\tau_1;\mathbf{r}_2,\tau_2) = \begin{cases} \dfrac{1}{\hbar}\langle\hat{\bar{\psi}}_\alpha(\mathbf{r}_1,\tau_1)\hat{\bar{\psi}}_\beta(\mathbf{r}_2,\tau_2)\rangle, & \tau_1 > \tau_2, \\ -\dfrac{1}{\hbar}\langle\hat{\bar{\psi}}_\beta(\mathbf{r}_2,\tau_2)\hat{\bar{\psi}}_\alpha(\mathbf{r}_1,\tau_1)\rangle, & \tau_1 < \tau_2; \end{cases} \tag{55.82b}$$

For systems with no net spin polarization we again have $\mathscr{G}_{\uparrow\uparrow} = \mathscr{G}_{\downarrow\downarrow} = \mathscr{G}$ and for singlet superconductivity $\mathscr{F}_{\alpha\beta}$ and $\bar{\mathscr{F}}_{\alpha\beta}$ obey

$$\mathscr{F}_{\alpha\beta}(\mathbf{r}_1,\tau_1;\mathbf{r}_2,\tau_2) = -\mathscr{F}_{\beta\alpha}(\mathbf{r}_1,\tau_1;\mathbf{r}_2,\tau_2) = \mathscr{F}_{\alpha\beta}(\mathbf{r}_2,\tau_2;\mathbf{r}_1,\tau_1) \equiv \mathrm{i}(\underset{\sim}{\sigma}_2)_{\alpha\beta}\mathscr{F}(\mathbf{r}_1,\tau_1;\mathbf{r}_2,\tau_2)$$

and

$$\bar{\mathscr{F}}_{\alpha\beta}(\mathbf{r}_1,\tau_1;\mathbf{r}_2,\tau_2) = -\mathrm{i}(\underset{\sim}{\sigma}_2)_{\alpha\beta}\bar{\mathscr{F}}(\mathbf{r}_1,\tau_1;\mathbf{r}_2,\tau_2),$$

where

$$\mathrm{i}\underset{\sim}{\sigma}_2 = \begin{pmatrix} 0 & 1 \\ -1 & 0 \end{pmatrix}.$$

In terms of the mean field expectation values for the $\hat{\psi}$ operator products, the unpolarized singlet equations analogous to (55.4), (55.7), and (55.9) are

$$\mathscr{G}(\mathbf{r}_1,\tau_1;\mathbf{r}_2,\tau_2) = \begin{cases} -\dfrac{1}{\hbar}\sum\limits_n [\bar{u}_n(\mathbf{r}_2,\tau_2)u_n(\mathbf{r}_1,\tau_1)(1-f_n) + v_n(\mathbf{r}_2,\tau_2)\bar{v}_n(\mathbf{r}_1,\tau_1)f_n], & \tau_1 > \tau_2, \\ \dfrac{1}{\hbar}\sum\limits_n [\bar{u}_n(\mathbf{r}_2,\tau_2)u_n(\mathbf{r}_1,\tau_1)f_n + v_n(\mathbf{r}_2,\tau_2)\bar{v}_n(\mathbf{r}_1,\tau_1)(1-f_n)], & \tau_1 < \tau_2, \end{cases} \quad (55.83)$$

$$\mathscr{F}(\mathbf{r}_1,\tau_1;\mathbf{r}_2,\tau_2) = \begin{cases} \dfrac{1}{\hbar}\sum\limits_n [\bar{v}_n(\mathbf{r}_2,\tau_2)u_n(\mathbf{r}_1,\tau_1)(1-f_n) - u_n(\mathbf{r}_2,\tau_2)\bar{v}_n(\mathbf{r}_1,\tau_1)f_n], & \tau_1 > \tau_2, \\ -\dfrac{1}{\hbar}\sum\limits_n [\bar{v}_n(\mathbf{r}_2,\tau_2)u_n(\mathbf{r}_1,\tau_1)f_n - u_n(\mathbf{r}_2,\tau_2)\bar{v}_n(\mathbf{r}_1,\tau_1)(1-f_n)], & \tau_1 < \tau_2; \end{cases} \quad (55.84a)$$

and

$$\bar{\mathscr{F}}(\mathbf{r}_1,\tau_1;\mathbf{r}_2,\tau_2) = \begin{cases} -\dfrac{1}{\hbar}\sum\limits_n [v_n(\mathbf{r}_2,\tau_2)\bar{u}_n(\mathbf{r}_1,\tau_1)f_n - \bar{u}_n(\mathbf{r}_2,\tau_2)v_n(\mathbf{r}_1,\tau_1)(1-f_n)], & \tau_1 > \tau_2, \\ \dfrac{1}{\hbar}\sum\limits_n [v_n(\mathbf{r}_2,\tau_2)\bar{u}_n(\mathbf{r}_1,\tau_1)(1-f_n) - \bar{u}_n(\mathbf{r}_2,\tau_2)v_n(\mathbf{r}_1,\tau_1)f_n], & \tau_1 < \tau_2. \end{cases} \quad (55.84b)$$

Using the imaginary time Bogoliubov equations (55.78a) and (55.78b) it is straightforward to show that the quantities $\mathscr{G}$, $\mathscr{F}$, and $\bar{\mathscr{F}}$ satisfy the following matrix equation:

$$\begin{pmatrix} -\hbar\dfrac{\partial}{\partial\tau_1} - \mathscr{H}'_1 & \Delta(\mathbf{r}_1) \\ -\Delta^*(\mathbf{r}_1) & \hbar\dfrac{\partial}{\partial\tau_1} - \mathscr{H}'_1 \end{pmatrix} \begin{pmatrix} \mathscr{G}(\mathbf{r}_1,\tau_1;\mathbf{r}_2,\tau_2) & \mathscr{F}(\mathbf{r}_1,\tau_1;\mathbf{r}_2,\tau_2) \\ \bar{\mathscr{F}}(\mathbf{r}_1,\tau_1;\mathbf{r}_2,\tau_2) & -\mathscr{G}(\mathbf{r}_2,\tau_2;\mathbf{r}_1,\tau_1) \end{pmatrix}$$

$$= \delta^{(3)}(\mathbf{r}_1 - \mathbf{r}_2)\delta(\tau_1 - \tau_2)\begin{pmatrix} 1 & 0 \\ 0 & 1 \end{pmatrix}. \quad (55.85)$$

In obtaining these equations we must recognize that when reversing the arguments $\mathbf{r}_1,\tau_1;\mathbf{r}_2,\tau_2$ to form $\mathscr{G}(\mathbf{r}_2,\tau_2;\mathbf{r}_1,\tau_1)$, the inequalities involving τ_1 and τ_2 must also be reversed. This reversal has the additional effect of changing the sign of the δ function involving $\tau_1 - \tau_2$. Note that in the general case $\mathscr{F}$ and $\bar{\mathscr{F}}$ are independent functions that are not trivially related to each other. The equations embodied in (55.85) are also referred to as the Nambu–Gorkov equations.

As $n \to \infty$ in the sums defining $\mathscr{G}$, $\mathscr{F}$, and $\bar{\mathscr{F}}$, the Fermi occupation factors $f_n \to \mathrm{e}^{-\varepsilon_n/k_B T}$. Assuming Bogoliubov wavefunctions of the forms (55.79) we see that as $n \to \infty$ the terms $\bar{u}_n(\tau_2)u_n(\tau_1)(1-f_n)$ and $v_n(\tau_2)\bar{v}_n(\tau_1)f_n$ in Eq. (55.83) are proportional to

$$\exp\left[\frac{\varepsilon_n}{\hbar}(\tau_2 - \tau_1) - \frac{\varepsilon_n}{k_B T}\right]$$

and

$$\exp\left[\frac{\varepsilon_n}{\hbar}(\tau_1 - \tau_2) - \frac{\varepsilon_n}{k_B T}\right],$$

respectively (where we neglected the 1 in the first form since we are here interested in the case where the sum can diverge). If the sums are to converge we must have[9]

$$-\frac{\hbar}{k_B T} \leq \tau_1 - \tau_2 \leq \frac{\hbar}{k_B T}. \tag{55.86}$$

Introducing the Matsubara time difference, $\tau = \tau_1 - \tau_2$ and using the properties $e^{-\varepsilon_n/k_B T} \times (1 - f_n) = f_n$ and $e^{\varepsilon_n/k_B T} f_n = 1 - f_n$, it is easy to show that $\mathscr{G}(\tau < 0)$ is related to $\mathscr{G}(\tau > 0)$ by

$$\mathscr{G}(\tau) = -\mathscr{G}\left(\tau + \frac{\hbar}{k_B T}\right) \quad (\tau < 0) \tag{55.87}$$

and similarly for $\bar{\mathscr{F}}$; i.e., these functions are *antisymmetric* in the variable τ (which, of course, arises from Fermi statistics); (55.87) is referred to as a *boundary condition*, which the Matsubara Green's functions must satisfy.

As with the real time Green's functions it is often useful to perform a Fourier transform on the variables τ and/or $\mathbf{r}$. Since τ is restricted to the interval (55.86), the function $\mathscr{G}(\tau)$ must be represented by a *Fourier series* rather than a Fourier integral and can be written as

$$\mathscr{G}(\mathbf{r}, \tau) = k_B T \sum_m e^{-i\omega_m \tau} \mathscr{G}(\mathbf{r}, \omega_m), \tag{55.88}$$

where

$$\omega_m = (2m + 1)\frac{\pi k_B T}{\hbar}; \tag{55.89}$$

the ω_m are referred to as *Matsubara frequencies* and the form selected builds in the antisymmetric property.[10] The Fourier transform of $\mathbf{r}$ would be performed in the usual way and the relations between fully transformed Matsubara Green's functions are

$$\mathscr{G}(\mathbf{r}, \tau) = k_B T \sum_m \int_{-\infty}^{+\infty} e^{i(\mathbf{k}\cdot\mathbf{r} - \omega_m \tau)} \mathscr{G}(\mathbf{k}, \omega_m) \frac{d^3k}{(2\pi)^3} \tag{55.90a}$$

and

$$\mathscr{G}(\mathbf{k}, \omega_m) = \int_0^{\hbar/k_B T} \int_{-\infty}^{+\infty} e^{-i(\mathbf{k}\cdot\mathbf{r} - \omega_m \tau)} \mathscr{G}(\mathbf{r}, \tau) d^3r\, d\tau, \tag{55.90b}$$

where we used the property (55.87) in restricting the limits of the τ integration in (55.90b).

9. It can be shown that the restriction (55.86) is a general property and is not dependent on the explicit forms given by (55.79).
10. The trick of expanding in a Fourier series was introduced by Abrikosov, Gorkov, and Dzyaloshinskii (1959) and Fredkin (1959).

We now establish a connection between the Matsubara Green's function and the retarded and advanced Green's functions defined by Eqs. (55.52a) and (55.52b); we will establish this connection using the mean field forms, although the relations we will obtain are true in general.[11] The expressions obtained earlier for the retarded Green's functions are

$$G^{\mathrm{R}}(\mathbf{r}_1, \tau_1; \mathbf{r}_2, \tau_2) = -\frac{\mathrm{i}}{\hbar}\theta(t_1 - t_2)\sum_n [u_n^*(\mathbf{r}_2, t_2)u_n(\mathbf{r}_1, t_1) + v_n(\mathbf{r}_2, t_2)v_n^*(\mathbf{r}_1, t_1)] \quad (55.91\mathrm{a})$$

and

$$F^{\mathrm{R}*}(\mathbf{r}_1, \tau_1; \mathbf{r}_2, \tau_2) = -\frac{\mathrm{i}}{\hbar}\theta(t_1 - t_2)\sum_n [v_n(\mathbf{r}_2, t_2)u_n^*(\mathbf{r}_1, t_1) - u_n^*(\mathbf{r}_2, t_2)v_n(\mathbf{r}_1, t_1)] \quad (55.91\mathrm{b})$$

with similar expressions for G^{A} and $F^{\mathrm{A}*}$; as noted earlier, Eqs. (55.91) are independent of temperature. For the case of a superfluid Fermi gas the space-time Fourier transforms of these expressions (which are analytic in the upper half plane) are

$$G^{\mathrm{R}}(\mathbf{k}, \omega) = \frac{u_k^2}{\hbar\omega - \varepsilon_k + \mathrm{i}0} + \frac{v_k^2}{\hbar\omega + \varepsilon_k + \mathrm{i}0} \quad (55.92\mathrm{a})$$

$$= \frac{\hbar\omega + \xi_k}{(\hbar\omega - \varepsilon_k + \mathrm{i}0)(\hbar\omega + \varepsilon_k + \mathrm{i}0)} \quad (55.92\mathrm{b})$$

and

$$F^{\mathrm{R}*}(\mathbf{k}, \omega) = \frac{-\Delta^*}{(\hbar\omega - \varepsilon_k + \mathrm{i}0)(\hbar\omega + \varepsilon_k + \mathrm{i}0)}; \quad (55.93)$$

to obtain G^{A} and F^{A} we substitute $\hbar\omega - \mathrm{i}0_+$ for $\hbar\omega + \mathrm{i}0_+$ with the result that the Green's functions are then analytic in the lower half plane. We next perform the space-time transform of the mean field Matsubara Green's functions (55.83), (55.84a), and (55.84b). Using plane wave forms for the spatial dependence of the wavefunctions (55.79a) and (55.79b) and evaluating the resulting Matsubara time integrals

$$\int_0^{\hbar/k_{\mathrm{B}}T} \mathrm{e}^{(\pm(\varepsilon_k/\hbar) + \mathrm{i}\omega_m)\tau}\mathrm{d}\tau = -\frac{\mathrm{e}^{\pm\varepsilon_k/k_{\mathrm{B}}T} + 1}{\pm\dfrac{\varepsilon_k}{\hbar} + \mathrm{i}\omega_m},$$

where the numerators combine with corresponding factors of f_n and $1 - f_n$ to produce unity, we obtain

$$\mathscr{G}(\mathbf{k}, \omega_m) = -\frac{\mathrm{i}\hbar\omega_m + \xi_k}{\hbar^2\omega_m^2 + \varepsilon_k^2}, \quad (55.94)$$

$$\mathscr{F}(\mathbf{k}, \omega_m) = \frac{\Delta}{\hbar^2\omega_m^2 + \varepsilon_k^2}, \quad (55.95\mathrm{a})$$

and

11. To establish the connection in general one must develop a so-called Källén–Lehmann spectral representation; see Lifshitz and Pitaevskii (1980).

$$\mathscr{F}(\mathbf{k}, \omega_m) = \frac{\Delta^*}{\hbar^2\omega_m^2 + \varepsilon_k^2}. \tag{55.95b}$$

Comparing (55.94) with (55.92b) and (55.95b) with (55.93) we obtain the relations

$$\mathscr{G}(\mathbf{k}, \omega_m) = G^{\mathrm{R}}(\mathbf{k}, \mathrm{i}\omega_m) \tag{55.96a}$$

and

$$\mathscr{F}(\mathbf{k}, \omega_m) = F^{\mathrm{R}*}(\mathbf{k}, \mathrm{i}\omega_m). \tag{55.96b}$$

Let us return to the full Green's function (55.5) for the case of the superfluid Fermi gas at finite temperature. We now show that the space-time transform of (55.5) and (55.9) are given by

$$G(\mathbf{k}, \omega) = \frac{u_k^2}{\hbar\omega - \varepsilon_k + \mathrm{i}0} + \frac{v_k^2}{\hbar\omega + \varepsilon_k - \mathrm{i}0} + 2\pi\mathrm{i}f_k[u_k^2\delta(\hbar\omega - \varepsilon_k) - v_k^2\delta(\hbar\omega + \varepsilon_k)] \tag{55.97}$$

and

$$F^*(\mathbf{k}, \omega) = \frac{-\Delta^*}{(\hbar\omega - \varepsilon_k + \mathrm{i}0)(\hbar\omega + \varepsilon_k - \mathrm{i}0)} - \pi\mathrm{i}\frac{\Delta^* f_k}{\varepsilon_k}[\delta(\hbar\omega - \varepsilon_k) + \delta(\hbar\omega + \varepsilon_k)]. \tag{55.98}$$

Consider the inverse Fourier transform of (55.97)

$$G(\mathbf{r}, t) = \int \mathrm{e}^{\mathrm{i}(\mathbf{k}\cdot\mathbf{r} - \omega t)} \frac{\mathrm{d}^3k}{(2\pi)^3} \frac{\mathrm{d}\omega}{2\pi}. \tag{55.99}$$

When performing the integration on $\mathrm{d}\omega/2\pi$ associated with the first two terms in (55.97) we must close the contour in the lower half plane for $t > 0$ and in the upper half plane for $t < 0$. For $t > 0$ this leads to

$$-\frac{\mathrm{i}}{\hbar}[u_k^2(1 - f_k)\mathrm{e}^{-(\mathrm{i}/\hbar)\varepsilon_k t} + v_k^2 f_k \mathrm{e}^{(\mathrm{i}/\hbar)\varepsilon_k t}],$$

while for $t < 0$ we have

$$\frac{\mathrm{i}}{\hbar}[u_k^2 f_k \mathrm{e}^{-(\mathrm{i}/\hbar)\varepsilon_k t} + v_k^2(1 - f_k)\mathrm{e}^{(\mathrm{i}/\hbar)\varepsilon_k t}];$$

performing the integration on $\mathrm{d}^3k/(2\pi)^3$, we obtain the Fermi gas form of (55.2). Performing similar manipulations we can verify (55.98).

Finally, let us determine the expression for the gap function $\Delta^*(\mathbf{r})$ in terms of $\mathscr{F}(\mathbf{r}_1, \mathbf{r}_2, \tau)$ for an inhomogeneous system in thermal equilibrium. Recalling our earlier expression (37.11) for $\Delta(\mathbf{r})$ and comparing it with (55.84b) we have immediately

$$\Delta^*(\mathbf{r}) = V\mathscr{F}(\tau_1 = \tau_2 + 0_+, \mathbf{r}_1 = \mathbf{r}_2 = \mathbf{r}). \tag{55.100}$$

For a uniform Fermi system we can write this as

$$\Delta^* = k_B TV \sum_{m=-\infty}^{+\infty} \int \bar{\mathscr{F}}(\mathbf{k}, \omega_m) \frac{d^3k}{(2\pi)^3}. \tag{55.101}$$

Substituting (55.95) we obtain the gap equation as

$$\frac{k_B TV}{(2\pi)^3} \sum_{m=-\infty}^{+\infty} \int \frac{d^3k}{\hbar^2\omega_m^2 + \varepsilon_k^2} = 1 \tag{55.102}$$

To perform the sum we note that our earlier expression (38.9) can be written

$$\tanh\frac{\varepsilon_k}{2k_B T} = k_B T \sum_m \left(\frac{1}{\varepsilon_k + i\hbar\omega_m} + \frac{1}{\varepsilon_k - i\hbar\omega_m} \right)$$
$$= 2k_B T \varepsilon_k \sum_m \frac{1}{\hbar^2\omega_m^2 + \varepsilon_k^2}.$$

We may then write (55.102) as

$$\frac{V}{(2\pi)^3} \int \frac{1}{2\varepsilon_k} \tanh\left(\frac{\varepsilon_k}{2k_B T}\right) d^3k = 1, \tag{55.103}$$

which is identical to Eq. (28.12) obtained earlier by more elementary means.

55.5 Dirty superconductors

In Secs. 38–40 we developed a microscopic theory of an inhomogeneous superconductor; however, this theory involved a linearized form of the self-consistency condition and is therefore valid only in the immediate vicinity of the transition temperature. The Green's function methods (and their quasiclassical variants), developed in the previous subsections, can be used to describe the properties at an arbitrary temperature. Since many experiments involve thin-film materials (which generally consist of small crystals leading to scattering of electrons by grain boundaries), or alloys (where scattering from potential fluctuations is present), it is first necessary to develop a Green's function formalism which incorporates the effects of electron scattering.

In what follows immediately, we limit ourselves to a model involving a free electron gas in which scattering centers (impurities) characterized by a potential $u(\mathbf{r})$ are present. The total potential in the presence of N_i impurities is then

$$U(\mathbf{r}, \mathbf{r}_1, \ldots, \mathbf{r}_{N_i}) = \sum_{a=1}^{N_i} u(\mathbf{r} - \mathbf{r}_a), \tag{55.104}$$

where the $\mathbf{r}_a$ denote the positions of the impurities.

55.5.1 The dirty-normal-metal Green's function

For orientation purposes we begin our discussion with the case of a normal metal in which the pairing potential, V, vanishes. The equation of motion for the Fourier time transform of the Green's function in the presence of impurity scattering is

$$\left[\hbar\omega + \frac{\hbar^2}{2m}\nabla^2 + \mu - \sum_a u(\mathbf{r} - \mathbf{r}_a)\right] G(\mathbf{r}, \mathbf{r}', \mathbf{r}_1, \ldots \mathbf{r}_{N_i}, \omega) = \delta^{(3)}(\mathbf{r} - \mathbf{r}'). \qquad (55.105)$$

As a first step we Fourier transform from the variables $\mathbf{r}, \mathbf{r}'$ to $\mathbf{k}, \mathbf{k}'$:

$$\int d^3r d^3r' e^{-i\mathbf{k}\cdot\mathbf{r} + i\mathbf{k}'\cdot\mathbf{r}'} \left[\hbar\omega + \frac{\hbar^2}{2m}\nabla^2 + \mu - \sum_a \int \frac{d^3q}{(2\pi)^3} u(\mathbf{q}) e^{i\mathbf{q}\cdot(\mathbf{r} - \mathbf{r}_a)}\right] G(\mathbf{r}, \mathbf{r}', \mathbf{r}_1, \ldots, \mathbf{r}_{N_i}, \omega)$$

$$= \int d^3(\mathbf{r} - \mathbf{r}') d^3r' e^{i\mathbf{k}\cdot(\mathbf{r} - \mathbf{r}') + i(\mathbf{k} - \mathbf{k}')\cdot\mathbf{r}'} \delta^{(3)}(\mathbf{r} - \mathbf{r}'), \qquad (55.106)$$

where we introduced the Fourier transform $u(\mathbf{r} - \mathbf{r}_a) = \int [d^3q/(2\pi)^3] u(\mathbf{q}) e^{i\mathbf{q}\cdot(\mathbf{r} - \mathbf{r}_a)}$. Carrying out the two $\mathbf{r}$-space integrations, using the property $\int d^3r' e^{i(\mathbf{k} - \mathbf{k}')\cdot\mathbf{r}'} = (2\pi)^3 \delta^{(3)}(\mathbf{k} - \mathbf{k}')$ on the right-hand side, and integrating the term involving ∇^2 twice by parts, Eq. (55.106) becomes

$$[\hbar\omega - \xi(k)] G(\mathbf{k}, \mathbf{k}', \mathbf{r}_1 \ldots \mathbf{r}_{N_i}, \omega) - \sum_a \int \frac{d^3q}{(2\pi)^3} e^{-i\mathbf{q}\cdot\mathbf{r}_a} u(\mathbf{q}) G(\mathbf{k} - \mathbf{q}\ \mathbf{k}', \mathbf{r}_1, \ldots, \mathbf{r}_{N_i}, \omega)$$

$$= (2\pi)^3 \delta^{(3)}(\mathbf{k} - \mathbf{k}'), \qquad (55.107)$$

where we wrote $\xi(k) = (\hbar^2 k^2/2m) - \mu$. Noting that $\hbar\omega - \xi(k) = [G^{(0)}(\omega, \mathbf{k})]^{-1}$ we may rewrite (55.107) as

$$G(\mathbf{k}, \mathbf{k}', \mathbf{r}_1, \ldots, \mathbf{r}_{N_i}, \omega) = (2\pi)^3 \delta^{(3)}(\mathbf{k} - \mathbf{k}') G^{(0)}(\mathbf{k}, \omega)$$

$$+ G^{(0)}(\mathbf{k}, \omega) \sum_a \int \frac{d^3q}{(2\pi)^3} e^{-i\mathbf{q}\cdot\mathbf{r}_a} u(\mathbf{q}) G(\mathbf{k} - \mathbf{q}, \mathbf{k}', \mathbf{r}_1, \ldots, \mathbf{r}_{N_i}, \omega). \qquad (55.108)$$

To solve the integral equation (55.108) iteratively we write $u(\mathbf{q})$ as $\lambda u(\mathbf{q})$ and expand G in the form

$$G = G^{(0)} + \lambda G^{(1)} + \lambda^2 G^{(2)} + \ldots;$$

substituting these forms into Eq. (55.108) and associating equal powers of λ we have

$$G^{(0)}(\mathbf{k}, \mathbf{k}', [\mathbf{r}_{N_i}], \omega) = G^{(0)}(\omega, \mathbf{k})(2\pi)^3 \delta^{(3)}(\mathbf{k} - \mathbf{k}')$$

$$= \frac{(2\pi)^3 \delta^{(3)}(\mathbf{k} - \mathbf{k}')}{\hbar\omega - \xi(k) + i0 \operatorname{sgn} \omega} \qquad (55.109a)$$

and

$$G^{(n)}(\mathbf{k}, \mathbf{k}', [\mathbf{r}_{N_i}], \omega) = G^{(0)}(\mathbf{k}, \omega) \sum_a \int \frac{d^3q}{(2\pi)^3} e^{-i\mathbf{q}\cdot\mathbf{r}_a} u(\mathbf{q}) G^{(n-1)}(\mathbf{k} - \mathbf{q}, \mathbf{k}', [\mathbf{r}_{N_i}], \omega), \qquad (55.109b)$$

where we denoted the set of impurity positions $\mathbf{r}_1, \ldots, \mathbf{r}_{N_i}$ as $[\mathbf{r}_{N_i}]$ and we have inserted the appropriate infinitesimal in the denominator of $G^{(0)}$. Carrying out the integration over $\mathbf{q}$, Eq. (55.109b) for $G^{(1)}$ becomes

$$G^{(1)}(\mathbf{k}, \mathbf{k}', [\mathbf{r}_{N_i}], \omega) = G^{(0)}(\mathbf{k}, \omega) \sum_a e^{-i(\mathbf{k} - \mathbf{k}')\cdot\mathbf{r}_a} u(\mathbf{k} - \mathbf{k}') G^{(0)}(\mathbf{k}', \omega). \qquad (55.110)$$

With the availability of $G^{(1)}$ we may go on to calculate $G^{(2)}$ as

$$G^{(2)}(\mathbf{k},\mathbf{k}',[\mathbf{r}_{N_i}],\omega) = G^{(0)}(\mathbf{k},\omega)\sum_a \int \frac{\mathrm{d}^3 q}{(2\pi)^3} \mathrm{e}^{-\mathrm{i}\mathbf{q}\cdot\mathbf{r}_a} u(\mathbf{q})$$

$$\times\, G^{(0)}(\mathbf{k}-\mathbf{q},\omega)\sum_b \mathrm{e}^{-\mathrm{i}(\mathbf{k}-\mathbf{q}-\mathbf{k}')\cdot\mathbf{r}_b\Omega^{\mathbf{r}_b}} u(\mathbf{k}-\mathbf{q}-\mathbf{k}')G^{(0)}(\mathbf{k}',\omega). \tag{55.111}$$

Before writing any higher order terms we first examine $G^{(1)}$ and $G^{(2)}$. We begin by noting that the positions $\mathbf{r}_a$ are in general not known. Since we seek a description which is not dependent on any specific configuration of impurities, it is appropriate to average successive terms in the expansion of G over the position of each impurity coordinate, $\mathbf{r}_a$, using the prescription $(1/V) \int \mathrm{d}^3 r_a$, where V is the volume of the specimen. Carrying out this procedure on (55.110) yields

$$\overline{G^{(1)}(\mathbf{k},\mathbf{k}',[\mathbf{r}_{N_i}],\omega)} = n_i G^{(0)}(\mathbf{k},\omega)u(0)G^{(0)}(\mathbf{k}',\omega)(2\pi)^3\delta^{(3)}(\mathbf{k}-\mathbf{k}'), \tag{55.112}$$

where $n_i \equiv N_i/V$ is the number of impurities per unit volume. Note that $\overline{G^{(1)}}$ is proportional to a momentum conserving δ function. This follows immediately from the fact that the configuration averaged Green's function must be translationally invariant. Thus we may write

$$\overline{G^{(n)}(\mathbf{k},\mathbf{k}',[\mathbf{r}_{N_i}],\omega)} = G^{(n)}(\mathbf{k},\omega)\delta^{(3)}(\mathbf{k}-\mathbf{k}'). \tag{55.113}$$

The configuration average of the right-hand side of (55.111) yields two terms; these correspond to the cases $a \neq b$ and $a = b$ which are given by

$$a \neq b\text{:}\quad G^{(0)}(\mathbf{k},\omega)n_i u(0)G^{(0)}(\mathbf{k},\omega)n_i u(0)G^{(0)}(\mathbf{k}',\omega)(2\pi)^3\delta^{(3)}(\mathbf{k}-\mathbf{k}') \tag{55.114a}$$

and

$$a = b\text{:}\quad G^{(0)}(\mathbf{k},\omega)\left[n_i \int \frac{\mathrm{d}^3 q}{(2\pi)^3} u(\mathbf{q})G^{(0)}(\mathbf{k}-\mathbf{q},\omega)u(-\mathbf{q})\right]G^{(0)}(\mathbf{k}',\omega)(2\pi)^3\delta^{(3)}(\mathbf{k}-\mathbf{k}'). \tag{55.114b}$$

Considered together with (55.109a) the contributions (55.112) and (55.114a) represent the first three terms in the expansion of the quantity $1/[\hbar\omega - \xi(k) - u(0) + \mathrm{i}0\,\mathrm{sgn}(\omega)]$. We may therefore regard $u(0)$ as a correction to the chemical potential, μ; since the latter is a parameter anyway we will simply absorb $u(0)$ into the definition of μ and henceforth disregard such corrections. In nth order of perturbation theory, n sums over the impurity atoms occur; that contribution for which $a \neq b \neq c\ldots$ (i.e., all the atoms are different) yields the shift arising from $u(0)$ in that order of perturbation theory.

The contribution (55.114b) is the first of a series of terms occurring for even values of n which arise when pairs of indices $a, b, c, d, \ldots$ in successive summations are equal. In second order we have one such term: $a = b$. In fourth order there are three cases:[12] (1) $a = b, c = d$; (2) $a = d, b = c$; and (3) $a = c, b = d$. Note that the earlier 'chemical potential-shifting' terms were proportional to n_i^n in each order; however, the terms considered now yield a contribution proportional to $n_i^{(n/2)}$.

We now examine the integral defined by the square brackets in (55.114b) which we denote as Σ. Defining a new variable of integration by $\mathbf{k}_1 = \mathbf{k} - \mathbf{q}$ this integral takes the form

12. For the nth order there are $(n-1)(n-3)(n-5)\ldots 1$ terms where the same atom appears twice in the summations. It can be shown that terms in which the same atom appears three or more times in the summations are small and may be neglected (see Abrikosov and Gorkov (1959)).

$$\Sigma = n_i \int \frac{d^3 k_1}{(2\pi)^3} \frac{|u(\mathbf{k} - \mathbf{k}_1)|^2}{\hbar\omega - \hbar v_F(k_1 - k_F) + i0 \operatorname{sgn} \omega} = \Sigma_1 + \Sigma_2.$$

We will primarily be interested in the behavior of this integral for values of k_1 near the Fermi wave vector, k_F, and for small ω. Therefore we have split the integral (55.114b) into two parts: a real part Σ_1 involving the two regions 0 to $k_F - \delta k$ and $k_F + \delta k$ to ∞, and a part Σ_2, covering the region in the immediate vicinity of the pole at $k_1 = (\omega/v_F) + k_F - i0 \operatorname{sgn} \omega$ with the limits $k_1 = k_F \pm \delta k$ (where $\delta k << k_F$). The part Σ_1 and the real part of Σ_2 may again be regarded as contributions to the chemical potential and will be discarded. Using the integral prescription $1/(x + i0) = P(1/x) - i\pi\delta(x)$, and assuming u is smoothly varying near the Fermi surface, our expression for Σ_2 is then

$$\Sigma_2 = -\frac{i}{2} \frac{k_F^2}{2\hbar v_F} \frac{n_i}{(2\pi)^2} \operatorname{sgn}(\omega) \int d\Omega \, |u(\theta)|^2, \tag{55.115}$$

where θ is the polar angle measured from $\hat{\mathbf{k}}$. Defining a mean scattering time τ by

$$\frac{1}{\tau} = \frac{n_i m k_F}{(2\pi)^2 \hbar^3} \int |u(\theta)|^2 d\Omega, \tag{55.116a}$$

we have

$$\Sigma = -i\hbar \frac{\operatorname{sgn} \omega}{2\tau}, \tag{55.116b}$$

where we have dropped the subscript from Σ_2.

In the fourth order of our perturbation theory the (1) term involving $a = b$ and $c = d$ discussed above turns out to be $(\Sigma)^2$ and likewise a term $(\Sigma)^{n/2}$ occurs in successive (even order) terms in the expansion. Including only these terms the perturbative series would take the form

$$G(\mathbf{k}, \omega) = G^{(0)}(\mathbf{k}, \omega) + G^{(0)}(\mathbf{k}, \omega)\Sigma(\omega)G^{(0)}(\mathbf{k}, \omega) + G^{(0)}(\mathbf{k}, \omega)\Sigma(\omega)G^{(0)}(\mathbf{k}, \omega)\Sigma(\omega)G^{(0)}(\mathbf{k}, \omega) + \ldots. \tag{55.117}$$

We note that if we factor a term $G^{(0)}(\mathbf{k}, \omega)\Sigma(\omega)$ out of all terms starting with the second term on the right we obtain the original series for $G(\mathbf{k}, \omega)$. We may therefore write

$$G(\mathbf{k}, \omega) = G^{(0)}(\mathbf{k}, \omega) + G^{(0)}(\mathbf{k}, \omega)\Sigma(\omega)G(\mathbf{k}, \omega). \tag{55.118}$$

Multiplying both sides by $[G(\mathbf{k}, \omega)]^{-1}[G^{(0)}(\mathbf{k}, \omega)]^{-1}$ we obtain

$$[G(\mathbf{k}, \omega)]^{-1} = [G^{(0)}(\mathbf{k}, \omega)]^{-1} - \Sigma(\omega) \tag{55.119}$$

or

$$G(\mathbf{k}, \omega) = \frac{1}{\hbar\omega - \xi - \Sigma(\omega)}. \tag{55.120}$$

Inserting our expression for Σ, (55.116b), we obtain $G(\mathbf{k}, \omega)$ as

$$G(\mathbf{k}, \omega) = \frac{1}{\hbar\omega - \xi + i\hbar \dfrac{\operatorname{sgn} \omega}{2\tau}}. \tag{55.121}$$

An analysis of the remaining two fourth order terms shows that (2) is included (in the approximation leading to (55.121)) and that (3) may be neglected in dilute alloys[13] (for a discussion see Abrikosov and Gorkov (1959)).

55.5.2 *The dirty superconductor Green's function*

The approach used in Sec. 55.5.1 will now be generalized to the case of a dirty superconductor. Since it will be applied first to the thermodynamics of a dirty superconductor, we will start with the Nambu–Gorkov equations in the Matsubara representation as embodied in the matrix equations (55.85). To obtain an equation analogous to (55.107) we must perform a Fourier transform from the variables $(\mathbf{r}_1, \mathbf{r}_2, \tau)$ to $(\mathbf{k}, \mathbf{k}', \omega_n)$. Now a position dependence of $\Delta(\mathbf{r})$ introduced by the impurities would complicate the analysis. However, we will assume that $\Delta(\mathbf{r})$ does not respond at the scale of an impurity potential fluctuation[14] and may therefore be accurately represented by some average value Δ (we are assuming no external potentials are present at this point in our discussion). With this assumption, and noting that the Fourier Matsubara time transform of $\mathscr{G}(-\tau)$ is $\mathscr{G}(-\omega_m)$, the transform of Eq. (55.85) may be written[15]

$$\begin{pmatrix} i\hbar\omega_m - \xi_k & \Delta \\ -\Delta^* & -i\hbar\omega_m - \xi_k \end{pmatrix} \begin{pmatrix} \mathscr{G}(\mathbf{k}, \mathbf{k}', [\mathbf{r}_{N_i}], \omega_m) \\ \mathscr{F}(\mathbf{k}, \mathbf{k}', [\mathbf{r}_{N_i}], \omega_m) \end{pmatrix}$$
$$- \sum_a \int \frac{d^3q}{(2\pi)^3} e^{-i\mathbf{q}\cdot\mathbf{r}_a} u(\mathbf{q}) \begin{pmatrix} \mathscr{G}(\mathbf{k}-\mathbf{q}, \mathbf{k}', [\mathbf{r}_{N_i}], \omega_m) \\ \mathscr{F}(\mathbf{k}-\mathbf{q}, \mathbf{k}', [\mathbf{r}_{N_i}], \omega_m) \end{pmatrix}$$
$$= (2\pi)^3 \delta^{(3)}(\mathbf{k}-\mathbf{k}') \begin{pmatrix} 1 \\ 0 \end{pmatrix}. \tag{55.122}$$

The matrix operator in (55.122) may be regarded as the inverse of a zeroth order Green's function matrix formed using Eqs. (55.94) and (55.95); i.e.,

$$\begin{pmatrix} \mathscr{G}^{(0)}(\mathbf{k}, \omega_m) & -\mathscr{F}^{(0)}(\mathbf{k}, \omega_m) \\ \mathscr{F}^{(0)}(\mathbf{k}, \omega_m) & \mathscr{G}^{(0)}(-\mathbf{k}, -\omega_m) \end{pmatrix} \begin{pmatrix} i\hbar\omega_m - \xi_k & \Delta \\ -\Delta^* & -i\hbar\omega_m - \xi_k \end{pmatrix} = \begin{pmatrix} 1 & 0 \\ 0 & 1 \end{pmatrix}, \tag{55.123a}$$

where the matrix on the left of (55.123a) is

$$\vec{\mathscr{G}}^{(0)}(\mathbf{k}, \omega_m) = \frac{1}{\hbar^2\omega_m^2 + \varepsilon_k^2} \begin{pmatrix} -i\hbar\omega_m - \xi_k & -\Delta \\ \Delta^* & i\hbar\omega_m - \xi_k \end{pmatrix}. \tag{55.123b}$$

We now rewrite Eq. (55.122) in a more compact form by writing the associated column vector as $\vec{\mathscr{G}}$:

$$\vec{\mathscr{G}}(\mathbf{k}, \mathbf{k}', [\mathbf{r}_{N_i}], \omega_m) = (2\pi)^3 \delta^{(3)}(\mathbf{k}-\mathbf{k}') \vec{\mathscr{G}}^{(0)}(\mathbf{k}, \omega_m)$$

13. The process (3) for $n = 4$ discussed above is the first in a series of terms in higher orders which are referred to as the 'maximally crossed diagrams'. An examination of the associated integral shows that such terms are smaller than those associated with processes (1) and (2) by a factor of order $(k_F\ell)^{-1}$, where ℓ is the mean free path. The inclusion of such terms, which are important for very dirty alloys, leads to electron localization effects at very low temperatures.
14. It will turn out that the average value of Δ is unchanged by scattering, assuming the pair potential remains fixed. Our assumption is valid only for nonmagnetic impurities.
15. With our assumption we need only two of the four equations embodied in (55.85).

$$+\sum_a \int \frac{d^3q}{(2\pi)^3} e^{-i\mathbf{q}\cdot\mathbf{r}_a} u(\mathbf{q}) \overleftrightarrow{\mathscr{G}}^{(0)}(\mathbf{k}, \omega_m) \overleftrightarrow{\mathscr{G}}(\mathbf{k}-\mathbf{q}, \mathbf{k}', [\mathbf{r}_{N_i}], \omega_m). \tag{55.124}$$

Written explicitly the two equations embodied in (55.124) are

$$\mathscr{G}(\mathbf{k}, \mathbf{k}', [\mathbf{r}_{N_i}], \omega_m) = (2\pi)^3 \delta^{(3)}(\mathbf{k}-\mathbf{k}') \mathscr{G}^{(0)}(\mathbf{k}, \omega_m)$$
$$+\sum_a \int \frac{d^3q}{(2\pi)^3} e^{-i\mathbf{q}\cdot\mathbf{r}_a} u(\mathbf{q}) [\mathscr{G}^{(0)}(\mathbf{k}, \omega_m) \mathscr{G}(\mathbf{k}-\mathbf{q}, \mathbf{k}', [\mathbf{r}_{N_i}], \omega_m)$$
$$-\mathscr{F}^{(0)}(\mathbf{k}, \omega_m) \bar{\mathscr{F}}(\mathbf{k}-\mathbf{q}, \mathbf{k}', [\mathbf{r}_{N_i}], \omega_m)]. \tag{55.125a}$$

and

$$\bar{\mathscr{F}}(\mathbf{k}, \mathbf{k}', [\mathbf{r}_{N_i}], \omega_m) = (2\pi)^3 \delta^{(3)}(\mathbf{k}-\mathbf{k}') \bar{\mathscr{F}}^{(0)}(\mathbf{k}, \omega_m) + \sum_a \int \frac{d^3q}{(2\pi)^3} e^{-iq\cdot r_a} u(q) [\bar{\mathscr{F}}^{(0)}_{(\mathbf{k}, w_m)} \mathscr{G}(\mathbf{k}-\mathrm{q}, \mathbf{k}', [\mathbf{r}_{N_i}], \omega_m)$$
$$+\mathscr{G}^{(0)}(-\mathbf{k}, -\omega_m) \bar{\mathscr{F}}(\mathbf{k}-\mathbf{q}, \mathbf{k}', [\mathbf{r}_{N_i}], \omega_m]. \tag{55.125b}$$

These equations are the generalization of Eq. (55.108) for a superconductor.

As with (55.108), we resort to an iterative solution of Eqs. (55.125a) and (55.125b). Writing u as λu, $\overleftrightarrow{\mathscr{G}}$ as $\overleftrightarrow{\mathscr{G}}^{(0)} + \lambda \overleftrightarrow{\mathscr{G}}^{(1)} \ldots$, and collecting equal powers of λ we obtain

$$\overleftrightarrow{\mathscr{G}}^{(0)}(\mathbf{k}, \mathbf{k}', [\mathbf{r}_{N_i}], \omega_m) = (2\pi)^3 \delta^{(3)}(\mathbf{k}-\mathbf{k}') \overleftrightarrow{\mathscr{G}}^{(0)}(\mathbf{k}, \omega_m) \tag{55.126a}$$

and

$$\overleftrightarrow{\mathscr{G}}^{(n)}(\mathbf{k}, \mathbf{k}', [\mathbf{r}_{N_i}], \omega_m) = \overleftrightarrow{\mathscr{G}}^{(0)}(\mathbf{k}, \omega_m) \sum_a \int \frac{d^3q}{(2\pi)^3} e^{-i\mathbf{q}\cdot\mathbf{r}_a} u(\mathbf{q}) \overleftrightarrow{\mathscr{G}}^{(n-1)}(\mathbf{k}-\mathbf{q}, \mathbf{k}', [\mathbf{r}_{N_i}], \omega). \tag{55.126b}$$

Performing a configuration average over $\mathbf{r}_a$ for the first order correction to $\overleftrightarrow{\mathscr{G}}$ yields

$$\overleftrightarrow{\mathscr{G}}^{(1)}(\mathbf{k}, \mathbf{k}', [\mathbf{r}_{N_i}], \omega_m) = \overleftrightarrow{\mathscr{G}}^{(0)}(\mathbf{k}, \omega_m) n_i u(0) \overleftrightarrow{\mathscr{G}}^{(0)}(\mathbf{k}, \omega_m) (2\pi)^3 \delta^{(3)}(\mathbf{k}-\mathbf{k}') \tag{55.127}$$

which corresponds to the thermal Green's function generalization of (55.112) to the case $\Delta \neq 0$. In second order we would have the two contributions corresponding to the generalization of the forms (55.114a) and (55.114b):

$$a \neq b: \quad \overleftrightarrow{\mathscr{G}}^{(0)}(\mathbf{k}, \omega_m) n_i u(0) \overleftrightarrow{\mathscr{G}}^{(0)}(\mathbf{k}, \omega_m) n_i u(0) \overleftrightarrow{\mathscr{G}}^{(0)}(\mathbf{k}', \omega_m) (2\pi)^3 \delta^{(3)}(\mathbf{k}-\mathbf{k}'), \tag{55.128a}$$

$$a = b: \quad \overleftrightarrow{\mathscr{G}}^{(0)}(\mathbf{k}, \omega_m) \left(n_i \int \frac{d^3q}{(2\pi)^3} |u(\mathbf{q})|^2 \overleftrightarrow{\mathscr{G}}^{(0)}(\mathbf{k}-\mathbf{q}, \omega_m) \right] \overleftrightarrow{\mathscr{G}}^{(0)}(\mathbf{k}', \omega_m) (2\pi)^3 \delta^{(3)}(\mathbf{k}-\mathbf{k}'). \tag{55.128b}$$

We next define a matrix $\overleftrightarrow{\Sigma}$ as

$$\overleftrightarrow{\Sigma} = n_i u(0) \overleftrightarrow{1} + \frac{1}{(2\pi)^3} \int d^3q\, |u(\mathbf{q})|^2 \overleftrightarrow{\mathscr{G}}^{(0)}(\mathbf{k}-\mathbf{q}, \omega_m). \tag{55.129}$$

If we neglect higher order terms involving 'nested' and 'intersecting' pairings (i.e., those not involving a serial repetition of (55.128b)), as well as other processes argued earlier to be small for dilute alloys, we may write our expansion of $\overleftrightarrow{\mathscr{G}}$ as

$$\overleftrightarrow{\mathscr{G}}(\mathbf{k}, \omega_m) = \overleftrightarrow{\mathscr{G}}^{(0)}(\mathbf{k}, \omega_m) + \overleftrightarrow{\mathscr{G}}^{0}(\mathbf{k}, \omega_m) \overleftrightarrow{\Sigma}(\omega) \overleftrightarrow{\mathscr{G}}^{(0)}(\mathbf{k}, \omega_m)$$
$$+ \overleftrightarrow{\mathscr{G}}^{(0)}(\mathbf{k}, \omega_m) \overleftrightarrow{\Sigma}(\omega) \overleftrightarrow{\mathscr{G}}^{(0)}(\mathbf{k}, \omega_m) \overleftrightarrow{\Sigma}(\omega) \overleftrightarrow{\mathscr{G}}^{(0)}(\mathbf{k}, \omega_m) + \ldots. \tag{55.130}$$

Factoring out a term $\overleftrightarrow{\Sigma}\overleftrightarrow{\mathscr{G}}^{(0)}$ starting with the second term on the right and again identifying the remaining series as $\overleftrightarrow{\mathscr{G}}$, Eq. (55.130) becomes

$$\overleftrightarrow{\mathscr{G}}(\mathbf{k}, \omega_m) = \overleftrightarrow{\mathscr{G}}^{(0)}(\mathbf{k}, \omega_m) + \overleftrightarrow{\mathscr{G}}^{(0)}(\mathbf{k}, \omega_m)\overleftrightarrow{\Sigma}(\omega)\overleftrightarrow{\mathscr{G}}(\mathbf{k}, \omega_m). \tag{55.131}$$

Multiplying both sides of (55.131) by the matrix

$$(\overleftrightarrow{\mathscr{G}}^{(0)}(\mathbf{k}, \omega_m))^{-1} = \begin{pmatrix} i\hbar\omega_m - \xi_k & \Delta \\ -\Delta^* & -i\hbar\omega_m - \xi_k \end{pmatrix} \tag{55.132}$$

(which we identify as the second matrix on the left-hand side of (55.123a)) we obtain the set of equations

$$\begin{pmatrix} i\hbar\omega_m - \xi_k - \Sigma_{11} & \Delta - \Sigma_{12} \\ -\Delta^* - \Sigma_{21} & -i\hbar\omega_m - \xi_k - \Sigma_{22} \end{pmatrix}\begin{pmatrix} \mathscr{G}(\mathbf{k}, \omega_m) \\ \overline{\mathscr{F}}(\mathbf{k}, \omega_m) \end{pmatrix} = \begin{pmatrix} 1 \\ 0 \end{pmatrix}. \tag{55.133}$$

The term $n_i u(0)\overleftrightarrow{1}$ on the right-hand side of Eq. (55.129) may be absorbed into the definition of the chemical potential as we did earlier. Again defining $\mathbf{k}_1 = \mathbf{k} - \mathbf{q}$, the contributions arising from the second term on the right-hand side of Eq. (55.131) may be written explicitly[16]

$$\Sigma_{11}(\omega_m) = \frac{n_i}{(2\pi)^3}\int \mathrm{d}^3k_1\,|u(\mathbf{k} - \mathbf{k}_1)|^2\mathscr{G}^{(0)}(\mathbf{k}_1, \omega_m), \tag{55.134a}$$

$$\Sigma_{21}(\omega_m) = \frac{n_i}{(2\pi)^3}\int \mathrm{d}^3k_1\,|u(\mathbf{k} - \mathbf{k}_1)|^2\overline{\mathscr{F}}^{(0)}(\mathbf{k}_1, \omega_m), \tag{55.134b}$$

and

$$\Sigma_{22}(\omega_m) = \Sigma_{11}(-\omega_m). \tag{55.134c}$$

The solution of (55.133) for $\mathscr{G}$ and $\overline{\mathscr{F}}$ is

$$\mathscr{G}(\mathbf{k}, \omega_m) = -\frac{i\hbar\omega_m - \Sigma_{11}(\omega_m) + \xi}{|i\hbar\omega_m - \Sigma_{11}(\omega_m)|^2 + \xi^2 + |\Delta^* + \Sigma_{21}(\omega_m)|^2} \tag{55.135a}$$

and

$$\overline{\mathscr{F}}(\mathbf{k}, \omega_m) = \frac{\Delta^* + \Sigma_{21}(\omega_m)}{|i\hbar\omega_m - \Sigma_{11}(\omega_m)|^2 + \xi^2 + |\Delta^* + \Sigma_{21}(\omega_m)|^2}, \tag{55.135b}$$

where we used the fact that $\Sigma_{11}(-\omega_M) = -\Sigma_{11}(\omega_m)$ (see below).

In evaluating Σ_{11}, the second term in the numerator of (55.94) for $\mathscr{G}^{(0)}$ is odd in the vicinity of μ and hence its contribution involves energies far from μ. At large energies ($\xi >> \Delta$), $\mathscr{G}^{(0)}$ approaches the normal-metal form and therefore the resulting contribution becomes the same one encountered earlier in evaluating $G(\mathbf{k}, \omega)$, which we absorbed into the definition of the chemical potential. What remains is the contribution from the $-i\hbar\omega_m$ term and we may then write

$$\Sigma_{11}(\omega_m) = \frac{-i\hbar\omega_m n_i}{(2\pi)^3}\int \mathrm{d}^3k_1 \frac{|u(\mathbf{k} - \mathbf{k}_1)|^2}{\hbar^2\omega_m^2 + \varepsilon_{k_1}^2}. \tag{55.136a}$$

16. More accurate expressions for Σ_{11} and Σ_{21} are obtained by replacing $\mathscr{G}^{(0)}$ and $\overline{\mathscr{F}}^{(0)}$ by $\mathscr{G}$ and $\overline{\mathscr{F}}$ in Eqs. (55.134a) and (55.134b), respectively. However, this does not affect the conclusion which follows.

On the other hand, using (55.95b) in Eq. (55.134b) we have

$$\Sigma_{21}(\omega_m) = \frac{\Delta^* n_i}{(2\pi)^3} \int d^3k_1 \frac{|u(\mathbf{k} - \mathbf{k}_1)|^2}{\hbar^2\omega_m^2 + \varepsilon_{k_1}^2}. \tag{55.136b}$$

We see that Eqs. (55.136) involve the same integral and therefore

$$-\frac{\Sigma_{11}(\omega_m)}{i\hbar\omega_m} = \frac{\Sigma_{21}(\omega_m)}{\Delta^*} \tag{55.137}$$

(from which we obtain the property $\Sigma_{11}(\omega) = -\Sigma_{11}(-\omega)$ used above). Eq. (55.137) allows us to 'rescale' Eqs. (55.135) by writing

$$\tilde{\Delta}^*(\omega_m) = \Delta^* + \Sigma_{21}(\omega_m) = \eta(\omega_m)\Delta^* \tag{55.138a}$$

and

$$i\hbar\tilde{\omega}_m = i\hbar\omega_m - \Sigma_{11}(\omega_m) = \eta(\omega_m) i\hbar\omega_m, \tag{55.138b}$$

where we defined the common factor as

$$\begin{aligned}\eta(\omega_m) &\equiv 1 + \frac{\Sigma_{21}(\omega_m)}{\Delta^*} \\ &= 1 + \frac{n_i}{(2\pi)^3} \int d^3k_1 \frac{|u(\mathbf{k} - \mathbf{k}_1)|^2}{\hbar^2\omega_m^2 + \varepsilon_{k_1}^2}.\end{aligned} \tag{55.138c}$$

To calculate Σ_{21} we will assume $|u(\mathbf{k} - \mathbf{k}_1)|^2$ is slowly varying on the energy scales associated with superconductivity (e.g., $k_B T_c$ or Δ). We then factor the denominator in Eq. (55.136c) as $[\xi_k + i(\hbar^2\omega_m^2 + \Delta^2)^{1/2}][\xi_k - i(\hbar^2\omega_m^2 + \Delta^2)^{1/2}]$; i.e., we have poles at $\xi_k = \pm i(\hbar^2\omega_m^2 + \Delta^2)^{1/2}$. Using Cauchy's theorem and closing the contour with an infinite semicircle in either half plane yields

$$\Sigma_{21} = \frac{\hbar\Delta^*}{2\tau(\hbar^2\omega_m^2 + \Delta^2)^{1/2}}, \tag{55.139}$$

where τ is given by Eq. (55.116a). The common factor occurring in (55.138a,b) is thus

$$\eta(\omega_m) = 1 + \frac{\hbar}{2\tau(\hbar^2\omega_m^2 + \Delta^2)^{1/2}}. \tag{55.140}$$

From the above it is clear that we may obtain the impurity averaged forms for $\mathscr{G}$ and $\mathscr{F}$ by making the substitutions $\omega_m \to \eta(\omega_m)\omega_m$ and $\Delta \to \eta(\omega_m)\Delta$ into the pure metal forms. The resulting gap equation, (55.102), is then

$$\frac{k_B T V}{(2\pi)^3} \sum_{m=-\infty}^{+\infty} \int \frac{\eta d^3k}{\eta^2(\hbar^2\omega_m^2 + \Delta^2) + \xi^2} = 1.$$

If we define a new energy variable by $\xi = \eta\xi'$, which requires we write $d^3k = \eta d^3k'$, this equation becomes identical to the gap equation in the absence of impurities, Eq. (55.102). This is the formal proof of the conclusion reached by Anderson on the basis of a time reversal symmetry argument. The agrument used here follows that of Abrikosov *et al.* (1963).

55.6 G–L theory revisited

In Sec. 45 we obtained the equations of the G–L theory. The treatment involved regarding the gap function as a perturbation in the Bogoliubov equations. We will now sketch how a similar approach can be applied to the Gorkov Green's function equation of motion; in fact this approach is actually easier and is the one used originally by Gorkov (1959). We start with the Fourier Matsubara time transform of (55.85)

$$(i\hbar\omega_m - \mathscr{H}'_1)\mathscr{G}(\mathbf{r}_1,\mathbf{r}_2,\omega_m) + \Delta(\mathbf{r}_1)\bar{\mathscr{F}}(\mathbf{r}_1,\mathbf{r}_2,\omega_m) = \delta^{(3)}(\mathbf{r}_1 - \mathbf{r}_2) \tag{55.141a}$$

and

$$\Delta^*(\mathbf{r}_1)\mathscr{G}(\mathbf{r}_1,\mathbf{r}_2,\omega_m) + (i\hbar\omega_m + \mathscr{H}'_1)\bar{\mathscr{F}}(\mathbf{r}_1,\mathbf{r}_2,\omega_m) = 0. \tag{55.141b}$$

We will denote the Green's function in the normal metal by $\mathscr{G}^{(0)}$ which satisfies the equations

$$(i\hbar\omega_m - \mathscr{H}'_1)\mathscr{G}^{(0)}(\mathbf{r}_1,\mathbf{r}_2,\omega_m) = \delta^{(3)}(\mathbf{r}_1 - \mathbf{r}_2) \tag{55.142a}$$

and

$$(-i\hbar\omega_m - \mathscr{H}'_1)\mathscr{G}^{(0)}(\mathbf{r}_1,\mathbf{r}_2,-\omega_m) = \delta^{(3)}(\mathbf{r}_1 - \mathbf{r}_2). \tag{55.142b}$$

We next multiply (55.141a) by $\mathscr{G}^{(0)}(\mathbf{r}_1,\mathbf{r}_3,\omega_m)$ and (55.141b) by $\mathscr{G}^{(0)}(\mathbf{r}_3,\mathbf{r}_1,-\omega_m)$ and then integrate over d^3r_1. After integrating twice by parts (so that the $\mathscr{H}'_1$ operator acts on $\mathscr{G}^{(0)}$ rather than $\mathscr{G}$ or $\bar{\mathscr{F}}$) and using Eq. (55.142) we obtain (on interchanging the labels on $\mathbf{r}_1$ and $\mathbf{r}_3$)

$$\mathscr{G}(\mathbf{r}_1,\mathbf{r}_2,\omega_m) = \mathscr{G}^{(0)}(\mathbf{r}_1,\mathbf{r}_2,\omega_m) - \int d^3r_3\mathscr{G}^{(0)}(\mathbf{r}_1,\mathbf{r}_3,\omega_m)\Delta(\mathbf{r}_3)\bar{\mathscr{F}}(\mathbf{r}_3,\mathbf{r}_2,\omega_m) \tag{55.143a}$$

and

$$\bar{\mathscr{F}}(\mathbf{r}_1,\mathbf{r}_2,\omega_m) = \int d^3r_3\mathscr{G}^{(0)}(\mathbf{r}_3,\mathbf{r}_1,-\omega_m)\Delta^*(\mathbf{r}_3)\mathscr{G}(\mathbf{r}_3,\mathbf{r}_2,\omega_m). \tag{55.143b}$$

These equations may be solved iteratively in terms of multiple integrals involving the functions $\Delta(\mathbf{r})$ and $\mathscr{G}^{(0)}(\pm\omega_m)$. We write $\Delta \to \lambda\Delta$ and expand our Green's function as $\mathscr{G} = \mathscr{G}^{(0)} + \lambda^2\mathscr{G}^{(2)}$ and $\bar{\mathscr{F}} = \bar{\mathscr{F}}^{(1)} + \lambda^3\bar{\mathscr{F}}^{(3)} + \ldots$; in successive orders of λ we have

$$\lambda^1:\ \bar{\mathscr{F}}^{(1)}(\mathbf{r}_1,\mathbf{r}_2,\omega_m) = \int d^3r_3\mathscr{G}^{(0)}(\mathbf{r}_3,\mathbf{r}_1,-\omega_m)\Delta^*(\mathbf{r}_3)\mathscr{G}^{(0)}(\mathbf{r}_3,\mathbf{r}_2,\omega_m); \tag{55.144a}$$

$$\begin{aligned}\lambda^2:\ \mathscr{G}^{(2)}(\mathbf{r}_1,\mathbf{r}_2,\omega_m) &= -\int d^3r_3\mathscr{G}^{(0)}(\mathbf{r}_1,\mathbf{r}_3,\omega_m)\Delta(\mathbf{r}_3)\bar{\mathscr{F}}^{(0)}(\mathbf{r}_3,\mathbf{r}_2,\omega_m)\\ &= -\int d^3r_3d^3r_4\mathscr{G}^{(0)}(\mathbf{r}_1,\mathbf{r}_3,\omega_m)\Delta(\mathbf{r}_3)\mathscr{G}^{(0)}(\mathbf{r}_4,\mathbf{r}_3,-\omega_m)\Delta^*(\mathbf{r}_4)\mathscr{G}^{(0)}(\mathbf{r}_4,\mathbf{r}_2,\omega_m);\end{aligned} \tag{55.144b}$$

$$\begin{aligned}\lambda^3:\ \bar{\mathscr{F}}^{(3)}(\mathbf{r}_1,\mathbf{r}_2,\omega_m) &= -\int d^3r_3\mathscr{G}^{(0)}(\mathbf{r}_3,\mathbf{r}_1,-\omega_m)\Delta^*(\mathbf{r}_3)\mathscr{G}^{(2)}(\mathbf{r}_3,\mathbf{r}_2,\omega_m)\\ &= -\int d^3r_3d^3r_4d^3r_5\mathscr{G}^{(0)}(\mathbf{r}_3,\mathbf{r}_1,-\omega_m)\Delta^*(\mathbf{r}_3)\mathscr{G}^{(0)}(\mathbf{r}_3,\mathbf{r}_4,\omega_m)\Delta(\mathbf{r}_4)\mathscr{G}^{(0)}(\mathbf{r}_5,\mathbf{r}_4,-\omega_m)\\ &\quad\times\Delta^*(\mathbf{r}_5)\mathscr{G}^{(0)}(\mathbf{r}_5,\mathbf{r}_2,\omega_m).\end{aligned} \tag{55.144c}$$

Recalling Eq. (55.101) for $\Delta^*(\mathbf{r})$ and retaining terms through third order in $\mathscr{F}$ we have

$$\Delta^*(\mathbf{r}) = k_B TV \sum_{m=-\infty}^{+\infty} \Bigg[\int d^3 r_1 \mathscr{G}^{(0)}(\mathbf{r}_1, \mathbf{r}, -\omega_m)\Delta^*(\mathbf{r}_1)\mathscr{G}^{(0)}(\mathbf{r}_1, \mathbf{r}, \omega_m)$$
$$- \int d^3 r_1 d^3 r_2 d^3 r_3 \mathscr{G}^{(0)}(\mathbf{r}_1, \mathbf{r}, -\omega_m)\Delta^*(\mathbf{r}_1)\mathscr{G}^{(0)}(\mathbf{r}_1, \mathbf{r}_2, \omega_m)\Delta(\mathbf{r}_2)\mathscr{G}^{(0)}(\mathbf{r}_3, \mathbf{r}_2, -\omega_m)$$
$$\times \Delta^*(\mathbf{r}_3)\mathscr{G}^{(0)}(\mathbf{r}_3, \mathbf{r}, \omega_m) \Bigg], \tag{55.145}$$

where we have again relabeled the variables.

The linearized theory is a special case of (55.145) where we neglect the term involving $\Delta^* |\Delta|^2$. It is immediately apparent that the complex conjugate of the kernel $K(\mathbf{r}, \mathbf{r}')$ introduced in Eq. (38.6) is given by

$$K^*(\mathbf{r}, \mathbf{r}') = k_B TV \sum_m \mathscr{G}^{(0)}(\mathbf{r}', \mathbf{r}, -\omega_m)\mathscr{G}^{(0)}(\mathbf{r}', \mathbf{r}, \omega_m). \tag{55.146}$$

The clean limit form of $\mathscr{G}^{(0)}$ follows from the $\Delta = 0$ limit of Eq. (55.94) as

$$\mathscr{G}^{(0)}(\mathbf{k}, \omega_m) = \frac{1}{i\hbar\omega_m - \xi_k}. \tag{55.147}$$

Performing a Fourier transform of $\mathscr{G}^{(0)}(\mathbf{k}, \omega_m)$ and $\mathscr{G}^{(0)}(\mathbf{k}', -\omega_m)$ to form $\mathscr{G}^{(0)}(\mathbf{r}, \omega_m)\mathscr{G}^{(0)}(\mathbf{r}, -\omega_m)$ in Eq. (55.146) leads us immediately to Eq. (45.5) from which the clean-limit, linearized, G–L theory follows (including the effects of a vector potential, $A(\mathbf{r})$), as discussed in Subsec. 45.2. If we ignore the position dependence of $\Delta^*(\mathbf{r})$ and $\Delta(\mathbf{r})$ in the last term in (55.145) and take them from under the integral sign (which is justified due to the fact that $\mathscr{G}^{(0)}(\mathbf{r}, \omega_m)$ falls off as $e^{-|\omega_m|r/v_F}$) we may carry out the remaining integration to obtain the coefficient of the $\Delta^* |\Delta|^2$ term. Lastly, the effects of impurities may be added to using the $\Delta = 0$ form of $\mathscr{G}(0)$ for a dirty superconductor to evaluate the linear term in Δ^*. Thus $\mathscr{G}^{(0)}$ follows from Eq. (55.135a) as

$$\mathscr{G}^{(0)}(\mathbf{k}, \omega_m) = \frac{1}{i\hbar\omega_m - \xi_k - \Sigma_{11}(\omega_m)}, \tag{55.148}$$

where $\Sigma_{11}(\omega_m)$ follows from (55.139) and (55.137) as

$$\Sigma_{11}(\omega_m) = -\frac{i\hbar^2\omega_m}{2\tau(\hbar^2\omega_m^2 + \Delta^2)^{1/2}} \tag{55.149}$$

with τ given by (55.116a). For details, including the derivation of the second G–L equation, the reader is referred to Gorkov (1959).

55.7 The Usadel equations

Much of Part III of the book has been devoted to discussing the properties of an inhomogeneous superconductor in the immediate vicinity of the transition temperature via the linearized self-consistency condition of deGennes. Our discussions have primarily emphasized dirty superconductors. Clean superconductors always have a highly-anisotropic and material-specific Fermi surface. (Only the alkali metals have nearly spherical Fermi surfaces and they are not superconducting.) In a dirty superconductor electron scattering tends to average out the underlying electronic anisotropy and this fact permits the construction of a more universal description (e.g., in terms of a diffusion constant, transition temperature, and a few other parameters).

The Nambu–Gorkov Green's function formalism is very powerful and is, in principle, capable of describing an inhomogeneous superconductor at all temperatures. The fact that the Green's functions depend on two space-time points makes the application of the Green's function method difficult in practice. The quasiclassical Green's function formalisms, either the generalized Boltzmann equations or the Eilenberger equations, are easier to work with in practice since they have less variables ($(\mathbf{r}, \mathbf{k}, t)$ or $(\mathbf{r}, \hat{\mathbf{k}}, \omega, t)$ for the two formalisms, respectively). For the case of a dirty superconductor, the quasiclassical formalism may be further simplified by observing that in the dirty limit the quasiclassical Green's functions are nearly isotropic. A description in which these functions are expanded in spherical harmonics and only the $\ell = 0$ and $\ell = 1$ terms are retained is adequate for most situations and the resulting equations, first obtained by Usadel (1970), are simpler to work with.

We start with the Eilenberger equations (Eqs. (55.69a)–(55.69c)) in the Matsubara representation. Onto these equations we will 'graft' the effects of impurity scattering;[17] however, the additional terms follow naturally from the procedures of Abrikosov and Gorkov for configuration averaging the effects of the impurity potential. Because of the normalization condition, (55.77b), we only require the equations for $f(\mathbf{r}, \hat{\mathbf{k}}, \omega_m)$ and $\bar{f}(\mathbf{r}, \hat{\mathbf{k}}, \omega_m)$ which are

$$\left\{2\hbar\omega_m + \hbar\mathbf{v}_F \cdot \left[\nabla - \frac{2ie}{\hbar c}\mathbf{A}(\mathbf{r})\right]\right\} f(\mathbf{r}, \hat{\mathbf{k}}, \omega_m) = 2\Delta(\mathbf{r}) g(\mathbf{r}, \hat{\mathbf{k}}, \omega_m)$$

$$+ \frac{n_i}{(2\pi)^2\hbar}\int \frac{d^2k'}{v'_F} |u(\hat{\mathbf{k}}, \hat{\mathbf{k}}')|^2 [g(\mathbf{r}, \hat{\mathbf{k}}, \omega_m) f(\mathbf{r}, \hat{\mathbf{k}}', \omega_m) - f(\mathbf{r}, \hat{\mathbf{k}}, \omega_m) g(\mathbf{r}, \hat{\mathbf{k}}', \omega_m)] \quad (55.150a)$$

and

$$2\hbar\omega_m \bar{f}(\mathbf{r}, \hat{\mathbf{k}}, \omega_m) - \hbar\mathbf{v}_F \cdot \left[\nabla + \frac{2ie}{\hbar c}\mathbf{A}(\mathbf{r})\right] \bar{f}(\mathbf{r}, \hat{\mathbf{k}}, \omega_m) = 2\Delta^*(\mathbf{r}) g(\mathbf{r}, \hat{\mathbf{k}}, \omega_m)$$

$$+ \frac{n_i}{(2\pi)^2\hbar}\int \frac{d^2k'}{v'_F} |u(\hat{\mathbf{k}}, \hat{\mathbf{k}}')|^2 [g(\mathbf{r}, \hat{\mathbf{k}}, \omega_m) \bar{f}(\mathbf{r}, \hat{\mathbf{k}}', \omega_m) - \bar{f}(\mathbf{r}, \hat{\mathbf{k}}, \omega_m) g(\mathbf{r}, \hat{\mathbf{k}}', \omega_m)]; \quad (55.150b)$$

here f, $\bar{f}$ and g are the energy integrated forms of $\mathscr{F}$, $\bar{\mathscr{F}}$ and $\mathscr{G}$, and $\hat{\mathbf{k}}$ and $\hat{\mathbf{k}}'$ denote vector directions on the Fermi surface.[18] This equation must be supplemented by the normalization condition (55.77b) which we rewrite as

$$g(\mathbf{r}, \hat{\mathbf{k}}, \omega_m) = [1 - |f(\mathbf{r}, \hat{\mathbf{k}}, \omega_m)|^2]^{1/2} \quad (55.151)$$

17. The Eilenberger equations were originally written in the Matsubara representation and the effects of impurity scattering were included.
18. It can be seen that the structure of this equation is basically equivalent to the equation of motion (55.69b) to which we have added the correction due to the configuration-averaged impurity potential, $\vec{\Sigma}\vec{\mathscr{G}}$ (see Eq. (55.133)) which we then integrate over the energy variable. However, the integration over $d\Omega'$ cannot be carried out at this point due to the nontrivial dependence of Green's function on $\hat{\mathbf{k}}'$ in the nonequilibrium case.

as well as the self-consistency conditions for the gap function, (55.101), and the Matsubara form for the current density, (55.65):

$$\Delta(\mathbf{r})\ell n\left(\frac{T}{T_c}\right) + 2\pi k_B T \sum_{m>0}\left[\frac{\Delta(\mathbf{r})}{\hbar\omega_m} - \int\frac{d\Omega_{\hat{k}}}{4\pi}f(\mathbf{r},\hat{\mathbf{k}},\omega_m)\right] = 0, \qquad (55.152)$$

$$\mathbf{j}(\mathbf{r}) = 2ie\mathcal{N}(0)2\pi k_B T\sum_{m>0}\frac{d\Omega_{\hat{k}}}{4\pi}\mathbf{v}g(\mathbf{r},\hat{\mathbf{k}},\omega_m); \qquad (55.153)$$

here we removed the large ω_m divergence of the sum on $f(\omega_m)$, as we did earlier in Sec. 40, by using the BCS expression

$$\mathcal{N}(0)V\ell n\frac{1.14\hbar\omega_D}{k_B T_c} = 1$$

and adding and subtracting our prescription for cutting off the divergent sum

$$2\pi k_B T\sum_{m>0}\frac{1}{\hbar\omega_m} = \ell n\left(\frac{1.14\hbar\omega_D}{k_B T}\right).$$

As discussed above we will approximate the angular dependence of $f(\mathbf{r},\hat{\mathbf{k}},\omega_m)$ by the $\ell = 0$ and $\ell = 1$ terms in an expansion in spherical harmonics which we write as

$$f(\mathbf{r},\hat{\mathbf{k}},\omega_m) = f_0(\mathbf{r},\omega_m) + \hat{\mathbf{k}}\cdot\mathbf{f}(\mathbf{r},\omega_m) + \ldots, \qquad (55.154)$$

where f_0 and $\mathbf{f}$ are scalar and vector functions, respectively, and $\hat{\mathbf{k}}$ is short hand for the three $\ell = 1$ spherical harmonics

$$\hat{k}_x = \sin\theta\cos\phi; \quad \hat{k}_y = \sin\theta\sin\phi; \quad \hat{k}_z = \cos\theta; \qquad (55.155)$$

we expand the function g in a similar manner:

$$g(\mathbf{r},\mathbf{k},\omega_m) = g_0(\mathbf{r},\omega_m) + \hat{\mathbf{k}}\cdot\mathbf{g}(\mathbf{r},\omega_m) + \ldots \qquad (55.156)$$

If the expansions (55.154) and (55.156) are rapidly convergent then we may assume $g_0 >> \hat{\mathbf{k}}\cdot\mathbf{g}$ and $f_0 >> \hat{\mathbf{k}}\cdot\mathbf{f}$. If we insert (55.154) and (55.156) into the normalization condition (55.151), we have to order $\ell = 1$ in our expansion,

$$g_0^2 + 2g_0\hat{\mathbf{k}}\cdot\mathbf{g} = 1 - |f_0|^2 - f_0\hat{\mathbf{k}}\cdot\mathbf{f}^* - f_0^*\hat{\mathbf{k}}\cdot\mathbf{f}$$

from which it follows that

$$g_0^2(\mathbf{r},\omega_m) = 1 - |f_0(\mathbf{r},\omega_m)|^2 \qquad (55.157a)$$

and

$$\mathbf{g}(\mathbf{r},\omega_m) = -\frac{f_0^*(\mathbf{r},\omega_m)\mathbf{f}(\mathbf{r},\omega_m) + f_0(\mathbf{r},\omega_m)\mathbf{f}^*(\mathbf{r},\omega_m)}{2g_0(\mathbf{r},\omega_m)}. \qquad (55.157b)$$

If we insert the expansions (55.154) and (55.156) into Eq. (55.150) we obtain in our approximation

$$\left\{2\hbar\omega_m + \hbar\mathbf{v}_F\cdot\left[\nabla - \frac{2ie}{\hbar c}\mathbf{A}(\mathbf{r})\right]\right\}[f_0(\mathbf{r},\omega_m) + \hat{\mathbf{k}}\cdot\mathbf{f}(\mathbf{r},\omega_m)] = 2\Delta(\mathbf{r})[g_0(\mathbf{r},\omega_m) + \hat{\mathbf{k}}\cdot\mathbf{g}(\mathbf{r},\omega_m)]$$

$$+\frac{n_i}{(2\pi)^2\hbar}\int\frac{d^3k'}{v_F'}|u(\hat{\mathbf{k}},\hat{\mathbf{k}}')|^2[\mathscr{g}(\mathbf{r},\omega_m)\mathscr{f}_0(\mathbf{r},\omega_m)-\mathscr{f}(\mathbf{r},\omega_m)\mathscr{g}_0(\mathbf{r},\omega_m)]\cdot(\hat{\mathbf{k}}-\hat{\mathbf{k}}'). \quad (55.158)$$

Integrating (55.158) over all solid angles yields[19]

$$2\hbar\omega_m\mathscr{f}_0(\mathbf{r},\omega_m)+\frac{\hbar}{3}v_F\left[\nabla-\frac{2ie}{\hbar c}\mathbf{A}(\mathbf{r})\right]\cdot\mathscr{f}(\mathbf{r},\omega_m)=2\Delta(\mathbf{r})\mathscr{g}_0(\mathbf{r},\omega_m). \quad (55.159a)$$

If we first multiply Eq. (55.150) by $\hat{\mathbf{k}}$ and again integrate over all angles we get

$$2\hbar\omega\mathscr{f}(\mathbf{r},\omega_m)+\hbar v_F\left[\nabla-\frac{2ie}{\hbar c}\mathbf{A}(\mathbf{r})\right]\mathscr{f}_0(\mathbf{r},\omega_m)=2\Delta(\mathbf{r})\mathscr{g}_0(\mathbf{r},\omega_m)$$

$$+\frac{\hbar}{\tau_{tr}}[\mathscr{g}(\mathbf{r},\omega_m)\mathscr{f}_0(\mathbf{r},\omega_m)-\mathscr{f}(\mathbf{r},\omega_m)\mathscr{g}_0(\mathbf{r},\omega_m)]; \quad (55.159b)$$

here τ_{tr} is the transport scattering time and is given by

$$\frac{1}{\tau_{tr}}\equiv\frac{n_i}{(2\pi)^2\hbar^2}\int\frac{d\Omega}{4\pi}\int\frac{d^2k'}{v_F'}|u(\hat{\mathbf{k}},\hat{\mathbf{k}}')|^2\hat{k}_i(\hat{k}_i-\hat{k}_i'), \quad (55.160)$$

where i denotes the x, y, or z component of $\hat{\mathbf{k}}$. For a free electron metal the expression becomes

$$\frac{1}{\tau_{tr}}=\frac{n_imk_F}{(2\pi)^2\hbar^3}\int|u(\theta)|^2(1-\cos\theta)d\Omega. \quad (55.161)$$

Eqs. (55.157) and (55.159) constitute a set of four equations in the four variables $\mathscr{f}_0(\mathbf{r},\omega_m)$, $\mathscr{f}(\mathbf{r},\omega_m)$, $\mathscr{g}_0(\mathbf{r},\omega_m)$, and $\mathscr{g}(\mathbf{r},\omega_m)$. We can reduce this to two equations by eliminating the vector functions $\mathscr{f}(\mathbf{r},\omega_m)$ and $\mathscr{g}(\mathbf{r},\omega_m)$. We will do this for the extreme dirty limit in which the following approximations are made:

$$\omega_m\tau_{tr}<<1 \quad (55.162a)$$

and

$$\frac{\Delta(\mathbf{r})\tau_{tr}}{\hbar}<<1. \quad (55.162b)$$

Eq. (55.159b) then has the approximate solution

$$-\mathscr{f}(\mathbf{r},\omega_m)=+\frac{v_F\tau_{tr}}{\mathscr{g}_0(\mathbf{r},\omega_m)}\left[\nabla-\frac{2ie}{\hbar c}\mathbf{A}(\mathbf{r})\right]\mathscr{f}_0(\mathbf{r},\omega_m)$$

$$-\frac{\mathscr{f}_0(\mathbf{r},\omega_m)}{\mathscr{g}_0(\mathbf{r},\omega_m)}\mathscr{g}(\mathbf{r},\omega_m), \quad (55.163a)$$

while its adjoint is given by

$$-\bar{\mathscr{f}}(\mathbf{r},\omega_m)=-\frac{v_F\tau_{tr}}{\mathscr{g}_0(\mathbf{r},\omega_m)}\left[\nabla+\frac{2ie}{\hbar c}\mathbf{A}(\mathbf{r})\right]\bar{\mathscr{f}}_0(\mathbf{r},\omega_m)$$

19. Here we will assume $\mathbf{v}_F$ is parallel to $\hat{\mathbf{k}}$ which is true only for a spherical Fermi surface. However, our final expressions will contain phenomenological parameters into which it is assumed we may absorb Fermi surface anisotropy effects.

$$-\frac{\bar{f}_0(\mathbf{r},\omega_m)}{g_0(\mathbf{r},\omega_m)}g(\mathbf{r},\omega_m). \tag{55.163b}$$

Multiplying (55.163a) by $\bar{f}_0$ and (55.163b) by f_0, adding, and dividing by $2g_0$ yields

$$g(\mathbf{r},\omega_m)=\frac{v_F\tau_{tr}}{2g_0^2(\mathbf{r},\omega_m)}\Big[\bar{f}_0(\mathbf{r},\omega_m)\nabla f_0(\mathbf{r},\omega_m)-f_0(\mathbf{r},\omega_m)\nabla\bar{f}_0(\mathbf{r},\omega_m)$$
$$-\frac{4ie}{\hbar c}\mathbf{A}(\mathbf{r})|f_0(\mathbf{r},\omega_m)|^2\Big]-\frac{|f_0(\mathbf{r},\omega_m)|^2}{g_0^2(\mathbf{r},\omega_m)}g(\mathbf{r},\omega_m), \tag{55.164}$$

where we wrote the left-hand side as g using Eq. (55.157b). Combining the terms in g and using Eq. (55.157a) gives

$$g(\mathbf{r},\omega_m)=\frac{v_F\tau_{tr}}{2}\Big[\bar{f}_0(\mathbf{r},\omega_m)\nabla f_0(\mathbf{r},\omega_m)-f_0(\mathbf{r},\omega_m)\nabla\bar{f}_0(\mathbf{r},\omega_m)$$
$$-\frac{4ie}{\hbar c}\mathbf{A}(\mathbf{r})|f_0(\mathbf{r},\omega_m)|^2\Big]. \tag{55.165}$$

Inserting (55.165) into (55.163), using (55.157a), and collecting terms gives

$$f(\mathbf{r},\omega_m)=-\ v_F\tau_{tr}\Big\{g_0(\mathbf{r},\omega_m)\Big[\nabla-\frac{2ie}{\hbar c}\mathbf{A}(\mathbf{r})\Big]f_0(\mathbf{r},\omega_m)$$
$$+\frac{f_0(\mathbf{r},\omega_m)}{2g_0(\mathbf{r},\omega_m)}\nabla|f_0(\mathbf{r},\omega_m)|^2\Big\}. \tag{55.166}$$

Inserting (55.166) into Eq. (55.159a) we obtain an equation involving only the quantities f_0 and g_0:

$$2\hbar\omega_m f_0(\mathbf{r},\omega_m)-\hbar D\Big[\nabla-\frac{2ie}{\hbar c}\mathbf{A}(\mathbf{r})\Big]\cdot\Big\{g_0(\mathbf{r},\omega_m)\Big[\nabla-\frac{2ie}{\hbar c}\mathbf{A}(\mathbf{r})\Big]f_0(\mathbf{r},\omega_m)$$
$$+\frac{1}{2}\frac{f_0(\mathbf{r},\omega_m)}{g_0(\mathbf{r},\omega_m)}\nabla|f_0(\mathbf{r},\omega_m)|^2\Big\}=2\Delta(\mathbf{r})g_0(\mathbf{r},\omega_m), \tag{55.167}$$

where we have defined a diffusion coefficient $D=\frac{1}{3}\tau_{tr}v_F^2$. Eq. (55.167) is the Usadel equation (Usadel 1970). To complete the formalism we require (55.157a)[20] to relate g_0 to f_0 and the self-consistency conditions which now take the form

$$\Delta(\mathbf{r})\ln\Big(\frac{T}{T_c}\Big)+2\pi k_B T\sum_{m>0}\Big[\frac{\Delta(\mathbf{r})}{\hbar\omega_m}-f_0(\mathbf{r},\omega_m)\Big]=0 \tag{55.168}$$

and

$$\mathbf{j}(\mathbf{r})=2ie\mathcal{N}_0\pi k_B TD\sum_{m>0}\Big\{\bar{f}_0(\mathbf{r},\omega_m)\Big[\nabla-\frac{2ie}{\hbar c}\mathbf{A}(\mathbf{r})\Big]f_0(\mathbf{r},\omega_m)$$
$$-f_0(\mathbf{r},\omega_m)\Big[\nabla-\frac{2ie}{\hbar c}\mathbf{A}(\mathbf{r})\Big]\bar{f}_0(\mathbf{r},\omega_m)\Big\}. \tag{55.169}$$

20. For cases where f_0 may be chosen to be real it is customary to write $f_0=\sin\theta$ and $g_0=\cos\theta$ which automatically satisfies Eq. (55.157a).

Near T_c, where $\mathscr{f}_0 \to 0$ and $\mathscr{g} \to 1$, we obtain a linearized form of (55.167) as

$$\left\{2\hbar\omega_m - \hbar D\left[\nabla - \frac{2ie}{\hbar c}\mathbf{A}(\mathbf{r})\right]^2\right\}\mathscr{f}_0(\mathbf{r}, \omega_m) = 2\Delta(\mathbf{r}). \tag{55.170}$$

We can now make contact with deGennes linearized gap equation. Disregarding **A** and Fourier transforming in **r** we obtain

$$\mathscr{f}_0(\mathbf{q}, \omega_m) = \frac{2}{\hbar}\frac{\Delta(\mathbf{q})}{2\omega_m + Dq^2}; \tag{55.171}$$

when this is inserted in the gap equation (55.168) the latter may be rewritten as

$$\Delta(\mathbf{q}) = \frac{2\pi}{\hbar}k_B T\mathscr{N}_0 V\sum_m \frac{\Delta(\mathbf{q})}{2\omega_m + Dq^2}, \tag{55.172}$$

where we absorbed a factor 2 by summing over positive and negative ω_m. We therefore recover the deGennes–Gorkov kernel

$$K(\mathbf{q}) = \frac{2\pi}{\hbar}k_B T\mathscr{N}_0 V\sum_m \frac{1}{2\omega_m + Dq^2} \tag{55.173}$$

which is identical to Eq. (40.2) obtained earlier by more elementary means. Eq. (55.167) is now seen to be a generalization of the deGennes theory of a dirty superconductor to all temperatures. Applications generally require numerical solutions. The theory may be generalized to include the effects of paramagnetic impurities and spin–orbit coupling; however, we will not take up these topics here. A rigorous discussion of boundary conditions has not been carried out to date; it is generally assumed that the deGennes boundary conditions (derived in the linear regime, see Sec. 42) are a reasonable choice for all temperatures.

Since much of this book involved the development and application of the linearized gap equation, it is perhaps appropriate to end here.

Appendix A

Identical particles and spin: the occupation number representation

A.1 The symmetry of many-particle wavefunctions

We must first discuss what the word identical means in connection with a system of particles obeying quantum mechanics. In classical mechanics we could imagine that two particles have identical physical properties but that by observing them continuously we could *distinguish* them at all future times. However, in quantum mechanics the act of observing adds uncertainty, and therefore an experiment which would distinguish them at all future times is not possible.

A way to build the property of indistinguishability formally into the quantum theory of identical particles is to impose a permutation (or particle interchange) symmetry on the associated many-particle wavefunction. We introduce a permutation operator, $\hat{P}$, which for a two-electron wavefunction has the property

$$\hat{P}\psi(\mathbf{r}_1, \mathbf{r}_2) = \psi(\mathbf{r}_2, \mathbf{r}_1). \tag{A.1}$$

We impose the physically reasonable condition that if the particles are identical, the probability density must be unchanged by this operation and therefore

$$|\psi(\mathbf{r}_1, \mathbf{r}_2)|^2 = |\psi(\mathbf{r}_2, \mathbf{r}_1)|^2. \tag{A.2}$$

From this requirement it follows that the only effect the permutation operation can have is to alter the phase of the original wavefunction; hence

$$\hat{P}\psi(\mathbf{r}_1, \mathbf{r}_2) = \psi(\mathbf{r}_2, \mathbf{r}_1) = e^{i\phi_P}\psi(\mathbf{r}_1, \mathbf{r}_2). \tag{A.3}$$

Applying the permutation operator twice yields

$$\hat{P}^2\psi(\mathbf{r}_1, \mathbf{r}_2) = \hat{P}e^{i\phi_P}\psi(\mathbf{r}_1, \mathbf{r}_2) = e^{2i\phi_P}\psi(\mathbf{r}_1, \mathbf{r}_2). \tag{A.4}$$

But from (A.1) two permutations completely restore the original wavefunction;[1] thus $e^{2i\phi_P} = 1$, $2\phi_P = 0, 2\pi$, $\phi_P = 0, \pi$ and therefore

$$\hat{P}\psi(\mathbf{r}_1, \mathbf{r}_2) = \psi(\mathbf{r}_2, \mathbf{r}_1) = \begin{cases} +\,\psi(\mathbf{r}_1, \mathbf{r}_2) & \text{symmetric case,} \\ -\,\psi(\mathbf{r}_1, \mathbf{r}_2) & \text{antisymmetric case.} \end{cases} \tag{A.5}$$

1. The argument given here follows that of Landau and Lifshitz (1977).

Experimentally, the symmetric and antisymmetric cases are found to apply to particles with integer ($s = 0, 1, 2, \ldots$) or half inter ($s = 1/2, 3/2, \ldots$) spin, respectively; the two kinds of particles are called Bose particles (bosons) and Fermi particles (fermions).[2]

If our two particles are weakly interacting we may seek an approximate description by writing our two-electron wavefunction as a product of two one-electron wavefunctions. In order to satisfy (A.5) these 'product' wavefunctions must have the forms

$$\psi_B(\mathbf{r}_1, \mathbf{r}_2) = \frac{1}{\sqrt{2}}[\psi_n(\mathbf{r}_1)\psi_m(\mathbf{r}_2) + \psi_n(\mathbf{r}_2)\psi_m(\mathbf{r}_1)] \tag{A.6}$$

and

$$\psi_F(\mathbf{r}_1, \mathbf{r}_2) = \frac{1}{\sqrt{2}}[\psi_n(\mathbf{r}_1)\psi_m(\mathbf{r}_2) - \psi_n(\mathbf{r}_2)\psi_m(\mathbf{r}_1)], \tag{A.7}$$

where the subscripts B and F identify the Bose and Fermi cases, the factor $(\sqrt{2})^{-1}$ insures normalization, and the subscripts n and m denote the (full set of) quantum numbers of the two one-electron states used to construct the product state. For the case $n = m$, (A.7) vanishes identically; this property is called the *Pauli exclusion principle*, which states that no two fermions may have the same quantum numbers

In constructing (A.6) and (A.7) we did not explicitly write the spin degrees of freedom. Formally we may do this by writing $\mathbf{r}_i, s_i$ in place of $\mathbf{r}_i$: $\psi_n(\mathbf{r}_i) \to \psi_n(\mathbf{r}_i, s_i)$. If we neglect any interactions which couple the translational (also called orbital) degrees of freedom with the spin degrees of freedom, we may write the total wave function as a product of a translational part and a spin part. For the Fermi case we write this as

$$\psi_F = \psi(\mathbf{r}_1, \mathbf{r}_2)\chi(s_1, s_2). \tag{A.8}$$

We require the total wavefunction, ψ_F, to be antisymmetric under the interchange of all coordinates defining the two electrons, $(\mathbf{r}_1, s_1)$ and $(\mathbf{r}_2, s_2)$. If we quantize the total spin along the z axis there are two possibilities: the singlet case, where ψ is symmetric and χ is antisymmetric, and the triplet case, where ψ is antisymmetric and χ is symmetric. We write these two possibilities explicitly below

$$\psi^{\text{singlet}} = \frac{1}{\sqrt{2}}[\psi_n(\mathbf{r}_1)\psi_m(\mathbf{r}_2) + \psi_n(\mathbf{r}_2)\psi_m(\mathbf{r}_1)]\frac{1}{\sqrt{2}}\left[\begin{pmatrix}1\\0\end{pmatrix}^{(1)}\begin{pmatrix}0\\1\end{pmatrix}^{(2)} - \begin{pmatrix}0\\1\end{pmatrix}^{(1)}\begin{pmatrix}1\\0\end{pmatrix}^{(2)}\right] \tag{A.9}$$

and

2. The permutation symmetry has a profound effect on the statistical properties of ensembles of particles, particularly when the interaction is weak and the temperature is low. The associated statistics are called Bose–Einstein and Fermi–Dirac and were discovered before the connection with wavefunction permutation symmetry was identified.

$$\psi^{\text{triplet}} = \frac{1}{\sqrt{2}}[\psi_n(\mathbf{r}_1)\psi_m(\mathbf{r}_2) - \psi_n(\mathbf{r}_2)\psi_m(\mathbf{r}_1)]\begin{cases} \begin{pmatrix}1\\0\end{pmatrix}^{(1)}\begin{pmatrix}1\\0\end{pmatrix}^{(2)} \\ \frac{1}{\sqrt{2}}\left[\begin{pmatrix}1\\0\end{pmatrix}^{(1)}\begin{pmatrix}0\\1\end{pmatrix}^{(2)} + \begin{pmatrix}0\\1\end{pmatrix}^{(1)}\begin{pmatrix}1\\0\end{pmatrix}^{(2)}\right]. \\ \begin{pmatrix}0\\1\end{pmatrix}^{(1)}\begin{pmatrix}0\\1\end{pmatrix}^{(2)} \end{cases} \tag{A.10}$$

We must now extend our discussion to the N-particle case. Let $\psi_1, \psi_2, \ldots$ etc. number the possible states that the particles *may* occupy. Let $p_1, p_2, \ldots p_N$ be the numbers of states *actually* occupied by the N particles. For example, we might have a string of numbers 6, 8, 8, 10. This means we have four particles, with the first in state 6, the second and third in state 8, and the fourth in state 10. The wavefunction of a system of *noninteracting* particles is made up of products of the form $\psi_{p_1}(\mathbf{r}_1, s_1)\psi_{p_2}(\mathbf{r}_2, s_2)\psi_{p_3}(\mathbf{r}_3, s_3)\ldots\psi_{pN}(\mathbf{r}_N, s_N)$. The interchange of any two particles i, j is denoted by a permutation operator $\hat{P}_{ij}$; thus[3]

$$\hat{P}_{12}[\psi_{p_1}(\mathbf{r}_1, s_1)\psi_{p_2}(\mathbf{r}_2, s_2)\psi_{p_3}(\mathbf{r}_3, s_3)\ldots\psi_{pN}(\mathbf{r}_N, s_N)]$$
$$= \psi_{p_1}(\mathbf{r}_2, s_2)\psi_{p_2}(\mathbf{r}_1, s_1)\psi_{p_3}(\mathbf{r}_3, s_3)\ldots\psi_{pN}(\mathbf{r}_N, s_N).$$

For the Bose case we may write the total symmetrized and normalized wave function as:

$$\Psi_B(\mathbf{r}_1, s_1 \ldots \mathbf{r}_N, s_N) = \left[\frac{N_1!N_2!\ldots}{N!}\right]^{1/2} \sum_{i>j} \hat{P}_{ij}[\psi_{p_1}(\mathbf{r}_1, s_1)\psi_{p_2}(\mathbf{r}_2, s_2)\ldots\psi_{p_N}(\mathbf{r}_n, s_n)], \tag{A.11}$$

where the N_ℓ are the number of particles in each of the basis states (which may be 0 since $0! \equiv 1$).

The Fermi case may be conveniently written in the form

$$\Psi_F(\mathbf{r}_1, s_1 \ldots \mathbf{r}_N, s_N) = \frac{1}{(N!)^{1/2}}\begin{vmatrix} \psi_{p_1}(\mathbf{r}_1, s_1) & \psi_{p_1}(\mathbf{r}_2, s_2)\ldots\psi_{p_1}(\mathbf{r}_N, s_N) \\ \psi_{p_2}(\mathbf{r}_1, s_1) & \psi_{p_2}(\mathbf{r}_2, s_2)\ldots\psi_{p_2}(\mathbf{r}_N, s_N) \\ \vdots & \\ \psi_{p_N}(\mathbf{r}_1, s_1) & \psi_{p_N}(\mathbf{r}_2, s_2)\ldots\psi_{p_N}(\mathbf{r}_N, s_N) \end{vmatrix}. \tag{A.12}$$

This form of wavefunction is referred to as a *Slater determinant*. It is automatically antisymmetric and vanishes if any two p_i are the same, hence satisfying of the Pauli principle.

In the presence of spin it is common to write the basis functions as a product of **r**-space, $\psi(\mathbf{r}_i)$, and spin-space, $\chi(s_i)$, states; i.e.,

$$\psi_n(\mathbf{r}_1, s_i) \rightarrow \psi_n(\mathbf{r}_1)\chi_\nu(s_i).$$

The basis functions are now designated by two quantum numbers (n, ν) and the p_i now refer to occupied pairs.

3. In our example above the wavefunction is $\psi = \psi_6(\mathbf{r}_1, s_1)\,\psi_8(\mathbf{r}_2, s_2)\,\psi_8(\mathbf{r}_3, s_3)\,\psi_{10}(\mathbf{r}_4, s_4)$. Then $\hat{P}_{12}\Psi = \psi_6(\mathbf{r}_2, s_2)\,\psi_8(\mathbf{r}_1, s_1)\,\psi_8(\mathbf{r}_3, s_3)\,\psi_{10}(\mathbf{r}_4, s_4)$.

A.2 Occupation number representation: Bose statistics

Having described the general form of the wavefunction given in Eqs. (A.11) and (A.12) for a system of noninteracting Bose and Fermi particles, one could go on to consider the effect of interactions among the particles using perturbation theory or some other scheme. Such calculations are said to be performed in the 'coordinate representation'. However, most many-body calculations in the literature are carried out using a different representation called the 'occupation number' representation or the 'second quantized' representation (the authors prefer the first terminology). The occupation number representation is used extensively because:

(i) it automatically *guarantees* that the relevant states have, at all steps in a calculation, the appropriate symmetry (Bose/symmetric or Fermi/antisymmetric); and
(ii) it allows a convenient description for the appearance or disappearance of excitations (e.g., photon or phonon emission or absorption).

To start our discussion consider a quantum mechanical operator that involves only the coordinates of the particles individually. An example is the total dipole moment operator for a collection of charged particles interacting with an external field: $e\Sigma_{a=1}^{N}(\mathbf{E}\cdot\mathbf{r}_a)$; we call this a *single-particle operator* and write it formally as[4]

$$\hat{F}^{(1)} = \sum_{a=1}^{N} \hat{f}_a^{(1)}. \tag{A.13}$$

Here $\hat{F}^{(1)}$ is the operator for all N particles and $\hat{f}^{(1)}$ operates on the coordinates of the individual particles. Consider the Bose case. If one evaluates the matrix elements of (A.13) between the wavefunction given in (A.11) one obtains a matrix in *occupation number space*. The off diagonal elements (corresponding to 'transitions' between states i and k) are given by $f_{ik}^{(1)}(N_iN_k)^{1/2}$ (where N_i and N_k are the number of particles in states i and k) with

$$f_{ik}^{(1)} \equiv \int \psi_i^* \hat{f}^{(1)} \psi_k \mathrm{d}^3r. \tag{A.14}$$

The diagonal elements are given by $f_{ii}^{(1)}N_i$.

These results may be written in a more elegant form by introducing the *occupation number representation.* This involves a new kind of state function $\phi(N_1, N_2, \ldots)$; here N_1 = number of particles in state 1, N_2 = number of particles in state 2, etc. Hence occupation numbers, N_i, specify which of a (generally infinite) set of symmetrized one-particles states are actually present. There is then a one-to-one correspondence between the form $\phi(N_1, N_2, \ldots)$ and the form (A.11).

For a Bose system $N_i = 0, 1, 2, \ldots$ and we write $\phi = \phi_{\mathrm{B}}$; ϕ_{B} is assumed to be normalized and states with different N_i are orthogonal. We next introduce new (Bose) operators $\hat{a}_i$ and $\hat{a}_i^\dagger$, which act in this occupation number space, and are called *destruction and creation operators*, respectively. They are defined to have the properties

$$\hat{a}_i\phi_{\mathrm{B}}(N_1, N_2, \ldots, N_i, \ldots) = (N_i)^{1/2}\phi_{\mathrm{B}}(N_1, N_2, \ldots, N_i - 1, \ldots) \tag{A.15}$$

and

4. We use letters at the beginning of the alphabet to denote individual particles.

$$\hat{a}_i^\dagger \phi_{\rm B}(N_1, N_2, \ldots) = (N_i + 1)^{1/2} \phi_{\rm B}(N_1, \ldots, N_i + 1, \ldots); \tag{A.16}$$

we then say that $\hat{a}_i$ destroys (removes) one particle from state i and $\hat{a}_i^\dagger$ creates (adds) one particle in state i. Note that

$$\langle N_i - 1 | \hat{a}_i | N_i \rangle^* = \langle N_i | \hat{a}_i^\dagger | N_i - 1 \rangle = N_i^{1/2}.$$

From the above we verify that

$$\hat{a}_i^\dagger \hat{a}_i | \phi_{\rm B} \rangle = N_i | \phi_{\rm B} \rangle; \tag{A.17a}$$

and

$$\hat{a}_i \hat{a}_i^\dagger | \phi_{\rm B} \rangle = (N_i + 1) | \phi_{\rm B} \rangle. \tag{A.17b}$$

From these we construct the commutation relations for $\hat{a}_i$:

$$[\hat{a}_i \hat{a}_i^\dagger] \equiv (\hat{a}_i \hat{a}_i^\dagger - \hat{a}_i^\dagger \hat{a}_i) = 1; \tag{A.18}$$

we verify directly that

$$[\hat{a}_i, \hat{a}_k] = 0, \tag{A.19a}$$

$$[\hat{a}_i^\dagger, \hat{a}_k^\dagger] = 0, \tag{A.19b}$$

and

$$[\hat{a}_i^\dagger, \hat{a}_k] = 0 \quad (i \neq k). \tag{A.19c}$$

In terms of the $\hat{a}_i$ operators we may write the total *one-particle operator* $\hat{F}^{(1)}$ in the *occupation number representation* as

$$\hat{F}^{(1)} = \sum_{i,k} f_{ik}^{(1)} \hat{a}_i^\dagger \hat{a}_k. \tag{A.20}$$

We also introduce *two-particle operators*, e.g., the total interparticle Coulomb potential $\hat{V}^{(2)} = \Sigma_{a>b}^N \hat{V}_{ab}^{(2)} = \Sigma_{a>b}^N e^2/|\mathbf{r}_a - \mathbf{r}_b|$. Formally, we write the total two-particle operator $\hat{\mathrm{F}}^{(2)}$ as

$$\hat{F}^{(2)} = \sum_{a>b}^{N} \hat{f}_{ab}^{(2)}. \tag{A.21}$$

If we take the matrix element of (A.21) with the Bose wavefunction (A.11) and use the operators $\hat{a}_i$, we find that $\hat{F}^{(2)}$ in the occupation number representation is given by

$$\hat{F}^{(2)} = \frac{1}{2} \sum_{ik\ell m} f_{ik\ell m}^{(2)} \hat{a}_i^\dagger \hat{a}_k^\dagger \hat{a}_m \hat{a}_\ell \tag{A.22}$$

where

$$f_{ik\ell m}^{(2)} = \langle ik | f^{(2)} | \ell m \rangle = \iint \psi_i^*(\mathbf{r}_1) \psi_k^*(\mathbf{r}_2) \hat{f}^{(2)} \psi_\ell(\mathbf{r}_1) \psi_m(\mathbf{r}_2) \mathrm{d}^3 r_1 \mathrm{d}^3 r_2. \tag{A.23}$$

An important special case of the above operators is the Hamiltonian

$$\hat{H} = \sum_a \hat{H}_a^{(1)} + \sum_{a>b} U^{(2)}(\mathbf{r}_a, \mathbf{r}_b) + \sum_{a>b>c} U^{(3)}(\mathbf{r}_a, \mathbf{r}_b, \mathbf{r}_c)\ldots, \tag{A.24}$$

where

$$\sum_{a=1}^{N} \hat{H}_a^{(1)} = \sum_{a=1}^{N} \left(-\frac{\hbar^2}{2m}\right)\nabla_a^2 + U^{(1)}(\mathbf{r}_a). \tag{A.25}$$

Here $U_a^{(1)}, U_{ab}^{(2)}, U_{abc}^{(3)}$ are one-particle, two-particle, etc. potentials. In the occupation number representation $\hat{H}$ is given by

$$\hat{H} = \sum_{ik} H_{ik}^{(1)} \hat{a}_i^\dagger \hat{a}_k + \frac{1}{2} \sum_{ik\ell m} U_{ik\ell m}^{(2)} \hat{a}_i^\dagger \hat{a}_k^\dagger \hat{a}_m \hat{a}_\ell + \ldots. \tag{A.26}$$

For noninteracting particles

$$\hat{H} = \sum_{ik} H_{ik}^{(1)} \hat{a}_i^\dagger \hat{a}_k. \tag{A.27}$$

In a representation where $H_{ik}^{(1)}$ is diagonal with elements ε_i we have

$$\hat{H} = \sum_i \varepsilon_i \hat{a}_i^\dagger \hat{a}_i. \tag{A.28}$$

Since $\hat{a}_i^\dagger \hat{a}_i$ yields N_i when operating on ϕ_B, we have for the expectation value of $\hat{H}$, which is the total energy E, the expression

$$E = \sum_i \varepsilon_i N_i. \tag{A.29}$$

We now introduce the concept of '$\hat{\psi}$ operators' defined by

$$\hat{\psi}(\mathbf{r}) \equiv \sum_i \psi_i(\mathbf{r}) \hat{a}_i \tag{A.30a}$$

and

$$\hat{\psi}^\dagger(\mathbf{r}) \equiv \sum_i \psi_i^*(\mathbf{r}) \hat{a}_i^\dagger. \tag{A.30b}$$

Here $\mathbf{r}$ is a *parameter* as opposed to a coordinate (the coordinates of the occupation number representation are the occupation numbers N_i) and $\psi_i(\mathbf{r})$ are any complete set of one-particle wavefunctions appropriate to the space under consideration, i.e., those used to define (A.11).

We have suppressed the spin coordinates in the definitions (A.30a) and (A.30b). When these are included we would write $\hat{\psi}_\alpha(\mathbf{r})$ and $\hat{\psi}_\alpha^\dagger(\mathbf{r})$ in place of $\hat{\psi}(\mathbf{r})$ and $\hat{\psi}^\dagger(\mathbf{r})$, where α denotes the spin projection. In addition the basis functions $\psi_k(\mathbf{r})$ would be written as $\psi_k(\mathbf{r})\chi_\kappa(\alpha)$, where ψ_k and χ_κ denote the real space and spin space basis functions, respectively, and k and κ are the associated quantum numbers.

We can rewrite our one- and two-particle operators in terms of $\hat{\psi}$ operators using (A.30) as

$$\hat{F}^{(1)} = \sum_{ik} f_{ik}^{(1)} \hat{a}_i^\dagger \hat{a}_k = \sum_{ik} \left[\int \psi_i^*(\mathbf{r}) f^{(1)} \psi_k(\mathbf{r}) d^3r \right] \hat{a}_i^\dagger \hat{a}_k$$

$$= \int d^3r \hat{\psi}^\dagger(\mathbf{r}) f^{(1)} \hat{\psi}(\mathbf{r}); \tag{A.31a}$$

similarly

$$\hat{F}^{(2)} = \frac{1}{2} \int d^3r d^3r' \hat{\psi}^\dagger(\mathbf{r}) \hat{\psi}^\dagger(\mathbf{r}') f^{(2)} \hat{\psi}(\mathbf{r}') \hat{\psi}(\mathbf{r}). \tag{A.31b}$$

The Hamiltonian can be rewritten as

$$\hat{H} = \int d^3r \hat{\psi}^\dagger(\mathbf{r}) \left[-\frac{\hbar^2}{2m} \nabla^2 + U^{(1)}(\mathbf{r}) \right] \hat{\psi}(\mathbf{r})$$

$$+ \frac{1}{2} \int d^3r d^3r' \hat{\psi}^\dagger(\mathbf{r}) \hat{\psi}^\dagger(\mathbf{r}') U^{(2)}(\mathbf{r}, \mathbf{r}') \hat{\psi}(\mathbf{r}') \hat{\psi}(\mathbf{r})$$

$$+ (\text{three-body terms}) + \ldots; \tag{A.32}$$

$U^{(1)}(\mathbf{r})$ and $U^{(2)}(\mathbf{r}, \mathbf{r}')$ are the one- and two-body potentials, respectively. The particle density operator is

$$\hat{\rho}(\mathbf{r}) \equiv \hat{\psi}^\dagger(\mathbf{r}) \hat{\psi}(\mathbf{r}) \tag{A.33}$$

and the number operator is defined by $\hat{N} \equiv \int d^3r \hat{\rho}(\mathbf{r})$ or

$$\hat{N} \equiv \int d^3r \hat{\psi}^\dagger(\mathbf{r}) \hat{\psi}(\mathbf{r}). \tag{A.34}$$

Substituting (A.30) in (A.34) and using the orthonormality of the ψ_i we have

$$\hat{N} = \sum_i a_i^\dagger \hat{a}_i. \tag{A.35}$$

Using the definition (A.30) and the commutation relations for the $\hat{a}_i$ given by (A.18) and (A.19) we obtain the commutation relations for the $\hat{\psi}$ operators

$$\hat{\psi}(\mathbf{r}) \hat{\psi}(\mathbf{r}') - \hat{\psi}(\mathbf{r}') \hat{\psi}(\mathbf{r}) = 0$$

or

$$[\hat{\psi}(\mathbf{r}), \hat{\psi}(\mathbf{r}')] = 0, \tag{A.36}$$

and

$$\hat{\psi}(\mathbf{r}) \hat{\psi}^\dagger(\mathbf{r}') - \hat{\psi}^\dagger(\mathbf{r}') \hat{\psi}(\mathbf{r}) = \delta^{(3)}(\mathbf{r} - \mathbf{r}')$$

or

$$[\hat{\psi}(\mathbf{r}), \hat{\psi}^\dagger(\mathbf{r}')] = \delta^{(3)}(\mathbf{r} - \mathbf{r}'), \tag{A.37}$$

where we have used the completeness relation $\Sigma_i \psi_i^*(\mathbf{r}) \psi_i(\mathbf{r}') = \delta^{(3)}(\mathbf{r} - \mathbf{r}')$.

A.3 Occupation number representation: Fermi statistics

We now treat the case of Fermi statistics; we write the wavefunction in the occupation number representation as $\phi = \phi_F(N_i)$. To fix the sign of ϕ_F we: (1) number the basis states ψ_i once and for all, and (2) complete the rows by ordering p_i such that

$$p_1 < p_2 < p_3 < \ldots < p_N. \tag{A.38}$$

This prescription is unique since no two p_i can be the same or the determinant will vanish; this is equivalent to the occupation numbers, N_i, being 0 or 1. As in the Bose case, explicit (but tedious) calculation shows that $F_i^{(1)}$ has nonzero matrix elements in occupation number space between the wavefunction ϕ_F given in (A.12).[5] The off diagonal elements between states i and k are given by:

$$\langle 1_i, 0_k \,|\, F^{(1)} \,|\, 0_i, 1_k\rangle = f_{ik}^{(1)}(-1)^{\Sigma(i+1,k-1)},\ i < k \tag{A.39a}$$

or

$$\langle 1_i, 0_k \,|\, F^{(1)} \,|\, 0_i, 1_k\rangle = f_{ik}^{(1)}(-1)^{\Sigma(k+1,i-1)},\ i > k \tag{A.39b}$$

where 0_i, 1_i signify $N_i = 0$, $N_i = 1$, respectively, and we *define*

$$\sum(k,\ell) \equiv \sum_{n=k}^{n=\ell} N_n, \tag{A.40}$$

where the sum is defined as zero when $\ell < k$.

The diagonal elements are $f_{ii}^{(1)}N_i$ as in the Bose case. We define new Fermi destruction and creation operators $\hat{c}_i$ and $\hat{c}_i^\dagger$ by

$$\langle 0_i \,|\, \hat{c}_i \,|\, 1_i\rangle = \langle 1_i \,|\, \hat{c}_i^\dagger \,|\, 0_i\rangle = (-1)^{\Sigma(1,i-1)}. \tag{A.41}$$

Multiplying these matrices we obtain the elements (when $i < k$)

$$\begin{aligned}\langle 1_i, 0_k \,|\, \hat{c}_i^\dagger \hat{c}_k \,|\, 0_i, 1_k\rangle &= \langle 1_i, 0_k \,|\, \hat{c}_i^\dagger \,|\, 0_i, 0_k\rangle\langle 0_i, 0_k \,|\, \hat{c}_k \,|\, 0_i, 1_k\rangle \\ &= (-1)^{\Sigma(1,i-1)}(-1)^{\Sigma(1,k-1)} \\ &= (-1)^{\Sigma(1,i-1)}(-1)^{\Sigma(1,i-1)+\Sigma(i+1,k-1)}.\end{aligned}$$

Now since $N_i = 0$, the ith term in the second sum may be omitted and since $(-1)^{2\Sigma(1,i-1)} = 1$ we are left with

$$\langle 1_i, 0_k \,|\, \hat{c}_i^\dagger \hat{c}_k \,|\, 0_i, 1_k\rangle = (-1)^{\Sigma(i+1,k-1)}. \tag{A.42}$$

In the case $i = k$, the matrix $\hat{c}_i^\dagger \hat{c}_i$ is diagonal and the elements are 1 for $N_i = 1$ and 0 for $N_i = 0$. Thus we can write

$$\hat{c}_i^\dagger \hat{c}_i = N_i. \tag{A.43}$$

Comparison of (A.42) with (A.39) yields

$$\hat{F}^{(1)} = \sum_{i,k} f_{ik}^{(1)} \hat{c}_i^\dagger \hat{c}_k,$$

5. For the reader who is not interested in the details of the calculations the results are identical to the Bose case with $\hat{a} \to \hat{c}$ but where the $\hat{c}$ obey the *anticommutation* relations (A.45); the $\hat{\psi}$ obey the anticommutation relations (A.46).

which is the same form as the Bose case, Eq. (A.20). To examine the commutation properties we evaluate ($k > i$)

$$\begin{aligned}\langle 1_i, 0_k | \hat{c}_k \hat{c}_i^\dagger | 0_i, 1_k \rangle &= \langle 1_i, 0_k | \hat{c}_k | 1_i, 1_k \rangle \langle 1_i, 1_k | \hat{c}_i^\dagger | 0_i, 1_k \rangle \\ &= (-1)^{\Sigma(1,k-1)}(-1)^{\Sigma(1,i-1)} \\ &= (-1)^{\Sigma(1,i-1)+\Sigma(i,k-1)}(-1)^{\Sigma(1,i-1)};\end{aligned}$$

in the first sum $N_i = 1$ so we can rewrite the expression as

$$\langle (1_i, 0_k | \hat{c}_k \hat{c}_i^\dagger | 0_i, 1_k) \rangle = (-1)^{\Sigma(1,i-1)+\Sigma(i+1,k-1)+1}(-1)^{\Sigma(1,i-1)} = -(-1)^{\Sigma(i+1,k-1)}. \tag{A.44}$$

Comparing (A.42) and (A.44) we have

$$\hat{c}_i^\dagger \hat{c}_k + \hat{c}_k \hat{c}_i^\dagger = 0, \quad i \neq k.$$

Evaluating the diagonal element in the opposite order of (A.43) gives

$$\hat{c}_i \hat{c}_k^\dagger + \hat{c}_k^\dagger \hat{c}_i = \delta_{ik}. \tag{A.45a}$$

Thus fermion creation and destruction operators *anticommute*. Similar calculations show that

$$\hat{c}_i \hat{c}_k + \hat{c}_k \hat{c}_i = 0. \tag{A.45b}$$

Calculations similar to the Bose case show that fermion $\hat{\psi}$ operators obey the *anticommutation* relations

$$\hat{\psi}^\dagger(\mathbf{r}')\hat{\psi}(\mathbf{r}) + \hat{\psi}(\mathbf{r})\hat{\psi}^\dagger(\mathbf{r}') = \delta^{(3)}(\mathbf{r} - \mathbf{r}'), \tag{A.46a}$$

$$\hat{\psi}(\mathbf{r})\hat{\psi}(\mathbf{r}') + \hat{\psi}(\mathbf{r}')\hat{\psi}(\mathbf{r}) = 0. \tag{A.46b}$$

Formally, $\mathbf{r}$ here includes both the real space and spin coordinates. If we write the spin coordinate explicitly as a subscript α (having projections $\pm\frac{1}{2}$ for a spin $\frac{1}{2}$ particle corresponding to, say, $\alpha = \pm 1$) then Eqs. (A.46) take the forms

$$\hat{\psi}_\alpha^\dagger(\mathbf{r}')\hat{\psi}_\beta(\mathbf{r}) + \hat{\psi}_\beta(\mathbf{r})\hat{\psi}_\alpha^\dagger(\mathbf{r}') = \delta_{\alpha\beta}\delta^{(3)}(\mathbf{r} - \mathbf{r}') \tag{A.47a}$$

and

$$\hat{\psi}_\alpha(\mathbf{r}')\hat{\psi}_\beta(\mathbf{r}) + \hat{\psi}_\beta(\mathbf{r})\hat{\psi}_\alpha(\mathbf{r}') = 0. \tag{A.47b}$$

Analogous to the Bose case we may write the two-particle operators as

$$\hat{F}^{(2)} = \frac{1}{2}\sum_{ik\ell m} f^{(2)}_{ik\ell m} \hat{c}_i^\dagger \hat{c}_k^\dagger \hat{c}_m \hat{c}_\ell, \tag{A.48}$$

with $f^{(2)}_{ik\ell m}$ given by (A.23) and the Hamiltonian as

$$\hat{H} = \sum_{ik} H_{ik} \hat{c}_i^\dagger \hat{c}_k + \frac{1}{2}\sum_{ik\ell m} U^{(2)}_{ik\ell m} \hat{c}_i^\dagger \hat{c}_k^\dagger \hat{c}_m \hat{c}_\ell + \ldots \tag{A.49}$$

Eq. (A.32) also holds for the Fermi case.

Appendix B

Some calculations involving the BCS wavefunction

BCS introduced *pairing operators* defined as

$$\hat{b}_{\mathbf{k}}^{\dagger} = \hat{c}_{\mathbf{k}\uparrow}^{\dagger}\hat{c}_{-\mathbf{k}\downarrow}^{\dagger} \tag{B.1a}$$

and

$$\hat{b}_{\mathbf{k}} = \hat{c}_{-\mathbf{k}\downarrow}\,\hat{c}_{\mathbf{k}\uparrow}, \tag{B.1b}$$

where $\hat{c}_{\mathbf{k}\sigma}^{\dagger}$ and $\hat{c}_{\mathbf{k}\sigma}$ are the usual fermion creation and destruction operators obeying the anticommutation relations (26.5a)–(26.5c). The pairing operators, on the other hand, obey commutation relations (which are verified by direct calculation) given by

$$\hat{b}_{\mathbf{k}}\hat{b}_{\mathbf{k}'}^{\dagger} - \hat{b}_{\mathbf{k}'}^{\dagger}\hat{b}_{\mathbf{k}} = 0, \quad \mathbf{k} \neq \mathbf{k}' \tag{B.2a}$$

$$\hat{b}_{\mathbf{k}}\hat{b}_{\mathbf{k}}^{\dagger} - \hat{b}_{\mathbf{k}}^{\dagger}\hat{b}_{\mathbf{k}} = 1 - \hat{n}_{\mathbf{k}\uparrow} - \hat{n}_{-\mathbf{k}\downarrow}, \tag{B.2b}$$

and

$$\hat{b}_{\mathbf{k}}\hat{b}_{\mathbf{k}'} - \hat{b}_{\mathbf{k}'}\hat{b}_{\mathbf{k}} = 0, \tag{B.2c}$$

where $\hat{n}_{\mathbf{k}\uparrow}$ and $\hat{n}_{\mathbf{k}\downarrow}$ are the number operators for spin-up and spin-down electrons, respectively. We may write the total number operator (for both spins) as

$$\hat{n}_{\mathbf{k}\uparrow} + \hat{n}_{\mathbf{k}\downarrow} = 2\hat{b}_{\mathbf{k}}^{\dagger}\hat{b}_{\mathbf{k}}. \tag{B.3}$$

In terms of the pairing operators the reduced Hamiltonian, $\hat{H}_{\mathrm{R}}$, can be written as

$$\hat{H}_{\mathrm{R}} = 2\sum_{\mathbf{k}} \xi_{\mathbf{k}}\hat{b}_{\mathbf{k}}^{\dagger}\hat{b}_{\mathbf{k}} - \sum_{\mathbf{k},\mathbf{k}'} V_{\mathbf{k}\mathbf{k}'}\hat{b}_{\mathbf{k}}^{\dagger}\hat{b}_{\mathbf{k}'}. \tag{B.4}$$

To calculate the ground-state energy we must take the expectation value of $\hat{H}_{\mathrm{R}}$ with the BCS wavefunction

$$\Psi = \prod_{\mathbf{k}} (1 + g_{\mathbf{k}} b_{\mathbf{k}}^{\dagger})\,|\,0\rangle, \tag{B.5}$$

subject to the constraint that the expectation value of the total particle number operator is fixed at the average particle number, N. Let us first normalize Ψ:

$$\begin{aligned}\langle\Psi\,|\,\Psi\rangle &= \prod_{\mathbf{k},\mathbf{k}'} \langle 0\,|\,(1 + g_{\mathbf{k}}^{*}\hat{b}_{\mathbf{k}})(1 + g_{\mathbf{k}'}\hat{b}_{\mathbf{k}'}^{\dagger})\,|\,0\rangle \\ &= \prod_{\mathbf{k},\mathbf{k}'} \langle 0\,|\,1 + g_{\mathbf{k}}^{*}\hat{b}_{\mathbf{k}} + g_{\mathbf{k}'}\hat{b}_{\mathbf{k}'}^{\dagger} + g_{\mathbf{k}}^{*}g_{\mathbf{k}'}\hat{b}_{\mathbf{k}}\hat{b}_{\mathbf{k}'}^{\dagger}\,|\,0\rangle.\end{aligned}$$

The terms with an odd number of operators vanish; using (B.2b) to commute the operators in the last term gives

$$\langle\Psi|\Psi\rangle = \prod_{\mathbf{k},\mathbf{k}'}\langle 0|1 + g^*_{\mathbf{k}}g_{\mathbf{k}'}(1 - \hat{n}_{\mathbf{k}\uparrow} - \hat{n}_{\mathbf{k}\downarrow})\delta_{\mathbf{k}\mathbf{k}'} + \hat{b}^\dagger_{\mathbf{k}}\hat{b}_{\mathbf{k}'}|0\rangle.$$

The operator $\hat{b}_{\mathbf{k}'}$ annihilates the vacuum; furthermore $\hat{n}_{\mathbf{k}}|0\rangle$ vanishes since the vacuum state has no particles. Therefore

$$\langle\Psi|\Psi\rangle = \prod_{\mathbf{k}}(1 + g^*_{\mathbf{k}}g_{\mathbf{k}})$$

and the normalized ground-state wavefunction is

$$|\Psi\rangle = \prod_{\mathbf{k}}\frac{(1 + g_{\mathbf{k}}b^\dagger_{\mathbf{k}})|0\rangle}{(1 + g^*_{\mathbf{k}}g_{\mathbf{k}})^{1/2}}. \tag{B.6}$$

It is useful to introduce the quantities

$$u_{\mathbf{k}} \equiv \frac{1}{(1 + |g_{\mathbf{k}}|^2)^{1/2}} \tag{B.7a}$$

and

$$v_{\mathbf{k}} \equiv \frac{g_{\mathbf{k}}}{(1 + |g_{\mathbf{k}}|^2)^{1/2}}, \tag{B.7b}$$

whereupon (B.6) becomes

$$|\Psi\rangle = \prod_{\mathbf{k}}(u_{\mathbf{k}} + v_{\mathbf{k}}b^\dagger_{\mathbf{k}})|0\rangle \tag{B.8}$$

where $u_{\mathbf{k}}$ and $v_{\mathbf{k}}$ satisfy

$$|u_{\mathbf{k}}|^2 + |v_{\mathbf{k}}|^2 = 1. \tag{B.9}$$

To evaluate the ground-state energy we require the expectation value of

$$\langle\Psi|\hat{H}_{\mathrm{R}}|\Psi\rangle = \langle\Psi|\hat{H}_0|\Psi\rangle + \langle\Psi|\hat{H}_{\mathrm{IR}}|\Psi\rangle \tag{B.10}$$

where

$$\hat{H}_0 = 2\sum_{\mathbf{k}}\xi_{\mathbf{k}}\hat{b}^\dagger_{\mathbf{k}}\hat{b}_{\mathbf{k}} \tag{B.11a}$$

and

$$\hat{H}_{\mathrm{IR}} = -\sum_{\mathbf{k},\mathbf{k}'}V_{\mathbf{k}\mathbf{k}'}\hat{b}^\dagger_{\mathbf{k}}\hat{b}_{\mathbf{k}'}. \tag{B.11b}$$

Let us write the product of the terms contributing to the wavefunction (B.8) explicitly:

$$|\Psi\rangle = (u_{\mathbf{k}_1} + v_{\mathbf{k}_1}b^\dagger_{\mathbf{k}_1})(u_{\mathbf{k}_2} + v_{\mathbf{k}_2}b^\dagger_{\mathbf{k}_2})\ldots(u_{\mathbf{k}_i} + (v_{\mathbf{k}_i}b^\dagger_{\mathbf{k}_i})\ldots|0\rangle,$$

where $\mathbf{k}_1, \mathbf{k}_2, \ldots$ correspond to the set of normal-state eigenfunctions of $\hat{H}_0$. In evaluating the expectation values in Eq. (B.10) those states which are not affected by $\hat{H}_0$ or $\hat{H}_{IR}$ will result in successive factors of unity due to the normalization condition (B.9). Only terms involving the pairing of a $\hat{b}_k$ and $\hat{b}_k^\dagger$ (associated with the *same* configuration space eigenfunction) gives a nonzero contribution to the expectation value. The expectation value $\langle\Psi|\hat{H}_{IR}|\Psi\rangle$ involves terms of the form

$$\langle 0|[\ldots(u_{k_i}^* + v_{k_i}^*\hat{b}_{k_i})\ldots(u_{k_j}^* + v_{k_j}^*\hat{b}_{k_j})]V_{qq'}\hat{b}_q^\dagger\hat{b}_{q'}[\ldots(u_{k_i'} + v_{k_i'}\hat{b}_{k_i'}^\dagger)\ldots(u_{k_j'} + v_{k_j'}\hat{b}_{k_j'}^\dagger)\ldots]|0\rangle.$$

The terms involving $\hat{b}_k^\dagger\hat{b}_k$ or $\hat{b}_k\hat{b}_k^\dagger$ pairs are

$$\langle 0|u_q^*u_qV_{qq}\hat{b}_q^\dagger\hat{b}_q + v_q^*v_{q'}V_{qq'}\hat{b}_q\hat{b}_q^\dagger\hat{b}_{q'}\hat{b}_{q'}^\dagger + u_q^*v_{q'}^*u_{q'}v_qV_{qq'}\hat{b}_{q'}\hat{b}_q^\dagger\hat{b}_{q'}\hat{b}_q^\dagger(1-\delta_{qq'})|0\rangle. \quad \text{(B.12)}$$

The first term vanishes since $\hat{b}|0\rangle = 0$. Using Eq. (B.2b) the second term may be written as $(1-\hat{n}_q)(1-\hat{n}_{q'})v_q^*v_{q'}V_{qq'}$; it corresponds to the Hartree–Fock contribution to the total energy and may be discarded if we assume that the ξ_k are quasiparticle energies (which already include electron–electron interaction effects). The last term contributes to the pairing energy (and vanishes in the normal state where $u_kv_k = 0$); again using (B.2b) and $\hat{n}_k|0\rangle = 0$ we have

$$\langle\Psi|\hat{H}_{IR}|\Psi\rangle = \sum_{k,k'} u_k^*v_ku_{k'}v_{k'}^*V_{kk'}. \quad \text{(B.13)}$$

We may use (B.12) to evaluate the contribution of $\hat{H}_0$ to (B.10) by replacing $V_{kk'}$ by $2\xi_k\delta_{kk'}$ to obtain

$$\langle\Psi|\hat{H}_0|\Psi\rangle = 2\sum_k \xi_k|v_k|^2. \quad \text{(B.14)}$$

Appendix C

The gap as a perturbation through third order

In Sec. 38 we regarded the gap as a perturbation in the Bogoliubov equations and calculated the first order corrections $u^{(1)}$ and $v^{(1)}$. As we now show, these calculations can be extended to higher order. We recall the expansions

$$u(\mathbf{r}) = u^{(0)}(\mathbf{r}) + u^{(1)}(\mathbf{r}) + u^{(2)}(\mathbf{r}) + u^{(3)}(\mathbf{r}) + \ldots, \tag{C.1a}$$

$$v(\mathbf{r}) = v^{(0)}(\mathbf{r}) + v^{(1)}(\mathbf{r}) + v^{(2)}(\mathbf{r}) + v^{(3)}(\mathbf{r}) + \ldots, \tag{C.1b}$$

and

$$\varepsilon = \varepsilon^{(0)} + \varepsilon^{(1)} + \varepsilon^{(2)} + \varepsilon^{(3)} + \ldots. \tag{C.1c}$$

Here

$$u_n^{(0)} = \phi_n \theta(\xi_n), \quad v_n^{(0)} = \phi_n^* \theta(-\xi_n), \tag{C.2}$$

where the eigenfunctions, ϕ_n, and the associated eigenvalues, ξ_n, are the exact solutions to Eq. (38.1) with a vector potential which is determined self-consistently from Maxwell's fourth equation

$$\nabla \times \mathbf{H}(\mathbf{r}) = \frac{4\pi}{c}\mathbf{j}(\mathbf{r}) \tag{C.3}$$

(with $\mathbf{H}(\mathbf{r}) = \nabla \times \mathbf{A}(\mathbf{r})$ and $\mathbf{j}(\mathbf{r})$ calculated from Eq. (37.17) with perturbed Bogoliubov wavefunctions expanded to the required order). Inserting Eqs. (C.1) into the Bogoliubov equations ((36.17a) and (36.17b)), and collecting terms of the same order (keeping in mind that Δ is inherently first order), we obtain

$$\varepsilon_n^{(1)} u_n^{(0)} + \varepsilon_n^{(0)} u_n^{(1)} = \mathscr{H}'_0 u_n^{(1)} + \Delta v_n^{(0)} \tag{C.4a}$$

and

$$\varepsilon_n^{(1)} v_n^{(0)} + \varepsilon_n^{(0)} v_n^{(1)} = -\mathscr{H}'^*_0 v_n^{(1)} + \Delta^* u_n^{(0)} \tag{C.4b}$$

in the first order. In second order we have

$$\varepsilon_n^{(2)} u_n^{(0)} + \varepsilon_n^{(1)} u_n^{(1)} + \varepsilon_n^{(0)} u_n^{(2)} = \mathscr{H}'_0 u_n^{(2)} + \Delta v_n^{(1)} \tag{C.5a}$$

and

$$\varepsilon_n^{(2)} v_n^{(0)} + \varepsilon_n^{(1)} v_n^{(1)} + \varepsilon_n^{(0)} v_n^{(2)} = -\mathscr{H}'^*_0 v_n^{(2)} + \Delta^* u_n^{(1)} \tag{C.5b}$$

while the third order equations are

$$\varepsilon_n^{(3)} u_n^{(0)} + \varepsilon_n^{(2)} u_n^{(1)} + \varepsilon_n^{(1)} u_n^{(2)} + \varepsilon_n^{(0)} u_n^{(3)} = \mathscr{H}'_0 u_n^{(3)} + \Delta v_n^{(2)} \tag{C.6a}$$

and

$$\varepsilon_n^{(3)} v_n^{(0)} + \varepsilon_n^{(2)} v_n^{(1)} + \varepsilon_n^{(1)} v_n^{(2)} + \varepsilon_n^{(0)} v_n^{(3)} = -\mathcal{H}_0'^{*} v_n^{(3)} + \Delta^* u_n^{(2)}. \tag{C.6b}$$

We expand the various orders of u and v in the form

$$u_n^{(i)} = \sum_m a_{nm}^{(i)} \phi_m; \tag{C.7a}$$

$$v_n^{(i)} = \sum_m b_{nm}^{(i)} \phi_m^*. \tag{C.7b}$$

Inserting Eqs. (C.7) into Eqs. (C.4), multiplying by ϕ_n^* and ϕ_n, respectively, and integrating over d^3r gives

$$\varepsilon^{(1)}\delta_{nm} + (|\xi_n| - \xi_m)a_{nm}^{(1)} = \int d^3r' \phi_m^*(\mathbf{r}')\Delta(\mathbf{r}')\phi_n^*(\mathbf{r}')\theta(-\xi_n) \tag{C.8a}$$

$$\varepsilon^{(1)}\delta_{nm} + (|\xi_n| + \xi_m)b_{nm}^{(1)} = \int d^3r' \phi_m(\mathbf{r}')\Delta^*(\mathbf{r}')\phi_n(\mathbf{r}')\theta(\xi_n). \tag{C.8b}$$

For the case where $n \neq m$, the coefficients are

$$a_{nm}^{(1)} = \frac{\theta(-\xi_n)}{|\xi_n| - \xi_m} \int \phi_m^*(\mathbf{r})\Delta(\mathbf{r})\phi_n^*(\mathbf{r})d^3r \tag{C.9a}$$

and

$$b_{nm}^{(1)} = \frac{\theta(\xi_n)}{|\xi_n| + \xi_m} \int \phi_m(\mathbf{r})\Delta^*(\mathbf{r})\phi_n(\mathbf{r})d^3r, \tag{C.9b}$$

as given earlier in Eqs. (38.5). For the case when $n = m$, Eqs. (C.8) become

$$\varepsilon_n^{(1)}\theta(\xi_n) + (|\xi_n| - \xi_n)a_{nn}^{(1)} = \int d^3r' \phi_n^*(\mathbf{r}')\Delta(\mathbf{r}')\phi_n(\mathbf{r}')\theta(-\xi_n) \tag{C.10a}$$

and

$$\varepsilon_n^{(1)}\theta(-\xi_n) + (|\xi_n| + \xi_n)b_{nn}^{(1)} = \int d^3r' \phi_n(\mathbf{r}')\Delta^*(\mathbf{r}')\phi_n(\mathbf{r}')\theta(\xi_n). \tag{C.10b}$$

These have the solutions

$$\varepsilon_n^{(1)} = 0, \tag{C.11a}$$

$$a_{nn}^{(1)} = \int d^3r' \frac{\phi_n^*(\mathbf{r}')\Delta(\mathbf{r}')\phi_n^*(\mathbf{r}')}{2|\xi_n|}\theta(-\xi_n), \tag{C.11b}$$

$$b_{nn}^{(1)} = \int d^3r' \frac{\phi_n(\mathbf{r}')\Delta(\mathbf{r}')\phi_n(\mathbf{r}')}{2|\xi_n|}\theta(\xi_n). \tag{C.11c}$$

However from Eqs. (C.10) $a_{nn}(\xi_n > 0)$ and $b_{nn}(\xi_n < 0)$ are arbitrary; they are fixed by the normalization condition (36.23b)

$$\int d^3r \phi_n^*(\mathbf{r})\phi(\mathbf{r})[(a_{nn} + a_{nn}^*)\theta(\xi_n) + (b_{nn} + b_{nn}^*)\theta(-\xi_n)] = 0.$$

If we set arbitrary phase factors to zero we obtain $a_{nn}(\xi > 0) = b_{nn}(\xi < 0) = 0$.

We now discuss the second order expansion terms. Starting with Eqs. (C.5) we insert the expressions for the first order corrections, with coefficients given by Eqs. (C.9), obtaining

$$\begin{aligned}&\varepsilon_n^{(2)}\theta(\xi_n)\delta_{nm} + (|\xi_n| - \xi_m)a_{nm}^{(2)}\\&= \sum_\ell{}' \int d^3r_1 d^3r_2 \frac{\phi_m^*(\mathbf{r}_1)\Delta(\mathbf{r}_1)\phi_\ell^*(\mathbf{r}_1)\phi_\ell(\mathbf{r}_2)\Delta^*(\mathbf{r}_2)\phi_n(\mathbf{r}_2)\theta(\xi_n)}{|\xi_n| + \xi_\ell}\end{aligned} \tag{C.12a}$$

and

$$\begin{aligned}&\varepsilon_n^{(2)}\theta(-\xi_n)\delta_{nm} + (|\xi_n| + \xi_m)b_{nm}^{(2)}\\&= \sum_\ell{}' \int d^3r_1 d^3r_2 \frac{\phi_m(\mathbf{r}_1)\Delta^*(\mathbf{r}_1)\phi_\ell(\mathbf{r}_1)\phi_\ell^*(\mathbf{r}_2)\Delta(\mathbf{r}_2)\phi_n^*(\mathbf{r}_2)\theta(-\xi_n)}{|\xi_n| - \xi_\ell},\end{aligned} \tag{C.12b}$$

where the prime on the summation implies the divergent terms, $\xi_\ell = -|\xi_n|$ and $\xi_\ell = +|\xi_n|$ in (C.12a) and (C.12b), respectively, are omitted from the sums. For $m \neq n$ we obtain immediately

$$a_{nm}^{(2)} = \sum_\ell{}' \int d^3r_1 d^3r_2 \frac{\phi_m^*(\mathbf{r}_1)\Delta(\mathbf{r}_1)\phi_\ell^*(\mathbf{r}_1)\phi_\ell(\mathbf{r}_2)\Delta^*(\mathbf{r}_2)\phi_n(\mathbf{r}_2)}{(|\xi_n| - \xi_m)(|\xi_n| + \xi_\ell)}\theta(\xi_n) \tag{C.13a}$$

and

$$b_{nm}^{(2)} = \sum_\ell{}' \int d^3r_1 d^3r_2 \frac{\phi_m(\mathbf{r}_1)\Delta^*(\mathbf{r}_1)\phi_\ell(\mathbf{r}_1)\phi_\ell^*(\mathbf{r}_2)\Delta(\mathbf{r}_2)\phi_n^*(\mathbf{r}_2)}{(|\xi_n| + \xi_m)(|\xi_n| - \xi_\ell)}\theta(-\xi_n). \tag{C.13b}$$

For the case $m = n$, $a_{nn}^{(2)}(\xi_n < 0)$ and $b_{nn}^{(2)}(\xi_n > 0)$ vanish; however, from Eqs. (C.12) $a_{nn}^{(2)}(\xi_n > 0)$ and $b_{nn}^{(2)}(\xi_n < 0)$ are arbitrary and must be fixed by the normalization condition (which we will consider shortly). The $m = n$ case also yields

$$\varepsilon_n^{(2)}(\xi_n > 0) = \sum_\ell{}' \int d^3r_1 d^3r_2 \frac{\phi_n^*(\mathbf{r}_1)\Delta(\mathbf{r}_1)\phi_\ell^*(\mathbf{r}_1)\phi_\ell(\mathbf{r}_2)\Delta^*(\mathbf{r}_2)\phi_n(\mathbf{r}_2)}{|\xi_n| + \xi_\ell}\theta(\xi_n) \tag{C.14a}$$

and

$$\varepsilon_n^{(2)}(\xi_n < 0) = \sum_\ell{}' \int d^3r_1 d^3r_2 \frac{\phi_n(\mathbf{r}_1)\Delta^*(\mathbf{r}_1)\phi_\ell(\mathbf{r}_1)\phi_\ell^*(\mathbf{r}_2)\Delta(\mathbf{r}_2)\phi_n^*(\mathbf{r}_2)}{|\xi_n| - \xi_\ell}\theta(-\xi_n). \tag{C.14b}$$

Note that in general $\varepsilon_n^{(2)}(\xi > 0) \neq \varepsilon_n^{(2)}(\xi < 0)$, which is referred to as particle–hole asymmetry. However, it is usually a very good approximation to regard ε_n as a symmetric function of ξ_n. Note that for the uniform case (where we may take the ϕ as real and where it is obvious the off-diagonal elements vanish) this expression reduces to $\varepsilon_n^{(2)} = (1/2)\Delta^2/|\xi_n|$ which is the leading order in the expansion of the BCS expression $\varepsilon_n = (\xi_n^2 + \Delta^2)^{1/2}$ in powers of Δ^2; clearly the summation in Eqs. (C.14) involves a singularity in the limit $|\xi_n| \to 0$. Maintaining the wavefunction normalization through second order requires

$$\begin{aligned}&\int d^3r[u_n^{(0)*}u_n^{(2)} + u_n^{(2)*}u_n^{(0)} + u_n^{(1)*}u_n^{(1)} + v_n^{(0)*}v_n^{(2)} + v_n^{(2)*}v_n^{(0)} + v_n^{(1)*}v_n^{(1)}]\\&= a_{nn}^{(2)} + a_{nn}^{(2)*} + \sum_m |a_{nm}^{(1)}|^2 + b_{nn}^{(2)} + b_{nn}^{(2)*} + \sum_m |b_{nm}^{(1)}|^2 = 0.\end{aligned} \tag{C.15}$$

Setting the arbitrary phases associated with this condition to zero we obtain the following expressions for the (real) diagonal second order expansion coefficient

$$a_{nn}^{(2)} = -\frac{1}{2}\sum_m{}' \left| \int d^3r \frac{\Delta(\mathbf{r})\phi_m^*(\mathbf{r})\phi_n^*(\mathbf{r})}{|\xi_n| - \xi_m} \theta(\xi_n) \right|^2 \tag{C.16a}$$

and

$$b_{nn}^{(2)} = -\frac{1}{2}\sum_m{}' \left| \int d^3r \frac{\Delta^*(\mathbf{r})\phi_m(\mathbf{r})\phi_n(\mathbf{r})}{|\xi_n| + \xi_m} \theta(-\xi_n) \right|^2, \tag{C.16b}$$

where the arbitrariness in the assignment of the terms $a_{nn}^{(1)}$ and $b_{nn}^{(1)}$ to the second order coefficients $a_{nn}^{(2)}$ and $b_{nn}^{(2)}$ is resolved by the energy restriction on the latter which was discussed above.

Finally, we examine the third order wavefunction corrections; substituting Eqs. (C.7) into Eqs. (C.6) gives

$$\varepsilon^{(3)}\theta(\xi_n)\delta_{nm} + \varepsilon_n^{(2)}a_{nm}^{(1)} + (|\xi_n| - \xi_m)a_{nm}^{(3)} = \sum_k \int d^3r_1 \phi_m^*(\mathbf{r}_1)\Delta(\mathbf{r}_1)\phi_k^*(\mathbf{r}_1)b_{nk}^{(2)} \tag{C.17a}$$

and

$$\varepsilon^{(3)}\theta(-\xi_n)\delta_{nm} + \varepsilon_n^{(2)}b_{nm}^{(1)} + (|\xi_n| + \xi_m)b_{nm}^{(3)} = \sum_k \int d^3r_1 \phi_m(\mathbf{r}_1)\Delta^*(\mathbf{r}_1)\phi_k(\mathbf{r}_1)a_{nk}^{(2)}. \tag{C.17b}$$

On substituting our earlier expressions from $a_{nm}^{(1)}$, $b_{nm}^{(1)}$, $a_{n\ell}^{(2)}$, $b_{n\ell}^{(2)}$, and $\varepsilon_n^{(2)}$ we obtain

$$\begin{aligned}
a_{nm}^{(3)} = {} & \sum_k{}'\sum_\ell{}' \int d^3r_1 d^3r_2 d^3r_3 \frac{\phi_m^*(\mathbf{r}_1)\Delta(\mathbf{r}_1)\phi_k^*(\mathbf{r}_1)\phi_k(\mathbf{r}_2)\Delta^*(\mathbf{r}_2)\phi_\ell(\mathbf{r}_2)\phi_\ell^*(\mathbf{r}_3)\Delta(\mathbf{r}_3)\phi_n^*(\mathbf{r}_3)}{(|\xi_n| - \xi_m)(|\xi_n| + \xi_k)(|\xi_n| - \xi_\ell)} \theta(-\xi_n) \\
& - \frac{1}{2}\int d^3r_1 \frac{\phi_m^*(\mathbf{r}_1)\Delta(\mathbf{r}_1)\phi_n^*(\mathbf{r}_1)}{|\xi_n| - \xi_m} \sum_\ell{}' \left| \int d^3r_2 \frac{\phi_\ell^*(\mathbf{r}_2)\Delta(\mathbf{r}_2)\phi_n^*(\mathbf{r}_2)}{|\xi_n| + \xi_\ell} \theta(-\xi_n) \right|^2 \\
& - \frac{1}{2}\sum_\ell{}' \int d^3r_1 d^3r_2 d^3r_3 \frac{\phi_m^*(\mathbf{r}_1)\Delta(\mathbf{r}_1)\phi_n^*(\mathbf{r}_1)\phi_n(\mathbf{r}_2)\Delta^*(\mathbf{r}_2)\phi_\ell(\mathbf{r}_2)\phi_\ell^*(\mathbf{r}_3)\Delta(\mathbf{r}_3)\phi_n^*(\mathbf{r}_3)}{(|\xi_n| - \xi_m)^2(|\xi_n| - \xi_\ell)} \theta(-\xi_n)
\end{aligned} \tag{C.18a}$$

and

$$\begin{aligned}
b_{nm}^{(3)} = {} & \sum_k{}'\sum_\ell{}' \int d^3r_1 d^3r_2 d^3r_3 \frac{\phi_m(\mathbf{r}_1)\Delta^*(\mathbf{r}_1)\phi_k(\mathbf{r}_1)\phi_k^*(\mathbf{r}_2)\Delta(\mathbf{r}_2)\phi_\ell^*(\mathbf{r}_2)\phi_\ell(\mathbf{r}_3)\Delta^*(\mathbf{r}_3)\phi_n(\mathbf{r}_3)}{(|\xi_n| + \xi_m)(|\xi_n| - \xi_k)(|\xi_n| + \xi_\ell)} \theta(\xi_n) \\
& - \frac{1}{2}\int d^3r_1 \frac{\phi_m(\mathbf{r}_1)\Delta^*(\mathbf{r}_1)\phi_n(\mathbf{r}_1)}{|\xi_n| + \xi_m} \sum_\ell{}' \left| \int d^3r_2 \frac{\phi_\ell(\mathbf{r}_2)\Delta^*(\mathbf{r}_2)\phi_n(\mathbf{r}_2)}{|\xi_n| - \xi_\ell} \theta(\xi_n) \right|^2 \\
& - \frac{1}{2}\sum_\ell{}' \int d^3r_1 d^3r_2 d^3r_3 \frac{\phi_m(\mathbf{r}_1)\Delta^*(\mathbf{r}_1)\phi_n(\mathbf{r}_1)\phi_n^*(\mathbf{r}_2)\Delta(\mathbf{r}_2)\phi_\ell^*(\mathbf{r}_2)\phi_\ell(\mathbf{r}_3)\Delta^*(\mathbf{r}_3)\phi_n(\mathbf{r}_3)}{(|\xi_n| + \xi_m)^2(|\xi_n| + \xi_\ell)} \theta(\xi_n)
\end{aligned} \tag{C.18b}$$

for the case $n \neq m$. We next examine the diagonal terms, $m = n$. For $\xi_n > 0$, Eq. (C.18a) requires $\varepsilon^{(3)}(\xi_n > 0) = 0$ and leaves $a_{nn}^{(3)}(\xi_n > 0)$ arbitrary; (C.18b) yields

$$b_{nn}^{(3)} = \frac{1}{2|\xi_n|}\sum_k\sum_\ell{}' \int d^3r_1 d^3r_2 d^3r_3 \frac{\phi_n(\mathbf{r}_1)\Delta^*(\mathbf{r}_1)\phi_k(\mathbf{r}_1)\phi_k^*(\mathbf{r}_2)\Delta(\mathbf{r}_2)\phi_\ell^*(\mathbf{r}_2)\phi_\ell(\mathbf{r}_3)\Delta^*(\mathbf{r}_3)\phi_\ell(\mathbf{r}_3)}{(|\xi_n| - \xi_k)(|\xi_n| + \xi_\ell)} \theta(\xi_n)$$

$$-\frac{1}{4\xi_n^2}\sum_{\ell}{}' \int d^3r_1 d^3r_2 d^3r_3 \frac{\phi_n^*(\mathbf{r}_1)\Delta(\mathbf{r}_1)\phi_\ell^*(\mathbf{r}_1)\phi_\ell(\mathbf{r}_2)\Delta^*(\mathbf{r}_2)\phi_n(\mathbf{r}_2)\phi_n(\mathbf{r}_3)\Delta^*(\mathbf{r}_3)\phi_n(\mathbf{r}_3)}{|\xi_n| + \xi_\ell}\theta(\xi_n). \tag{C.19a}$$

For $\xi_n < 0$, (C.18b) requires $\varepsilon^{(3)}(\xi_n < 0) = 0$ and leaves $b_{nn}^{(3)}(\xi_n < 0)$ arbitrary; (C.18a) yields

$$a_{nn}^{(3)} = \frac{1}{2|\xi_n|}\sum_k \sum_\ell{}' \int d^3r_1 d^3r_2 d^3r_3 \frac{\phi_n^*(\mathbf{r}_1)\Delta(\mathbf{r}_1)\phi_k^*(\mathbf{r}_1)\phi_k(\mathbf{r}_2)\Delta^*(\mathbf{r}_2)\phi_\ell(\mathbf{r}_2)\phi_\ell^*(\mathbf{r}_3)\Delta(\mathbf{r}_3)\phi_\ell^*(\mathbf{r}_3)}{(|\xi_n| + \xi_k)(|\xi_n| - \xi_\ell)}\theta(-\xi_n)$$

$$-\frac{1}{4\xi_n^2}\sum_{\ell}{}' \int d^3r_1 d^3r_2 d^3r_3 \frac{\phi_n(\mathbf{r}_1)\Delta^*(\mathbf{r}_1)\phi_\ell(\mathbf{r}_1)\phi_\ell^*(\mathbf{r}_2)\Delta(\mathbf{r}_2)\phi_n^*(\mathbf{r}_2)\phi_n^*(\mathbf{r}_3)\Delta(\mathbf{r}_3)\phi_n^*(\mathbf{r}_3)}{|\xi_n| - \xi_\ell}\theta(-\xi_n); \tag{C.19b}$$

$a_{nn}^{(3)}(\xi_n > 0)$ and $b_{nn}(\xi_n < 0)$ are determined from the normalization condition

$$\int d^3r(u^{(0)*}u^{(3)} + u^{(0)}u^{(3)*} + u^{(1)*}u^{(2)} + u^{(1)}u^{(2)*} + v^{(0)*}v^{(3)} + v^{(0)}v^{(3)*} + v^{(1)*}v^{(2)} + v^{(1)}v^{(2)*}) = 0,$$

leading to

$$a_{nn}^{(3)}(\xi_n > 0) + a_{nn}^{(3)*}(\xi_n > 0) + b_{nn}^{(3)}(\xi_n < 0) + b_{nn}^{(3)*}(\xi_n < 0) = 0; \tag{C.20}$$

again setting the arbitrary phases to zero we may take

$$a_{nn}^{(3)}(\xi_n > 0) = 0; \tag{C.21a}$$

$$b_{nn}^{(3)}(\xi_n < 0) = 0. \tag{C.21b}$$

Superconducting transition temperature, thermodynamic critical field, Debye temperature and specific heat coefficient for the elements

Element	T_c(K)	H_c(Oe)	θ_D(K)	γ(mJ mol^{-1} K^{-1})
Al	1.75 ± 0.002	104.9 ± 3	420	1.35
Am*(α, ?)	0.6			
Am*(β, ?)	1.0			
Be	0.026			0.21
Cd	0.517 ± 0.002	28 ± 1	209	0.69
Ga	1.083 ± 0.001	58.3 ± 0.2	325	0.60
Ga(β)	5.9, 6.2	560		
Ga(γ)	7	950		
Ga(Δ)	7.85	815		
Hf	0.128	12.7		2.21
Hg(α)	4.154 ± 0.001	411 ± 2	87, 71.9	1.81
Hg(β)	3.949	339	93	1.37
In	3.408 ± 0.001	281.5 ± 2	109	1.672
Ir	0.1125 ± 0.001	16 ± 0.05	425	3.19
La(α)	4.88 ± 0.02	800 ± 10	151	9.8
La(β)	6.00 ± 0.1	1096, 1600	139	11.3
Lu	0.1 ± 0.03	350 ± 50		
Mo	0.915 ± 0.005	96 ± 3	460	1.83
Nb	9.25 ± 0.02	2060 ± 50	276	7.80
Os	0.66 ± 0.03	70	500	2.35
Pa	1.4			
Pb	7.196 ± 0.006	803 ± 1	96	3.1
Re	1.697 ± 0.006	200 ± 5	4.5	2.35
Ru	0.49 ± 0.015	69 ± 2	580	2.8
Sn	3.722 ± 0.001	305 ± 2	195	1.78
Ta	4.47 ± 0.04	829 ± 6	258	6.15
Tc	7.8 ± 0.1	1410	411	6.28
Th	1.38 ± 0.02	160 ± 3	165	4.32
Ti	0.40 ± 0.04	56	415	3.3
Tl	2.38 ± 0.02	178 ± 2	78.5	1.47
V	5.40 ± 0.05	1408	383	9.82
W	0.0154 ± 0.0005	1.15 ± 0.03	383	0.90
Zn	0.85 ± 0.01	54 ± 0.3	310	0.66
Zr	0.61 ± 0.15	47	290	2.77
Zr(ω)	0.65, 0.95			

From Roberts, B. W., *Properties of Selected Superconducting Materials* (1978) Supplement, NBS Technical Note 983, US Government Printing Office, Washington DC.

References

Abragam, A., (1961) *Principles of Nuclear Magnetism*, Oxford: Oxford University Press, Oxford.

Abramowitz, M. and **Stegun, I. A.**, (1970) *Handbook of Mathematical Functions*, New York: Dover Publications.

Abrikosov, A. A., (1957) *Zh. Eksp. Teor. Fiz.* **32**, 1442; (1957) *Sov. Phys. JETP* **5**, 1174.

Abrikosov, A. A., and **Gorkov, L. P.**, (1959) *Sov. Phys. JETP* **8**, 1090.

Abrikosov, A. A., and **Gorkov, L. P.**, (1961) *Sov. Phys. JETP* **12**, 1243.

Abrikosov, A. A., and **Gorkov, L. P.**, (1962) *Sov. Phys. JETP* **15**, 752.

Abrikosov, A. A., **Gorkov, L. P.**, and **Dzyaloshinski, I. E.**, (1959) *Sov. Phys. JETP* **9**, 636.

Abrikosov, A. A., **Gorkov, L. P.**, and **Dzyaloskinski, I. E.**, (1963) *Methods of Quantum Field Theory in Statistical Physics*, New York: Prentice-Hall.

Abrikosov, A. A., **Gorkov, L. P.**, and **Khalatnikov, I. M.**, (1959) *Sov. Phys. JETP* **8**, 182.

Adenwalla, S., **Lin, S. W.**, **Ran, Q. Z.**, **Zhao, Z.**, **Ketterson, J. B.**, **Sauls, J. A.**, **Taillefer, L.**, **Hinks, D.**, **Levy, M.**, and **Sarma, B. K.**, (1990) *Phys. Rev. Lett.* **65**, 2298.

Alexander, J. A. X., **Orlando, T. P.**, **Rainer, D.**, and **Tedrow, P. M.**, (1985) *Phys. Rev.* **B31**, 5811.

Ambegaokar, V., and **Baratoff, A.**, (1963) *Phys. Rev. Lett.* **10**, 486 (1963); Errata, *ibid.* **11**, 104.

Ambegaokar, V., **deGennes, P. G.**, and **Rainer, D.**, (1974) *Phys. Rev.* **A9**, 2676; *ibid.* **A12**, E 345.

Anderson, P. W., (1959) *Phys. Rev. Lett.* **3**, 325.

Anderson, P. W., (1961) in *Proceedings of the 8th Conference on Low Temperature Physics*, Toronto: University of Toronto Press.

Anderson, P. W., (1962) *Phys. Rev. Lett.* **9**, 309.

Anderson, P. W., (1964) in *Lectures on the Many Body Problem*, ed. E. R. Caianello, New York: Academic.

Anderson, P. W., and **Brinkman, W. F.**, (1973) *Phys. Rev. Lett.* **30**, 1108.

Anderson, P. W., and **Brinkman, W. F.**, (1978) in *Theory of Anisotropic Superfluidity in ^3He* in *The Physics of Liquid and Solid Helium*, Part II, edited by K. H. Bennemann and J. B. Ketterson, New York: John Wiley, Inc., p. 177.

Anderson, P. W., and **Kim, Y. B.**, (1964) *Rev. Mod. Phys.* **36**, 39.

Anderson, P. W., and **Morel, P.**, (1960) *Phys. Rev. Lett.* **5**, 136; 282(E).

Andreev, A. F., (1964) *Sov. Phys. JETP* **19**, 1228.

Aslamasov, L., and **Larkin, A.**, (1968) *Phys. Lett.* **26A**, 238.

Auvil, P. R., and **Ketterson, J. B.**, (1987) *J. Appl. Phys.* **61**, 1957.

Auvil, P. R., and **Ketterson, J. B.**, (1988) *Solid State Commun.* **67**, 1003.

Auvil, P. R., **Ketterson, J. B.**, and **Song, S. N.**, (1989) *J. Low Temp. Phys.* **74**, 103.

Balian, R., and **Werthamer, N. R.**, (1963) *Phys. Rev.* **131**, 1553.

Balsamo, E. P., **Paterno, G.**, **Barone, A.**, **Rissman, P.**, and **Russo, M.**, (1974) *Phys. Rev.* **B10**, 1881.

Bardeen, J., (1956) *Encyclopedia of Physics*, Vol. 15, Berlin: Springer-Verlag, p. 274.

Bardeen, J., (1961) *Phys. Rev. Lett.* **6**, 57.

Bardeen, J., (1962) *Phys. Rev. Lett.* **9**, 147.

Bardeen, J., and **Stephen, M. J.**, (1965) *Phys. Rev.* **140**, A1197.

Bardeen, J., **Cooper, L. N.**, and **Schrieffer, J. R.**, (1957) *Phys. Rev.* **108**, 1175.

Bardeen, J., **Kummel, R.**, **Jacobs, A. E.**, **Tewordt, L.**, (1969) *Phys. Rev.* **187**, 556.

Barone, A., and **Paterno, G.**, (1982) *Physics and Applications of the Josephson Effect*, New York: John Wiley & Sons.

Barton, G., and **Moore, M. A.**, (1974) *J. Low Temp. Phys.*, **21**, 489; (1975) *J. Phys.* **C8**, 970.

Bean, C. P., (1962) *Phys. Rev. Lett.* **8**, 250.

Bean, C. P., (1964) *Rev. Mod. Phys.* **36**, 31.

Beasley, M. R., Labusch, R., and **Webb, W. W.,** (1969) *Phys. Rev.* **181**, 682.

Belykh, V. N., Pedersen, N. F., and **Soerensen, O. H.,** (1977) *Phys. Rev.* **B16**, 4853.

Berk, N. F., and **Schrieffer, J. R.,** (1966) *Phys. Rev. Lett.* **17**, 433.

Blackford, B. L., and **March, R. H.,** (1968) *Can. J. Phys.* **46**, 141.

Blatter, G., Feigelman, M. V., Goshhenbein, V. B., Larkin, A. I., and **Vinokur, V. M.,** (1994) *Rev. Mod. Phys.* **66**, 1128.

Bogoliubov, N. N., (1958) *Zh. Eksp. Teor. Fiz.* **34**, 58; (1958) *Sov. Phys. JETP* **7**, 41; (1958) *Nuovo Cimento* **7**, 794.

Bogoliubov, N. N., Tolmachev, V. V., and **Shirkov, D. V.,** (1959) *A New Method in the Theory of Superconductivity*, New York: Consultants Bureau.

Brandt, E. H., (1977) *J. Low Temp. Phys.* **26**, 709; (1977) *ibid.* **26**, 735; (1977) *ibid.* **28**, 263; (1977) *ibid.* **28**, 291.

Brandt, E. H., (1986a) *J. Low Temp. Phys.* **64**, 375.

Brandt, E. H., (1986b) *Phys. Rev.* **B34**, 6514.

Brandt, E. H., and **Essmann, U.,** (1987) *Phys. Stat. Sol.* **B144**, 13.

Brinkman, W. F., and **Anderson, P. W.,** (1973) *Phys. Rev.* **A8**, 2732.

Campbell, A. M., and **Evetts, J. E.,** (1972) *Adv. Phys.* **21**, 199.

Carlson, R. V., (1975) Thesis, University of Minnesota (available through University Microfilms, Ann Arbor, Michigan).

Carlson, R. V., and **Goldman, A. M.,** (1975) *Phys. Rev. Lett.* **34**, 11.

Caroli, C., and **Maki, K.,** (1967a) *Phys. Rev.* **159**, 306.

Caroli, C., and **Maki, K.,** (1967b) *Phys. Rev.* **164**, 591.

Caroli, C., and **Maki, K.,** (1968) *Phys. Rev.* **169**, 381.

Caroli, C., deGennes, P. G., and **Matricon, J.,** (1964) *Phys. Lett.* **9**, 307.

Chambers, R. G., (1952) *Proc. Roy. Soc. London* **A215**, 481.

Chandrasekhar, B. S., (1962) *Appl. Phys. Lett.* **1**, 7.

Chang, J. J., and **Scalapino, D. J.,** (1978) *J. Low Temp. Phys.* **31**, 1.

Clem, J. R., (1991) *Phys. Rev.* **B43**, 7837.

Clem, J. R., and **Coffey, M. W.,** (1990) *Phys. Rev.* **B42**, 6209.

Clogston, A. M., (1962) *Phys. Rev. Lett.* **9**, 266.

Cohen, M. H., Falicov, L. M., and **Phillips, J. C.,** (1962) *Phys. Rev. Lett.* **8**, 316.

Cooper, L. N., (1956) *Phys. Rev.* **104**, 1189.

Cribier, D., Jacrot, B., Madhav Rao, L., and **Farnoux, B.,** (1964) *Phys. Lett.* **9**, 106.

Cyrot, M., (1973) *Rep. Prog. Phys.* **36**, 103.

deGennes, P. G., (1964) *Phys. Kondens, Mater.* **3**, 79; (1964) *Rev. Mod. Phys.* **36**, 225.

deGennes, P. G., (1966) *Superconductivity of Metals and Alloys*, New York: Benjamin.

deGennes, P. G., and **Matricon, J.,** (1964) *Rev. Mod. Phys.* **36**, 45.

deGennes, P. G., and **Sarma, G.,** (1966) *Solid State Commun.* **4**, 449.

Deutcher, G., and **Entin-Wohlman, O.,** (1978) *Phys. Rev.* **B17**, 1249.

Doniach, S., and **Engelsberg, S.,** (1966) *Phys. Rev. Lett.* **17**, 750.

Efetof, K. B., (1980) *Zh. Eksp. Teor. Fiz.* **78**, 2017; *Sov. Phys. JETP* **51**, 1015.

Eilenberger, G., (1968) *Z. Phys.* **214**, 195.

Eliashberg, G. M., (1960) *JETP* **11**, 696.

Ferrel, R. A., (1959) *Phys. Rev. Lett.* **3**, 262.

Ferrel, R. A., (1969) *J. Low Temp. Phys.* **1**, 241.

Feynman, R. P., and Hibbs, A. R., (1965) *Quantum Mechanics and Path Integrals*, New York: McGraw-Hill.

Fisher, D. S., Fisher, M. P. A., and **Huse, D. A.,** (1991) *Phys. Rev.* **B43**, 130.

Fiske, M. D., (1964) *Rev. Mod. Phys.* **36**, 221; this journal issue is the proceedings of a conference held in 1963 and a qualitative explanation of the origin of the steps was given by A. B. Pippard during the discussion following the presentation of Fiske's paper.

Fredkin, E. S. (1959) *Sov. Phys. JETP* **9**, 912.

Friedel, J., deGennes, P. G., and **Matricon, J.,** (1963) *Appl. Phys. Lett.* **2**, 119.

Fröhlich, H., (1950) *Phys. Rev.* **79**, 845.

Fulde, P., and **Ferrell, R. A.,** (1964) *Phys. Rev.* **135A**, 550.

Fulde, P., and **Maki, K.,** (1966) *Phys. Rev.* **141**, 275; (1966) *ibid.* **147**, 414.

Gammel, P. L., Schneemeyer, L. F., and **Bishop, D. J.,** (1991) *Phys. Rev. Lett.* **66**, 953.

Garland, J. W., (1963) *Phys. Rev. Lett.* **11**, 114.

Ginzburg, V. L., and **Landau, L. D.**, (1950) *Zh. Eksp. Teor. Fiz.* **20**, 1064.

Goldstone, J., (1961) *Nuovo Cimento* **19**, 154.

Gorkov, L. P., (1958) *Sov. Phys. JETP* **7**, 505.

Gorkov, L. P., (1959) *Sov. Phys. JETP* **9**, 1364.

Gorkov, L. P., and **Eliashberg, G. M.**, (1968) *Sov. Phys. JETP* **27**, 328; (1968) *Sov. Phys. JETP Lett.* **8**, 202.

Gorter, C. J., and **Casimir, H. B. G.**, (1934a) *Phys. Z.* **35**, 963.

Gorter, C. J., and **Casimir, H. B. G.**, (1934b) *Physica* **1**, 306.

Gradshteyn, I. S., and **Ryzhik, I. M.**, (1994) *Tables of Integrals, Series and Products*, Boston: Academic Press.

Hebel, L. C., (1959) *Phys. Rev.* **116**, 1504.

Hebel, L. C., and **Slichter, C. P.**, (1957) *Phys. Rev.* **113**, 901.

Hess, D. W., **Riseborough, P. S.**, and **Smith, J. L.**, (1993) *Encyclopedia of Applied Physics*, New York: VCH Publishers.

Houghton, A., **Pelcovits, R. A.**, and **Sudbo, S.**, (1989) *Phys. Rev.* **B40**, 6763.

Hu, C. R., and **Thompson, R. S.**, (1972) *Phys. Rev.* **B6**, 110.

Imry, Y., and **Strongin, M.**, (1981) *Phys. Rev.* **B24**, 6353.

Johnson, W. J., (1968) Ph.D. thesis, University of Wisconsin, Madison.

Josephson, B. D., (1962) *Phys. Lett.* **1**, 251.

Josephson, B. D., (1964) *Rev. Mod. Phys.* **36**, 216.

Josephson, B. D., (1965) *Adv. Phys.* **14**, 419.

Josephson, B. D., (1969) in *Superconductivity*, ed. R. D. Parks, Vol. 1, New York: Marcel Dekker.

Kamerlingh-Onnes, H., (1911) *Leiden Commun.* 120b, 122b, 124c.

Kaplan, S., **Chi, C. C.**, **Langenberg, D. N.**, **Chang, J. J.**, **Jafarey, S.**, and **Scalapino, D. J.**, (1976) *Phys. Rev.* **B14**, 4854.

Keesom, W. H., and **van Laer, P. H.**, (1938) *Physica* **5**, 193.

Keldysh, L. V., (1965) *Sov. Phys. JETP* **20**, 1018.

Kim, Y. B., and **Stephen, M. J.**, (1969) in *Superconductivity*, ed. R. D. Parks, Vol. 2, New York: Marcel Dekker.

Kim, Y. B., **Hempstead, C. F.**, and **Strand, A. R.**, (1963) *Phys. Rev.* **129**, 528.

Kim, Y. B., **Hempstead, C. F.**, and **Strand, A. R.**, (1965) *Phys. Rev.* **139**, A1163.

Kleiner, W. M., **Roth, L. M.**, and **Autler, S. H.**, (1964) *Phys. Rev.* **133A**, 1226.

Klemm, R., (1974) Thesis, Harvard University.

Klemm, R. A., (1983) *Solid State Commun.* **46**, 705.

Klemm, R. A., **Luther, A.**, and **Beasley, M. R.**, (1975) *Phys. Rev.* **B12**, 877.

Koch, V. E., and **Wölfle, P.**, (1981) *Phys. Rev. Lett.* **46**, 486.

Kogan, V. G., (1981) *Phys. Rev.* **B24**, 1572.

Koperdraad, R., (1995) Ph.D. thesis, Vrije Universiteit, Amsterdam.

Kulik, I. O., (1965a) *Sov. Phys. JETP Lett.* **2**, 85.

Kulik, I. O., (1965b) *Zh. Eskp. Teor. Fiz.* **49**, 1212.

Kulik, I. O., (1966) *Sov. Phys. JETP* **22**, 841.

Labusch, R., (1967) *Phys. Status Solidi* **19**, 715.

Labusch, R., (1969a) *Cryst. Lattice Defects* **1**, 1.

Labusch, R., (1969b) *Phys. Status Solidi* **32**, 439.

Landau, L. D., (1937) *Phys. Z. Sowjet* **11**, 129.

Landau, L. D., (1941) *J. Phys. (USSR)* **5**, 71.

Landau, L. D., (1957a) *Sov. Phys. JETP* **3**, 920.

Landau, L. D., (1957b) *Sov. Phys. JETP* **5**, 101.

Landau, L. D., (1959) *Sov. Phys. JETP* **8**, 70.

Landau, L. D., and **Khalatnikov, I. M.**, (1954) *Dokl. Adad. Nauk S.* **96**, 469.

Landau, L. D., and **Lifshitz, E. M.**, (1977) *Quantum Mechanics*, Oxford: Pergamon.

Landau, L. D. and **Lifshitz, E. M.**, (1987) *Fluid Mechanics*, Oxford: Pergamon.

Landau, L. D., **Lifshitz, E. M.**, and **Pitaevskii, L. P.**, (1984) *Electrodynamics of Continuous Media*, New York: Pergamon.

Larkin, A. I., and **Ovchinnikov, Yu. N.**, (1978) *Sov. Phys. JETP* **27**, 280.

Larkin, A. I., and **Ovchinnikov, Yu. N.**, (1979) *J. Low Temp. Phys.* **34**, 409.

Lawrence, W. E., and **Doniach, S.**, (1971) in *Proceedings of the 16th International Conference on Low Temperature Physics*, ed. E. Kanda, Kyoto: Academic Press of Japan, p. 361.

Leggett, A. J., (1966) *Prog. Theor. Phys.* **36**, 901.

Leggett, A. J., (1975) *Rev. Mod. Phys.* **47**, 331.

Lifshitz, E. M., and **Pitaevskii, L. P.**, (1980) *Statistical Physics*, Part 2, Oxford: Pergamon, Ch. 5.

Lifshitz, E. M., and **Pitaevskii, L. P.**, (1981) *Physical Kinetics*, New York: Pergamon.

Livingston, J. D., (1963) *Phys. Rev.* **129**, 1943.

London, F., (1950) *Superfluids*, Vol. 1, New York: John Wiley & Sons. Reprinted (1961) by Dover Publications, Inc., New York.

London, F., and **London, H.**, (1935) *Proc. Roy. Soc.* (*London*) **A149**, 71.

Luders, G., and **Usadel, K.**, (1971) *The Method of Correlation Functions in Superconductivity Theory*, Berlin: Springer.

Maki, K., (1966) *Phys. Rev.* **148**, 362.

Mast, D. B., Sarma, B. K., Owers-Bradley, J. R., Calder, I. D., Ketterson, J. B., and **Halperin, W. P.**, (1980) *Phys. Rev. Lett.* **45**, 266.

Mattis, D. C., and **Bardeen, J.**, (1958) *Phys. Rev.* **111**, 412.

McLaughlin, D. W., and **Scott, A. C.**, (1977) *Appl. Phys. Lett.* **30**, 545.

McLaughlin, D. W., and **Scott, A. C.**, (1978) *Phys. Rev.* **A18**, 1652.

Meissner, W., and **Ochsenfeld, R.**, (1933) *Naturwissen* **21**, 787.

Menon, M., and **Arnold, G. B.**, (1985) *Superlattices and Microstructures* **1**, 451.

Mermin, N. D., (1974) *Phys. Rev.* **A9**, 865.

Mermin, N. D., and **Stare, G.**, (1973) *Phys. Rev. Lett.* **30**, 1135.

Minenko, E. V., (1983) *Fiz. Nizk. Temp.* **9**, 1036; (1983) *Sov. J. Low Temp. Phys.* **9**, 535.

Miyano, K., and **Ketterson, J. B.**, (1979) *Physical Acoustics*, eds. W. P. Mason and R. N. Thurston, New York: Academic Press.

Monthouk, P., Balatsky, A., and **Pines, D.**, (1992) *Phys. Rev.* **B46**, 14803.

Morse, P. M., and **Feshbach, H.**, (1953) *Methods of Theoretical Physics*, New York: McGraw-Hill.

Morse, R. W., and **Bohm, H. V.**, (1957) *Phys. Rev.* **108**, 1094.

Morse, R. W., Olsen, T., and **Gavenda, J. D.**, (1959) *Phys. Rev. Lett.* **3**, 15.

Nambu, Y., (1960) *Phys. Rev.* **117**, 648.

Nakajima, S., (1973) *Prog. Theor. Phys.* **50**, 1101.

Nozières, P., (1964) *Theory of Interacting Fermi Systems*, New York: W. A. Benjamin, Inc.

Osheroff, D. D., Richardson, R. C., and **Lee, D. M.**, (1972) *Phys. Rev. Lett.* **28**, 885.

Patashinskii, A. Z., and **Pokrovskii, V. L.**, (1979) *Fluctuation Theory of Phase Transitions*, Oxford: Pergamon.

Pearl, J., (1964) *Appl. Phys. Lett.* **5**, 65.

Pines, D., and **Nozières, P.**, (1966) *The Theory of Quantum Liquids*, New York: Benjamin.

Pippard, A. B., (1953) *Proc. Roy. Soc. London* **A216**, 547.

Radović, Z., Ledvij, M., and **Dobrosavljević Grujić, L.**, (1991) *Phys. Rev.* **B43**, 8613.

Rothwarf, A., and **Taylor, B. N.**, (1967) *Phys. Rev. Lett.* **19**, 27.

Rutgers, J. (1933) in Ehrenfest, P., (1933) *Leiden Commun. Suppl.* 75b.

Saint-James, D., (1965) *Phys. Lett.* **16**, 218.

Saint-James, D., and **deGennes, P. G.**, (1963) *Phys. Lett.* **7**, 306.

Saint-James, D., Thomas, E. J., and **Sarma, G.**, (1969) *Type II Superconductivity*, New York: Pergamon.

Sakurai, J., (1985) *Modern Quantum Mechanics*, Menlo Park, California: Benjamin Cummings.

Salomaa, M. M., and **Volovik, G. E.**, (1989) *J. Low Temp. Phys.* **77**, 17.

Sarma, B. K., Levy, M., Adenwalla, S., and **Ketterson, J. B.**, (1992) *Physical Acoustics*, Vol. XX, ed. M. Levy, Boston: Academic Press, Inc., p. 107.

Sarma, G., (1963) *J. Phys. Chem. Solids* **24**, 1029.

Sarma, G., and **Saint-James, D.**, (1969) unpublished; reproduced in Saint-James, Thomas, and Sarma (1969).

Schmid, A., (1966) *Phys. Kondens. Mater.* **5**, 302.

Schmid, A., (1969) *Phys. Rev.* **180**, 527.

Schmidt, H., (1968) *Z. Phys.* **216**, 336.

Serene, J., and **Rainer, D.**, (1983) *Phys. Rep.* **101**, 223.

Shapiro, S., (1963) *Phys. Rev. Lett.* **11**, 34; for a review see Shapiro, S., Janus, A. R., and **Holly, S.**, (1964) *Rev. Mod. Phys.* **36**, 223.

Shelankov, A. L., (1985) *J. Low Temp. Phys.* **60**, 29.

Shubnikov, L. V., Khotkevich, V. I., Shepelev, Y. D., and **Riabinin, Y. N.**, (1937) *Zh. Eskp. Teor. Fiz.* **7**, 221.

Shul, H., Matthias, B. T., and **Walker, L. R.**, (1959) *Phys. Rev. Lett.*, **3**, 552.

Silin, V. P., (1958) *Sov. Phys. JETP* **6**, 387.

Silver, A. H., and **Zimmerman, J. E.**, (1967) *Phys.*

Rev. **157**, 317.

Simáneck, E., (1979) *Solid State Commun.* **31**, 419.

Singwi, K. S., (1964) in *Phonons and Phonon Interactions*, ed. T. Bak, New York: Benjamin, p. 167.

Stewart, W. C., (1968) *Appl. Phys. Lett.* **12**, 277.

Stewart, W. C., (1974) *J. Appl. Phys.* **45**, 452.

Stoner, E. C., (1938) *Proc. Roy. Soc. London* **A165**, 372.

Swihart, J., (1961) *J. Appl. Phys.* **32**, 451.

Takahashi, S., and **Tachiki, M.**, (1985) *Physica* **135B**, 178.

Takahashi, S., and **Tachiki, M.**, (1986a) *Phys. Rev.* **B33**, 4620.

Takahashi, S., and **Tachiki, M.**, (1986b) *Phys. Rev.* **B34** 3162.

Takahashi, S., and **Tachiki, M.**, (1987) *Phys. Rev.* **B35**, 145.

Taylor, B. N., **Parker, W. H.**, and **Langenberg, D. N.**, (1969) *Rev. Mod. Phys.* **41**, 375.

Theodorakis, S., (1988) *Phys. Rev.* **B37**, 3318.

Thompson, R. S., and **Hu, C. R.**, (1971) *Phys. Rev. Lett.* **27**, 1352.

Tinkham, M., (1963) *Phys. Rev.* **129**, 2413.

Tinkham, M., (1964) *Phys. Rev. Lett.* **13**, 804.

Tinkham, M., (1975) *Introduction to Superconductivity*, New York: McGraw-Hill.

Tinkham, M., (1988) *Phys. Rev. Lett.* **61**, 1658.

Tokuyasu, T. A., (1990) Vortex states in unconventional superconductors, thesis, Princeton University.

Tokuyasu, T. A., **Hess, D. W.**, and **Sauls, J. A.**, (1990) *Phys. Rev.* **B41**, 8891.

Trauble, H., and **Essmann, U.**, (1968) *J. Appl. Phs.* **39**, 4052.

Usadel, K. D., (1970) *Phys. Rev. Lett.* **25**, 507.

Valatin, J. G., (1958) *Nuovo Cimento* **7**, 843.

Vollhardt, D., and **Wölfle, P.**, (1990) *The Superfluid Phase of Helium* 3, London: Taylor and Francis.

Volovik, G. E., and **Gorkov, L. P.**, (1985) *Sov. Phys. JETP* **61**, 843.

Volovik, G. E. and **Khazan, M. V.**, (1984) *Sov. Phys. JETP* **60**, 276.

Werthamer, N. R., **Helfand, E.**, and **Hohenberg, P. C.**, (1966) *Phys. Rev.* **147**, 1295.

Wohlforth, E. P., (1953) *Rev. Mod. Phys.* **24**, 211.

Wolf, E. L., (1985) *Principles of Electron Tunneling Spectroscopy*, New York: Oxford University Press.

Wollman, D. A., **vanHarlingen, D. J.**, **Lee, W. C.**, **Ginsburg, D. M.**, and **Leggett, A. J.**, (1993) *Phys. Rev. Lett.* **71**, 2134.

Wong, H. K., and **Ketterson, J. B.**, (1986) *J. Low Temp. Phys.* **63**, 307.

Yamafuji, K., **Kawashima, T.**, and **Irie, F.**, (1966) *Phys. Lett.* **20**, 123.

Yeshurun, Y., and **Malozemoff, A. P.**, (1988) *Phys. Rev. Lett.* **60**, 2202.

Yoshida, H., **Katada, F.**, **Sakamoto, Y.**, **Iwasa, A.**, and **Endo, T.**, (1993) *Jpn. J. Appl. Phys.* **32**, Pt. 1, 3643.

Zhao, Z., **Adenwalla, S.**, **Sarma, B.**, and **Ketterson, J. B.**, (1992) *Adv. Phys.* **41**, 147.

Ziman, J. M., (1964) *Principles of the Theory of Solids*, New York: Cambridge University Press, p. 242.

Zimmerman, J. E., (1972) *Cryogenics* **12**, 19.

Additional reading

Fundamentals of Superconductivity, **V. Kresin** and **S. Wolf**, New York: Plenum (1980). This book is authored by an active theorist and experimentalist, respectively.

Introduction to Superconductivity, **M. Tinkham**, New York: McGraw-Hill (1975); Krieger (1980). Written by one of the leading experimentalists, the book contains very readable accounts of BCS and G–L theory. The chapters on magnetic properties, Josephson effect and fluctuations are quite valuable.

The Many Body Problem, **D. Pines**, New York: Benjamin (1962). The book begins with a brief summary discussion by Pines of a number of topics and methods in many-body theory, followed by reprints of classic papers in the field. One comes to appreciate Pines' advice to learn several many-body-theory 'languages' and one is constantly referring to various reprints.

Many Body Physics, **G. D. Mahan**, New York: Plenum Press (1981). An exhaustive book on many-body theory which one can learn from due to the fact the author is kind enough to put in the needed mathematical discussions (and steps). It contains much material on superconductivity and the discussion of Eliashburg theory (not treated in the present book) is very useful.

The Method of Correlation Functions in Superconductivity Theory, **G. Luders** and **K. D. Usadel**, Berlin: Springer-Verlag (1971). An interesting book that develops correlation function and quasiclassical methods along lines initiated by deGennes' approach.

Methods of Quantum Field Theory in Statistical Physics, **A. A. Abrikosov, L. P. Gorkov,** and **I. E. Dzyaloshinski**, Englewood Cliffs, New Jersey: Prentice Hall (1963). The authors introduced Green's function methods into the theory of superconductivity. The book is the primary reference for the theory of superconducting alloys.

A New Method in the Theory of Superconductivity, **N. N. Bogoliubov, V. V. Tolmachev** and **D. V. Shirkov**, New York: Consultants Bureau (1959). A somewhat difficult book to read with extended discussions of the justification of, and possible shortcomings of, the pairing theory.

Physics and Applications of the Josephson Effect, **A. Barone** and **G. Paterno**, New York: John Wiley and Sons (1982). The Josephson effect and superconducting magnets are the presently important applications of superconductivity. This book provides an exhaustive account of the former and is recommended to those specializing in this area.

Statistical Physics Part II, **E. M. Lifshitz** and **L. P. Pitaevskii**, New York: Pergamon (1980). Written in the Landau and Lifshitz style this book wastes no words but simultaneously is very clear. The discussion of Green's functions in many-body physics is the clearest exposition in print in the opinion of the present authors. The discussions of the Landau Fermi liquid theory and the G–L theory are excellent.

Superconductivity of Metals and Alloys, **P. G. deGennes**, New York: W. A. Benjamin (1966). An excellent text (and a motivating force behind the present book) from which much can be extracted, particularly with respect to using the Bogoliubov approach to derive a broad range of superconducting and superfluid phenomena.

Superconductivity Parts I and II, **R. Parks**, Ed., New York: Marcel Dekker (1969). Exhaustive treatments, by acknowledged experts in the field, intended to summarize the state of the field of superconductivity as it existed in 1969. Although a primary reference for many subjects, the level of sophistication needed to utilize the material varies widely among sections.

Superfluidity and Superconductivity, **O. R. Tilley** and **J. Tilley**, Bristol: Adam Hilger Ltd. (1986). This book, which is intended for graduate and undergraduate students, treats many of the topics discussed here, but at a somewhat simpler level. It is useful as an introduction to various topics in superconductivity.

Superfluids, Vol. I and II, **F. London**, New York: John Wiley and Sons (1950); New York: Dover (1961). Although now quite dated, this book has useful discussions of the macroscopic aspects of superconductivity. The book was of great historical importance and flux quantization was first proposed there.

Theory of Superconductivity, **J. R. Schrieffer**, New York: Benjamin (1964). An excellent exposition of the BCS pairing theory itself including a fascinating discussion of the motivation behind Schrieffer's choice of the many-body ground-state wavefunction.

The Theory of Quantum Liquids, Vol. I. *Normal Fermi Liquids*, **D. Pines** and **P. Nozières**, New York: W. A. Benjamin, Inc. (1966). The book begins with a discussion of the Landau Fermi liquid theory and contains some insights not found elsewhere. Most of the remainder of the book is devoted to developing methods involving correlation functions.

Type II Superconductivity, **D. Saint-James, E. J. Thomas** and **G. Sarma**, New York: Pergammon (1969). Likely the best discussion of the high field properties of Type II superconductors. The Bogoliubov–deGennes approach is adopted for the microscopic properties and the discussions of paramagnetic effects and gapless superconductivity are particularly useful. The G–L treatment of surface superconductivity is also recommended.

List of mathematical and physical symbols

Symbol	Explanation	Introductory section
a	defined by $\alpha = a(T - T_c)$, where α is G–L parameter	10
a	vortex lattice periodicity	14
a	insulator layer thickness	19
$\hat{a}_k^\dagger, \hat{a}_k$	Bose creation/destruction operator for state k	A
a_H	Landau-level magnetic length	10
$a_{nm}, a_{nm}(t)$	Bogoliubov wavefunction expansion coefficients	38
α	amplitude of G–L wavefunction	9
$\hat{A}$	antisymmetrizing operator	26
$\mathbf{A}(\mathbf{r}), \mathbf{A}(\mathbf{r}, t)$	magnetic vector potential	3, 50
$A(\mathbf{k}, \mathbf{k}')$	scattering amplitude	55
$\mathscr{A}$	interface area	6
b	vortex lattice periodicity	14
b	superconductor layer thickness	19
$b_{nm}, b_{nm}(t)$	Bogoliubov wavefunction expansion coefficients	38
$\hat{b}_k^\dagger, \hat{b}_k$	BCS pair creation/destruction operators	B
$\mathbf{B}$	thermodynamic magnetic induction	1
$\mathbf{B}(\mathbf{r})$	macroscopic magnetic induction	20
c	velocity of light	2
$\bar{c}$	mode velocity in a junction	19
$\hat{c}_{\mathbf{k}\sigma}$	electron destruction operator for a state $\vert\mathbf{k}\sigma\rangle$	26
$\hat{c}_{k\sigma}^\dagger$	electron creation operator for a state $\vert\mathbf{k}\sigma\rangle$	26
c_{ij}	vortex lattice elastic constants	20
C	total heat capacity	4
C	electrical capacity	18
$\mathscr{C}$	charge conjugation operator	38
d	vortex spacing	8

Symbol	Explanation	Introductory section
d	superconductor film thickness	12
$d_0(\mathbf{k})$	singlet BCS gap function	52
$\mathbf{d}(\mathbf{k})$	triplet vector BCS gap function	52
D	superlattice layer thickness defined by $D = d + s$	16
D	diffusion coefficient	39
$\mathbf{D}(\mathbf{r})$	electric displacement field	31
$D_{\alpha\beta}$	matrix defining p and d paired gaps	23, 53
e	charge of the electron	2
$\mathbf{e}(\mathbf{p})$	quasiparticle spin energy function	24, 52
e^*	G–L charge ($= 2e$)	2
$\mathbf{E}(\mathbf{r}), \mathbf{E}(\mathbf{r}, t)$	electric field	2, 50
E	total energy	2
E_J	Josephson coupling energy	18
E_{0S}	superconducting ground-state energy	36
$f(\mathbf{p}, \mathbf{p}')$	Landau quasiparticle interaction function	24
$f^{(s)}(\mathbf{p}, \mathbf{p}'), f^{(a)}(\mathbf{p}, \mathbf{p}')$	symmetric and antisymmetic Landau interaction functions	52
$f_{\mathbf{k}\sigma}$	Fermi distribution function for state $\vert \mathbf{k}\sigma\rangle$	28
$f(\mathbf{r}, t; \hat{\mathbf{k}}, \omega)$	quasiclassical (energy integrated) anomalous Green's function	55
$f(\mathbf{r}, t; \hat{\mathbf{k}}, \omega_m)$	quasiclassical (energy-intergrated) anomolous Matsubara Green's function	55
$\mathbf{f}_\eta$	viscous damping force density	20
$\mathbf{f}_p$	pinning force density	20
$\mathbf{f}_{el}$	elastic force density	20
$\mathbf{f}_L$	Lorentz force density	20
F	total Helmholtz free energy	2
$F(\mathbf{r})$	pair amplitude	44, 55
$F_{\alpha\beta}(\mathbf{r}_1, t_1; \mathbf{r}_2, t_2)$	anomalous one-particle Green's function	55
$F_{\alpha\beta}(\mathbf{k}, \omega)$	Fourier transform of anomalous one-particle Green's function	55
F_p	pinning free energy	20
$F(\Omega)$	phonon density of states	35
F_e	effective free energy	37
$\mathscr{F}(\mathbf{r})$	Helmholtz free energy density	9
$\mathscr{F}_p(\mathbf{r})$	Helmholtz pinning free energy density	20
$\mathscr{F}_{\alpha\beta}(\mathbf{r}_1, t_1; \mathbf{r}_2, t_2)$	Matsubara anomalous Green's function	55
$g(\mathbf{k})$	Fourier transform of center-of-mass wavefunction	25
$g_\xi(\mathbf{r}, \mathbf{r}', t)$	correlation function	38
$g_\xi(\mathbf{r}, \mathbf{r}', \Omega)$	Fourier time transform of correlation function	38
$g(\mathbf{r}, t; \hat{\mathbf{k}}, \omega)$	quasiclassical (energy integrated) Green's function	55
$g(\mathbf{r}, \hat{\mathbf{k}}, \omega_m)$	quasiclassical Matsubara anomalous Green's function	55

Symbol	Explanation	Introductory section
$g_\xi(\mathbf{r}, \mathbf{r}', t)$	unnormalized correlation function	42
$g_\xi(\mathbf{r}, \mathbf{r}', \Omega)$	Fourier transform of unnormalized correlation function	42
$G_{\alpha\beta}(\mathbf{r}_1, t_1; \mathbf{r}_2, t_2)$	one-particle Green's function	55
$G_{\alpha\beta}(\mathbf{k}, \omega)$	Fourier transform of one-particle Green's function	55
$G^{\mathrm{R}}, G^{\mathrm{A}}, G^{\mathrm{K}}$	retarded, advanced, and Keldysh Green's functions, respectively	55
G	total Gibbs free energy	4
$\mathcal{G}_{\alpha\beta}(\mathbf{r}_1, t_1; \mathbf{r}_2, t_2)$	Matsubara Green's function	55
$\hbar$	Planck's constant ($h/2\pi$)	3
$\underset{\sim}{h}_m, \underset{\sim}{h}_{m\mu\mu'}$	energy matrices defined by Eqs. (50.7) and (54.12)	50, 54
$\underset{\sim}{h}_{mn\mu\nu}(t)$	matrix element of total mean field	54
$\mathbf{H}(\mathbf{r}), \mathbf{H}(\mathbf{r}, t)$	local magnetic field	2, 50
H_{c}	thermodynamic critical field	1
H_{c1}	lower critical field	8
H_{c2}	upper critical field	10
H_{c3}	upper critical field for surface superconductivity	13
$H_{\mathrm{c2}\parallel}$	parallel upper critical field	11
$H_{\mathrm{c2}\perp}$	perpendicular upper critical field	11
$\hat{H}$	second quantized Hamiltonian	26
$\hat{H}_0$	zeroth order second quantized Hamiltonian	26
$\hat{H}_{\mathrm{I}}$	interaction second quantized Hamiltonian	26
$\hat{H}_{\mathrm{IR}}$	reduced second quantized interaction Hamiltonian	26
$\hat{H}_{\mathrm{R}}$	BCS reduced second quantized Hamiltonian	26
H_1	perturbing Hamiltonian	35
$\hat{H}_{\mathrm{T}}$	second quantized tunneling Hamiltonian	29
$\hat{H}^{(\mathrm{e})}$	second quantized effective Hamiltonian	36
$H_\mu^{(1),(2)}(z)$	Hankel functions of the first and second kinds	48
$\mathcal{H}$	granular superconductor Hamiltonian	18
$\mathcal{H}'_0$	one-electron Hamiltonian	35
I	contact interaction strength	31
I	electric current	29
j_{max}	maximum Josephson supercurrent	15
$\hat{\mathbf{j}}(\mathbf{r})$	second quantized current operator	37
$\mathbf{j}(\mathbf{r}), \mathbf{j}_{\mathrm{e}}(\mathbf{r})$	electron current density	2, 31
$\mathbf{j}_{\mathrm{i}}(\mathbf{r})$	ion current density	31
$\mathbf{j}_{\mathrm{s}}(\mathbf{r})$	microscopic supercurrent density	20
$\mathbf{J}_{\mathrm{s}}(\mathbf{r})$	macroscopic supercurrent density	20
$\mathbf{J}(\mathbf{r})$	electron current density	19
J_{m}	maximum Josephson current	20
J_{c}	critical current density	24
$\mathbf{k}$	wave vector	10

Symbol	Explanation	Introductory section
k_B	Boltzmann's constant	3
K_n	modified Bessel function	7
$\mathbf{K}$	surface current density	17
$K(\mathbf{r},\mathbf{r}')$	kernel of linearized self-consistency condition	38
$K(\mathbf{r},t;\mathbf{r}',t')$	time-dependent kernel of linearized self-consistency condition	49
$K(\mathbf{q})$	Fourier transform of $K(\mathbf{r}-\mathbf{r}')$	40
$\mathscr{K}(t)$	Wigner time reversal operator	41
ℓ	electron mean free path	3
ℓ	vortex hopping range	20
ℓ	element of line integration	7
L	junction length	19
$L(\xi_m, \xi_n)$	BCS function entering diamagnetic response	46
$L(\omega, \xi_m, \xi_n)$	function entering finite frequency diamagnetic response	50
$\mathscr{L}$	granular superconductor Hamiltonian	18
$\mathscr{L}$	differential operator defined by $-D(\mathbf{r})\left[\mathbf{\nabla}-\frac{2ie}{\hbar c}\mathbf{A}(\mathbf{r})\right]^2$	44
m	mass of the electron	2
m^*	G–L mass ($=2m$)	9
m^*	quasiparticle effective mass	24
$\mathbf{m}(\mathbf{p})$	vector spin distribution function	24, 52
$m(t)$	time-dependent nuclear magnetization	34
m_i	diagonal normalized mass tensor elements; $m_i \equiv M_i/\bar{M}$	17
m_{ij}	mass tensor elements	17
M	ion mass	31
M_i	diagonal mass tensor elements	17
$\bar{M}$	geometric mean mass	17
$\mathbf{M}(\mathbf{r})$	magnetization per unit volume (magnetization)	8
M_{ij}	matrix element entering deGennes–Ginzburg–Landau boundary condition	9
$M_{nm}, M_{nm}(t)$	matrix elements entering diamagnetic current	46, 50
n	electron number density; density of superconducting electrons	2
$n(\mathbf{p}), n(\mathbf{r},\mathbf{p},t)$	quasiparticle distribution function	24, 52, 55
$n_{\alpha\beta}(\mathbf{k})$	spin-generalized distribution function	41, 52
$n^{\mathrm{A}}_{\alpha\beta}(\mathbf{k})$	spin-generalized anomalous distribution function	41, 52
$n^{\mathrm{s}}_{\alpha\beta}(\mathbf{k}), n^{\mathrm{s}}(\mathbf{k}), n^{\mathrm{A}}(\mathbf{k},\mathbf{r},t)$	singlet anomalous distribution functions	41, 52, 55
$n^{\mathrm{t}}_{\alpha\beta}(\mathbf{k}), \mathbf{n}^{\mathrm{t}}(\mathbf{k})$	triplet anomalous distribution functions	41, 52
$n(\Omega)$	Bose distribution function for phonons	35
$\hat{n}$	granular superconductor number operator	18
n_{L}	number of flux lines per unit area	8

Symbol	Explanation	Introductory section
$\bar{N}$	average number of electrons	26
N	number of electrons	26
$\hat{N}$	number operator	26
N_{qp}	number of quasiparticles	35
N_{ph}	number of phonons	35
$\mathcal{N}(\xi, \mathbf{r})$	local electronic density of states	42
$\mathcal{N}(\xi)$	electronic density of states per spin	24
$\mathcal{N}_{\mathrm{F}}$	Fermi surface density of states for both spins	24
$\mathbf{p}$	particle momentum	9
$\hat{p}$	operator conjugate to $\hat{n}$	18
P_{ij}	permutation operator	A
$\mathbf{q}$	momentum transfer; wave vector	25
$Q_\omega(\mathbf{r}, \mathbf{r}')$	a quantity entering kernel of linearized self-consistency condition	42
R	electrical resistance	17
$\mathbf{R}$	coordinate of center of mass	25
$\mathbf{R}(\psi)$	Euler rotation matrix	53
R	quasiparticle recombination strength	35
$R_\pm(r)$	radial contributions to vortex wavefunction	48
$R(t-\tau)$	tunneling correlation function	25
$R_\omega(\mathbf{r}, \mathbf{r}')$	a quantity analogous to $Q_\omega(\mathbf{r}, \mathbf{r}')$	44
s	Josephson lattice spacing; insulator thickness	16
s	action relative to value at Fermi surface	47
s_u	particle-like action relative to value at Fermi surface	47
s_v	hole-like action relative to value at Fermi surface	47
$\bar{s}$	$\equiv \frac{1}{2}(s_u + s_v)$	47
s	$\equiv \frac{1}{2}(s_u - s_v)$	47
S	total entropy	4
S	time-independent action	47
S_{F}	action evaluated at Fermi surface	47
$S(t-\tau)$	tunneling correlation function	30
$S(\mathbf{r}, t)$	probability flux for Bogoliubov equations	47
$S_{ij}(\mathbf{r}, \mathbf{r}')$	kernel of nonlocal London equation	46
$S_{ij}(\mathbf{r}, \mathbf{r}', \omega)$	finite frequency kernel of nonlocal London equation	50
T	absolute temperature	1
T	transmission coefficient	47
$T_{r\rho\ell\lambda}$	tunneling matrix element	29
T_{c}	transition temperature	1
u	reduced soliton velocity	19
$u_{N\alpha}(\mathbf{r}), u_{N\alpha}(\mathbf{r}, t)$	spin-generalized Bogoliubov particle-like wavefunction for state $\vert N\rangle$	41. 54

Symbol	Explanation	Introductory section
$u_n(\mathbf{r}), u_n(\mathbf{r}, t)$	Bogoliubov particle-like wavefunction for state $\|n\rangle$	35
$u_\mathbf{k}$	BCS amplitude for state $\|\mathbf{k}\rangle$	26
$U(z)$	'universal function' $\left(\equiv \psi\left(\frac{z}{2}+\frac{1}{2}\right)-\psi\left(\frac{1}{2}\right)\right)$	40
U	single-electron charging energy: $U = e^2/C$	18
$U^{(1)}_{\alpha\beta}(\mathbf{r})$	one-body potential	36
$U^{(2)}_{\delta\gamma,\ \alpha\beta}(\mathbf{r}, \mathbf{r}'')$	two-body potential	36
$U(\mathbf{r}), U(\mathbf{r}, t)$	mean one-electron potential	36
$U_{\alpha\beta}(\mathbf{r}, \mathbf{r}'), U_{\alpha\beta}(\mathbf{r}, \mathbf{r}', t)$	nonlocal spin-generalized one-electron potential	51, 54
$U_{\alpha\beta}(\mathbf{k})$	spin-generalized mean field contribution to one-electron energies	51
$\mathbf{v}$	electron velocity	2
$\mathbf{v}$	vortex velocity	20
$v_n(\mathbf{r}), v_n(\mathbf{r}, t)$	Bogoliubov hole-like wavefunction for state $\|n\rangle$	26, 49
$v_{N\alpha}(\mathbf{r}), v_{N\alpha}(\mathbf{r}, t)$	spin-generalized Bogoliubov hole-like wavefunction for state $\|N\rangle$	41, 54
v_F	Fermi velocity	
$v_\mathbf{k}$	BCS amplitude for state $\|\mathbf{k}\rangle$	26
V	Josephson barrier energy	18
$V(\mathbf{r}_1, \mathbf{r}_2)$	pairing potential	25
$V_{\mathbf{k},\mathbf{k}'}$	Fourier transform of pairing potential	25
V	BCS model potential strength	25
$V^{(s)}, V^{(t)}$	singlet–triplet BCS model potential strengths	52
$V_{0,1}$	singlet and triplet interaction functions	31
$\hat{V}_2$	second quantized interaction Hamiltonian	
V	junction voltage	19
V_i	ion velocity	31
w	transition probability per unit time	29
w_{nm}	transition probability per unit time	46
$w_n(\mathbf{r})$	one-electron wavefunctions (real)	36
W	power dissipation	46
$\alpha(T)$	coefficient of quadratic G–L order parameter expansion	9
α	spin index	26
α	electron–phonon matrix element	35
β	coefficient of quartic G–L order parameter expansion	9
β	spin index	26
β	phonon pair-breaking parameter	35
β_A	Abrikosov parameter	14
γ	surface tension of normal–superconducting boundary	6
γ	defined by $\gamma = \xi_\parallel/\xi_\perp$	17

Symbol	Explanation	Introductory section
γ	coefficient of gradient term in G–L expansion	20
$\Gamma_{s,\rho}(\mathbf{q}), \Gamma_{s,\rho}(\mathbf{p}, \mathbf{p}')$	generalized quasiparticle interaction functions	31, 52
$\Gamma_{s,\rho}(\mathbf{r}, \mathbf{r}')$	**r**-dependent generalized quasiparticle interaction functions	52
Γ, γ	propagation constants	19
$\hat{\gamma}_{\mathbf{k}\sigma}$	Bogoliubov destruction operator for state $\|\mathbf{k}\sigma\rangle$	27
$\hat{\gamma}^{\dagger}_{-\mathbf{k}-\sigma}$	Bogoliubov creation operator for state $\|-\mathbf{k}-\sigma\rangle$	27
δ_{mn}	Kronecker delta function	36
$\delta^{(2)}(\mathbf{r})$	two-dimensional Dirac delta function	7
$\delta^{(3)}(\mathbf{r})$	three-dimensional Dirac delta function	
$\Delta, \Delta_{\mathbf{k}}$	BCS superconducting energy gap	3, 26
$\Delta_{\alpha\beta}(\mathbf{k})$	spin-generalized energy gap	52
$\Delta(\mathbf{r}), \Delta(\mathbf{r}, t)$	Bogoliubov pair potential	36, 49
$\Delta_{\alpha\beta}(\mathbf{r}, \mathbf{r}', t)$	nonlocal spin generalized Bogoliubov pair potential	51, 54
ε	dielectric constant	2
$\varepsilon(\mathbf{k})$	excitation energy	24
ε_n	Landau level excitation energy	10
$\varepsilon(\mathbf{q}, \omega)$	Fourier transform of dielectric function	35
$\zeta(n)$	Riemann zeta function	45
η	BCS coherence factor parameter ($= \pm 1$)	34
η	acoustic impedance factor	35
θ_{D}	Debye temperature	26
$\theta_{\mathbf{k}}$	Bogoliubov rotation angle	26
$\theta(\xi)$	Heaviside step function	38
κ	G–L parameter $\equiv \lambda_{\mathrm{L}}(T)/\xi(T)$	10
λ	effective magnetic field penetration depth	3
λ	Josephson coupling strength	9
λ	BCS interaction strength ($= \mathcal{N}_0 V$)	32
λ	electron–acoustic wave coupling parameter	34
λ_{L}	London magnetic field penetration depth	2
λ_{J}	Josephson penetration depth	19
Λ	vortex screening length	17
μ	chemical potential; Fermi energy	2
μ_3	magnetic moment of ^{3}He	24
μ_{B}	Bohr magneton	41
μ^*	Coulomb pseudopotential	32
ν	Matsubara integers	38
$\nu(\hat{\mathbf{p}})$	Fermi surface distortion function	24, 55
$\xi(\mathbf{k})$	single-particle excitation energy	9
$\xi_{\alpha\beta}(\mathbf{k})$	spin-generalized single-particle excitation energies	52
ξ_0	BCS coherence length	3
$\xi, \xi(T)$	G–L coherence length	9

Symbol	Explanation	Introductory section
ξ_d	dirty limit coherence length	40
$\xi_{\parallel}$	parallel coherence length	17
$\xi_{\perp}$	perpendicular coherence length	17
ξ_ω	defined by $\xi_\omega = (D/2\lvert\omega\rvert)^{1/2}$ where ω are the Matsubara frequencies	12
$\hat{\Pi}$	canonical momentum operator	14
ρ	electron mass density	2
$\rho(\mathbf{r}, t)$	probability density for Bogoliubov equations	47
$\rho_{\pm}(\mathbf{r})$	quasiclassical envelope wavefunctions for a vortex	48
ρ	coordinate in center of mass	25
ρ	$\equiv \dfrac{\mathrm{d}\xi}{\mathrm{d}\varepsilon}$	35
$\rho(\mathbf{r}, \mathbf{r}')$, $\rho_{\alpha\beta}(\mathbf{r}, \mathbf{r}')$, $\rho(\mathbf{r}, \mathbf{r}', t)$	one-particle density matrix	41, 52, 55
$\rho^{\mathrm{A}}(\mathbf{r}, \mathbf{r}')$, $\rho^{\mathrm{A}}_{\alpha\beta}(\mathbf{r}, \mathbf{r}')$, $\rho^{\mathrm{A}}(\mathbf{r}, \mathbf{r}', t)$	anomalous density matrix	41, 52, 55
$\rho_{\xi}(\mathbf{r}, \mathbf{r}')$	energy shell averaged density matrix	46
$\underset{\sim}{\rho}$	matrix entering Wigner time reversal operator ($\equiv \mathrm{i}\underset{\sim}{\sigma}_y$)	2
$\underset{\sim}{\sigma}_i$	Pauli spin matrices (in spin space)	41
$\sigma_{ij}(\mathbf{r}, \mathbf{r}', \Omega)$	nonlocal frequency-dependent conductivity tensor	46
Σ	self-energy	55
τ, τ_e	electron relaxation (scattering) time	2
τ_{ph}	phonon relaxation time	35
τ_{es}	phonon escape lifetime	35
τ_s	spin scattering time	41
τ'	spin–orbit scattering time	41
$\mathrm{d}\tau$	phase space volume element ($= [2/(2\pi\hbar)^3]\mathrm{d}^3p$)	24
$\underset{\sim}{\tau}_i$	Pauli matrices (in particle–hole space)	36
ϕ	magnetic flux density	7
$\lvert\phi_0\rangle$	second quantized vacuum state	26
ϕ_0	London flux quantum	7
$\phi_{\mathbf{q}}(\rho)$	center of mass wavefunction	25
$\phi_n(\mathbf{r})$	one-electron normal-state wavefunction (complex)	38
Φ	G–L phase	9
$\chi(\rho)$	a function entering the G–L gradient energy	45
$\chi(\sigma_1, \sigma_2), \chi_0, \chi_1$	two-particle spin wavefunctions	25, 42
$\chi_{\mu\alpha}$	spin 1/2 wavefunction for state μ and component α	54
$\chi(\mathbf{r})$	a gauge function	46
$\chi(\mathbf{q}, \omega)$	generalized susceptibililty	31
χ_n	normal state susceptibility	41
χ_s	superconducting state susceptibility	41
$\psi, \psi(\mathbf{r})$	G–L order parameter	9

Symbol	Explanation	Introductory section
$\psi(\mathbf{r}_1, \mathbf{r}_2)$	pair wavefunction	25
$\hat{\psi}^{\dagger}_{\alpha}(\mathbf{r})$	second quantized creation operator at point **r** for spin α	36
$\hat{\psi}_{\alpha}(\mathbf{r})$	second quantized destruction operator at point **r** for spin α	36
$\psi_{\mathrm{B,F}}(\mathbf{r}_1 \ldots \mathbf{r}_N)$	Bose/Fermi symmetrized N-particle wavefunction	A
$\psi_n(z)$	eigenfunctions of the Fourier time transformed diffusion equation	43
$\psi(z)$	digamma function	40
Ψ_N	many-body wavefunction of a system of N electrons	26
Ψ_{BCS}	BCS ground-state wavefunction	26
ω	Matsubara frequencies ($= (2\pi k_{\mathrm{B}}T/\hbar)(\nu + \frac{1}{2})$)	38
ω_{c}	cyclotron frequency ($= e^*H/m^*c$)	10
ω_{c}	Coulomb cut-off frequency	32
ω_{D}	Debye cut-off frequency	25
$\omega_{\mathrm{p}}, \omega_{\mathrm{ep}}$	electron plasma frequency	31
ω_{ip}	ion plasma frequency	31
Ω	frequency associated with Fourier time transform of correlation function	38
Ω_{L}	Larmor frequency	41
Ω_{n}	eigenvalues of the Fourier time transformed diffusion equation	43

Index

Printed in the United States
By Bookmasters